Series on the Foundations of Natural Science and Technology

ISSN: 2010-1961

Published:

Vol. 17 *Quantum Capacitance in Quantized Transistors*
by K. P. Ghatak and J. Pal

Vol. 16 *Electron Statistics in Quantum Confined Superlattices*
by K. P. Ghatak and A. Biswas

Vol. 15 *Topics in Nanoscience*
In 2 Parts
Part I: Basic Views, Complex Nanosystems: Typical Results and Future
Part II: Quantized Structures, Nanoelectronics, Thin Films
edited by W. Schommers

Vol. 14 *Elastic Constants in Heavily Doped Low Dimensional Materials*
by K. P. Ghatak and M. Mitra

Vol. 13 *Scalar Field Cosmology*
by S. Chervon, I. Fomin, V. Yurov and A. Yurov

Vol. 12 *The Origin of Natural Order: An Axiomatic Theory of Biology*
by Q. Zhao

Vol. 9 *The Timeless Approach: Frontier Perspectives in 21st Century Physics*
by D. Fiscaletti

Vol. 8 *Quantum Effects, Heavy Doping, and the Effective Mass*
by K. P. Ghatak

*For further details, please visit: http://www.worldscientific.com/series/sfnst

(Continued at end of book)

Quantum Capacitance in Quantized Transistors

Series on the Foundations of Natural Science and Technology — Vol. 17

Quantum Capacitance in Quantized Transistors

Kamakhya Prasad Ghatak

University of Engineering and Management, India

Jayita Pal

Meghnad Saha Institute of Technology, India

World Scientific

NEW JERSEY • LONDON • SINGAPORE • BEIJING • SHANGHAI • HONG KONG • TAIPEI • CHENNAI • TOKYO

Published by

World Scientific Publishing Co. Pte. Ltd.
5 Toh Tuck Link, Singapore 596224
USA office: 27 Warren Street, Suite 401-402, Hackensack, NJ 07601
UK office: 57 Shelton Street, Covent Garden, London WC2H 9HE

British Library Cataloguing-in-Publication Data
A catalogue record for this book is available from the British Library.

Series on the Foundations of Natural Science and Technology — Vol. 17
QUANTUM CAPACITANCE IN QUANTIZED TRANSISTORS

ISBN 978-981-12-7939-3 (hardcover)
ISBN 978-981-12-7940-9 (ebook for institutions)
ISBN 978-981-12-7941-6 (ebook for individuals)

For any available supplementary material, please visit
https://www.worldscientific.com/worldscibooks/10.1142/13500#t=suppl

Desk Editor: Rhaimie Wahap

Typeset by Stallion Press
Email: enquiries@stallionpress.com

An academician is measured by not what he says or does
but by what he becomes.

Dedication

From the Desk of the First Author

This book is dedicated to Sir Konstantin Novoselov (Nosel Laureate in Physics, 2010) *for presenting an eye-opening crystal clear Lecture on 'Graphene' for the students and family members of IEM UEM Group, Kolkata, in 2022 (which was, in turn, arranged by the Hon'ble President and the respected Director of the said group) for instilling the seed in the mind of this author which with time have been grown into a mighty tree with the name*

"Quantum Capacitance in Quantized Transistors"

From the Desk of the Second Author

The author dedicates this book to her loving parents and her sweet family members.

Preface

Don't try to get, but try to become.

It is well known that the electrons in bulk semiconductors in general have three-dimensional (3D) freedom of motion and the electrons are moving on the closed 3D wave vector (k) space where the total electron energy (E) is constant. The system asymmetry can produce 2D electron motion when these electrons are confined in a one-dimensional (1D) potential well whose width is of the order of de Broglie wavelength, the motion in that particular direction gets quantized while that along the other two directions remains as free. Thus, the energy spectrum appears in the shape of discrete levels for the 1D quantization, each of which has a continuum for the two-dimensional (2D) free motion. The physical properties of such 1D quantum confined carriers have been studied extensively [1–20]. For the metal-oxide-semiconductor (MOS) structures, the work functions of the metal and the semiconductor substrate are different and the application of an external gate voltage at the metal-gate surface causes the change in the charge density of the carriers at the oxide semiconductor interface creates a bending of the energy bands of the semiconductor near the surface. As a result, a triangular potential well is being created at the surface of the semiconductor. The spatial variation of the potential profile is so sharp that for considerably large values of the electric field, the width of the potential well becomes of the order of the de Broglie wavelength of the carriers as discussed already. The Fermi energy, which is near the edge of the conduction band in the bulk, becomes nearer to the edge of the valance band at the surface creating inversion layers. The energy levels of the carriers bound within the potential well get quantized and form electric sub-bands. Each of the sub-bands corresponds to a quantized level in a plane perpendicular to the surface leading to a quasi-2D electron gas. Thus, the extreme band bending at low temperatures allows us to observe the quantum effects at the surface. Though considerable work has already been done, nevertheless it appears from the literature that there remains scope for the investigations made while the interest for further research of some other aspects of such 2D systems is becoming increasingly important.

One such vital quantity which generates considerable interest is the quantum capacitance (QC) of metal oxide semiconductor quantum field effect transistors (QMOSFETs) under the presence of a large gate voltage which creates a large surface electric field [21–25]. The 2D electron statistics in the inversion layers of such devices can rather easily be varied by changing the gate voltage which, in turn, brings a change in the surface electric field, the QC depends on the gate-voltage. This variation has been investigated for n-channel inversion layers on p-type Si which shows few important oscillatory characteristics due to the 2D density-of-states function of the electrons in the quantized inversion layers. There has been considerable progress [26–33] in the investigations of the physical properties of the 2D QMOSFET devices operated under inversion mode made of different materials whose bulk dispersion relations (DR) are defined by the two-band model of Kane. It appears then that there remains considerable scope of investigations on the QC for other types of technologically important FETs made of various vital compounds of non-standard energy bands.

This book contains 27 chapters is partially based on our ongoing research on the QC for the last 40+ years and an attempt has been made to present a cross-section of the same for wide ranges of metal oxide semiconductor field effect transistors (MOSFETs), quantum well field effect transistors (QWFETs) and nano wire field effect transistors (NWFETs), made of nonlinear optical, III–V, II–VI, IV–VI, stressed Kane type, Bi_2Te_3, Ge, GaSb, GaAs, GaP, InSb, $PtSb_2$ and II–V materials, respectively, under various physical conditions. We have also considered the heavily doped MOSFETs, QWFETs and NWFETs, respectively. In Chapter 1, we study the QC in 2D QMOSFETs of tetragonal and nonlinear optical materials on the basis of newly formulated 2D electron statistics in both weak and strong electric field limits. Besides, the QC in 2D QMOSFETs of III–V, ternary and quaternary materials have been obtained as special cases of our generalized analysis. The said materials find extensive applications in various fields of surface and allied technologies as given in the literature [34, 35]. Besides, as a collateral study, we have also investigated in the same chapter the effective electron mass (EEM), the sub-band energy, the experimental determinations of Einstein's relation (ER) and the elastic constants together with the fact the surface concentration is connected with 25 important transport topics of quantum effect devices, namely *the Effective Electron Mass* [36], *the Thermoelectric Power* [37], *the experimental determinations of* ΔC_{44} *and* ΔC_{456} [38], *the Debye screening length* [39], *Einstein relation* [40], *the diffusion coefficient of the minority carriers* [41], *the normalized optical response* [42], *the third-order nonlinear optical susceptibility* [43], *the generalized Raman gain* [44], *the plasma frequency* [45], *the activity coefficient* [46], *the magneto-thermal effect in quantized structures* [47], *the normalized Hall coefficient* [48], *the reflection coefficient* [49], *the heat capacity* [50], *the magnetic susceptibilities* [51], *the faraday rotation* [52], *the Fowler–Nordheim field emission* [53], *the optical effective mass* [54], *Einstein's Photoemission* [55], *the Righi-Leduc coefficient* [56], *the Electric Susceptibility* [57], *the Electric Susceptibility Mass* [58], *the Electron Diffusion Thermo-power* [59] *and the Hydrostatic Piezo-resistance Coefficient* [60], *respectively.*

In Chapter 2, we study the QC in 2D QMOSFETs of II–VI, IV–VI, stressed Kane type compounds, Bi_2Te_3, Ge, GaAs, GaP, GaSb, InSb and Te, respectively. The quantum well transistors (QWTs) are recently been studied in the literature [34] and in Chapter 3, we study the QC in QWTs of nonlinear optical, III–V, ternary, quaternary, II–VI, IV–VI, stressed Kane type, Te, GaP, $PtSb_2$, Bi_2Te_3, Ge, GaAs, GaSb and InSb, respectively.

It is well known that the effects of quantizing magnetic field (B) on the band structures of compound semiconductors are most striking than that of the parabolic one and are easily observed in experiments. A number of interesting physical features originate from the significant changes in the basic energy wave vector relation of the carriers caused by the magnetic field. The valuable information could also be obtained from experiments under magnetic quantization regarding the important physical properties such as Fermi energy and effective masses of the carriers, which affect almost all the transport properties of the 2D devices made of various materials having different carrier DRs. *In Chapter 4, we also study the QC of 2D MOSFETs and 2D QWTs made of the aforementioned materials in the presence of a strong quantizing magnetic field, respectively.*

It is worth remarking that the effects of crossed electric and quantizing magnetic fields on the transport properties of 2D materials having various band structures have relatively been less investigated as compared with the corresponding magnetic quantization, although, the study of the cross-fields are of fundamental importance with respect to the addition of new physics and the related experimental findings in modern quantum effect devices. It is well known that in the presence of electric field (E_o) along x-axis and the quantizing magnetic field (B) along z-axis, the dispersion laws of the carriers in materials become modified and for which the carriers move in both the z and y directions, respectively. The motion along y-direction is purely due to the presence of E_0 along x-axis and in the absence of electric field, the EEM along y-axis tends to infinity indicating the fact that the electron motion along y-axis is forbidden. The EEM of the isotropic, bulk materials having parabolic energy bands exhibits mass anisotropy in the presence of cross-fields and this anisotropy depends on the electron energy, the magnetic quantum number, the electric and the magnetic fields, respectively, although, the EEM along z-axis is a constant quantity. In 1966, Zawadzki and Lax [61] derived the expression of the DR of the conduction electrons for III–V materials in accordance with the two-band model of Kane under cross-fields configuration which generates the interest to study this particular topic in low-dimensional materials in general [62]. In Chapter 5, we study the QC in 2D QWTs made of nonlinear optical, III–V, ternary, quaternary, II–VI, IV–VI and stressed Kane type materials under cross-fields configurations. In Chapter 6, we study the QC in 2D MOSFETs of all the materials as stated in Chapter 5 under cross-fields configurations. In Chapter 7, we study the QC in NWFETs of nonlinear optical, III–V, ternary, quaternary, II–VI, IV–VI, stressed Kane type, Bi_2Te_3, Ge, GaAs, GaP, $PtSb_2$, II–V materials and CNTFETs on the basis of newly formulated surface concentration in each case.

The study of MOSFETs operated under accumulation layer mode has relatively been less investigated as compared with the corresponding inversion mode of operation. In this book, in Chapter 8, we study the QC in 2D MOSFETs made of the same materials as stated in Chapter 5 under accumulation mode by formulating the respective 2D electron statistics. The influence of magnetic quantization of the QC of 2D MOSFETs made of the said materials under accumulation mode has been studied in Chapter 9. In Chapter 10, we study the QC in heavily doped (HD) MOSFETs operated under accumulation mode in the presence of crossed fields configuration under accumulation mode of all the materials as studied in Chapter 5. The QC in HD QWTs made of nonlinear optical, III–V, ternary, quaternary, II–VI, IV–VI, stressed Kane type, Te, GaP, $PtSb_2$, Bi_2Te_3, Ge and GaSb has been studied in Chapter 11. The QC under magnetic quantization in HDQWTs made of all the materials of Chapter 7 together with Te and lead germanium telluride has been investigated in Chapter 12. In Chapter 13, the QC in HD QWTs' made of all the materials of Chapter 8 have been studied under cross-fields configurations. In Chapter 14, the

QC in HD NWFETs made of all the materials of Chapter 7 has been investigated.

With the advent of nanophotonics, there has been considerable interest in studying the optical processes in semiconductors and their nanostructures in the presence of intense light waves [56]. It appears from the literature, that the investigations of the physical properties of different materials in the presence of external intense photo-excitation have been carried out on the assumption that the DRs are invariant quantities under strong external light waves, which **is not fundamentally true**. The physical properties of semiconductors in the presence of strong light waves which alter the basic DRs have relatively been much less investigated [57–60] as compared with the cases of other external fields and in optoelectronics the influence of strong light-waves is needed for the characterization of the low dimensional optoelectronic devices. In solo Chapter 15, we have formulated the DR of III–V materials in the presence of strong radiations and also used it to study the QC in the 2D MOSFETs and also studied the same in the presence of magnetic quantization together with the cross-fields configuration. The QC in QWFETs and NWFETs for the aforementioned physical conditions has also been studied in this context. The QC in CNTFETs under strong radiation has been studied in this chapter. Finally, we have investigated the QC for all types of FETs as written above in the presence of heavy doping by formulating appropriate DR in each case.

With the advent of nanodevices, the built-in electric field becomes so large that the DR changes fundamentally [61] instead of being invariant. The solo Chapter 16 investigates the QC in QWFETs, NWFETs and CNTFETs in the presence of electric field, on the basis of newly formulated appropriate DR. Chapter 17 contains conclusion and scope for future research. Chapter 18 contains the values of the energy band constants of the said materials in this context.

In Chapters 19 and 20, we have explored the I_D–V_D equations for the ballistic NWFETs and ballistic HDNWFETs of all the materials of Chapter 14, respectively. In Chapters 21 and 22, we study the I_D–V_D equations for the ballistic MOSFETs and ballistic QWFETs made of all the materials of Chapter 4, respectively. In Chapter 23, we study the I_D–V_D equations for ballistic HDQWFETs made of all the materials of Chapter 22. In Chapters 24 and 25, we study the I_D–V_D equations for ballistic magneto QWFETs and ballistic magneto HDQWFETs made of all the materials of Chapter 4 respectively. In Chapter 26, the I_D–V_D equations for magneto MOSFETs made of the said materials of Chapter 4 have been derived.

It may be noted that the $I_D - V_D$ *equations as derived from Chapters 19–26 are new and reported for the first time in a book.*

The generalized distribution function for heavily doped materials have been derived in Chapter 27 together with few new plots and associated discussions in this regard. In this last chapter of this book, we have proposed the complex distribution function which may be used for studying the transport properties of heavily doped quantized structures to generate new physics and inferences from them which all together is a sea in itself. The book contains circa 200 open research problems and forms the most important integral part of the book.

It is needless to say that this book is based on the iceberg principle [63] and the rest of which will be explored by the researchers of different appropriate fields. Since, there is no existing report devoted solely to the study of the QC in 2D MOSFETs, 2D QWTFETs, 1D NWFETs made of different materials together with the respective $I_D - V_D$ equations in the presence of external quantizing fields under various physical conditions to the best of our knowledge, we earnestly hope that the present book will a useful reference source for the present and the next generation of readers and researchers of low dimensional device electronics and allied sciences in general. We have discussed enough regarding the QCs for low-dimensional transistors and their respective $I_D - V_D$ equations made of various materials having different DRs and lots of new computer-oriented numerical analyses are being left for the purpose of being computed by our alert readers, to generate the new graphs and the inferences from them which all together is a sea in itself. Since the production of error-free first edition of any book from every point of view is a permanent member of impossibility theorems, therefore in spite of our joint concentrated efforts together with the seasoned team of Ms Nur Syarfeena Binte Mohd Fauzi, the Physics Editor, of World Scientific Publishing Co, the same stands very true for this book also. *Various expressions and many chapters of this book have been appearing for the first time in printed form.* The suggestions from the readers for the development of the book will be highly appreciated for the purpose of inclusion in the future edition, if any. In this book, we have presented circa 200 multidimensional open research problems for graduate students, PhD aspirants, researchers, engineers in this pinpointed research topic. We emphatically hope that the readers of this book will not only solve the said open research problems by removing all the mathematical approximations and establishing the appropriate uniqueness conditions but also generate new research problems both theoretical and experimental. Incidentally, our readers after reading this book will easily understand how little is presented and how much more is yet to be investigated in this exciting topic which is the signature of coexistence of new physics, advanced mathematics combined with the inner fire for performing creative researches in this context from the young scientists since like Kikoin [64] we strongly assert that "A young scientist is no good if his teacher learns nothing **new** from him and gives his teacher nothing to be proud of". The open research problems will be useful for the readers to initiate their own contributions on the QC in quantized transistors and like Sakurai [65] we firmly believe "The reader who has read the book but cannot do the exercise has *learned nothing*".

In this book, as stated already, we investigate not only the QC in quantized transistors (QTs) but also many important electronic properties *for the purpose of in-depth investigations of the said important pinpointed research topics.* Thus, in this book, the readers will get much information regarding the influence of quantization on the electronic properties of various QTs made of different materials having different band structures. Although the name of this book is an example of extremely high Q-factor, from the various chapters, one can easily infer that it should be useful in graduate courses on surface science, materials science, condensed matter physics, solid-states electronics, nanoscience and technology and solid-state sciences and devices in many Universities and the Institutions. *Last but not least, the authors hope that their humble effort will kindle the desire to delve deeper into this fascinating and deep topic by any one engaged in materials research and modern low-dimensional electronic device development either in academics or in industries.*

References

1. T. Ando, H. Fowler and F. Stern, *Rev. Mod. Phys.* **54**, 437 (1982); C. Hoon Choi, Y. Wu, J. Suk Goo, Z. Yu and R. W. Dutton, *IEEE Trans. Electron Dev.* **47**, 1843 (2000); L. Larcher, P. Pavan, F. Pellizzer and G. Ghidini, *IEEE Trans. Electron Dev.* **48**, 935 (2001); J.-M. Sallese, W. Grabinski, V. Meyer, C. Bassin and P. Fazan, *Sensors Actuators A: Phys.* **94**, 53 (2001).
2. J. J. Quinn and P. J. Styles. (ed.) *Electronic Properties of Quasi Two Dimensional Systems* (North Holland, Amsterdam, 1976); Y. Yu Fan, R. E. Nieh, J. C. Lee, G. Lucovsky, G. A. Brown, L. F. Register and S. K. Banerjee, *IEEE Trans. Electr. Dev.* **49**, 1969 (2002); A. Haque and M. Z. Kauser, *IEEE Trans. Electr. Dev.* **49**, 1580 (2002).

3. G. A. Antcliffe, R. T. Bate and R. A. Reynolds, *Proceedings of the International Conference, Physics of Semi-metals and Narrow-Gap Semiconductors*, D. L. Carter and R. T. Bate (ed.) (Pergamon Press, Oxford, 1971).
4. Z. A. Weinberg, *J. Solid-State Electr.* **20**, 11 (1977); J. Guang Su, S. Chih Wong, C. Yen Chang and T. Yuan Huang, *Solid State Electron*, **46**, 1163 (2002); C. Hoon Choi, Z. Yu and R. W. Dutton, *IEEE Electron Dev. Lett.* **24**, 330 (2003); P. A. Kraus, K. Z. Ahmed and J. S. Williamson, *IEEE Trans. Electron Devi.* **51**, 1350 (2004).
5. G. Paasch, T. Fiedler, M. Kolar and I. Bartos, *J. Physica Status Solidi (b)* **118**, 641 (1983).
6. S. Lamari, *J. Physical Review B* **64**, 245340 (2001).
7. T. Matsuyama, R. Kürsten, C. Meißner and U. Merkt, *J. of Physical Review B* **61**, 15588 (2000).
8. P. V. Santos and M. Cardona, *J. Phys. Revi. Lett.* **72**, 432 (1994); M. M. A. Hakim and A. Haque, *Solid State Electronics* 48, 1095 (2004); J. Joon Kim and K. Roy, *IEEE Trans. Electron Devi.* **51**, 1468 (2004).
9. L. Bu, Y. Zhang, B. A. Mason, R. E. Doezema and J. A. Slinkman, *J. Phys. Rev. B* **45**, 11336 (1992).
10. P. D. Dresselhaus, C. M. Papavassiliou, R. G. Wheeler and R. N. Sacks, *J. Phys. Rev. Lett.* **68**, 106 (1992).
11. U. Kunze, *J. Phys. Rev. B* **41**, 1707 (1990); M. J. Kumar, V. Venkataraman and S. K. Gupta, *IEEE Trans. Electr. Dev.* **52**, 1676 (2005); L. Ge, F. Gamiz, G. O. Workman and S. Veeraraghavan, *IEEE Trans. Electr. Dev.* **53**, 753 (2006); A. E. Islam and A. Haque, *IEEE Trans. Electr. Dev.* **53**, 1364 (2006).
12. E. Yamaguchi, *J. Phys. Rev. B* **32**, 5280 (1985); O. Wunnicke, *Appl. Phys. Lett.* **89**, 083102 (2006); Y. Wen Chang, H. Wen Chang, T. Cheng Lu, Y. Chin King, W. Ting, Y. Hui Joseph Ku and C. Yuan Lu, *IEEE Electr. Dev. Lett.* **27**, 390 (2006); J. Deng and H.-S. Philip Wong, *IEEE Trans. Electr. Dev.* **54**, 2377 (2007).
13. Th. Lindner and G. Paasch, *J. Appl. Phys.* **102**, 054514 (2007).
14. S. Lamari, *J. Appl. Phys.* **91**, 1698 (2002); B. Bindu, N. DasGupta and A. DasGupta, *IEEE Trans. Electr. Dev.* **54**, 1889 (2007); A. Khakifirooz and D. A. Antoniadis, *IEEE Trans. Electr. Dev.* **55**, 1401 (2008); S. Ho Hong, G. Bok Choi, R. Hyun Baek, H. Sung Kang, S. Woo Jung and Y. Ha Jeong, *IEEE Electr. Devi. Lett.* **29**, 775 (2008).
15. K. P. Ghatak and M. Mondal, *J. Appl. Phys.* **70**, 299 (1991).
16. K. P. Ghatak and S. N. Biswas, *J. Vac. Sci. Technol. B* **7**, 104 (1989); S. Mookerjea, R. Krishnan, S. Datta and V. Narayanan, *IEEE Trans. Electr. Devi.* **56**, 2092 (2009); R. Granzner, S. Thiele, C. Schippel and F. Schwierz, *IEEE Trans. Electr. Dev.* **57**, 3231 (2010); F. J. García Ruiz, I. María Tienda-Luna, A. Godoy, L. Donetti and F. Gamiz, *IEEE Trans. Electron. Dev.* **57**, 2477 (2010).
17. B. Mitra and K. P. Ghatak, *J. Solid-State Electr.* **32**, 177 (1989).
18. K. P. Ghatak and M. Mondal, *J. Appl. Phys.* **62**, 922 (1987).

19. M. Mondal and K. P. Ghatak, *J. Magn. Magn. Mater.* **62**, 115 (1986); M. Mondal and K. P. Ghatak, *J. Phys. Scri.* **31**, 613 (1985); K. P. Ghatak and M. Mondal, *J. Zeitschrift für Physik B Condensed Matter* **64**, 223 (1986).
20. K. P. Ghatak and S. N. Biswas, *J. Solid-State Electr.* **37**, 1437 (1994); K. P. Ghatak, N. Chattopadhyay and M. Mondal, *J. Appl. Phys. A* **48**, 365 (1989).
21. A. S. Grove and D. J. Fitzgerald, *J. Solid State Electr.* **9**, 783 (1966); K. P. Ghatak and S. N. Banik, *Fizika A* **5**, 31 (1996); B. Mitra, A. Ghoshal and K. P Ghatak, *Phys. Stat. Soli. (b)*, **153**, K209 (1989).
22. A. M. Voshchenkov and J. N. Zemel, *J. Phys. Rev. B* **9**, 4410 (1974); K. P. Ghatak and M. Mondal, *J. Appl. Phys.* **64**, 918 (1988); S. N. Biswas and K. P. Ghatak, *Advan. Proce. Semico. Dev.* **797**, 286 (1987).
23. Z. Djurid, Z. Spasojevid and D. Tjapkin, *J. Solid State Electr.* **19**, 931 (1976); S. N. Biswas and K. P. Ghatak, *Megaga. Tech. Puls. Power. Appl* **8**, 219 (1987); K. P. Ghatak, N. Chattopadhyay and S. N. Biswas, *Optoelectr. Mater. Devi. Packa. Interconn.* **836**, 203 (1987).
24. M. Kaplit and Z. N. Zemel, *J. Phys. Rev. Lett.* **21**, 212 (1963).
25. A. Daerr, J. P. Kotthaus and J. F. Koch, *J. Solid State Commun.* **17**, 455 (1975).
26. A. Daerr and J. P. Kotthaus, *J. Surf. Sci.* **73**, 549 (1978).
27. H. A. Washburn, and J. R. Sites, *J. Surf. Sci.* **73**, 537 (1978).
28. J. W. Rowe and J. L. Shay, *J. Phys. Rev. B* **3**, 451 (1973).
29. H. Kildal, *J. Phys. Rev. B* **10**, 5082 (1974).
30. J. Bodnar, Proceedings of the International Conference on Physics of Narrow-gap Semiconductors (Polish Science Publishers, Warsaw, 1978).
31. G. P. Chuiko and N. N. Chuiko, *J. Sov. Phys.: Semiconductors.* **15**, 739 (1981).
32. K. P. Ghatak and S. N. Biswas, *J. Proc. SPIE.* **1484**, 149 (1991).
33. G. Paasch, T. Fiedler, M. Kolar and I. Bartos, *J. Physica Status Solidi B* **118**, 641 (1983).
34. M. S. Laundstrom and J. Guo, *Nanoscale Transistors: Device Physics, Modeling and Simulation* (Springer), pp. 1–217 and the reference cited therein, 2006.
35. D. M. Kin and Y. H. Jeong (ed.), *Nanowire Field Effect Transistors: Principles and Applications* (Springer), pp. 1–290 and the reference cited therein, 2014.
36. S. Bhattacharya and K. P. Ghatak, *Effective Electron Mass in Low Dimensional Semiconductors* Springer Series in Materials Sciences, Vol. 167 (Springer 2013).
37. K. P. Ghatak, *Magneto Thermoelectric Power in Heavily Doped Quantum Structures*, Series on the Foundations of Natural Science and Technology, Vol. 7 (World Scientific, 2016); K. V. Klitzing, G. Dorda and M. Pepper, *Phys. Rev. Lett.* **45**, 494 (1980); K. V. Klitzing, *Rev. Mod. Phys.* **58**, 519 (1986);

    W. Zawadzki, in *Two-Dimensional Systems, Heterostructures and Superlattices*, Springer Series in Solid State Sciences, Vol. 53, G. Bauer, F. Kuchar and H. Heinrich (eds.) (Springer, Berlin, Heidelberg, Germany, 1984).

38. K. P. Ghatak and M. Mitra, *Elastic Constants in Heavily Doped Low Dimensional Materials*, Series on the Foundations of Natural Science and Technology, Vol. 14, (2021), P. 1–1036.
39. K. P. Ghatak and S. Bhattacharya, *Debye Screening Length: Effects of Nanostructured Materials*, Springer Tracts in Modern Physics, Vol. 255 (Springer, 2014); K. P. Ghatak and S. Bhattacharya, *J. Appl. Phys.* **102**, 073704 (2007); E. O. Kane, *Solid-State Electron.* **28**, 3 (1985); W. Zawadzki, in *Handbook on Semiconductors*, W. Paul (ed.), North Holland, New York, **1**, p. 715 (1982).
40. K. P. Ghatak, S. Bhattacharya and D. De, *Einstein Relation in Compound Semiconductors and their Nano Structures*, Springer series in Materials Science, Vol. 116, (Springer, 2009); A. N. Chakravarti and B. R. *Nag, Phys. Stat. Sol. (a)* **14**, K55 (1972); A. N. Chakravarti and D. P. Parui, *Canad. J. Phys.* **51**, 451 (1973).
41. K. P. Ghatak and S. Bhattacharya, *Heavily Doped 2D-Quantized Structures and the Einstein Relation*, Springer Tracts in Modern Physics, Vol. 260 (Springer, Germany, 2015).
42. A. S. Filipchenko, I. G. Lang, D. N. Nasledov, S. T. Pavlov and L. N. Radaikine, *Phys. Stat. Sol. (b)* **66**, 417 (1974).
43. M. Wegener, *Extreme Nonlinear Optics* (Springer-Verlag, Germany, 2005).
44. O. Madelung, *Physics of III–V Compounds* (John Wiley and Sons, Inc. New York, (1966).
45. B. S. Wherreff, W. Wolland, C. R. Pidgeon, R. B. Dennis and S. D. Smith, Proceedings of the 12th International Conference of the Physics of the Semiconductors, M. H. Pilkahn, R. G. Tenbner (eds.) (Staffgard, 1978), p.793.
46. C. H. Peschke, *Phys. Stat. Sol. (b)* **191**, 325 (1995).
47. C. J. Hwang and J. R. Brews, *J. Phys. Chem. Solids* **32**, 837 (1971); P. T Landsberg and H. C Cheng, *Phys. Rev. B* **32**, 8021 (1985).
48. A. Isihara and Y. Shiwa, *J. Phys. C* **18**, 4703 (1985) V. G. I. Ravish, B. A. Efimova and I. A. Smirnov, *Semiconducting Lead Chalocogenides* (Plenum Press, New York, 1970), p.163.
49. M. Kohomoto, L. P. Kadanoff and C. Tang, *Phys. Rev. Lett.* **50**, 1870 (1983).
50. D. K. Roy, *Quantum Mechanical Tunneling and Its Applications* (World Scientific Publication Co.Ptc.Ltd, 1986).
51. M. Mondal and K. P. Ghatak, *Acta Physica Polonica. A* **66**, 539 (1984); M. Mondal and K. P. Ghatak, *J. Phys. Status Solidi (b)* **128**, K81 (1985); K. P. Ghatak, A. Ghoshal, S. Bhattacharyya and M. Mondal, Thermo sense XII: An International Conference on Thermal Sensing and Imaging Diagnostic Applications, (*Int. Soc. Opt. Photon. 1313*, 225–237, 1990); W. Zawadzki, *Phys.* **127**, 388 (1980).
52. J. Todd, R. Merlin, R. Clarke, M. K. Mohanty and J. D. Axe, *Phys. Rev. Lett.* **57**, 1157 (1986).
53. H. Piller, *Semiconductors and Semimetals*, R. K. Willardson and A. C. Beer, (eds.) **124**, 8 (1972).

54. S. Bhattacharya and K. P. Ghatak, *Fowler-Nordheim Field Emission*, Springer Series in Solid-State Sciences, Vol. 170 (Springer, 2012); R. H. Fowler and L. Nordheim, *Proc. Roy. Soc. Lond., Seri-A* **119**, 173 (1928); A. Van Der Ziel, *Solid State Physical Electronics* (Prentice-Hall Inc., Englewood Cliffs, USA, 1957); B. Mitra and K. P. Ghatak, *Phys. Lett. A* **357**, 146 (1990).
55. K. P. Ghatak, D. De and S. Bhattacharya, *Photoemission from Optoelectronic Materials and Their Nanostructures*, Springer Series in Nanostructure Science and Technology (Springer, 2009); R. K. Pathria, Statistical Mechanics, 2nd edition (Butterworth-Heinemann, Oxford 1996); L. Torres, L. Lopez-Diaz, O. Alejos and J. Iniguez, *J. Appl. Phys.* **85**, 6208 (1999); S. Tiwari, *Cryst. Res. Technol.* **41**, 78 (2006); R. Houdré, C. Hermann, G. Lampel, P. M. Frijlink and A. C. Gossard, *Phys. Rev. Lett.* **55**, 734 (1985).
56. B. M. Askerov, N. F. Gushimzade and M. M. Panakhov, *Sov. Phys. Sol. Stat.* **29**, 465 (1987).
57. B. Mitra, A. Ghoshal and K. P. Ghatak, *Phys. Stat Sol. (b)*, **155**, K23 (1989).
58. S. N. Biswas, N. Chattopadhyay and K. P. Ghatak, *Phys. Stat Sol. (b)* **141**, K47 (1987).
59. J. A. Woollam, *Phys. Rev.* **3**, 1148 (1971).
60. V. V. Kaminoskii, N. N. Stepanav and l. M. Smirnov, *Sov. Phys. Sol. Stat.* **27**, 1295 (1985); K. P. Ghatak, S. Chakrabarti, B. Chatterjee, P. K. Das, P. Dutta and A. Halder, *J. Mater. Focus* **7**, 390 (2018).
61. W. Zawadzki and B. Lax, *Phys. Rev. Lette.* **16**, 1001 (1966).
62. B. Mitra and K. P. Ghatak, *Phys. Stat. Sol. (b)* **164**, K13 (1991); K. P. Ghatak and B. Mitra, *Int. J. Electron.* **70**, 345 (1991); S.M. Adhikari, D. De, J.K. Baruah, S. Chowdhury and K.P. Ghatak, *Adv. Sci. Focus* **1**, 57 (2013); S. Pahari, S. Bhattacharya, D. De, S. M. Adhikari, A. Niyogi, A. Dey, N. Paitya, S. C. Saha, K. P. Ghatak and P. K. Bose, *Physica B: Condensed Matter.* **405**, 4064 (2010); S. Bhattaacharya, S. Choudhary, S. Ghoshal, S. K. Bishwas, D. De and K. P. Ghatak, *J. Comp. Theo. Nanosci.* **3**, 423 (2006); M. Mondal and K. P. Ghatak, *Ann. der Physik.* **46**, 502 (1989); K. P. Ghatak and B. De, *MRS Proceedings* **242**, 377 (1992); K. P. Ghatak and B. Mitra, *Int. J. Electron. Theo. Exp.* **70**, 343 (1991); B. Mitra and K. P. Ghatak, *Phys. Lett. A* **141**, 81 (1989); B. Mitra and K. P. Ghatak, *Phys. Lett. A* **137**, 413 (1989); M. Mondal, N. Chattopadhyay and K. P. Ghatak, *J. low Temp. Phys.* **73**, 321 (1988); K. P. Ghatak, N. Chattopadhyay and S. N. Biswas, *International Society for Optics and Photonics, Proc. Soc. Photo-Optical Instru. Engg.*, USA, 203 (1987).
63. A. Pais, J. Robert, *Oppenheimer* (Oxford University Press, UK, 2006) (p. xviii).
64. I. K. Kikoin, *Science for Everyone: Encounters with Physicists and Physics* (Mir Publishers, Russia, 1989).
65. J. J. Sakurai, *Advanced Quantum Mechanics* (Addison–Wesley Publishing Company, USA, 1967).

Acknowledgments

I must be ready to give a little more and take a little less.

Acknowledgment by Kamakhya Prasad Ghatak

I am grateful to Late A.N. Chakravarti, formerly Head of the Department of the Institute of Radio Physics and Electronics of the University of Calcutta, my mentor and an *ardent advocator of the concept of theoretical minimum of Landau*, who convinced a 21 years old Circuit theorist that the Quantized Structures, in general, is the hidden dual dance of quantum mechanics, statistical mechanics together with advanced mathematical techniques, and even to appreciate the incredible beauty, he not only placed a stiff note for me to derive all the equations in the Monumental Course of Theoretical Physics, the Classics of Landau–Lifshitz together with the two-volume classics of Morse-Feshbach 50 years ago but also forced me to stay in the *creative critical zone of research till date*. I express my gratitude to Late B.R. Nag, formerly Head of the Departments of Radio Physics and Electronics and Electronic Science of the University of Calcutta, to whom I am ever grateful as a student and research worker, the Semiconductor Grandmaster of India, for his three books on semiconductor science in general and more than 200 research papers (***many of them are absolutely in Honor's Class***) which still fire my imagination. I consider myself to be rather fortunate to study Mathematics under the direct influence of Late S.C. Dasgupta, formerly Head of the Department of Mathematics of the then Bengal Engineering College, Shibpur (presently Indian Institute of Engineering Science and Technology) and Late M. Mitra (both of them were formidable Applied Mathematicians with deep physical insight and could solve any problem from the two-volume classics of Morse-Feshbach and the famous Whittaker and Watson instantly on the blackboard) of the said Department of Mathematics for teaching me deeply the various methods of different branches of Applied Mathematics with special emphasis on ***analytic number theory*** when I was in my late teens, pursuing the bachelor's degree in the branch of Electronics and Telecommunication Engineering 55 years ago. I offer my thanks to Late S.S. Boral, formerly Founding Head of the Department of Electronics and Telecommunication Engineering of the then Bengal Engineering College, Shibpur, for teaching me the *Theoretical Acoustics from the Classic of Morse and Ingard* by urging me to solve the problems. I am grateful to Late S. K. Sen, formerly Head of the Department of Electrical Engineering of Bengal Engineering College, Shibpur, Ex-Vice-Chancellor of Jadavpur University and Ex-Power Minister of West Bengal State for teaching me deeply *Non-linear network analysis and synthesis and Non-linear control systems in general.* I am indebted to Late C.K. Majumdar of the Department of Physics of the University of Calcutta to light the fire for Theoretical Physics.

I am grateful to all my friends and colleagues for the last 55 years from my school days till date for forming my inner fire to solve independently the problems from three volumes Sakurai, 10 volumes of Addison-Wesley Modular Series on Solid State Devices, two volumes Cohen-Tannoudji *et al.*, three volumes Carslaw, six volumes Guillemin, three volumes Tuttle, two volumes Scott, two volumes Budak, five volumes Chen.... It is curious to note that they are insisting me in the real sense of the term to verify all the entries of Gradshteyn-Ryzhik and three volumes Bateman manuscript project for the last 50 years. It may be noted that academic output of a nice scholar is the response of a good quality RC coupled amplifier in the mid-band zone whereas the same for a fine research worker is the Dirac's delta function. Incidentally, I can safely say that I belong to neither as a consequence of the academic surroundings that this life presents to me. I express my warm gratitude to J. S. Blakemore, H. L. Hartnagel, D. Bimberg, W. L. Freeman and W. Schommers for various academic interactions spanning over the last four decades. I remember the sweet memories of P.N. Robson, I.M. Stephenson, V.S. Letokhov, J. Bodnar and K.F. Brennan with true reverence for inspiring me with the ebullient idea that the publications of the research monographs and advanced books from internationally reputed publishers containing innovative concepts are of prime importance to excel in creative research activity.

From the flash back of memory, I wish to note that the Late P. T. Landsberg (popularly known as PTL to his scientific friends) visited the Institute of Radio Physics and Electronics of the University of Calcutta in the late seventies where I was a PhD student of the said Institute.

I was very impressed by his crystal-clear lecture on semiconductor Physics and I formed permanent covalent bonds with him. Whatever small I have learned with respect to the fundamentals of semiconductor science is mainly due to him. Incidentally, his wife Mrs. Sophia Landsberg in her letter of desperate sadness dated January 12, 2009, informed me that from the last phase of 2008 onward, her husband had to give up scientific works due to Alzheimer's disease. The disappearance of P.T.L from my research scenario is a big shock to me which I have to bear until my own physical disappearance. I still remember his power-packed words that "The definition of Kamakhya Prasad Ghatak (K.P.G) = Ph.D. + D. Engg. + more than 300 research chapters in nanoscience and technology in SCOPUS enlisted journals, + Ph.D. guide of more than a few dozen PhD students + many research books + HOD + Dean.... is a *big zero to P.T.L if K.P.G. cannot live in a world full with new creative concepts and not merely repeating the past successes, since past success is a dead concept at least to P.T.L.*

In December 21, **1974**, A. N. Chakravarti, my mentor in my first interaction with him emphatically energized me by making myself acquainted with the ***famous Fermi-Dirac Integrals*** and telling me that ***I must master*** *Semiconductor Statistics* (Pergamon Press, London, 1962) by J. S. Blakemore for my initiation in semiconductor physics. Later on, I came in deep touch with K. Seeger, the well-known author of the book *Semiconductor Physics* (Springer Series in Solid State Sciences, vol. 40, Springer-Verlag, Germany, 1982). The solid Mathematical Physicist Late P. N. Butcher has been a steady hidden force since 1983 before his sad demise with respect to our scripting the series in band structure-dependent properties of nano-structured materials. He insisted me repeatedly regarding it and to tune with his high rigorous academic standard, myself with the prominent members of my research group wrote [1] as the first one, [2] as the second one, [3] as the third one, [4] as the fourth one, [5] as the fifth one, [6] as the sixth one, [7] as the seventh one, [8] as the eighth one, [9] as the ninth one, [10] as the tenth one, [11] as the eleventh one, [12] as the twelfth one, [13] as the thirteenth one, [14] as the fourteenth one, [15] as the fifteenth one, [16] as the sixteenth one, [17] as the seventeenth one, [18] as the eighteenth one, [19] as the nineteenth one and the present book as the twentieth one. Both P. T. L and P. N. Butcher visited the Institute of Radio Physics and Electronics of the University of Calcutta, my ALMA MATER where I started my research as M.Tech. student and later on as a faculty member. I formed permanent covalent bonds with J. S. Blakemore, K. Seeger, P. N. Butcher and P. T. L through letters (Pre email era) and these four first-class semiconductor physicists, in turn, shared with pleasure their vast creative knowledge of Semiconductors and related sciences with a novice like me.

I offer special thanks to Late N. Guhachoudhury of Jadavpur University for instilling in me the thought that the *academic output = ((desire X determination X dedication) – (false enhanced self-ego pretending like a true friend although a real unrecognizable foe))*, although a thank you falls in the band gap regime for my beloved better half See who really formed the backbone of my long unperturbed research career, since in accordance with "Sanatan" Hindu Dharma, the fusion of marriage has transformed us to form a single entity, where the individuality is being lost. I am grateful to all the members of my research group (from 1983 till date) for not only quantum confining me in the infinitely deep quantum wells of *Ramanujan and Rabindranath* but also for inspiring me to teach quantum mechanics and related topics from the classical volumes of Greiner *et al.*

In this context, I wish to offer my indebtedness to M. Mondal, the first member of my research team who in 1983 joined with me to complete his PhD work under my guidance when R. K. Poddar, the then Vice-chancellor of the University of Calcutta selected me as a Lecturer (presently Assistant Professor) of the famous Department of Radio Physics and Electronics. In **1987**, S. K. Sen, the then Vice-chancellor of Jadavpur University accepted me as the Reader (presently Associate Professor) in the Department of Electronics and Telecommunication Engineering and since then a flood of young researchers (more than 12 in numbers consisting of B. Mitra, A. Ghoshal, D. Bhattacharya, A. R. Ghatak, *et al.*) around me created my research team and insisted me to work with them at the @ 16 hours per day including holidays in different directions of research for the purpose of my creative conversion from an ordinary engineer to a 360° research scientist and consequently I enjoyed the high rate of research publications in different reputed International journals in various aspects of band structure-dependent properties of quantized structures. It is nice to note that the said young talented researchers obtained their respective PhD degree under my direct supervision. Incidentally in **1994**, R. N. Basu, the then Vice-chancellor of the University of Calcutta selected me as a Professor in the Department of Electronic Science and another flood of research over helmed me in a new direction of materials science. The persons responsible for this change include S. Datta, S. Sengupta, A. Ali, *et al.* The 11th and 12th names of this golden series are S. Bhattacharya and D. De respectively who, in turn formed permanent covalent bonds with me not only with respect to research (S. Bhattacharya and D. De are, respectively, the co-authors of seven and two books in different series of Springer) but also in all aspects of life in general.

It is curious to note that after serving 18 years as a Professor in the Department of Electronic Science, in 2012, P. K. Bose, and the then Director of the National Institute of Technology, Agartala requested me to join as a Professor and Departmental Head in Electronics and Communication Engineering. Being my life-long friend, I accepted his offer (and later on as a DEAN) and more than ten young scholars around me again directed my research in an altogether new direction. In 2015 the respected Director Professor S. Chakrabarti of the Institute of Engineering and Management in Salt lake City Kolkata kindly offered me the position of Research Director and Senior Professor in his famous Institute in the academic fag end of my life to complete my last run towards the creative knowledge temple with my new young research workers and I am grateful to him for his creative gesture. In my 50+ years of academic life (I have the wide experience of teaching quantum mechanics (from non-relativistic quantum mechanics up to string theory), Engineering Mathematics, Applied Mechanics (from engineering statics up to nonlinear mechanics including indeterminate structures) and 50% of the courses in Electronic Engineering) and 50+ years of research life (mostly in Materials Science and Quantized Structures), *I have finally realized that the teaching without research is body without brain and research without teaching is body without blood although my all-time hero, creatively prolific number theorist Godfrey Harold Hardy in his classic book entitled "A Mathematician's Apology" (Cambridge University Press, 1990) tells us "I hate teaching".*

Incidentally, one young theoretician friend of mine often tells me that many works in theoretical semiconductor science are based on the following seven principles:

1. Principles of placing the necessary and sufficient conditions of a proof in the band gap regime.
2. Principles of interchange of the summation and the integration processes and unconditioned convergences of the series and the integrals.
3. Principles of random applications of one electron theory and superposition theorem in studying the properties of materials, although the many-body effects are very important together with the fact that the nature is fundamentally nonlinear in nature.
4. Principles of using the invariant band structure concept of materials even in the presence of strong external fields (magnetic, electric, heavy

doping, etc.) and the random applications of perturbation theory, which is in a sense quantum mechanical Taylor series without considering the related consequences.
5. Principle of random applications of the binomial theorem without considering the important concept of branch cut.
6. Principle of little discussion regarding the whole set of materials science comprising of different compounds having various band structures under different physical conditions as compared with the simplified two-band model of Kane for III–V materials.
7. Principle of using Fermi's golden rule, the band structure, and the related features, which are valid for non-degenerate materials to materials having degenerate carrier concentrations directly.

Although my young friend is a purist in his conjecture, there are no doubt certain elements of truth inside his beautiful comments. I can only hope that our readers will present their intricate and advanced theories after paying due weightage to the aforementioned seven principles of my friend.

I am grateful to Prof. Dr. S. Chakrabarti the Director of the Institute of Engineering & Management for his encouragement and enthusiastic support, his keen academic insight and the creation of congenial atmosphere to promote research and higher learning among the faculty members. I must express my gratitude to Prof. Dr. B. Chatterjee, presently Vice-Chancellor of the University of Engineering & Management (UEM), Jaipur for offering important suggestions for the condensed presentation of this book. I am extremely grateful to Prof. R. Paul, Director IQAC and Deputy Dean Research of UEM, Kolkata for really helping me in front of academic turbulence. I offer special thanks to P. K. Das, M. Ghosh and other members of my research team for placing their combined effort towards the development of this book in the DO-LOOP of a computer and critically reading the manuscript in its last phase before sending it to *Mrs. Nur Syarfeena Mohd Fauzi, Physics Editor, WSPC.* I am grateful to M. Ghosh, my personal assistant, for typing the manuscript of this book three times very diligently. I am also grateful to D. Bhattacharjee for the overall supervision of this book. Last but not least, I am grateful for ever to our life long time tested friend S. Sanyal, Ex-Principal, Lake School for Girls, Kolkata for not only motivating me at rather turbulent moments of my academic carrier but also instilling in me the concept that *the ratio of total accumulated knowledge in my chosen field of research to my own contribution in my topic of research tends to infinity at any time and is the definition of a non-removable essential pole in the transfer function of my life.*

Acknowledgment by Jayita Pal

It is a great pleasure to express my gratitude to Professor K.P. Ghatak for instigating me to carry out extensive research in the field of nano science and nano technology. My family members especially my husband and son deserve a very special mention for their constant support, cooperation and also for forming the backbone of my long research carrier. Lastly, I wish to offer special thanks to my parents for their motivation and guidance in the midst of all difficulties.

Joint Acknowledgements by Kamakhya Prasad Ghatak and Jayita Pal

We are grateful to Mrs. Nur Syarfeena Binte Mohd Fauzi, Physics Editor, WSPC, who single-handedly carried out this project starting from the very inception of the duly filled, in book proposal form for this book. She arranged for referring of the book proposal by the referees for critical analysis. In spite of her 360 degrees heavy commitment and other academic duties, she helped us from every point of view. Besides, she extended our last date of submission to 3rd April 2023 and we are grateful to her for this very act of kindness. Without her solid help, this book would be a mere dream and we can only hope that every author should get an Editor like Syarfeena. Incidentally, at the fag end of this project she took a new carrier path and from then Mr. Rhaimie Wahap, editor WSPC completed her unfinished project with extreme care and we thank him in the real sense of the term. We offer special thanks to Professor K. K. Phua (D.Sc), Chairman and Editor-in-Chief, WSPC and Ms. Lakshmi Narayanan, Deputy Director (Physics) at WSPC for further necessary action in this context. The authors are responsible for non-imaginative shortcomings. *We firmly believe that our mother nature has propelled this project in her own unseen way in spite of several insurmountable obstacles.*

References

1. K. P. Ghatak, S. Bhattacharya and D. De, *Einstein Relation in Compound Semiconductors and their Heterostructures*, Springer Series in Materials Science, Vol. 116 (Springer, 2009).
2. K. P. Ghatak, D. De and S. Bhattacharya, *Photoemission from Optoelectronic Materials and Their Nanostructures*, Springer Series in Nanostructure Science and Technology (Springer, 2009).
3. K. P. Ghatak and S. Bhattacharya, *Thermo Electric Power In Nano structured Materials Strong Magnetic Fields*, Springer Series in Materials Science, Vol. 137 (Springer, 2010).

4. S. Bhattacharya and K. P. Ghatak, *Fowler-Nordheim Field Emission: Effects in semiconductor nanostructures*, Springer Series in Solid state Sciences, Vol. 170 (Springer, 2012).
5. K. P. Ghatak and S. Bhattacharaya (Eds), *Bismuth: Characteristics, Production and Applications*, Nova Series in Materials Science and Technologies (NOVA Science Publishers, USA, 2012).
6. D. De, S. Bhattacharya and K. P. Ghatak (Eds), *Quantum Dots and Quantum Cellular Automata*, Series in Physics Research and Technology (NOVA Science Publishers, USA, 2013).
7. S. Bhattacharya and K. P. Ghatak, *Effective Electron Mass in Low Dimensional Semiconductors*, Springer Series in Materials Sciences, Vol. 167 (Springer, 2013).
8. K. P. Ghatak and S. Bhattacharya, *Debye Screening Length: Effects of Nanostructured Materials*; Springer Tracts in Modern Physics, Vol. 255 (Springer, 2014).
9. K. P. Ghatak and S. Bhattacharya, *Heavily Doped 2D Quantized Structures and the Einstein Relation*; Springer Tracts in Modern Physics, Vol. 260 (Springer, 2015).
10. K. P. Ghatak, *Einstein's Photo-emission: Emission from Heavily-Doped Quantized Structures*; Springer Tracts in Modern Physics, Vol. 262 (Springer, 2015).
11. K. P. Ghatak, *Dispersion Relations in Heavily-Doped Nanostructures*; Springer Tracts in Modern Physics, Vol. 265 (Springer, 2016).
12. K. P. Ghatak, *Magneto Thermoelectric Power in Heavily Doped Quantized Structures*; Series on the Foundations of Natural Science and Technology, Vol. 7 (World Scientific, 2016).
13. K. P. Ghatak, *Quantum Effects, Heavy Doping, and The Effective Mass*; Series on the Foundations of Natural Science and Technology, Vol. 8 (World Scientific, 2017).
14. K. P. Ghatak (ed.), *Quantum Wires: An Overview*, Series on the NanoTechnology Science and Technology (NOVA Science Publishers, USA, 2020).
15. K. P. Ghatak and M. Mitra, *Electronic Properties, Series in Nanomaterials*, Vol. 1 (De Guyter, Germany, 2020).
16. K. P. Ghatak and M. Mitra, *Quantization and Entropy*, Series in Nanomaterials, Vol. 2 (De Guyter, Germany, 2020).
17. K. P. Ghatak and M. Mitra, *Elastic Constants in Heavily Doped Low Dimensional Materials*, Series on the Foundations of Natural Science and Technology, Vol. 14 (World Scientific, 2021).
18. K. P. Ghatak, M. Mitra and A. Biswas, *Heisenberg's Uncertainty Principle and the Electron Statistics in Quantized Structures* (Springer, Germany, 2022).
19. K. P. Ghatak and A. Biswas, *Electron Statistics in Quantum Confined Superlattices*, Series on the Foundations of Natural Science and Technology, Vol. 16 (World Scientific, 2023).

About the Authors

Professor Kamakhya Prasad Ghatak (h-index-37, i-10 index-182, maximum citation of a research paper = 325 & total citations-5757) obtained his D. Phil (Tech) Degree (on the basis of 27 published research papers in eminent SCI journals which is still a record in the Institute of Radio Physics and Electronics) of the University of Calcutta, India, in 1988.He is the first recipient of the Degree of Doctor of Engineering of Jadavpur University, Kolkata in 1991 since the University inception in 1955 and in the same year he received the INSA visiting fellowship to IIT-Kharagpur. In accordance with the analysis of World Ranking of top 2% Scientists as prepared by Stanford University in 2020, USA, (shorturl.at/qHIJ4), Professor K. P. Ghatak of IEM UEM group stays within less than 1% band width in the field of Applied Physics. Prof. Ghatak is the principal co-author of more than 200 research papers in SCI Journals and 385 research publications in SCOPUS respectively. From the position of Assistant Professor in the Department of Radio physics and Electronics of the University of Calcutta in 1983 up to the position of senior professor and research Director in nano technology center of excellence of the University of Engineering and Management, Kolkata, he was at the top of the Merit lists in all the above cases. Prof. Ghatak is the Principal author of 20 research monographs of high selling cross-sections on various topics of Nano Technology from Springer-Verlag Germany, De Gryuter, World Scientific and NOVA respectively. The score in Vidwan portal unit, of Central Government of India, of Prof. K. P. Ghatak is 8.9 out of 10, the highest score among the nano-technologists of private engineering colleges and Universities in West Bengal. Prof. Ghatak has successfully supervised more than 100 PhD students in different topics of Nanotechnology and the list includes Director, Vice-Chancellor, Principal, Professor and CEO of different reputed Institutes, Universities, Colleges and private organizations respectively.

The AICTE has selected the first R&D project in his life for the best project award in Electronics and second best research project award considering all the branches of Engineering for the year 2006. In 2011, the UGC awarded a research project to him and placed him at the top in the list of awardees. He is the first recipient of the academic excellence award of the University of Engineering & Management, Kolkata from the then Governor of West Bengal in its first convocation in 2017. He is the referee, Editor and Guest Editor of different reputed International Journals. He is the main referee of International and Intercontinental experimental research projects of very high value between Austria and France. His present research interests are Nano Materials and Genetic Electronics respectively. Incidentally, from 2015-2017, in accordance with Google Scholar, the number of research publications of Prof. Ghatak in consecutive three years have exceeded one hundred. His brief CV has been enlisted in many biographical references of USA and UK. For more details please visit https://www.amazon.com/Kamakhya-Prasad-Ghatak/e/B003B09OEY%3Fref=dbs_a_mng_rwt_scns_share

Jayita Pal is an Assistant Professor in the Department of Physics of Meghnad Saha Institute of Technology, Kolkata. She obtained her Ph.D. degree in 2010 and is the author of 20 research papers and chapters in eminent journals and edited monographs published by Springer. She has 20 years of teaching experience and her present teaching and research interests include quantum mechanics and quantized structures, respectively. She is a prominent member in the research group of Prof. Ghatak.

Contents

Chapter 1

The Quantum Capacitance (QC) in 2D Quantum Metal-Oxide-Semiconductor-Field Effect Transistors (QMOSFET) of Nonlinear Optical, Tetragonal, III–V and Optoelectronic Materials

The collapse of education is the real disaster

1.1 Introduction

The importance of studying the Quantum Capacitance (QC) (C_G) has been described in detail in the preface. In this chapter, we investigate the QC in quantum metal-oxidi-semiconductor-field effect transistors (QMOSFETs) operated under inversion mode made of tetragonal and nonlinear optical materials. For tetragonal materials, we shall take Cd_3As_2 as an example. As Dirac semimetal Cd_3As_2, exhibits very high mobility and large thermoelectric power, we might think of its possible thermoelectric applications. The values of thermoelectric figure of merit in Cd_3As_2 can be enhanced further to reach the practical limit for the application by increasing the temperature and sample modification (doping, nanostructuring, inclusion of nanocomposite, etc.). Besides Cd_3As_2 finds extensive applications in magnetic random access memory and magnetic sensors. We shall also study the QC in QMOSFETs made of nonlinear optical compounds operated under inversion mode (taking $CdGeAs_2$ as an example) which are being used in nonlinear optics and light emitting diodes [1–28]. The quasi-cubic model can be used to investigate the symmetric properties of both the bands at the zone center of wave vector space of the same compound. Including the anisotropic crystal potential in the Hamiltonian and special features of the nonlinear optical compounds, Kildal [29] formulated the electron dispersion law under the assumptions of isotropic momentum matrix element and the isotropic spin-orbit splitting constant, respectively, although the anisotropies in the two aforementioned band constants are the significant physical features of the said materials [30–32]. $CdGeAs_2$ is known for having one of the highest nonlinear optical coefficients of any inorganic compound (236 pm/V), making it attractive for CO_2-doubling as well as other mid- to far-IR laser frequency conversion applications. It finds applications in the family of ternary pnictides and the third harmonic of CO_2-laser radiation has been obtained by passing the laser beam through two $CdGeAs_2$ crystals in tandem, with the first crystal oriented for second-harmonic generation and the second for sum mixing. In Section 1.2.1, the QC in QMOSFETs operated under inversion mode made of tetragonal compounds and nonlinear optical materials has been investigated both under weak and strong electric field limits on the basis of newly formulated DR of the said compounds by considering the combined influence of the anisotropies of the aforementioned energy band constants together with the inclusion of the crystal field splitting, respectively within the framework of $\vec{k}.\vec{p}$ formalism.

In Section 1.2.2, we shall study the QC in QMOSFETs operated under inversion mode made of III–V, ternary and quaternary compounds for both the weak and strong electric field limits as a special case of our generalized analysis as presented in the same section. It is important to note that the III–V compounds find applications in infrared detectors [33], quantum dot light emitting diodes [34], quantum cascade lasers [35], NWs [36], optoelectronic sensors [37], high electron mobility transistors [38], etc. The DR of III–V materials can be described by the three- and two-band models of Kane [39, 40]. In this context, it may be noted that the ternary and quaternary compounds enjoy the singular position in the entire spectrum of optoelectronic materials. The ternary alloy $Hg_{1-x}Cd_xTe$ is a classic narrow gap compound. The band gap of this ternary alloy can be varied to cover the spectral range from 0.8 to over 30 μm [41] by adjusting the alloy composition. $Hg_{1-x}Cd_xTe$ finds extensive applications in infrared detector materials and photovoltaic detector arrays in the 8–12 μm wave bands [42]. The above uses have generated the $Hg_{1-x}Cd_xTe$ technology for the experimental realization of high-mobility single crystals with specially prepared surfaces. The same compound has emerged to be the optimum choice for illuminating the narrow sub-band physics because the relevant material constants can easily be experimentally measured [43]. Besides, the quaternary alloy $In_{1-x}Ga_xAs_yP_{1-y}$ lattice matched to InP, also finds wide use in the fabrication of avalanche photo-detectors [44], hetero-junction lasers [45], light emitting diodes [46] and avalanche photodiodes [47], field effect transistors, detectors, switches, modulators, solar cells, filters, and new types of integrated optical devices are made from the quaternary systems [48].

It is well-known that the concept of EEM in quantum materials is one of the basic pillars in the realm of modern nanomaterials [49–123]. Among the various definitions of the effective mass (e.g. effective acceleration mass, density-of-state effective mass, concentration effective mass, conductivity effective mass, Faraday rotation effective mass, etc.) [93], it is the effective momentum mass that should be regarded as the basic quantity [94], since it is this mass which appears in the description of transport phenomena and all other properties of the carriers in materials having arbitrary band structures [94] *together with the fact that only the basic definition of mass generates Einstein's* $E= mc^2$ *where E, m and c are the energy, mass and velocity of light respectively, for photons and under the condition of stationary frame of reference* [96]. It can be shown that it is the effective momentum mass that enters various transport coefficients and plays the most dominant role in explaining the experimental results of different scattering mechanisms through Boltzmann's transport equation [96]. The carrier degeneracy in materials influences the effective mass when it is energy dependent. Under degenerate conditions, only the electrons at the Fermi surface of n-type materials participate in the conduction process and hence, the effective momentum mass of the electrons corresponding to the Fermi level would be of interest in electron transport under such conditions. The nature of variations of the EEM has widely been studied in the literature [109–122] and in this chapter, we shall study the EEM in inversion layers of different materials as stated in the content of Chapter 1.

It is worth remarking that Einstein relation (ER) for the diffusivity–mobility ratio of the carriers in semiconductors occupies a central position in the whole field of semiconductor science and technology [124] since the diffusion constant (a quantity very useful for device analysis where the exact experimental determination is rather difficult) can be obtained from this ratio by knowing the experimental values of the mobility. The classical value of the ER is equal to $(k_BT/|e|)$, (where k_B, T, and $|e|$ are Boltzmann's constant, temperature and the magnitude of the carrier charge, respectively). This relation in this form was first introduced to the study of the diffusion of gas particles and is known as the ER [125, 126].

Therefore, it appears that the ER increases linearly with increasing T and is independent of electron concentration. This relation holds for both types of charge carriers only under non-degenerate carrier concentration although its validity has been suggested erroneously for degenerate materials [127]. Landsberg first pointed out that the ER for semiconductors having degenerate electron concentration is essentially determined by their energy band structures [128–133]. This relation is useful for semiconductor homo-structures [127, 131], semiconductor–semiconductor hetero-structures [132, 133], metals–semiconductor hetero-structures [134–142] and insulator–semiconductor hetero-structures [143–149]. We shall also investigate the ER in this context whose *experimental determination* has been studied here. It is worth remarking that the second- and third-order elastic constants (ΔC_{44} and ΔC_{456}) are very important mechanical properties which have been described in this context. Section 1.3, contains a few related applications. Section 1.4 contains the result and discussions of this chapter. Section 1.5 contains few open research problems.

1.2 Theoretical Background

1.2.1 *The QC in QMOSFETs of nonlinear optical and tetragonal materials*

In MOS structure, the QC is formed by the series combination of a fixed capacitance C_{OX} due to the oxide layer and a variable capacitance C_s (surface capacitance) due to the space charge layer which can be controlled by the gate voltage.

The relation among these capacitances is given by

$$\frac{1}{C_G} = \frac{1}{C_s} + \frac{1}{C_{OX}} \tag{1.1}$$

where $C_{OX} = \varepsilon_{OX}/d_{OX}, \varepsilon_{OX}$ and d_{OX} being the permittivity and the thickness of the oxide layer, respectively.

The C_s can also be expressed as

$$C_s = e\frac{dN_s}{dV_0} = e\frac{dN_s}{dV_g}\left[\left\{1 - e\frac{d_{OX}}{\varepsilon_{OX}}\frac{dN_s}{dV_g}\right\}\right]^{-1} \tag{1.2}$$

where e is the electron charge, N_S is defined as the surface concentration per unit area, V_g is the gate voltage and V_0 defined as the surface potential and can be written as

$$V_0 = V_g - eN_s\frac{d_{OX}}{\varepsilon_{OX}} \tag{1.3}$$

Using (1.1), (1.2) and (1.3) the QC per unit area can be written as

$$C_G = e\frac{dN_s}{dV_g} \tag{1.4}$$

It appears, therefore, that we should express the surface concentration as a function of the surface electric field in order to study the QC. This in turn would require DRs for two-dimensional electron motion as appropriate in this case.

The form of **k. p** matrix for tetragonal and nonlinear optical compounds can be expressed extending Bodnar [30] as

$$H = \begin{bmatrix} H_1 & H_2 \\ H_2^+ & H_1 \end{bmatrix} \tag{1.5}$$

where

$$H_1 = \begin{bmatrix} E_g & 0 & P_{||}k_z & 0 \\ 0 & (-2\Delta_{||}/3) & (\sqrt{2}\Delta_\perp/3) & 0 \\ P_{||}k_z & (\sqrt{2}\Delta_\perp/3) & -\left(\delta + \frac{1}{3}\Delta_{||}\right) & 0 \\ 0 & 0 & 0 & 0 \end{bmatrix},$$

$$H = \begin{bmatrix} 0 & -f_{,+} & 0 & f_{,-} \\ f_{,+} & 0 & 0 & 0 \\ 0 & 0 & 0 & 0 \\ f_{,+} & 0 & 0 & 0 \end{bmatrix}$$

in which E_g is the band gap in the absence of any field, $P_{||}$ and $P_\perp$ are the momentum matrix elements parallel and perpendicular to the direction of crystal axis, respectively, δ is the crystal field splitting constant, $\Delta_{||}$ and $\Delta_\perp$ are the spin-orbit splitting constants parallel and perpendicular to the C-axis, respectively, $f_{,\pm} \equiv (P_\perp/\sqrt{2})(k_x \pm ik_y)$ and $i = \sqrt{-1}$. Thus, neglecting the contribution of the higher bands and the free electron term, the diagonalization of the above matrix leads to the DR of the conduction electrons in bulk specimens of nonlinear optical and tetragonal materials as

$$\gamma(\bar{E}) = f_1(\bar{E})k_s^2 + f_2(\bar{E})k_z^2 \tag{1.6}$$

where

$$\gamma(\bar{E}) = \bar{E}(\bar{E} + E_{g_0})\left[(\bar{E} + E_{g_0})(\bar{E} + E_{g_0} + \Delta_{||}) + \delta\left(\bar{E} + E_{g_0} + \frac{2}{3}\Delta_{||}\right) + \frac{2}{3}(\Delta_{||}^2 - \Delta_\perp^2)\right]$$

$\bar{E}$ is the total energy of the electron as measured from the edge of the conduction band in the vertically upward direction in the absence of any quantization and the symbols $\bar{E}$ and E are being used interchangeably, E_{g_0} is the band gap in the absence of any field, δ is the crystal field splitting constant, $\Delta_{||}$ and $\Delta_{\perp}$ are the spin-orbit splitting constants parallel and perpendicular to the C-axis, respectively, $k_s^2 = k_x^2 + k_y^2$,

$$f_1(\bar{E}) = \frac{\hbar^2 E_{g_0}(E_{g_0} + \Delta_\perp)}{\left[2m_\perp^*\left(E_{g_0} + \frac{2}{3}\Delta_\perp\right)\right]}\left[\delta\left(\bar{E} + E_{g_0} + \frac{1}{3}\Delta_{||}\right) + (\bar{E} + E_{g_0}) \times \left(\bar{E} + E_{g_0} + \frac{2}{3}\Delta_{||}\right) + \frac{1}{9}(\Delta_{||}^2 - \Delta_\perp^2)\right],$$

$$f_2(\bar{E}) = \frac{\hbar^2 E_{g_0}(E_{g_0} + \Delta_{||})}{\left[2m_{||}^*\left(E_{g_0} + \frac{2}{3}\Delta_{||}\right)\right]}\left[(\bar{E} + E_{g_0})\left(\bar{E} + E_{g_0} + \frac{2}{3}\Delta_{||}\right)\right],$$

$\hbar$ is the Dirac constant, $m_{||}^*$ and $m_\perp^*$ are the longitudinal and transverse effective electron masses at the edge of the conduction band, respectively.

The quantization rule for inversion layers is given by [149, 150]

$$\int_0^{z_t} k_z dz = \frac{2}{3}(S_i)^{3/2} \tag{1.7}$$

where z_t is the classical turning point and S_i is the zeros of the Airy function $(Ai(-S_i) = 0)$.

(a) Weak Electric Field Limit

In the presence of a surface electric field F_s along z direction and perpendicular to the surface, (1.6) assumes the form

$$\psi_1(E - |e|F_s z) = \psi_2(E - |e|F_s z)k_s^2 + \psi_3(E - |e|F_s z)k_z^2 \tag{1.8}$$

where $\psi_1(E) \equiv \gamma(E), \psi_2(E) \equiv f_1(E)$ and $\psi_3(E) \equiv f_2(E)$,

Using (1.7) and (1.8), under the weak electric field limit $(\frac{eF_s z}{E} \ll 1)$, we can write

$$\int_0^{\bar{z}_t} (\sqrt{A_7(E) - |e|F_s z D_7(E)})dz = \frac{2}{3}(S_i)^{3/2} \tag{1.9}$$

in which

$$z_t \equiv \frac{A_7(E)}{[|e|F_s D_7(E)]}, \quad A_7(E) \equiv \left[\frac{\psi_1(E) - \psi_2(E)k_s^2}{\psi_3(E)}\right],$$

$$D_7(E) \equiv [B_7(E) - A_7(E)C_7(E)],$$

$$B_7(E) \equiv \left[\frac{(\psi_1(E))' - (\psi_2(E))'k_s^2}{\psi_3(E)}\right], \quad C_7(E) \equiv \left[\frac{(\psi_3(E))'}{\psi_3(E)}\right]$$

and the primes indicate the differentiation of the differentiable functions with respect to E.

Thus, the 2D electron dispersion law in QMOSFETs operated under inversion mode and made of nonlinear optical and tetragonal materials under the weak electric field limit can approximately be written as

$$\psi_1(E) = P_7(E,i)k_s^2 + Q_7(E,i) \tag{1.10}$$

where

$$P_7(E,i) \equiv [\psi_2(E) - \left(\frac{2t_2(E)}{3[t_2(E)]^{1/3}}\right)\psi_3(E)\bar{S}_i(|\bar{e}|\bar{F}_s)^{2/3}],$$

$$t_2(E) \equiv \left[\frac{[\psi_2(E)]'}{\psi_3(E)} - \left(\frac{\psi_2(E)[\psi_3(E)]'}{[\psi_3(E)]^2}\right)\right]$$

$$t_1(E) \equiv \left[\frac{[\psi_1(E)]'}{\psi_3(E)} - \left(\frac{\psi_1(E)[\psi_3(E)]'}{[\psi_3(E)]^2}\right)\right],$$

and

$$Q_7(E,i) \equiv S_i\psi_3(E)[|e|F_s t_1(E)]^{2/3}$$

The general expression of the 2D total DOS function in this case is given by

$$N_{2D_i}(E) = \frac{2g_v}{(2\pi)^2}\sum_{i=0}^{i_{\max}}\frac{\partial}{\partial E}[A(E,i)H(E-E_{n_i})] \tag{1.11}$$

where g_v is the valley degeneracy, $A(E,i)$ is the area of the constant energy 2D wave vector space for inversion layers, $H(E-E_{n_i})$ is the Heaviside step function and E_{n_i} is the corresponding sub-band energy.

Using (1.11) and (1.10), the total 2D DOS function under the weak electric field limit can be expressed as

$$N_{2D_i}(E) = \frac{g_v}{(2\pi)}\sum_{i=0}^{i_{\max}}[G_7(E,i)H(E-E_{n_{iw1}})] \tag{1.12}$$

where $G_7(E,i) \equiv [P_7(E,i)]^{-2}[P_7(E,i)\{(\psi_1(E))' - (Q_7(E,i))'\} - \{\psi_1(E) - (Q_7(E,i))\}(P_7(E,i))']$ and sub-band energy $(E_{n_{iw1}})$ in this case can be obtained from (1.10) as

$$\psi_1(E_{n_{iw1}}) = Q_7(E_{n_{iw1}}, i) \tag{1.13}$$

Using (1.12) and the Fermi–Dirac occupation probability factor, the 2D surface electron concentration in this case under the weak electric field limit (n_{2Dw}) can be written as

$$n_{2Dw} = g_v(2\pi)^{-1}\sum_{i=0}^{i_{\max}}[P_{7w}(E_{Fiw},i) + Q_{7w}(E_{Fiw},i)] \tag{1.14}$$

where $P_{7w}(E_{Fiw},i) \equiv [\psi_1(E_{Fiw},i) - Q_7(E_{Fiw},i)]\{P_7(E_{Fiw},i)\}^{-1}, E_{Fiw}$ is the Fermi energy under the weak electric field limit as measured from the edge of the conduction band at the surface in the vertically upward direction and is given by, $E_{Fiw} = (eV_g - \frac{e^2 n_{2Dw}}{C_{OX}} - E_{fb})$, C_{OX} is the capacitance per unit area of the oxide layer and E_{fb} is the energy separation between the Fermi level and the edge of the conduction band in the bulk of the substrate material,

$$Q_{7w}(E_{Fiw},i) \equiv \sum_{r=1}^{s}\{L(r)[P_7(E_{Fiw},i)]\},$$

$$L(r) = [2(k_BT)^{2r}(1-2^{1-2r})\xi(2r)]\left[\frac{\partial^{2r}}{\partial E_F^{2r}}\right],$$

where E_F represents the Fermi energy in general, $F_s \equiv \frac{|e|n_{2Dw}}{\varepsilon_{sc}}$ and ε_{sc} is the semiconductor permittivity, k_B is Boltzmann constant, T is temperature and $\xi(2r)$ is the Zeta function of order 2r.

The expression of 2D surface electron concentration per unit area under the condition of extreme carrier degeneracy can be written as

$$n_s = \frac{g_v}{2\pi}\sum_{\bar{i}=0}^{\bar{i}_{\max}}\left[\frac{\gamma(E_{Fiw}) - Q(E_{Fiw},i)}{P(E_{Fiw},i)}\right] \tag{1.15}$$

Therefore, the QC per unit area of the said materials at the low electric field limit can be expressed as

$$C_G = \left[e^2\sum_{i=0}^{i_{\max}} R_7(E_{Fiw},i)\right]\left[\frac{2\pi}{\bar{g}_v} + \sum_{i=0}^{i_{\max}}\left(\bar{R}_8(E_{Fiw},i)\right.\right.$$
$$\left.\left.+\bar{R}_7(E_{Fiw},i)\frac{e^2 d_{ox}}{\epsilon_{ox}}\right)\right]^{-1} \tag{1.16}$$

where

$$R_7(E_{Fiw},i) = \bar{R}_3(E_{Fiw},i) - \bar{R}_5(E_{Fiw},i),$$
$$\bar{R}_3(E_{Fiw},i) = [\bar{t}_1(E_{Fiw}) - \bar{R}_2(E_{Fiw},i)(n_{2Dw})^{2/3}]^{-2}[\{\bar{t}_1(E_{Fiw})$$
$$- \bar{R}_2(E_{Fiw},i)\bar{n}_s^{2/3}\}\{\gamma'(E_{Fiw}) - \bar{R}'_1(E_{Fiw},i)(n_{2Dw})^{2/3}\}]$$

$$\bar{R}_2(E_{Fiw}, i) = \frac{2}{3}\frac{t_2(E_{Fiw})}{t_1^{1/3}(E_{Fiw})} t_2(E_{Fiw}) S_i \left(\frac{e^2}{\in_{sc}}\right)^{2/3},$$

$$t_2(E_{Fiw}) = \left[\frac{t_1'(E_{Fiw})}{f_2(E_{Fiw})} - \frac{t_1(E_{Fiw})t_2'(E_{Fiw})}{f_2^2(E_{Fiw})}\right],$$

$$t_1'(E_{Fiw}) = [\hbar^2 E_{g_0}(E_{g_0} + \Delta_\perp)] \left[2m_\perp^* \left(E_{g_0} + \frac{2}{3}\Delta_\perp\right)\right]^{-1} \times \left[2E_{Fiw} + 2E_{g_0} + \frac{2}{3}\Delta_{||} + \delta\right]$$

$$t_2'(E_{Fiw}) = [\hbar^2 E_{g_0}(E_{g_0} + \Delta_{||})] \left[2m_{||}^* \left(E_{g_0} + \frac{2}{3}\Delta_{||}\right)\right]^{-1} \times \left[2E_{Fiw} + 2E_{g_0} + \frac{2}{3}\Delta_{||}\right]$$

$$t_1(E_{Fiw}) = \left[\frac{\gamma'(E_{Fiw})}{f_2(E_{Fiw})} - \frac{\gamma(E_{Fiw})f_2'(E_{Fiw})}{f_2^2(E_{Fiw})}\right],$$

$$\gamma(E_{Fiw}) = (4E_{Fiw}^3 + 3L_1 E_{Fiw}^2 + 2L_2 E_{Fiw} + L_3),$$

$$L_1 = [3E_{g_0} + \Delta_{||} + \delta], \quad L_2 = \left[3E_{g_0}^2 + 2E_{g_0}\Delta_{||} + 2E_{g_0}\Delta_{||} + \frac{2}{3}\Delta_{||}\delta + \frac{2}{9}(\Delta_{||} - \Delta_\perp)\right]$$

$$L_3 = \left[E_{g_0}^3 + E_{g_0}^2\Delta_{||} + \delta E_{g_0}^2 + \frac{2}{3}\Delta_{||}\delta E_{g_0} + \frac{2}{9}(\Delta_{||}^2 - \Delta_\perp^2)E_{g_0}\right]$$

$$R_1'(E_{Fiw}, i) = S_i \left(\frac{e^2}{\in_{sc}}\right)^{2/3} \left[f_2'(E_{Fiw}) \times t_1(E_{Fiw})^{2/3} + \frac{2}{3}\{t_1(E_{Fiw})\}^{-1/3} t_1'(E_{Fiw}) f_2(E_{Fiw})\right]$$

$$t_1'(E_{Fiw}) = \left[\frac{\gamma''(E_{Fiw})}{f_2(E_{Fiw})} - \frac{2\gamma'(E_{Fiw})f_2'(E_{Fiw})}{f_2^2(E_{Fiw})} - \frac{\gamma(E_{Fiw})f_2''(E_{Fiw})}{f_2^2(E_{Fiw})} + 2\frac{\gamma(E_{Fiw})[f_2'(E_{Fiw})]^2}{f_2^3(E_{Fiw})}\right]$$

$$\gamma''(E_{Fiw}) = (12E_{Fiw}^2 + 6L_1 E_{Fiw} + 2L_2),$$

$$t_2''(E_{Fiw}) = 2[\hbar^2 E_{g_0}(E_{g_0} + \Delta_{||})]\left[2m_{||}^*\left(E_{g_0} + \frac{2}{3}\Delta_{||}\right)^{-1}\right]$$

$$\begin{aligned} R_5(E_{Fiw}, i)) &= [t_1(E_{Fiw}) - R_2(E_{Fiw}, i)(n_{2Dw})^{2/3}]^{-2}[\{\gamma(E_{Fiw}) \\ &\quad - R_1(E_{Fiw}, i)n_s^{2/3}\}\{t_1'(E_{Fiw}) - R_2'(E_{Fiw}, i)(n_{2Dw})^{2/3}\}] \\ R_8(E_{Fiw}, i) &= R_4(E_{Fiw}, i) - R_6(E_{Fiw}, i), \\ R_4(E_{Fiw}, i) &= [t_1(E_{Fiw}) - R_2(E_{Fiw}, i)(n_{2Dw})^{2/3}]^{-2}\,[\{t_1(E_{Fiw}) \\ &\quad - R_2(E_{Fiw}, i)(n_{2Dw})^{2/3}\}\left\{R_1(E_{Fiw}, i)\frac{2}{3}(n_{2Dw})^{-1/3}\right\}] \\ R_1(E_{Fiw}, i) &= [S_i t_2(E_{Fiw})[t_1(E_{Fiw})]^{2/3}\left(\frac{e^2}{\in_{sc}}\right)^{2/3}], \\ R_6(E_{Fiw}, i) &= [t_1(E_{Fiw}) - R_2(E_{Fiw}, i)(n_{2Dw})^{2/3}]^{-2} \\ &\quad \times \left[R_2(E_{Fiw}, i)\frac{2}{3}(n_{2Dw})^{-1/3}\right] \end{aligned}$$

Incidentally, it may be noted that under electric quantum limit as in inversion layers, the ER can be written as

$$\frac{D}{\mu} = \left(\frac{\bar{n}_{2Dw}}{e}\right) \Big/ \left(\frac{d\bar{n}_{2Dw}}{d(\bar{E}_{Fw} - \bar{E}_{n0w1})}\right) \tag{1.17}$$

where $\bar{n}_{2Dw}$ is the surface electron concentration under weak electric field quantum limit at low temperatures, $\bar{E}_{Fw}$ is the Fermi energy under weak electric field quantum limit as measured from the edge of the conduction band at the surface and $\bar{E}_{n0w1}$ is the sub-band energy for the weak electric field quantum limit, respectively.

The surface electron concentration under the weak electric field quantum limit assumes the form

$$\bar{n}_{2Dw} = g_v(2\pi)^{-1}[P_{7w}(\bar{E}_{Fw}, 0)] \tag{1.18}$$

where $\bar{E}_{Fw}$ is the Fermi energy under the weak electric field quantum limit as measured from the edge of the conduction band at the surface.

Using (1.17) and (1.18), the ER in this case can be expressed as

$$\frac{D}{\mu} = \frac{\bar{n}_{2Dw}}{|e|}(x_{71}(\bar{E}_{Fw}, 0) - y_{71}(\bar{E}_{Fw}, 0)) \tag{1.19}$$

where

$$x_{71}(\bar{E}_{Fw}, 0) \equiv \frac{\omega_{71}(\bar{E}_{Fw}, 0)}{\omega_{72}(\bar{E}_{Fw}, 0)},$$

$$\omega_{71}(\bar{E}_{Fw}, 0) \equiv \left[1 - \left(\frac{(\bar{n}_{2Dw})^{-1/3}}{3P_7(\bar{E}_{Fw}, 0)}\right)\left\{2\bar{n}_{2Dw}V_{72}(\bar{E}_{Fw}) - \frac{g_v}{\pi}V_{71}(\bar{E}_{Fw})\right\}\right],$$

$$V_{72}(\bar{E}_{Fw}) \equiv \frac{2}{3}\left[\frac{t_2(\bar{E}_{Fw})}{[t_1(\bar{E}_{Fw})]^{1/3}}\right]\psi_3(\bar{E}_{Fw})S_0\left(\frac{|e|^2}{\varepsilon_{sc}}\right)^{2/3},$$

$$V_{71}(\bar{E}_{Fw}) \equiv \psi_3(\bar{E}_{Fw})S_0\left(\frac{|e|^2}{\varepsilon_{sc}}t_1(\bar{E}_{Fw})\right)^{2/3},$$

$$y_{71}(\bar{E}_{0w1}, 0) \equiv \left[\frac{2}{3}V_{71}(\bar{E}_{01w})(\bar{n}_{2Dw})^{-1/3}\right][\{\psi_1(\bar{E}_{01w})\}' - \{V_{71}(\bar{E}_{01w})\}'(\bar{n}_{2Dw})^{2/3}]^{-1}$$

$\bar{E}_{0w1}$ is the electron sub-band energy and can be determined from the equation

$$\psi_1(\bar{E}_{0w1}) = V_{71}(\bar{E}_{0w1})(n_{2Dw})^{2/3}$$

The EEM in the $x - y$ plane can be expressed as

$$m^*(E_{Fiw}, i) = \left(\frac{\hbar^2}{2}\right)G_7(E, i)\Big|_{E=E_{Fiw}} \tag{1.20}$$

Thus, we observe that the EEM is the function of sub-band index, the Fermi energy and other band constants due to the combined influence of the crystal filed splitting constant and the anisotropic spin-orbit splitting constants, respectively.

The second- and third-order elastic constants (denoted by ΔC_{44} and ΔC_{456}, respectively) can be expressed as

$$\Delta C_{44} = \left(\frac{-G_0^2}{9}\right)\left(\frac{dn_{2Dw}}{d(E_{Fiw} - E_{niw1})}\right) \tag{1.21}$$

$$\Delta C_{456} = \left(\frac{G_0^3}{27}\right)\left(\frac{d^2n_{2Dw}}{d(E_{Fiw} - E_{niw1})^2}\right) \tag{1.22}$$

where G_0 is the deformation potential constant.

Using (1.21), (1.22) and (1.14) we can study by ΔC_{44} and ΔC_{456} in this case.

(b) Strong Electric Field Limit

The 2D DR of the carriers in QMOSFETs operated under inversion mode and made of tetragonal and nonlinear optical and materials under the strong electric field limit can approximately be written as

$$k_s^2 = P_2(E,i) \tag{1.23}$$

where

$$P_2(E,i) \equiv [F_7(E)]^{-1}[F_6(E) - F_8(E,i)],$$

$$F_6(E) \equiv \left[\frac{\gamma(E)}{t_2(E)}\right]\left[1 + \frac{\gamma'(E)\bar{t}'_2(E)}{t_2(E)\gamma''(E)} + \frac{t_2''(E)\gamma(E)}{2\gamma''(E)\bar{t}_2(E)}\right],$$

$$F_7(E) \equiv \left[\frac{t_1(E)}{\bar{t}_2(E)} - \frac{\gamma(E)t_1''(E)}{2t_2(E)\gamma''(E)} + \frac{t_1(E)}{t_2(E)}\left[\frac{\gamma'(E)t_2'(E)}{t_2(E)\gamma''(E)} + \frac{t_2''(E)\gamma(E)}{2\gamma''(E)\bar{t}_2(E)}\right]\right.$$
$$\left. + \frac{\gamma(E)}{t_2(E)}\left[\frac{t_1'(E)t_2'(E)}{\gamma''(E)\bar{t}_2(E)} + \frac{t_1(E)t_2''(E)}{2\gamma''(E)t_2(E)}\right]\right]$$

and

$$F_8(E,i) \equiv \left[\frac{2\sqrt{2}}{3}(S_i)^{3/2}(|e|\bar{F}_s)\sqrt{[\gamma(E)]''}\right]$$

The sub-band energy ($E_{n_{is1}}$) in this case can be obtained from (1.23) as

$$P_2(E_{n_{is1}}, i) = 0 \tag{1.24}$$

The total 2D DOS function under the strong electric field limit can be written as

$$N_{2D_i}(E) = \frac{g_v}{(2\pi)}\sum_{i=0}^{i_{\max}}[(P_2(E,i))H(E - E_{n_{is1}}, i)] \tag{1.25}$$

Using (1.25) and the Fermi–Dirac occupation probability factor, the 2D surface electron concentration (n_{2Ds}) in this case can be expressed as

$$n_{2Ds} = g_v(2\pi)^{-1}\sum_{i=0}^{i_{\max}}[P_2(E_{Fis}, i) + Q_2(E_{Fis}, i)] \tag{1.26}$$

where $Q_2(E_{Fis}, i) \equiv \sum_{r=1}^{s}\{L(r)[P_2(E_{Fis}, i)]\}$

The surface electron concentration under the condition of very low temperatures where the quantum effects become prominent, can be expressed as

$$n_{2Ds} = \frac{g_v}{2\pi}\sum_{\bar{i}=0}^{\bar{i}_{\max}}\left[\frac{R_9(E_{Fis})}{R_{10}(E_{Fis}, i)n_{2Ds}}\right] \tag{1.27}$$

where

$$R_9(E_{Fis}) = F_6(E_{Fis}) - F_7(E_{Fis}) \quad \text{and}$$

$$R_{10}(E_{Fis}, i) = \frac{2\sqrt{2}}{3}\left(\frac{e^2}{\in_{sc}}\right) S_i^{3/2}[\gamma''(E_{Fis})]^{1/2}$$

Therefore, the QC of the said materials at the high electric field limit can be expressed as

$$C_G = \frac{e^2 \sum_{i=0}^{i_{\max}} R_{11}(E_{Fis}, i)}{\left[\frac{2\pi}{g_v} + \frac{e^2 d_{ox}}{\in_{ox}} \sum_{i=0}^{i_{\max}} R_{11}(E_{Fis}, i) + \frac{R_9(E_{Fis})}{R_{10}(E_{Fis}, i)(n_{2Ds})^2}\right]} \tag{1.28}$$

where

$$R_{11}(E_{Fis}, i) = \left[(E_{Fis}) - \frac{R_9(E_{Fis})R'_{10}(E_{Fis})}{R_{10}^2(E_{Fis}, i)(n_{2Ds})^2}\right]$$

Thus, the surface electron concentration under the strong electric quantum limit assumes the form

$$\bar{n}_{2Ds} = g_v(2\pi)^{-1}[P_2(\bar{E}_{Fis}, 0)] \tag{1.29}$$

where $\bar{E}_{Fs}$ is the Fermi energy in the n-channel inversion layer under the strong electric field quantum limit as measured from the edge of the conduction band at the surface.

Using (1.29) and (1.17), the ER under the strong electric field quantum limit, in this case, can be expressed as

$$\frac{D}{\mu} = \frac{\bar{n}_{2Ds}}{|e|}(x_{72}(\bar{E}_{Fis}, 0) - y_{72}(\bar{E}_{0s1}, 0)) \tag{1.30}$$

where

$$x_{72}(\bar{E}_{Fs}, 0) = \frac{\omega_{73}(\bar{E}_{Fs}, 0)}{\omega_{74}(\bar{E}_{Fs}, 0)}, \quad \omega_{73}(\bar{E}_{Fs}, 0) \equiv \left[\frac{2\pi}{g_v} + \beta_{71}(\bar{E}_{Fs})\right],$$

$$\beta_{71}(\bar{E}_{Fs}) \equiv \left[\frac{2\sqrt{2}}{3}\left(\frac{|e|^2}{\varepsilon_{sc}}\right)(S_0)^{3/2}\frac{\sqrt{\{\psi_1(\bar{E}_{Fs})\}''}}{F_7(\bar{E}_{Fs})}\right]$$

$$\omega_{74}(\bar{E}_{Fs}) \equiv \left[\frac{\{F_6(\bar{E}_{Fs})\}'}{F_7(\bar{E}_{Fs})} - \left[\frac{\{F_6(\bar{E}_{Fs})\}}{\{F_7(\bar{E}_{Fs})\}^2}\right][F_7(\bar{E}_{Fs})]' - \{[\beta_{71}(\bar{E}_{Fs})]'\bar{n}_{2Ds}\}\right]$$

$$y_{72}(\bar{E}_{0s1}, 0) \equiv \frac{\beta_{72}(\bar{E}_{0s1})}{[\{F_6(\bar{E}_{0s1})\}' - \{[\beta_{72}(\bar{E}_{0s1})]'\bar{n}_{2Ds}\}]} \text{ and}$$

$$\beta_{72}(\bar{E}_{0s1}) \equiv \frac{2\sqrt{2}}{3}\left(\frac{|e|^2}{\varepsilon_{sc}}\right)(S_0)^{3/2}\sqrt{\{\psi_1(\bar{E}_{0s1})\}''}$$

$\bar{E}_{0s1}$ is the sub-band energy under the strong electric field quantum limit in this case and is given by the equation $F_6(\bar{E}_{0s1}) = \beta_{72}(\bar{E}_{0s1})\bar{n}_{2Ds}$

The EEM in the $x - y$ plane can be written in this case as

$$m^*(E_{Fis}, i) = \left(\frac{\hbar^2}{2}\right)[P_2(E, i)]'\bigg|_{E=E_{Fis}} \tag{1.31}$$

where E_{Fis} is the Fermi energy under the strong electric field limit as measured from the edge of the conduction band at the surface. Thus, we note that the EEM is the function of sub-band index and the Fermi energy due to the combined influence of the crystal filed splitting constant and the anisotropic spin-orbit splitting constants, respectively.

1.2.2 *The QC in QMOSFETs of III–V, ternary and quaternary materials*

(**A**) Under the conditions $\delta = 0, \Delta_{||} = \Delta_{\perp} = \Delta_0$ (the isotropic spin-orbit splitting), $m^*_{||} = m^*_{\perp} = m_c$ (the isotropic effective electron mass at the edge of the conduction band), the generalized dispersion law as given by (1.6) gets reduced into the form as

$$\frac{\hbar^2 k^2}{2m_c} = I_{11}(\bar{E}), \quad I_{11}(\bar{E}) \equiv \frac{\bar{E}(\bar{E} + E_{g_0})(\bar{E} + E_{g_0} + \Delta_0)\left(E_{g_0} + \frac{2}{3}\Delta_0\right)}{E_{g_0}(E_{g_0} + \Delta_0)\left(\bar{E} + E_{g_0} + \frac{2}{3}\Delta_0\right)} \tag{1.32}$$

which is the well-known three band model of Kane [39,40]. (1.32) is the DR of the conduction electrons of III–V, ternary and quaternary materials and should be used as such for studying the electron transport in n-InAs where the spin-orbit splitting constant is of the order of band gap.

(a) Weak Electric Field Limit

Under the condition of weak electric field limit ($\frac{eF_s z}{E} \ll 1$), the 2D electron dispersion law assumes the form

$$I_{11}(E) = \frac{\hbar^2 k_s^2}{2m_c} + S_i\left[\frac{\hbar\,|e|\,F_s[I_{11}(E)]'}{\sqrt{2m_c}}\right]^{2/3} \tag{1.33}$$

(1.33) represents the DR of the 2D electrons in QMOSFETs of III–V, ternary and quaternary materials operated under inversion layer mode in the presence of the weak electric field limit whose bulk electrons obey the three-band model of Kane.

The sub-band energy ($E_{n_{iw2}}$) in this case can be obtained from (1.33) as

$$I(E_{n_{iw2}}) = S_i \left[\frac{\hbar\,|e|\,F_s[I(E_{n_{iw2}})]'}{\sqrt{2m_c}} \right]^{2/3} \tag{1.34}$$

Using the appropriate equations the 2D total DOS function in weak electric field limit can be expressed as

$$N_{2D_i}(E) = \frac{m_c g_v}{\pi\hbar^2} \sum_{i=0}^{i_{\max}} [P_3(E,i) H(E - E_{n_{iw2}})] \tag{1.35}$$

where

$$P_3(E,i) \equiv \left\{ [I(E)]' - \left\{ \frac{2}{3} S_i \left[\frac{\hbar\,|e|\,F_s}{\sqrt{2m^*}} \right]^{2/3} \{[I(E)]'\}^{-1/3} [I(E)]'' \right\} \right\}$$

Using (1.35) and the occupation probability, the n_{2Dw} in the present case can be written as

$$n_{2Dw} = \frac{g_v m_c}{\pi\hbar^2} \sum_{i=0}^{i_{\max}} [P_{4w}(E_{Fiw}, i) + Q_{4w}(E_{Fiw}, i)] \tag{1.36}$$

where

$$P_{4w}(E_{Fiw}, i) \equiv \left\{ I(E_{Fiw}) - S_i \left[\frac{\hbar e F_s [I(E_{Fiw})]'}{\sqrt{2m_c}} \right]^{2/3} \right\}$$

and

$$Q_4(E_{Fiw}, i) \equiv \sum_{r=1}^{s} \{L(r)[P_4(E_{Fiw}, i)]\}$$

The surface electron concentration under the condition of very low temperatures and extreme degeneracy where the quantum effects become prominent can be expressed as

$$n_{2Dw} = \frac{m_c g_v}{\pi\hbar^2} \sum_{i=0}^{i_{\max}} \left[I_{11}(E_{Fiw}) - S_i \left[\frac{\hbar\,|e|\,F_s [I_{11}(E_{Fiw})]'}{\sqrt{2m_c}} \right]^{2/3} \right] \tag{1.37}$$

The QC in QMOSFETs under low electric field limit and in the presence of extreme carrier degeneracy at low temperatures where the quantum effects become prominent for materials whose bulk electrons obey the three-band model of Kane can be written as

$$C_G = e^2 \left[\sum_{\bar{i}=0}^{\bar{i}_{\max}} \left[I'_{11}(E_{Fiw}) - \frac{2}{3} S_i \left(\frac{\hbar e F_s}{\sqrt{2m_c}} \right)^{2/3} \right.\right.$$

$$\left.\left. \times [I'_{11}(E_{Fiw})]^{-1/3} [I''_{11}(E_{Fiw})] \right] \left[\frac{\pi\hbar^2}{m_c} + \sum_{\bar{i}=0}^{\bar{i}_{\max}} \left[I'_{11}(E_{Fiw}) \frac{e^2 d_{ox}}{\varepsilon_{ox}} \right.\right.\right.$$

$$+ \frac{2}{3} \frac{S_i}{n_{2Dw}} \left(\frac{\hbar e^2 n_{2Dw} I'_{11}(E_{Fiw})}{\varepsilon_{sc}\sqrt{2m_c}} \right)^{2/3}$$

$$\left.\left.\left. - \frac{e^2 d_{ox}}{\varepsilon_{ox}} \frac{2 S_i I'_{11}(E_{Fiw})}{3 I'_{11}(E_{Fiw})} \left(\frac{\hbar e F_s I''_{11}(E_{Fiw})}{\sqrt{2m_c}} \right)^{2/3} \right]\right]^{-1} \right] \tag{1.38}$$

The electron concentration under the weak electric field quantum limit assumes the form

$$\bar{n}_{2Dw} = \frac{g_v m_c}{\pi\hbar^2} [P_{4w}(\bar{E}_{Fw}, 0)] \tag{1.39}$$

Using (1.39) and (1.17), the ER under the electric field quantum limit, in this case, can be expressed as

$$\frac{D}{\mu} = \frac{\bar{n}_{2Dw}}{|e|} (x_{73}(\bar{E}_{Fw}, 0) - y_{73}(\bar{E}_{0w2}, 0)) \tag{1.40}$$

where

$$x_{73}(\bar{E}_{Fw}, 0) \equiv \left[\frac{\left(\frac{\pi\hbar^2}{m_c g_v} \right) + \frac{2}{3} (\bar{n}_{2Dw})^{-1/3} V_{73}(\bar{E}_{Fw})}{\{I(\bar{E}_{Fw})\}' - \{V_{73}(\bar{E}_{Fw})\}' (\bar{n}_{2Dw})^{2/3}} \right],$$

$$V_{73}(\bar{E}_{Fw}) \equiv S_0 \left[\frac{\hbar |e|^2}{\varepsilon_{sc}\sqrt{2m_c}} \{I(\bar{E}_{Fw})\}' \right]^{2/3},$$

$$y_{73}(\bar{E}_{0w2}, 0) \equiv \frac{\frac{2}{3} V_{73}(\bar{E}_{0w2}, 0)(\bar{n}_{2Dw})^{-1/3}}{[\{I(\bar{E}_{0w2})\}' - \{V_{73}(\bar{E}_{0w2}, 0)\}' (\bar{n}_{2Dw})^{2/3}]}$$

$\bar{E}_{0w2}$ is the sub-band energy in QMOSFETs of III–V, ternary and quaternary materials operated under inversion layer mode under the weak electric field quantum limit and is determined by the equation

$$I(\bar{E}_{0w2}) = V_{73}(\bar{E}_{0w2})(\bar{n}_{2Dw})^{2/3}.$$

The EEM can be expressed as

$$m^*(E_{Fiw}, i) = m_c \left[P_3(E, i)\right]\big|_{E=E_{Fiw}} \tag{1.41}$$

Thus, one can observe that the EEM is a function of the sub-band index, surface electric field, the Fermi energy and the other spectrum constants due to the combined influence of E_g and Δ.

Using (1.21), (1.22) and (1.37) we can study by ΔC_{44} and ΔC_{456} in this case

(B) Using the constraints $\Delta \gg E_g$ or $\Delta \ll E_g$, the (1.32) and (1.33) assume the forms

$$\bar{E}(1 + \alpha\bar{E}) = \frac{\hbar^2 k^2}{2m_c} \tag{1.42a}$$

$$E(1 + \alpha E) = \frac{\hbar^2 k_s^2}{2m_c} + S_i \left[\frac{\hbar|e|F_s(1 + 2\alpha E)}{\sqrt{2m_c}}\right]^{2/3} \tag{1.42b}$$

It may be noted that (1.42a) *is the well-known two-band model of Kane and is used in the literature to study the physical properties of those III–V and optoelectronic materials whose energy band structures obey the aforementioned inequalities.*

For large values of $i, S_i \to [\frac{3\pi}{2}(i + \frac{3}{4})]^{2/3}$ [150], and the (1.42b) gets simplified as

$$E(1 + \alpha E) = \frac{\hbar^2 k_s^2}{2m_c} + \left[\frac{3\pi\hbar\,|e|\,F_s}{2}\left(i + \frac{3}{4}\right)\frac{(1 + 2\alpha E)}{\sqrt{2m_c}}\right]^{2/3} \tag{1.43}$$

(1.43) was derived for the first time by Antcliffe *et al.* [3].

The sub-band energies ($E_{n_{iw3}}$) are given by

$$E_{n_{iw3}}(1 + \alpha E_{n_{iw3}}) = S_i \left[\frac{\hbar\,|e|\,F_s(1 + 2\alpha E_{n_{iw3}})}{\sqrt{2m_c}}\right]^{2/3} \tag{1.44}$$

The total 2D density-of-states function can be written as

$$N_{2D}(E) = \frac{m_c g_v}{\pi\hbar^2}\sum_{i=0}^{i_{\max}}\left\{\left[1 + 2\alpha E - \frac{4\alpha}{3}S_i\left[\frac{\hbar\,|e|\,F_s}{\sqrt{2m_c}}\right]^{2/3}(1 + 2\alpha E)^{-1/3}\right] \times H(E - E_{n_{iw3}})\right\} \tag{1.45}$$

Under the condition $\alpha E \ll 1$, the use of (1.45) and the Fermi–Dirac integral leads to the expression of n_{2Dw} as

$$n_{2Dw} = \left(\frac{g_v m_c k_B T}{\pi\hbar^2}\right)\sum_{i=0}^{i_{\max}} \{[1 + D_i + 2\alpha E_{n_{iw3}}]F_0(\eta_{iw}) + 2\alpha k_B T F_1(\eta_{iw})\} \tag{1.46}$$

where

$$D_i \equiv \frac{4\alpha S_i}{3}\left(\frac{\hbar\,|e|\,F_s}{\sqrt{2m_c}}\right)^{2/3}, \quad \eta_{iw} \equiv \left[\frac{E_{Fiw} - E_{n_{iw3}}}{k_B T}\right],$$

$F_j(\eta)$ is the one parameter Fermi–Dirac integral of order j which can be written as [60]

$$F_j(\eta) = \left(\frac{1}{\Gamma(j+1)}\right)\int_0^\infty y^j(1+\exp(y-\eta)^{-1}\,dy, \quad j > -1 \tag{1.47}$$

where $\Gamma(j+1)$ is the complete Gamma function or for all j, analytically continued as a complex integral around the negative axis

$$F_j(\eta) = A_j\int_{-\infty}^{(0+)} y^j(1+\exp(-y-\eta))^{-1}\,dy, \tag{1.48}$$

in which $A_j \equiv \frac{\Gamma(-j)}{2\pi\sqrt{-1}}$

For all values of αE_{Fiw}, the n_{2Dw} can be written by using (1.46) as

$$n_{2Dw} = \left(\frac{g_v m_c}{\pi\hbar^2}\right)\sum_{i=0}^{i_{\max}}[P_{5w}(E_{Fiw}, i) + Q_{5w}(E_{Fiw}, i)] \tag{1.49}$$

where

$$P_{5w}(E_{Fiw}, i) \equiv \left[E_{Fiw}(1+\alpha E_{Fiw}) - S_i\left[\frac{\hbar|e|F_s}{\sqrt{2m_c}}(1+2\alpha E_{Fiw})\right]^{2/3}\right] \text{ and}$$

$$Q_{5w}(E_{Fiw}, i) \equiv \sum_{r=1}^{s} L(r)P_{5w}(E_{Fiw}, i)$$

The n_{2Dw} under the condition $T \to 0$ can be expressed as

$$n_{2Dw} = \left(\frac{g_v m_c}{\pi\hbar^2}\right)\sum_{i=0}^{i_{\max}}[P_{5w}(E_{Fiw}, i)] \tag{1.50}$$

Using (1.4) and (1.50), the QC under the low electric field limit for materials whose bulk electrons obey the two-band model of Kane can, respectively be expressed as

$$C_G = e^2 \left[\sum_{i=0}^{i_{\max}} \left[(1+2\alpha E_F) - \frac{2}{3} S_i \left(\frac{\hbar e F_s}{\sqrt{2m_c}} \right)^{2/3} \times [(1+2\alpha E_F)]^{-1/3} [(2\alpha)] \right] \right.$$
$$\times \left[\frac{\pi\hbar^2}{m_c} + \sum_{i=0}^{i_{\max}} \left[(1+2\alpha E_F) \frac{e^2 d_{ox}}{\varepsilon_{ox}} + \frac{2}{3}\frac{S_i}{n_s} \left(\frac{\hbar e^2 \bar{n}_s (1+2\alpha E_F)}{\varepsilon_{sc}\sqrt{2m_c}} \right)^{2/3} \right. \right.$$
$$\left. \left. \left. - \frac{e^2 d_{ox}}{\varepsilon_{ox}} \frac{2S_i(1+2\alpha E_F)}{3 I_{11}'(E_F)} \left(\frac{\hbar e F_s (2\alpha)}{\sqrt{2m_c}} \right)^{2/3} \right] \right]^{-1} \right] \tag{1.51}$$

where $\alpha = \frac{1}{E_{g0}}$

The electron concentration under the weak electric field quantum limit assumes the form

$$\bar{n}_{2Dw} = \frac{g_v m_c}{\pi\hbar^2} \left\{ \bar{E}_{Fw}(1+\alpha \bar{E}_{Fw}) - S_0 \left(\frac{\hbar |e| F_s}{\sqrt{2m_c}} \right)^{2/3} (1+2\alpha \bar{E}_{Fw})^{2/3} \right\} \tag{1.52}$$

Using (1.52) and (1.17), the ER in the present case can be expressed as

$$\frac{D}{\mu} = \frac{\bar{n}_{2Dw}}{|e|} (x_{75}(\bar{E}_{0w3}, 0) - y_{75}(\bar{E}_{0w3}, 0)) \tag{1.53a}$$

where

$$x_{75}(\bar{E}_{Fw}, 0) \equiv \frac{\omega_{75}(\bar{E}_{Fw}, 0)}{\omega_{76}(\bar{E}_{Fw}, 0)},$$

$$\omega_{75}(\bar{E}_{Fw}, 0) \equiv \left[\frac{\pi\hbar^2}{m^* g_v} + \frac{2}{3} S_0 (\bar{n}_{2Dw})^{-1/3} \left(\frac{\hbar |e|^2}{\varepsilon_{sc}\sqrt{2m_c}} (1+2\alpha \bar{E}_{Fw}) \right)^{2/3} \right],$$

$$\omega_{76}(\bar{E}_{Fw}, 0) \equiv \left[1 + 2\alpha \bar{E}_{Fw} - \left\{ \frac{4 S_0 \alpha}{3} (1+2\alpha \bar{E}_{Fw})^{-1/3} \right\} \right.$$
$$\left. \times \left(\frac{\bar{n}_{2Dw} \hbar |e|^2}{\varepsilon_{sc}\sqrt{2m_c}} \right)^{2/3} \right],$$

$$y_{75}(\bar{E}_{0w3}, 0) \equiv \left[\frac{2}{3} \frac{\bar{E}_{0w3}(1+\alpha \bar{E}_{0w3})}{\bar{n}_{2Dw}} \right] \left[1 + 2\alpha \bar{E}_{0w3} \right.$$
$$\left. - \left\{ \frac{4\alpha \bar{E}_{0w3}(1+\alpha \bar{E}_{0w3})}{3(1+2\alpha \bar{E}_{0w3})} \right\} \right]^{-1}$$

$\bar{E}_{0w3}$ is the sub-band energy in the present case and is determined by (1.43)

$$\bar{E}_{0w3}(1+\alpha\bar{E}_{0w3}) - S_0\left(\frac{\hbar\,|e|\,F_s}{\sqrt{2m_c}}\right)^{2/3}(1+2\alpha\bar{E}_{0w3})^{2/3} = 0 \qquad (1.53b)$$

The EEM in this case is given by

$$m^*(E_{Fiw}, i) = m_c\left[P_6(E,i)\right]|_{E=E_{Fiw}} \qquad (1.54)$$

where

$$P_6(E,i) \equiv \left\{1+2\alpha E - \frac{4\alpha}{3}S_i\left[\frac{\hbar|e|F_s}{\sqrt{2m^*}}\right]^{2/3}\{1+2\alpha E\}^{-1/3}\right\}$$

Thus, one can observe that the EEM is a function of the sub-band index, surface electric field and the Fermi energy due to the presence of band non-parabolicity only.

(C) For $\alpha \to 0$, as for inversion layers, whose bulk electrons are defined by the parabolic energy bands, from (1.42b), we can write,

$$E = \frac{\hbar^2 k_s^2}{2m_c} + S_i\left[\frac{\hbar\,|e|\,F_s}{\sqrt{2m_c}}\right]^{2/3} \qquad (1.55)$$

(1.55) is valid for all values of the surface electric field [1].

The electric sub-band energy ($E_{n_{i4}}$) assumes the form, from (1.55) as

$$E_{n_{i4}} = S_i\left[\frac{\hbar|e|F_s}{\sqrt{2m_c}}\right]^{2/3} \qquad (1.56)$$

The total DOS function can be written using Equation (1.45) as

$$N_{2D}(E) = \frac{m_c g_v}{\pi\hbar^2}\sum_{i=0}^{i_{\max}} H(E - E_{n_{i4}}) \qquad (1.57)$$

The use of (1.57) leads to the expression of n_{2Di} as [1]

$$n_{2Di} = \frac{g_v m_c k_B T}{\pi\hbar^2}\sum_{i=0}^{i_{\max}} F_0(\eta_i) \qquad (1.58)$$

where $\eta_i \equiv (k_B T)^{-1}\left[E_{Fi} - S_i\left[\frac{\hbar|e|F_s}{\sqrt{2m_c}}\right]^{2/3}\right]$,

E_{Fi} is the Fermi energy as measured from the edge of the conduction band at the surface.

Under the condition $T \to 0$, the 2D electron statistics assumes the form

$$n_s = \frac{m_c g_v}{\pi\hbar^2}\sum_{i=0}^{i_{\max}}\left[E_{Fi} - S_i\left[\frac{\hbar|e|F_s}{\sqrt{2m_c}}\right]^{2/3}\right] \tag{1.59}$$

The QC can be written using (1.4) and (1.59) as

$$C_G = e^2\left[\sum_{i=0}^{i_{\max}}\left[1 - \left[\frac{\pi\hbar^2}{m_c} + \sum_{i=0}^{i_{\max}}\left[\frac{e^2 d_{ox}}{\varepsilon_{ox}} + \frac{2}{3}\frac{S_i}{n_s}\left(\frac{\hbar e^2 n_s}{\varepsilon_{sc}\sqrt{2m_c}}\right)^{2/3}\right]\right]^{-1}\right]\right] \tag{1.60}$$

The ER in this case for $i = 0$ and $T \neq 0$ assumes the form

$$\frac{D}{\mu} = \frac{k_B T}{|e|}\left(\frac{F_0(\eta_0)}{F_{-1}(\eta_0)}\right) \tag{1.61}$$

where

$$\eta_0 \equiv \frac{(\bar{E}_{F0} - \bar{E}_0)}{k_B T}$$

$\bar{E}_{F0}$ and $\bar{E}_0$ are the Fermi energy and also the electric sub-band energy under the electric field quantum limit, respectively.

At low temperatures, (1.61) gets simplified as

$$\frac{D}{\mu} = \frac{1}{|e|}(\bar{E}_{F0} - \bar{E}_0) \tag{1.62}$$

(b) Strong Electric Field Limit

Using the substitutions $\delta = 0, \Delta_{||} = \Delta_{\perp} = \Delta$ and $m^*_{||} = m^*_{\perp} = m_c$, the 2D $E - k_s$ relation under the condition of strong electric field limit can be written from (1.54) as

$$\left[I_{11}(E) - \left\{\frac{|e|F_s\hbar}{\sqrt{2m_c}}\left(\frac{2\sqrt{2}(S_i)^{3/2}}{3}\right)\sqrt{[I_{11}(E)]''}\right\}\right] = \frac{\hbar^2 k_s^2}{2m_c} \tag{1.63}$$

(1.63) represents the dispersion relation of the 2D electrons in QMOSFETs made of III–V, ternary and quaternary materials operated in the presence of inversion layer mode and under the strong electric field limit whose bulk conduction electrons are defined by the three-band model of Kane.

The sub-band energy ($E_{n_{is2}}$) in this case can be obtained from (1.63) as

$$I_{11}(E_{n_{is2}}) - \left\{\frac{|e|F_s\hbar}{\sqrt{2m_c}}\left(\frac{2\sqrt{2}(S_i)^{3/2}}{3}\right)\sqrt{[I_{11}(E_{n_{is2}})]''}\right\} = 0 \tag{1.64}$$

Using (1.63) and (1.11), the total 2D DOS function under the strong electric field limit can be expressed as

$$N_{2D_i}(E) = \frac{m_c g_v}{\pi\hbar^2}\sum_{i=0}^{i_{\max}}[P_5(E,i)H(E-E_{n_{is2}})] \tag{1.65}$$

Using (1.65) and the Fermi–Dirac occupation probability factor, the n_{2Ds} in the present case under the strong electric field can be written as

$$n_{2Ds} = \frac{g_v m_c}{\pi\hbar^2}\sum_{i=0}^{i_{\max}}[P_{6s}(E_{Fis},i)+Q_{6s}(E_{Fis},i)] \tag{1.66}$$

where

$$P_{6s}(E_{Fis},i) \equiv \left\{I_{11}(E_{Fis}) - \left[\frac{2\sqrt{2}}{3}(S_i)^{3/2}\frac{\hbar|e|F_s[I_{11}(E_{Fis})]''}{\sqrt{2m_c}}\right]\right\}$$

and

$$Q_6(E_{Fis},i) \equiv \sum_{r=1}^{s}\{L(r)[P_6(E_{Fis},i)]\}$$

The surface electron concentration, under the condition of very low temperatures where the quantum effects become prominent, can be expressed as

$$n_s = \frac{m_c g_v}{\pi\hbar^2}\sum_{i=0}^{i_{\max}}\left[I_{11}(E_{Fis}) - \frac{2}{3}\frac{eF_s}{\sqrt{m_c}}S_i^{3/2}\hbar[I''_{11}(E_{Fis})]^{1/2}\right] \tag{1.67}$$

Using (1.4) and (1.67) the QC under the constraints of low electric field limit and $T \to 0$ where the quantum effects become prominent for QMOSFETs whose unperturbed bulk electrons obey the three-band model of Kane can be written as

$$\begin{aligned} C_G = e^2 &\left[\sum_{i=0}^{i_{\max}}\left[I'_{11}(E_{Fis}) - \frac{n_{2Ds}\hbar e^2 S_i^{3/2}}{3\varepsilon_{sc}\sqrt{m_c}}\frac{I'''_{11}(E_{Fis})}{\sqrt{I''_{11}(E_{Fis})}}\right]\right]\left[\left[\frac{\pi\hbar^2}{m_c}\right.\right. \\ &+ \sum_{i=0}^{i_{\max}}\frac{e^2 d_{ox}}{\varepsilon_{ox}}I'_{11}(E_{Fis}) + \frac{2}{3}\frac{e^2 S_i^{3/2}\hbar}{\varepsilon_{sc}\sqrt{\bar{m}_c}}(I''_{11}(E_{Fis}))^{1/2} \\ &\left.\left. - \frac{n_{2Ds}\hbar e^2 S_i^{3/2}}{3\varepsilon_{sc}\sqrt{\bar{m}_c}}, \frac{e^2 d_{ox}}{\varepsilon_{ox}}\left(\frac{I'''_{11}(E_{Fis})}{\sqrt{I''_{11}(E_{Fis})}}\right)\right]\right]^{-1} \end{aligned} \tag{1.68}$$

The electron concentration under the strong electric field quantum limit assumes the form

$$\bar{n}_{2Ds} = \frac{g_v m_c}{\pi\hbar^2}[P_{6s}(\bar{E}_{Fs},0)] \tag{1.69}$$

Using (1.69) and (1.17), the ER under the electric field quantum limit, in this case, can be expressed as

$$\frac{D}{\mu} = \frac{\bar{n}_{2Ds}}{|e|}(x_{74}(\bar{E}_{Fs},0) - y_{74}(\bar{E}_{0s2},0)) \tag{1.70}$$

where, $x_{74}(\bar{E}_{Fs},0) \equiv [\frac{(\frac{\pi\hbar^2}{m_c g_v})+V_{74}(\bar{E}_{Fs})}{\{I_{11}(\bar{E}_{Fs})\}'-\{V_{74}(\bar{E}_{Fs})\}'(\bar{n}_{2Ds})}]$, $V_{74}(\bar{E}_{Fs}) \equiv (\frac{|e|^2}{\varepsilon_{sc}\sqrt{2m_c}})(\frac{2\sqrt{2}}{3}S_0)\sqrt{\{I_{11}(\bar{E}_{Fs})\}''}$

$y_{74}(\bar{E}_{0s2},0) \equiv \frac{V_{74}(\bar{E}_{0s2})}{[\{I_{11}(\bar{E}_{0s2})\}'-\bar{n}_{2Ds}\{y_{72}(\bar{E}_{0s2})\}']}$

$\bar{E}_{0s2}$ is the sub-band energy in the present case and is determined by the equation

$$I_{11}(\bar{E}_{0s2}) - \left[\frac{2\sqrt{2}}{3}(S_i)^{3/2}\frac{\hbar|e|F_s[I_{11}(\bar{E}_{0s2})]''}{\sqrt{2m_c}}\right] = 0.$$

The EEM can be expressed as

$$m^*(E_{Fis},i) = m_c\,[P_5(E,i)]|_{E=E_{Fis}} \tag{1.71}$$

where

$$P_5(E,i) \equiv \left\{\{I_{11}(E)\}' - \left\{\frac{|e|\,F_s\hbar}{\sqrt{2m_c}}\left(\frac{\sqrt{2}(S_i)^{3/2}}{3}\right) \times ([I_{11}(E)]'')^{-1/2}[I_{11}(E)]'''\right\}\right\}$$

Thus, one can observe that the EEM is a function of the sub-band index, surface electric field, the Fermi energy and the other spectrum constants due to the combined influence of E_g and Δ.

Using the constraints $\Delta \gg E_g$ or $\Delta \ll E_g$, (1.63) under the strong electric field limit assumes the form

$$E(1+\alpha E) = \frac{\hbar^2 k_s^2}{2m_c} + \frac{2\sqrt{2}}{3}(S_i)^{3/2}\left[\frac{\hbar|e|F_s}{\sqrt{2m_c E_g}}\right] \tag{1.72}$$

For large values of i, $S_i \to \left[\frac{3\pi}{2}\left(i+\frac{3}{4}\right)\right]^{2/3}$ [150] and (1.72) gets simplified as

$$E(1+\alpha E) = \frac{\hbar^2 k_s^2}{2m_c} + \left[\frac{\pi\hbar\,|e|\,F_s\sqrt{2}}{\sqrt{m_c E_g}}\left(i+\frac{3}{4}\right)\right] \tag{1.73}$$

(1.73) was derived for the first time by Antcliffe *et al.* [3].

From (1.73), we observe that under the condition $E_g \to 0$, one cannot obtain the corresponding parabolic case, since under high electric field limit, the band becomes permanently non-parabolic.

The electric sub-band energy ($E_{n_{is5}}$) in the high electric field limit is given by

$$E_{n_{iw5}}(1+\alpha E_{n_{is5}}) = \left[\frac{\pi\hbar\,|e|\,F_s\sqrt{2}}{\sqrt{m_c E_g}}\left(i+\frac{3}{4}\right)\right] \tag{1.74}$$

The 2D total DOS function in this case can be written as

$$N_{2D_i}(E) = \frac{m_c g_v}{\pi\hbar^2}\sum_{i=0}^{i_{\max}}\{[1+2\alpha E]H(E-E_{n_{is5}})\} \tag{1.75}$$

The surface electron concentration for all values of αE_{Fis} in this case assumes the form

$$n_{2Dw} = \left(\frac{g_v m_c k_B T}{\pi\hbar^2}\right)\sum_{i=0}^{i_{\max}}\{[1+2\alpha k_B T]F_0(\eta_{is}) + 2\alpha k_B T E_{n_{is5}} F_1(\eta_{is})\} \tag{1.76}$$

where

$$\eta_{is} \equiv \left[\frac{E_{Fis}-E_{n_{is5}}}{k_B T}\right]$$

The QC under the high electric field limit at $T \to 0$ in this case can be expressed as

$$C_G = e^2\left[\sum_{i=0}^{i_{\max}}[E_{Fis}(1+\alpha E_{Fis})]\left[\frac{\pi\hbar^2}{m_c} + \sum_{i=0}^{i_{\max}}\frac{e^2 d_{ox}}{\varepsilon_{ox}}E_{Fis}(1+\alpha E_{Fis}) + \frac{2}{3}\frac{e^2 S_i^{3/2}\hbar}{\varepsilon_{sc}\sqrt{m_c}}(2\alpha)^{1/2}\right]\right]^{-1} \tag{1.77}$$

The surface electron concentration under high electric field quantum limit can be written as

$$\bar{n}_{2Ds} = \frac{g_v m_c}{\pi\hbar^2}\{\bar{E}_{Fs}(1+\alpha\bar{E}_{Fs}) - N_7\bar{n}_{2Ds}\} \tag{1.78a}$$

where

$$N_7 \equiv \frac{2\sqrt{2}}{3}(S_0)^{3/2}\left(\frac{\hbar|e|^2}{\varepsilon_{sc}\sqrt{m_c E_g}}\right)$$

Using (1.78a) and (1.17), the ER in this case can be expressed as

$$\frac{D}{\mu}=\frac{\bar{n}_{2Ds}}{|e|}\left(\frac{\pi\hbar^2}{m_c g_v(1+2\alpha\bar{E}_{Fs})}\right) \tag{1.78b}$$

where

$$t_{77}=\left[\left(\frac{1}{m_c g_v}+\frac{N_7}{\pi\hbar^2}\right)-\frac{N_7}{\pi\hbar^2}\frac{(1+2\alpha\bar{E}_{Fis})}{(1+2\alpha\bar{E}_{F0s})}\right]$$

in which

$$\bar{E}_{F0s}(1+\alpha\bar{E}_{F0s})=\left(\frac{\pi\hbar\,|e|^2\,\sqrt{2}\bar{n}_{2Ds}}{\varepsilon_{SC}\sqrt{m_c E_g}}\right)$$

The EEM is given by

$$m^*(E_{Fis},i)=m_c(1+2\alpha E_{Fis}) \tag{1.79}$$

Thus, in the high electric field limit, the EEM is a function of Fermi energy due to the presence of band non-parabolicity only and is independent of the sub-band index.

Using (1.18), (1.19) and (1.76) we can study by ΔC_{44} and ΔC_{456} in this case

1.3 Few Applications

1.3.1 *Introduction*

In this monograph, we investigate different QCs' in various QFETs made of different technologically important materials and its implications in low-dimensional electronics in general. The concept of QCs' is one of the main keys for investigating the transport properties of different QFETs. Incidentally, in the introduction and theoretical background, we have discussed a lot about the EEM, a cardinal transport property. Besides, in this section, we shall discuss *25 different applications* in this context.

1.3.2 *Thermoelectric power*

In recent years, with the advent of quantum Hall effect (QHE) [151–158], there has been considerable interest in studying the thermoelectric power under strong magnetic field (TPSM) in various types of nanostructured materials having quantum confinement of their charge carriers in one, two and three dimensions of the respective wave vector space by using different carrier energy spectra [159–163]. The classical TPSM equation

$G = \pi^2 k_B/3e$ (which is a function of three fundamental constants of nature) is valid only under the condition of carrier non-degeneracy, being independent of carrier concentration and reflects the fact that the signature of the band structure of any material is totally absent in the same.

Zawadzki [157] demonstrated that the TPSM for electronic materials having degenerate electron concentration is essentially determined by their respective energy band structures. It has, therefore, different values in different materials and changes with the doping, magnitude of the reciprocal quantizing magnetic field under magnetic quantization, quantizing electric field as in inversion layers, nanothickness as in quantum wells, wires and dots, with super-lattice period as in quantum confined semiconductor super-lattices with graded interfaces having various carrier energy spectra and also in other types of field-assisted nanostructured materials.

The magnitude of the thermoelectric power G can be written as [158]

$$G = \frac{1}{|e|\, T n_0} \int_{-\infty}^{\infty} (E - E_F) R(E) \left[-\frac{\partial f_0}{\partial E} \right] dE \tag{1.80}$$

where $R(E)$ is the total number of states.

The (1.80) can be written under the condition of carrier degeneracy [159]

$$G = \left(\frac{\pi^2 k_B^2 T}{3|e| n_0} \right) \left(\frac{\partial n_0}{\partial E_F} \right) \tag{1.81}$$

For inversion layers, (1.80) assumes the form

$$G = \frac{\pi^2 k_B^2 T}{3 e n_0} \left[\frac{\partial n_0}{\partial (E_F - E_{SB})} \right] \tag{1.82}$$

where E_F is the corresponding Fermi energy and E_{SB} is the corresponding sub-band energy. These two quantities are band structure dependent and we can use the carrier statistics for different low-dimensional materials to investigate the TPSM in such cases.

1.3.3 *Experimental determinations of ΔC_{44} and ΔC_{456}*

One important mechanical property of the materials is the carrier contribution to the elastic constants (CECs) which has been studied in the last 30 years mainly by the group of Ghatak *et al.* [164–181]. It has been shown that the carrier contribution of the second and third-order elastic constants depends on the (DOS) function [164]. It has therefore different values in various materials and varies with the electron concentration, with doping, with the thickness of ultra-thin films and with temperature for materials and their hetero-structures having various carrier energy spectra.

Using (1.21), (1.22) and (1.82), with the appropriate changes in this case) we get

$$\Delta C_{44} = \left[-n_0 (G_0)^2 \, |e| \, \frac{G}{3\pi^2 k_B^2 T} \right] \tag{1.83}$$

and

$$\Delta C_{456} = \left[n_0 |e| (G_0)^3 \frac{G^2}{3\pi^4 k_B^3 T} \right] \left(1 + \frac{n_0}{G} \frac{\partial G}{\partial n_0} \right) \tag{1.84}$$

Thus, the experimental graph of G versus n_0 allows us to determine ΔC_{44} and ΔC_{456} for materials having arbitrary DRs.

1.3.4 *Experimental determination of Debye screening length*

It is well known that the Debye screening length (DSL) of the carriers in materials is a very important quantity characterizing the screening of the Coulomb field of the ionized impurity centers by the free carriers [182–194]. It affects many of the special features of modern nanodevices, the carrier motilities under different mechanisms of scattering, and the carrier plasmas in materials [183]. The DSL is a very good approximation to the accurate self-consistent screening in the presence of band tails and is also used to illustrate the interaction between the colliding carriers in Auger effect in solids [182]. The classical value of the DSL is equal to $[\varepsilon_{sc} k_D T/(e^2 n_0)]^{1/2}$ where ε_{sc} is the semiconductor permittivity which is valid for both the carriers. In this conventional form, the DSL decreases with increasing carrier concentration at a constant temperature and this relation holds only under the condition of carrier non-degeneracy. It is interesting to note that the under the condition of extreme degeneracy, the expression of DSL for materials having parabolic energy bands can be written as

$$L_D = (\pi^{2/3} \hbar \sqrt{\varepsilon_{sc}})(e g_v^{1/3} 3^{1/6} n_0^{1/6} \sqrt{m_c})^{-1} \tag{1.85}$$

Thus, we observed that in this case the result is independent of temperature, but depends on n_0, g_v and m_c. Besides, the indices of inverse electron variation change from half in the former case to one-sixth in the latter case. Since, the performance of the electron devices at the device terminals and the speed of operation of modern switching transistors are significantly influenced by the degree of carrier degeneracy present in these devices, the simplest way of analyzing such devices taking into account of the degeneracy of the band is to use the appropriate DSL to express the performance at the device terminal and switching speed in terms of the carrier concentration [184].

The DSL depends on the DOS function which, in turn, is significantly affected by the different carrier energy spectra of different materials having various band structures [182, 184]. It is well known, that the DSL for electronic materials having degenerate electron concentration is essentially determined by their respective energy band structures. It has, therefore, different values in different materials and varies with the electron concentration, with the magnitude of the reciprocal quantizing magnetic field under magnetic quantization, with the quantizing electric field as in inversion layers, with the nanothickness as in quantum wells, with super-lattice period as in the quantum confined super-lattices of small gap compounds with graded interfaces having various carrier energy spectra.

The 3D DSL can, in general, be expressed as [184]

$$L_{3D} = \left[\frac{e^2}{\varepsilon_{sc}} \cdot \frac{\partial n_0}{\partial E_F}\right]^{-1/2} \tag{1.86}$$

where n_0 is the electron concentration per unit volume in bulk materials and E_F is the corresponding Fermi energy.

Thus, we can write

$$L_{3D} = \left[\frac{\pi^2 k_B^2 \varepsilon_{sc}}{3e^3 n_0 G}\right]^{1/2} \tag{1.87}$$

For inversion layers the 2D Debye screening length is given by

$$L_{2D} = \left[\frac{e^2}{2\varepsilon_{sc}} \frac{\partial \bar{n}_0}{\partial(\bar{E}_{F0} - \bar{E}_0)}\right]^{-1} \tag{1.88}$$

where $\bar{n}_0$, $\bar{E}_{F0}$ and $\bar{E}_0$ are the 2D electron concentration, Fermi energy and the sub-band energy in the electric quantum limit, respectively.

Using (1.82) and (1.88) we get,

$$L_{2D} = \frac{2\pi^2 k_B^2 T \varepsilon_{sc}}{3e^3 \bar{n}_0 G} \tag{1.89}$$

From the suggestion for the experimental determination of the 3D DSL for degenerate materials having arbitrary dispersion laws as given by (1.87), we observe that for a constant T, the DSL varies inversely with the square root of $(Gn_0)^{-1}$. Only the experimental values of G for any material as a function of electron concentration will generate the experimental values of the 3D DSL for that range of $\mathrm{n_0}$ for that material. Since $(Gn_0)^{-1/2}$ decreases with increasing n_0 for constant T, from (1.87), we can conclude that the 3D DSL will decrease with increasing n_0. For 2D DSL L_{2D} at a constant temperature varies inversely with $\bar{n}_0 G$ as appears from (1.89). Since $(\bar{n}_0 G)^{-1}$ decreases with increasing surface concentration, we can infer that 2D Debye screening Length will decrease with decreasing $\bar{n}_0$ for the appropriate 2D systems. This statement provides a compatibility test of our theoretical analysis. *Thus,* (1.87) *and* (1.89) *provide experimental checks of both 3D and 2D DSL s and also a technique for probing the band structures of the degenerate materials having arbitrary band structures.*

1.3.5 *Experimental determination of Einstein relation (ER)*

Using (1.82) and (1.17) we get,

$$\frac{D}{\mu} = \left(\frac{\pi^2 k_B^2 T}{3 \left| e \right|^2 G} \right) \tag{1.90}$$

Thus, the ER for degenerate materials can be determined by knowing the experimental values of G. The suggestion for the experimental determination of the ER for degenerate materials having arbitrary dispersion laws as given by (1.90) does not contain any energy band constants. For a fixed temperature, the D/μ varies inversely as G. Only the experimental values of G for any material as a function of electron concentration will generate the experimental values of the ER for that range of n_0 for that system. Since G decreases with increasing $\mathrm{n_0}$, from (1.90), one can infer that the ER will increase with an increase in n_0. *This statement is the compatibility test so far as the suggestion for the experimental determination of ER for degenerate materials is concerned* [185, 186].

1.3.6 *Diffusion coefficient of the minority carriers*

This particular coefficient in quantum-confined lasers can be expressed as

$$\frac{D_i}{D_0} = \frac{dE_{Fi}}{dE_F} \tag{1.91}$$

where D_i and D_0 are the diffusion coefficients of the minority carriers both in the presence and absence of quantum confinements. It appears then that the formulation of the above ratio requires a relation between E_{Fi} and E_F, which, in turn, can be formed through the respective carrier statistics.

1.3.7 *Nonlinear optical response*

The nonlinear response from the optical excitation of the free carriers is given by [187]

$$Z_0 = \frac{e^2}{\omega^2\hbar^2}\int_0^\infty \left(k_x\frac{\partial k_x}{\partial E}\right)^{-1} f_0 N(E)dE \tag{1.92}$$

where ω is the optical angular frequency, $N(E)$ is the DOS function. From the various $E-k$ relations of different materials under different physical conditions, we can formulate the expression of $N(E)$ and from band structure we can derive the term $(k_x\frac{\partial k_x}{\partial E})$ and thus by using the DOS function as formulated, we can study the Z_0 for all types of materials as considered in this book.

1.3.8 *Third-order nonlinear optical susceptibility*

This particular susceptibility can be written as [188]

$$\chi_{NP}(\omega_1,\omega_2,\omega_3) = \frac{n_0 e^4\left\langle\varepsilon^4\right\rangle}{24\omega_1\omega_2\omega_3(\omega_1+\omega_2+\omega_3)\hbar^4} \tag{1.93}$$

where $n_0\langle\varepsilon^4\rangle = \int_0^\infty \frac{\partial^4 E}{\partial k_z^4}N(E)f_0 dE$ and the other notations are defined in [187–190]. The term $(\frac{\partial^4 E}{\partial k_z^4})$ can be formulated by using the respective DR of different materials. Thus, one can investigate the $\chi_{NP}(\omega_1,\omega_2,\omega_3)$ for all materials as considered in this book.

1.3.9 *Generalized Raman gain*

The generalized Raman gain can be expressed as [188]

$$R_G = \bar{I}\left(\frac{16\pi^2c^2}{\hbar\omega\rho g\omega_s^2 n_s n_p}\right)\left(\frac{\Gamma_\rho}{\Gamma}\right)\left(\left(\frac{e^2}{mc^2}\right)^2 m^2R^2\right) \tag{1.94}$$

where $\bar{I} = \sum_{n,t_z}^{[} f_0(n,k_z,\uparrow) - f_0(n,k_z,\downarrow)]$, $f_0(n,k_z,\uparrow)$ is the Fermi factor for spin-up Landau levels, $f_0(n,k_z,\uparrow)$ is the Fermi factor for spin-down

Landau levels, n is the Landau quantum number and the other notations are defined in [189]. It appears then the formulation of R_G is determined by the appropriate derivation requires the magneto-dispersion relations. By using the different appropriate formulas as formulated in various chapters of this book R_G can, in general, be investigated.

1.3.10 *The plasma frequency*

The plasma frequency ω_p can be expressed as [190]

$$\omega_p^2 = \frac{e^2}{\varepsilon_{sc}} \int_{E'}^{\infty} N(E) v_x^2 \frac{\partial f_0(E)}{\partial E} dE \tag{1.95}$$

where E' can be obtained from $N(E') = 0$.

Thus, by using the expression of $N(E)$ and formulating the expression of v_x, we can study the plasma frequency from equation (1.95) for all types of materials having arbitrary band structures.

1.3.11 *The activity coefficient*

Dilute solution transport equations introduced by many authors [191] have been very effective in describing the fluxes of electrons and holes in semiconductors. These expressions, however, are inaccurate at high carrier concentrations due to no idealities associated with carrier degeneracy and band gap narrowing. Within the framework of dilute solution theory, these no idealities can be included by the introduction of concentration-dependent activity coefficients. Expressions for activity coefficients should also be useful in the area of chemical kinetics where mass-action principles are applicable. Activity coefficients were derived under the assumption of complete ionization.

The activity coefficient can, in general, be expressed as [191]

$$f_e^* = \frac{k_B T}{n_0} \int_0^{\infty} \frac{N(E)}{(1 - f_0)^2} \left[-\frac{\partial f_0}{\partial E} \right] dE \tag{1.96}$$

Thus, by using the expression of $N(E)$, we can study the activity coefficient from equation (1.96) for all types of materials having arbitrary band structures.

1.3.12 *Magneto-thermal effect in quantized structures*

The magneto-thermal effect can be expressed by the dimensionless quantity as [192]

$$\phi = \frac{B}{T}\frac{\partial T}{\partial B} = -\frac{B}{T}\left[\left(\frac{\partial S}{\partial B}\right)_{(T,n)}\right]\left[\left(\frac{\partial S}{\partial T}\right)_{(\beta,n)}\right]^{-1} \tag{1.97}$$

where S is the entropy which can be written as [197]

$$S = k_B \int_0^{\infty} -\left(\frac{\partial f_0}{\partial E}\right)\left(\frac{E-E_F}{k_BT}\right)[\int_0^E N(E)dE]dE \tag{1.98}$$

Under magnetic quantization we know, the DOS function assumes the form

$$N(E) = \left(\frac{eBg_v}{2\pi^2\hbar}\right)\sum_{n=0}^{n_{\max}}\frac{\partial k_z'(E)}{\partial E}H(E-E') \tag{1.99}$$

where E' is the Landau sub-band energy.

Therefore,

$$S = k_B \sum_{n=0}^{n_{\max}}\int_{E'}^{\alpha} -\left(\frac{\partial f_0}{\partial E}\right)\left(\frac{E-E_F}{k_BT}\right)\frac{eBg_v}{2\pi^2\hbar}k_z dE \tag{1.100}$$

Since

$$\int N(E)dE = \frac{eBg_v}{2\pi^2\hbar}\sum_{n=0}^{n_{\max}} k_z(E) \tag{1.101}$$

$$S = c_0 \sum_{n=0}^{n_{\max}}\left[\int_{E'}^{\alpha}\left\{k_z E\left(\frac{\partial f_0}{\partial E}\right)dE\right\} - E_F\int_{E'}^{\alpha}\left\{k_z E\left(\frac{-\partial f_0}{\partial E}\right)dE\right\}\right] \tag{1.102}$$

where $c_0 = (eBg_v/2\pi^2\hbar T)$

Using the concept of 3D quantization (magneto doping super-lattices, quantum dot super-lattices, magneto-size quantized super-lattices and the respective heavily doped cases together with the influence of light waves and electric fields on the 3D quantized structures), the DOS functions can respectively be written as

$$N(E) = \frac{g_v eB}{2\pi\hbar} \sum_{n=0}^{n_{z\max}} \sum_{n=0}^{n_{\max}} \delta'(E - E'') \tag{1.103}$$

$$N(E) = \frac{g_v eB}{2\pi\hbar} \sum_{i=0}^{i_{\max}} \sum_{n=0}^{n_{\max}} \delta'(E - E'') \tag{1.104}$$

and

$$N(E) = \frac{2g_v}{(d_x d_y d_z)} \sum_{n_x=1}^{n_{x\max}} \sum_{n_y=1}^{n_{y\max}} \sum_{n_z=1}^{n_{z\max}} \delta'(E - E''') \tag{1.105}$$

where E'' is the totally quantized energy in 2D systems in the presence of applied quantizing magnetic field along z-direction and E''' is the totally quantized energy in 0D systems in the absence of quantizing magnetic field along z-direction, (1.103) represents all the cases corresponding to E'', (1.104) represents all the cases corresponding to magneto inversion layers and magneto accumulation layers and (1.105) represents the rest cases.

In the three cases we can show

$$S = c_0'' \sum \sum E''[F_0(\eta'') - (\eta'')F_{-1}(\eta'')] \tag{1.106}$$

$$S = c'''_0 \sum \sum E'''[F_0(\eta'') - (\eta'')F_{-1}(\eta'')] \tag{1.107}$$

where

$$c_0'' = \frac{g_v eBk_B}{2\pi\hbar},\ c'''_0 = \frac{2g_v}{d_x d_y d_z}, \eta'' = \frac{E_F - E''}{k_B T}, \eta''' = \frac{E_F - E'''}{k_B T}$$

and E_F in this case is the Fermi energy for all the previous respective cases of 3D quantization of wave vector space.

Thus, by using the (1.97), (1.106) and (1.107), we can study the magneto thermal effect in 3D quantized structures for all types of materials having arbitrary band structures.

1.3.13 *Normalized Hall coefficient*

The normalized Hall coefficient in a semiconductor can be expressed as [193, 194]

$$\frac{R(T)}{R(0)} = 1 + \frac{(\pi k_B T)^2}{3} \left[\frac{1}{N(E)} \frac{\partial N(E)}{\partial E} \right]^2 \Bigg|_{E=E_{FT}} \tag{1.108}$$

where $R(T)$ is the Hall coefficient at T, $R(0)$ is the same at $T \to 0$ and E_{FT} is the Fermi energy at T.

Thus, by using the expression of $N(E)$, we can study the normalized Hall coefficient from equation (1.108) for all types of materials having arbitrary band structures.

1.3.14 *Reflection coefficient*

In the presence of light waves, the reflection coefficient (R) is given by [195]

$$R = \frac{(x-1)^2}{(x+1)^2} \tag{1.109}$$

where

$$x = \sqrt{\varepsilon_{sc} - \frac{n_0 e^2 \lambda^2}{m^* \pi c^2}},$$

m^* is the effective electron mass and c is the velocity of light.

As effective mass is a DR-dependent concept thus we can study the reflection coefficient by using (1.109) for all types of materials having arbitrary band structures.

1.3.15 *Heat capacity*

The heat capacity of the carriers in electronic materials is a very important thermal quantity.

The total heat capacity c_t is given by

$$c_t = c_e + c_l \tag{1.110}$$

where c_e is the electronic heat capacity and is given by [196]

$$c_e = \int_0^\infty -\frac{\partial f_0}{\partial E}\left[\frac{E-E_F}{T} + \frac{\partial E_F}{\partial T}\right](E-E_F)N(E)dE \tag{1.111}$$

and c_l is the lattice heat capacity.

Thus, (1.110) can be written as

$$c_t = \frac{\pi^2 k_B^2 T N(E_F)}{3}\left[1 + T\left(\frac{\partial E_F}{\partial T}\right)\left(\frac{N'(E_F)}{N(E_F)}\right)\right] + c_l \tag{1.112}$$

Thus, by using the expression of $N(E)$, we can study the Heat capacity from equation (1.112) for all types of materials having arbitrary band structures.

1.3.16 *Magnetic susceptibilities*

The Landau diamagnetic susceptibility (χ_d) can be written as

$$\chi_d = -\mu_0 \frac{\partial^2 F(B)}{\partial B^2} \tag{1.113}$$

where μ_0 is the free space permeability and $F(B)$ is the free energy which can be written as

$$F(B) = n_0 E_{FB} + \int_0^\infty D_2(E) \frac{\partial f_0}{\partial E} dE \tag{1.114}$$

where $D_2(E)$ is twice the integration of the DOS function with respect to E.

Pauli's paramagnetic susceptibility (χ_p) can be expressed as [197]

$$\chi_p = -2\mu_B^2 \int_{E_0}^\infty N(E) \frac{\partial f_0}{\partial E} dE \tag{1.115}$$

where μ_B is the Bohr magnetic field.

Thus, by using the expression of $N(E)$, we can study the magnetic susceptibilities by using (1.114) and (1.115), respectively for all types of materials having arbitrary band structures.

1.3.17 *Faraday rotation*

The Faraday rotation of the electrons in electronic materials can generally be expressed under the condition $\omega\tau \gg 1$, as [198]

$$\theta = AI \tag{1.116}$$

where ω is the angular frequency of the incident light, τ is the momentum relaxation time,

$$A \equiv \frac{e^3 l}{8\pi^3 \hbar^4 \omega^2} \left(\frac{1}{2n_r \varepsilon_0 c} \right),$$

l is the thickness of the specimen, n_r is the real part of the refractive index at that frequency and

$$I = \frac{(2\pi)^3}{2g_v} \int_0^\infty \frac{\partial E}{\partial k_x} \left[\frac{\partial E}{\partial k_y} \cdot \frac{\partial^2 E}{\partial k_x \partial k_y} - \frac{\partial E}{\partial k_x} \cdot \frac{\partial^2 E}{\partial k_y^2} \right] \frac{\partial f_0}{\partial E} N(E) dE \tag{1.117}$$

Thus, from different DRs as formulated in this book, we can formulate the partial differentials and from the expressions of $N(E)$ as derived in different chapters, we can study the Faraday rotations for various materials.

1.3.18 *Fowler–Nordheim field emission*

The Fowler–Nordheim field emission (FNFE) is a well-known quantum-mechanical phenomenon that involves tunneling of electrons through a surface barrier due to the application of an intense external electric field. Normally, at field strengths of the order of 10^8 V/m (below the electrical breakdown), the potential barriers at the surfaces of metals and semiconductors usually become very thin and result in field emission of electrons due to the tunnel effect [199]. This has been well-investigated with reference to 3D electron gases in metals and semiconductors and the FNFE from quantum-confined structures has also been studied in this context [200–206]

The current (I) due to FNFE can be written as

$$I_F = \frac{1}{2}\sum_{n_x=1}^{n_{x\max}}\sum_{n_y=1}^{n_{y\max}}\left[\int_{E_{11}}^{i}[e.vn_1t_{11}]\right] \tag{1.118}$$

where is the magnitude of the electron charge, v is the velocity of the electron and given by $v = \frac{1}{\hbar}\frac{\partial E}{\partial k_z}n_1 = N_{1D}(E).f_0dE$, $N_{1D}(E)$ is the 1D DOS function per sub-bands and can be written as $N_{1D}(E) = \frac{2g_v}{\pi}\frac{\partial k_z}{\partial E}$, t_{11} is the transmission coefficient which can be expressed as $t_{11} = \exp[-2\int_i |k_i|di]$ in which $i = x, i, z$

Thus, by using (1.118), we can study the FNFE for all types of materials having arbitrary band structures.

1.3.19 *Optical effective mass*

The investigation of the energy band structure of noble metals is one of the standard problems in solid-state physics. The topology and geometry of the Fermi surfaces of the noble metals are very well known in detail from the de Hass–van Alphen measurements of Shoenberg. Very accurate values for the cyclotron masses of the noble metals were experimentally obtained from de Hass–van Alphen investigations [207]. The optical effective mass being used to study the different optical properties of semiconductors can be expressed following Madelung [213] as

$$m_{opt} = 3\hbar^2 n_0\left[\int_0^\infty N(E)\frac{\partial E}{\partial k}f_0 dE\right]^{=1} \tag{1.119}$$

Thus, by using the expression of $N(E)$ and $\frac{\partial E}{\partial k}$, we can study the optical effective mass from (1.109a) for all types of materials having arbitrary band structures.

1.3.20 *Einstein's photoemission*

Einstein's photoemission (EP) is a physical phenomenon and occupies a singular position in the whole arena of materials science and related disciplines in general. The EP in recent years finds extensive applications in modern optoelectronics, characterization and investigation of condensed matter systems, photoemission spectroscopy and related aspects in connection with the investigations of the optical properties of nanostructures [207–211]. Interest in low-dimensional silicon nanostructures also grew up and gained momentum, after the discovery of room-temperature photoluminescence and electroluminescence of silicon nanowires in porous silicon [207]. Work on ultrathin layers of $SiSiO_2$ super-lattices resulting in visible light emission at room temperature clearly exhibited low-dimensional quantum confinement effect [208] and one of the most popular techniques for analyzing the low-dimensional structures is to employ photoemission techniques. Recent observation of room temperature photoluminescence and electro luminescence in porous silicon has stimulated vigorous research activities in silicon nanostructures [209].

It is well-known that the classical equation of the photo-emitted current density is [210, 211]

$$J = [4\pi\alpha_0 e m_c g_v (k_B T)^2/h^3]\exp[(hv - \phi)/(k_B T)],$$

where α_0, hv, and ϕ are the probability of photoemission, incident photon energy along z-axis and work function, respectively. The afore-mentioned equation is valid for both the charge carriers and in this conventional form it appears that, the photoemission changes with the effective mass, temperature, work function and the incident photon energy, respectively. This relation holds only under the condition of carrier non-degeneracy. The EP has different values for different materials and varies with doping and with external fields which creates quantization of the wave-vector space of the carriers leading to various types of quantized structures.

The J from bulk specimens can be written as

$$J = \frac{\alpha_0 e}{4}(P_I) \tag{1.120}$$

in which

$$P_I = \int_{E_0}^{\infty} N(E')\nu_z(E')f(E)dE' \tag{1.121}$$

where $E_0 \equiv W - hv, W$ is the electron affinity, $E' \equiv E - E_0, E$ is the total energy of the electron as measured from the edge of the conduction band in the vertically upward direction in the absence of any quantization, $N(E')$ is the DOS function at $E = E', \nu_z(E')$ is the velocity of the emitted electron along z-axis when, $E = E'$ and $f(E)$ is the Fermi–Dirac occupation probability factor.

The photo electric current density under magnetic quantization can be written as

$$J_{BP} = \frac{\alpha_0 g_v e}{2} \sum_{n=0}^{n_{\max}} \int_{E'_n}^{\infty} N'_B(E)\nu_{zB}(E)f(E)dE \tag{1.122}$$

where n is the Landau quantum number, $E'_n \equiv (E_n + W - h\nu)$, E_n is the nth Landau level energy, $\nu_{zB}(E)$ is the velocity of the photo emitted electron along z-direction under quantizing magnetic field B which is also applied parallel to x-axis and $N'_B(E)$ is the DOS function under magnetic quantization.

Einstein's photoelectric current density in quantum wells can be written as

$$J_{2D} = \frac{\alpha_o e g_v}{2d_z} \sum_{n_{z\min}}^{n_{z\max}} \int_{E_{n_z}}^{\infty} N_{2D}(E)f(E)v_z(E_{n_z})dE \tag{1.123}$$

where d_z is the nanothickness along z-direction, E_{n_z} is the energy of the nth sub-band, $N_{2D}(E)$ is the DOS function per sub-band and $v_z(E_{n_z})$ is the velocity of the electron in the n_zth sub-band.

Thus, using (1.109b), (1.109c), (1.109d) and (1.109e), we can determine the EP from different materials under various physical conditions.

1.3.21 *Righi–Leduc coefficient*

It is well known that some transport effects under strong magnetic field are independent of relaxation mechanism and are determined by the DRs only. Among them, one is thermoelectric power and the other one is Righi–Leduc coefficient, which can in turn, be written as [216–218]

$$R = \frac{cn_0}{eB^2\chi_{ph}}[\langle(E - E_F)^2\rangle - \langle E - E_F\rangle^2] \tag{1.124}$$

where χ_{ph} is the phonon thermal conductivity,

$$\langle(E - E_F)^2\rangle = \frac{1}{n}\int (E - E_F)^2 G(E)\left[-\frac{\partial f_0}{\partial E}\right] dE \tag{1.125}$$

where $G(E)$ is the total number of states and is given by $G(E) = \int_{E_{30}}^{E} N(\zeta)d\zeta$ in which ζ is the variable of integration, E_{30} is the lower limit of integration and

$$\langle E - E_F \rangle = \frac{1}{n}\int (E - E_F)G(E)\left[-\frac{\partial f_0}{\partial E}\right] dE \tag{1.126}$$

Thus by using (1.124), (1.125) and (1.126), we can study the Righi–Leduc coefficient for all types of materials having arbitrary band structures.

1.3.22 *Electric susceptibility*

In semiconductors, the effect of free carriers on the optical properties becomes important at wavelengths longer than the intrinsic absorption edge. Free carriers produce absorption and affect also the dispersion at sufficiently long wave lengths. Electric susceptibility can be written as [213]

$$\chi_c = \frac{-e^2}{3\omega^2\varepsilon_0\hbar^2}\int_0^\infty N(E)\frac{\partial E}{\partial k} f_0 dE \tag{1.127}$$

Thus, by using the expression of $N(E)$ and $\frac{\partial E}{\partial k}$, we can study the electric susceptibility from (1.127) for all types of materials having arbitrary band structures.

1.3.23 *Electric susceptibility mass*

In recent years, there has been considerable interest in studying the electric susceptibility mass of the carriers in narrow gap materials because of its direct influence in the study of carrier scattering mechanisms in semiconductor. Electric susceptibility mass can be determined from measurements in the infrared region of the frequency dependence of the spectral reflectivity at normal incidence and provides useful information regarding band structures. Electric susceptibility mass can be written as [214]

$$m_s^* = 3A_{20}\left[\int_{E_{20}}^\infty N(E)\frac{\partial E}{\partial k} f_0 dE\right]^{-1} \tag{1.128}$$

where $A_{20} = n_0\varepsilon_0\hbar^2$ and E_{20} is the lower limit of integration.

Thus, by using the expression of $N(E)$ and $\frac{\partial E}{\partial k}$, we can study the electric susceptibility mass from (1.128) for all types of materials having arbitrary band structures.

1.3.24 *Electron diffusion thermopower*

The electron diffusion thermopower (S_e) is an important property of AlGaN/GaN QLD structures since it yields useful information about the carrier transport mechanism. S_e can be written as [215]

$$S_e = -\frac{1}{n_0 e}\frac{\partial \bar{f}}{\partial T} \tag{1.129}$$

where

$$\bar{f} = n_0 E_F - \sum \int_{E_{21}}^{\infty} G(E) f_0 dE \tag{1.130}$$

In which

$$G(E) = \int_{E_{30}}^{E} N(\zeta) d\zeta \tag{1.131}$$

where E_{21} and E_{30} are the lower limits of integrations of $\bar{f}$ and $G(E)$, respectively.

Thus, by using (1.129), (1.30) and (1.131), we can study the electron diffusion thermopower for all types of materials having arbitrary band structures.

1.3.25 *Hydrostatic piezo-resistance coefficient*

The importance of hydrostatic piezo-resistance coefficient (H_C) is well known in the sensor technology and H_c can be expressed as [216]

$$H_c = \frac{-\partial n_0}{n_0 \partial P} \tag{1.132}$$

where P is the pressure in kilobar.

Thus, by using (1.132), we can study the hydrostatic piezo-resistance coefficient for all types of materials having arbitrary band structures.

1.4 Result and Discussions

Using the appropriate equations and the values of the energy band constants from Chapter 18, we have plotted in Fig. 1.1, in the electric quantum limit ($i = 0$) the QC in QMOSFETs of Cd_3As_2 operated under inversion layer mode as a function of surface electric field in the low electric field limit in accordance with (a) generalized band model (red line), (b) generalized band model with $\delta = 0$ (green line), (c) three-band model of Kane (brown line),

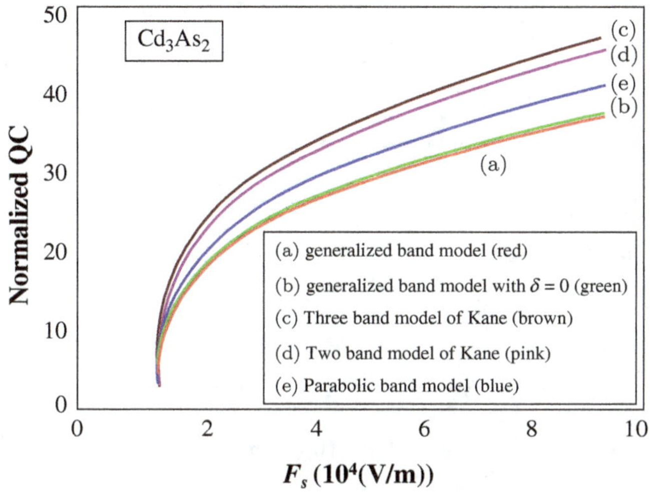

Fig. 1.1. The normalized QC in QMOSFETs of Cd_3As_2 operated under inversion mode as a function of surface electric field has been plotted in the low electric field limit in accordance with (a) generalized band model (red line, (1.16)), (b) generalized band model with $\delta = 0$ (green line), (c) three-band model of Kane (brown line, (1.38)), (d) two-band model of Kane (pink line, (1.51)) and (e) parabolic energy bands (blue line, (1.60)). We have used the electric quantum limit condition in obtaining all the plots.

(d) two-band model of Kane (pink line) and (e) parabolic energy bands (blue line). In Fig. 1.2, the QC in QMOSFETs of Cd_3As_2 operated under inversion layer mode as a function of surface electric field for $i = 0$ in the high electric field limit has been plotted for all cases of Fig. 1.1. In Fig. 1.3, the QC in QMOSFETs of Cd_3As_2 operated under inversion layer mode as a function of gate voltage has been plotted for $i = 0$ in the low electric field limit for all cases of Fig. 1.1. In Fig. 1.4, the QC in QMOSFETs of Cd_3As_2 operated under inversion layer mode as a function of gate voltage has been plotted in the high electric field limit for all cases of Fig. 1.1. In Fig. 1.5, the QC in QMOSFETs of $CdGeAs_2$ operated under inversion layer mode as a function of surface electric field has been plotted for $i = 0$ in the low electric field limit in accordance with (a) generalized band model (brown line), (b) generalized band model with $\delta = 0$ (yellow line), (c) three-band model of Kane (green line), (d) two-band model of Kane (pink line) and (e) parabolic energy bands (blue line). In Fig. 1.6, the same has been plotted as a function of surface electric field for $i = 0$ in the high electric field limit for all cases of Fig. 1.5. In Fig. 1.7, the QC in QMOSFETs of $CdGeAs_2$

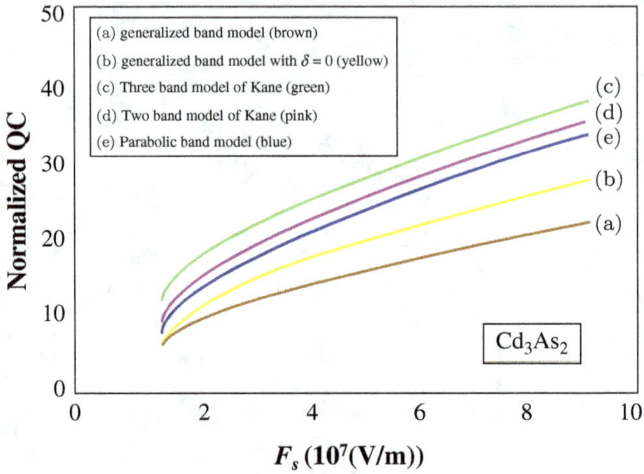

Fig. 1.2. The normalized QC in QMOSFETs of Cd_3As_2 operated under inversion mode as a function of surface electric field has been plotted in the high electric field limit in accordance with (a) generalized band model (brown line, (1.28)), (b) generalized band model with $\delta = 0$ (yellow line), (c) three-band model of Kane (green line, (1.68)), (d) two-band model of Kane (pink line, (1.77)) and (e) parabolic energy bands (blue line, (1.60)). We have used the electric quantum limit condition in obtaining all the plots.

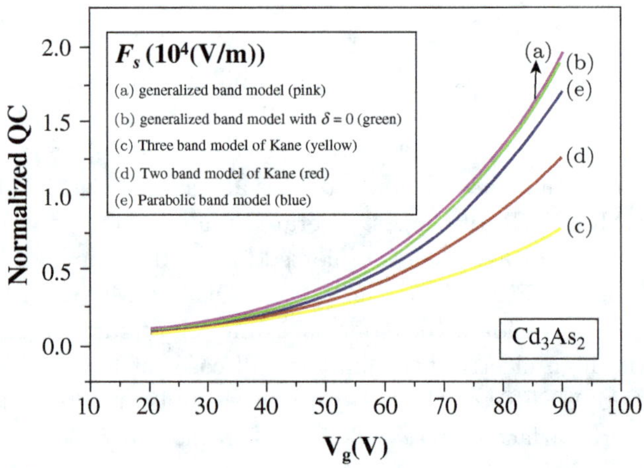

Fig. 1.3. The normalized QC in QMOSFETs of Cd_3As_2 operated under inversion mode as a function of gate voltage has been plotted in the low electric field limit in accordance with (a) generalized band model (pink line, (1.16)), (b) generalized band model with $\delta = 0$ (green line), (c) three-band model of Kane (yellow line, (1.38)), (d) two-band model of Kane (red line, (1.51)) and (e) parabolic energy bands (blue line, (1.60)). We have used the electric quantum limit condition in obtaining all the plots.

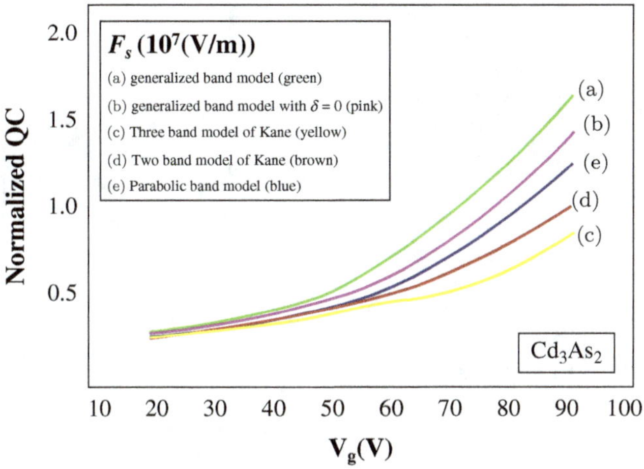

Fig. 1.4. The normalized QC in QMOSFETs of Cd_3As_2 operated under inversion mode as a function of gate voltage has been plotted in the high electric field limit in accordance with (a) generalized band model (green line, (1.28)), (b) generalized band model with $\delta = 0$ (pink line), (c) three-band model of Kane (yellow line, (1.68)), (d) two-band model of Kane (brown line, (1.77)) and (e) parabolic energy bands (blue line, (1.60)). We have used the electric quantum limit condition in obtaining all the plots.

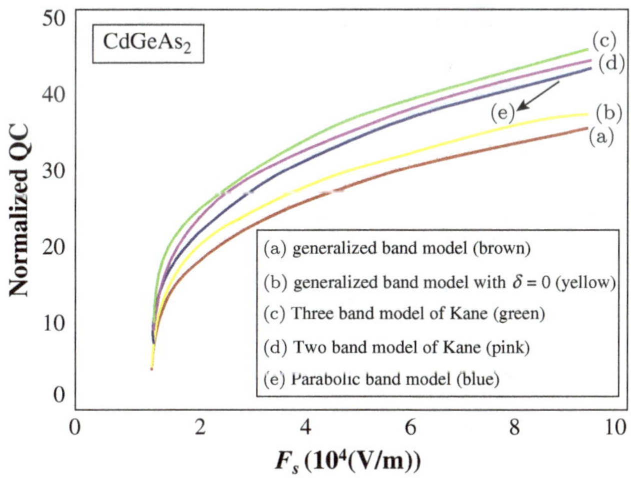

Fig. 1.5. The normalized QC in QMOSFETs of $CdGeAs_2$ operated under inversion mode as a function of surface electric field has been plotted in the low electric field limit in accordance with (a) generalized band model (brown line, (1.16)), (b) generalized band model with $\delta = 0$ (yellow line), (c) three-band model of Kane (green line, (1.38)), (d) two-band model of Kane (pink line, (1.51)) and (e) parabolic energy bands (blue line, (1.60)). We have used the electric quantum limit condition in obtaining all the plots.

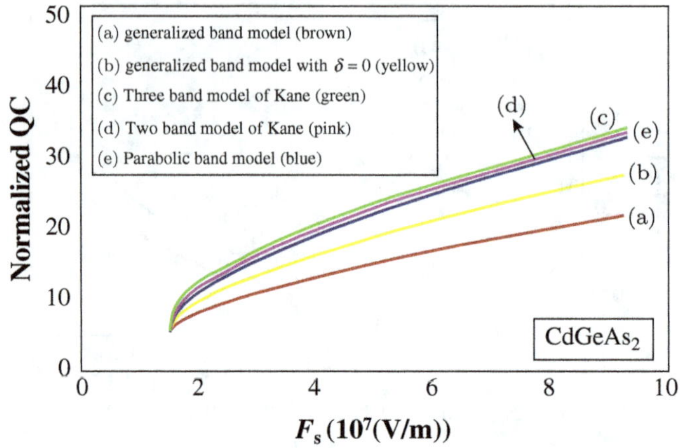

Fig. 1.6. The normalized QC in QMOSFETs of $CdGeAs_2$ operated under inversion mode as a function of surface electric field has been plotted in the high electric field limit in accordance with (a) generalized band model (brown line, (1.28)), (b) generalized band model with $\delta = 0$ (yellow line), (c) three-band model of Kane (green line, (1.68)), (d) two-band model of Kane (pink line, (1.77)) and (e) parabolic energy bands (blue line, (1.60)). We have used the electric quantum limit condition in obtaining all the plots.

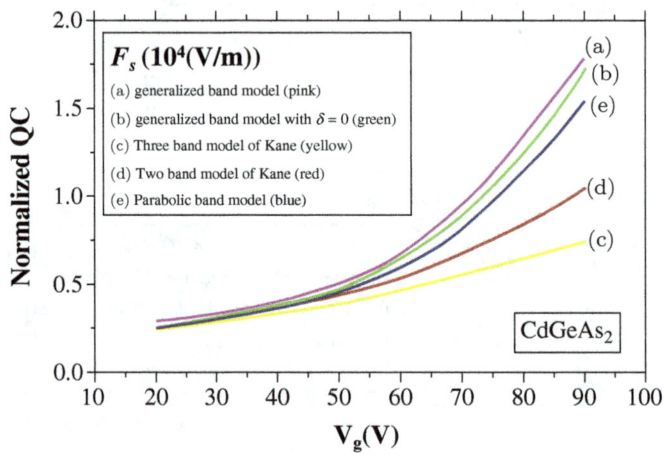

Fig. 1.7. The normalized QC in QMOSFETs of $CdGeAs_2$ operated under inversion mode as a function of gate voltage has been plotted in the low electric field limit in accordance with (a) generalized band model (pink line (1.16)), (b) generalized band model with $\delta = 0$ (green line), (c) three-band model of Kane (yellow line, (1.38)), (d) two-band model of Kane (red line, (1.51)) and (e) parabolic energy bands (blue line, (1.60)). We have used the electric quantum limit condition in obtaining all the plots.

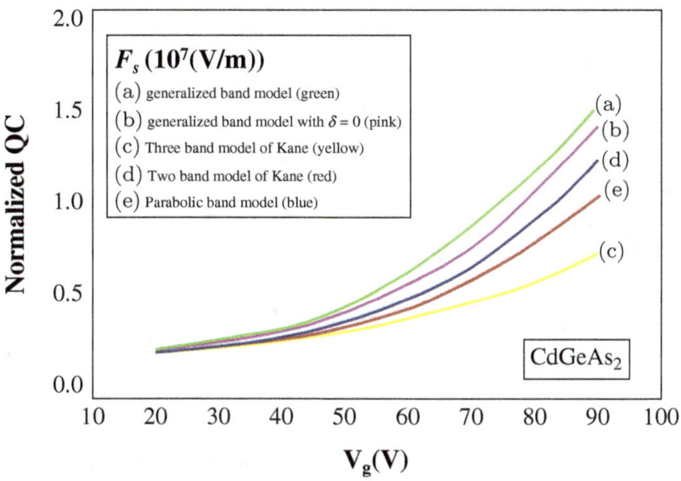

Fig. 1.8. The normalized QC in QMOSFETs of $CdGeAs_2$ operated under inversion mode as a function of gate voltage has been plotted in the high electric field limit in accordance with (a) generalized band model (green line, (1.28)), (b) generalized band model with $\delta = 0$ (pink line), (c) three-band model of Kane (yellow line, (1.68)), (d) two-band model of Kane (red line, (1.77)) and (e) parabolic energy bands (blue line, (1.60)). We have used the electric quantum limit condition in obtaining all the plots.

operated under inversion layer mode as a function of gate voltage has been plotted for $i = 0$ in the low electric field limit for all cases of Fig. 1.5. In Fig. 1.8, the same has been plotted as a function of gate voltage for $i = 0$ in the high electric field limit for all cases of Fig. 1.5. In Fig. 1.9, the QC in QMOSFETs of InSb operated under inversion layer mode as a function of surface electric field has been plotted for $i = 0$ in the low electric field limit in accordance with (a) three-band model of Kane (blue line) (b) two-band model of Kane (red line) and (c) parabolic energy bands (green line). In Fig. 1.10, the QC in QMOSFETs of InSb operated under inversion layer mode as a function of surface electric field has been plotted for $i = 0$ in the high electric field limit for all cases of Fig. 1.9. In Fig. 1.11, the QC in this case as a function of gate voltage has been plotted for $i = 0$ in the low electric field limit for all cases of Fig. 1.9. In Fig. 1.12, the QC in QMOSFETs of InSb as a function of gate voltage has been plotted for $i = 0$ in the high electric field limit for all cases of Fig. 1.9. In Fig. 1.13, the QC in QMOSFETs of InAs as a function of surface electric field has been plotted for $i = 0$ in the low electric field limit in accordance with (a) three-band model of Kane (blue line), (b) two-band model of Kane (red line) and (c)

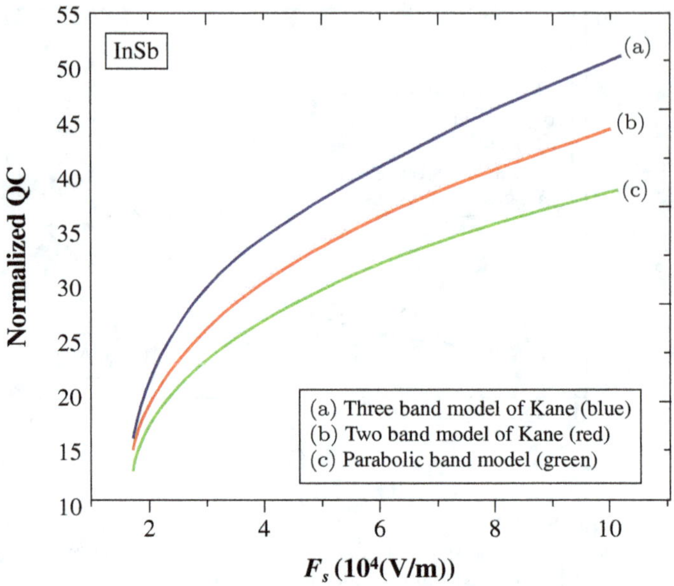

Fig. 1.9. The normalized QC in QMOSFETs of InSb operated under inversion mode as a function of surface electric field has been plotted in the low electric field limit in accordance with (a) three-band model of Kane (blue line, (1.38)), (b) two-band model of Kane (red line, (1.51)) and (c) parabolic energy bands (green line, (1.60)). We have used the electric quantum limit condition in obtaining all the plots.

parabolic energy bands (green line). In Fig. 1.14, the QC in QMOSFETs of InAs as a function of surface electric field has been plotted for $i = 0$ in the high electric field limit for all cases of Fig. 1.13. In Fig. 1.17, the QC in QMOSFETs of InAs as a function of gate voltage field has been plotted for $i = 0$ in the low electric field limit for all cases of Fig. 1.13. In Fig. 1.16, the QC in QMOSFETs of InAs as a function of gate voltage field has been plotted for $i = 0$ in the high electric field limit for all cases of Fig. 1.13. In Fig. 1.17, the QC in QMOSFETs of $Hg_{1-x}Cd_xTe$ as a function of surface electric field has been plotted for $i = 0$ in the low electric field limit in accordance with (a) three-band model of Kane (blue line), (b) two-band model of Kane (red line) and (c) parabolic energy bands (green line). In Fig. 1.18, the QC in QMOSFETs of $Hg_{1-x}Cd_xTe$ as a function of surface electric field has been plotted for $i = 0$ in the high electric field limit for all cases of Fig. 1.17. In Fig. 1.19, the QC in QMOSFETs of $Hg_{1-x}Cd_xTe$ as a function of gate voltage has been plotted for $i = 0$ in the low electric

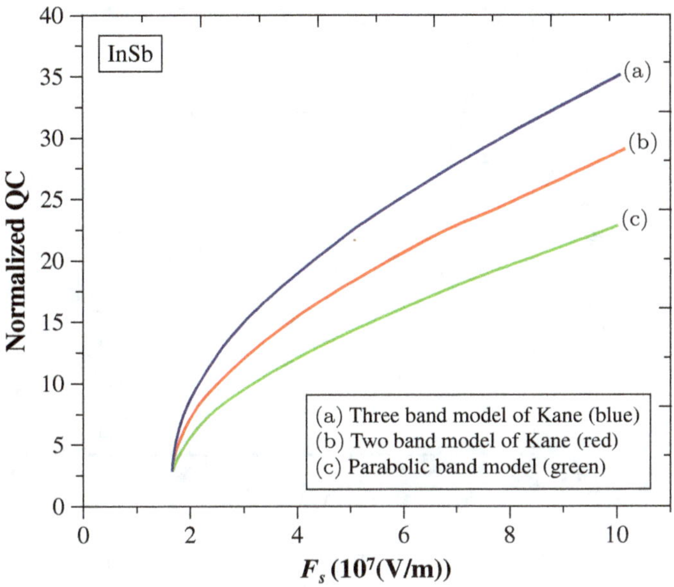

Fig. 1.10. The normalized QC in QMOSFETs of InSb operated under inversion mode as a function of surface electric field has been plotted in the high electric field limit in accordance with (a) three-band model of Kane (blue line, (1.68)), (b) two-band model of Kane (red line, (1.77)) and (c) parabolic energy bands (green line, (1.60)). We have used the electric quantum limit condition in obtaining all the plots.

field limit for all cases of Fig. 1.17. In Fig. 1.20, the QC in QMOSFETs of $Hg_{1-x}Cd_xTe$ as a function of gate voltage has been plotted for $i = 0$ in the high electric field limit for all cases of Fig. 1.17. In Fig. 1.21, the QC in QMOSFETs of $Hg_{1-x}Cd_xTe$ as a function of alloy composition has been plotted for $i = 0$ in the low electric field limit with $V_g = 30\,\text{V}$ in accordance with (a) three-band model of Kane (green line), (b) two-band model of Kane (red line) and (c) parabolic energy bands (blue line). In Fig. 1.22, the QC in QMOSFETs of $Hg_{1-x}Cd_xTe$ as a function of alloy composition has been plotted for $i = 0$ in the low electric field limit with $V_g = 80\,\text{V}$ for all cases of Fig. 1.21. In Fig. 1.23, the QC in QMOSFETs of $Hg_{1-x}Cd_xTe$ as a function of alloy composition has been plotted in the high electric field limit with $V_g = 30\,\text{V}$ for all cases of Fig. 1.21. In Fig. 1.24, the QC in QMOSFETs of $Hg_{1-x}Cd_xTe$ as a function of alloy composition has been plotted for $i = 0$ in the high electric field limit with $V_g = 80\,\text{V}$ for all cases of Fig. 1.21. In Fig. 1.25, the QC in QMOSFETs of $Ga_{1-x}Al_xAs$ as a function of surface

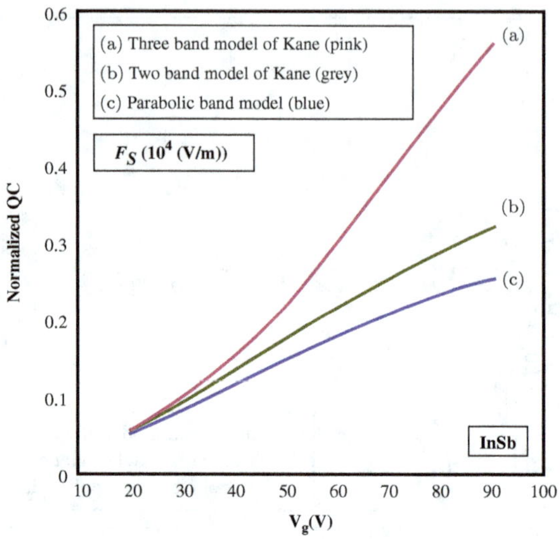

Fig. 1.11. The normalized QC in QMOSFETs of InSb operated under inversion mode as a function of gate voltage has been plotted in the low electric field limit in accordance with (a) three-band model of Kane (pink line, (1.38)), (b) two-band model of Kane (green line, (1.51)) and (c) parabolic energy bands (blue line, (1.60)). We have used the electric quantum limit condition in obtaining all the plots.

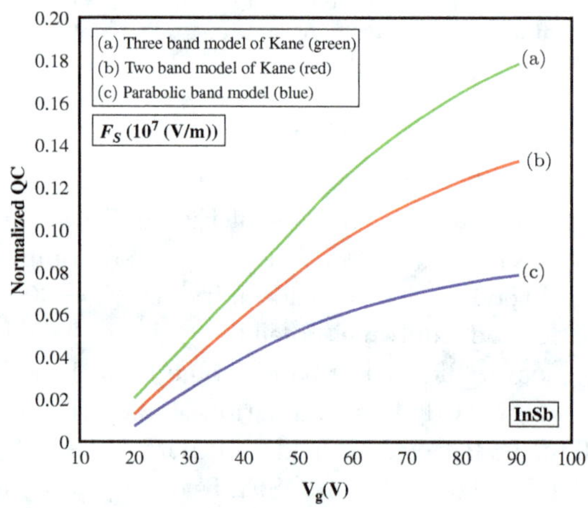

Fig. 1.12. The normalized QC in QMOSFETs of InSb operated under inversion mode as a function of gate voltage has been plotted in the high electric field limit in accordance with (a) three-band model of Kane (green line, (1.68)), (b) two-band model of Kane (red line, (1.77)) and (c) parabolic energy bands (blue line, (1.60)). We have used the electric quantum limit condition in obtaining all the plots.

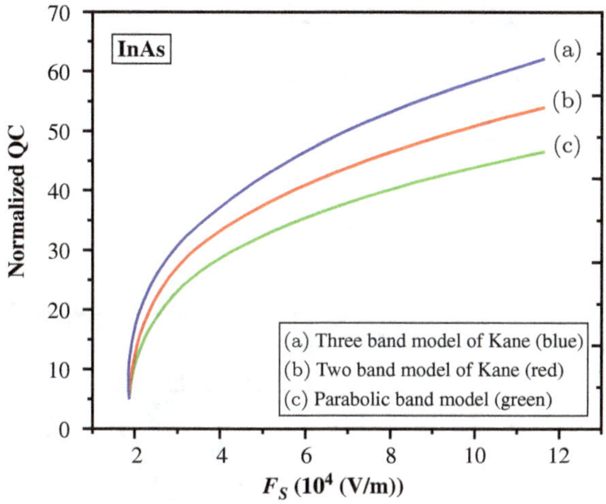

Fig. 1.13. The normalized QC in QMOSFETs of InAs operated under inversion mode as a function of surface electric field has been plotted in the low electric field limit in accordance with (a) three-band model of Kane (blue line, (1.38)), (b) two-band model of Kane (red line, (1.51)) and (c) parabolic energy bands (green line, (1.60)). We have used the electric quantum limit condition in obtaining all the plots.

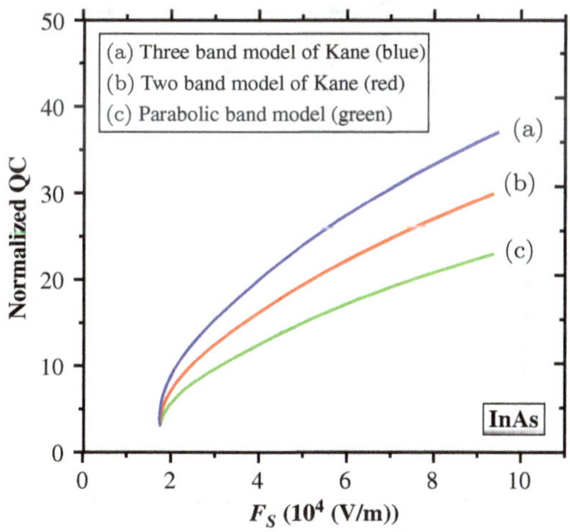

Fig. 1.14. The normalized QC in QMOSFETs of InAs operated under inversion mode as a function of surface electric field has been plotted in the high electric field limit in accordance with (a) three-band model of Kane (blue line, (1.68)), (b) two-band model of Kane (red line, (1.77)) and (c) parabolic energy bands (green line, (1.60)). We have used the electric quantum limit condition in obtaining all the plots.

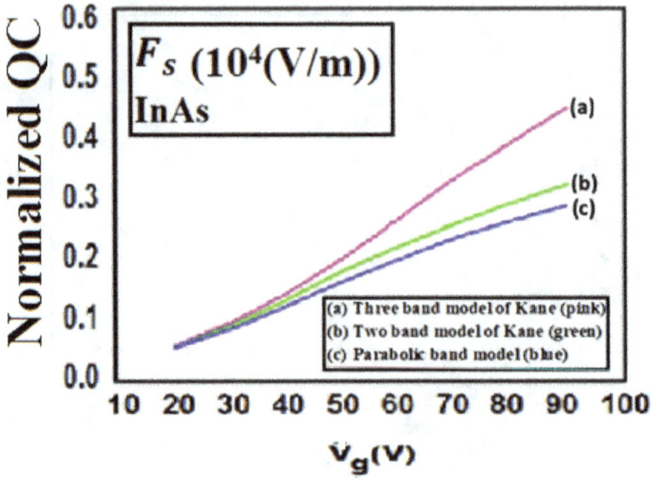

Fig. 1.15. The normalized QC in QMOSFETs of InAs operated under inversion mode as a function of gate voltage field has been plotted in the low electric field limit in accordance with (a) three-band model of Kane (pink line, (1.38)), (b) two-band model of Kane (green line, (1.51)) and (c) parabolic energy bands (blue line, (1.60)). We have used the electric quantum limit condition in obtaining all the plots.

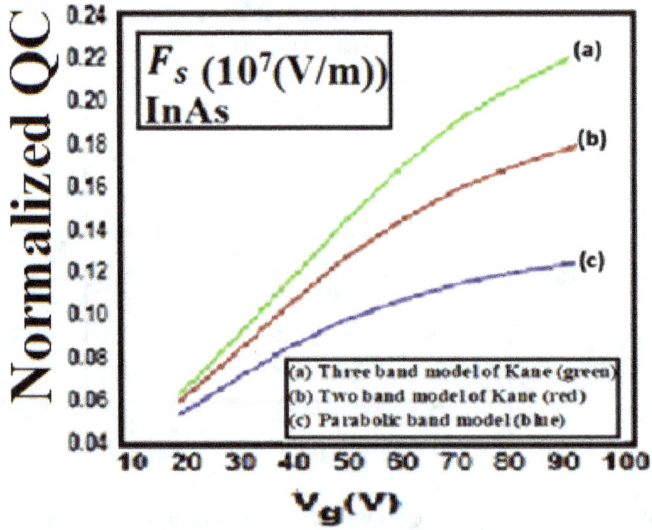

Fig. 1.16. The QC in QMOSFETs of InAs operated under inversion mode as a function of gate voltage field has been plotted in the high electric field limit in accordance with (a) three-band model of Kane (green line, (1.68)), (b) two-band model of Kane (red line, (1.77)) and (c) parabolic energy bands (blue line, (1.60)). We have used the electric quantum limit condition in obtaining all the plots.

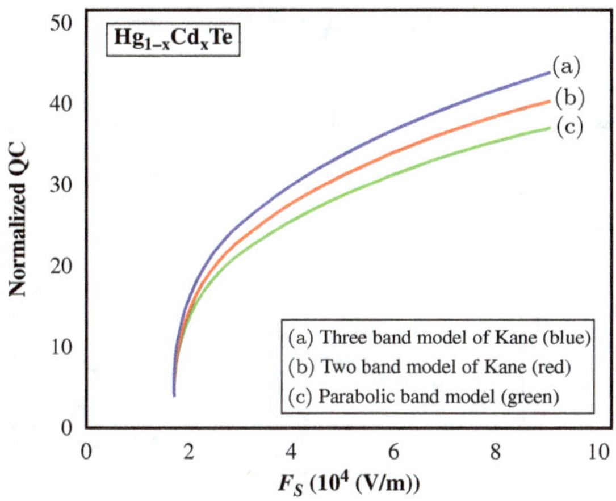

Fig. 1.17. The normalized QC in QMOSFETs of $Hg_{1-x}Cd_xTe$ operated under inversion mode as a function of surface electric field has been plotted in the low electric field limit in accordance with (a) three-band model of Kane (blue line, (1.38)), (b) two-band model of Kane (red line, (1.51)) and (c) parabolic energy bands (green line, (1.60)). We have used the electric quantum limit condition in obtaining all the plots.

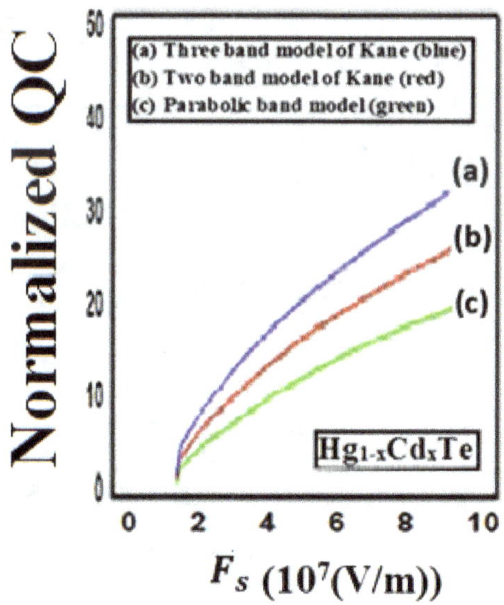

Fig. 1.18. The normalized QC in QMOSFETs of $Hg_{1-x}Cd_xTe$ operated under inversion mode as a function of surface electric field has been plotted in the high electric field limit in accordance with (a) three-band model of Kane (blue line, (1.68)), (b) two-band model of Kane (red line, (1.77)) and (c) parabolic energy bands (green line, (1.60)). We have used the electric quantum limit condition in obtaining all the plots.

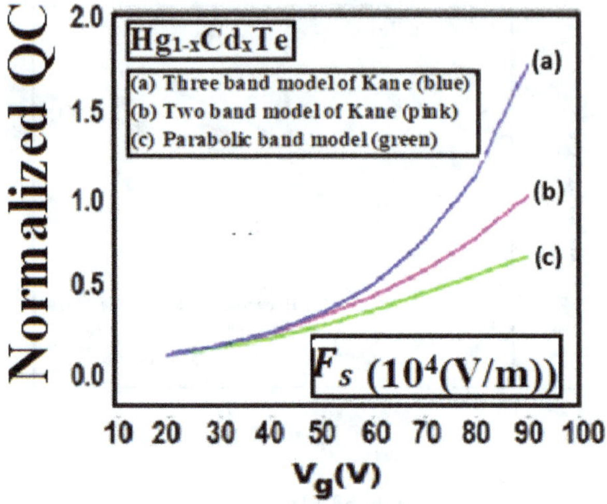

Fig. 1.19. The normalized QC in QMOSFETs of $Hg_{1-x}Cd_xTe$ operated under inversion mode as a function of gate voltage has been plotted in the low electric field limit in accordance with (a) three-band model of Kane (blue line, (1.38)), (b) two-band model of Kane (pink line, (1.51)), and (c) parabolic energy bands (green line, (1.60)). We have used the electric quantum limit condition in obtaining all the plots.

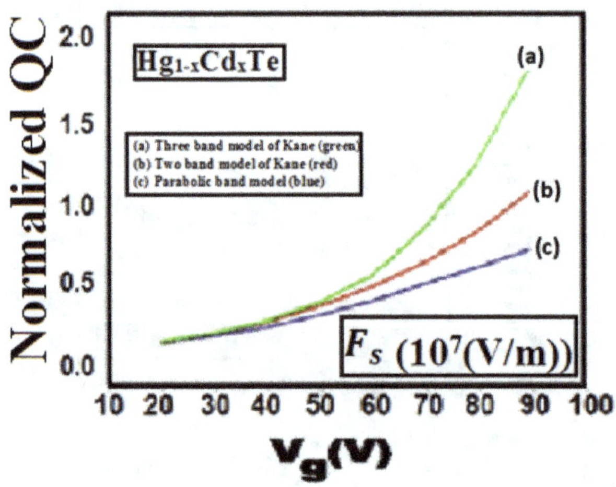

Fig. 1.20. The normalized QC in QMOSFETs of $Hg_{1-x}Cd_xTe$ operated under inversion mode as a function of gate voltage has been plotted in the high electric field limit in accordance with (a) three-band model of Kane (green line, (1.68)), (b) two-band model of Kane (red line, (1.77)) and (c) parabolic energy bands (blue line, (1.60)). We have used the electric quantum limit condition in obtaining all the plots.

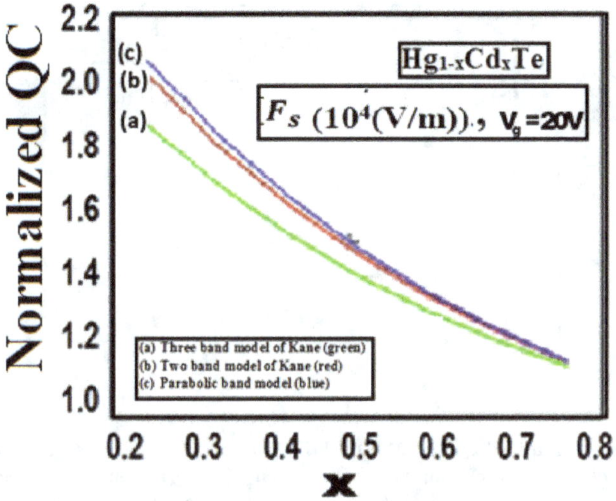

Fig. 1.21. The normalized QC in QMOSFETs of $Hg_{1-x}Cd_xTe$ operated under inversion mode as a function of alloy composition has been plotted in the low electric field limit with $V_g = 20$ V in accordance with (a) three-band model of Kane (green line, (1.38)), (b) two-band model of Kane (red line, (1.51)) and (c) parabolic energy bands (blue line, (1.60)). We have used the electric quantum limit condition in obtaining all the plots.

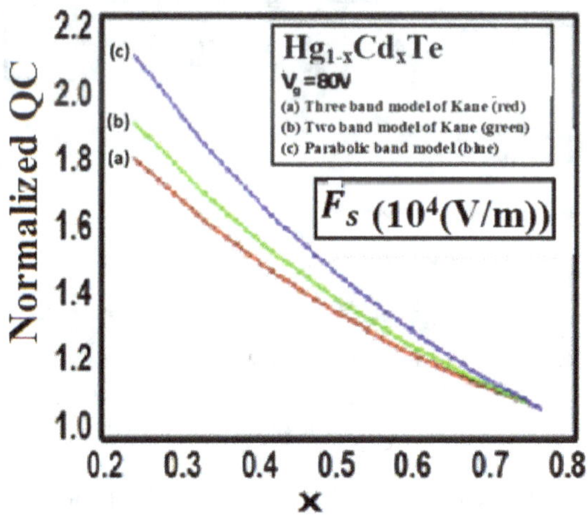

Fig. 1.22. The normalized QC in QMOSFETs of $Hg_{1-x}Cd_xTe$ operated under inversion mode as a function of alloy composition has been plotted in the low electric field limit with $V_g = 80$ V in accordance with (a) three-band model of Kane (red line, (1.38)), (b) two-band model of Kane (green line, (1.51)) and (c) parabolic energy bands (blue line, (1.60)). We have used the electric quantum limit condition in obtaining all the plots.

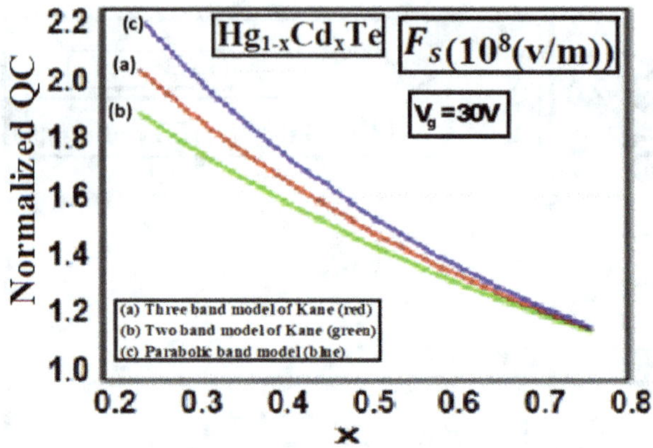

Fig. 1.23. The normalized QC in QMOSFETs of $Hg_{1-x}Cd_xTe$ operated under inversion mode as a function of alloy composition has been plotted in the high electric field limit with $V_g = 30$ V in accordance with (a) three-band model of Kane (red line, (1.68)), (b) two-band model of Kane (green line, (1.77)) and (c) parabolic energy bands (blue line, (1.60)). We have used the electric quantum limit condition in obtaining all the plots.

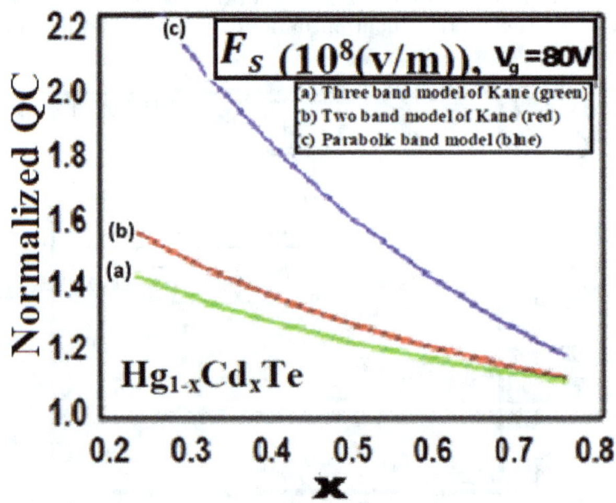

Fig. 1.24. The normalized QC in QMOSFETs of $Hg_{1-x}Cd_xTe$ operated under inversion mode as a function of alloy composition has been plotted in the high electric field limit with $V_g = 80$ V in accordance with (a) three-band model of Kane (green line, (1.68)), (b) two-band model of Kane (red line, (1.77)) and (c) parabolic energy bands (blue line, (1.60)). We have used the electric quantum limit condition in obtaining all the plots.

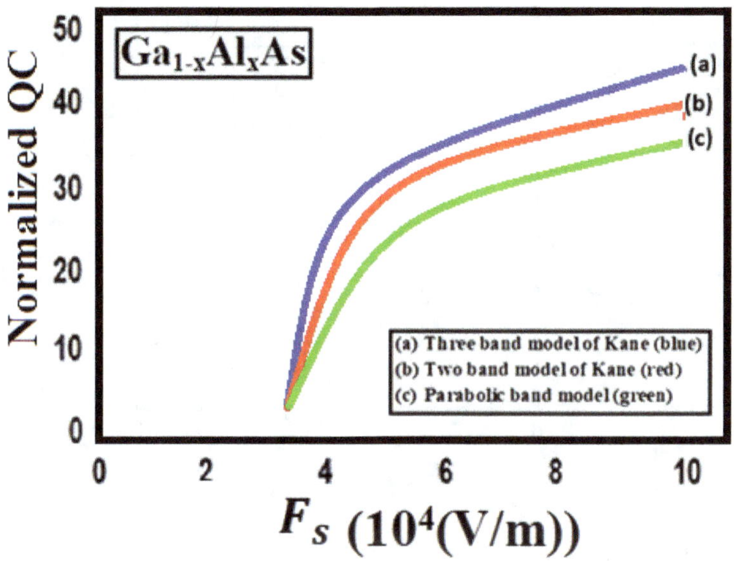

Fig. 1.25. The normalized QC in QMOSFETs of $Ga_{1-x}Al_xAs$ operated under inversion mode as a function of surface electric field has been plotted in the low electric field limit in accordance with (a) three-band model of Kane (blue line, (1.38)), (b) two-band model of Kane (brown line, (1.51)) and (c) parabolic energy bands (green line, (1.60)). We have used the electric quantum limit condition in obtaining all the plots.

electric field has been plotted for $i = 0$ in the low electric field limit in accordance with (a) three-band model of Kane (blue line), (b) two-band model of Kane (brown line) and (c) parabolic energy bands (green line). In Fig. 1.26, the QC in QMOSFETs of $Ga_{1-x}Al_xAs$ as a function of surface electric field has been plotted for $i = 0$ in the high electric field limit for all cases of Fig. 1.25. In Fig. 1.27, the QC in QMOSFETs of $Ga_{1-x}Al_xAs$ as a function of gate voltage has been plotted for $i = 0$ in the low electric field limit for all cases of Fig. 1.25. In Fig. 1.28, the QC in QMOSFETs of $Ga_{1-x}Al_xAs$ as a function of gate voltage has been plotted for $i = 0$ in the high electric field limit for all cases of Fig. 1.25. In Fig. 1.29, the QC in QMOSFETs of $Ga_{1-x}Al_xAs$ as a function of alloy composition has been plotted for $i = 0$ in the high electric field limit for all cases of Fig. 1.25. In Fig. 1.30, the QC in QMOSFETs of $In_{1-x}Ga_xAs_yP_{1-y}$ as a function of gate voltage has been plotted for $i = 0$ in the low electric field limit in accordance with (a) three-band model of Kane (green line), (b) two-band model of Kane (red line) and (c) parabolic energy bands (blue line). In

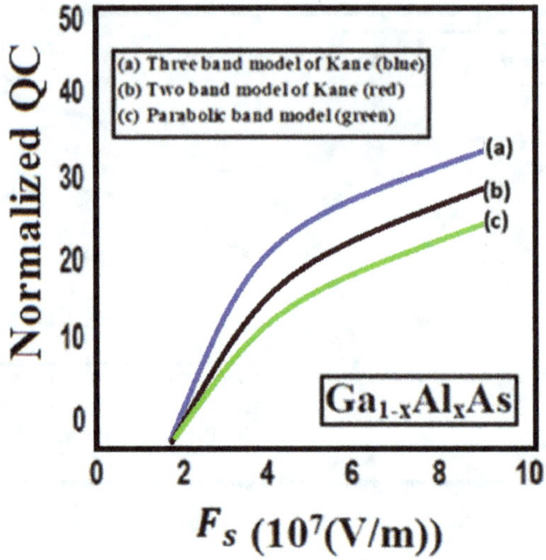

Fig. 1.26. The normalized QC in QMOSFETs of $Ga_{1-x}Al_xAs$ operated under inversion mode as a function of surface electric field has been plotted in the high electric field limit in accordance with (a) three-band model of Kane (blue line, (1.68)), (b) two-band model of Kane (brown line, (1.77)) and (c) parabolic energy bands (green line, (1.60)). We have used the electric quantum limit condition in obtaining all the plots.

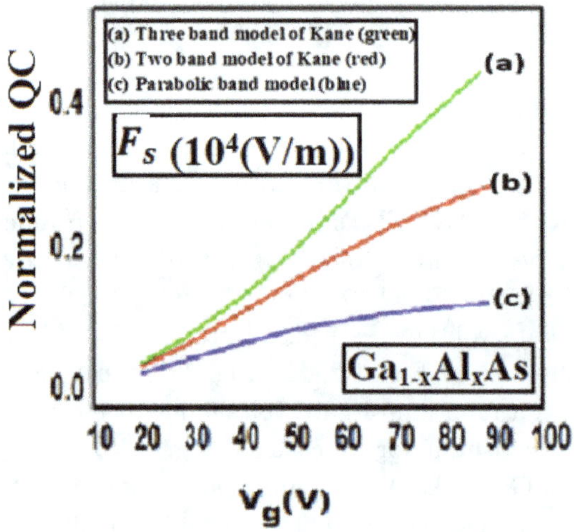

Fig. 1.27. The normalized QC in QMOSFETs of $Ga_{1-x}Al_xAs$ operated under inversion mode as a function of gate voltage has been plotted in the low electric field limit in accordance with (a) three-band model of Kane (green line, (1.38)), (b) two-band model of Kane (red line, (1.51)) and (c) parabolic energy bands (blue line, (1.60)). We have used the electric quantum limit condition in obtaining all the plots.

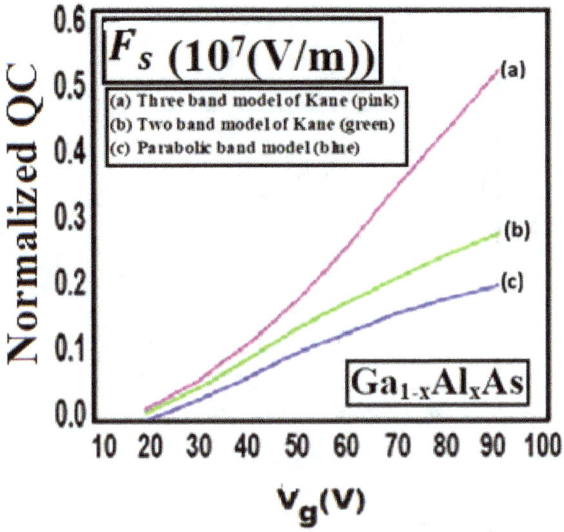

Fig. 1.28. The normalized QC in QMOSFETs of $Ga_{1-x}Al_xAs$ operated under inversion mode as a function of gate voltage has been plotted in the high electric field limit in accordance with (a) three-band model of Kane (pink line, (1.68)), (b) two-band model of Kane (green line, (1.77)) and (c) parabolic energy bands (blue line, (1.60)). We have used the electric quantum limit condition in obtaining all the plots.

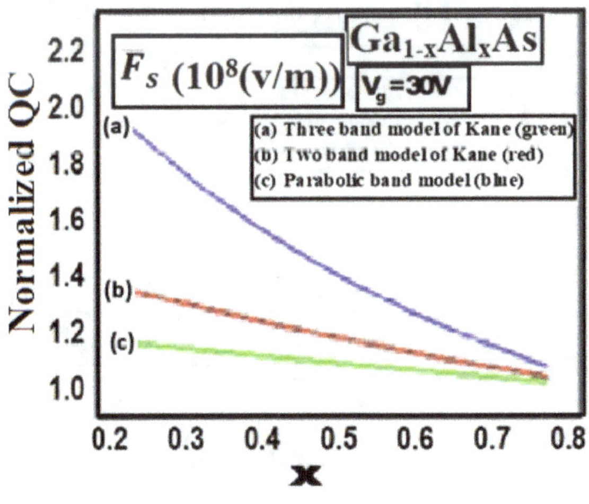

Fig. 1.29. The normalized QC in QMOSFETs of $Ga_{1-x}Al_xAs$ operated under inversion mode as a function of alloy composition has been plotted in the high electric field limit in accordance with (a) three-band model of Kane (green line, (1.68)), (b) two-band model of Kane (red line, (1.77)) and (c) parabolic energy bands (blue line, (1.60)). We have used the electric quantum limit condition in obtaining all the plots.

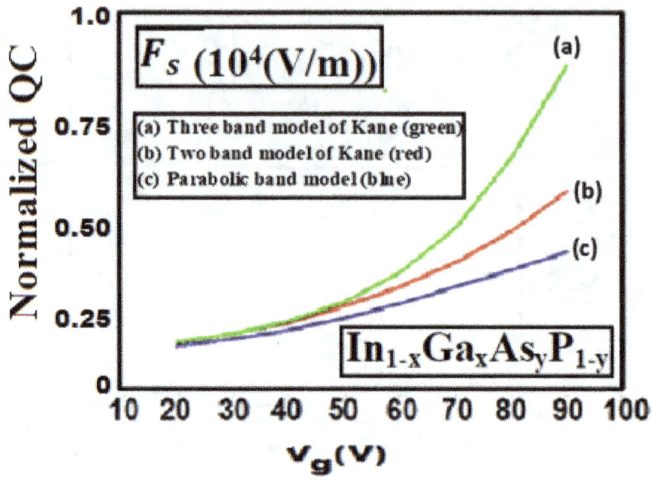

Fig. 1.30. The normalized QC in QMOSFETs of $In_{1-x}Ga_xAs_yP_{1-y}$ operated under inversion mode as a function of gate voltage has been plotted in the low electric field limit in accordance with (a) three-band model of Kane (green line, (1.68)), (b) two-band model of Kane (red line, (1.77)) and (c) parabolic energy bands (blue line, (1.60)). We have used the electric quantum limit condition in obtaining all the plots.

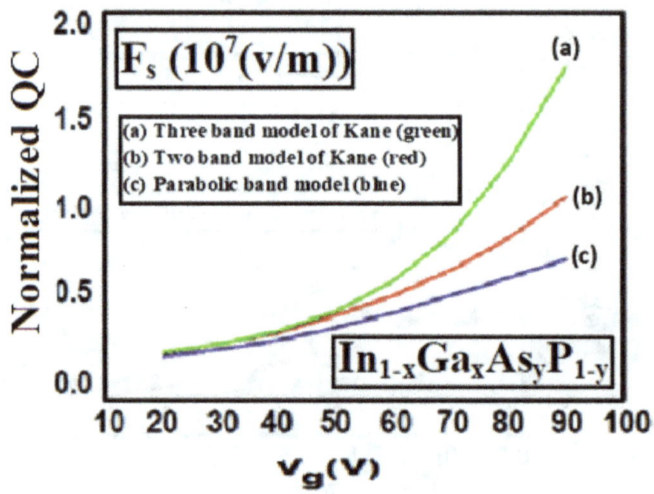

Fig. 1.31. The normalized QC in QMOSFETs of $In_{1-x}Ga_xAs_yP_{1-y}$ operated under inversion mode as a function of gate voltage has been plotted in the high electric field limit in accordance with (a) three-band model of Kane (green line, (1.68)), (b) two-band model of Kane (red line, (1.77)) and (c) parabolic energy bands (blue line, (1.60)). We have used the electric quantum limit condition in obtaining all the plots.

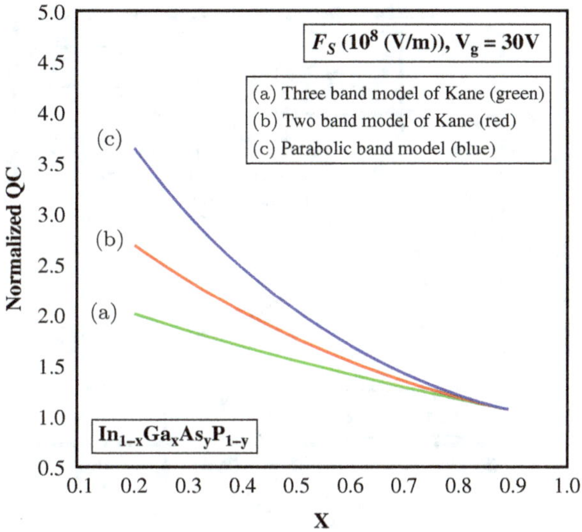

Fig. 1.32. The normalized QC in QMOSFETs of $In_{1-x}Ga_xAs_yP_{1-y}$ operated under inversion mode as a function of alloy composition has been plotted in the high electric field limit with $V_g = 30$ V in accordance with (a) three-band model of Kane (green line, (1.68)), (b) two-band model of Kane (red line, (1.77)) and (c) parabolic energy bands (blue line, (1.60)). We have used the electric quantum limit condition in obtaining all the plots.

Fig. 1.31, the QC in QMOSFETs of $In_{1-x}Ga_xAs_yP_{1-y}$ as a function of gate voltage has been plotted for $i = 0$ in the high electric field limit for all cases of Fig. 1.30. In Fig. 1.32, the QC in QMOSFETs of $In_{1-x}Ga_xAs_yP_{1-y}$ as a function of alloy composition has been plotted for $i = 0$ in the high electric field limit with $V_g = 30\,\mathrm{V}$ for all cases of Fig. 1.30. In Fig. 1.33, the QC in QMOSFETs of $In_{1-x}Ga_xAs_yP_{1-y}$ as a function of alloy composition has been plotted for $i = 0$ in the high electric field limit with $V_g = 80\,\mathrm{V}$ for all cases of Fig. 1.30.

It is apparent from Fig. 1.1 that the QC increases with increasing surface field in the low electric field limit showing a tendency of saturation at relatively low values of the surface field for all types of band models as considered here. The rate of increase in all the cases depends on the values of the energy band constant which differ from one type of band model to another type. Thus, the influence of the energy band models on the QC is also apparent from all the figures of this chapter. From Fig. 1.2, we observe that the QC increases with increasing surface electric field at the

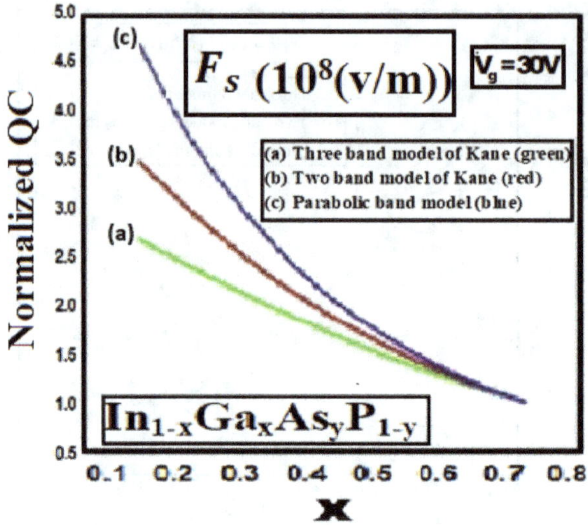

Fig. 1.33. The normalized QC in QMOSFETs of $In_{1-x}Ga_xAs_yP_{1-y}$ operated under inversion mode as a function of alloy composition has been plotted in the high electric field limit with $V_g = 30$ V in accordance with (a) three-band model of Kane (green line, (1.68)), (b) two-band model of Kane (red line, (1.77)) and (c) parabolic energy bands (blue line, (1.60)). We have used the electric quantum limit condition in obtaining all the plots.

high electric field limit with widely different numerical values for all types of band models. From Figs. 1.3 and 1.4, it appear that the QC increases with increasing gate voltage for both the limits for QMOSFETs of $CdGeAs_2$ operated under inversion layer mode. From Figs. 1.5 and 1.6, we note that the QC increases with increasing surface electric field for both weak and strong electric field limits for QMOSFETs of $CdGeAs_2$ operated under inversion layer mode. From Figs. 1.7 and 1.8, we can write that the QC in the present case increases with increasing gate voltage for both low and high electric field limits where the numerical values depend on the values of the energy band constants.

From Figs. 1.9 and 1.10, we observe that the QC in QMOSFETs of InSb operated under inversion layer mode increases with increasing surface electric field for both the limits in accordance with the three and two band models of Kane together with parabolic energy bands and they exhibit wide difference with each other. From Figs. 1.11 and 1.12, we note that the QC in the present case increases with increasing gate voltage for both the limits and as gate voltage increases the difference among the respective curves also increases.

From Figs. 1.13 and 1.14, we note that the QC in QMOSFETs of InAs operated under inversion layer mode increases with increasing surface electric field for both the limits in accordance with the three- and two-band models of Kane together with parabolic energy bands and they exhibit wide difference with each other. From Figs. 1.15 and 1.16, we observe that the QC in the present case increases with increasing gate voltage for both the limits and as gate voltage increases the difference among the respective curves also increases. From Figs. 1.17 and 1.18, we note that the QC in QMOSFETs of $Hg_{1-x}Cd_xTe$ operated under inversion layer mode increases with increasing surface electric field for both the limits in accordance with the three and two band models of Kane together with parabolic energy bands. From Figs. 1.19 and 1.20, we observe that the QC in the present case increases with increasing gate voltage for both the limits and as gate voltage increases the difference among the respective curves also increases. From Figs. 1.21 to 1.24, we observe that the QC in QMOSFETs of $Hg_{1-x}Cd_xTe$ operated under inversion layer mode decreases with increasing alloy composition for both high and low electric field limits for all types of band models for different values of gate voltage.

From Figs. 1.25 and 1.26 we note that the QC in QMOSFETs of $Ga_{1-x}Al_xAs$ increases with increasing surface electric field for both the limits in accordance with the three and two-band models of Kane together with parabolic energy bands. From Figs. 1.27 and 1.28, we observe that the QC in QMOSFETs of $Ga_{1-x}Al_xAs$ increases with increasing gate voltage for both the limits and as gate voltage increases the difference among the respective curves also increases. From Fig. 1.29, we observe that the QC in QMOSFETs of $Ga_{1-x}Al_xAs$ decreases with increasing alloy composition for all types of band models.

From Figs. 1.30 and 1.31, we observe that the QC in QMOSFETs of $In_{1-x}Ga_xAs_yP_{1-y}$ increases with increasing gate voltage for both the limits and as gate voltage increases the difference among the respective curves also increases. From Figs. 1.32 and 1.33, we observe that the QC in QMOSFETs of $In_{1-x}Ga_xAs_yP_{1-y}$ decreases with increasing alloy composition for all types of band models. It may be noted that the thickness of the oxide layer is a critical parameter that determines the nature of dependence of the QC on different external variables. For thin oxide layer, C_{ox} will be very large and the QC will be determined mainly by the capacitance of the surface space charge layer and, consequently, will be dependent on the surface concentration which increases with increasing gate voltage. On the other hand, when the oxide layer is relatively thick, C_{ox} and, hence, will be fairly independent of the gate voltage.

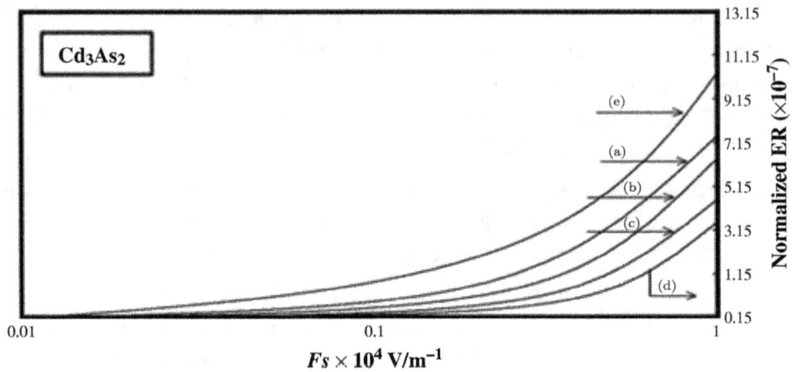

Fig. 1.34. The plot of the normalized ER in QMOSFETs of Cd_3As_2 operated under inversion layer mode and in the presence of weak electric field quantum limit as a function of surface electric field in accordance with (a) the generalized band model (1.19), (b)$\delta = 0$, (c) the three-band model of Kane (1.40), (d) the two-band model of Kane (1.53a) and (e) the parabolic energy bands.

It is felt that the theoretical investigation presented here would be of much significance, as the interest in QC has been growing very much in recent years from the point of view of technical application and of exploration of other fundamental aspect of semiconductor surface in MOS structure.

Using the appropriate equations and the values of the energy band constants from Chapter 18, the plot for the normalized ER for n-channel inversion layers of Cd_3As_2 under weak electric field limit as a function of surface electric field has been shown in curve (a) of Fig. 1.34. The curve (b) corresponds to $\delta = 0$ and the curve (c) exhibits the dependence of the ER on the surface electric field in accordance with the three-band model of Kane, respectively. The plots (d) and (e) correspond to the two-band model of Kane and that of parabolic energy bands, respectively. By comparing curves (a) and (b) of Fig. 1.34, one can assess the influence of crystal field splitting of the ER in inversion layers of tetragonal compounds. Figure 1.35 represents all cases of Fig. 1.34 under high electric field limit.

It appears from Fig. 1.35 that, the ER increases with increasing surface electric field and the numerical values of the ER in the high electric field are much greater as compared with the corresponding low electric field limit. In Figs. 1.36 and 1.37, the ER for n-channel inversion layers of $CdGeAs_2$ has been drawn as a function of surface electric field for both weak and strong electric field limits respectively for all the cases of Fig. 1.34.

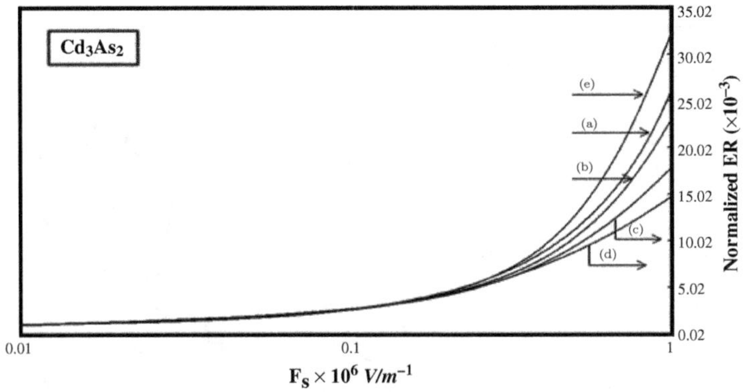

Fig. 1.35. The plot of the normalized ER in QMOSFETs of Cd_3As_2 operated under inversion layer mode and in the presence of strong electric field quantum limit as a function of surface electric field in accordance with (a) the generalized band model (1.30), (b) $\delta = 0$, (c) the three-band model of Kane (1.70), (d) the two-band model of Kane (1.78b) and (e) the parabolic energy bands.

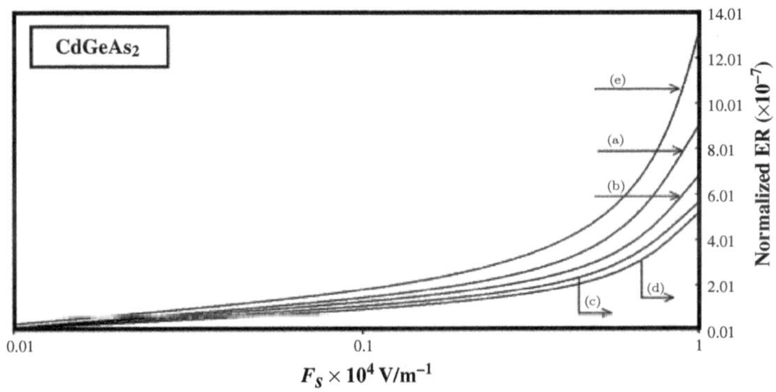

Fig. 1.36. The plot of the normalized ER in QMOSFETs of $CdGeAs_2$ operated under inversion layer mode and in the presence of weak electric field quantum limit as a function of surface electric field in accordance with (a) the generalized band model (1.19), (b) $\delta = 0$, (c) the three-band model of Kane (1.40), (d) the two-band model of Kane (1.53a) and (e) the parabolic energy bands.

The trend of variation of the ER for n-channel inversion layers of $CdGeAs_2$ is more or less the same with different numerical magnitudes as compared with n-channel inversion layers of Cd_3As_2 for both the limits.

Figures 1.38 and 1.39 exhibit the normalized ER in n-channel inversion layers of InAs for both weak and strong electric field quantum limits have

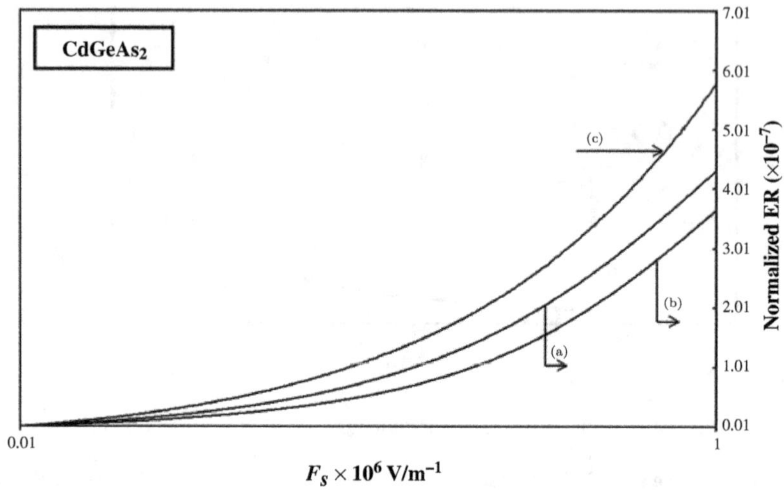

Fig. 1.37. The plot of the normalized ER in QMOSFETs of $CdGeAs_2$ operated under inversion layer mode and in the presence of strong electric field quantum limit as a function of surface electric field in accordance with (a) the generalized band model (1.30), (b) $\delta = 0$, (c) the three-band model of Kane (1.70), (d) the two-band model of Kane (1.78b) and (e) the parabolic energy bands.

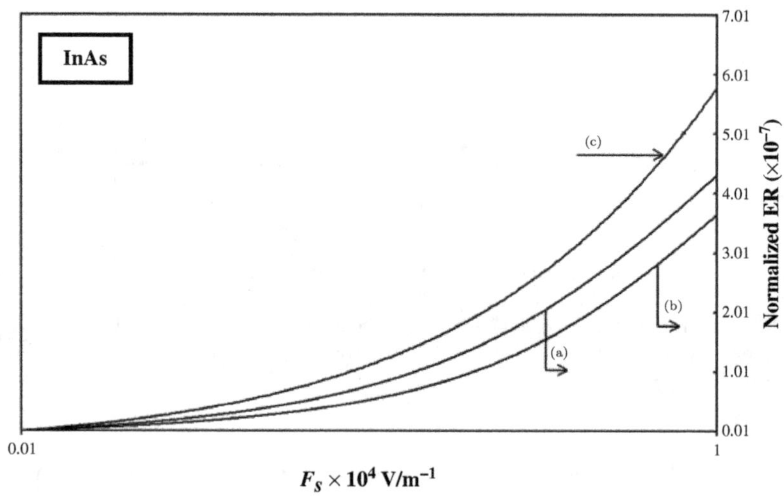

Fig. 1.38. The plot of the normalized ER in QMOSFETs of InAs operated under inversion layer mode and in the presence of weak electric field quantum limit as a function of surface electric field in accordance with (a) the three-band model of Kane (1.40), (b) the two-band model of Kane (1.53a) and (c) the parabolic energy bands.

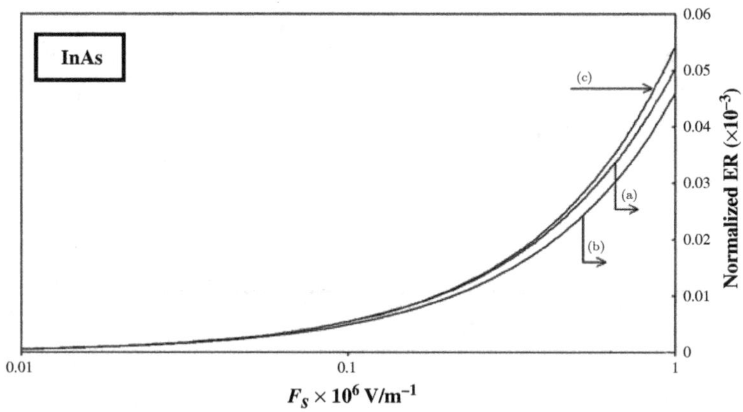

Fig. 1.39. The plot of the normalized ER in QMOSFETs of InAs operated under inversion layer mode and in the presence of strong electric field quantum limit as a function of surface electric field in accordance with (a) the three-band model of Kane (1.70), (b) the two-band model of Kane (1.78b) and (c) the parabolic energy bands.

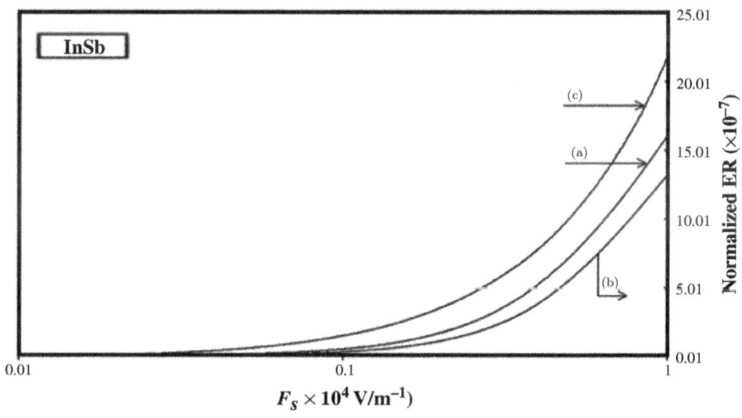

Fig. 1.40. The plot of the normalized ER in QMOSFETs of InSb operated under inversion layer mode and in the presence of weak electric field quantum limit as a function of surface electric field in accordance with (a) the three-band model of Kane (1.40), (b) the two-band model of Kane (1.53a) and (c) the parabolic energy bands.

been drawn as a function of surface electric field in accordance with (a) the three-band model of Kane, (b) the two-band model of Kane and (c) the parabolic energy bands, respectively. In Figs. 1.40 and 1.41, all cases of Figs. 1.38 and 1.39 have been drawn as a function of surface electric field for n-channel inversion layers of InSb for both the limits. In Figs. 1.42 and

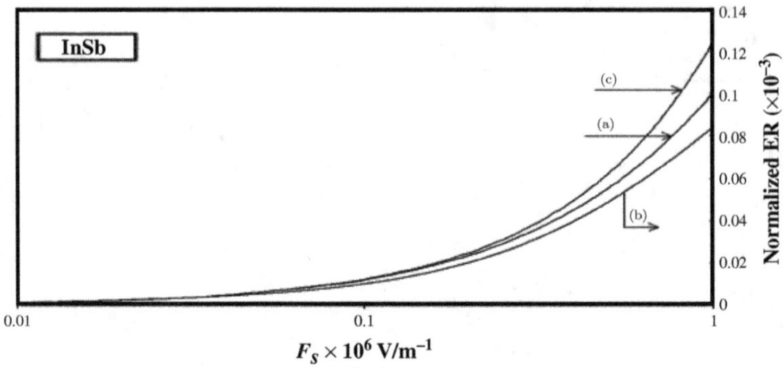

Fig. 1.41. The plot of the normalized ER in QMOSFETs of InSb operated under inversion layer mode and in the presence of strong electric field quantum limit as a function of surface electric field in accordance with (a) the three-band model of Kane (1.70), (b) the two-band model of Kane (1.78b) and (c) the parabolic energy bands.

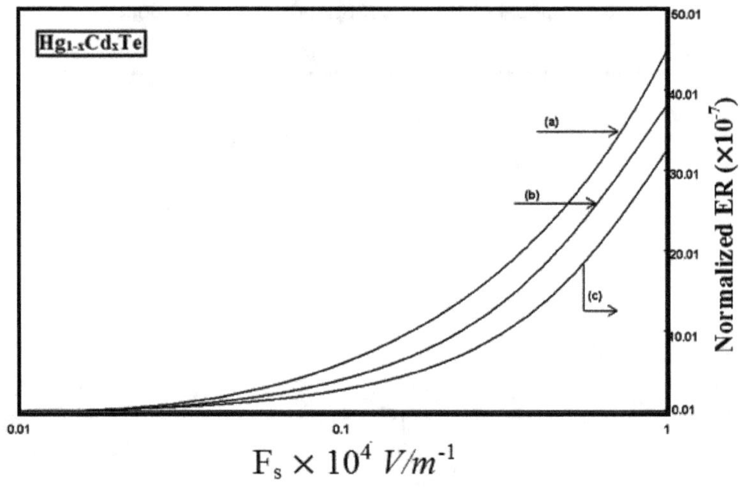

Fig. 1.42. The plot of the normalized ER in QMOSFETs of $Hg_{1-x}Cd_xTe$ operated under inversion layer mode and in the presence of weak electric field quantum limit as a function of surface electric field in accordance with (a) the three-band model of Kane (1.40), (b) the two-band model of Kane (1.53a) and (c) the parabolic energy bands.

1.43, the normalized ER in n-channel inversion layers of $Hg_{1-x}Cd_xTe$ have been drawn for both the weak and strong electric field limits as functions of surface electric field for all the cases of Fig. 1.38. In Figs. 1.44 and 1.45, the normalized ER in n-channel inversion layers of $In_{1-x}Ga_xAs_yP_{1-y}$ lattice matched to InP have been drawn for both the limits as functions of surface

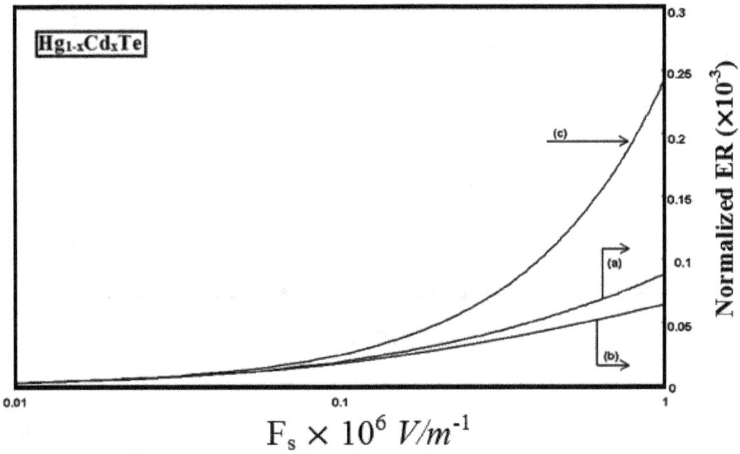

Fig. 1.43. The plot of the normalized ER in QMOSFETs of $Hg_{1-x}Cd_xTe$ operated under inversion layer mode and in the presence of strong electric field quantum limit as a function of surface electric field in accordance with (a) the three-band model of Kane (1.70), (b) the two-band model of Kane (1.78b) and (c) the parabolic energy bands.

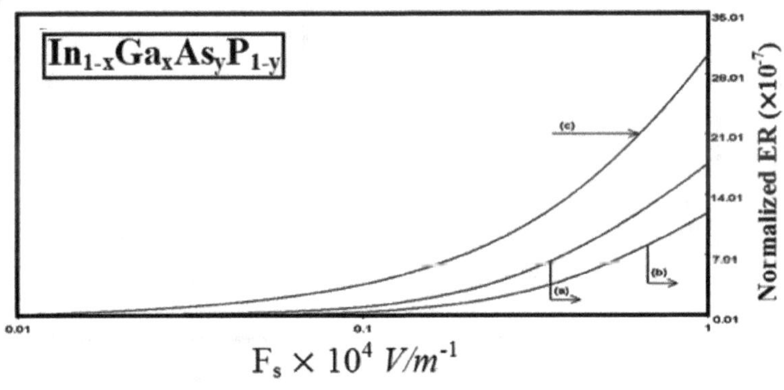

Fig. 1.44. The plot of the normalized ER in QMOSFETs of $In_{1-x}Ga_xAs_yP_{1-y}$ operated under inversion layer mode and in the presence of weak electric field quantum limit as a function of surface electric field in accordance with (a) the three-band model of Kane (1.40), (b) the two-band model of Kane (1.53a) and (c) the parabolic energy bands.

electric field for all the cases of Fig. 1.38. It appears from Figs. 1.38 to 1.45 that the ER for n-channel inversion layers of InAs, InSb, $Hg_{1-x}Cd_xTe$ and $In_{1-x}Ga_xAs_yP_{1-y}$ lattice matched to InP increases with increase in surface electric field for both weak and strong electric field limits with different numerical values and the influence of the energy band constants can also

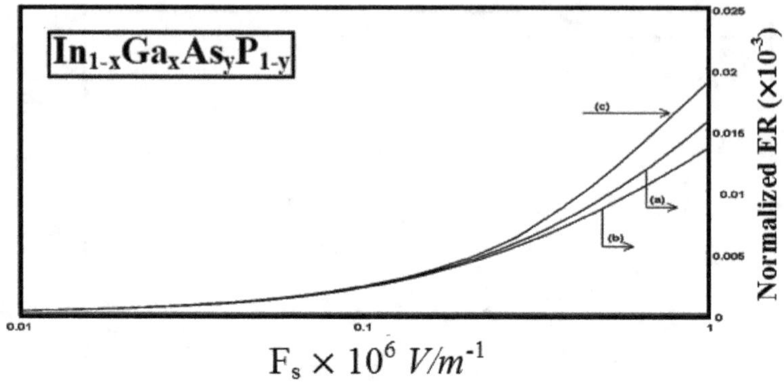

Fig. 1.45. The plot of the normalized ER in QMOSFETs of $In_{1-x}Ga_xAs_yP_{1-y}$ operated under inversion layer mode and in the presence of strong electric field quantum limit as a function of surface electric field in accordance with (a) the three-band model of Kane (1.70), (b) the two-band model of Kane (1.78b) and (c) the parabolic energy bands.

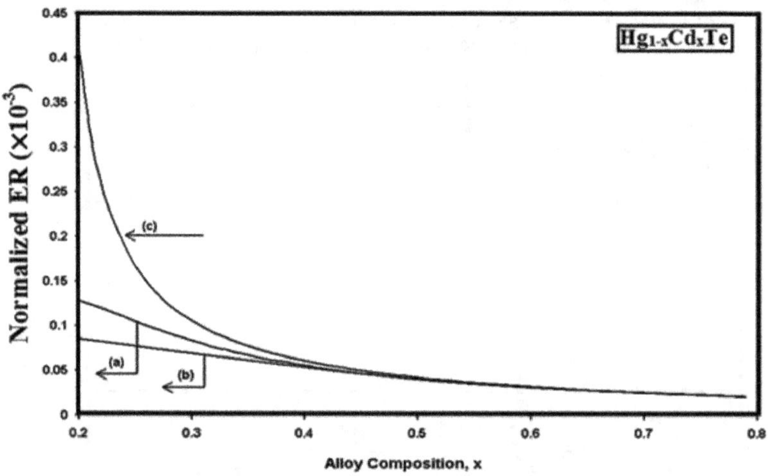

Fig. 1.46. The plot of the normalized ER in QMOSFETs of $Hg_{1-x}Cd_xTe$ operated under inversion layer mode and in the presence of strong electric field quantum limit as a function of alloy composition in accordance with (a) the three-band model of Kane (1.70), (b) the two-band model of Kane (1.78b) and (c) the parabolic energy bands.

be assessed from the said figures. In Figs. 1.46 and 1.47, the normalized ER for n-channel inversion layers of $Hg_{1-x}Cd_xTe$ and $In_{1-x}Ga_xAs_yP_{1-y}$ lattice matched to InP have been drawn as a function of alloy composition under strong electric field limit in accordance with the three- and two-band models of Kane together with parabolic energy bands, respectively. It appears from

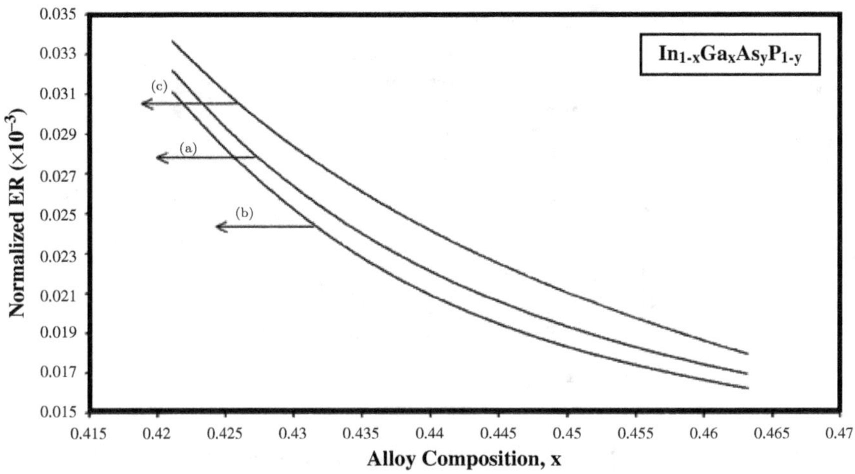

Fig. 1.47. The plot of the normalized ER in QMOSFETs of $In_{1-x}Ga_xAs_yP_{1-y}$ operated under inversion layer mode and in the presence of strong electric field quantum limit as a function of alloy composition in accordance with (a) the three-band model of Kane (1.70), (b) the two-band model of Kane (1.78b) and (c) the parabolic energy bands.

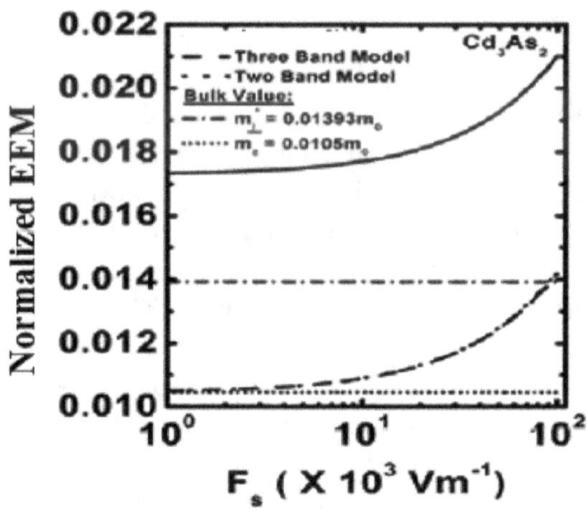

Fig. 1.48. Plot of the normalized EEM at weak electric field limit as the function of surface electric field for QMOSFETs of Cd_3As_2 operated under inversion mode in accordance with the generalized theory (1.20). The simplified results for three (1.41) and two-band (1.54) models of Kane have also been exhibited in which, $m_\perp^* = 0.01393m_0$ and $m_c = \frac{1}{2}(m_\perp^* + m_\parallel^*) = 0.0105m_0$ are the corresponding bulk values.

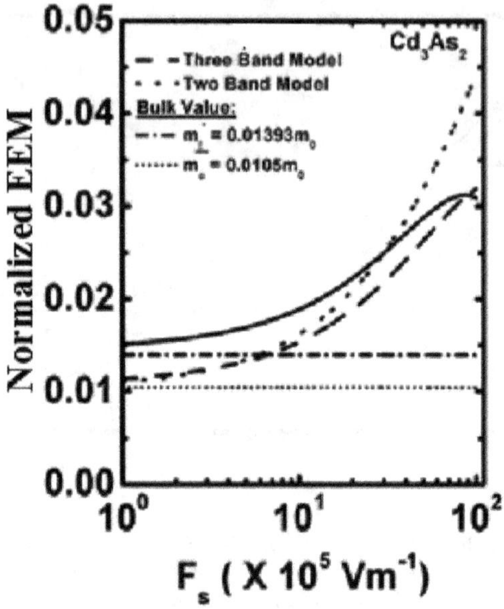

Fig. 1.49. Plot of the normalized EEM as a function of surface electric field under weak electric field limit for QMOSFETs of Cd_3As_2 operated under inversion mode for all cases of Fig. 1.48.

Figs. 1.46 and 1.47 that the ER decreases with increasing alloy composition although the rate of decrease is determined by the respective energy band constants of the ternary and quaternary materials.

The effect of surface electric field on the normalized EEM at the quantum limit in n-channel inversion layers of Cd_3As_2 and $CdGeAs_2$ has been exhibited in Figs. 1.48–1.51. In Figs. 1.48 and 1.49, we have demonstrated the variation of the normalized EEM with electric field in the weak inversion regime which was extended up to 10^5 Vm^{-1}. It appears that with the increase in the electric field, the normalized EEM in Fig. 1.48 increases considering the generalized energy band model and the three- and the two-band models of Kane which are the special cases of our generalized analysis. It appears that in the weak field regime, the deviation between the three- and two-band models of Kane is less however significant difference is with the consideration of the crystal field. It should be noted that it is these two models which tend to the isotropic bulk effective mass value $0.0105m_0$, rather than the generalized model. In the high field regime, Fig. 1.49, the difference in the three- and two-band models of Kane appears which marks a significant variation in the value of the normalized EEM. The effect of

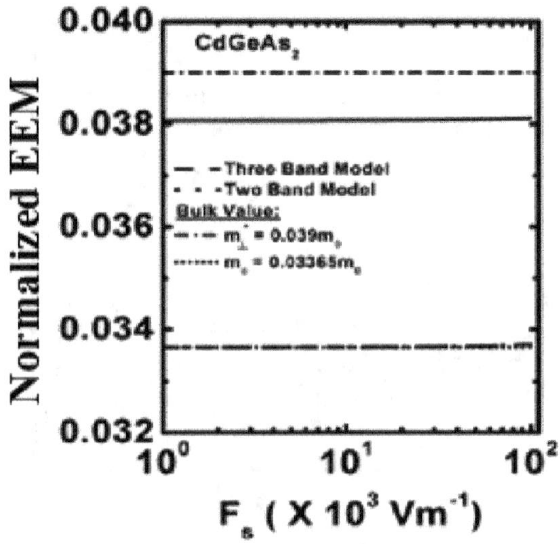

Fig. 1.50. Plot of the normalized EEM as a function of surface electric field under weak electric field limit for QMOSFETs of $CdGeAs_2$ operated under inversion mode in accordance with the generalized theory (1.20). The simplified results for three (1.41) and two-band (1.54) models of Kane have also been exhibited in which, $m_{\perp}^* = 0.039m_0$ and $m_c = \frac{1}{2}(m_{\perp}^* + m_{\parallel}^*) = 0.03365m_0$ are the corresponding bulk values.

crystal field splitting tends to decrease the normalized EEM considering the generalized energy band model. A closer look at the two figures reveals more interesting features of the continuity of the weak inversion energy band model in the strong field and strong inversion energy band model in the weak field. It is due to this non-convergence there is a slight mismatch of the normalized EEM at the boundary of 10^5 Vm^{-1} in both Figs. 1.48 and 1.49 and needs more attention towards the development of the generalized theory valid for all values of electric field is still a formidable problem in this case.

Figures 1.50 and 1.51 exhibit the normalized EEM in n-$CdGeAs_2$ for all the cases of Figs. 1.48 and 1.49, respectively. It appears that in the weak inversion regime, the normalized EEM is almost invariant of the electric field; however, with the increase in the field, the normalized EEM sharply decreases and tends to take negative values which challenge the applicability of the quantization condition (1.7) at strong electric field for n-channel $CdGeAs_2$. The effect of electric field on the normalized EEM of n-channel InAs has been exhibited in Figs. 1.52 and 1.53. Almost no variation of the normalized EEM in weak field appears for n-InAs channel while for higher

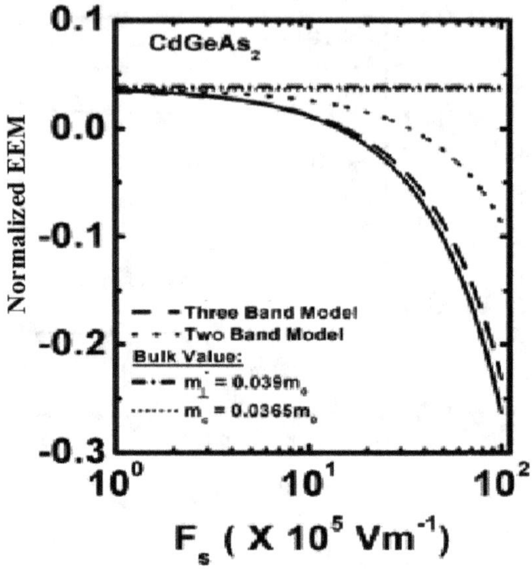

Fig. 1.51. Plot of the normalized EEM as a function of surface electric field under strong electric field limit for QMOSFETs of $CdGeAs_2$ operated under inversion mode for all cases of Fig. 1.50.

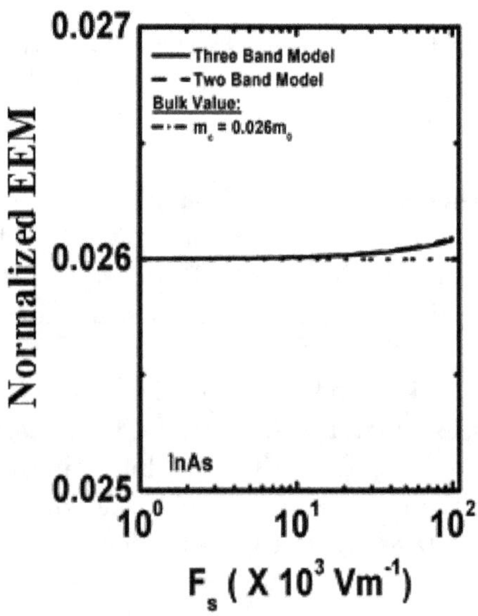

Fig. 1.52. Plot of the normalized EEM at low electric field limit as a function of surface electric field for QMOSFETs of InAs operated under inversion mode.

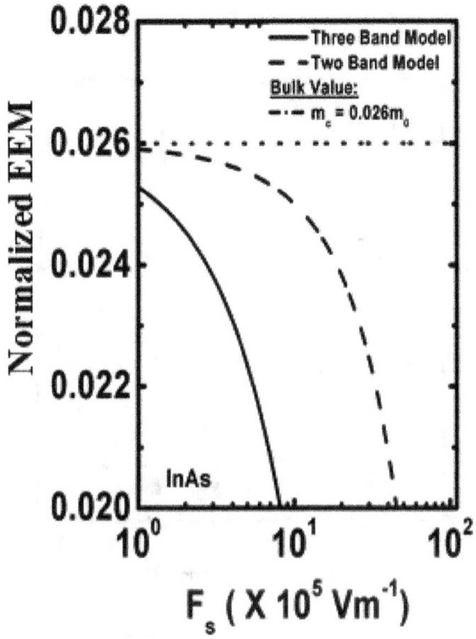

Fig. 1.53. Plot of the normalized EEM at high electric field limit as a function of surface electric field for QMOSFETs of InAs operated under inversion mode.

fields, the normalized EEM tends to decrease. The effect of surface electric field on n-channel GaAs and InSb at weak and strong electric field has been exhibited in Fig. 1.54.

Using the appropriate equations, in Fig. 1.55 we have plotted the normalized ΔC_{44} as a function of low electric field in the quantum limit in n-channel inversion layers of Cd_3As_2 in accordance with the model of (a) generalized band model with $\delta \neq 0$, (b) generalized band model with $\delta = 0$, (c) three-band model of Kane and (d) two-band model of Kane. In Fig. 1.56, we have plotted all cases of Fig. 1.55 for high electric field limit. From both the figures we note that ΔC_{44} increases with increasing electric field and the numerical magnitude of ΔC_{44} in each case depends on the values of the energy band constants of each material, respectively. In Fig. 1.57, we have plotted the normalized ΔC_{44} as a function of electric field in n-channel inversion layers of $CdGeAs_2$ in accordance with the generalized band model under weak field limit at the electric quantum limit. In Fig. 1.60, we have plotted the normalized ΔC_{44} for n-$CdGeAs_2$ under strong electric field quantum limit as a function of surface electric field in accordance with (a) generalized band model with $\delta \neq 0$, (b) generalized band model with

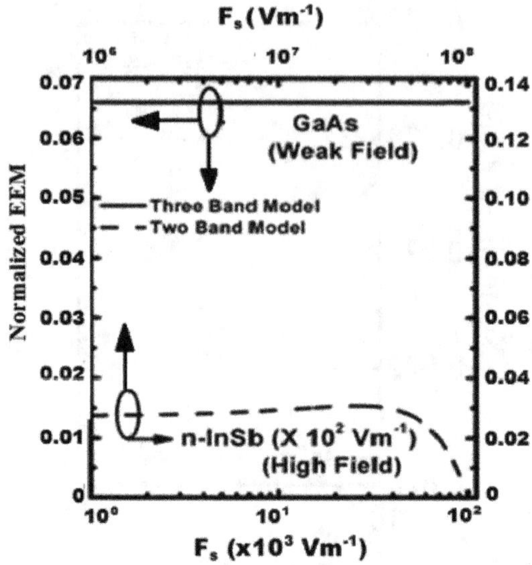

Fig. 1.54. Plot of the normalized EEM at low and high electric field limits as a function of surface electric field for QMOSFETs of GaAs and InSb, respectively, operated under inversion mode, respectively.

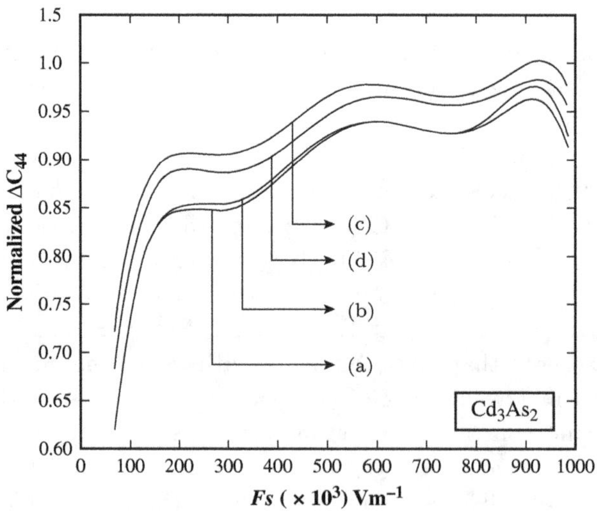

Fig. 1.55. Plot of the normalized ΔC_{44} for QMOSFETs of Cd_3As_2 under weak electric field quantum limit operated under inversion mode as a function of surface electric field in accordance with (a) generalized band model with $\delta \neq 0$, (b) generalized band model with $\delta = 0$, (c) three-band model of Kane and (d) two-band model of Kane.

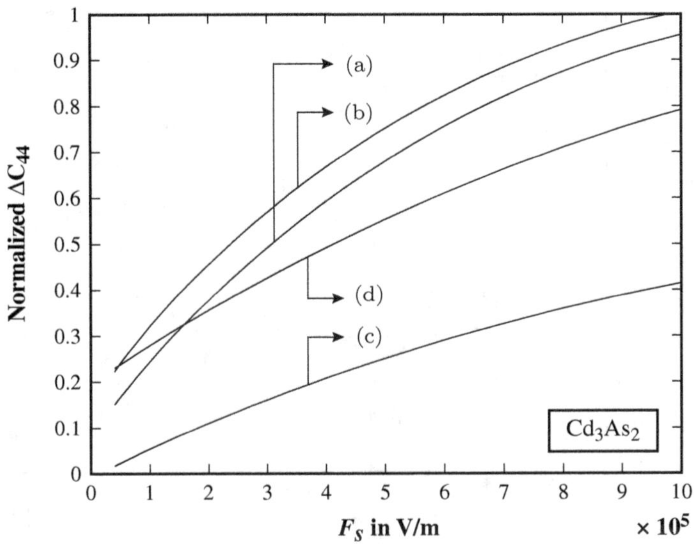

Fig. 1.56. Plot of the normalized ΔC_{44} for QMOSFETs of Cd_3As_2 under strong electric field quantum limit operated under inversion mode as a function of surface electric field in accordance with (a) generalized band model with $\delta \neq 0$, (b) generalized band model with $\delta = 0$, (c) three-band model of Kane and (d) two-band model of Kane.

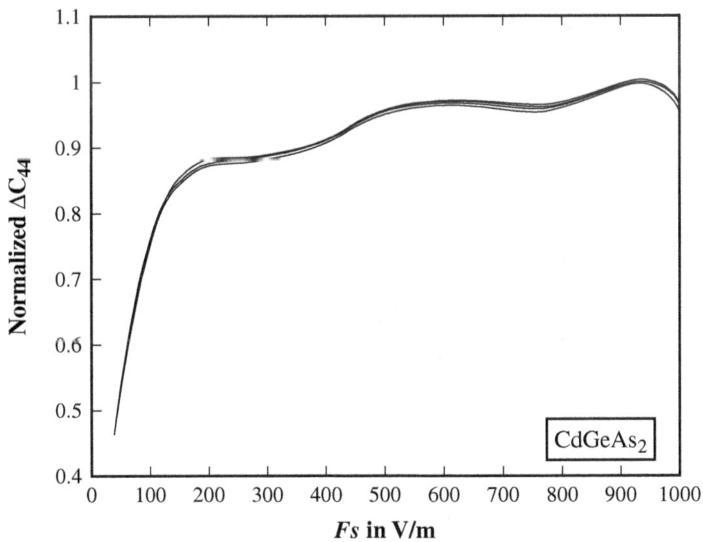

Fig. 1.57. Plot of the normalized ΔC_{44} for QMOSFETs of $CdGeAs_2$ under weak electric field quantum limit operated under inversion mode as a function of surface electric field in accordance with the generalized band model.

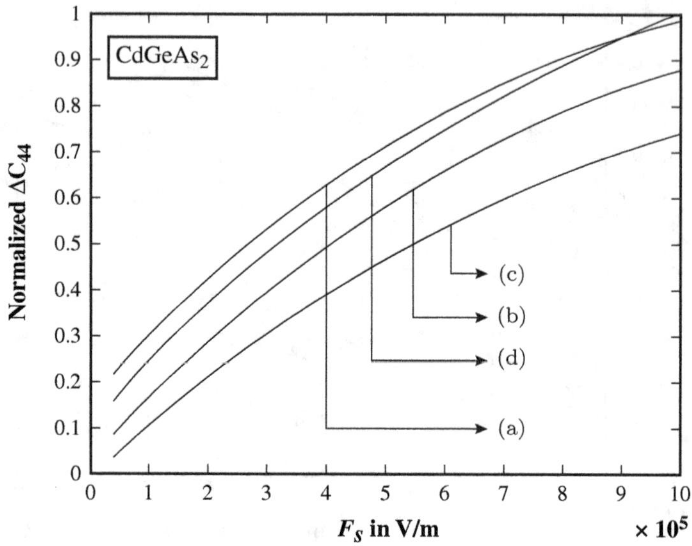

Fig. 1.58. Plot of the normalized ΔC_{44} for QMOSFETs of $CdGeAs_2$ under strong electric field quantum limit operated under inversion mode as a function of surface electric field in accordance with (a) generalized band model with $\delta \neq 0$, (b) generalized band model with $\delta = 0$, (c) three-band model of Kane and (d) two-band model of Kane.

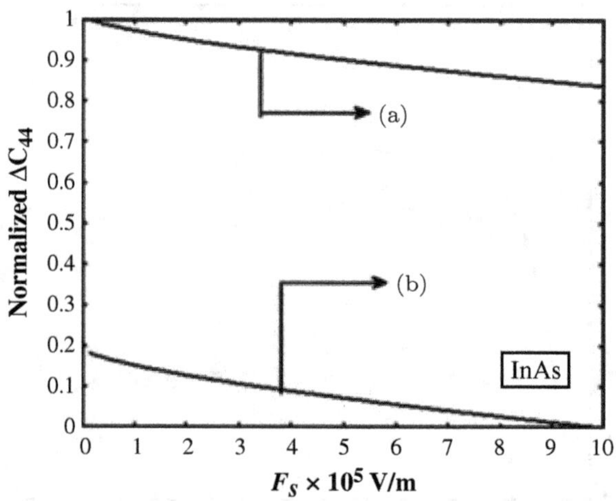

Fig. 1.59. Plot of the normalized ΔC_{44} in QMOSFETs of InAs under weak electric field quantum limit operated under inversion mode as a function of surface electric field in accordance with (a) three-band model of Kane and (b) two-band model of Kane, respectively.

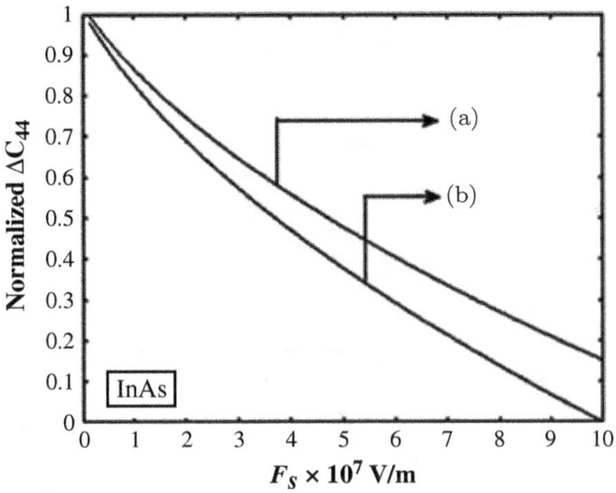

Fig. 1.60. Plot of the normalized ΔC_{44} in QMOSFETs of InAs under strong electric field quantum limit operated under inversion mode as a function of surface electric field in accordance with (a) three-band model of Kane and (b) two-band model of Kane, respectively.

$\delta = 0$, (c) three-band model of Kane and (d) two-band model of Kane. We note that the nature of variations of ΔC_{44} with electric field has not altered much, the numerical values of ΔC_{44} for all the band models are different which is the direct influence of the energy band constants in both cases and for all types of dispersion relations. In Figs. 1.61 and 1.62, we have plotted of the normalized ΔC_{44} in n-channel inversion layers of InAs under both weak and strong electric field quantum limits as a function of surface electric field in accordance with (a) three-band model of Kane and (b) two-band model of Kane, respectively. It appears from both the figures that ΔC_{44} decreases with increasing surface electric field in both the cases although for low electric field limit the curves for both the models of ΔC_{44} are more wide as compared with the corresponding high electric field limit for the whole range of the surface electric field as considered here for Figs. 1.59 and 1.60, respectively. In Figs. 1.61 and 1.62, we have plotted of the normalized ΔC_{44} in n-channel inversion layers of InSb under both weak and strong electric field quantum limits as a function of surface electric field in accordance with (a) three-band model of Kane and (b) two-band model of Kane, respectively. In Figs. 1.63 and 1.64, we have plotted the normalized ΔC_{44} in n-channel inversion layers of $Hg_{1-x}Cd_xTe$ under both weak and strong electric field quantum limits as a function of surface electric

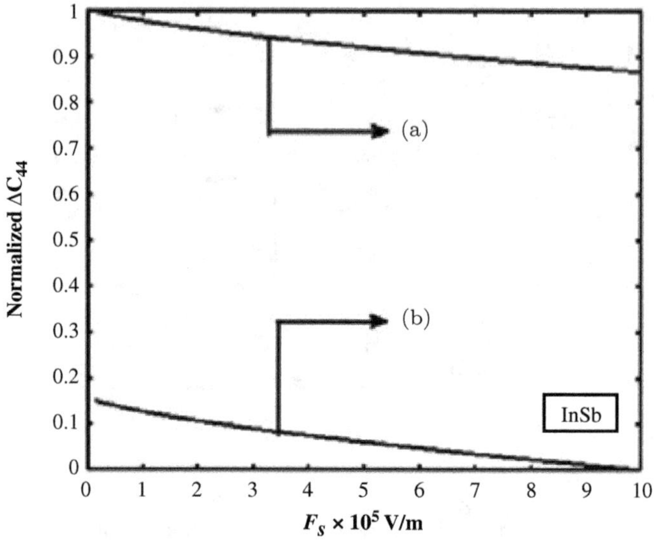

Fig. 1.61. Plot of the normalized ΔC_{44} in QMOSFETs of InSb under weak electric field quantum limit operated under inversion mode as a function of surface electric field in accordance with (a) three-band model of Kane and (b) two-band model of Kane, respectively.

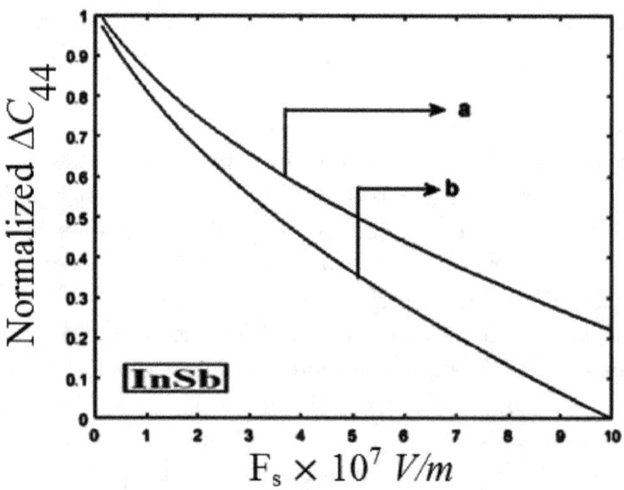

Fig. 1.62. Plot of the normalized ΔC_{44} in QMOSFETs InSb under strong electric field quantum limit operated under inversion mode as a function of surface electric field in accordance with (a) three-band model of Kane and (b) two-band model of Kane, respectively.

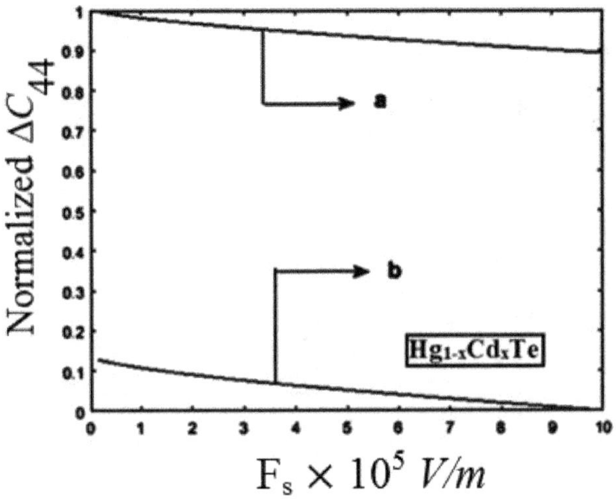

Fig. 1.63. Plot of the normalized ΔC_{44} in QMOSFETs of $Hg_{1-x}Cd_xTe$ under weak electric field quantum limit operated under inversion mode as a function of surface electric field in accordance with (a) three-band model of Kane and (b) two-band model of Kane, respectively.

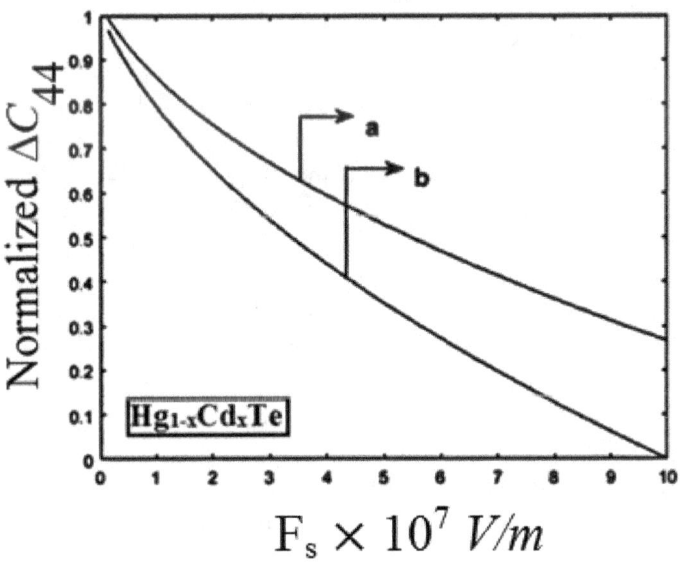

Fig. 1.64. Plot of the normalized ΔC_{44} in QMOSFETs of $Hg_{1-x}Cd_xTe$ under strong electric field quantum limit operated under inversion mode as a function of surface electric field in accordance with (a) three-band model of Kane and (b) two-band model of Kane, respectively.

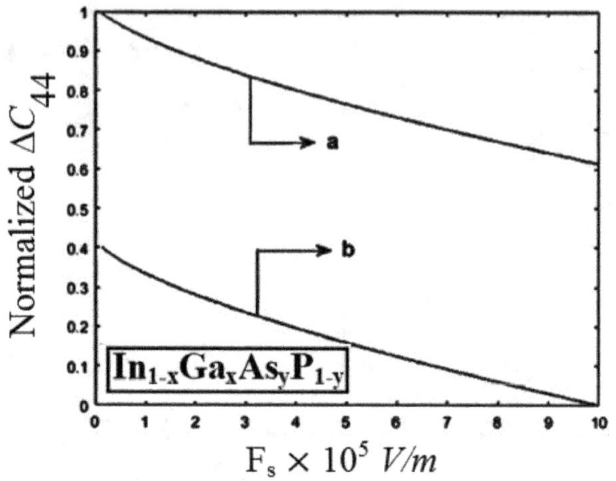

Fig. 1.65. Plot of the normalized ΔC_{44} in QMOSFETs of $In_{1-x}Ga_xAs_yP_{1-x}$ lattice matched to InP under weak electric field quantum limit operated under inversion mode as a function of surface electric field in accordance with (a) three-band model of Kane and (b) two-band model of Kane, respectively.

field in accordance with (a) three-band model of Kane and (b) two-band model of Kane, respectively. In Figs. 1.65 and 1.66, we have plotted of the normalized ΔC_{44} in n-channel inversion layers of $In_{1-x}Ga_xAs_yP_{1-y}$ lattice matched to InP under both weak and strong electric field quantum limits as a function of surface electric field in accordance with (a) three-band model of Kane and (b) two-band model of Kane, respectively. From Figs. 1.59 up to 1.66, we observe that ΔC_{44} decreases with increasing electric field with different gradients in all the cases and numerical values of ΔC_{44} are different in each case for both three- and two-band models of Kane. In Figs. 1.67 and 1.68, we have plotted of the normalized ΔC_{44} in n-channel inversion layers of $Hg_{1-x}Cd_xTe$ under both weak and strong electric field quantum limits as a function of alloy composition in accordance with (a) three-band model of Kane and (b) two-band model of Kane, respectively. In Figs. 1.69 and 1.70, we have plotted of the normalized ΔC_{44} in n-channel inversion layers of $In_{1-x}Ga_xAs_yP_{1-y}$ lattice matched to InP under both weak and strong electric field quantum limits as a function of alloy composition in accordance with (a) three-band model of Kane and (b) two-band model of Kane, respectively. It appears from Figs. 1.67 up to 1.70 that ΔC_{44} increases with decreasing alloy composition for all the cases.

In Fig. 1.71, we have plotted the normalized ΔC_{456} for n-Cd_3As_2 under weak electric field quantum limit as a function of surface electric field

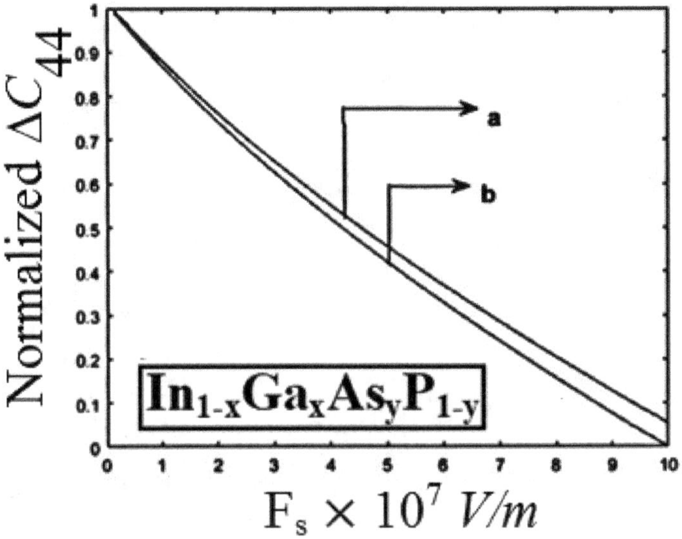

Fig. 1.66. Plot of the normalized ΔC_{44} in QMOSFETs of $In_{1-x}Ga_xAs_yP_{1-x}$ lattice matched to InP under strong electric field quantum limit operated under inversion mode as a function of surface electric field in accordance with (a) three-band model of Kane and (b) two-band model of Kane, respectively.

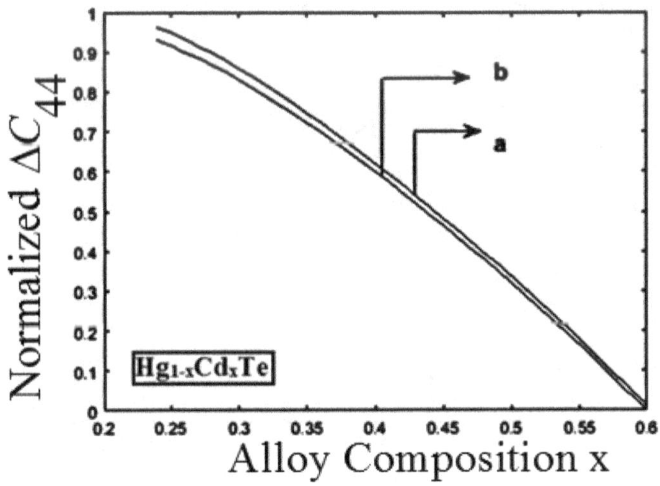

Fig. 1.67. Plot of the normalized ΔC_{44} in QMOSFETs of $Hg_{1-x}Cd_xTe$ under weak electric field quantum limit operated under inversion mode as a function of alloy composition in accordance with (a) three-band model of Kane and (b) two-band model of Kane, respectively.

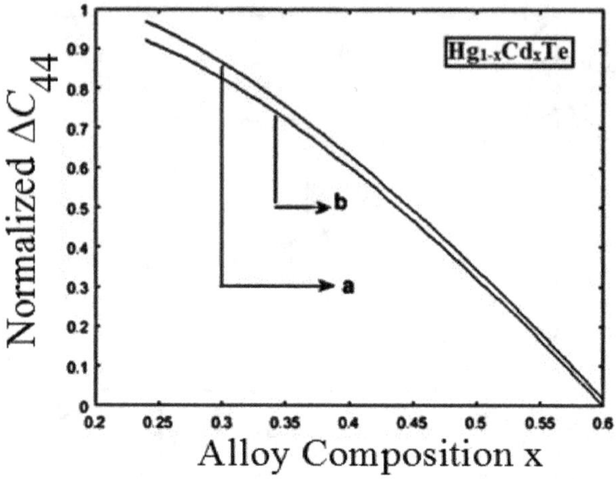

Fig. 1.68. Plot of the normalized ΔC_{44} in QMOSFETs of $Hg_{1-x}Cd_xTe$ under strong electric field quantum limit operated under inversion mode as a function of alloy composition in accordance with (a) three-band model of Kane and (b) two-band model of Kane, respectively.

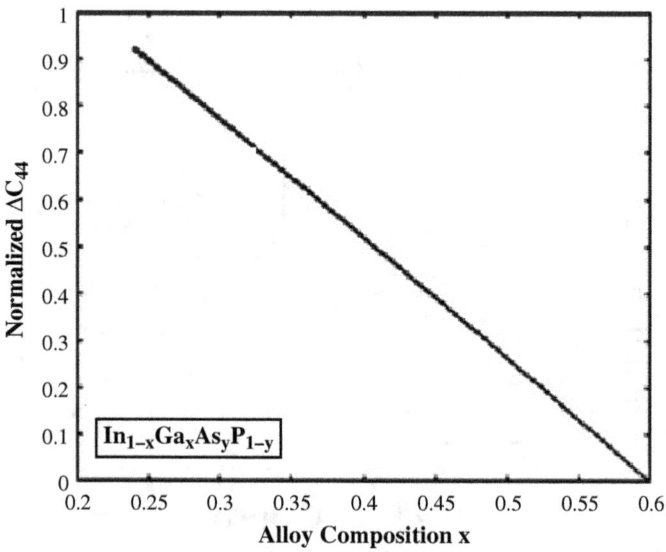

Fig. 1.69. Plot of the normalized ΔC_{44} in QMOSFETs of $In_{1-x}Ga_xAs_yP_{1-x}$ lattice matched to InP under weak electric field quantum limit operated under inversion mode as a function of alloy composition in accordance with three-band model of Kane.

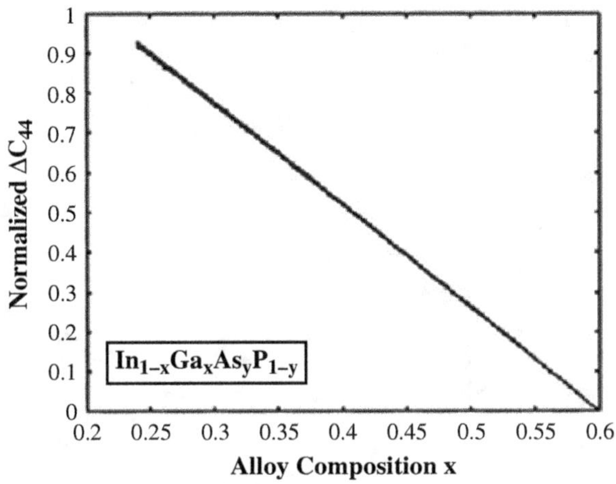

Fig. 1.70. Plot of the normalized ΔC_{44} in QMOSFETs of $In_{1-x}Ga_xAs_yP_{1-x}$ lattice matched to InP under strong electric field quantum limit operated under inversion mode as a function of alloy composition in accordance with three-band model of Kane.

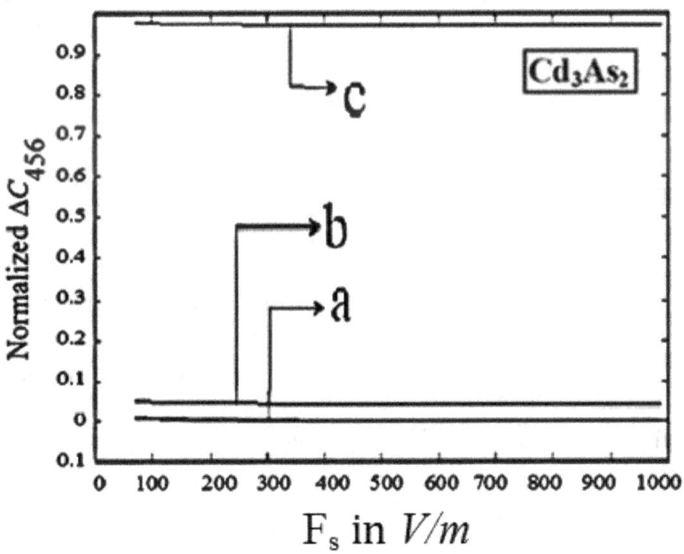

Fig. 1.71. Plot of the normalized ΔC_{456} for QMOSFETs of Cd_3As_2 under weak electric field quantum limit operated under inversion mode as a function of surface electric field in accordance with (a) generalized band model with $\delta \neq 0$, (b) three-band model of Kane and (c) two-band model of Kane.

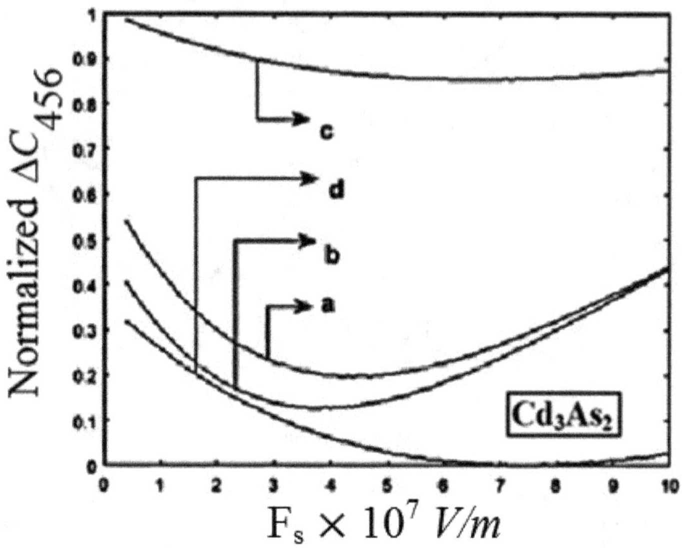

Fig. 1.72. Plot of the normalized ΔC_{456} for QMOSFETs of Cd_3As_2 under strong electric field quantum limit operated under inversion mode as a function of surface electric field in accordance with (a) generalized band model with $\delta \neq 0$, (b) generalized band model with $\delta = 0$, (c) three-band model of Kane and (d) two-band model of Kane.

in accordance with (a) generalized band model with $\delta \neq 0$, (b) three-band model of Kane and (c) two-band model of Kane. In Fig. 1.72, we have plotted the normalized ΔC_{456} for Cd_3As_2 under strong electric field quantum limit as a function of surface electric field in accordance with (a) generalized band model with $\delta \neq 0$, (b) generalized band model with $\delta = 0$, (c) three-band model of Kane and (d) two-band model of Kane. From both the figures we note that ΔC_{456} changes with increasing electric field and the numerical magnitude of ΔC_{456} in each case depends on the values of the energy band constants of each materials, respectively. In Fig. 1.73, we have plotted the normalized ΔC_{456} for n-$CdGeAs_2$ under weak electric field quantum limit as a function of surface electric field in accordance with (a) generalized band model with $\delta \neq 0$, (b) generalized band model with $\delta = 0$, (c) three-band model of Kane.

In Fig. 1.74, we have plotted the normalized ΔC_{456} for $CdGeAs_2$ under strong electric field quantum limit as a function of surface electric field in accordance with (a) generalized band model with $\delta \neq 0$, (b) generalized band model with $\delta - 0$, (c) three-band model of Kane and (d) two-band model of Kane. The nature of variations of $\Delta C_{456}$with electric field is

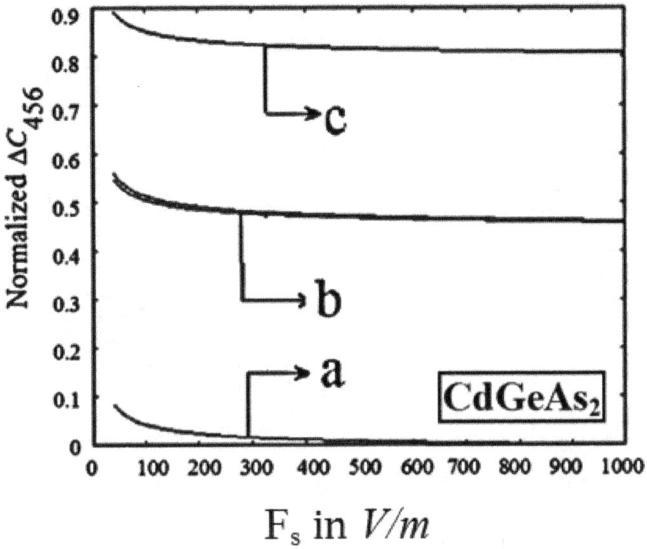

Fig. 1.73. Plot of the normalized ΔC_{456} for QMOSFETs of $CdGeAs_2$ under weak electric field quantum limit operated under inversion mode as a function of surface electric field in accordance with (a) generalized band model with $\delta \neq 0$, (b) generalized band model with $\delta = 0$, (c) three-band model of Kane.

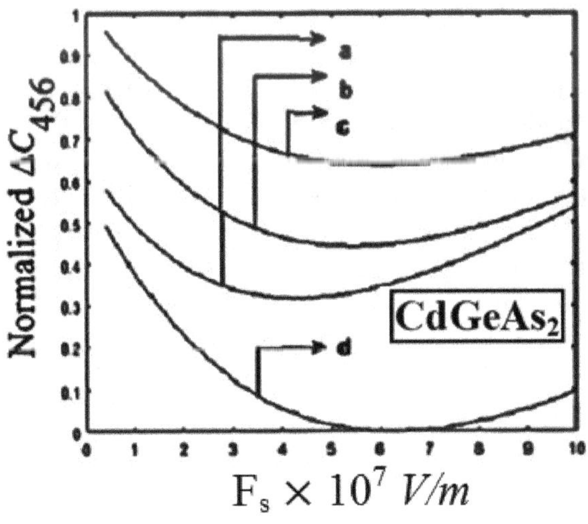

Fig. 1.74. Plot of the normalized ΔC_{456} for QMOSFETs of $CdGeAs_2$ under strong electric field quantum limit operated under inversion mode as a function of surface electric field in accordance with (a) generalized band model with $\delta \neq 0$, (b) generalized band model with $\delta = 0$, (c) three-band model of Kane and (d) two-band model of Kane.

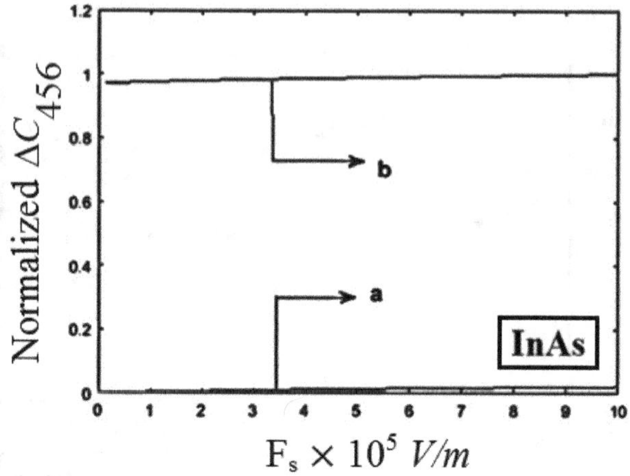

Fig. 1.75. Plot of the normalized ΔC_{456} in QMOSFETs of InAs under weak electric field quantum limit operated under inversion mode as a function of surface electric field in accordance with (a) three-band model of Kane and (b) two-band model of Kane, respectively.

apparent from both the figures and the numerical values of ΔC_{456} for all the band models are different which is the direct influence of the energy band constants in both cases and for all types of dispersion relations. In Fig. 1.75, we have plotted of the normalized ΔC_{456} in n-channel inversion layers of InAs under weak electric field quantum limit as a function of surface electric field in accordance with (a) three-band model of Kane and (b) two-band model of Kane, respectively. We note that the influence of spin-orbit splitting constant on the ΔC_{456} is to reduce its value drastically as compared with the two-band model of Kane of the whole range of electric field as considered here. In Fig. 1.76, we have plotted of the normalized ΔC_{456} in n-channel inversion layers of InAs under strong electric field quantum limit as a function of surface electric field in accordance with three-band model of Kane. In Fig. 1.77, we have plotted of the normalized ΔC_{456} in n-channel inversion layers of InSb under weak electric field quantum limit as a function of surface electric field in accordance with (a) three-band model of Kane and (b) two-band model of Kane, respectively. Our observation regarding the influence of spin-orbit splitting constant on the ΔC_{456} as apparent from Fig. 1.75 is also applicable in this case. In Fig. 1.78, we have plotted of the normalized ΔC_{456} in n-channel inversion layers of InSb under strong electric field quantum limit as a function of surface electric field in accordance with three-band model of Kane. From Figs. 1.76

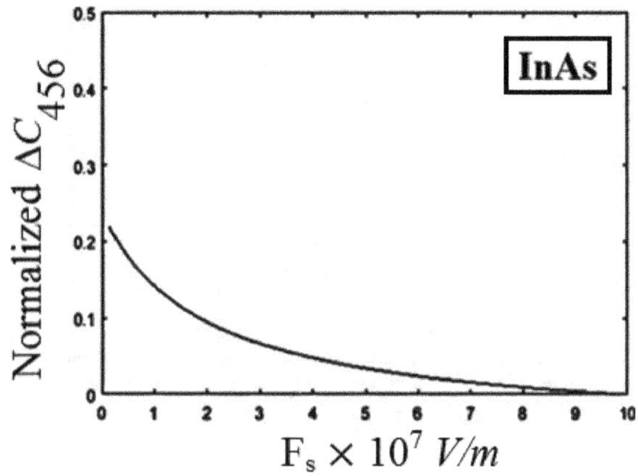

Fig. 1.76. Plot of the normalized ΔC_{456} in QMOSFETs of InAs under strong electric field quantum limit operated under inversion mode as a function of surface electric field in accordance with three-band model of Kane.

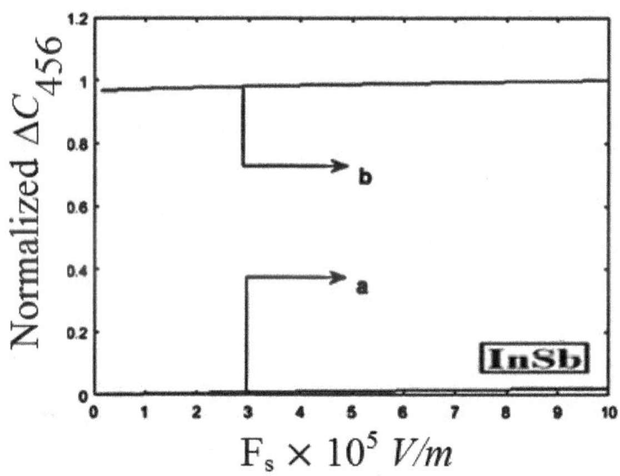

Fig. 1.77. Plot of the normalized ΔC_{456} in QMOSFETs of InSb under weak electric field quantum limit operated under inversion mode as a function of surface electric field in accordance with (a) three-band model of Kane and (b) two-band model of Kane, respectively.

and 1.78, we note that ΔC_{456} decreases with increasing surface electric field in both the cases.

In Fig. 1.79, we have plotted of the normalized ΔC_{456} in n-channel inversion layers of $Hg_{1-x}Cd_xTe$ under weak electric field quantum limit as a function of surface electric field in accordance with (a) three-band model

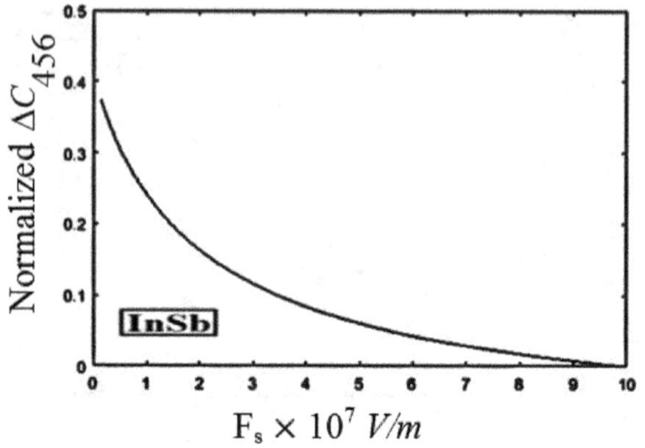

Fig. 1.78. Plot of the normalized ΔC_{456} in QMOSFETs of InSb under strong electric field quantum limit operated under inversion mode as a function of surface electric field in accordance with three-band model of Kane.

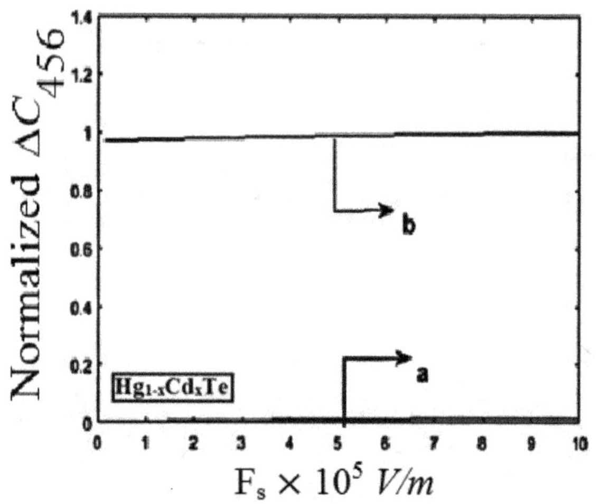

Fig. 1.79. Plot of the normalized ΔC_{456} in QMOSFETs of $Hg_{1-x}Cd_xTe$ under weak electric field quantum limit operated under inversion mode as a function of surface electric field in accordance with (a) three-band model of Kane and (b) two-band model of Kane, respectively.

of Kane and (b) two-band model of Kane, respectively. Our observation regarding the influence of spin-orbit splitting constant on the ΔC_{456} as apparent from Fig. 1.75 is also applicable in this case. In Fig. 1.80, we have plotted of the normalized ΔC_{456} in n-channel inversion layers of $Hg_{1-x}Cd_xTe$ under strong electric field quantum limit as a function of

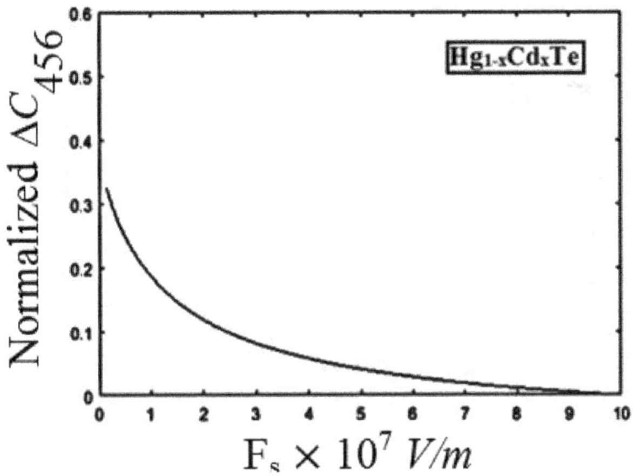

Fig. 1.80. Plot of the normalized ΔC_{456} in QMOSFETs of $Hg_{1-x}Cd_xTe$ under strong electric field quantum limit operated under inversion mode as a function of surface electric field in accordance with three-band model of Kane.

surface electric field in accordance with three-band model of Kane. From Figs. 1.80 we note that ΔC_{456} decreases with increasing surface electric field.

In Fig. 1.81, we have plotted the normalized ΔC_{456} in n-channel inversion layers of $In_{1-x}Ga_xAs_yP_{1-x}$ lattice matched to InP under weak electric field quantum limit as a function of surface electric field in accordance with (a) three-band model of Kane and (b) two-band model of Kane, respectively. Our observation regarding the influence of spin-orbit splitting constant on the ΔC_{456} as apparent from Fig. 1.81 is also applicable in this case. In Fig. 1.82, we have plotted of the normalized ΔC_{456} in n-channel inversion layers of $In_{1-x}Ga_x\ As_yP_{1-x}$ lattice matched to InP under strong electric field quantum limit as a function of surface electric field in accordance with three-band model of Kane. From Fig. 1.82, we note that ΔC_{456} increases with decreasing surface electric field.

In Fig. 1.83, we have plotted of the normalized ΔC_{456} in n-channel inversion layers of $Hg_{1-x}Cd_xTe$ under weak electric field quantum limit as a function of alloy composition in accordance with (a) three-band model of Kane and (b) two-band model of Kane, respectively. The ΔC_{456} changes with alloy composition in accordance with two-band model of Kane whereas for three-band model of Kane, the ΔC_{456} becomes invariant with alloy composition. In Fig. 1.84, we have plotted the normalized ΔC_{456} in n-channel inversion layers of $Hg_{1-x}Cd_xTe$ under strong electric field quantum

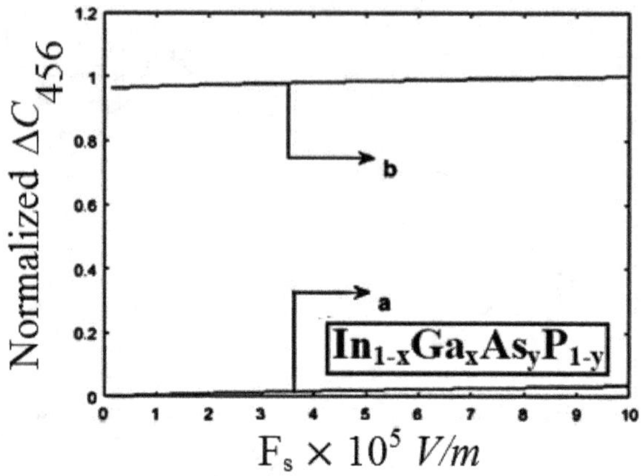

Fig. 1.81. Plot of the normalized ΔC_{456} in QMOSFETs of $In_{1-x}Ga_xAs_yP_{1-x}$ lattice matched to InP under weak electric field quantum limit operated under inversion mode as a function of surface electric field in accordance with (a) three-band model of Kane and (b) two-band model of Kane, respectively.

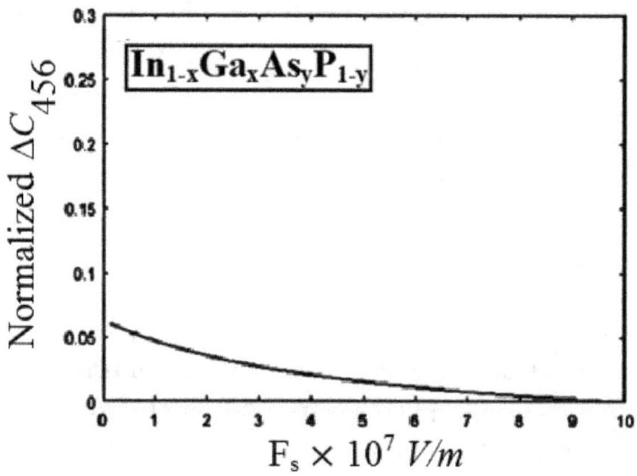

Fig. 1.82. Plot of the normalized ΔC_{456} in n-channel inversion layers of $In_{1-x}Ga_xAs_yP_{1-x}$ lattice matched to InP under strong electric field quantum limit operated under inversion mode as a function of surface electric field in accordance with three-band model of Kane.

limit as a function of alloy composition in accordance with three-band model of Kane. From Fig. 1.84, we note that ΔC_{456} increases with increasing alloy composition for the whole range of x as considered here. In Fig. 1.85, we have plotted of the normalized ΔC_{456} in n-channel inversion layers of

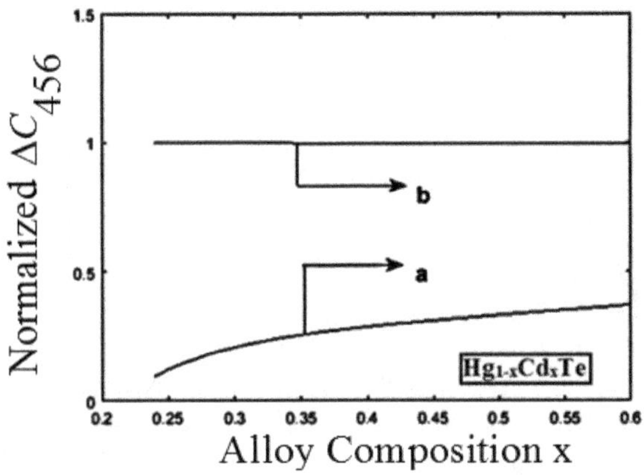

Fig. 1.83. Plot of the normalized ΔC_{456} in QMOSFETs of $Hg_{1-x}Cd_xTe$ under weak electric field quantum limit operated under inversion mode as a function of alloy composition in accordance with (a) three-band model of Kane and (b) two-band model of Kane, respectively.

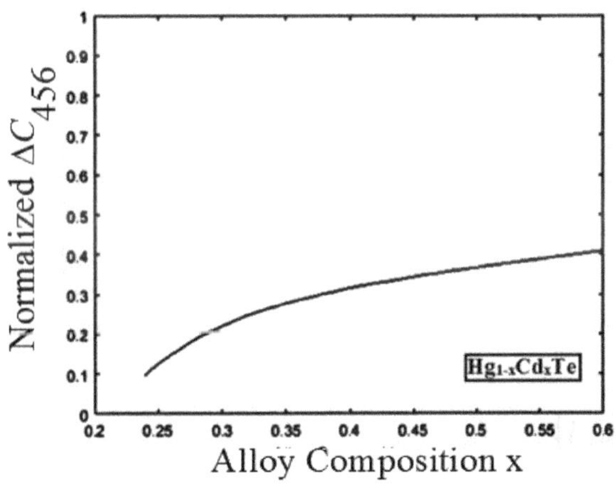

Fig. 1.84. Plot of the normalized ΔC_{456} in QMOSFETs of $Hg_{1-x}Cd_xTe$ under strong electric field quantum limit operated under inversion mode as a function of alloy composition in accordance with three-band model of Kane.

$In_{1-x}Ga_xAs_yP_{1-x}$ lattice matched to InP under weak electric field quantum limit as a function of alloy composition in accordance with (a) three-band model of Kane and (b) two-band model of Kane, respectively. The ΔC_{456} changes radically with changing alloy composition in accordance with both type of band models. In Fig. 1.86, we have plotted the normalized ΔC_{456}

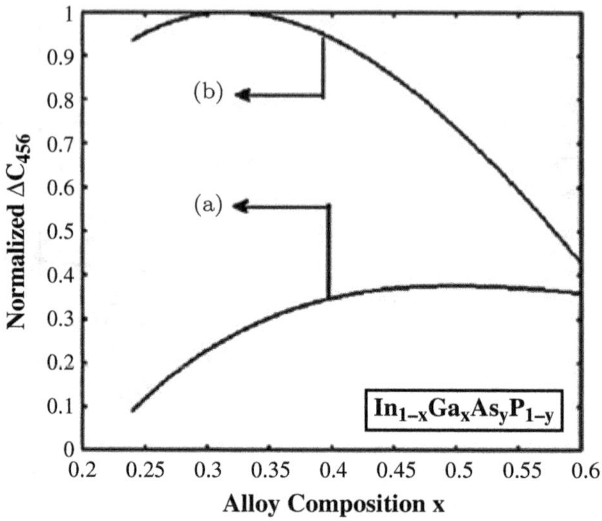

Fig. 1.85. Plot of the normalized ΔC_{456} in QMOSFETs of $In_{1-x}Ga_xAs_yP_{1-y}$ lattice matched to InP under weak electric field quantum limit operated under inversion mode as a function of alloy composition in accordance with (a) three-band model of Kane and (b) two-band model of Kane, respectively.

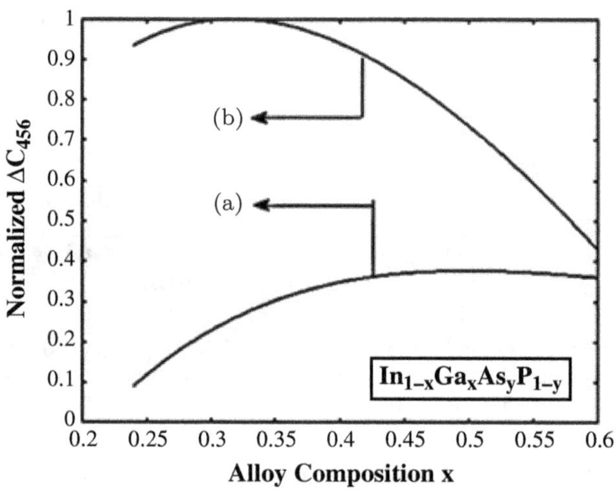

Fig. 1.86. Plot of the normalized ΔC_{456} in QMOSFETs of $In_{1-x}Ga_xAs_yP_{1-y}$ lattice matched to InP under strong electric field quantum limit operated under inversion mode as a function of alloy composition in accordance with (a) three-band model of Kane and (b) two-band model of Kane, respectively.

in n-channel inversion layers of $In_{1-x}Ga_xAs_yP_{1-x}$ lattice matched to InP under strong electric field quantum limit as a function of alloy composition in accordance with (a) three-band model of Kane and (b) two-band model of Kane, respectively. The influence of band models on the variation of the ΔC_{456} under strong electric field quantum limit is rather prominent in this case which is the consequence of different values of the energy band constants for $In_{1-x}Ga_xAs_yP_{1-x}$ lattice matched to InP and $Hg_{1-x}Cd_x$ Te, respectively.

It may be noted that if the direction of application of the surface electric field applied perpendicular to the surface be taken as either k_x or k_y and not as k_z as assumed in the present work, all the expressions of QC, ER, EEM, ΔC_{44} and ΔC_{456} would be different analytically for both the limits. Nevertheless, the arbitrary choice of the direction normal to the surface would not result in a change of the basic qualitative features of the said transport quantities in QMOSFETs of different materials operated under inversion mode. The approximation of the potential well at the surface by a triangular well introduces some errors, as for instance the omission of the free charge contribution to the potential. This kind of approach is reasonable if there are only a few charge carriers in the QMOSFETs of different materials operated under inversion mode, but is responsible for an overestimation of the splitting when the accumulation carrier density exceeds that of the depletion layer. It has been observed that the maximum error due to the triangular potential well is tolerable in the practical sense because for actual calculations, one needs a self-consistent solution which is a formidable problem, for the present generalized systems due to the non-availability of the proper analytical techniques, without exhibiting a widely different qualitative behavior. The second assumption of electric quantum limit in the numerical calculation is valid in the range of low temperatures, where the quantum effects become prominent. The errors which are being introduced for these assumptions are found not to be serious enough at low temperatures. Thus, whenever the condition of the electric quantum limit has been applied, the temperature has been assumed to be low enough so that the assumption becomes well grounded because at low temperature, one can assume that nearly all electrons are at the lowest electric sub-band. We wish to note that the many-body effects, the hot electron effects, the formation of band tails, arbitrary orientation of the direction of the electric quantization and the effects of surface of states have been neglected in our simplified theoretical formalism due to the lack of availability in the literature of the proper analytical techniques for including them for the generalized systems as considered in this chapter.

Our simplified approach will be useful for the purpose of comparison, when, the methods of tackling the aforementioned formidable problems for the present generalized system appear. The inclusion of the said effects would certainly increase the accuracy of our results, and the qualitative features of all the aforementioned electronic quantities discussed in this chapter would not change in the presence of the aforementioned influences.

1.5 Open Research Problems

The problems under these sections of this book are by far the most important part for the alert readers and few open research problems are presented from this chapter till the end. The numerical values of the energy band constants of the different materials as considered here are given in Chapter 18 for the purpose of numerical computations.

(R.1.1) *Investigate and explain with the help of appropriate graphs the DR, the ES, the sub-band energy, the effective mass, the density-of-states function, the electron statistics,* the Burstein Moss shift, Debye screening length, Diffusion coefficient of the minority carriers, Nonlinear optical response, Third-order nonlinear optical susceptibility, generalized Raman gain, the plasma frequency, the activity coefficient, the Magneto-Thermal effect in quantized structures, normalized Hall coefficient, reflection coefficient, heat capacity, magnetic susceptibilities, Faraday rotation, Fowler–Nordheim Field Emission, optical effective mass, Einstein's photoemission, Righi–Leduc coefficient, electric susceptibility, electric susceptibility mass, electron diffusion thermo-power, hydrostatic piezo-resistance coefficient, *relaxation time and the corresponding mobility for various scattering mechanisms both the low and high electric field limits of the QMOSFETs whose constitution bulk materials having the respective DRs as given below:*

(A) The conduction electrons of n-GaP obey two different dispersion laws as given literature.

(i) In accordance with Rees [217], the electron energy spectrum is given by

$$E = \frac{\hbar^2 k_s^2}{2m_\perp^*} + \frac{\hbar^2}{2m_{||}^*}[\wp k_s^2 + k_z^2] - \left[\frac{\hbar^4 k_0^2}{m_{||}^{*2}}(k_s^2 + k_z^2) + |V_G|^2\right]^{1/2} - |V_G| \tag{R1.1a}$$

where k_0 and $|V_G|$ are constants of the energy spectrum with $m_{||}^* = 0.92m_0$, $m_\perp^* = 0.25m_0$,

$$k_0 = 1.7 \times 10^{19}m^{-1}, |V_G| = 0.21\,\text{eV}, g_v = 6, g_s = 2 \text{ and } \wp = 1$$

(ii) In accordance with Ivchenko and Pikus [219], the electron dispersion law can be written as

$$E = \frac{\hbar^2 k_s^2}{2m_{||}^*} + \frac{\hbar^2 k_s^2}{2m_\perp^*} \mp \frac{\bar{\Delta}}{2} \pm \left[\left(\frac{\bar{\Delta}}{2}\right)^2 + P_1 k_z^2 + D_1 k_x^2 k_y^2\right]^{1/2} \quad \text{(R1.1b)}$$

where $\bar{\Delta} = 335$ meV, $P_1 = 2 \times 10^{-10}$ eVm, $D_1 = P_1 a_1$ and $a_1 = 5.4 \times 10^{-10}$m.

(B) In addition to the Cohen model, the dispersion relation for the conduction electrons for IV–VI compounds can also be described by the models of Dimmock [220], Bangert *et al.* [221] and Foley *et al.* [222], respectively.

(i) In accordance with Dimmock model [220], the carrier energy spectrum of IV–VI materials assumes the form

$$\left[\in - \frac{E_g}{2} - \frac{\hbar^2 k_s^2}{2m_t^-} - \frac{\hbar^2 k_z^2}{2m_l^-}\right]\left[\in + \frac{E_g}{2} + \frac{\hbar^2 k_s^2}{2m_t^+} + \frac{\hbar^2 k_z^2}{2m_l^+}\right]$$
$$= P_\perp^2 k_s^2 + P_{||}^2 k_z^2 \quad \text{(R1.2)}$$

where $\in$ is the energy as measured from the center of the band gap E_g, $m_t^\pm$ and $m_l^\pm$ represent the contribution of the transverse and longitudinal effective masses of the external L_6^+ and L_6^- bands arising from the **k.p** perturbations with the other bands taken to the second-order and $g_v = 4$.

(ii) In accordance with Bangert *et al.* [221] the dispersion relation is given by

$$\Gamma(E) = F_1(E)k_s^2 + F_2(E)k_z^2 \quad \text{(R1.3)}$$

where

$$\Gamma(E) \equiv 2E, F_1(E) \equiv \frac{R_1^2}{E+E_g} + \frac{S_1^2}{E+\Delta_c'} + \frac{Q_1^2}{E+E_g},$$

$$F_2(E) \equiv \frac{2C_5^2}{E+E_g} + \frac{(S_1+Q_1)^2}{E+\Delta_c''}$$

$$R_1^2 = 2.3 \times 10^{-19}(\text{eVm})^2, C_5^2 = 0.83 \times 10^{-19}(\text{eVm})^2,$$

$$Q_1^2 = 1.3R_1^2, S_1^2 = 4.6R_1^2, \Delta_c' = 3.07\,\text{eV}, \Delta_c'' = 3.28\,\text{eV}$$

and $g_v = 4$. It may be noted that under the substitution $S_1 = 0$, $Q_1 = 0$, $R_1^2 \equiv \frac{\hbar^2 E_g}{m_\perp^*}$, $C_5^2 \equiv \frac{\hbar^2 E_g}{2m_{||}^*}$, equation (R1.3) assumes the form $E(1+\alpha E) = \frac{\hbar^2 k_s^2}{2m_\perp^*} + \frac{\hbar^2 k_z^2}{2m_{||}^*}$ which is the simplified Lax model.

(iii) The carrier energy spectrum of IV–VI semiconductors in accordance with Foley *et al.* [222] can be written as

$$E + \frac{E_g}{2} = E_-(k) + \left[\left[E_+(k) + \frac{E_g}{2}\right]^2 + P_\perp^2 k_s^2 + P_{||}^2 k_z^2\right]^{1/2} \quad \text{(R1.4)}$$

where

$$E_+(k) = \frac{\hbar^2 k_s^2}{2m_\perp^+} + \frac{\hbar^2 k_z^2}{2m_{||}^+}, E_-(k) = \frac{\hbar^2 k_s^2}{2m_\perp^-} + \frac{\hbar^2 k_z^2}{2m_{||}^-}$$

represents the contribution from the interaction of the conduction and the valance band edge states with the more distant bands and the free electron term,

$$\frac{1}{m_\perp^\pm} = \frac{1}{2}\left[\frac{1}{m_{tc}} \pm \frac{1}{m_{tv}}\right], \frac{1}{m_{||}^\pm} = \frac{1}{2}\left[\frac{1}{m_{1c}} \pm \frac{1}{m_{1v}}\right]$$

For n-PbTe,

$$P_\perp = 4.61 \times 10^{-10}\,\text{eVm}, P_{||} = 1.48 \times 10^{-10}\,\text{eVm}, \frac{m_0}{m_{tv}} = 10.36,$$

$$\frac{m_0}{m_{1v}} = 0.75, \frac{m_0}{m_{tc}} = 11.36 \frac{m_0}{m_{1c}} = 1.20 \text{ and } g_v = 4$$

(C) The importance of Germanium is well known since the inception of semiconductor physics. The conduction electrons of n-Ge obey two different dispersion laws since band non-parabolicity has been included in two different ways as given in literature [223,224]. In accordance with Cardona *et al.* [223] and Wang *et al.* [224] the electron dispersion laws in Ge can, respectively, be expressed as

$$E = -\frac{E_g}{2} + \frac{\hbar^2 k_z^2}{2m_{||}^*} + \left[\frac{E_g^2}{4} + E_g k_s^2 \left(\frac{\hbar^2}{2m_\perp^*}\right)\right]^{1/2} \quad \text{(R1.5)}$$

and

$$E = a_9 k_z^2 + l_9 k_s^2 - c_9 k_s^4 - d_9 k_s^2 k_z^2 - e_9 k_z^4 \quad \text{(R1.6)}$$

where

$a_9 = (\hbar^2/2m_{||}^*), m_{||}^* = 1.588 m_0, l_9 = (\hbar^2/2m_\perp^*), m_\perp^* = 0.0815 m_0,$

$c_9 = 1.4A_9$,

$$A_9 \equiv \frac{1}{4}(\hbar^4/E_g m_\perp^{*2})\left[1 - \frac{m_\perp^*}{m_0}\right]^2, E_g = 2.2eV, d_9 = 0.8A_9,$$

$e_9 = 0.005A_9, g_v = 4$ and $g_s = 2$

(D) The DR of the conduction electrons of zero-gap materials (e.g. HgTe) is given by [225]

$$E = \frac{\hbar^2 k^2}{2m^*} + \frac{3e^2}{128\varepsilon_\infty}k - \left(\frac{2E_B}{\pi}\right)\ln\left|\frac{k}{k_0}\right| \tag{R1.7}$$

where ε_∞ is the semiconductor permittivity in the high-frequency limit,

$$E_B \equiv \frac{m_0 e^2}{2\hbar^2\varepsilon_\infty^2} \text{ and } k_0 \equiv \frac{m_0 e^2}{\hbar^2\varepsilon_\infty}$$

(E) The conduction electrons of n-GaSb obey the following three dispersion relations:

(i) In accordance with the model of Seiler *et al.* [226]

$$E = \left[-\frac{E_g}{2} + \frac{E_g}{2}[1+\alpha_4 k^2]^{1/2} + \frac{\bar{\varsigma}_0\hbar^2 k^2}{2m_o} + \frac{\bar{v}_0 f_1(k)\hbar^2}{2m_o}\right.$$
$$\left.\pm \frac{\bar{\omega}_0 f_2(k)\hbar^2}{2m_o}\right] \tag{R1.8a}$$

where $\alpha_4 \equiv 4P^2(E_g + \frac{2}{3}\Delta)[E_g^2(E_g + \Delta)]^{-1}$, P is the isotropic momentum matrix element, $f_1(k) \equiv k^{-2}[k_x^2 k_y^2 + k_y^2 k_z^2 + k_z^2 k_x^2]$ represents the warping of the Fermi surface, $f_2(k) \equiv [\{k^2(k_x^2 k_y^2 + k_y^2 k_z^2 + k_z^2 k_x^2) - 9k_x^2 k_y^2 k_z^2\}^{1/2} k^{-1}]$ represents the inversion asymmetry splitting of the conduction band and $\bar{\varsigma}_0, \bar{v}_0$ and $\bar{\omega}_0$ represent the constants of the electron spectrum in this case.

It should be noted that under the substitutions,

$$\bar{\varsigma}_0 = 0, \bar{v}_0 = 0, \bar{\omega}_0 = 0 \text{ and } P^2 \equiv \frac{\hbar^2}{2m^*}\frac{E_g(E_g+\Delta)}{(E_g + \frac{2}{3}\Delta)}$$

equation (R1.8a) assumes the well-known two-band model of Kane which is given by

$$E\left(1 + \frac{E}{E_g}\right) = \frac{\hbar^2 k^2}{2m^*} \tag{R1.8b}$$

(ii) In accordance with the model of Mathur *et al.* [227],

$$E = \frac{\hbar^2 k^2}{2m_0} + \frac{E_{g1}}{2}\left[\left(1 + \frac{2\hbar^2 k^2}{E_{g1}}\left\{\frac{1}{m^*} - \frac{1}{m_0}\right\}\right)^{1/2} - 1\right] \quad \text{(R1.9)}$$

where $E_{g1} = \{E_g + [(5 \times 10^{-5}T^2)(2(112+T)^{-1})]\}\text{eV}$.

(iii) In accordance with the model of Zhang. [228]

$$E = [E_2^{(1)} + E_2^{(2)}K_{4,1}]k^2 + [E_4^{(1)} + E_4^{(2)}K_{4,1}]k^4 + k^6[E_6^{(1)} + E_6^{(2)}K_{4,1} + E_6^{(3)}K_{6,1}] \quad \text{(R1.10)}$$

where

$$K_{4,1} \equiv \frac{5}{4}\sqrt{21}\left[\frac{k_x^4 + k_y^4 + k_z^4}{k^4} - \frac{3}{5}\right],$$

$$K_{6,1} \equiv \sqrt{\frac{639639}{32}}\left[\frac{k_x^2 k_y^2 k_z^2}{k^6} + \frac{1}{22}\left(\frac{k_x^4 + k_y^4 + k_z^4}{k^4} - \frac{3}{5}\right) - \frac{1}{105}\right]$$

the coefficients are in eV, the values of k are $10(\frac{a}{2\pi})$ times those of k in atomic units (a is the lattice constant), $E_2^{(1)} = 1.0239620$, $E_2^{(2)} = 0$, $E_4^{(1)} = -1.1320772$, $E_4^{(2)} = 0.05658$, $E_6^{(1)} = 1.1072073$, $E_6^{(2)} = -0.1134024$ and $E_6^{(3)} = -0.0072275$

(F) The dispersion relation of the carriers in p-type Platinum antimonide ($PtSb_2$) following Emtage [229] can be written as

$$\left[E + \overline{\lambda_0}\frac{(\bar{a})^2}{4}k^2 - \bar{l}\frac{(\bar{a})^2}{4}k_s^2\right]\left[E + \overline{\delta_0} - \bar{\nu}\frac{(\bar{a})^2}{4}k^2 - \bar{n}\frac{(\bar{a})^2}{4}k_s^2\right] = \frac{I(\bar{a})^4}{16}k^4 \quad \text{(R1.11)}$$

where $\overline{\lambda_0}, \bar{a}, \bar{l}, \overline{\delta_0}, \bar{\nu}, \bar{n}$ and I are system constants and $\lambda_1, l_1, \delta_0', \nu_1, n_1$ and I_0 are the energy band constants and a is the lattice constant.

(G) In addition to the well-known three-band model of Kane, the conduction electrons of n-GaAs obey the following three dispersion relations:

(i) In accordance with the model of Stillman *et al.* [230]

$$E = \frac{\hbar^2 k^2}{2m^*} - \left(1 - \frac{m^*}{m_0}\right)^2\left(\frac{\hbar^2 k^2}{2m^*}\right)^2\left[\frac{3E_g + 4\Delta + \left(\frac{2\Delta^2}{E_g}\right)}{(E_g + \Delta)(2\Delta + 3E_g)}\right] \quad \text{(R1.12)}$$

(ii) In accordance with the model of Newson *et al.* [231]

$$E = \alpha_{13}k_z^4 + \left[\frac{\hbar^2}{2m_c}a_{14}k_s^2\right]k_z^2 + \frac{\hbar^2}{2m_c}k_s^2 + a_{11}k_x^2k_y^2 + a_{13}(k_x^4 + k_y^4) \tag{R1.13}$$

where a_{13} is the non-parabolicity constant and $a_{14}(\equiv 2a_{13} + a_{15})$ and a_{15} is the wrapping constant.

(iii) In accordance with the model of Rossler [232]

$$E = \frac{\hbar^2k^2}{2m^*} + \bar{\alpha}_{10}k^4 + \bar{\beta}_{10}[k_x^2k_y^2 + k_y^2k_z^2 + k_z^2k_x^2] \pm \bar{\gamma}_{10}[k^2(k_x^2k_y^2 + k_y^2k_z^2 + k_z^2k_x^2) - 9k_x^2k_y^2k_z^2]^{1/2} \tag{R1.14}$$

where $\bar{\alpha}_{10} = \bar{\alpha}_{11} + \bar{\alpha}_{12}k$, $\bar{\beta}_{10} = \bar{\beta}_{11} + \bar{\beta}_{12}k$ and $\bar{\gamma}_{10} = \bar{\gamma}_{11} + \bar{\gamma}_{12}k$, in which, $\bar{\alpha}_{11} = -2132 \times 10^{-40}\text{eVm}^4, \bar{\alpha}_{12} = 9030 \times 10^{-50}\text{eVm}^5, \bar{\beta}_{11} = -2493 \times 10^{-40}\text{eVm}^4 \bar{\beta}_{12} = 12594 \times 10^{-50}\text{eVm}^5, \bar{\gamma}_{11} = 30 \times 10^{-30}\text{eVm}^3$ and $\bar{\gamma}_{12} = -154 \times 10^{-42}\text{eVm}^4$

(H) In addition to the well-known three-band model of Kane, the conduction electrons of n-InSb obey the following three dispersion relations:

(i) To the fourth-order effective mass theory, and taking into account the interactions of the conduction, the heavy hole, the light hole and the split-off bands, the electron energy spectrum in n-InSb is given by [233]

$$E = \frac{\hbar^2k^2}{2m^*} + \bar{b}_1k^4 \tag{R1.15}$$

where

$$\bar{b}_1 \equiv \frac{K_2\hbar^4}{4E_g(m^*)^2}, \quad K_2 \equiv -\left[\frac{\left(1 + \frac{1}{2}x_1^2\right)}{1 + \frac{1}{2}x_1}\right](1 - y_1)^2,$$

$$x_1 \equiv \left[1 + \left(\frac{\Delta}{E_g}\right)\right]^{-1} \quad \text{and} \quad y_1 \equiv \frac{m^*}{m_0}$$

(ii) In accordance with Johnson and Dickey [234], the electron energy spectrum assumes the form

$$E = -\frac{E_g}{2} + \frac{\hbar^2 k^2}{2}\left[\frac{1}{m_0} + \frac{1}{m_{\gamma b}}\right] + \frac{E_g}{2}\left[1 + 4\frac{\hbar^2 k^2}{2m_c'}\frac{\bar{f}_1(E)}{E_g}\right]^{1/2} \tag{R1.16}$$

where

$$\frac{m_0}{m_c'} \equiv P^2\left[\frac{(E_g + \frac{2\Delta}{3})}{E_g(E_g + \Delta)}\right], \bar{f}_1(E) \equiv \frac{(E_g + \Delta)(E + E_g + \frac{2\Delta}{3})}{(E_g + \frac{2\Delta}{3})(E + E_g + \Delta)},$$

$$m_c' = 0.139 m_0 \text{ and } m_{\gamma b} = \left[\frac{1}{m_c'} - \frac{2}{m_0}\right]^{-1}$$

(iii) In accordance with Agafonov *et al.* [235], the electron energy spectrum can be written as

$$E = \frac{\bar{\eta} - E_g}{2}\left[1 - \frac{\hbar^2 k^2}{2\bar{\eta} m^*}\left\{\frac{D\sqrt{3} - 3\bar{B}}{2\left(\frac{\hbar^2}{2m^*}\right)}\right\}\left[\frac{k_x^4 + k_y^4 + k_z^4}{k^4}\right]\right] \tag{R1.17}$$

where

$$\bar{\eta} \equiv \left(E_g^2 + \frac{8}{3}P^2 k^2\right)^{1/2}, \bar{B} \equiv -21\frac{\hbar^2}{2m_0} \quad \text{and} \quad D \equiv -40\left(\frac{\hbar^2}{2m_0}\right)$$

(I) The dispersion relation of the carriers in n-type $\mathrm{Pb}_{1-x}\mathrm{Ga}_x\mathrm{Te}$ with $x = 0.01$ following Vassilev [236] can be written as

$$\begin{aligned}&[E - 0.606k_s^2 - 0.0722k_z^2][E + \bar{E}_g + 0.411k_s^2 + 0.0377k_z^2]\\ &= 0.23k_s^2 + 0.02k_z^2 \pm [0.06\bar{E}_g + 0.061k_s^2 + 0.0066k_z^2]k_s\end{aligned} \tag{R1.18}$$

where $\bar{E}_g (= 0.21\mathrm{eV})$ is the energy gap for the transition point, the zero of the energy E is at the edge of the conduction band of the Γ point of the Brillouin zone and is measured positively upwards, k_x, k_y and k_z are in the units of $10^9\mathrm{m}^{-1}$.

(J) The dispersion relation of the holes in the valance bands of II–V compounds in accordance with Yamada [237] can be written as

$$E = \theta_1 k_x^2 + \theta_2 k_y^2 + \theta_3 k_z^2 + \delta_4 k_x \mp [\{\theta_5 k_x^2 + \theta_6 k_y^2 + \theta_7 k_z^2 + \delta_5 k_x\}^2 + G_3^2 k_y^2 + \Delta_3^2]^{\frac{1}{2}} \pm \Delta_3 \quad \text{(R1.19)}$$

where k_x, k_y and k_z are expressed in the units of $10^{10}\ \text{m}^{-1}$,

$$\theta_1 = \tfrac{1}{2}(a_1 + b_1), \theta_2 = \tfrac{1}{2}(a_2 + b_2), \theta_3 = \tfrac{1}{2}(a_3 + b_3), \delta_4 = \tfrac{1}{2}(A + B)$$

$$\theta_5 = \tfrac{1}{2}(a_1 - b_1), \theta_6 = \tfrac{1}{2}(a_2 - b_2), \theta_7 = \tfrac{1}{2}(a_3 - b_3), \delta_5 = \tfrac{1}{2}(A - B)$$

$a_i (i = 1, 2, 3, 4), b_i, A, B, G_3$ and Δ_3 are system constants.

(R.1.2) Investigate the said electronic quantities in the presence of an arbitrarily oriented electric field for MOSFETs of tetragonal semiconductors. Study all the special cases for MOSFETs of III–V, ternary and quaternary materials in this context. Draw the corresponding graphs and explain the results.

(R.1.3) Investigate the said electronic quantities for MOSFETs of IV–VI, II–VI and stressed Kane type compounds. Draw the corresponding graphs and explain the results.

(R.1.4) Investigate the said electronic quantities for MOSFETs of IV–VI, II–VI and stressed Kane type compounds in the presence of an arbitrarily oriented electric field. Draw the corresponding graphs and explain the results.

References

1. T. Ando., A. B. Fowler and F. Stern, *J. Revi. Modern Phys.* **54**, 437 (1982).
2. J. JJ. Quinn and P. J. Styles. (ed.) *Electronic Properties of Quasi Two Dimensional Systems* (North Holland, Amsterdam, 1976).
3. G. A. Antcliffe, R. T. Bate and R. A. Reynolds, *Proceedings of the International Conference, Physics of Semi-metals and Narrow-Gap Semiconductors*, D. L. Carter and R. T. Bate (eds.) (Pergamon Press, Oxford, 1971), pp. 499.
4. Z. A. Weinberg, *J. Solid-State Electron.* **20**, 11 (1977).
5. G. Paasch., T. Fiedler, M. Kolar and I. Bartos., *J. Phys. Status Solidi (b)* **118**, 641 (1983).
6. S. Lamari., *J. Phys. Rev. B* **64**, 245340 (2001).
7. T. Matsuyama., R. Kürsten, C. Meißner and U. Merkt, *J. Phys. Rev. B* **61**, 15588 (2000).
8. P. V. Santos and M. Cardona., *J. Phys. Rev. Lett.* **72**, 432 (1994).

9. L. Bu., Y. Zhang, B. A. Mason, R. E. Doezema and J. A. Slinkman, *J. Phys. Rev. B* **45**, 11336 (1992).
10. P. D. Dresselhaus, C. M. Papavassiliou, R. G. Wheeler, R. N. and Sacks, *J. Phys. Review Letters.* **68**, 106 (1992).
11. U. Kunze, *Journal of Physical Review B* **41**, 1707 (1990).
12. E. Yamaguchi, *Journal of Physical Review B* **32**, 5280 (1985).
13. Th. Lindner and G. Paasch, *Journal of Applied Physics* **102**, 054514 (2007).
14. S. Lamari, *Journal of Applied Physics* **91**, 1698 (2002); M. Mondal and K. P. Ghatak, *Journal of Physica Status Solidi (a)* **93**, 377 (1986); D. R. Choudhury, A. K. Choudhurhy and A. N. Chakravarti, *Physica Status Solidi (a)* **56**, K69 (1980).
15. K. P. Ghatak and M. Mondal, *Journal of Applied Physics.* **70**, 299 (1991); D. R. Choudhury, A. K. Choudhurhy and A. N. Chakravarti, Czechoslovak, Journal of Physica, **30**, 1157 (1980); K. P. Ghatak, and M. Mondal, *J. Appl. Phys.* **62**, 922 (1987).
16. K. P. Ghatak and S. N. Biswas, *Journal of Vacuum Science & Technology* **7B**, 104 (1989); K. P. Ghatak, N. Chattopadhyay and S. N. Biswas, *Optoelectr. Mater, Devi, Packa, and Interconn.* **836**, 203 (1987).
17. B. Mitra and K. P. Ghatak, *Journal of Solid-State Electronics* **32**, 177 (1989); S. N. Biswas and K.P. Ghatak, *Advan. Proce. of Semico. Dev.* **797**, 286 (1987); K.P. Ghatak and M. Mondal, *J. Appl. Phys.* **64**, 918 (1988).
18. K. P. Ghatak and M. Mondal, *Journal of Applied Physics.* **62**, 922 (1987);); S. N. Biswas and K. P. Ghatak, *Megaga. Tech. and Puls. Power. Appl* 8, 219 (1987); B. Mitra, A. Ghoshal and K. P Ghatak, *Phys. stat. soli. (b)* 153, K209 (1989).
19. M. Mondal and K. P. Ghatak, *Journal of Magnetism and Magnetic Materials* **62**, 115 (1986); M. Mondal and K. P. Ghatak, *Journal of Physica Scripta* **31**, 613 (1985).
20. K. P. Ghatak and M. Mondal, *Journal of Zeitschrift für Physik B Condensed Matter*, **64**, 223 (1986); K. P. Ghatak and S.N. Biswas, *Journal of Solid-State Electronics*, **37**, 1437 (1994); K. P. Ghatak, N. Chattopadhyay and M. Mondal, *Journal of Applied Physics A*, **48**, 365 (1989); R. Paul, P. K. Das, M. Mitra and K. P. Ghatak, *Advance Science Engineering and Medicine.* **11**, 1 (2019); K. P. Ghatak, S. N. Biswas and S. N. Banik, *Nonlin. Opt. II*, **1409**, 240 (1991).
21. A. S. Grove and D. J. Fitzgerald, *J. Solid State Electron.* **9**, 783 (1966).
22. A. M. Voshchenkov and J. N. Zemel, *J. Phys. Rev. B* **9**, 4410 (1974).
23. Z. Djurid, Z. Spasojevid and D. Tjapkin, *J. Solid State Electron.* **19**, 931 (1976).
24. M. Kaplit and Z. N. Zemel, *J. Phys. Rev. Lett.* **21**, 212 (1963).
25. A. Daerr, J. P. Kotthaus and J. F. Koch, *J. Solid State Commun.* **17**, 455 (1975).
26. A. Daerr and J. P. Kotthaus, *J. Surf. Sci.* **73**, 549 (1978).
27. H. A. Washburn and J. R. Sites, *J. Surf. Sci.* **73**, 537 (1978).
28. J. W. Rowe and J. L. Shay, *J. Phys. Rev. B* **3**, 451 (1973).

29. H. Kildal, *J. Phys. Rev. B* **10**, 5082 (1974).
30. J. Bodnar, in *Proceedings of the International Conference on Physics of Narrow-gap Semiconductors* (Polish Science Publishers, Warsaw, 1978).
31. G. P. Chuiko and N. N. Chuiko, *J. Sov. Phys. Semiconduct.* **15**, 739 (1981).
32. K. P. Ghatak and S. N. Biswas, *J. Proce. SPIE* **1484**, 149 (1991).
33. A. Rogalski, *J. Alloys Comp.* **371**, 53 (2004).
34. A. Baumgartner, A. Chaggar, A. Patanè, L. Eaves and M. Henini, *Appl. Phys. Lett.* **92**, 091121 (2008).
35. J. Devenson, R. Teissier, O. Cathabard and Baranov, *A.N. Proc. SPIE* **6909**, 69090 (2008).
36. B. S. Passmore, J. Wu, M. O. Manasreh and G. J. Salamo, *Appl. Phys. Lett.* **91**, 233508 (2007).
37. M. Mikhailova, N. Stoyanov, I. Andreev, B. Zhurtanov, S. Kizhaev, Kunitsyna E., K. Salikhov and Y. Yakovlev, *Proc. SPIE* **6585**, 658526 (2007).
38. W. Kruppa, J. B. Boos, B. R. Bennett, N. A. Papanicolaou, D. Park and R. Bass, *Electron. Lett.* **42**, **688** (2006).
39. E. O. Kane, in *Semi-conductors and Semimetals*, vol. 1, R. K. Willardson and A. C. Beer (eds) (Academic Press, New York, 1966).
40. B. R. Nag, *Electron Transport in Compound Semi-conductors* (Springer, Heidelberg, 1980).
41. P. Y. Lu, C. H. Wung, C. M. Williams, S. N. G. Chu and C. M. Stiles, *Appl. Phys. Lett.* **49**, 1372 (1986).
42. N. R. Taskar, I. B. Bhat, K. K. Prat, D. Terry, H. Ehasani and S. K. Ghandhi, *J. Vac. Sci. Tech.* **7A**, 281 (1989).
43. F. Koch, *Springer Series in Solid States Sciences* (Springer, Germany, 1984).
44. L. R. Tomasetta, H. D. Law, R. C. Eden, I. Reyhimy and K. Nakano, *IEEE J. Quant. Electron.* **14**, 800 (1978).
45. T. Yamato, K. Sakai, S. Akiba and Y. Suematsu, *IEEE J. Quantum Electron.* **14**, 95 (1978).
46. T. P. Pearsall, B. I. Miller and R. J. Capik, *Appl. Phys. Lett.* **28**, 499 (1976).
47. M. A. Washington, R. E. Nahory, M. A. Pollack and E. D. Beeke, *Appl. Phys. Lett.* **33**, 854 (1978).
48. M. I. Timmons, S. M. Bedair, R. J. Markunas and J. A. Hutchby, *Proceedings of the 16th IEEE Photovoltaic Specialist Conference* (IEEE, San Diego, California, 1982).
49. I. M. Tsidilkovski., *Band Structures of Semiconductors* (Pergamon Press, London 1982); L. Landau and E. M. Liftshitz, *Statistical Physics, Part-II* (Pergamon Press, Oxford, 1980).
50. P. R. Wallace, *Phys. Stat. Sol.* (b) **92**, 49 (1979).
51. S. J. Adachi, *J. Appl. Phys.* **58**, 11 (1985).
52. R. Dornhaus and G. Nimtz, *Springer Tracts in Modern Physics*, Vol. 78 (Berlin-Heidelberg, Springer, 1, 1976).

53. W. Zawadzki, *Handbook of Semiconductors*, Vol. 1 W. Paul (ed.) (Amsterdam, North Holland, 719, 1982).
54. I. M. Tsidilkovski, *Cand. Thesis Leningrad University USSR* (1955).
55. F. G. Bass and I. M. Tsidilkovski, *Ivz. Acad. Nauk Azerb USSR* **10**, 3 (1966).
56. B Mitra and K.P Ghatak, *Physica Scripta* **40**, 776 (1989).
57. K. P. Ghatak and S. N. Biswas, *Nonlin. Opt. Quant. Opts.* **4**, 347 (1993).
58. A. N. Chakravarti, A. K. Choudhury, K. P. Ghatak, S. Ghosh and A. Dhar, *Appl. Phys.* **25**, 105 (1981); K. P. Ghatak and M. Mondal, Z. F. Physik B, **69**, 471 (1988); M. Mondal and K. P. Ghatak, *Phys. Letts. A*, **131**, 529 (1988).
59. K. P. Ghatak, A. Ghoshal and B. Mitra, *Nouvo Cimento D* **14**, 903 (1992).
60. B. Mitra, A. Ghoshal and K. P. Ghatak, *Nouvo Cimento D* **12**, 891 (1990); K. P. Ghatak and S. N. Biswas, *Nonlin. Opt. Quant. Opts.* **12**, 83 (1995).
61. M. Mondal and K. P. Ghatak, *Graphite Intercalation Compounds: Science and Applications, MRS Proceedings*, M. Endo, M. S. Dresselhaus and G. Dresselhaus (eds.), MRS Fall Meeting, EA **16**, 173 (1988).
62. M. Mondal, N. Chattapadhyay and K. P. Ghatak, *J. Low Temp. Phys.* **66**, 131 (1987); A. N. Chakravarti, K. P. Ghatak, K. K. Ghosh, S. Ghosh and A. Dhar, *Z. Physik B.* **47**, 149 (1982).
63. V. K. Arora and H. Jeafarian, *Phys. Rev. B.* **13**, 4457 (1976).
64. M. Singh, P. R. Wallace, S. D. Jog and J. J. Erushanov, *J. Phys. Chem. Solids.* **45**, 409 (1984).
65. W. Zawadski, *Adv. Phys.* **23**, 435 (1974).
66. K. P. Ghatak and M. Mondal, *Z. fur Nature A* **41**, 881 (1986).
67. T. Ando, A. H. Fowler and F. Stern, *Rev. Modern Phys.* **54**, 437 (1982).
68. R. K. Pathria, *Statistical Mechanics*, 2nd Edition (Butterworth-Heinmann, Oxford, 1996).
69. R. A. Smith, *Wave Mechanics of Crystalline Solids* (Chapman and Hall, London, 1969).
70. O. Madelung, *Semiconductors: Data handbook*, 3rd edition, (Springer, Germany, 2004); M. Krieehbaum, P. Kocevar, H. Pascher and G. Bauer, *IEEE QE* **24** 1727 (1988); A. N. Chakravarti and B. R. Nag, *Int. J. Elect.* **37**, 281 (1974).
71. B. A. Aronzon and E. Z. Meilikhov, *Phys. Stat. Sol. (a)* **19**, 313 (1973).
72. C. C. Wu and C. J. Lin, *J. Low Temp. Phys.* **57**, 469 (1984); M. H. Chen, C. C. Wu and C. J. Lin, *J. Low Temp. Phys.* **55**, 127 (1984).
73. N. G. Anderson, W. D. Laidig, R. M. Kolbas and Y. C. Lo, *J. Appl. Phys.* **60**, 2361 (1986).
74. N. Paitya and K. P Ghatak, *J. Adv. Phys.* **1**, 1 (2012).
75. F. Capasso, K. Mohammed, A.Y. Cho, R. Hull and A.L. Hutchinson, *Appl. Phys. Letts.* **47**, 420 (1985).
76. S. Bhattacharya, D. De and K. P Ghatak, *J. Sci. Engg. Medi.* **6**, 5 (2012).
77. N. Paitya and K. P. Ghatak, *J. Nanoengg. Nanomanufac.* **2**, 1 (2012).
78. N. Paitya, S. Bhattacharya, D. De and K. P. Ghatak, *Adv. Sc. Engg. Medi.* **4**, 96 (2012).

79. R. Grill, C. Metzner and G. H. Döhler, *Phys. Rev. B* **63**, 235316 (2001); J. Z. Wang, Z. G. Wang, Z. M. Wang, S. L. Feng and Z. Yang, *Phys. Rev. B* **62**, 6956 (2000); *Phys. Rev. B* **61**, 15614 (2000).
80. A. R. Kost, M. H. Jupina, T. C. Hasenberg and E. M. Garmire, *J. Appl. Phys.* **99**, 023501 (2006).
81. A. G. Smirnov, D. V. Ushakov and V. K. Kononenko, *Proc. SPIE* **4706**, 70 (2002).
82. D. V. Ushakov, V. K. Kononenko and I. S. Manak, *Proc. SPIE* **4358**, 171 (2001).
83. J. Z. Wang, Z. G. Wang, Z. M. Wang, S. L. Feng and Z. Yang, *Phys. Rev. B* **62**, 6956 (2000).
84. A. R. Kost, L. West, T. C. Hasenberg, J. O. White, M. Matloubian and G. C. Valley, *Appl. Phys. Lett.* **63**, 3494 (1993).
85. S. Bastola, S. J. Chua and S. J. Xu, *J. Appl. Phys.* **83**, 1476 (1998).
86. Z. J. Yang, E. M. Garmire and D. Doctor, *J. Appl. Phys.* **82**, 3874 (1997).
87. G. H. Avetisyan, V. B. Kulikov, I. D. Zalevsky and P. V. Bulaev, *Proc. SPIE* **2694**, 216 (1996).
88. U. Pfeiffer, M. Kneissl, B. Knüpfer, N. Müller, P. Kiesel, G. H. Döhler and J. S. Smith, *Appl. Phys. Lett.* **68**, 1838 (1996).
89. H. L. Vaghjiani, E. A. Johnson, M. J. Kane, R. Grey and C. C. Phillips, *J. Appl. Phys.* **76**, 4407 (1994).
90. P. Kiesel, K. H. Gulden, A. Hoefler, M. Kneissl, B. Knuepfer, S. U. Dankowski, P. Riel, X. X. Wu, J. S. Smith and G. H. Doehler, *Proc. SPIE* **1985**, 278 (1993).
91. G.H. Doheler, *Phys. Script.* **24**, 430 (1981).
92. S. Mukherjee, S. N. Mitra, P. K. Bose, A. R. Ghatak, A. Neoigi, J. P. Banerjee, A. Sinha, M. Pal, S. Bhattacharya and K. P. Ghatak, *J. Comput. Theor. Nanosc.* **4**, 550 (2007).
93. T. Ono, Y. Fujimoto and S. Tsukamoto, *Quan. Matt.* **1**, 4 (2012); H. De Raedt, F. Jin and K. Michielsen, *Quan. Matt.* **1**, 20 (2012); R. Hofmann and D. Kaviani, *Quan. Matt.* **1**, 41 (2012); M. Narayanan and A. J. Peter, *Quan. Matt.* **1**, 53 (2012); S. Olszewski, *Quan. Matt.* **1**, 59 (2012); A. Heidari, N. Heidari and M. Ghorbani, *Quan. Matt.* **1**, 86 (2012); S. Olszewski, *Quan. Matt.* **1**, 127 (2012); V. Sajfert, P. Mali, N. Bednar, N. Pop, D. Popov and B. Tosic, *Quan. Matt.* **1**, 134 (2012); B. Tuzun and S. Erkoc, *Quan. Matt.* **1**, 136 (2012); A. Heidari, N. Heidari and M. Ghorbani, *Quan. Matt.* **1**, 149 (2012); C. Falquez, R. Hofmann and T. Baumbach, *Quan. Matt.* **1**, 153 (2012); I. S Amiri, A. Nikoukar, J. Ali and P. P Yupapin, *Quan. Matt.* 1, 159 (2012); N. Paitya, S. Bhattacharya, D. De and K. P. Ghatak, *Quan. Matt.* **1**, 63 (2012); S. Bhattacharya, D. De, S. Ghosh and K. P. Ghatak, *J. Comp. Theo. NanoSc.* **10**, 1 (2013); S. Bhattacharya, D. De, N. Paitya, S. M Adhikari and K. P Ghatak, *Adv. Sc. Lett.* **16**, 348 (2012).
94. R. Dornhaus and G. Nimtz, *Springer Tracts in Modern Physics*, Vol. 78 (Berlin-Heidelberg, N., Springer, 1976).

95. D. De, S. Bhattacharya, S. Ghosh and K. P Ghatak, *Adv. Sc. Engg. Med.* **4**, 211 (2012).
96. T. N. Sen and K. P. Ghatak, *Quant. Matt.* **5**, 1 (2016).
97. S. K. Biswas, A. R. Ghatak, A. Neogi, A. Sharma, S. Bhattacharya and K. P. Ghatak, *Physica E* **36**, 163 (2007).
98. N. Paitya, S. Bhattacharya, D. De, S. Ghosh and K. P Ghatak, *J. Nanoengg. Nanomanufac.* **2**, 211 (2012).
99. M. Mondal, S. Banik and K. P. Ghatak, *J. Low Temp. Phys.* **74**, 423 (1989).
100. K. P. Ghatak, S. Bhattacharya, H. Saikia and A. Sinha, *J. Comp. Theor. Nanosci.* **3**, 1 (2006).
101. K. P. Ghatak and S. N. Biswas, J. Vac. Sc. Tech. **7B**, 104 (1989); P. K. Chakraborty, S.Choudhury and K. P. Ghatak, *Phys. B* **387**, 333 (2007).
102. S. Bhattacharya, S. Pahari, D. K. Basu and K. P. Ghatak, *J. Comp. Theo. Nanosci.* **3**, 280 (2006); A. Sinha, A. K Sharma, R. Barui, A. R. Ghatak, S. Bhattacharya, and K. P. Ghatak, *Phys. B* **391**, 141 (2007).
103. S. Choudhury, L. J. Singh and K. P. Ghatak, *Nanotechnology* **15**, 180 (2004).
104. P. K. Chakraborty, A. Sinha, S. Bhattacharya and K. P. Ghatak, *Phys. B* **390**, 325 (2007); K. P. Ghatak, J. P. Banerjee and D. Bhattacharyya, *Nanotechnology* **7**, 110 (1996)
105. P. K. Chakraborty, B. Nag and K. P. Ghatak. *J. Phys. Chem. Solids* **64**, 2191 (2003).
106. K. P. Ghatak, J. P. Banerjee and B. Nag, *J. Appl. Phys.* **83**, 1420 (1998).
107. B. Nag and K. P. Ghatak, *J. Phys. Chem. Solids* **59**, 713 (1998). B. Nag and K. P. Ghatak, *J. Phys. Chem. Sol.* **58**, 427 (1997).
108. K. P. Ghatak, D. K. Basu and B. Nag. *J. Phys. Chem. Solids* **58**, 133 (1997); B. Nag, K. P. Ghatak, Nonlin. *Opt. Quan. Opt.* **19**, 1 (1998);
109. K. P. Ghatak and B. Nag, *Nanostruct. Mat.* **10**, 923 (1998).
110. P. K. Bose, N. Paitya, S. Bhattacharya, D. De, S. Saha, K. M. Chatterjee, S. Pahari and K. P. Ghatak, *Quant. Matt.* **1**, 89 (2012); K. P. Ghatak, B. Mitra, *Int. J. Electron.* **72**, 541 (1992); M. Mondal, N. Chattopadhyay and K. P. Ghatak, *J. Low Temp. Phys.* **66**, 131 (1987); P. K. Chakraborty, G. C. Datta and K. P. Ghatak, *Physica Scripta*, **68**, 368 (2003); K. P. Ghatak and M. Mondal, *Zeitschrift für Physik B Condensed Matter*, **69**, 471 (1988); B. Mitra and K. P. Ghatak, *Solid-state Electron.* **32**, 177 (1989).
111. S. M. Adhikari, A. Sakar and K. P. Ghatak, *Quant. Mat.* **2**, 455 (2013); K. P. Ghatak, S. Bhattacharya, S. K. Biswas, A. Dey and A. K. Dasgupta, *Physica Scripta*, **75**, 820 (2007); M. Mondal and K. P. Ghatak, *Phys. Lett. A*, **131**, 529 (1988).
112. A. N. Chakravarti, K. P. Ghatak, K. K. Ghosh, S. Ghosh and A. Dhar, *Zeitschrift für Physik B Condensed Matt.* **47**, 149 (1982); A. Ghoshal, B. Mitra and K. P. Ghatak, *Il Nuovo Cimento D*, **12**, 891 (1990); A. N. Chakravarti, A. K. Chowdhury, K. P. Ghatak, S. Ghosh and A. Dhar, *Appli. Phys.*, **25** 105 (1981).
113. M. Mondal and K. P. Ghatak, *Physica Status Solidi (b)* **135**, K21 (1986); M. Mondal and K. P. Ghatak, *Physica Status Solidi (b)*, **123**, K143 (1984);

K. P. Ghatak, A. Ghoshal and B. Mitra, *Il Nuovo Cimento D,* **13**, 867 (1991).

114. K. P. Ghatak and B. Mitra, *Int. J. Electron. Theoret. Exp.* **70**, 343 (1991); K. P. Ghatak, B. Mitra and A. Ghoshal, *Physica Status Solidi (b)* **154**, K121 (1989); K. P. Ghatak and M. Mondal, *Zeitschrift für Naturforschung A* **41**, 881 (1986).
115. M. Mondal and K. P. Ghatak, *Physica Status Solidi (b)* **129**, 745 (1985); B. Mitra and K. P. Ghatak, *Solid-state Electron.* **32**, 515 (1989).
116. K. P. Ghatak, N. Chatterjee and M. Mondal, *Physica Status Solidi (b)* **139**, K25 (1987); A. N. Chakravarti and K. P. Ghatak, K. K. Ghosh and H. M. Mukherjee, *Physica Status Solidi (b)* **116**, 17 (1983); A. N. Chakravarti, K. P. Ghatak, A. Dhar and K. K. Ghosh, *Czechoslovak J. Phys.* **33**, 65 (1983).
117. K. P. Ghatak and A. Ghosal, *Physica Status Solidi (b)* **151**, K135 (1989); S. Bhattacharya, D. De, S. M. Adhikari and K. P. Ghatak, *Super-lattices Microstruct.*, **51**, 203 (2012); K. P. Ghatak and M. Mondal, *Physica Status Solidi (b)*, **175**, 113 (1993).
118. M. Mondal and K. P. Ghatak, *Czechoslovak J. Phys. B* **36**, 1389 (1986); M. Mondal and K. P. Ghatak, *Physica Status Solidi (b)* **126**, K47 (1984); D.De, S. Bhattacharya, S. M. Adhikari, A. Kumar, P. K. Bose and K. P. Ghatak, *Beilstein J. Nanotechnolo.* **2**, 339 (2011).
119. M. Mondal and K. P. Ghatak, *Physica Status Solidi (b)* **139**, 185 (1987); S. Debbarma, A. Bhattacharjee, S. Bhattacharyya, A. Mondal and K. P. Ghatak, *J. Adv. Phys.* **1**, 84 (2012); D. Bhattacharyya and K. P. Ghatak, *Physica Status Solidi (b)* **187**, 523 (1995).
120. S. M. Adhikari, D. De, J. K. Baruah, S. Chowdhury and K. P. Ghatak, *Adv. Sci. Focus* **1**, 57 (2013); P. K. Bose, S. Bhattacharya, D. De, N. Paitya and K. P. Ghatak, *Adv. Sci. Engi. Med.* **5**, 245 (2013).
121. K. P. Ghatak, *Acta Physica Hungarica* **74**, 257 (1994); M. Mondal and K. P. Ghatak, *Acta Physica Slovaca,* **36**, 325 (1986); M. Mondal, K. P. Ghatak and *Acta physica Polonica.* A, **66**, 47 (1984).
122. S. Bhattachrya, N. Paitya and K. P. Ghatak, *J. Comput. Theoret. Nanosci.* **10**, 1999 (2013); N. Paitya and K. P. Ghatak, *J. Nanosci. Nanotechnol.* **12**, 8985 (2012); N. Chattopadhyay, K. P. Ghatak and M. Mondal, *Acta Physica Hungarica,* **67**, 163 (1990); M. Mondal and K. P. Ghatak, *Acta Physica Polonica. A* **67**, 983 (1985).
123. P. K. Chakraborty, G. C. Datta and K. P. Ghatak, *Physica Scripta* **68**, 368 (2003); B. Nag, P. K. Chakrabarty and K. P. Ghatak, *J. Wave Mater. Interact.* **11**, 211 (1996).
124. K. P. Ghatak, D. K. Basu and B. Nag, *J. Wave Mater. Interact.* **10**, 29 (1995); K. P. Ghatak and S. N Banik, *FIZIKA A,* **4**, 33 (1995); K. P. Ghatak and D. Bhattacharyya, *Physica Status Solidi (b)* **179**, 383 (1993); K. P. Ghatak and D. Bhattacharyya, *J. Wave Mater. Interact.* **8**, 233 (1993).
125. A. Ghoshal and K. P. Ghatak, *Acta Physica Slovaca, Czechoslovakia* **42**, 370 (1992); K. P. Ghatak and A. Ghoshal, *Physica Status Solidi (a)* **126**, K53 (1991); K. P. Ghatak, *Acta Physica Hungarica* **70**, 77 (1991).

126. H. Kroemer, *IEEE Trans. Electron Devi.* **25**, 850 (1978); M. Mondal and K. P. Ghatak, *Physica Status Solidi (b)* **128**, K133 (1985).
127. R. W.Lade, *Proc. IEEE* **52**, 743 (1965).
128. A. Einstein, *Ann. der Physik* **17**, 549 (1905); W. Nernst, *Z. Phys. Chem. Leipzig* **2**, 613 (1888); J. S. Townsend, *Trans. Roy. Soc.* 193A, 129 (1900).
129. P. T. Landsberg, *Thermodynamics and Statistical Mechanics* (Oxford University Press, Oxford, 1978); *In Recombination in Semiconductors* (Cambridge University Press, U.K., 1991); C. Wagner, *Z. Physik, Chem. B* **21**, 24 (1933); C. Herring and M. H. Nichols, *Rev. Mod. Phys.* **21**, 185 (1949).
130. P. T. Landsberg, *Proc R. Soc. A* **213**, 226 (1952); *Proc. Phys. Soc. A* **62**, 806 (1949).
131. P. T. Landsberg, *Eur. J. Phys.* **2**, 213 (1981).
132. C. H. Wang and A. Neugroschel, *IEEE Electron. Dev. Lett.* **ED-11**, 576 (1990).
133. I.-Y. Leu and A. Neugroschel, *IEEE Trans. Electron. Dev.* **ED-40**, 1872 (1993).
134. F. Stengel, S. N. Mohammad and H. Morkoç, *J. Appl. Phys.* **80**, 3031 (1996).
135. H.-J. Pan, W.-C. Wang, K.-B. Thai, C.-C. Cheng, K.-H. Yu, K.-W. Lin, C.-Z. Wu and W.-C. Liu, *Semicond. Sci. Technol.* **15**, 1101 (2000).
136. S. N. Mohammad, *J. Appl. Phys.* **95**, 4856 (2004).
137. V. K. Arora, *Appl. Phys. Lett.* **80**, 3763 (2002).
138. S. N. Mohammad, *J. Appl. Phys.* **95**, 7940 (2004).
139. S. N. Mohammad, *Philos. Mag.* **84**, 2559 (2004).
140. S. N. Mohammad, *J. Appl. Phys.* **97**, 063703 (2005).
141. K. Suzue, S. N. Mohammad, Z. F. Fan, W. Kim, O. Aktas, A. E. Botchkarev and H. Morkoç, *J. Appl. Phys.* **80**, 4467 (1996).
142. S. N. Mohammad, Z. F. Fan, W. Kim, O. Aktas, A. E. Botchkarev, A. Salvador and H. Morkoç, *Electron. Lett.* **32**, 598 (1996).
143. Z. Fan, S. N. Mohammad, W. Kim, O. Aktas, A. E. Botchkarev, K. Suzue and H. Morkoç, *J. Electron. Mater.* **25**, 1703 (1996).
144. C. Lu, H. Chen, X. Lv, X. Xia and S. N. Mohammad, *J. Appl. Phys.* **91**, 9216 (2002).
145. S. G. Dmitriev and Yu V. Markin, *Semiconductors* **34**, 931 (2000).
146. M. Tao, D. Park, S. N. Mohammad, D. Li, A. E. Botchkerav and H. Morkoç, *Philos. Mag. B* **73**, 723 (1996).
147. D. G. Park, M. Tao, D. Li, A. E. Botchkarev, Z. Fan, S. N. Mohammad and H. Morkoç, *J. Vac. Sci. Technol. B* **14**, 2674 (1996).
148. Z. Chen, D. G. Park, S. N. Mohammad and H. Morkoç, *Appl. Phys. Lett.* **69**, 230 (1996).
149. G. Paasch, T, Fiedler, M. Kolar and I. Bartos, *J. Physica Status Solidi B* **118**, 641 (1983).
150. K. P. Ghatak and M. Mitra, *Elastic Constants in Heavily Doped Low Dimensional Materials*, Series on the Foundations of Natural Science and Technology, Vol. 14 (World Scientific, 2021).

151. K. v. Klitzing, G. Dorda and M. Pepper, *Phys. Rev. Lett.* **45**, 494 (1980); K. v. Klitzing, *Rev. Mod. Phys.* **58**, 519 (1986).
152. J. Hajdu and G. Landwehr, In: *Strong and Ultrastrong Magnetic Fields and Their Applications,* F. Herlach (ed.) (Springer, 1985).
153. I. M. Tsidilkovskii, *Band Structure of Semi-conductors* (Pergamon Press, Oxford, 1982).
154. E. A. Arushanov, A. F. Knyazev, A. N. Natepov and S. I. Radautsan, *Sov. Phys. Semi.* **15**, 828 (1981).
155. S. P. Zelenim, A. S. Kondratev and A. E. Kuchma, *Sov. Phys. Semi.* **16**, 355 (1982).
156. F. M. Peeters and P. Vasilopoulos, *Phys. Rev. B* **46**, 4667 (1992).
157. W. Zawadzki, In: *Two-Dimensional Systems, Heterostructures and Superlattices*, Springer Series in Solid State Sciences, Vol. 53, G. Bauer, F. Kuchar, H. Heinrich (eds.) (Springer, Berlin, Heidelberg, 1984).
158. B. M. Askerov, N. F. Gashimzcede and M. M. Panakhov, *Sov. Phys. Sol. State* **29**, 465 (1987).
159. G. P. Chuiko, *Sov. Phys. Semi.* **19**, 1279 (1985).
160. S. Pahari, S. Bhattacharya and K. P. Ghatak, *J. Comput. Theor. Nanosci.* **6**, 2088 (2009).
161. K. P. Ghatak, S. Bhattacharya, S. Bhowmik, R. Benedictus and S. Choudhury, *J. Appl. Phys.* **103**, 034303 (2008).
162. K. P. Ghatak, S. Bhattacharya, S. Pahari, D. De, S. Ghosh and M. Mitra, *Annalen der Physik* **17**, 195 (2008).
163. K. P. Ghatak and S. N. Biswas, *J. Appl. Phys.* **70**, 299 (1991).
164. K. P. Ghatak and S. N. Biswas, *J. Low Temp. Phys.* **78**, 219 (1990).
165. K. P. Ghatak and M. Mondal, *J. Appl. Phys.* **65**, 3480 (1989).
166. K. P. Ghatak and B. Nag, *Nanostruct. Materi.* **5**, 769 (1995).
167. K. P. Ghatak and M. Mondal, *Phys. Stat. Sol. (b)* **185**, K5 (1994).
168. K. P. Ghatak and B. Mitra, *IlNuovoCimento D* **15**, 97 (1993).
169. K. P. Ghatak and S. N. Biswas, *Phys. Stat. Solidi (b)* **140**, K107 (1987).
170. K. P. Ghatak, *IlNuovoCimentoD* **13**, 1321 (1991).
171. K. P. Ghatak and A. Ghoshal, *Phys. Stat. Sol. (b)***170**, K27 (1992).
172. K. P. Ghatak, B. De, B. Nag and P. K. Chakraborty, *Mol. Cryst. Liquid Cryst. Sci. Technolo. Sect. B: Nonlin. Opt.* **16**, 221 (1996).
173. K. P. Ghatak, M. Mitra, B. Goswami and B. Nag, *Mol. Cryst. Liquid Cryst. Sci. Technolo. Sect. B: Nonlin. Opt.* **16**, 167 (1996).
174. K. P. Ghatak, D. K. Basu, D. Basu and B. Nag, *NuovoCimentodellaSocietaItaliana di Fisica D — Condensed Matter, Atomic, Mol. Chem. Phys., Biophys.* **18**, 947 (1996).
175. B. Mitra and K. P. Ghatak, *Phys. Letts. A* **141**, 81 (1989); S. K. Biswas, A. R. Ghatak, A. Neogi, A. Sharma, S. Bhattacharya and K. P. Ghatak, *Physica E: Low-Dimen. Syst. Nanostruct.* **36**, 163 (2007).
176. M. Mondal, A. Ghoshal and K. P. Ghatak, *Il NuovoCimento D***14**, 63 (1992); K. P. Ghatak and M. Mondal, *Phys. Stat. Sol. (b)*, **135**, 819 (1986); L. J. Singh, S. Choudhury, D. Baruah, S. K. Biswas, S. Pahari and K. P. Ghatak, *Phys. B* **368**, 188 (2005).

177. K. P. Ghatak, B. De, M. Mondal and S. N. Biswas, *Mater. Rese. Soc. Symp. — Proce.* **198**, 327 (1990); K. P. Ghatak, *Proc. SPIE — The Int. Soci. Opti. Eng.* **1584**, 435 (1992).
178. K. P. Ghatak and B. De, *Mater. Res. Soc. Symp. — Proc.* **234**, 55 and 59 (1991).
179. K. P. Ghatak, B. De, M. Mondal and S. N. Biswas, *Mater. Rese. Soci. Sympo. — Proc.* **184**, 261 (1990).
180. M. Mondal and K. P. Ghatak, *Physica Scripta* **31**, 613 (1985).
181. K. P. Ghatak and S. N. Biswas, *Mater. Rese. Soci. Symp. — Proc.* **216**, 465 (1990).
182. P. T. Landsberg, *Eur. J. Phys.* **2**, 213 (1981).
183. R. B. Dingle, *Philos. Mag.* **46**, 813 (1955); D. Redfield, M. A. Afromowitz, *Philos. Mag.* **19**, 831 (1969); H. C. Casey, F. Stern, *J. Appl. Phys.* **47**, 631 (1976).
184. K. P. Ghatak, S. Bhattacharya and D. De, *Einstein Relation in Compound semiconductors and their Nano Structures*, Springer series in Materials Science, Vol. 116 (Springer, Germany, 2009); A. N. Chakravarti and B. R. Nag, *Phys. Stat. Sol. (a)* **14**, K55 (1972); A. N. Chakravarti and D. P. Parui, *Canad. J. Phys.* **51**, 451 (1973).
185. K. P. Ghatak and S. Bhattacharya, *Heavily Doped 2D-Quantized Structures and the Einstein Relation, Tracts in Modern Physics*, Vol. 260 (Springer, Germany, 2015).
186. A. S. Filipchenko, I. G. Lang, D. N. Nasledov, S. T. Pavlov and L. N. Radaikine, *Phys. Stat. Sol. (b)* **66**, 417 (1974).
187. M. Wegner, *Extreme Nonlinear Optics* (Springer-Verlag, Germany, 2005).
188. B. S. Wherreff, W. Wolland, C. R. Pidgeon, R. B. Dennis and S. D. Smith, in *Proceedings of the 12th International Conference of the Physics of the Materials*, M. H. Pilkahn, and R. G. Tenbner (eds.) (Staffgard,1978).
189. C. H. Peschke, *Phys. Stat. Sol. (b)* **191**, 325 (1995).
190. M. E. Orazem and J. Newman, *J. Electrochem. Soc.* **2**, 2715 (1984), P. T. Landsberg and H.C Cheng, *Phys. Rev.* **32B**, 8021 (1985); H. Reiss, C. S. Fuller and F. J. Morin, *Bell Syst. Tech. J.* **35**, 535 (1956); P. T. Landsberg and A. G. Guy, *Phys. Rev. B* **28**, 1187 (1983); *Handbook on Materials*, W. Paul (ed.), (North-Holland Publishing Co., 1982).
191. A. Isihara and Y. Shiwa, *J. Phys. C* **18**, 4703 (1985).
192. V. G.I Ravish, B. A. Efimova and I.A Smirnov, *Semiconducting Lead Chalocogenides* (Plenum Press, New York, 1970).
193. F. J Blatt, *J. Phys. Chem. Solids* **17**, 177 (1961).
194. D. K Roy, *Quantum Mechanical Tunnelling and Its Applications* (World Scientific Publication Co. Ptc. Ltd, 1986).
195. W. Zawadzki, *Physica* **127**, 388 (1980).
196. E. V. Rozenfeld, Yu. P. Irkhin and P. G. Guletskii, *Sov. Phys. Solid State* **29**, 1134 (1987).
197. H. Piller, *Materials and Semimetals* 124, **8** (1972).

198. R. H. Fowler and L. Nordheim, *Proc. Roy. Soci. Lond. Series A* **119**, 173 (1928); A.Van Der Ziel, *SolidState Physical Electronics* (Prentice-Hall, Inc., Englewood Cliffs U. S.A., 1957).
199. B. Mitra and K. P. Ghatak, *Phys. Lett. A* **357**, 146 (1990); K. P. Ghatak and M. Mondal, *J. Mag. Mag. Mat.* **74**, 203 (1988); K. P. Ghatak and B. Mitra, *Phys. Lett.* **156A**, 233 (1991); K. P. Ghatak, A. Ghosal, S. N. Biswas and M. Mondal, *Proc. SPIE, USA* **1308**, 356 (1990).
200. V. T. Binh and Ch. Adessi, *Phys. Rev. Lett.* **85**, 864 (2000); R. G. Forbes, *Ultramicroscopy* **79**, 11 (1999); J. W. Gadzuk and E. W. Plummer, *Rev. Mod. Phys* **45**, 487 (1973); J. M. Beebe, B. Kim, J. W. Gadzuk, C. D. Frisbie and J. G. Kushmerick, *Phys. Rev. Lett* **97**, 026801 (1999); Y. Feng and J. P. Verboncoeur, *Phys. Plasmas* **12**, 103301 (2005); W. S. Koh and L. K. Ang, *Nanotechnology* **19**, 235402 (2008); M. Razavy, *Quantum Theory of Tunneling* (World Scientific Publishing Co. Pte. Ltd, Singapore 2003).
201. S. I. Baranchuk and N. V. Mileshkina, *Soviet Phys. Solid State* **23**, 1715 (1981); P. G. Borzyak and A. A. Dadykin, *Soviet Phys. Docklady* **27**, 335 (1982); S. Bono and R. H. Good, Jr., *Surface Sci.* **134**, 272 (1983); S. M. Lyth and S. R. P. Silva, *Appl. Phys. Letts.* **90**, 173124 (2007); C. Xu and X. Sun, *Int. J. Nanotech.* **1**, 452 (2004); S. D. Liang, L. Chen, *Phys. Rev. Letts.* **101**, 027602 (2008).
202. E. C. Heeres, E. P. A. M. Bakkers, A. L. Roest, M. Kaiser, T. H, Oosterkamp and N. de Jonge, *Nano Lett.* **7** 536 (2007); L. Dong, J. Jiao, D. W. Tuggle, J. M. Petty, S. A. Elliff and M. Coulter, *Appl. Phys. Lett.* **82**, 1096 (2003); S. Y. Li, P.Lin, C. Y. Lee and T. Y. Tseng, *J. Appl. Phys.*, **95**, 3711 (2004); N. N. Kulkarni, J. Bae, C. K. Shih, S. K. Stanley, S. S. Coffee and J. G. Ekerdt, *Appl. Phys. Lett.* **87**, 213115 (2005).
203. K. Senthil and K. Yong, *Mat. Chem. Phys.* **112**, 88, (2008); R. Zhou, H. C. Chang, V. Protasenko, M. Kuno, A. K. Singh, D. Jena and H. Xing, *J. Appl. Phys.* **101**, 073704 (2007); K. S. Yeong, J. T. L. Thong, *J. Appl. Phys.* **100**, 114325 (2006); C. H. Oon, S. H. Khong, C. B. Boothroyd and J. T. L. Thong, *J. Appl. Phys.* **99**, 064309 (2006).
204. B. H. Kim, M. S. Kim, K. T. Park, J. K. Lee, D. H. Park, J. Joo, S. G. Yu and S. H. Lee, *Appl. Phys. Lett.* **83**, 539 (2003); Z. S. Wu, S. Z. Deng, N. S. Xu, J. Chen, J. Zhou and J. Chen, *Appl. Phys. Lett.* **80**, 3829 (2002); Y. W. Zhu, T. Yu, F. C. Cheong, X. J. Xu, C. T. Lim, V. B. C. Tan, J. T. L. Thong and C. H. Sow, *Nanotechnology* **16**, 88 (2005); Y. W. Zhu, H. Z. Zhang, X. C. Sun, S. Q. Feng, J. Xu, Q. Zhao, B. Xiang, R. M. Wang and D. P. Yu, *Appl. Phys. Lett.* **83**, 144 (2003); S. Bhattacharjee and T. Chowdhury, *Appl. Phys. Lett.* **95**, 061501 (2009); S. Kher, A. Dixit, D. N. Rawat and M. S. Sodha, *Appl. Phys. Lett.* **96**, 044101 (2010).
205. I. Shigeo, W. Teruo, O. Kazuyoshi, T. Masateru, U. Satoshi and N. Norio, *J. Vac. Sc. and Tech. B*, **13**, 487 (2009); C. A. Spindt, I. Brodie, L. Humphrey and E. R. Westerberg, *J. Appl. Phys.* **47**, 5248 (2009); Q. Fu, A. V. Nurmikko, L. A. Kolodziejski, R. L. Gunshor and J. W. Wu, *Appl. Phys. Lett.* **51**, 578 (2009).

206. I. Mertig, E. Mrosan, V. N Antonov, V. I Antonov and P. Ziesche, *Phys. Stat. sol. (b)* **135**, K13 (1986); O. *Madelung, Physics of III-V Compounds* (John Wiley and Sons, Inc. New York, 1966).
207. K. P. Ghatak and B. Mitra, *IlNuovoCimento D* **15**, 97 (1993).
208. K. P. Ghatak and S. N. Biswas, *Phys. Stat. Solidi (b)* **140**, K107 (1987).
209. K. P. Ghatak, *IlNuovoCimentoD* **13**, 1321 (1991).
210. K. P. Ghatak and A. Ghoshal, *Phys. Stat. Sol. (b)* **170**, K27 (1992).
211. K. P. Ghatak, B. De, B. Nag and P. K. Chakraborty, *Mol. Cryst. Liquid Cryst. Sci. Technolo. Sect. B: Nonlin. Opt.* **16**, 221 (1996).
212. K. P. Ghatak, M. Mitra, B. Goswami and B. Nag, *Mol. Cryst. Liquid Cryst. Sci. Technolo. Sect. B: Nonlin. Opt.* **16**, 167 (1996).
213. K. P. Ghatak, D. K. Basu, D. Basu and B. Nag, *NuovoCimentodellaSocietaItaliana di Fisica D — Condensed Matter, Atomic, Mol. Chem. Phys., Biophys.* **18**, 947 (1996).
214. V. V. Kaminoskii, N. N. Stepanav and l. M. Smirnov, *Sov. Phys. Sol. Stat.* **27**, 1295 (1985); K. P. Ghatak, S. Chakrabarti, B. Chatterjee, P. K. Das, P. Dutta, A. Halder, *J. Mater. Focus*, **7**, 390 (2018).
215. K. S. Hong, R. F. Speyer and R. A. Condrate, *J. Phys. Chem. Solids* **51**, 969 (1990).
216. K. P. Ghatak, *Dispersion Relation in Heavily-Doped Nanostuructures* (Springer Tracts in Modern Physics, Vol. 265, 2016).
217. G. J. Rees, Phys. of Compounds, *Proccedings of the 13th International Nature Conference* F. G. Fumi, (ed.), (1166, North Holland Company, 1976).
218. K. P. Ghatak and M. Mitra, *Nano-materials Vol 2: Quantization and Entropy* (De Gruyter, Germany, 2020).
219. E. L. Ivchenko and G. E. Pikus, *Sov. Phys. Semicond.* **13**, 579 (1979).
220. J. O. Dimmock, *the Physics of Semimetals and Narrow Gap Compounds*, D. L. Carter and R. T. Bates (eds.), (Pergamon Press, Oxford 1971).
221. E. Bangert and P. Kastner, *Phys. Stat. Sol. (b)* **61**, 503 (1974).
222. G. M. T. Foley and P. N. Langenberg, *Phys. Rev. B* **15B**, 4850 (1977).
223. M. Cardona, W. Paul and H. Brooks Helv, *Acta Physica* **33**, 329 (1960) and A. F. Gibson in *Proceeding of International School of Physics "ENRICO FERMI"* course XIII, R.A Smith (ed.) (Academic Press, 1963).
224. C. C. Wang and N. W. Ressler, *Phys. Rev.* **2**, 1827 (1970).
225. V. I. Ivanov-Omskii, A. Sh. Mekhtisev, S. A. Rustambekova and E. N. Ukraintsev, *Phys. Stat. Sol. (b)* **119**, 159 (1983).
226. D. G. Seiler, W. M. Beeker and L. M. Roth, *Phys. Rev.* **1**, 764 (1970).
227. P. C. Mathur and S. Jain, *Phys. Rev.* **19**, 3159 (1979).
228. H. I. Zhang, *Phys. Rev.* **1B**, 3450 (1970)
229. P. R. Emtage, *Phys. Rev. A* **246**, 138 (1965).
230. G. E. Stillman, C. M. Wolfe and J. C. Dinmockin, in *Semiconductors and Semimetals*, R. K. Willardson and A. C. Beer (eds.), **12**, 169 (1977).
231. D. J. Newson and A. Kurobe, *Semicond. Sci. Tech.* **3**, 786 (1988).
232. U. Rossler, *Solid State Communi.* **49**, 943 (1984).

233. E. D. Palik, G. S. Picus, S. Teither and R. F. Wallis, *Phys. Rev.* **122**, 475 (1961).
234. E. J. Johnson and D. H. Dicley, *Phys. Rev.* **1**, 2676 (1970).
235. V. G. Agafonov, P. M. Valov, B. S. Ryvkin and I. D. Yaroshetskii, *Sov. Phys. Semicond.* **12**, 1182 (1978).
236. L. A. Vassilev, *Phys. Stat. Sol. (b)* **121**, 203 (1984).
237. Y. Yamada, *J. Phys. Soc. Japan* **37**, 606 (1974).

Chapter 2

The QC in QMOSFETs of Other Important Materials

I must try to get the permanent passport for living in the wonderful world known as solitude

2.1 Introduction

In this chapter, an attempt is made to investigate the QC in QMOSFETs of II–VI, IV–VI, stressed Kane type, Bi_2Te_3, Ge, GaAs, GaP, GaSb, InSb and Te, respectively. The II–VI compounds are being extensively used in nanoribbons, blue-green diode lasers, photosensitive thin films, infrared detectors, ultra-high-speed bipolar transistors, fiber optic communications, microwave devices, photovoltaic and solar cells, semiconductor gamma-ray detector arrays, semiconductor detector gamma camera and allow for a greater density of data storage on optically addressed compact discs [1–3]. The carrier energy spectra in II–VI materials are defined by the Hopfield model [4–8] where the splitting of the two-spin states by the spin-orbit coupling and the crystalline field has been taken into account. Section 2.2.1 contains the investigation of the QC in QMOSFETs of II–VI compounds.

Lead chalcogenides (PbTe, PbSe and PbS) are IV–VI compounds whose studies over several decades have been motivated by their importance in infrared IR detectors, lasers, light-emitting devices, photovoltaics and high temperature thermoelectric [9–13]. PbTe, in particular, is the end compound of several ternary and quaternary high performance high temperature thermoelectric materials [14–18]. It has been used not only as bulk but also as films [19–22], quantum wells [23], superlattices [24, 25], nanowires [26], and colloidal and embedded nanocrystals [27–30]. PbTe films doped with various impurities have also been investigated [31–34]. These studies revealed some of the interesting features that have been observed in bulk PbTe, such as Fermi level pinning and, in the case of superconductivity [35]. Section 2.2.2 contains an investigation of QC in QMOSFETs of IV–VI compounds.

It may be noted that the stressed materials are being widely investigated for strained silicon transistors, quantum cascade lasers, semiconductor strain gages, thermal detectors and strained-layer structures [36–39]. Section 2.2.3 contains the investigation of the QC in QMOSFETs of stressed Kane type compounds.

Bi_2Te_3 was first identified as a material for thermoelectric refrigeration in 1954 [40] and its physical properties were later improved by the addition of bismuth selenide and antimony telluride to form solid solutions [41–45]. The alloys of Bi_2Te_3 are very important compounds for the thermoelectric industry and have extensively been investigated in the literature [41–45]. Section 2.2.4 contains the investigation of the QC in QMOSFETs of Bismuth telluride compounds.

The importance of Germanium is already well-known since the inception of transistor technology and in recent years, memory circuits, single-photon detectors, single photon avalanche diode, ultrafast all-optical switches, THz lasers and THz spectrometers [46–49] are made of Ge. The Section 2.2.5 contains the investigation of the QC in QMOSFETs of Ge. In Section 2.2.6 we have studied the QC in QMOSFETs of GaAs compounds.

The n-Gallium Phosphide (n-GaP) is being used in quantum dots light emitting diode [50], high efficiency yellow solid state lamps, light sources and high peak current pulse for high gain tubes. The green and yellow light emitting diodes made of nitrogen-doped n-GaP possess a longer device life at high drive currents [51–53]. Section 2.2.7 contains the investigation of the QC in QMOSFETs of GaP.

Gallium antimonide (GaSb) finds applications in the fiber optic transmission window, heterojunctions and quantum wells. A complementary hetero-junction field effect transistor (CHFET) in which the channels for the p-FET device and the FET device forming the complementary FET are formed from GaSb. The band gap energy of GaSb makes it suitable for low-power operation [54–59]. Section 2.2.8 contains the investigation of the QC in QMOSFETs of InSb.

Tellurium (Te) is also an elemental semiconductor that has been used as the semiconductor layer in thin-film transistors (TFT) [60]. Te also finds extensive applications in CO_2 laser detectors [61], electronic imaging, strain-sensitive devices [62] and multichannel Bragg cell [63]. Section 1.2.6 contains the investigation of TPSM in QDs of Tellurium. Section 2.2.9 contains the investigation of the QC in QMOSFETs of Te. Section 2.3, contains the summary and conclusions of this chapter and Section 2.4 contains few open research problem.

2.2 Theoretical Background

2.2.1 *The QC in QMOSFETs of II–VI materials*

The energy spectrum of the conduction electron in bulk specimens of II–VI semiconductors is given by [8]

$$E = \frac{\hbar^2 k_s^2}{2m_\perp^*} + \frac{\hbar^2 k_z^2}{2m_\parallel^*} \pm C_0 k_s \tag{2.1}$$

where E is the total energy as measured from the edge of the conduction band in vertically upward direction and C_0 represents the splitting of the two-spin states by the spin-orbit coupling and the crystalline field.

In the presence of the surface electric field F_S along z direction, (2.1) assumes the form

$$E - eF_S z = \frac{\hbar^2 k_s^2}{2m_\perp^*} + \frac{\hbar^2 k_z^2}{2m_\parallel^*} \pm C_0 k_s \tag{2.2}$$

Using (1.7) and (2.2) we get

$$\frac{\hbar^2 k_s^2}{2m_\perp^*} \pm C_0 k_s + S_i \left(\frac{\hbar e F_S}{\sqrt{2m_\parallel^*}} \right)^{2/3} - E = 0 \tag{2.3a}$$

(2.3a) is a general result and is valid for all values of F_S

The EEM is given by

$$m^*(E, i) = m_\perp^* \left[1 - \frac{\rho_{71}}{\sqrt{E + \rho_{72}}} \right] \tag{2.3b}$$

where

$$\rho_{71} \equiv \frac{C_0}{2\sqrt{a_0'}} \quad \text{and} \quad \rho_{72} \equiv \left[(\rho_{71})^2 - \left(\frac{\hbar |e| F_s}{\sqrt{2m_\parallel^*}} \right)^{2/3} \right]$$

Thus, the EMM depends on both the Fermi energy and the sub-band index due to the presence of the term C_0.

The use of (2.3a) leads to the expression of the density-of-states function as

$$N_{2D}(E) = \frac{2g_v m_\perp^*}{\pi\hbar^2}\sum_{i=0}^{i_{\max}} H(E - E_{i1}) \tag{2.4a}$$

In which the sub-band energy (E_{i1}) is given by

$$E_{i1} = \left[S_i\left(\frac{\hbar e F_S}{\sqrt{2m_\parallel^*}}\right)^{2/3} - \frac{m_\perp^* C_0^2}{\hbar^2}\right] \tag{2.4b}$$

The surface electron concentration assumes the form

$$N_S = N_{C_{2D}}\sum_{i=0}^{i_{\max}} F_0(n_i) \tag{2.5}$$

where

$$N_{C_{2D}} = \frac{2m_\perp^* k_B T}{\pi\hbar^2}, n_i = \frac{\bar{E}_F - E_{i1}}{k_B T} \quad \text{and} \quad \bar{E}_F = eV_g - \frac{e^2 N_s}{C_{OX}} - E_{fb}$$

Thus using (1.4) and (2.5), the expression of QC can be written as

$$C_G = \frac{\frac{e^2 N_{C_{2D}}}{k_B T}\sum_{i=0}^{i_{\max}} F_{-1}(\eta_i)}{\left[1 + \frac{N_{C_{2D}}}{k_B T}\sum_{i=0}^{i_{\max}} F_{-1}(\eta_i)\left[\frac{e^2}{C_{OX}} + \frac{2}{3}N_S^{-1/3} S_i\left(\frac{\hbar e^2}{\varepsilon_{SC}\sqrt{2m_\parallel^*}}\right)^{2/3}\right]\right]} \tag{2.6}$$

This is the general expression of QC for both the limits.

2.2.2 *The QC in MOSFETs of IV–VI materials*

The DR of the conduction electrons in bulk specimens of IV–VI materials can be written in accordance with Cohen [64] as

$$E(1+\alpha E) = \frac{p_x^2}{2m_1} + \frac{p_z^2}{2m_3} - \frac{\alpha E p_y^2}{2m_2'} + \left(\frac{\alpha p_y^4}{4m_2 m_2'}\right) + \frac{p_y^2}{2m_2}(1+\alpha E) \tag{2.7}$$

where m_1, m_2 and m_3 are the effective carrier masses at the band-edge along x, y and z directions, respectively and m_2' is the effective-mass tensor component at the top of the valence band (for electrons) or at the bottom of the conduction band (for holes)..

(a) Weak Electric Field Limit

In the low electric field limit the (2.7) assumes the form

$$E(1+\alpha E)-|e|F_s z(1+2\alpha E) = \frac{p_x^2}{2M_1}+\frac{p_z^2}{2M_3}+\left(\frac{\alpha p_y^4}{4M_2M_2'}\right)+\frac{p_y^2}{2M_2}+\frac{p_y^2}{2M_2}\alpha(E-|e|F_s z)\left(1-\frac{M_2}{M_2'}\right) \tag{2.8}$$

where $M_1 = m_{\perp c}$, $m_{\perp c}$ is the transverse effective electron mass at the edge of the conduction band at $k = 0$, $M_2 = \left(\frac{m_{\perp c}+2m_{||c}}{3}\right)$, $m_{||c}$ is the longitudinal effective electron mass at the edge of the conduction band at $k = 0$, $M_3 = \left(\frac{3m_{\perp c}m_{||c}}{2m_{||c}+m_{\perp c}}\right)$, $M_2' = \left(\frac{m_{\perp v}+2m_{||v}}{3}\right)$, $m_{\perp v}$ and $m_{||v}$ are the effective transverse and longitudinal hole masses at the edge of the valance band at $k = 0$.

The use of (2.8) and (1.7) leads to the simplified expression of the 2D electron dispersion law in n-channel inversion layers of IV–VI materials under the weak electric field limit as

$$\gamma_{71}(E,i) = p_{71}k_x^2 + q_{71}(E,i)k_y^2 + r_{71}k_y^4 \tag{2.9}$$

where

$$\gamma_{71}(E,i) \equiv \left[E(1+\alpha E) - S_i\left(\frac{\hbar|e|F_s}{\sqrt{2M_3}}\right)^{2/3}\left(1+\frac{4}{3}\alpha E\right)\right],\quad p_{71} \equiv \frac{\hbar^2}{2M_1},$$

$$q_{71}(E,i) \equiv \left(\frac{\hbar^2}{2M_2}\right)\left[1+\alpha E\left(1-\frac{M_2}{M_2'}\right) - \frac{2\alpha S_i}{3}\left(\frac{\hbar|e|F_s}{\sqrt{2M_3}}\right)^{2/3}\left(1-\frac{M_2}{M_2'}\right)\right]$$

$$\text{and}\quad r_{71} \equiv \left(\frac{\alpha\hbar^4}{4M_2M_2'}\right)$$

The area enclosed by (2.9) is given by

$$A(E,i) = \frac{4}{3}\left(\frac{r_{71}}{p_{71}}\right)^{1/2}[\{a_{71w}(E,i)\}^2+\{b_{71w}(E,i)\}^2]^{1/2} \left[\{a_{71w}(E,i)\}^2 F\left[\frac{\pi}{2},\ell_{71w}(E,i)\right] - (\{a_{71w}(E,i)\}^2 + \{b_{71w}(E,i)\}^2)E\left[\frac{\pi}{2},\ell_{71w}(E,i)\right]\right] \tag{2.10}$$

in which,

$$\{a_{71w}(E,i)\}^2 \equiv \left[\frac{q_{71}(E,i)}{2r_{71}} + \frac{1}{2}\left[\frac{\{q_{71}(E,i)\}^2}{(r_{71})^2} + \frac{4\gamma_{71}(E,i)}{r_{71}}\right]^{1/2}\right]\{b_{71w}(E,i)\}^2$$

$$\equiv \left[\frac{1}{2}\left[\frac{\{q_{71}(E,i)\}^2}{(r_{71})^2} + \frac{4\gamma_{71}(E,i)}{r_{71}}\right]^{1/2} - \left(\frac{q_1(E,i)}{2r_{71}}\right)\right]$$

$$\ell_{71w}(E,i) \equiv \frac{b_{71w}(E,i)}{\sqrt{\{a_{71w}(E,i)\}^2 + \{b_{71w}(E,i)\}^2}}, \quad F\left[\frac{\pi}{2}, \ell_{71w}(E,i)\right] \quad \text{and}$$

$$E\left[\frac{\pi}{2}, \ell_{71w}(E,i)\right]$$

are the complete elliptic integral of the first and second kinds, respectively.

Using (2.10), the EMM in this case can be expressed as

$$m^*(E_{Fiw}, i) = \frac{\hbar^2}{2\pi} R_{71}(E,i)|_{E=E_{Fiw}} \tag{2.11}$$

where

$$R_{71}(E,i) \equiv \left[\frac{4}{3}\left(\frac{r_{71}}{p_{71}}\right)^{1/2} [\{a_{71w}(E,i)\}\{a_{71w}(E,i)\}' + b_{71w}(E,i)\{b_{71w}(E,i)\}'\}[\{a_{71w}(E,i)\}^2 + \{b_{71w}(E,i)\}^2\right]^{-1/2}$$

$$\left[\{a_{71w}(E,i)\}^2 F\left[\frac{\pi}{2}, \ell_{71w}(E,i)\right] - (\{a_{71w}(E,i)\}^2 - \{b_{71w}(E,i)\}^2) E\left[\frac{\pi}{2}, \ell_{71w}(E,i)\right]\right]$$

$$+ \frac{4}{3}\left(\frac{r_{71}}{p_{71}}\right)^{1/2} (\{a_{71w}(E,i)\}^2 + \{b_{71w}(E,i)\}^2)^{1/2}$$

$$2a_{71w}(E,i)\{a_{71w}(E,i)\}' F\left[\frac{\pi}{2}, \ell_{71w}(E,i)\right] + \{a_{71w}(E,i)\}^2 \left\{F\left[\frac{\pi}{2}, \ell_{71w}(E,i)\right]\right\}' - \left\{E\left[\frac{\pi}{2}, \ell_{71w}(E,i)\right]\right\}' (\{a_{71w}(E,i)\}^2 - \{b_{71w}(E,i)\}^2)$$

$$E\left[\frac{\pi}{2}, \ell_{71w}(E,i)\right] [2a_{71w}(E,i)\{a_{71w}(E,i)\}' - 2b_{71w}(E,i)\{a_{71w}(E,i)\}']]$$

Thus, the EMM is a function of the sub-band index number and the Fermi energy due to the presence of band non-parabolicity only.

The sub-band energies ($E_{n_{iw7}}$) are given by

$$\left[E_{n_{iw7}}(1+\alpha E_{n_{iw7}}) - S_i\left(\frac{\hbar|e|F_s}{\sqrt{2M_3}}\right)^{2/3}\left(1+\frac{4}{3}\alpha E_{n_{iw7}}\right)\right] = 0 \tag{2.12}$$

The total 2D density-of-states function can be written as

$$N_{2D_i}(E) = \frac{g_v}{2\pi^2}\sum_{i=0}^{i_{\max}}\{R_{71}(E,i)H(E-E_{n_{iw7}})\} \tag{2.13}$$

The surface electron concentration assumes the form

$$n_{2Dw} = \frac{2g_v}{3\pi^2}\left(\frac{r_{71}}{p_{71}}\right)^{1/2}\left\{\sum_{i=0}^{i_{\max}}[P_{7w}(E_{Fwi},i)+Q_{7w}(E_{Fwi},i)]\right\} \tag{2.14}$$

where

$$P_{7w}(E_{Fwi},i) \equiv [\{a_{71w}(E_{Fwi},i)\}^2 + \{b_{71w}(E_{Fwi},i)\}^2]^{1/2}$$
$$\left[\{a_{71w}(E_{Fwi},i)\}^2 F\left[\frac{\pi}{2},\ell_{71w}(E_{Fwi},i)\right] - (\{a_{71w}(E_{Fwi},i)\}^2\right.$$
$$\left. - \{b_{71w}(E_{Fwi},i)\}^2)E\left[\frac{\pi}{2},\ell_{71w}(E_{Fwi},i)\right]\right]$$

and

$$Q_{7w}(E_{Fwi},i) \equiv \sum_{r=1}^{s} L(r)P_{7w}(E_{Fwi},i)$$

From the allied definitions of (7.75), it appears that $\ell_{71w}(E,i) \ll 1$. Retaining the first two terms of the expansions of $F\left[\frac{\pi}{2},\ell_{71w}(E,i)\right]$ and $E\left[\frac{\pi}{2},\ell_{71w}(E,i)\right]$ together with the simple algebraic manipulations, one can approximately write

$$A(E,i) = \pi\left[\frac{\gamma_{71}(E,i)}{\sqrt{p_{71}q_{71}(E,i)}}\right]\left[1+\frac{\gamma_{71}(E,i)r_{71}}{(q_{71}(E,i))^2}\right]^{-1} \tag{2.15}$$

Thus, the expression of the surface electron concentration under the weak electric field limit under the condition $T \to 0$ can be written as

$$\bar{n}_{2Dw} = \frac{g_v}{2\pi}\sum_{i=0}^{i_{\max}}\left[\frac{\gamma_{71}(E_{Fiw},i)}{\sqrt{p_{71}q_{71}(E_{Fiw},i)}}\right]\left[1+\frac{\gamma_{71}(E_{Fiw},i)r_{71}}{(q_{71}(E_{Fiw},i))^2}\right]^{-1} \tag{2.16}$$

The QC is given by

$$C_{Glow} = \frac{\left[e^2 \sum_{i=0}^{i_{\max}} A(E_{Fiw}, i) I_{10}(E_{Fiw}, i)\right]}{\left[\frac{2\pi}{g_v} + \sum_{i=0}^{i_{\max}} \left[\frac{A(E_{Fiw},i) I_{10}(E_{Fiw},i) e^2 d_{OX}}{C_{OX}} + A(E_{Fiw}, i) I_{11}(E_{Fiw}, i)\right]\right]} \tag{2.17}$$

where

$$I_{10}(E_{Fiw}, i) = [I_8(E_{Fiw}, i) I_5(E_{Fiw}, i) I_6(E_{Fiw}, i)],$$

$$I_8(E_{Fiw}, i) = \left[\frac{1}{\gamma_{71}(E_{Fiw}, i)}\left[1 + 2\alpha E_{Fiw} - I_1'(E_{Fiw}, i) N_s^{2/3}\right] - \frac{1}{2q_{71}(E_{Fiw}, i)} I_2'(E_{Fiw}, i)\right],$$

$$I_1(E_{Fiw}, i) = \left[S_i \left(\frac{\hbar e^2}{\varepsilon_{sc}\sqrt{2M_3}}\right)^{2/3} \left(1 + \frac{4}{3}\alpha E_{Fiw}\right)\right],$$

$$I_2(E_{Fiw}, i) = \left[\frac{\hbar^2}{2M_2}\left[1 + \alpha E_{Fiw}\left(1 - \frac{M_2}{M_2'}\right)\right]\right],$$

$$I_5(E_{Fiw}, i) = \left[\left[1 + \frac{r_{71}\gamma_{71}(E_{Fiw}, i)}{q_{71}^2(E_{Fiw}, i)}\right]^{-1} q_{71}^{-4}(E_{Fiw}, i)\right],$$

$$I_6(E_{Fiw}, i) = [\{q_{71}^2(E_{Fiw}, i) r_{71}[1 + 2\alpha E_{Fiw} - I_1'(E_{Fiw}, i) N_S^{2/3}] - 2r_{71}\gamma_{71}(E_{Fiw}, i) q_{71}(E_{Fiw}, i)\{I_2'(E_{Fiw}, i) - I_3'(E_{Fiw}, i) N_S^{2/3}\}\}],$$

$$I_{11}(E_{Fiw}, i) = [I_9(E_{Fiw}, i) I_5(E_{Fiw}, i) I_7(E_{Fiw}, i)]$$

$$I_9(E_{Fiw}, i) = \left[\frac{2I_1(E_{Fiw}, i)}{\gamma_{71}(E_{Fiw}, i)} N_s^{-1/3} - \frac{1}{2q_{71}(E_{Fiw}, i)} \frac{2}{3} N_s^{-1/3} I_3(E_{Fiw}, i)\right],$$

$$I_3(E_{Fiw}, i) = \left[\frac{\hbar^2}{2M_2}\frac{2\alpha S_i}{3}\left(1 - \frac{M_2}{M_2'}\right)\left(\frac{\hbar e^2}{\varepsilon_{sc}\sqrt{2M_3}}\right)^{2/3}\right]$$

$$I_7(E_{Fiw}, i) = \left[q_{71}^2(E_{Fiw}, i) r_{71} I_1(E_{Fiw}, i) \frac{2}{3} N_S^{-1/3} - 2r_{71}\gamma_{71}(E_{Fiw}, i) q_{71}(E_{Fiw}, i) \frac{2}{3} N_S^{-1/3} I_3(E_{Fiw}, i)\right]$$

(b) Strong Electric Field Limit

Under the strong electric field limit, the dispersion relation assumes the form

$$\gamma_{72}(E,i) = p_{72}p_x^2 + q_{72}(E)k_y^2 + r_{72}k_y^4 \tag{2.18a}$$

where

$$\gamma_{72}(E,i) \equiv \left[E(1+\alpha E) - \frac{2\sqrt{2\alpha}}{3}\frac{\hbar|e|F_s}{\sqrt{2M_3}}(S_i)^{3/2}\right], \quad p_{72} \equiv \left(\frac{\hbar^2}{2M_1}\right)$$

$$q_{72}(E) \equiv \left(\frac{\hbar^2}{2M_2}\right)\left[1+\alpha E\left(1-\frac{M_2}{M_2'}\right)\right] \quad \text{and} \quad r_{72} \equiv \left[\frac{\alpha\hbar^4}{4M_2M_2'}\right]$$

Following the method as used earlier, the area enclosed by (2.18a) is given by

$$\begin{aligned} A(E,i) = \frac{4}{3}\left(\frac{r_{72}}{p_{72}}\right)^{1/2} &[\{a_{72s}(E,i)\}^2 + \{b_{72s}(E,i)\}^2]^{1/2} \\ &\times \left[\{a_{72s}(E,i)\}^2 F\left[\frac{\pi}{2}, \ell_{72s}(E,i)\right] - (\{a_{72s}(E,i)\}^2\right. \\ &\left. - \{b_{72s}(E,i)\}^2)E\left[\frac{\pi}{2}, \ell_{72s}(E,i)\right]\right] \end{aligned} \tag{2.18b}$$

in which,

$$\{a_{72s}(E,i)\}^2 \equiv \left[\frac{q_{72}(E,i)}{2r_{72}} + \frac{1}{2}\left[\frac{\{q_{72}(E,i)\}^2}{(r_{72})^2} + \frac{4\gamma_{72}(E,i)}{r_{72}}\right]^{1/2}\right]$$

$$\{b_{72s}(E,i)\}^2 \equiv \left[\frac{1}{2}\left[\frac{\{q_{72}(E,i)\}^2}{(r_{72})^2} + \frac{4\gamma_{72}(E,i)}{r_{72}}\right]^{1/2} - \left(\frac{q_{72}(E,i)}{2r_{72}}\right)\right]$$

$$\ell_{72s}(E,i) \equiv \frac{b_{72s}(E,i)}{\sqrt{\{a_{72s}(E,i)\}^2 + \{b_{72s}(E,i)\}^2}}, \; F\left[\frac{\pi}{2}, \ell_{72s}(E,i)\right]$$

and $E[\frac{\pi}{2}, \ell_{72s}(E,i)]$ are the complete elliptic integral of the first and second kinds, respectively.

Using (2.18b), the EMM in this case can be expressed as

$$m^*(\bar{E}_{Fis}, i) = \frac{\hbar^2}{2\pi}R_{72}(E,i)|_{E=\bar{E}_{Fis}} \tag{2.18c}$$

where

$$R_{72}(E,i) \equiv \left[\frac{4}{3}\left(\frac{r_{72}}{p_{72}}\right)^{1/2} \left[\{a_{72s}(E,i)\}\{a_{72s}(E,i)\}' + b_{72s}(E,i)\{b_{72s}(E,i)\}'\right]\left[\{a_{72s}(E,i)\}^2 + \{b_{72s}(E,i)\}^2\right]^{-1/2}\right.$$

$$\left[\{a_{72s}(E,i)\}^2 F\left[\frac{\pi}{2}, \ell_{72s}(E,i)\right] - (\{a_{72s}(E,i)\}^2 - \{b_{72s}(E,i)\}^2) E\left[\frac{\pi}{2}, \ell_{72s}(E,i)\right]\right] + \frac{4}{3}\left(\frac{r_{72}}{p_{72}}\right)^{1/2} (\{a_{72s}(E,i)\}^2 + \{b_{72s}(E,i)\}^2)^{1/2}\Bigg]\Bigg]$$

$$2a_{72s}(E,i)\{a_{72s}(E,i)\}' F\left[\frac{\pi}{2}, \ell_{72s}(E,i)\right] + \{a_{72s}(E,i)\}^2 \left\{F\left[\frac{\pi}{2}, \ell_{72s}(E,i)\right]\right\}' - \left\{E\left[\frac{\pi}{2}, \ell_{72s}(E,i)\right]\right\}' (\{a_{72s}(E,i)\}^2 - \{b_{72s}(E,i)\}^2)$$

$$E\left[\frac{\pi}{2}, \ell_{72s}(E,i)\right] [2a_{72s}(E,i)\{a_{72s}(E,i)\}' - 2b_{72s}(E,i)\{a_{72s}(E,i)\}']]$$

Thus, the EMM is a function of the sub-band index number and the Fermi energy due to the presence of band nonparabolicity only.

The sub-band energies ($\bar{E}_{n_{is8}}$) are given by

$$\left[\bar{E}_{n_{is8}}(1 + \alpha\bar{E}_{n_{is8}}) - \frac{2\sqrt{2\alpha}}{3}(S_i)^{3/2}\left(\frac{\hbar|e|F_s}{\sqrt{M_3}}\right)\right] = 0 \tag{2.18d}$$

The total 2D density-of-states function can be written as

$$N_{2D_i}(E) = \frac{g_v}{2\pi^2}\sum_{i=0}^{i_{\max}} \{R_{72}(E,i)H(E - \bar{E}_{n_{is8}})\} \tag{2.18e}$$

The surface electron concentration assumes the form

$$n_{2Ds} = \frac{2g_v}{3\pi^2}\left(\frac{r_{72}}{p_{72}}\right)^{1/2}\left\{\sum_{i=0}^{i_{\max}} [P_{8s}(E_{F2.2s},i) + Q_{8s}(E_{F2.2s},i)]\right\} \tag{2.18f}$$

where

$$P_{8s}(E_{F2.2s},i) \equiv [\{a_{72s}(E_{F2.2s},i)\}^2 + \{b_{72s}(E_{F2.2s},i)\}^2]^{1/2}$$

$$\left[\{a_{72s}(E_{F2.2s},i)\}^2 F\left[\frac{\pi}{2},\ell_{72s}(E_{F2.2s},i)\right] - (\{a_{72s}(E_{F2.2s},i)\}^2 - \{b_{72s}(E_{F2.2s},i)\}^2)E\left[\frac{\pi}{2},\ell_{72s}(E_{F2.2s},i)\right]\right]$$

and

$$Q_{8s}(E_{F2.2s},i) \equiv \sum_{r=1}^{s} L(r)P_{8s}(E_{F2.2s},i)$$

From the allied definitions of (7.75), it appears that $\ell_{72s}(E_{F2.2s},i) \ll 1$. Retaining the first two terms of the expansions of $F\left[\frac{\pi}{2},\ell_{72s}(E_{F2.2s},i)\right]$ and $E\left[\frac{\pi}{2},\ell_{72s}(E_{F2.2s},i)\right]$ together with the simple algebraic manipulations, one can approximately write the surface electron concentration in the strong electric field limit in this case under the condition $T \to 0$ can be written as

$$\bar{n}_{2Ds} = \frac{g_v}{2\pi}\sum_{i=0}^{i_{\max}}\left[\frac{\gamma_{72}(E_{Fis},i)}{p_{72}q_{72}(E_{Fis})}\right]\left[1+\frac{\gamma_{72}(E_{Fis},i)r_{72}}{\{q_{72}(E_{Fis})\}^2}\right]^{-1} \tag{2.19}$$

The QC can be expressed as

$$C_{Ghigh} = \frac{\left[e^2 B_1 \sum_{i=0}^{i_{\max}} A'(E_{Fis},i)\right]}{\left[1+\sum_{i=0}^{i_{\max}}\left[\frac{B_1 e^2}{C_{OX}} + A'(E_{Fis},i)\right]\right]} \tag{2.20}$$

where

$$\begin{aligned} B_1 &= \frac{2g_v}{(2\pi)^2}, A'(E_{Fis},i) \\ &= \left[\frac{\gamma'_{72}(E_{Fis},i)}{\gamma_{72}(E_{Fis},i)} - \frac{q'_{72}(E_{Fis})}{2q_{72}(E_{Fis})} - \left[1+\frac{\gamma_{72}(E_{Fis},i)r_{72}}{q_{72}^2(E_{Fis})}\right]\right. \\ &\quad \times \left[r_{72}\{q_{72}^2(E_{Fis})\gamma'_{72}(E_{Fis},i) - 2q_{72}(E_{Fis})q'_{72}(E_{Fis})\gamma_{72}(E_{Fis},i)\}\right. \\ &\quad \left.\left.\times \{q_{72}(E_{Fis})\}^{-4}\right]\right] \end{aligned}$$

2.2.3 *The QC in MOSFETs of stressed Kane type materials*

The electron energy spectrum in stressed Kane type semiconductors can be written [65] as

$$\left(\frac{k_x}{\bar{a}_0(\bar{E})}\right)^2 + \left(\frac{k_y}{\bar{b}_0(\bar{E})}\right)^2 + \left(\frac{k_z}{\bar{c}_0(\bar{E})}\right)^2 = 1 \tag{2.21}$$

where

$$[\bar{a}_0(\bar{E})]^2 \equiv \frac{\bar{K}_0(\bar{E})}{\bar{A}_0(\bar{E}) + \frac{1}{2}\bar{D}_0(\bar{E})}, \quad \bar{K}_0(\bar{E}) \equiv \left[\bar{E} - C_1\varepsilon - \frac{2C_2^2\varepsilon_{xy}^2}{3E_g'}\right]\left(\frac{3E_g'}{2B_2^2}\right)$$

C_1 is the conduction band deformation potential, ε is the trace of the strain tensor $\hat{\varepsilon}$ which can be written as

$$\hat{\varepsilon} = \begin{bmatrix} \varepsilon_{xx} & \varepsilon_{xy} & 0 \\ \varepsilon_{xy} & \varepsilon_{yy} & 0 \\ 0 & 0 & \varepsilon_{zz} \end{bmatrix}$$

C_2 is a constant which describes the strain interaction between the conduction and valance bands, $E_g' \equiv E_g + \bar{E} - C_1\varepsilon$, B_2 is the momentum matrix element,

$$\bar{A}_0(\bar{E}) \equiv \left[1 - \frac{(\bar{a}_0 + C_1)}{E_g'} + \frac{3\bar{b}_0\varepsilon_{xx}}{2E_g'} - \frac{\bar{b}_0\varepsilon}{2E_g'}\right], \quad \bar{a}_0 \equiv -\frac{1}{3}(\bar{b}_0 + 2\bar{m}),$$

$$\bar{b}_0 \equiv \frac{1}{3}(\bar{l} - \bar{m}), \quad \bar{d}_0 \equiv \frac{2\bar{n}}{\sqrt{3}}\bar{l}, \bar{m}, \bar{n}$$

are the matrix elements of the strain perturbation operator, $\bar{D}_0(\bar{E}) \equiv (\bar{d}_0\sqrt{3})\frac{\varepsilon_{xy}}{E_g'}$,

$$[\bar{b}_0(\bar{E})]^2 \equiv \frac{\bar{K}_0(\bar{E})}{\bar{A}_0(\bar{E}) - \frac{1}{2}\bar{D}_0(\bar{E})}, \quad [\bar{c}_0(\bar{E})]^2 \equiv \frac{\bar{K}_0(\bar{E})}{\bar{L}_0(\bar{E})} \quad \text{and}$$

$$\bar{L}_0(\bar{E}) \equiv \left[1 - \frac{(\bar{a}_0 + C_1)}{E_g'} + \frac{3\bar{b}_0\varepsilon_{zz}}{E_g'} - \frac{\bar{b}_0\varepsilon}{2E_g'}\right]$$

The (2.21) can be written as

$$(\bar{E} - \alpha_1)k_x^2 + (\bar{E} - \alpha_2)k_y^2 + (\bar{E} - \alpha_3)k_z^2 = t_1(\bar{E})^3 - t_2(\bar{E})^2 + t_3\bar{E} + t_4 \quad (2.22)$$

where

$$\alpha_1 \equiv [E_g - C_1\varepsilon - (\bar{a}_0 + C_1)\varepsilon + \frac{3}{2}\bar{b}_0\varepsilon_{xx} - \frac{\bar{b}_0}{2}\varepsilon + \left(\frac{\sqrt{3}}{2}\right)\varepsilon_{xy}\bar{d}_0],$$

$$\alpha_2 \equiv \left[E_g - C_1\varepsilon - (\bar{a}_0 + C_1)\varepsilon + \frac{3}{2}\bar{b}_0\varepsilon_{xx} - \frac{\bar{b}_0}{2}\varepsilon - \left(\frac{\sqrt{3}}{2}\right)\varepsilon_{xy}\bar{d}_0\right],$$

$$\alpha_3 \equiv \left[E_g - C_1\varepsilon - (\bar{a}_0 + C_1)\varepsilon + \frac{3}{2}\bar{b}_0\varepsilon_{zz} - \frac{\bar{b}_0}{2}\varepsilon\right],$$

$$t_1 \equiv \left(\frac{3}{2B_2^2}\right), \quad t_2 \equiv \left(\frac{1}{2B_2^2}\right)[6(E_g - C_1\varepsilon) + 3C_1\varepsilon],$$

$$t_3 \equiv \left(\frac{1}{2B_2^2}\right)[3(E_g - C_1\varepsilon)^2 + 6C_1\varepsilon(E_g - C_1\varepsilon) - 2C_2^2\varepsilon_{xy}^2]$$

and

$$t_4 \equiv \left(\frac{1}{2B_2^2}\right)[-3C_1\varepsilon(E_g - C_1\varepsilon)^2 + 2C_2^2\varepsilon_{xy}^2].$$

We use E and $\bar{E}$ interchangeably throughout the book.

(a) Weak electric field limit

The use of (2.22) and (1.7) leads to the expression of the dispersion relation of the 2D electrons in n-channel inversion layers of stressed III–V materials under the low electric field limit as

$$[T_{57}(E,i)]k_x^2 + [T_{67}(E,i)]k_y^2 = T_{77}(E,i) \tag{2.23}$$

where

$$T_{57}(E,i) \equiv \left[E - \alpha_1 + \frac{2}{3}S_i\left(\frac{|e|^2}{\varepsilon_{sc}}\right)^{2/3}(n_{2Dw})^{2/3}L_{17}(E)\right],$$

$$L_{17}(E) \equiv \left[\frac{(E-\alpha_1)}{(E-\alpha_3)^{2/3}[\bar{T}_{47}(E)]^{1/3}} - (E-\alpha_3)^{1/3}[\bar{T}_{47}(E)]^{-1/3}\right],$$

$$[\bar{T}_{47}(E)] \equiv \left[\{\rho_5(E)\}' - \left(\frac{\rho_5(E)}{E-\alpha_3}\right)\right], \quad \rho_5(E) = [t_1E^3 - t_2E^2 + t_3E + t_4]$$

$$T_{67}(E,i) \equiv \left[E - T_2 + \frac{2}{3}S_i\left(\frac{|e|^2}{\varepsilon_{sc}}\right)^{2/3}(n_{2Dw})^{2/3}L_{27}(E)\right],$$

$$L_{27}(E) \equiv \left[\frac{(E-\alpha_2)}{(E-\alpha_3)^{2/3}[\bar{T}_{47}(E)]^{1/3}} - \left(\frac{(E-\alpha_3)^{1/3}}{[\bar{T}_{47}(E)]^{1/3}}\right)\right]$$

$$T_{77}(E,i) \equiv \left[\rho_5(E) - S_i\left(\frac{|e|^2}{\varepsilon_{sc}}\right)^{2/3}(n_{2Dw})^{2/3}L_{37}(E)\right] \quad \text{and}$$

$$L_{37}(E) \equiv (E-\alpha_3)^{1/3}[\bar{T}_{47}(E)]^{2/3}$$

The area of the 2D surface under the weak electric field limit can be written as

$$A(E,i) = \frac{\pi T_{77}(E,i)}{\sqrt{T_{57}(E,i)T_{67}(E,i)}} \tag{2.24}$$

The sub-band energies ($E_{n_{iw8}}$) in this case are defined by

$$T_{47}(E_{n_{iw8}}) = S_i \left(\frac{|e|^2}{\varepsilon_{sc}}\right)^{2/3} (n_{2Dw})^{2/3} L_{37}(E_{n_{iw8}}) \tag{2.25}$$

The expression of the EMM in this case can be written as

$$m^*(E_{Fiw}, i) = \frac{\hbar^2}{2} L_{47}(E, i)|_{E=E_{Fiw}} \tag{2.26}$$

where

$$L_{47}(E, i) \equiv \left[\frac{1}{T_{57}(E, i)T_{67}(E, i)}\right] \left[\{T_{77}(E, i)\}'[T_{57}(E, i)T_{67}(E, i)]^{1/2} - \left(\frac{T_{77}(E, i)}{2}\right)\left\{\{T_{57}(E, i)\}'\left[\frac{T_{67}(E, i)}{T_{57}(E, i)}\right]^{1/2} + \{T_{67}(E, i)\}'\left[\frac{T_{57}(E, i)}{T_{67}(E, i)}\right]^{1/2}\right\}\right]$$

The total 2D density-of-states function can be expressed as

$$N_{2D}(E) = \frac{g_v}{2\pi} \sum_{i=0}^{i_{\max}} \{L_{47}(E, i) H(E - E_{n_{iw8}})\} \tag{2.27}$$

The surface electron concentration under the weak electric field quantum limit assumes the form

$$n_{2Dw} = \frac{g_v}{(2\pi)} \left\{\sum_{i=0}^{i_{\max}} [P_{8w}(E_{Fwi}, i) + Q_{8w}(E_{Fwi}, i)]\right\} \tag{2.28}$$

where

$$P_{8w}(E_{Fwi}, i) \equiv \frac{T_{77}(E_{Fwi}, i)}{\sqrt{T_{57}(E_{Fwi}, i)T_{67}(E_{Fwi}, i)}} \quad \text{and}$$

$$Q_{8w}(E_{Fwi}, i) \equiv \sum_{r=1}^{s} L(r) P_{8w}(E_{Fwi}, i)$$

The surface electron concentration under weak electric field limit for $T \to 0$ can be written as

$$N_S = \frac{g_v}{2\pi} \sum_{i=0}^{i_{\max}} \frac{T_{77}(E_{Fwi}, i)}{\sqrt{T_{57}(E_{Fwi}, i)T_{67}(E_{Fwi}, i)}} \tag{2.29}$$

The QC can be written as

$$C_{Glow} = \frac{\left[e^2 \sum_{i=0}^{i_{\max}} h_{10}\right]}{\left[\frac{2\pi}{g_v} + \sum_{i=0}^{i_{\max}} \left[h_{11} + \frac{e^2 d_{OX}}{\epsilon_{OX}} h_{10}\right]\right]} \tag{2.30}$$

where

$$h_{10} = h_4 - h_6 - h_8, h_4 = [T_{57}(E_{Fiw}, i)T_{67}(E_{Fiw}, i)]^{-1/2}[P_5(E_{Fiw}, i) - h'_1(E_{Fiw}, i)N_S^{2/3}]$$

$$h_1(E_{Fiw}, i) = S_i \left(\frac{e^2}{\varepsilon_{SC}}\right)^{2/3} L_{37}(E_{Fiw}, i)$$

$$h_6 = T_{77}(E_{Fiw}, i)\frac{1}{2}[T_{57}(E_{Fiw}, i)T_{67}(E_{Fiw}, i)]^{-3/2}[1 + h'_2(E_{Fiw}, i)N_S^{2/3}]$$

$$h_2(E_{Fiw}, i) = \frac{2}{3}S_i \left(\frac{e^2}{\varepsilon_{SC}}\right)^{2/3} L_{17}(E_{Fiw}, i)$$

$$h_8 = T_{57}(E_{Fiw}, i)\frac{1}{2}[T_{57}(E_{Fiw}, i)T_{67}(E_{Fiw}, i)]^{-3/2}[1 + h'_3(E_{Fiw}, i)N_S^{2/3}]$$

$$h_3(E_{Fiw}, i) = \frac{2}{3}S_i \left(\frac{e^2}{\varepsilon_{SC}}\right)^{2/3} L_{27}(E_{Fiw}, i), h_{11} = h_5 - h_7 - h_9$$

$$h_5 = [T_{57}(E_{Fiw}, i)T_{67}(E_{Fiw}, i)]^{-1/2}\left[\frac{2}{3}N_S^{-1/3} - h_1(E_{Fiw}, i)\right]$$

$$h_7 = T_{77}(E_{Fiw}, i)\frac{1}{2}[T_{57}(E_{Fiw}, i)T_{67}(E_{Fiw}, i)]^{-3/2}\left[h_2(E_{Fiw}, i)\frac{2}{3}N_S^{2/3}\right]$$

$$h_9 = T_{57}(E_{Fiw}, i)\frac{1}{2}[T_{57}(E_{Fiw}, i)T_{67}(E_{Fiw}, i)]^{-3/2}\left[h_3(E_{Fiw}, i)\frac{2}{3}N_S^{2/3}\right]$$

(b) Strong electric field limit

The use of (2.22) and (1.4) leads to the simplified dispersion relation of the 2D electrons in n-channel inversion layers of stressed III–V materials under the high electric field limit can be expressed as

$$[T_{117}(E, i)]k_x^2 + [T_{127}(E, i)]k_y^2 = T_{137}(E, i) \tag{2.31}$$

where

$$T_{117}(E, i) \equiv \left[E - \alpha_1 + \frac{4}{3}|e|F_s(S_i)^{2/3}a_{77}(E)\right],$$

$$a_{77}(E) \equiv \frac{1}{2}\left[\frac{1}{\sqrt{(E-\alpha_3)[T_{97}(E)]}} - \left\{\frac{(E-\alpha_1)^2}{\sqrt{T_{97}(E)}(E-\alpha_3)^{3/2}}\right\}\right],$$

$$[T_{97}(E)] \equiv \left[\frac{\{\rho_5(E)\}''}{2} - \left(\frac{(\rho_5(E))'}{E-\alpha_3}\right) + \left(\frac{\rho_5(E)}{(E-\alpha_3)^2}\right)\right]$$

$$T_{127}(E,i) \equiv \left[E - \alpha_2 + \frac{4}{3}|e|F_s(S_i)^{2/3}a_{87}(E)\right]$$

$$a_{87}(E) \equiv \frac{1}{2}\left[\frac{1}{\sqrt{(E-a_3)\,T_{97}(E)}} - \left(\frac{(E-\alpha_2)^{1/3}}{\sqrt{T_{97}(E)}(E-T_3)^{3/2}}\right)\right]$$

$$T_{137}(E,i) \equiv [\rho_5(E) - (S_i)^{3/2}F_s a_{97}(E)] \quad \text{and}$$

$$a_{97}(E) \equiv \frac{4}{3}\sqrt{T_{97}(E)(E-\alpha_3)}$$

The area of the 2D surface in this case is given by

$$A(E,i) = \frac{\pi T_{137}(E,i)}{\sqrt{T_{117}(E,i)T_{127}(E,i)}} \tag{2.32}$$

The sub-band energies ($E_{n_{is8}}$) in this case can be written as

$$T_{137}(E_{n_{is8}}, i) = 0 \tag{2.33}$$

The EMM in this case assumes the form

$$m^*(E_{Fis}, i) = \frac{\hbar^2}{2}T_{147}(E,i)|_{E=E_{Fis}} \tag{2.34}$$

where

$$\begin{aligned} T_{147}(E,i) \equiv &\left[\frac{1}{T_{117}(E,i)T_{127}(E,i)}\right]\left[\{T_{137}(E,i)\}'[T_{117}(E,i)T_{127}(E,i)]^{1/2}\right. \\ &- \left(\frac{T_{137}(E,i)}{2}\right)\{T_{117}(E,i)\}'\left[\frac{T_{127}(E,i)}{T_{117}(E,i)}\right]^{1/2} \\ &\left. + \{T_{127}(E,i)\}'\left[\frac{T_{117}(E,i)}{T_{127}(E,i)}\right]^{1/2}\right] \end{aligned}$$

The expression of the total 2D density-of-states function is given by

$$N_{2D_i}(E) = \frac{g_v}{2\pi}\sum_{i=0}^{i_{\max}}\{T_{147}(E,i)H(E-E_{n_{is8}})\} \tag{2.35}$$

The surface electron concentration in the strong electric field limit can be expressed as

$$n_{2Ds} = \frac{g_v}{(2\pi)}\left\{\sum_{i=0}^{i_{\max}}[P_{9s}(E_{Fis},i) + Q_{9s}(E_{Fis},i)]\right\} \tag{2.36}$$

where

$$P_{9s}(E_{Fis},i) \equiv \frac{T_{137}(E_{Fis},i)}{\sqrt{T_{117}(E_{Fis},i)T_{127}(E_{Fis},i)}} \quad \text{and}$$

$$Q_{9s}(E_{Fis},i) \equiv \sum_{r=1}^{s} L(r)P_{9s}(E_{Fis},i)$$

Using (2.36), the expression of the 2D surface electron concentration under the strong electric field limit can be written as

$$N_S = \frac{g_v}{2\pi}\sum_{i=0}^{i_{\max}}\left(\frac{T_{137}(E_{Fis},i)}{\sqrt{T_{117}(E_{Fis},i)T_{127}(E_{Fis},i)}}\right) \tag{2.37}$$

The QC can be written as

$$C_{Ghigh} = \frac{\left[e^2\sum_{i=0}^{i_{\max}} g_6(E_{Fis},i)\right]}{\left[\frac{2\pi}{g_v} + \sum_{i=0}^{i_{\max}}[g_7(E_{Fis},i)]\right]} \tag{2.38}$$

where

$$g_6(E_{Fis},i) = \left[\frac{T_{137}(E_{Fis},i)}{\sqrt{T_{117}(E_{Fis},i)T_{127}(E_{Fis},i)}} \times g_4(E_{Fis},i)\right],$$

$$\begin{aligned} g_4(E_{Fis},i) &= \frac{1}{T_{137}(E_{Fis},i)}[\rho_5'(E_{Fis},i) - g_1'(E_{Fis},i)N_s \\ &\quad - \frac{1}{2}\frac{1}{T_{117}(E_{Fis},i)}[1 + g_2'(E_{Fis},i)N_s], \\ &\quad - \frac{1}{2}\frac{1}{T_{127}(E_{Fis},i)}[1 + g_3'(E_{Fis},i)N_s]], g_1(E_{Fis},i) \\ &= (S_i)^{2/3}\frac{e^2}{\varepsilon_{SC}}a_{97}(E_{Fis},i), \end{aligned}$$

$$g_2(E_{Fis},i) = \frac{4}{3}\frac{e^2}{\varepsilon_{sc}}S_i^{2/3}a_{77}(E_{Fis},i), \quad g_3(E_{Fis},i) = \frac{4}{3}\frac{e^2}{\varepsilon_{sc}}S_i^{2/3}a_{87}(E_{Fis},i)$$

$$g_7(E_{Fis},i) = \left[\frac{T_{137}(E_{Fis},i)g_5(E_{Fis},i) + g_4(E_{Fis},i)\frac{e^2 d_{OX}}{\varepsilon_{OX}}}{\sqrt{T_{117}(E_{Fis},i)T_{127}(E_{Fis},i)}}\right],$$

$$g_5(E_{Fis},i) = \left[\frac{1}{T_{137}(E_{Fis},i)}g_1(E_{Fis},i) + \frac{1}{2T_{117}(E_{Fis},i)}g_2(E_{Fis},i) + \frac{1}{2T_{127}(E_{Fis},i)}g_3(E_{Fis},i)\right]$$

2.2.4 *The QC in MOSFETs of bismuth telluride*

The DR of the valance bands in bulk p-type Bismuth telluride can be written as [66]

$$\bar{E}(1+\alpha\bar{E}) = (I_1k_x^2 + I_2k_y^2 + I_3k_z^2 + I_4k_yk_z) \tag{2.39}$$

where

$$I_1 = \frac{\hbar^2}{2m_0}\alpha_{11}, \quad I_2 = \frac{\hbar^2}{2m_0}\alpha_{22}, \quad I_3 = \frac{\hbar^2}{2m_0}\alpha_{33},$$

$$I_4 = \left(\frac{\hbar^2}{m_0}\alpha_{23}\right), \quad \alpha_{11},\alpha_{22},\alpha_{33} \quad \text{and} \quad \alpha_{23}$$

are constant band parameters m_0 is the free electron mass,

$$\alpha = E_g^{-1}, \quad E_g = 0.145eV, \quad g_s = 2, \quad g_v = 6, \quad \alpha_{11} = 32.5, \quad \alpha_{22} = 4.81,$$
$$\alpha_{33} = 9.02 \quad \text{and} \quad \alpha_{23} = 4.15$$

(a) Weak electric field limit

Using (1.7) and (2.39) the 2D dispersion relation can be written as

$$E(1+\alpha E) = (I_2k_y^2 + I_3k_z^2 + I_4k_yk_z) + S_i[(eF_S)^{2/3}(I_1)^{1/3}(1+2\alpha E)^{2/3}] \tag{2.40}$$

The density-of-state can be written as

$$N(E) = \frac{g_v}{2\pi}(I_2I_3 - I_4^2)^{-1/2}\sum_{i=0}^{i_{\max}} T'_{46}(E,i)H(E-E_{46}) \tag{2.41}$$

where $T_{46}(E,i) = [E(1+\alpha E) - S_i(eF_S(1+2\alpha E))^{2/3}(I_1)^{1/3}]$ and $E_{46}(1+\alpha E_{46}) = S_i[(eF_S)^{2/3}(I_1)^{1/3}(1+2\alpha E_{46})^{2/3}]$.

The surface electron concentration under the condition $T \to 0$ assumes the form

$$n_0 = \frac{g_v}{2\pi}(I_2 I_3 - I_4^2)^{-1/2} \sum_{i=0}^{i_{\max}} [E_{Fiw}(1 + \alpha E_{Fiw}) - S_i(eF_S(1 + 2\alpha E_{Fiw}))^{2/3}(I_1)^{1/3}] \tag{2.42}$$

Thus, using (1.4) and (2.42), we can study the QC in this case

(b) Strong electric field limit

Using (1.7) and (2.39) the 2D dispersion relation can be written as

$$E(1 + \alpha E) = \left[(I_2 k_y^2 + I_3 k_z^2 + I_4 k_y k_z) + \frac{4}{3} S_i^{3/2} eF_S (I_1 \alpha)^{1/2}\right] \tag{2.43}$$

The density-of-state can be written as

$$N(E) = \frac{g_v}{2\pi}(I_2 I_3 - I_4^2)^{-1/2} \sum_{i=0}^{i_{\max}} T_{47}'(E, i) H(E - E_{47}) \tag{2.44}$$

where

$$T_{47}(E, i) = \left[E(1 + \alpha E) - \frac{4}{3} S_i^{3/2} eF_S (I_1 \alpha)^{1/2}\right]$$

and

$$E_{47}(1 + \alpha E_{47}) = \left[\frac{4}{3} S_i^{3/2} eF_S (I_1 \alpha)^{1/2}\right]$$

The surface electron concentration under the condition $T \to 0$ assumes the form

$$n_0 = \frac{g_v}{2\pi}(I_2 I_3 - I_4^2)^{-1/2} \sum_{i=0}^{i_{\max}} \left[E_{Fiw}(1 + \alpha E_{Fiw}) - \frac{4}{3} S_i^{3/2} eF_S (I_1 \alpha)^{1/2}\right] \tag{2.45}$$

Thus, using (1.4) and (2.45), we can study the QC in this case

2.2.5 *The QC in MOSFETs of Germanium*

It is well known that the conduction electrons of n-Ge obey two different types of dispersion laws since band non-parabolicity has been included in two different ways as given in the literature [67, 68].

(i) Model of Cardona *et al.*

(a) The energy spectrum of the conduction electrons in bulk specimens of n-Ge can be expressed in accordance with Cardona *et al.* [67] as

$$\bar{E} = -\frac{E_{g_0}}{2} + \frac{\hbar^2 k_z^2}{2m_\parallel^*} + \left[\frac{E_{g_0}^2}{4} + E_{g_0} k_s^2 \left(\frac{\hbar^2}{2m_\perp^*}\right)\right]^{1/2} \tag{2.46}$$

where in this case $m_\parallel^*$ and $m_\perp^*$ are the longitudinal and transverse effective masses along $\langle 111 \rangle$ direction at the edge of the conduction band, respectively.

The (2.46) can be written as

$$\frac{\hbar^2 k_s^2}{2m_\perp^*} = \bar{E}(1+\alpha\bar{E}) + \alpha\left(\frac{\hbar^2 k_z^2}{2m_\parallel^*}\right)^2 - (1+2\alpha\bar{E})\left(\frac{\hbar^2 k_z^2}{2m_\parallel^*}\right) \tag{2.47}$$

Using (1.7) and (2.47), the 2D electron dispersion law in this case can be written as

$$\frac{\hbar^2 k_x^2}{2m_1} + \frac{\hbar^2 k_y^2}{2m_2} = [E(1+\alpha E) + \alpha E_{i2}^2 - E_{i2}(1+2\alpha E)] \tag{2.48}$$

where

$$m_1 = m_\perp^*, \quad m_2 = \frac{m_\perp^* + 2m_\parallel^*}{3}, \quad m_3 = \frac{3m_\parallel^* m_\perp^*}{2m_\parallel^* + m_\perp^*} \quad \text{and}$$

$$E_{i2} = S_i(\hbar e F_S/\sqrt{2m_3})^{2/3}$$

The area of 2D space is

$$A = \frac{2\pi\sqrt{m_1 m_2}}{\hbar^2}[E(1+\alpha E) + \alpha E_{i2}^2 - E_{i2}(1+2\alpha E)] \tag{2.49}$$

The EEM assumes the form

$$m^*(E_{Fiw}, i) = \sqrt{m_1 m_2}[1 + 2\alpha E_{Fiw} - E_{i2} 2\alpha] \tag{2.50}$$

Thus, the EEM is the function of both Fermi energy and quantum number due to band non-parabolicity.

The density-of-states function is given by

$$N_{2D}(E) = \frac{2g_v}{(2\pi)^2} \cdot \frac{2\pi\sqrt{m_1 m_2}}{\hbar^2} \sum_{i=0}^{i_{\max}} [1 + 2\alpha E - 2\alpha E_{i2}] H(E - E_{i2}) \tag{2.51}$$

The surface electron concentration is given by

$$n_{Ds} = \frac{g_v k_B T}{\pi} \cdot \frac{\sqrt{m_1 m_2}}{\hbar^2} \sum_{i=0}^{i_{\max}} [\tau_i F_0(\eta_{4i}) + 2\alpha k_B T F_1(\eta_{4i})] \tag{2.52}$$

where

$$\tau_i = (1 - 2\alpha E_{i2}) \quad \text{and} \quad \eta_{4i} = (E_{Fiw} - E_{i2})(k_B T)^{-1}$$

Thus, using (1.4) and (2.52), we can study the QC in this case

It may be noted that (2.52) is valid for all values of E

(ii) Model of Wang *et al.*

The dispersion relation of the conduction electron in bulk specimens of n-Ge can be expressed in accordance with the model of Wang and Ressler [68] can be written as

$$\bar{E} = \frac{\hbar^2 k_z^2}{2m_\parallel^*} + \frac{\hbar^2 k_s^2}{2m_\perp^*} - \bar{c}_1 \left(\frac{\hbar^2 k_s^2}{2m_\perp^*}\right)^2 - \bar{d}_1 \left(\frac{\hbar^2 k_s^2}{2m_\perp^*}\right)\left(\frac{\hbar^2 k_z^2}{2m_\parallel^*}\right) - \bar{e}_1 \left(\frac{\hbar^2 k_z^2}{2m_\parallel^*}\right)^2 \tag{2.53}$$

where

$$\bar{c}_1 = \bar{C}(2m_\perp^*/\hbar^2)^2, \quad \bar{C} = 1.4\bar{A}, \quad \bar{A} = \frac{1}{4}\left(\hbar^4/E_{g_0} m_\perp^{*2}\right)\left(1 - \frac{m_\perp^*}{m_0}\right)^2,$$

$$\bar{d}_1 = d\left(\frac{4m_\perp^* m_\parallel^*}{\hbar^4}\right), \quad \bar{d} = 0.8\bar{A}, \quad \bar{e}_1 = \bar{e}_0(2m_\parallel^*/\hbar^2)^2 \quad \text{and} \quad \bar{e}_0 = 0.005\bar{A}$$

The 2D dispersion law in this case can be written as

$$\left[\left[1 - d_1\left(\frac{\hbar^2 k_x^2}{2m_1} + \frac{\hbar^2 k_y^2}{2m_2}\right)\right] 3^{-1}\right]$$

$$\times \left[\left(1 - d_1\left(\frac{\hbar^2 k_x^2}{2m_1} + \frac{\hbar^2 k_y^2}{2m_2}\right)\right)\left\{\left(1 - d_1\left(\frac{\hbar^2 k_x^2}{2m_1} + \frac{\hbar^2 k_y^2}{2m_2}\right)\right)^2\right.\right.$$

$$- 4e_1\left(E - \left(\frac{\hbar^2 k_x^2}{2m_1} + \frac{\hbar^2 k_y^2}{2m_2}\right)\right) + 4\left(\frac{\hbar^2 k_x^2}{2m_1} + \frac{\hbar^2 k_y^2}{2m_2}\right)$$

$$+4\left(\frac{\hbar^2k_x^2}{2m_1}+\frac{\hbar^2k_y^2}{2m_2}\right)^2\Bigg)\Bigg\}^{1/2}\Bigg]^{3/2} - \frac{1}{5}\left[\left(1-d_1\left(\frac{\hbar^2k_x^2}{2m_1}+\frac{\hbar^2k_y^2}{2m_2}\right)\right)\right.$$

$$-\left[\left(1-d_1\left(\frac{\hbar^2k_x^2}{2m_1}+\frac{\hbar^2k_y^2}{2m_2}\right)\right)^2 - 4e_1\left(E-\left(\frac{\hbar^2k_x^2}{2m_1}+\frac{\hbar^2k_y^2}{2m_2}\right)\right)\right.$$

$$\left.\left.+C_1\left(\frac{\hbar^2k_x^2}{2m_1}+\frac{\hbar^2k_y^2}{2m_2}\right)^2\right]^{1/2}\right]^{5/2} = \frac{2}{3}\hbar eF_S(e_1S_i)^{3/2}(m_3)^{-1/2} \quad (2.54)$$

where

$$C_1 = 1.4A_1(2m_1/\hbar^2)^2, \quad m_1 = m_\perp^*, \quad A_1 = \frac{\hbar^4}{4E_g}\left(\frac{1}{m_1}-\frac{1}{m_0}\right)^2,$$

$$d_1 = 0.8A\left(\frac{4m_1m_3}{\hbar^4}\right), \quad m_2 = \frac{m_\perp^*+2m_\parallel^*}{3}, \quad m_3 = \frac{3m_\parallel^*m_\perp^*}{2m_\parallel^*+m_\perp^*},$$

$$e_1 = 0.005A_1(2m_3/\hbar^2)^2$$

The surface electron concentration in the electric quantum limit can be written as

$$\left[\frac{1}{3}\left(1-\frac{d_1n_0'\pi\hbar^2}{\sqrt{m_1m_2}}\right)\left[\left(1-\frac{d_1n_0'\pi\hbar^2}{\sqrt{m_1m_2}}\right)-\left[\left(1-\frac{d_1n_0'\pi\hbar^2}{\sqrt{m_1m_2}}\right)^2\right.\right.\right.$$

$$\left.\left.-4e_1\left\{E_F-\frac{n_0'\pi\hbar^2}{\sqrt{m_1m_2}}+c_1\left(\frac{n_0'\pi\hbar^2}{\sqrt{m_1m_2}}\right)^2\right\}\right]^{1/2}\right]^{3/2}$$

$$-\frac{1}{5}\left[\left(1-\frac{d_1n_0'\pi\hbar^2}{\sqrt{m_1m_2}}\right)-\left[\left(1-\frac{d_1n_0'\pi\hbar^2}{\sqrt{m_1m_2}}\right)\left[\left(1-\frac{d_1n_0'\pi\hbar^2}{\sqrt{m_1m_2}}\right)^2\right.\right.\right.$$

$$\left.\left.\left.\left.-4e_1\left\{E_F-\frac{n_0'\pi\hbar^2}{\sqrt{m_1m_2}}+c_1\left(\frac{n_0'\pi\hbar^2}{\sqrt{m_1m_2}}\right)^2\right\}\right]^{1/2}\right]^{3/2}\right]\right]$$

$$= \frac{2}{3}\hbar eF_S(e_1S_0)^{3/2}(m_3)^{-1/2} \quad (2.55)$$

where $n_0' = N_S/g_v$.

Thus using (1.4) and (2.55), we can study the QC in this case.

2.2.6 *The QC in MOSFETs of Gallium Aresenide*

It is well known that the conduction electrons of n-GaAs obey two different types of dispersion laws since band non-parabolicity has been included in two different ways as given in the literature [69, 70].

The dispersion relation in n-GaAs including band non-parabolacity, warping and spin-splitting can be written following Rossler [69] as

$$\bar{E} = \frac{\hbar^2 k^2}{2m_c} + \alpha_0 k^4 + \beta_0[k_x^2 k_y^2 + k_y^2 k_z^2 + k_z^2 k_x^2]$$
$$\pm \gamma_0[k^2(k_x^2 k_y^2 + k_y^2 k_z^2 + k_z^2 k_x^2) - 9k_x^2 k_y^2 k_z^2]^{1/2} \tag{2.56}$$

where

$$\alpha_0 = -1984 \times 10^{-40} eVm^4, \quad \beta_0 = -1380 \times 10^{-40} eVm^4 \quad \text{and}$$
$$\bar{\gamma}_0 = 19 \times 10^{-30} eVm^3$$

Using the method of spherical averaging [71] we can write

$$k^2 = \gamma_{4\pm}(\bar{E}) \tag{2.57}$$

where

$$\gamma_{4\pm}(\bar{E}) = (2e_2)^{-1}\left[+e_1 - \sqrt{e_1^2 - 4e_2[E \pm \Gamma_4(\bar{E})]}\right],$$

$$e_2 = \alpha_0, e_1 = \left(\frac{\hbar^2}{2m_c} + \frac{\beta_0}{\sqrt{5}}\right), \quad \Gamma_4(\bar{E}) = \left(e_3\left(\frac{1}{2e_2}\right)^{3/2}\right)$$
$$\times\left[+e_1 - \sqrt{e_1^2 - 4\bar{E}e_2}\right]^{3/2}$$

(a) Weak electric field limit

Under this condition using (1.7) and (2.57), the 2D electron dispersion law assumes the form

$$\gamma_{4\pm}(E) = k_s^2 + S_i[eF_s\gamma'_{4,\pm}(E)]^{2/3} \tag{2.58}$$

The density-of-states function can be written as

$$N(E) = \frac{\pi g_v}{(2\pi)^2}\sum_{i=0}^{i_{\max}} \frac{\partial}{\partial E}[T_{31,\pm}(E,i)H(E - \bar{E}_{31,\pm})] \tag{2.59}$$

where

$$T_{31,\pm}(E,i) = [\gamma_{4,\pm}(E) - S_i[eF_S\gamma'_{4,\pm}(E)]^{2/3}] \quad \text{and}$$
$$[\gamma_{4,\pm}(\bar{E}_{31,\pm}) - S_i[eF_S\gamma'_{4,\pm}(\bar{E}_{31,\pm})]^{2/3}] = 0$$

The electron concentration at $T \to 0$ assumes the form

$$N_S = \frac{g_v}{4\pi} \sum_{i=0}^{i_{\max}} [T_{31,\pm}(E_{Fiw}, i)] \tag{2.60}$$

Using (2.60) and (1.4) we can study the QC in this case.

(b) Strong electric field limit

Under this condition using (1.7) and (2.57), the 2D electron dispersion law assumes the form

$$\gamma_{4\pm}(E) = k_s^2 + \frac{2\sqrt{2}eF_s}{3} S_i^{3/2} [\gamma'_{4,\perp}(E)]^{1/2} \tag{2.61}$$

The density-of-states function can be written as

$$N(E) = \frac{\pi g_v}{(2\pi)^2} \sum_{i=0}^{i_{\max}} \frac{\partial}{\partial E} [T_{32,\pm}(E, i) H(E - \bar{E}_{32,\pm})] \tag{2.62}$$

where

$$T_{32,\pm}(E, i) = \left[\gamma_{4,\pm}(E) - \frac{2\sqrt{2}eF_s}{3} S_i^{3/2} [\gamma''_{4,\pm}(E)]^{1/2}\right] + \quad \text{and}$$

$$[\gamma_{4,\pm}(\bar{E}_{32,\pm}) - S_i [eF_S \gamma''_{4,\pm}(\bar{E}_{32,\pm})]^{2/3}] = 0$$

The electron concentration at $T \to 0$ assumes the form

$$N_S = \frac{g_v}{4\pi} \sum_{i=0}^{i_{\max}} [T_{32,\pm}(E_{Fis}, i)] \tag{2.63}$$

Using (2.63) and (1.4) we can study the QC in this case.

(ii) Stillman *et al.*

(i) The dispersion relation of the conduction electrons in bulk specimens of GaAs can be expressed in accordance with the model of Stillman *et al.* [70] as

$$\bar{E} = e_5 k^2 - e_6 k^4 \tag{2.64}$$

where

$$e_5 = \frac{\hbar^2}{2m_c} \quad \text{and} \quad e_6 = e_5^2 \left(1 - \frac{m_c}{m_0}\right)^2 \left[\frac{3E_g + 4\Delta + \left(\frac{2\Delta^2}{E_g}\right)}{(E_g + \Delta)(2\Delta + 3E_g)}\right]$$

The (2.64) can be written as

$$k^2 = \gamma_5(\bar{E}) \quad (2.65)$$

where

$$\gamma_5(\bar{E}) = [e_5 - \sqrt{e_5^2 - 4\bar{E}e_6}][2e_6]^{-1}$$

(a) Weak electric field limit

Using (2.65) and (1.7) the 2D electron dispersion law can be written as

$$\gamma_5(E) = k_s^2 + S_i[eF_s\gamma_5'(E)]^{2/3} \quad (2.66)$$

The density-of-states function assumes the form

$$N(E) = \frac{g_v}{2\pi}\sum_{i=0}^{i_{\max}} \frac{\partial}{\partial E}[T_{33}(E,i)H(E-\bar{E}_{33})] \quad (2.67)$$

where

$$T_{33}(E,i) = [\gamma_5(E) - S_i[eF_S\gamma_5'(E)]^{2/3}],$$
$$[\gamma_5(\bar{E}_{33}) - S_i[eF_S\gamma_5'(\bar{E}_{33})]^{2/3}] = 0$$

The electron concentration at $T \to 0$ assumes the form

$$N_S = \frac{g_v}{2\pi}\sum_{i=0}^{i_{\max}}[T_{33}(E_{Fis}, i)] \quad (2.68)$$

Using (2.68) and (1.4) we can study the QC in this case.

(b) Strong electric field limit

Under this condition using (1.7) and (2.65), the 2D electron dispersion law assumes the form

$$\gamma_5(E) = k_s^2 + \frac{2\sqrt{2}eF_s}{3}S_i^{3/2}[\gamma_5'(E)]^{1/2} \quad (2.69)$$

The density-of-states function can be written as

$$N(E) = \frac{2\pi g_v}{(2\pi)^2}\sum_{i=0}^{i_{\max}} \frac{\partial}{\partial E}[T_{34}(E,i)H(E-\bar{E}_{34})] \quad (2.70)$$

where

$$T_{34}(E,i) = \left[\gamma_5(E) - \frac{2\sqrt{2}eF_s}{3}S_i^{3/2}[\gamma_5''(E)]^{1/2}\right]$$

and

$$[\gamma_5(\bar{E}_{34}) - S_i[eF_S\gamma_5''(\bar{E}_{34})]^{2/3}] = 0$$

The electron concentration at $T \to 0$ assumes the form

$$N_S = \frac{g_v}{2\pi}\sum_{i=0}^{i_{\max}}[T_{34}(E_{Fis}, i)] \quad (2.71)$$

Using (2.71) and (1.4) we can study the QC in this case.

2.2.7 *The QC in MOSFETs of gallium phosphide*

The dispersion relation in bulk specimens of n-GaP can be written as

$$\bar{E} = \frac{\hbar^2 k_s^2}{2m_\perp^*} + \frac{\hbar^2}{2m_\parallel^*}[Ak_s^2 + k_z^2] - \left[\frac{\hbar^4 k_0^2}{m_\parallel^{*2}}(k_s^2 + k_z^2) + |V_G|^2\right]^{1/2} + |V_G| \quad (2.72)$$

where k_0 and $|V_G|$ are constants of the energy spectrum with $m_\parallel^* = 0.92m_0$, $m_\perp^* = 0.25m_0$, $k_0 = 1.7 \times 10^{19}m^{-1}$, $|V_G| = 0.21eV$, $g_v = 6$, $g_s = 2$ and $A = 1$.

The above equation can be expressed as

$$k_z^2 = \frac{\bar{E}}{b} - g_1 - g_2k_s^2 - g_3[\bar{E} + g_4 + g_5k_s^2]^{1/2} \quad (2.73)$$

where

$$g_1 = \frac{2bD + c}{2b^2}, \quad b = \frac{\hbar^2}{2m_\parallel^*}, \quad D = (V_G)^2, \quad c = \left[\frac{\hbar^4 k_0^2}{(m_\parallel^*)^2}\right],$$

$$g_2 = \frac{a}{b}, \quad a = \left[\frac{\hbar^2}{2m_\perp^*} + \frac{\hbar^2}{2m_\parallel^*}\right], \quad g_3 = \frac{\sqrt{c}}{b^{3/2}},$$

$$g_4 = (4b^2D^2 + c^2 + 4bcD)(4bc)^{-1} \quad \text{and} \quad g_5 = (4b^2 - ab)(b)^{-1}$$

The 2D dispersion relation can be expressed as

$$\left[\frac{2}{3}[(\alpha_1^2 - \alpha_1 t_2 + t_3)^{3/2} - (\alpha_2^2 - \alpha_2 t_2 + t_3)^{3/2}]\right.$$
$$+ \left(\alpha_1 - \frac{t_2}{2}\right) \cdot \frac{t_2}{2}(\alpha_1^2 + t_3 - \alpha_1 t_2)^{1/2}$$

$$
\begin{aligned}
& -\frac{t_2}{2}\left(\alpha_2-\frac{t_2}{2}\right)\cdot(\alpha_2^2+t_3-\alpha_2 t_2)^{1/2} \\
& +\left(t_2\cdot\frac{t_4^2}{2}\right)\ln\left[\left\{\left(\alpha_1-\frac{t_2}{2}\right)+(\alpha_1^2+t_3-\alpha_1 t_2)^{1/2}\right\}\right. \\
& \left.\times\left\{\alpha_2-\frac{t_2}{2}+.(\alpha_2^2+t_3-\alpha_2 t_2)^{1/2}\right\}^{-1}\right]\Bigg] \\
= {} & \frac{2}{3}(S_i d_1)^{3/2}(b_1)^{-1/2}
\end{aligned} \tag{2.74}
$$

where

$$
\begin{aligned}
& \alpha_1=(E+g_4+g_5k_s^2),\quad t_2=\frac{g_3d_1}{b_1},\quad b_1=\frac{eF_S}{b},\quad D_1=eF_s, \\
& t_3=\frac{d_1}{b_1}\left[a_1-\frac{b_1c_1}{d_1}\right],\quad a_1=\left[\frac{E}{b}-g_1-g_2k_s^2\right], \\
& \alpha_2=[a_1-b_1z_t][g_3]^{-1},\quad z_t=(2b_1^2)^{-1}[2a_1b_1-d_1g_3^2 \\
& \quad -[g_3^2d_1^2-2a_1d_1b_1g_3^2+4b_1^2c_1g_3^2]^{1/2}]\quad\text{and}\quad t_4^2=\left[t_3-\frac{t_2^2}{2}\right]
\end{aligned}
$$

The 2D electron statistics at the electric quantum limit can be written as

$$
\begin{aligned}
& \left[\frac{2}{3}[(\alpha_1^2(N_s)-\alpha_1(N_s)t_2+t_3(N_s))^{3/2}-(\alpha_2^2(N_s)-\alpha_2(N_s)t_2+t_3(N_s))^{3/2}]\right. \\
& +\left(\alpha_1(N_s)-\frac{t_2}{2}\right)\cdot\frac{t_2}{2}(\alpha_1^2(N_s)-\alpha_1(N_s)t_2+t_3(N_s))^{1/2} \\
& -\frac{t_2}{2}\left(\alpha_2(N_s)-\frac{t_2}{2}\right)\cdot(\alpha_2^2(N_s)+t_3(N_s)-\alpha_2(N_s)t_2)^{1/2} \\
& +\left(t_2\cdot\frac{t_4^2(N_s)}{2}\right)\ln\left[\left\{\left(\alpha_1(N_s)-\frac{t_2}{2}\right)+(\alpha_1^2(N_s)+t_3(N_s)\right.\right. \\
& \left.-\alpha_1(N_s)t_2)^{1/2}\right\}\left\{\alpha_2(N_s)-\frac{t_2}{2}+.(\alpha_2^2(N_s)+t_3(N_s)\right. \\
& \left.\left.-\alpha_2(N_s)t_2)^{1/2}\right\}^{-1}\right]\Bigg] \\
= {} & \frac{2}{3}(S_0d_1)^{3/2}(b_1)^{-1/2}
\end{aligned} \tag{2.75}
$$

where

$$\alpha_1(N_s) = \left(E_{Fiw} + g_4 + g_5\left(\frac{2N_s\pi}{g_v}\right)\right), \quad t_3(N_s) = \frac{d_1}{b_1}\left[a_1(N_S) - \frac{b_1c_1(N_S)}{d_1}\right],$$

$$a_1(N_S) = \left[\frac{E_{Fiw}}{b} - g_1 - g_2\left(\frac{2N_s\pi}{g_v}\right)\right],$$

$$\alpha_2(N_s) = [a_1(N_s) - b_1z_t(N_s)][g_3]^{-1},$$

$$z_t = (2b_1^2)^{-1}[2a_1(N_s)b_1 - d_1g_3^2 - [g_3^2d_1^2 - 2a_1(N_S)d_1b_1g_3^2 + 4b_1^2c_1g_3^2]^{1/2}]$$

$$\text{and} \quad t_4^2(N_S) = \left[t_3(N_S) - \frac{t_2^2}{2}\right]$$

Thus, using (2.73) and (1.4), we can study the QC in this case.

2.2.8 *The QC in MOSFETs of gallium antimonide*

The conduction electrons of n-GaSb obey the following three dispersion relations:

(i) In accordance with the model of Seiler *et al.* [72]

$$\bar{E} = \left[-\frac{E_g}{2} + \frac{E_g}{2}[1 + \alpha_4k^2]^{1/2} + \frac{\bar{\varsigma}_0\hbar^2k^2}{2m_o} + \frac{\bar{v}_0f_1(k)\hbar^2}{2m_o} \pm \frac{\bar{\omega}_0f_2(k)\hbar^2}{2m_o}\right] \tag{2.76}$$

where $\alpha_4 \equiv 4P^2(E_g + \frac{2}{3}\Delta)[E_g^2(E_g + \Delta)]^{-1}$, P is the isotropic momentum matrix element, $f_1(k) \equiv k^{-2}[k_x^2k_y^2 + k_y^2k_z^2 + k_z^2k_x^2]$ represents the warping of the Fermi surface, $f_2(k) \equiv [\{k^2(k_x^2k_y^2 + k_y^2k_z^2 + k_z^2k_x^2) - 9k_x^2k_y^2k_z^2\}^{1/2}k^{-1}]$, represents the inversion asymmetry splitting of the conduction band and $\bar{\varsigma}_0, \bar{v}_0$ and $\bar{\omega}_0$ represent the constants of the electron spectrum in this case.

After spherical averaging (2.76) assumes the form

$$k^2 = \gamma_6(\bar{E}) \tag{2.77}$$

where

$$\gamma_6(\bar{E}) = (2S_1^2)^{-1}[S_1(1 + 2\alpha\bar{E}) - \alpha_4 - [\alpha_4^2 - 4S_1\alpha_4(1 + 2\alpha\bar{E}) + 4S_1^2(1 - 20\alpha\bar{E} - 28\alpha^2(\bar{E})^2)]^{1/2}]$$

in which,

$$S_1 = \left(\frac{\alpha\hbar^2}{m_0}\right)[\bar{\xi}_0 + 0.1\bar{v}_0 + 0.856\bar{\omega}_0]$$

(a) Weak electric field limit

The 2D dispersion relation and the density-of-states function can be written as

$$k_s^2 = \gamma_6(E) - S_i[eF_s\gamma_6'(E)]^{2/3} \tag{2.78}$$

and

$$N(E) = \frac{g_v}{4\pi}\sum_{i=0}^{i_{\max}} T_{35}'(E,i)H(E-\bar{E}_{35}) \tag{2.79}$$

where $T_{35}(E,i) = [\gamma_6(E) - S_i[eF_s\gamma_6'(E)]^{2/3}]$ and $\bar{E}_{35}$ can be expressed through the equation

$$[\gamma_6(\bar{E}_{35}) - S_i[eF_s\gamma_6'(\bar{E}_{35})]^{2/3}] = 0$$

The surface electron concentration at $T \to 0$ can be expressed as

$$N_s = \frac{g_v}{4\pi}\sum_{i=0}^{i_{\max}} T_{35}(E_{Fiw},i) \tag{2.80}$$

Using (1.4) and (2.80) we can study the QC in this case.

(b) Strong electric field limit

The 2D dispersion relation and the density-of-states function can be written as

$$k_s^2 = \gamma_6(E) - \frac{2\sqrt{2}}{3}eF_sS_i^{3/2}[\gamma_6''(E)]^{1/2} \tag{2.81}$$

and

$$N(E) = \frac{g_v}{4\pi}\sum_{i=0}^{i_{\max}} T_{36}'(E,i)H(E-\bar{E}_{36}) \tag{2.82}$$

where

$$T_{36}(E,i) = \left[\gamma_6(E) - \frac{2\sqrt{2}}{3}eF_sS_i^{3/2}[\gamma_6''(E)]^{1/2}\right] \quad \text{and} \quad \bar{E}_{36}$$

can be expressed through the equation

$$\left[\gamma_6(\bar{E}_{36}) - \frac{2\sqrt{2}}{3}eF_sS_i^{3/2}[\gamma_6''(\bar{E}_{36})]^{1/2}\right] = 0 \tag{2.83}$$

The surface electron concentration at $T \to 0$ can be expressed as

$$N_s = \frac{g_v}{4\pi} \sum_{i=0}^{i_{\max}} T_{36}(E_{Fis}, i) \tag{2.84}$$

Thus, by using (1.4) and (2.84), we can study the QC in this case.

(ii) The dispersion relation of the conduction electrons in n-GaSb in accordance with the model of Mathur and Jain [73]

$$\bar{E} - \frac{\hbar^2 k^2}{2m_0} + \frac{E_{g1}}{2} = \frac{E_{g1}}{2}\left[\left(1 + \frac{2\hbar^2 k^2}{E_{g1}}\left\{\frac{1}{m_c} - \frac{1}{m_0}\right\}\right)^{1/2}\right] \tag{2.85}$$

where $E_{g1} = \{E_g + [(5 \times 10^{-5} T^2)(2(112 + T)^{-1})]\} eV$.

The (2.85) can be written as

$$k^2 = \gamma_7(\bar{E}) \tag{2.86}$$

where

$$\gamma_7(\bar{E}) = [a_7 + \bar{E} - [b_7\bar{E} + C_7]^{1/2}]\left[\frac{2m_c}{\hbar^2}\right], \quad a_7 = \left[E_{g1} - \frac{m_c}{m_0}\left(\frac{E_{g1}}{2}\right)\right],$$

$$b_7 = E_{g1}\left[\left(1 - \frac{m_c}{m_0}\right)\right] \quad \text{and} \quad C_7 = a_7^2$$

(a) Weak electric field limit

The 2D dispersion relation and the density-of-states function can be written as

$$k_s^2 = \gamma_7(E) - S_i[eF_s\gamma_7'(E)]^{2/3} \tag{2.87}$$

and

$$N(E) = \frac{g_v}{2\pi} \sum_{i=0}^{i_{\max}} T_{37}'(E, i) H(E - \bar{E}_{37}) \tag{2.88}$$

where $T_{37}(E, i) = [\gamma_7(E) - S_i[eF_s\gamma_7'(E)]^{2/3}]$ and $\bar{E}_{37}$ can be expressed through the equation

$$[\gamma_7(\bar{E}_{37}) - S_i[eF_s\gamma_7'(\bar{E}_{37})]^{2/3}] = 0$$

The surface electron concentration at $T \to 0$ can be expressed as

$$N_s = \frac{g_v}{2\pi} \sum_{i=0}^{i_{\max}} T_{37}(E_{Fiw}, i) \tag{2.89}$$

Using (1.4) and (2.89) we can study the QC in this case.

(b) Strong electric field limit

The 2D dispersion relation and the density-of-states function can be written as

$$k_s^2 = \gamma_7(E) - \frac{2\sqrt{2}}{3} e F_s S_i^{3/2} [\gamma_7''(E)]^{1/2} \tag{2.90}$$

and

$$N(E) = \frac{g_v}{2\pi} \sum_{i=0}^{i_{\max}} T_{38}'(E,i) H(E - \bar{E}_{38}) \tag{2.91}$$

where

$$T_{38}(E,i) = [\gamma_7(E) - \frac{2\sqrt{2}}{3} e F_s S_i^{3/2} [\gamma'{}_7'(E)]^{1/2}]$$

and $\bar{E}_{38}$ can be expressed through the equation

$$\left[\gamma_7(\bar{E}_{38}) - \frac{2\sqrt{2}}{3} e F_s S_i^{3/2} [\gamma_7''(\bar{E}_{38})]^{1/2}\right] = 0 \tag{2.92}$$

The surface electron concentration at $T \to 0$ can be expressed as

$$N_s = \frac{g_v}{2\pi} \sum_{i=0}^{i_{\max}} T_{38}(E_{Fis}, i) \tag{2.93}$$

Thus, by using (1.4) and (2.93), we can study the QC in this case.

(iii) The dispersion relation of the conduction electrons in n-GaSb in accordance with the model of Zhang [74] can be written as

$$\begin{aligned}\bar{E} = {} & [E_2^{(1)} + E_2^{(2)} K_{4,1}]k^2 + [E_4^{(1)} + E_4^{(2)} K_{4,1}]k^4 \\ & + k^6[E_6^{(1)} + E_6^{(2)} K_{4,1} + E_6^{(3)} K_{6,1}]\end{aligned} \tag{2.94}$$

where

$$K_{4,1} \equiv \frac{5}{4}\sqrt{21}\left[\frac{k_x^4 + k_y^4 + k_z^4}{k^4} - \frac{3}{5}\right],$$

$$K_{6,1} \equiv \sqrt{\frac{639639}{32}}\left[\frac{k_x^2 k_y^2 k_z^2}{k^6} + \frac{1}{22}\left(\frac{k_x^4 + k_y^4 + k_z^4}{k^4} - \frac{3}{5}\right) - \frac{1}{105}\right]$$

the coefficients are in eV, the values of k are $10\left(\frac{a}{2\pi}\right)$ times those of k in atomic units (a is the lattice constant), $E_2^{(1)} = 1.0239620$, $E_2^{(2)} = 0$, $E_4^{(1)} = -1.1320772$, $E_4^{(2)} = 0.05658$, $E_6^{(1)} = 1.1072073$, and $E_6^{(2)} = -0.1134024$, $E_6^{(3)} = -0.0072275$.

Using the method of spherical averaging and the method of successive approximation (2.94) can be written as

$$k^2 = \gamma_8(\bar{E}) \tag{2.95}$$

where

$$\gamma_8(\bar{E}) = (2\delta_2)^{-1}[\delta_1 - \sqrt{\delta_1^2 - 4\delta_2[\bar{E} - \delta_4(\bar{E})]}], \quad \delta_2 = E_4^{(1)}, \quad \delta_1 = E_2^{(1)},$$

$$\delta_4(\bar{E}) = \left[\frac{\delta_1 - \sqrt{\delta_1^2 - 4\bar{E}\delta_3}}{2\delta_3}\right]^3 \quad \text{and} \quad \delta_3 = [E_6^{(1)} + 0.094E_6^{(3)}]$$

(a) Weak electric field limit

The 2D dispersion relation and the density-of-states function can be written as

$$k_s^2 = \gamma_8(E) - S_i[eF_s\gamma_8'(E)]^{2/3} \tag{2.96}$$

and

$$N(E) = \frac{g_v}{2\pi}\sum_{i=0}^{i_{\max}} T_{39}'(E,i)H(E - \bar{E}_{39}) \tag{2.97}$$

where $T_{39}(E,i) = [\gamma_8(E) - S_i[eF_s\gamma_8'(E)]^{2/3}]$ and $\bar{E}_{39}$ can be expressed through the equation. The surface electron concentration at $T \to 0$ can be $[\gamma_8(\bar{E}_{39}) - S_i[eF_s\gamma_8'(\bar{E}_{39})]^{2/3}] = 0$ expressed as

$$N_s = \frac{g_v}{2\pi}\sum_{i=0}^{i_{\max}} T_{39}(E_{Fiw}, i) \tag{2.98}$$

Using (1.4) and (2.98) we can study the QC in this case.

(b) Strong electric field limit

The 2D dispersion relation and the density-of-states function can be written as

$$k_s^2 = \gamma_8(E) - \frac{2\sqrt{2}}{3}eF_sS_i^{3/2}[\gamma_8''(E)]^{1/2} \tag{2.99}$$

and

$$N(E) = \frac{g_v}{2\pi} \sum_{i=0}^{i_{\max}} T'_{40}(E,i) H(E - \bar{E}_{40}) \tag{2.100}$$

where

$$T_{40}(E,i) = \left[\gamma_8(E) - \frac{2\sqrt{2}}{3} eF_s S_i^{3/2} [\gamma_8''(E)]^{1/2}\right]$$

and $\bar{E}_{40}$ can be expressed through the equation

$$\left[\gamma_8(\bar{E}_{40}) - \frac{2\sqrt{2}}{3} eF_s S_i^{3/2} [\gamma_8''(\bar{E}_{40})]^{1/2}\right] = 0 \tag{2.101}$$

The surface electron concentration at $T \to 0$ can be expressed

$$N_s = \frac{g_v}{2\pi} \sum_{i=0}^{i_{\max}} T_{40}(E_{Fis}, i) \tag{2.102}$$

Thus, by using (1.4) and (2.102) we can study the QC in this case.

2.2.9 *The QC in MOSFETs of indium antimonide*

In addition to the well-known three-band model of Kane, the conduction electrons of n-InSb obey the following three dispersion relations:

(i) Model of Palik *et al.*

To the fourth-order effective mass theory, and taking into account the interactions of the conduction, the heavy hole, the light hole and the split-off bands, the electron energy spectrum in n-InSb is given by [75]

$$\bar{E} = \frac{\hbar^2 k^2}{2m_c} - \bar{B}_{11} k^4 \tag{2.103}$$

where

$$\bar{B}_{11} \equiv \frac{K_2 \hbar^4}{4E_g (m_c)^2}, \quad K_2 \equiv -\left[\frac{\left(1 + \frac{1}{2}x_1^2\right)}{1 + \frac{1}{2}x_1}\right](1 - y_1)^2,$$

$$x_1 \equiv \left[1 + \left(\frac{\Delta}{E_g}\right)\right]^{-1} \quad \text{and} \quad y_1 \equiv \frac{m_c}{m_0}$$

The (2.84) can be expressed as

$$k^2 = \gamma_9(\bar{E}) \tag{2.104}$$

where

$$\gamma_9(\bar{E}) = (2\bar{B}_{11})^{-1}\left[\frac{\hbar^2}{2m_c} - \sqrt{\left(\frac{\hbar^2}{2m_c}\right)^2 - 4\bar{B}_{11}\bar{E}}\right]$$

(a) Weak electric field limit

The 2D dispersion relation and the density-of-states function can be written as

$$k_s^2 = \gamma_9(E) - S_i[eF_s\gamma_9'(E)]^{2/3} \tag{2.105}$$

and

$$N(E) = \frac{g_v}{2\pi}\sum_{i=0}^{i_{\max}} T'_{401}(E,i)H(E - E_{401}) \tag{2.106}$$

where $T_{401}(E,i) = [\gamma_9(E) - S_i[eF_s\gamma_9'(E)]^{2/3}]$ and E_{401} can be expressed through the equation

$$[\gamma_9(E_{401}) - S_i[eF_s\gamma_9'(E_{401})]^{2/3}] = 0 \tag{2.107}$$

The surface electron concentration at $T \to 0$ can be expressed as

$$N_s = \frac{g_v}{2\pi}\sum_{i=0}^{i_{\max}} T_{401}(E_{Fiw}, i) \tag{2.108}$$

Using (1.4) and (2.108) we can study the QC in this case.

(b) Strong electric field limit

The 2D dispersion relation and the density-of-states function can be written as

$$k_s^2 = \gamma_9(E) - \frac{2\sqrt{2}}{3}eF_sS_i^{3/2}[\gamma_9''(E)]^{1/2} \tag{2.109}$$

and

$$N(E) = \frac{g_v}{2\pi}\sum_{i=0}^{i_{\max}} T'_{410}(E,i)H(E - E_{410}) \tag{2.110}$$

and E_{410} can be expressed through the equation

$$\left[\gamma_9(E_{410}) - \frac{2\sqrt{2}}{3}eF_sS_i^{3/2}[\gamma_9''(E_{410})]^{1/2}\right] = 0 \tag{2.111}$$

The surface electron concentration at $T \to 0$ can be expressed as

$$N_s = \frac{g_v}{2\pi} \sum_{i=0}^{i_{\max}} T_{410}(E_{Fis}, i) \tag{2.112}$$

Thus, by using (1.4) and (2.112), we can study the QC in this case.

(ii) Model of Johnson and Dickey

In accordance with Johnson and Dickey [76], the electron energy spectrum assumes the form

$$\bar{E} = -\frac{E_g}{2} + \frac{\hbar^2 k^2}{2}\left[\frac{1}{m_0} + \frac{1}{m_{\gamma b}}\right] + \frac{E_g}{2}\left[1 + 4\frac{\hbar^2 k^2}{2m_c'}\frac{\bar{f}_1(\bar{E})}{E_g}\right]^{1/2} \tag{2.113}$$

where

$$\frac{m_0}{m_c'} \equiv P^2\left[\frac{(E_g + \frac{2\Delta}{3})}{E_g(E_g+\Delta)}\right], \quad \bar{f}_1(\bar{E}) \equiv \frac{(E_g+\Delta)\,(\bar{E}+E_g+\frac{2\Delta}{3})}{(E_g+\frac{2\Delta}{3})\,(\bar{E}+E_g+\Delta)},$$

$$m_c' = 0.139 m_0 \quad \text{and} \quad m_{\gamma b} = \left[\frac{1}{m_c'} - \frac{2}{m_0}\right]^{-1}$$

The (2.93) can be expressed as

$$k^2 = \gamma_{10}(\bar{E}) \tag{2.114}$$

where

$$\gamma_{10}(\bar{E}) = \left[\left[E_g\bar{e}_7 + 2\bar{E}\bar{e}_7 + \frac{E_g^2}{4}\bar{e}_8(\bar{E})\right] - \left[E_g^2(\bar{e}_7)^2 + \frac{E_g^4}{16}(\bar{e}_8(\bar{E}))^2 + \bar{E}(\bar{e}_7)E_g^2(\bar{e}_8(\bar{E})) + \frac{\bar{e}_7}{2}E_g^3(\bar{e}_8(\bar{E})r\right]^{1/2}\right]$$

(a) Weak electric field limit

The 2D dispersion relation and the density-of-states function can be written as

$$k_s^2 = \gamma_{10}(E) - S_i[eF_s\gamma_{10}'(E)]^{2/3} \tag{2.115}$$

and

$$N(E) = \frac{g_v}{2\pi}\sum_{i=0}^{i_{\max}} T_{41}'(E,i)H(E-E_{41}) \tag{2.116}$$

where $T_{41}(E,i) = [\gamma_{10}(E) - S_i[eF_s\gamma'_{10}(E)]^{2/3}]$ and E_{41} can be expressed through the equation

$$[\gamma_{10}(E_{41}) - S_i[eF_s\gamma'_{10}(E_{41})]^{2/3}] = 0 \tag{2.117}$$

The surface electron concentration at $T \to 0$ can be expressed as

$$N_s = \frac{g_v}{2\pi}\sum_{i=0}^{i_{\max}} T_{41}(E_{Fiw}, i) \tag{2.118}$$

Using (1.4) and (2.118) we can study the QC in this case.

(b) Strong electric field limit

The 2D dispersion relation and the density-of-states function can be written as

$$k_s^2 = \gamma_{10}(E) - \frac{2\sqrt{2}}{3}eF_sS_i^{3/2}[\gamma''_{10}(E)]^{1/2} \tag{2.119}$$

and

$$N(E) = \frac{g_v}{2\pi}\sum_{i=0}^{i_{\max}} T'_{42}(E,i)H(E - E_{42}) \tag{2.120}$$

where

$$T_{42}(E,i) = [\gamma_{10}(E) - \frac{2\sqrt{2}}{3}eF_sS_i^{3/2}[\gamma''_{10}(E)]^{1/2}]$$

and E_{42} can be expressed through the equation

$$\left[\gamma_{10}(E_{42}) - \frac{2\sqrt{2}}{3}eF_sS_i^{3/2}[\gamma''_{10}(E_{42})]^{1/2}\right] = 0 \tag{2.121}$$

The surface electron concentration at $T \to 0$ can be expressed as

$$N_s = \frac{g_v}{2\pi}\sum_{i=0}^{i_{\max}} T_{42}(E_{Fis}, i) \tag{2.122}$$

Using (1.4) and (2.122) we can study the QC in this case.

(iii) Model of Agafonov *et al.*

In accordance with Agafonov *et al.* [77], the electron energy spectrum can be written as

$$\bar{E} = \frac{\bar{\eta} - E_g}{2}\left[1 - \frac{\hbar^2k^2}{2\bar{\eta}m_c}\left\{\frac{D\sqrt{3} - 3\bar{B}}{2\left(\frac{\hbar^2}{2m_c}\right)}\right\}\left[\frac{k_x^4 + k_y^4 + k_z^4}{k^4}\right]\right] \tag{2.123}$$

where

$$\bar{\eta} \equiv \left(E_g^2 + \frac{8}{3}P^2k^2\right)^{1/2}, \quad \bar{B} \equiv -21\frac{\hbar^2}{2m_0} \quad \text{and} \quad D \equiv -40\left(\frac{\hbar^2}{2m_0}\right)$$

Using the method of averaging [71] and the method of successive approximation we can write

$$k^2 = \gamma_{11}(\bar{E}) \tag{2.124}$$

where

$$\gamma_{11}(\bar{E}) = \left[\frac{8(\bar{E})^2N_{01}^2}{(2\bar{E}+E_g)^2}\right]^{-1}\left[-\left[4\bar{E}N_{01} - \frac{8P^2}{3}\right] + \left[\left(4\bar{E}N_{01} - \frac{8P^2}{3}\right)^2 - \frac{64(\bar{E})^2N_{01}^2(\bar{E})^2 + \bar{E}E_g)}{(2\bar{E}+E_g)^2}\right]^{1/2}\right] \quad \text{and} \quad N_{1s} = \frac{3}{10}(D\sqrt{3} - 3B)$$

(a) Weak electric field limit

The 2D dispersion relation and the density-of-states function can be written as

$$k_s^2 = \gamma_{11}(E) - S_i[eF_s\gamma_{11}'(E)]^{2/3} \tag{2.125}$$

and

$$N(E) = \frac{g_v}{2\pi}\sum_{i=0}^{i_{\max}} T_{43}'(E,i)H(E - E_{43}) \tag{2.126}$$

where $T_{43}(E,i) = [\gamma_{11}(E) - S_i[eF_s\gamma_{11}'(E)]^{2/3}]$ and E_{43} can be expressed through the equation

$$[\gamma_{11}(E_{43}) - S_i[eF_s\gamma_{11}'(E_{43})]^{2/3}] = 0 \tag{2.127}$$

The surface electron concentration at $T \to 0$ can be expressed as

$$N_s = \frac{g_v}{2\pi}\sum_{i=0}^{i_{\max}} T_{43}(E_{Fiw}, i) \tag{2.128}$$

Using (1.4) and (2.128) we can study the QC in this case.

(b) Strong electric field limit

The 2D dispersion relation and the density-of-states function can be written as

$$k_s^2 = \gamma_{11}(E) - \frac{2\sqrt{2}}{3}eF_sS_i^{3/2}[\gamma_{11}''(E)]^{1/2} \tag{2.129}$$

and

$$N(E) = \frac{g_v}{2\pi} \sum_{i=0}^{i_{\max}} T'_{45}(E,i) H(E - E_{45}) \tag{2.130}$$

where

$$T_{45}(E,i) = \left[\gamma_{11}(E) - \frac{2\sqrt{2}}{3} eF_s S_i^{3/2} [\gamma''_{11}(E)]^{1/2} \right]$$

and E_{45} can be expressed through the equation

$$\left[\gamma_{11}(E_{45}) - \frac{2\sqrt{2}}{3} eF_s S_i^{3/2} [\gamma''_{11}(E_{45})]^{1/2} \right] = 0 \tag{2.131}$$

The surface electron concentration at $T \to 0$ can be expressed as

$$N_s = \frac{g_v}{2\pi} \sum_{i=0}^{i_{\max}} T_{45}(E_{Fis}, i) \tag{2.132}$$

Using (1.4) and (2.132) we can study the QC in this case.

2.2.10 *The QC in MOSFETs of tellurium*

The DR of the conduction electrons in n-Te can be expressed as

$$\bar{E} = \psi_1 k_z^2 + \psi_2 k_s^2 \pm [\psi_3^2 k_z^2 + \psi_4^2 k_s^2]^{1/2} \tag{2.133}$$

where $\psi_1 = 6.7 \times 10^{-6}\,\text{meV} \cdot \text{m}^2$, $\psi_2 = 4.21 \times 10^{-16}\,\text{meV} \cdot \text{m}^2$, $\psi_3 = 6 \times 10^{-8}\,\text{meV} \cdot \text{m}$ and $\psi_4 = 3.6 \times 10^{-8}\,\text{meV} \cdot \text{m}$.

In the presence of a surface electric field (F_S) along z-direction, the (2.133) can be written as

$$k_z^2 = a_1(\bar{E}) - b_1 z \pm \sqrt{C_1(\bar{E}) - D_1 z} \tag{2.134}$$

where $a_1(\bar{E}) = (2\psi_1^2)^{-1}[\bar{E}\psi_1 + \psi_3^2 - 2\psi_1\psi_2 k_s^2]$, $b_1 = (eF_s/2\psi_1)$, $C_1(\bar{E}) = (4\psi_1^4)^{-1}[\psi_3^4 + 4\psi_1\psi_3^2\bar{E} - 4\psi_1\psi_2\psi_3^2 k_s^2 + 4\psi_1^2\psi_4^2 k_s^2]^{1/2}$ and $D_1 = [\psi_3^2 eF_s/\psi_1^3]$.

Using (2.116) and (1.7), the 2D dispersion law assumes the form

$$\left[\frac{a_1(E)D_1}{b_1} \pm \frac{D_1\sqrt{C_1(E)}}{b_1} \right]^{2/3} - \left[\frac{a_1(E)D_1}{b_1} - D_1 z_{t,\mp} \pm \left(\frac{D_1}{b_1} \right) \sqrt{C_1(E) - D_1 z_{t,\mp}} \right]^{2/3}$$

$$\begin{aligned}&\mp\frac{3}{4}\left(\frac{D_1}{b_1}\right)\left[\beta_1(E)\sqrt{\beta_1^2(E)-e_1^2(E)}-\beta_2(E)\sqrt{\beta_2^2(E)-e_1^2(E)}\right.\\&\left.+e_1^2(E)l_n\left|\frac{\beta_2(E)+\sqrt{\beta_2^2(E)-e_1^2(E)}}{\beta_1(E)+\sqrt{\beta_1^2(E)-e_1^2(E)}}\right|\right]\\&=\frac{(S_iD_1)^{3/2}}{\sqrt{b_1}}\end{aligned} \tag{2.135}$$

where

$$e_1^2(E)=\left[\frac{D_1^2}{4b_1}+C_1(E)-\frac{a_1(E)D_1}{b_1}\right],\quad \beta_1(E)=\sqrt{C_1(E)}\pm\frac{D_1}{2b_1},$$

$$\beta_2(E)=\sqrt{C_1(E)-D_1z_{t,\mp}}\pm\frac{D_1}{2b_1}z_{t,\mp}=(eF_s)^{-1}[E-\psi_2k_s^2\mp\psi_4k_s]$$

Therefore, the electron concentration at $T\to 0$ and $i=0$ can be written from (2.135) as

$$\begin{aligned}&\left[\frac{a_1(E_F)D_1}{b_1}\pm\frac{D_1\sqrt{C_1(E_F)}}{b_1}\right]^{2/3}\\&-\left[\frac{a_1(E_F)D_1}{b_1}-D_1z_{t,\mp}(N_s)\pm\left(\frac{D_1}{b_1}\right)\sqrt{C_1(E_F)-D_1z_{t,\mp}(N_s)}\right]^{2/3}\\&\mp\frac{3}{4}\left(\frac{D_1}{b_1}\right)\left[\beta_1(E_F)\sqrt{(\beta_1(E_F)^2)-e_1^2(E_F)}\right.\\&-\beta_2(E_F)\sqrt{(\beta_2(E_F))^2-e_1^2(E_F)}\\&\left.+e_1^2(E_F)l_n\left|\frac{\beta_2(E_F)+\sqrt{(\beta_2(E_F))^2-e_1^2(E_F)}}{\beta_1(E_F)+\sqrt{(\beta_1(E_F))^2-e_1^2(E_F)}}\right|\right]\\&=\frac{(S_0D_1)^{3/2}}{\sqrt{b_1}}\end{aligned} \tag{2.136}$$

where

$$a_1(E_F)=(2\psi_1^2)^{-1}\left[E_F\psi_1+\psi_3^2-(\psi_1\psi_2)\left(\frac{4\pi N_S}{g_v}\right)\right],$$

$$C_1(E_F) = (4\psi_1^4)^{-1}\left[\psi_3^4 + 4\psi_1\psi_3^2 E - (4\psi_1\psi_2\psi_3^2)\left(\frac{4\pi N_S}{g_v}\right) + (4\psi_1^2\psi_4^2)\left(\frac{4\pi N_S}{g_v}\right)\right]^{1/2}$$

$$e_1^2(E_F) = \left[\frac{D_1^2}{4b_1} + C_1(E_F) - \frac{a_1(E_F)D_1}{b_1}\right], \quad \beta_1(E_F) = \left[\sqrt{C_1(E_F)} \pm \frac{D_1}{2b_1}\right],$$

$$\beta_2(E_F) = \left[\sqrt{C_1(E_F) - D_1 z_{t,\mp}(N_s)} \pm \frac{D_1}{2b_1}\right]$$

$$z_{t,\mp}(N_s) = (eF_s)^{-1}\left[E_F - \psi_2\left(\frac{4\pi N_S}{g_v}\right) \mp \psi_4\left(\frac{4\pi N_S}{g_v}\right)^{1/2}\right]$$

Using (1.7) and (2.136) we can study the QC in this case.

2.3 Summary and Conclusion

In Section 2.2.1, we have formulated the electron statistics and the QC in inversion layers of II–VI as given by (2.5) and (2.6), respectively, in accordance with Hopfield model. The EEM and the sub-band energies are given by (2.3b) and (2.4b), respectively, together with the fact that all the results for this section are valid for all values of F_s. In Section 2.2.2, we have formulated N_s and QC for inversion layers of IV–VI materials in the weak electric field limit by using Cohen model as given by (2.16) and (2.17), respectively. The EEM and sub-band energies are given by (2.11) and (2.12), respectively. For high electric field limit the expressions of N_s and QC are given by (2.19) and (2.20), respectively. In Section 2.2.3, we have formulated N_s and QC for inversion layers of stressed Kane type materials in the weak electric field limit by using the model of Seiler *et al.* as given by (2.29) and (2.30), respectively. The EEM and sub-band energies are given by (2.26) and (2.25), respectively. For high electric field limit, the expressions of N_s and QC are given by (2.37) and (2.38) whereas the EEM and sub-band energies are given by (2.34) and (2.33), respectively.

In Section 2.2.4, we have formulated the expression of N_s for inversion layers of Bi_2Te_3, in the weak electric field limit by using the model of Kohlar as given by (2.42) and the expression of N_s for high electric field limit is given by (2.45), respectively. By using (1.4), (2.42) and (2.45) we can study the QC for both the limits in this case.

In Section 2.2.5, we have formulated the expression of N_s for inversion layers of Ge by using the model of Cardona *et al.* as given by (2.52) and the expression of N_s in accordance with the model of Wang *et al.* is given by (2.55), respectively. By using (1.4), (2.52) and (2.55) we can study the QC for both the models in this case. In Section 2.2.6, we have studied the N_s for n-channel inversion layers of GaAs for both weak and strong electric field limits as given by (2.60) and (2.63), respectively in accordance with the model of Rossler whereas for the model of Stillman *et al.* the N_s for both the limits are given by (2.68) and (2.71), respectively. Thus, by using (1.4), (2.60), (2.63), (2.68) and (2.71), we can study the QC for both the models and for both weak and strong electric field limits. In Section 2.2.7, we have formulated the N_s for n-channel inversion layers of GaP as given by (2.75) and by using (1.4) and (2.75) we can study the QC in this case. In Section 2.2.8, we have formulated the N_s in n-channel inversion layers of GaSb in the low electric field limit in accordance with the models of Seiler *et al.*, Mathur *et al.* and Zhang, respectively, as given by (2.80), (2.89) and (2.98), respectively and using the said equations with (1.4) we can study the QC in accordance with the said models. In the same section, we have also formulated the expressions of N_s in the high-electric field limit in accordance with the said models as given by Equations (2.84), (2.93) and (2.102), respectively, where by combining these equations we can study the QC in the high electric field case. In Section 2.2.9, we have formulated the N_s in n-channel inversion layers of InSb in the low-electric field limit in accordance with the models of Palik *et al.*, Johnson *et al.* and Agafonov *et al.*, respectively, as given by (2.108), (2.118) and (2.128), respectively, and using the said equations with (1.4) we can study the QC in accordance with the said models. In the same section, we have also formulated the expressions of N_s in the high-electric field limit in accordance with the said models as given by Equations (2.112), (2.122) and (2.132), respectively, where by combining these equations we can study the QC in the high electric field case. In Section 2.2.10, we have formulated the expression of N_s as given by (2.136) in n-channel inversion layers of Te and combining (2.136) and (1.4), we can study the QC in this case.

It may be noted that if the direction of application of the surface electric field applied perpendicular to the surface be taken as either k_x or k_y and not as k_z as assumed in the present work, the QCs' would be different analytically for both the limits. Nevertheless, the arbitrary choice of the direction normal to the surface would not result in a change of the basic qualitative feature of the QCs' in inversion layers of semiconductors. The approximation of the potential well at the surface by a triangular well introduces some errors, as for instance the omission of the free charge contribution to the potential. This kind of approach is reasonable if there are only few charge carriers in the accumulation layer, but is responsible for an overestimation of the splitting when the accumulation carrier density exceeds that of the depletion layer. It has been observed that the maximum error due to the triangular potential well is tolerable in the practical sense because, for actual calculations, one needs a self-consistent solution which is a formidable problem, for the present generalized systems due to the non-availability of the proper analytical techniques, without exhibiting a widely different qualitative behavior [79]. The second assumption of electric quantum limit in the numerical calculation is valid in the range of low temperatures, where the quantum effects become prominent. The errors which are being introduced for these assumptions are found not to be serious enough at low temperatures [82]. Thus, whenever the condition of the electric quantum limit has been applied, the temperature has been assumed to be low enough so that the assumption becomes well-grounded because, at low temperature, one can assume that nearly all electrons are at the lowest electric sub-band [82]. We wish to note that the many-body effects, the hot electron effects, the formation of band tails, the arbitrary orientation of the direction of the electric quantization and the effects of surface of states have been neglected in our simplified theoretical formalism due to the lack of availability in the literature of the proper analytical techniques for including them for the generalized systems as considered in this chapter. Our simplified approach will be useful for the purpose of comparison, when, the methods of tackling the aforementioned formidable problems for the present generalized system appear. The inclusion of the said effects would certainly increase the accuracy of our results, and the qualitative features of the QCs' as discussed in this chapter would not change in the presence of the aforementioned influences.

We have not given the graphs of the QCs for inversion layers since we feel that our readers will perform the intricate numerical computation to obtain the oscillatory plots for better assessment and joy of understanding. Incidentally, one can easily infer that how little is presented and how much more there is yet to be investigated in the research field of QCs' in this case for different MOSFETs having different band structures in general which is the signature of the co-existence of new physics, advanced mathematics combined with the inner fire for performing innovative researches in this context from the young scientists since like Kikoin [83], we firmly believe that "A young scientist is no good if his teacher learns nothing from him and gives his teacher nothing to be proud of". Finally, we can easily understand that it is the band structure that changes in a fundamental way and consequently all the physical properties of all the electronic materials change radically leading to new physical concepts.

2.4 Open Research Problems

R.2.1 Investigate the QCs in the presence of an arbitrarily oriented electric quantization for inversion layers of all the materials as studied in this chapter. Draw the corresponding graphs and explain the results.

R.2.2 Investigate the QCs in inversion layers of IV–VI, II–VI and stressed Kane type compounds in the presence of an arbitrarily oriented quantizing non-uniform electric field. Draw the corresponding graphs and explain the results.

R.2.3 Investigate the QCs in inversion layers of all the materials as stated in R.1.1 of Chapter 1 in the presence of an arbitrarily oriented quantizing electric field. Draw the corresponding graphs and explain the results.

R.2.4 Investigate the QCs in the presence of an arbitrarily oriented non-quantizing magnetic field in inversion layers of all the materials as studied in this chapter by including the electron spin. Draw the corresponding graphs and explain the results.

R.2.5 Investigate the QCs in inversion layers of IV–VI, II–VI and stressed Kane type compounds in the presence of an arbitrarily oriented non-quantizing magnetic field by including the electron spin. Draw the corresponding graphs and explain the results.

R.2.6 Investigate the QCs in inversion layers of all the materials as stated in R.1.1 of Chapter 1 in the presence of an arbitrarily oriented non-quantizing magnetic field by including electron spin. Draw the corresponding graphs and explain the results.

R.2.7 Investigate the QCs in inversion ion layers for all the problems from R.2.1 to R.2.6 in the presence of an additional arbitrarily oriented electric field. Draw the corresponding graphs and explain the results.

R.2.8 Investigate the QCs in inversion layers for all the problems from R.2.1 to R.2.3 in the presence of arbitrarily oriented crossed electric and magnetic fields. Draw the corresponding graphs and explain the results.

R.2.9 Investigate the QCs in inversion layers for all the problems from R.2.1 to R.2.8 in the presence of surface states. Draw the corresponding graphs and explain the results.

R.2.10 Investigate the QCs in inversion layers for all the problems from R.2.1 to R.2.8 in the presence of hot electron effects. Draw the corresponding graphs and explain the results.

R.2.11 Investigate the QCs in inversion layers for all the problems from R.2.1 to R.2.6 by including the occupancy of the electrons in various electric sub-bands. Draw the corresponding graphs and explain the results.

References

1. J. A. Zapien, Y. K. Liu, Y. Y. Shan, H. Tang, C. S. Lee and S. T. Lee, *Appl. Phys. Lett.* **90**, 213114 (2007).
2. R. M. Park, *Proc. SPIE* **2524**, 142 (1995).
3. H. Kroemer, *Rev. Mod. Phys.* **73**, 783 (2001).
4. T. Nguyen Duy, J. Meslage and G. Pichard, *J. Crys. Growth* **72**, 490 (1985).
5. T. Aramoto, F. Adurodija, Y. Nishiyama, T. Arita, A. Hanafusa, K. Omura and A. Morita, *Solar Energy Mat. & Solar Cells* **75**, 211 (2003).
6. H. B. Barber, *J. Elect. Mat.* **25**, 1232 (1996).
7. S. Taniguchi, T.Hino, S. Itoh, K. Nakano, N. Nakayama, A. Ishibashi and M. Ikeda, *Elect. Lett.* **32**, 552 (1996).
8. J. J. Hopfield, *J. Appl. Phys.* **32**, 2277 (1961).
9. G. P. Agrawal and N. K. Dutta, *Semiconductor Lasers* (Van Nostrand Reinhold, New York, 1993).
10. S. Chatterjee and U. Pal, *Opt. Eng.* **32**, 2923 (1993).
11. T. K. Chaudhuri, *Int. J. Energy Res.* **16**, 481 (1992).
12. J. H. Dughaish, *Physica B* **322**, 205 (2002).
13. C. Wood, *Rep. Prog. Phys.* **51**, 459 (1988).
14. K.-F. Hsu, S. Loo, F. Guo, W. Chen, J. S. Dyck, C. Uher, T. Hogan, E. K. Polychroniadis and M. G. Kanatzidis, *Science* **303**, 818 (2004).
15. J. Androulakis, K. F. Hsu, R. Pcionek, H. Kong, C. Uher, J. J. D'Angelo, A. Downey, T. Hogan and M. G. Kanatzidis, *Adv. Mater.* **18**, 1170 (2006).
16. P. F. P. Poudeu, J. D'Angelo, A. D. Downey, J. L. Short, T. P. Hogan and M. G. Kanatzidis, *Angew. Chem., Int. Ed.* **45**, 3835 (2006).
17. P. F. Poudeu, J. D'Angelo, H. Kong, A. Downey, J. L. Short, R. Pcionek, T. P. Hogan, C. Uher and M. G. Kanatzidis, *J. Am. Chem. Soc.* **128**, 14347 (2006).
18. J. R. Sootsman, R. J. Pcionek, H. Kong, C. Uher and M. G. Kanatzidis, *Chem. Mater.* **18**, 4993 (2006).
19. A. J. Mountvala and G. Abowitz, *J. Am. Ceram. Soc.* **48**, 651 (1965).
20. E. I. Rogacheva, I. M. Krivulkin, O. N. Nashchekina, A. Yu. Sipatov, V. A. Volobuev and M. S. Dresselhaus, *Appl. Phys. Lett.* **78**, 3238 (2001).

21. H. S. Lee, B. Cheong, T. S. Lee, K. S. Lee, W. M. Kim, J. W. Lee, S. H. Cho and J. Y. Huh, *Appl. Phys. Lett.* **85**, 2782 (2004).
22. K. Kishimoto, M. Tsukamoto and T. Koyanagi, *J. Appl. Phys.* **92**, 5331 (2002).
23. E. I. Rogacheva, O. N. Nashchekina, S. N. Grigorov, M. A. Us, M. S. Dresselhaus and S. B. Cronin, *Nanotechnology* **14**, 53 (2003).
24. E. I. Rogacheva, O. N. Nashchekina, A. V. Meriuts, S. G. Lyubchenko, M. S. Dresselhaus and G. Dresselhaus, *Appl. Phys. Lett.* **86**, 063103 (2005).
25. E. I. Rogacheva, S. N. Grigorov, O. N. Nashchekina, T. V. Tavrina, S. G. Lyubchenko, A. Yu. Sipatov, V. V. Volobuev, A. G. Fedorov and M. S. Dresselhaus, *Thin Solid Films* **493**, 41 (2005).
26. X. Qiu, Y. Lou, A. C. S. Samia, A. Devadoss, J. D. Burgess, S. Dayal and C. Burda, *Angew. Chem. Int. Ed.* **44**, 5855 (2005).
27. C. Wang, G. Zhang, S. Fan and Y. Li, *J. Phys. Chem. Solids* **62**, 1957 (2001).
28. B. Poudel, W. Z. Wang, D. Z. Wang, J. Y. Huang and Z. F. Ren, *J. Nanosci. Nanotechnol.* **6**, 1050 (2006).
29. B. Zhang, J. He and T. M. Tritt, *Appl. Phys. Lett.* **88**, 043119 (2006).
30. W. Heiss, H. Groiss, E. Kaufmann, G. Hesser, M. Böberl, G. Springholz, F. Schäffler, K. Koike, H. Harada and M. Yano, *Appl. Phys. Lett.* **88**, 192109 (2006).
31. C. Wang, G. Zhang and S. Fan, Y. Li, *J. Phys. Chem. Solids* **62**, 1957 (2001).
32. B. Poudel, W. Z. Wang, D. Z. Wang, J. Y. Huang and Z. F. Ren, *J. Nanosci. Nanotechnol.* **6**, 1050 (2006).
33. B. Zhang, J. He and T. M. Tritt, *Appl. Phys. Lett.* **88**, 043119 (2006).
34. W. Heiss, H. Groiss, E. Kaufmann, G. Hesser, M. Böberl, G. Springholz, F. Schäffler, K. Koike, H. Harada and M. Yano, *Appl. Phys. Lett.* **88**, 192109 (2006).
35. B. A. Volkov, L. I. Ryabova and D. R. Khokhlov, *Phys. Usp.* **45**, 819 (2002), and references therein.
36. F. Hüe, M. Hÿtch, H. Bender, F. Houdellier and A. Claverie, *Phys. Rev. Lett.* **100**, 156602 (2008).
37. S. Banerjee, K. A. Shore, C. J. Mitchell, J. L. Sly and M. Missous, *IEE Proc. Circuits Devices Syst.* **152**, 497 (2005).
38. M. Razeghi, A. Evans, S. Slivken, J. S. Yu, J. G. Zheng and V. P. Dravid, *Proc. SPIE* **5840**, 54 (2005).
39. R. A. Stradling, *Semicond. Sci. Technol.* **6**, C52 (1991).
40. H. J. Goldsmid and R. W. Douglas, *Brit. J. Appl. Phys.* **5**, 386 (1954).
41. F. D. Rosi, B. Abeles and R. V. Jensen, *J. Phys. Chem. Sol.* **10**, 191 (1959).
42. T. M. Tritt (ed.), *Semiconductors and Semimetals, Vols. 69, 70 and 71: Recent Trends in Thermoelectric Materials Research I, II and III* (Academic Press, USA, 2000).
43. D. M. Rowe (ed.), *CRC Handbook of Thermoelectrics* (CRC Press, USA, 1995).
44. D. M. Rowe and C. M. Bhandari, *Modern Thermoelectrics* (Reston Publishing Company, Virginia, 1983).

45. D. M. Rowe (ed.), *Thermoelectrics Handbook: Macro to Nano* (CRC Press, USA, 2006).
46. H. Choi, M. Chang, M. Jo, S. J. Jung and H. Hwang, *Electrochem. Solid-State Lett.* **11**, H154 (2008).
47. S. Cova, M. Ghioni, A. Lacaita, C. Samori and F. Zappa, *Appl. Opt.* **35**, 1956 (1996).
48. H. W. H. Lee, B. R. Taylor and S. M. Kauzlarich, *Nonlinear Optics: Materials, Fundamentals, and Applications*, (Technical Digest, 12, 2000).
49. E. Brundermann, U. Heugen, A. Bergner, R. Schiwon, G. W. Schwaab, S. Ebbinghaus, D. R. Chamberlin, E. E. Haller and M. Havenith, in *29th International Conference on Infrared and Millimeter Waves and 12th International Conference on Terahertz Electronics* 283 (2004).
50. F. Hatami, V. Lordi, J. S. Harris, H. Kostial and W. T. Masselink, *J. Appl. Phys.* **97**, 096106 (2005).
51. B. W. Wessels, *J. Electrochem. Soc.* **122**, 402 (1975).
52. D. W. L. Tolfree, *J. Sci. Instrum.* **41**, 788 (1964).
53. P. B. Hart, *Proc. the IEEE* **61**, 880 (1973).
54. A. N. Baranov, T. I. Voronina, N. S. Zimogorova, L. M. Kauskaya and Y. P. Yakoviev, *Sov. Phys. Semicond.* **19**, 1676 (1985).
55. M. Yano, Y. Suzuki, T. Ishii, Y. Matsushima and M. Kimata, *Jpn. J. Appl. Phys.* **17**, 2091 (1978).
56. F. S. Yuang, Y. K. Su and N. Y. Li, *Jpn. J. Appl. Phys.* **30**, 207 (1991).
57. F. S. Yuang, Y. K. Su, N. Y. Li and K. J. Gan, *J. Appl. Phys.* **68**, 6383 (1990).
58. Y. K. Su and S. M. Chen, *J. Appl. Phys.* **73**, 8349 (1993).
59. S. K. Haywood, A. B. Henriques, N. J. Mason, R. J. Nicholas and P. J. Walker, *Semicond. Sci. Technol.* **3**, 315 (1988).
60. P. K. Weimer, *Proc. IEEE* **52**, 608 (1964).
61. G. Ribakovs and A. A. Gundjian, *IEEE J. Quant. Electron.* **QE-14**, 42 (1978).
62. S. K. Dey, *J. Vac. Sc. Tech.* **10**, 227 (1973).
63. S. J. Lynch, *Thin Solid Films* **102**, 47 (1983).
64. M. H. Cohen, *Phys. Rev.* **121**, 387 (1961).
65. D. G. Seiler, B. D. Bajaj and A. E. Stephens, *Phys. Rev. B* **16**, 2822 (1977); A. V. Germaneko and G. M. Minkov, *Phys. Stat. Sol. (b)* **184**, 9 (1994); G. L. Bir and G. E. Pikus, *Symmetry and Strain-induced Effects in Semiconductors* (Nauka, Russia, 1972) (in Russian); M. Mondal and K. P. Ghatak, *Phys. Stat. Sol. (b)* **135**, K21 (1986).
66. H. Kohler, *Phys. Stat. Sol. B* **74**, 591 (1976); D. R. Lovett, *Semimetals and Narrow Band Gap Semiconductors* (Pion Limited, London, 1977), p. 185.
67. M. Cardona, W. Paul, H. Brooks Helv, *Acta Phys.* **33**, 329 (1960); A. F. Gibson in *Proceeding of International school of physics "ENRICO FERMI"* course XIII, R. A Smith (ed.) (Academic Press, 1963), p. 171.
68. C. C. Wang and N. W. Ressler, *Phys. Rev.* **2**, 1827 (1970).
69. U. Rossler, *Solid State Commun.* **49**, 943 (1984).

70. G. E. Stillman, C. M. Wolfe and J. C. Dinmockin, in *Semiconductors and Semimetals*, Vol. 12, R. K. Willardson and A. C. Beer (eds.) **12** (1977), p. 169.
71. W. Szwozkiewicz, *Phys. Stat. Sol. B*, **79**, 691 (1977).
72. D. G. Seiler, W. M. Beeker and L. M. Roth, *Phys. Rev.* **1**, 764 (1970).
73. P. C. Mathur and S. Jain, *Phys. Rev.* **19**, 3159 (1979).
74. P. R. Zhang, *Phys. Rev.* **1B**, 3450 (1970).
75. E. D. Palik, G. S. Picus, S. Teither and R. F. Wallis, *Phys. Rev.* **122**, 475 (1961).
76. E. J. Johnson and D. H. Dicley, *Phys. Rev.* **1**, 2676 (1970).
77. V. G. Agafonov, P. M. Valov, B. S. Ryvkin and I. D. Yaroshetskii, *Sov. Phys. Semicond.* **12**, 1182 (1978).
78. J. Bouat and J. C. Thuillier, *Surface Sci.* **73**, 528 (1978).
79. T. Ando, H. Fowler and F. Stern, *Rev. Mod. Phys.* **54**, 437 (1982).
80. J. J. Quinn and P. J. Styles (eds.), *Electronic Properties of Quasi Two Dimensional Systems* (North Holland, Amsterdam, 1976).
81. G. A. Antcliffe, R. T. Bate and R. A. *Reynolds, Proceedings of the International Conference, Physics of Semi-metals and Narrow-Gap semiconductors*, D. L. Carter and R. T. Bate (eds.) (Pergamon Press, Oxford, 1971), p. 499.
82. Z. A. Weinberg, *Sol. Stat. Electron.* **20**, 11 (1977).
83. I. K. Kikoin, *Science for Everyone: Encounters with Physicists and Physics* (Mir Publishers, Russia, 1989), p. 154.

Chapter 3

The QC in QWT of Important Materials

Not all readers are leaders, but all leaders are readers

3.1 Introduction

In this chapter from Sections 3.2.1 up to 3.2.7, we shall study the QC in QWTs made of tetragonal and nonlinear optical, III–V, II–VI, IV–VI, stressed Kane type, Te and GaP, respectively. The platinum antimonide ($PtSb_2$) finds application in device miniaturization, colloidal nanoparticle synthesis, sensors and detector materials and thermo-photovoltaic devices [1–3]. In Section 3.2.8, we study the QC in QWTs of $PtSb_2$. In Section 3.2.9, up to 3.2.13, we study the QC in QWT of Bi_2Te_3, Ge, GaAs, GaSb and InSb in accordance with various band models, respectively. Section 3.3 contains summary and conclusion and in Section 3.4, we present few open research problems.

3.2 Theoretical Background

3.2.1 *The QC in QWTs of tetragonal and nonlinear optical materials*

For dimensional quantization along z-direction, the dispersion relation of the 2D electrons in this case can be written following (1.6) as

$$\psi_1(E) = \psi_2(E)k_s^2 + \psi_3(E)(n_z\pi/d_z)^2 \tag{3.1}$$

where $\psi_1(E) = \gamma(E)$, $\psi_2(E) = f_1(E)$, $\psi_3(E) = f_2(E)$, $n_z(= 1, 2, 3, \ldots)$ and d_z are the size quantum number and the nanothickness along the z-direction, respectively.

The EEM at the Fermi level in the xy-plane can be written as

$$m^*(E_{Fs}, n_z) = \hbar^2 k_s \left.\frac{\partial k_s}{\partial E}\right|_{E=E_{Fs}} \tag{3.2}$$

where

$$E_{Fs} = \left(eV_g - \frac{e^2 n_{2D}}{C_{0x}} - E_{fb2D}\right)$$

E_{Fs} is the Fermi energy as measured from the edge of the conduction band at the surface in the presence of size quantization and E_{fb2D} is the energy separation between the Fermi level and the conduction band edge in the bulk of the substrate material under present condition.

From (3.1) and (3.2), the EEM in this case can be written as

$$\begin{aligned} m^*(E_{Fs}, n_z) = {} & \left(\frac{\hbar^2}{2}\right)[\psi_2(E_{Fs})]^{-2} \\ & \times \left[\psi_2(E_{Fs})\left\{\{\psi_1(E_{Fs})\}' - \{\psi_3(E_{Fs})\}'\left(\frac{n_z\pi}{d_z}\right)^2\right\}\right. \\ & \left. - \left\{\psi_1(E_{Fs}) - \psi_3(E_{Fs})\left(\frac{n_z\pi}{d_z}\right)^2\right\}\{\psi_2(E_{Fs})\}'\right] \end{aligned} \tag{3.3}$$

where the primes denote the differentiation of the differentiable functions with respect to Fermi energy. Thus, we observe that the EEM is the function of size quantum number and the Fermi energy due to the combined influence of the crystal filed splitting constant and the anisotropic spin-orbit splitting constants, respectively.

The general expression of the total 2D DOS ($N_{2DT}(\bar{E})$) in this case is given by

$$N_{2DT}(E) = \frac{2g_v}{(2\pi)^2} \sum_{n_z=1}^{n_{z\max}} \frac{\partial A(E, n_z)}{\partial E} H(E - E_{n_z}) \tag{3.4}$$

where $A(E, n_z)$ is the area of the constant energy 2D wave vector space for quantum wells, $H(E - E_{n_z})$ is the Heaviside step function and E_{n_z} is the corresponding sub-band energy. Using (3.1) and (3.4), the expression of the $N_{2DT}(\bar{E})$ for quantum wells of nonlinear optical semiconductors can be written as

$$N_{2DT}(E) = \left(\frac{g_v}{2\pi}\right) \sum_{n_z=1}^{n_{z\max}} [\psi_2(E)]^{-2} \left[\psi_2(E) \left\{ \{\psi_1(E)\}' - \{\psi_3(E)\}' \left(\frac{n_z\pi}{d_z}\right)^2 \right\} - \left\{ \psi_1(E) - \psi_3(E) \left(\frac{n_z\pi}{d_z}\right)^2 \right\} \{\psi_2(E)\}' \right] H(E - E_{n_{z_1}}) \quad (3.5)$$

where the sub-band energies ($E_{n_{z_1}}$) in this case given by

$$\psi_1(E_{n_{z_1}}) = \psi_2(E_{n_{z_1}})(n_z\pi/d_z)^2 \quad (3.6)$$

Combining (1.7) with the Fermi–Dirac occupation probability factor, integrating between $E_{n_{z_1}}$ to the Fermi energy, the 2D carrier statistics in QWTs of tetragonal and nonlinear materials at $T \to 0$ in this case assumes the form

$$n_{2D} = \frac{g_v}{2\pi} \sum_{n_z=1}^{n_{z\max}} [T_{51}(E_{Fs}, n_z)] \quad (3.7)$$

where

$$T_{51}(E_{Fs}, n_z) \equiv \left[\frac{\psi_1(E_{Fs}) - \psi_3(E_{Fs})(n_z\pi/d_z)^2}{\psi_2(E_{Fs})} \right]$$

Thus, using (1.4) and (3.7) we can study the QC in this case.

3.2.2 *The QC in QWTs of III–V, ternary and quaternary materials*

(a) The three-band model of Kane

The (1.32) in this case assumes the form

$$\frac{\hbar^2 k_s^2}{2m_c} + \frac{\hbar^2}{2m_c}(n_z\pi/d_z)^2 = I_{11}(E) \quad (3.8)$$

Using (3.2) and (3.8), the EEM in $x-y$ plane for this case can be written as

$$m^*(E_{Fs}) = m_c\{I_{11}(E_{Fs})\}' \quad (3.9)$$

It is worth noting that the EEM in this case is a function of Fermi energy alone and is independent of size quantum number.

The total 2D density-of-states function can be written as

$$N_{2DT}(E) = \left(\frac{m_c g_v}{\pi\hbar^2}\right) \sum_{n_z=1}^{n_{z\max}} \{[I_{11}(E)]' H(E - E_{n_{z2}})\} \tag{3.10}$$

where the sub-band energies $E_{n_{z2}}$ can be expressed as

$$I_{11}(E_{n_{z2}}) = \frac{\hbar^2}{2m_c}(n_z\pi/d_z)^2 \tag{3.11}$$

The 2D carrier concentration in QWTs of III–V ternary and quaternary materials at $T \to 0$ assumes the form

$$n_{2D} = \frac{m_c g_v}{\pi\hbar^2} \sum_{n_z=1}^{n_{z\max}} [T_{53}(E_{Fs}, n_z)] \tag{3.12}$$

where

$$T_{53}(E_{Fs}, n_z) \equiv \left[I_{11}(E_{Fs}) - \frac{\hbar^2}{2m_c}\left(\frac{n_z\pi}{d_z}\right)^2\right]$$

Thus, using (1.4) and (3.12) we can study the QC in this case.

(b) The two-band model of Kane

Under the inequalities $\Delta \gg E_g$ or $\Delta \ll E_g$, (3.8) can be expressed as

$$E(1 + \alpha E) = \frac{\hbar^2 k_s^2}{2m_c} + \frac{\hbar^2}{2m_c}\left(\frac{n_z\pi}{d_z}\right)^2 \tag{3.13}$$

The EEM in this case can be written as

$$m^*(E_{Fs}) = m_c(1 + 2\alpha E_{Fs}) \tag{3.14}$$

Thus, we see that the EEM in the present case is a function of Fermi energy only due to the presence of band non-parabolicity.

The total 2D density-of-states function assumes the form

$$N_{2DT}(E) = \frac{m_c g_v}{\pi\hbar^2} \sum_{n_z=1}^{n_{z\max}} (1 + 2\alpha E) H(E - E_{n_{z3}}) \tag{3.15}$$

where the sub-band energy $(E_{n_{z3}})$ can be expressed as

$$\frac{\hbar^2}{2m_c}(n_z\pi/d_z)^2 = E_{n_{z3}}(1 + \alpha E_{n_{z3}}) \tag{3.16}$$

The 2D electron statistics in QWTs of III–V ternary and quaternary materials whose energy band structures are defined by the two-band model of Kane can be written as

$$n_{2D} = \frac{m_c g_v}{\pi\hbar^2} \sum_{n_z=1}^{n_{z\max}} \int_{E_{n_{z3}}}^{\infty} \frac{(1+2\alpha E)dE}{1+\exp\left(\frac{E-E_{Fs}}{k_B T}\right)}$$

$$= \frac{m_c k_B T g_v}{\pi\hbar^2} \sum_{n_z=1}^{n_{z\max}} [(1+2\alpha E_{n_{z3}})F_0(\eta_{n_1}) + 2\alpha k_B T F_1(\eta_{n_1})] \quad (3.17)$$

where $\eta_{n_1} \equiv (E_{Fs} - E_{n_{z3}})/k_B T$ and $F_j(\eta)$ is the one parameter Fermi–Dirac integral of order j.

Thus by using (1.4) and (3.17) we can study the QC in this case.

(c) Parabolic energy bands

Under the condition $\alpha \to 0$, the expressions of the 2D DR, total 2D density-of-states, for semiconducting films whose bulk electrons are defined by the isotropic parabolic energy bands can, be written as

$$E = \frac{\hbar^2 k_s^2}{2m_c} + \frac{\hbar^2}{2m_c}\left(\frac{n_z \pi}{d_z}\right)^2 \quad (3.18a)$$

$$N_{2DT}(E) = \frac{m_c g_v}{\pi\hbar^2} \sum_{n_z=1}^{n_{z\max}} H(E - E_{n_{zp}}) \quad (3.18b)$$

The sub-band energy ($E_{n_{zp}}$), the EEM and the n_{2D} can, respectively, be expressed as

$$E_{n_{zp}} = \frac{\hbar^2}{2m_c}\left(\frac{n_z \pi}{d_z}\right)^2 \quad (3.19)$$

$$m^*(E_{Fs}) = m_c \quad (3.20)$$

and

$$n_{2D} = \frac{m_c k_B T g_v}{\pi\hbar^2} \sum_{n_z=1}^{n_{z\max}} F_0(\eta_{n_2}) \quad (3.21)$$

where

$$\eta_{n_2} \equiv \frac{1}{k_B T}\left[E_{Fs} - \frac{\hbar^2}{2m_c}\left(\frac{n_z \pi}{d_z}\right)^2\right]$$

It may be noted that the results of this section are already well known in the literature.

Using (1.4) and (3.21) we can study the QC in this case.

3.2.3 The QC in QWTs of II–VI materials

The DR of the conduction electrons in QWTs of II–VI materials for dimensional quantization along z-direction can be written in accordance with Hopfield model [4] and following (2.1) as

$$E = a_0' k_s^2 + b_0' \left(\frac{n_z \pi}{d_z}\right)^2 \pm C_0 k_s \tag{3.22}$$

where

$$a_0' = \frac{\hbar^2}{2m_\perp^*} \quad \text{and} \quad b_0' = \frac{\hbar^2}{2m_\parallel^*}$$

Using (3.22), the EEM in this case can be written as

$$m^*(E_{Fs}, n_z) = m_\perp^* \left[1 \mp \frac{(C_0)}{\left[(C_0)^2 - 4a_0' b_0' \left(\frac{n_z \pi}{d_z}\right)^2 + 4a_0' E_{Fs}\right]^{1/2}}\right] \tag{3.23}$$

where

$$E_{Fs} = eV_g - \frac{e^2 N_s}{C_{OX}} - E_{fB}$$

Thus, we can infer that the EEM in QWTs of II–VI compounds is a function of both the size quantum number and the Fermi energy due to the presence of the term C_0.

The sub-band energy $E_{n_{z5}}$ assumes the form

$$E_{n_{z5}} = b_o' (n_z \pi / d_z)^2 \tag{3.24}$$

From (3.22), one can write

$$N_{2DT}(E) = \frac{g_v m_\perp^*}{\pi \hbar^2} \sum_{n_z=1}^{n_{z\max}} \left[1 - \frac{\left(C_0 / 2\sqrt{a_0'}\right)}{\sqrt{E + \delta_{51}(n_z)}}\right] H(E - E_{n_{z5}}) \tag{3.25}$$

where

$$\delta_{51}(n_z) \equiv \frac{1}{4a_0'} \left[(C_0)^2 - 4a_0' b_0' \left(\frac{n_z \pi}{d_z}\right)^2\right]$$

Using (3.25), n_{2D} in QWTs of II–VI materials assumes the form

$$n_{2D} = \frac{g_v m_\perp^* k_B T}{\pi \hbar^2} \sum_{n_z=1}^{n_{z\max}} \left[F_0(\eta_{n_{z5}}) - \frac{C_0 f_s(E_{Fs}, n_z)}{2\sqrt{a_0' k_B T}}\right] \tag{3.26}$$

where

$$\eta_{n_{z5}} \equiv (k_BT)^{-1}[E_{Fs} - b_0'(n_z\pi/d_z)^2]$$

$$f_s(E_{Fs}, n_z) \equiv \Bigg[2\left(\sqrt{\eta_{n_{z5}} + \delta_{52}(n_z)} - \sqrt{\delta_{52}(n_z)}\right) + \sum_{r=1}^{s}\left[2(1-2^{1-2r})\zeta(2r)\frac{(-1)^{2r-1}(2r-1)!}{(\eta_{n_{z5}} + \delta_{52}(n_z))^{2r}}\right]\Bigg]$$

and

$$\delta_{52}(n_z) \equiv \left[\frac{b_0'(n_z\pi/d_z)^2 + \delta_{51}(n_z)}{k_BT}\right]$$

Using (1.4) and (3.26) we can study the QC in this case.

3.2.4 *The QC in QWTs of IV–VI materials*

The carrier dispersion laws in bulks specimens of IV–VI materials can be described by models of McClure and Choi [5], Hybrid [6], Cohen [7], Lax [8], Dimmock [9] and the ellipsoidal parabolic, respectively, and we study the QC chronologically.

(a) The McClure and Choi model

The DR of the carriers in IV–VI materials can be written, following the McClure and Choi [5], as

$$\bar{E}(1+\alpha\bar{E}) = \frac{p_x^2}{2m_1} + \frac{p_y^2}{2m_2} + \frac{p_z^2}{2m_3} + \frac{p_y^2}{2m_2}\alpha\bar{E}\left\{1 - \left(\frac{m_2}{m_2'}\right)\right\} + \frac{p_y^4\alpha}{4m_2m_2'} - \frac{\alpha p_x^2p_y^2}{4m_1m_2} - \frac{\alpha p_y^2p_z^2}{4m_2m_3} \tag{3.27}$$

where $p_i \equiv \hbar k_i$, $i = x, y, z$, m_1, m_2 and m_3 are the effective carrier masses at the band-edge along x, y and z directions, respectively, and m_2' is the effective-mass tensor component at the top of the valence band (for electrons) or at the bottom of the conduction band (for holes).

The DR of the conduction electrons in QWTs of IV–VI materials for dimensional quantization along k_z direction can be written following (3.27) for this model as

$$E(1+\alpha E) = \frac{p_x^2}{2m_1} + \frac{p_y^2}{2m_2} + \frac{\hbar^2}{2m_3}\left(\frac{n_z\pi}{d_z}\right)^2 + \frac{p_y^2}{2m_2}\alpha E\left\{1 - \left(\frac{m_2}{m_2'}\right)\right\} + \frac{p_y^4\alpha}{4m_2m_2'} - \frac{\alpha p_x^2p_y^2}{4m_1m_2} - \frac{\alpha p_y^2\hbar^2}{4m_2m_3}\left(\frac{n_z\pi}{d_z}\right)^2 \tag{3.28}$$

The (3.28) can approximately be written as

$$\gamma_1(E,n_z) = p_1 k_x^2 + q_1(E)k_y^2 + R_1(E,n_z)k_y^4 \tag{3.29}$$

where

$$\gamma_1(E,n_z) \equiv \left[E(1+\alpha E) - \frac{\hbar^2}{2m_3}\left(\frac{n_z\pi}{d_z}\right)^2\right], \quad p_1 \equiv \frac{\hbar^2}{2m_1},$$

$$q_1(E) \equiv \frac{\hbar^2}{2m_2}\left[1+\alpha E\left(1-\frac{m_2}{m_2'}\right) - \alpha E(1+\alpha E)\right]$$

and

$$R_1(E,n_z) \equiv \left[\frac{\alpha\hbar^4}{4m_2m_2'} + \alpha\left(\frac{\hbar^2}{2m_2}\right)^2\left\{1+\alpha E\left(1-\frac{m_2}{m_2'}\right) - \frac{\alpha\hbar^2}{2m_3}\left(\frac{n_z\pi}{d_z}\right)^2\right\}\right]$$

The area enclosed by (3.29) is defined by the following integral:

$$A(E,n_z) = 4\left[\frac{R_1(E,n_z)}{p_1}\right]^{1/2} J_1(E,n_z) \tag{3.30}$$

where

$$J_1(E,n_z) \equiv \int_0^{u_0(E,n_z)} \left[\frac{\gamma_1(E,n_z)}{R_1(E,n_z)} - \frac{q_1(E)k_y^2}{R_1(E,n_z)} - k_y^4\right]^{1/2} dk_y$$

and

$$u_0(E,n_z) \equiv \left[\sqrt{\frac{q_1^2(E)}{4R_1^2(E,n_z)} + \frac{\gamma_1(E,n_z)}{R_1(E,n_z)}} - q_1(E)\right]^{1/2}.$$

Thus, the area enclosed can be written as

$$\begin{aligned} A(E,n_z) = \frac{4}{3}\left[\frac{R_1(E,n_z)}{p_1}\right]^{1/2} & [a^2(E,n_z)+b^2(E,n_z)]^{1/2} \\ & \times \left[a^2(E,n_z)F\left[\frac{\pi}{2},\ell(E,n_z)\right] - [a^2(E,n_z)-b^2(E,n_z)]\right. \\ & \left.\times E\left[\frac{\pi}{2},\ell(E,n_z)\right]\right] \end{aligned} \tag{3.31}$$

where

$$a^2(E,n_z) \equiv \frac{q_1(E)}{2R_1(E,n_z)} + \frac{1}{2}\left[\frac{q_1^2(E)}{R_1^2(E,n_z)} + \frac{4\gamma_1(E,n_z)}{R_1(E,n_z)}\right]^{1/2}$$

$$b^2(E,n_z) \equiv \frac{1}{2}\left[\frac{q_1^2(E)}{R_1^2(E,n_z)} + \frac{4\gamma_1(E,n_z)}{R_1(E,n_z)}\right]^{1/2} - \left(\frac{q_1(E)}{2R_1(E,n_z)}\right),$$

$$\ell(E,n_z) \equiv \frac{b(E,n_z)}{\sqrt{a^2(E,n_z)+b^2(E,n_z)}}$$

$F\left[\frac{\pi}{2},\ell(E,n_z)\right]$ and $E\left[\frac{\pi}{2},\ell(E,n_z)\right]$ are the complete elliptic integral of the first and second kinds, respectively [10].

Using (3.31), the EEM can be written as

$$m^*(E_{Fs},n_z)=\left(\frac{2\hbar^2}{3\pi\sqrt{p_1}}\right)[R_3(E,n_z)]|_{E=E_{Fs}} \tag{3.32}$$

where

$$\begin{aligned}R_3(E_{Fs},n_z)\equiv\Bigg[&\frac{1}{2}[R_1(E_{Fs},n_z)]^{-1/2}[R_1(E_{Fs},n_z)]'[a^2(E_{Fs},n_z)\\&+b^2(E_{Fs},n_z)]^{1/2}\left[a^2(E_{Fs},n_z)F\left(\frac{\pi}{2},\ell(E_{Fs},n_z)\right)\right.\\&\left.-[a^2(E_{Fs},n_z)-b^2(E_{Fs},n_z)]E\left(\frac{\pi}{2},\ell(E_{Fs},n_z)\right)\right]\\&+\sqrt{R_1(E_{Fs},n_z)}[a^2(E_{Fs},n_z)+b^2(E_{Fs},n_z)]^{-1/2}\\&\times[a(E_{Fs},n_z)(a(E_{Fs},n_z))'+b(E_{Fs},n_z)\,(b(E_{Fs},n_z))']\\&\times\left[a^2(E_{Fs},n_z)F\left(\frac{\pi}{2},\ell(E_{Fs},n_z)\right)-[a^2(E_{Fs},n_z)\right.\\&\left.-b^2(E_{Fs},n_z)]E\left(\frac{\pi}{2},\ell(E_{Fs},n_z)\right)\right]\\&+\sqrt{R_1(E_{Fs},n_z)}[a^2(E_{Fs},n_z)+b^2(E_{Fs},n_z)]^{1/2}\\&\times\Bigg[2a(E_{Fs},n_z)(a(E_{Fs},n_z))'F\Big(\frac{\pi}{2},\ell(E_{Fs},n_z)\Big)+a^2(E_{Fs},n_z)\\&\times\left\{F\left(\frac{\pi}{2},\ell(E_{Fs},n_z\right)\right\}'-[2a(E_{Fs},n_z)[a(E_{Fs},n_z)]'\\&-2b(E_{Fs},n_z)(b(E_{Fs},n_z))']E\left(\frac{\pi}{2},\ell(E_{Fs},n_z)-[a^2(E_{Fs},n_z)\right.\\&\left.-b^2(E_{Fs},n_z)]\left(E\left(\frac{\pi}{2},\ell(E_{Fs},n_z)\right)'\right]\Bigg]\end{aligned}$$

Thus, the EEM in this case is a function of both the Fermi energy and the size quantum number due to the presence of band non-parabolicity only.

The total 2D density-of-states function can be written following (3.31), as

$$N_{2DT}(E)=\left(\frac{2g_v}{3\pi^2\sqrt{p_1}}\right)\sum_{n_z=1}^{n_{z\max}}R_3(E,n_z)H(E-E_{n_{z6}}) \tag{3.33}$$

where the sub-band energies $E_{n_{z6}}$ assumes the form

$$E_{n_{z6}}(1+\alpha E_{n_{z6}})=\frac{\hbar^2}{2m_3}\left(\frac{n_z\pi}{d_z}\right)^2 \tag{3.34}$$

Combining (3.33) with the Fermi–Dirac occupation probability factor, the 2D electron statistics in quantum wells of IV–VI materials in accordance with the McClure and Choi model can be expressed at $T \to 0$ as

$$n_{2D} = \left(\frac{2g_v}{3\pi^2\sqrt{p_1}}\right)\sum_{n_z=1}^{n_{z\max}}[\theta_1(E_{Fs}, n_z)] \tag{3.35}$$

where

$$\begin{aligned}\theta_1(E_{Fs}, n_z) \equiv \Big\{&\sqrt{R_1(E_{Fs}, n_z)}[a^2(E_{Fs}, n_z) + b^2(E_{Fs}, n_z)]^{1/2} \\ &\times \left[a^2(E_{Fs}, n_z)F\left(\frac{\pi}{2}, \ell(E_{Fs}, n_z)\right) - [a^2(E_{Fs}, n_z)\right. \\ &\left.-b^2(E_{Fs}, n_z)]F\left(\frac{\pi}{2}, \ell(E_{Fs}, n_z)\right)\right]\Big\}\end{aligned}$$

Using (1.4) and (3.35) we can study the QC in this case.

(b) The Hybrid model

The DR of the carriers in bulk specimens of IV–VI materials in accordance with the Hybrid model can be represented as [6]

$$E(1+\alpha E) = \frac{\theta_0(E)(\hbar k_y)^2}{2M_2} + \frac{\alpha\gamma_0\hbar^4 k_y^4}{4M_2^2} + \frac{\hbar^2 k_x^2}{2m_1} + \frac{\hbar^2 k_z^2}{2m_3} \tag{3.36}$$

in which $\theta_0(E) \equiv [1 + \alpha E(1-\gamma_0) + \bar{\delta}_0], \gamma_0 \equiv \frac{M_2}{m_2}, \bar{\delta}_0 \equiv \frac{M_2}{M_2'}$ and the other notations are defined in [5].

The 2D DR in QWTs of IV–VI materials in accordance with Hybrid model can be written as

$$\frac{\hbar^2 k_x^2}{2m_1} + \frac{\hbar^2 k_z^2}{2m_3} = E(1+\alpha E) - \frac{\theta_0(E)\hbar^2}{2M_2}\left(\frac{\pi n_y}{d_y}\right)^2 - \frac{\alpha\gamma_0\hbar^4}{4M_2^2}\left(\frac{\pi n_y}{d_y}\right)^4 \tag{3.37}$$

The 2D area is given by

$$A(E, n_y) = \frac{2\pi\sqrt{m_1 m_3}}{\hbar^2}t_{29}(E, n_y) \tag{3.38}$$

where

$$t_{29}(E, n_y) = \left[E(1+\alpha E) - \frac{\theta_0(E)\hbar^2}{2M_2}\left(\frac{\pi n_y}{d_y}\right)^2 - \frac{\alpha\gamma_0\hbar^4}{4M_2^4}\left(\frac{\pi n_y}{d_y}\right)^4\right]$$

The effective mass in the x–z plane can be written as

$$m^*(E_{Fs}, n_y) = [\sqrt{m_1 m_3}]\, t'_{29}(E_{Fs}, n_y) \tag{3.39}$$

Therefore, the effective mass in QWTs of IV–VI materials in accordance with Hybrid model is a function of Fermi energy and the size quantum number due to the presence of band non-parabolicity only.

The sub-band energy is given as

$$E_{n_y}(1+\alpha E_{n_y}) - \frac{\theta_0(E_{n_y})\hbar^2}{2M_2}\left(\frac{\pi n_y}{d_y}\right)^2 - \frac{\alpha\gamma_0\hbar^4}{4M_2^2}\left(\frac{\pi n_y}{d_y}\right)^4 = 0 \tag{3.40}$$

The total DOS function in this case can be written as

$$N_{2DT}(E) = \frac{g_v\sqrt{m_1 m_3}}{\pi\hbar^2}\sum_{n_y=1}^{n_{y\max}}\{t_{29}(E,n_y)\}' H(E-E_{n_y}) \tag{3.41}$$

The use of (3.41) leads to the 2D electron statistics in QWTs of IV–VI materials in this case at $T \to 0$ as

$$n_{2D} = \frac{g_v\sqrt{m_1 m_3}}{\pi\hbar^2}\sum_{n_y=1}^{n_{y\max}}[t_{29}(E_{Fs},n_y)] \tag{3.42}$$

Using (1.4) and (3.42) we can study the QC in this case.

(c) The Cohen model

In accordance with the Cohen model [7], the 2D electron dispersion law in QWTs of IV–VI in accordance with this model can be written following (2.7) as

$$E(1+\alpha E) = \frac{p_x^2}{2m_1} + \frac{\hbar^2}{2m_3}\left(\frac{n_z\pi}{d_z}\right)^2 - \frac{\alpha E p_y^2}{2m_2'} + \left(\frac{\alpha p_y^4}{4m_2 m_2'}\right) + \frac{p_y^2}{2m_2}(1+\alpha E) \tag{3.43}$$

The (3.43) can be written as

$$\gamma_1(E,n_z) = p_1 k_x^2 + q_2(E)k_y^2 + R_2 k_y^4 \tag{3.44}$$

where

$$q_2(E) \equiv \left[\frac{\hbar^2}{2m_2}(1+\alpha E) - \frac{\alpha E\hbar^2}{2m_2'}\right] \quad \text{and} \quad R_2 \equiv \left(\frac{\alpha\hbar^4}{4m_2 m_2'}\right)$$

The EEM in this case can be written as

$$m^*(E_{Fs},n_z) = \left(\frac{2\hbar^2}{3\pi\sqrt{p_1}}\right)[R_4(E,n_z)]|_{E=E_{Fs}} \tag{3.45}$$

in which

$$R_4(E_{Fs},n_z) \equiv \sqrt{R_2}[a_1^2(E_{Fs},n_z)+b_1^2(E_{Fs},n_z)]^{-1/2}[a_1(E_{Fs},n_z) \times (a_1(E_{Fs},n_z))' + b_1(E_{Fs},n_z)(b_1(E_{Fs},n_z))'] \times \left[a_1^2(E_{Fs},n_z)F\left(\frac{\pi}{2},\ell_1(E_{Fs},n_z)\right) - [a_1^2(E_{Fs},n_z) - b_1^2(E_{Fs},n_z)]E\left(\frac{\pi}{2},\ell_1(E_{Fs},n_z)\right)\right] + \sqrt{R_2}[a_1^2(E_{Fs},n_z) + b_1^2(E_{Fs},n_z)]^{1/2}\left[2a_1(E_{Fs},n_z)\,(a_1(E_{Fs},n_z))' \times F\left(\frac{\pi}{2},\ell_1(E_{Fs},n_z)\right) + a_1^2(E_{Fs},n_z)\left\{F\left(\frac{\pi}{2},\ell_1(E_{Fs},n_z)\right)\right\}' \times [2a_1(E_{Fs},n_z)(a_1(E_{Fs},n_z))' - 2b_1(E_{Fs},n_z)(b_1(E_{Fs},n_z))'] \times E\left(\frac{\pi}{2},\ell_1(E_{Fs},n_z)\right) - [a_1^2(E_{Fs},n_z) - b_1^2(E_{Fs},n_z)] \times E\left(\frac{\pi}{2},\ell_1(E_{Fs},n_z)\right)'\right]\Bigg]$$

$$a_1^2(E_{Fs},n_z) \equiv \frac{q_2(E_{Fs})}{2R_2} + \frac{1}{2}\left[\frac{q_2^2(E_{Fs})}{R_2^2} + \frac{4\gamma_1(E_{Fs},n_z)}{R_2}\right]^{1/2},$$

$$b_1^2(E_{Fs},n_z) \equiv \frac{1}{2}\left[\frac{q_2^2(E_{Fs})}{R_2^2} + \frac{4\gamma_1(E_{Fs},n_z)}{R_2}\right]^{1/2} - \left(\frac{q_2(E_{Fs})}{2R_2}\right) \quad \text{and}$$

$$\ell_1(E_{Fs},n_z) \equiv \frac{b_1(E_{Fs},n_z)}{\sqrt{a_1^2(E_{Fs},n_z)+b_1^2(E_{Fs},n_z)}}$$

which shows that the EEM in this present case is again a function of both the size quantum number and the Fermi energy due to the presence of the band non-parabolicity only.

The total density-of-states is given by

$$N_{2DT}(E) = \left(\frac{2g_v}{3\pi^2\sqrt{p_1}}\right)\sum_{n_z=1}^{n_{z\max}} R_4(E,n_z)H(E-E_{n_{z7}}) \tag{3.46}$$

where $E_{n_{z7}}$ is the lowest positive root of the equation

$$\gamma_1(E_{n_{z7}},n_z) = 0 \tag{3.47}$$

Combining (3.46) with the Fermi–Dirac occupation probability factor, the 2D electron statistics in QWTs of IV–VI materials in accordance with the

Cohen model can be written as

$$n_{2D} = \left(\frac{2g_v}{3\pi^2\sqrt{p_1}}\right)\sum_{n_z=1}^{n_{z\max}}[\theta_3(E_{Fs},n_z)] \tag{3.48}$$

where

$$\begin{aligned}\theta_3(E_{Fs},n_z) \equiv \Big\{&\sqrt{R_2}[a_1^2(E_{Fs},n_z)+b_1^2(E_{Fs},n_z)]^{1/2}\\ &\times\Big[a_1^2(E_{Fs},n_z)F\left(\frac{\pi}{2},\ell_1(E_{Fs},n_z)\right)\\ &-[a_1^2(E_{Fs},n_z)-b_1^2(E_{Fs},n_z)]F\left(\frac{\pi}{2},\ell_1(E_{Fs},n_z)\right)\Big]\Big\},\end{aligned}$$

Using (1.4) and (3.48) we can study the QC in this case.

(d) The Lax model

The electron energy spectra in bulk specimens of IV–VI materials in accordance with the Lax model can be written as [8]

$$E(1+\alpha E) = \frac{p_x^2}{2m_1}+\frac{p_y^2}{2m_2}+\frac{p_z^2}{2m_3} \tag{3.49}$$

The 2D electron dispersion law in this case can be written as

$$E(1+\alpha E) = \frac{\hbar^2k_x^2}{2m_1}+\frac{\hbar^2k_y^2}{2m_2}+\frac{\hbar^2}{2m_3}\left(\frac{n_z\pi}{d_z}\right)^2 \tag{3.50}$$

The EEM in this case assumes the form

$$m^*(E_{Fs}) = \sqrt{m_1m_2}(1+2\alpha E_{Fs}) \tag{3.51}$$

Thus, we see that the EEM for the Lax model is a function of the Fermi energy alone due to the band non-parabolicity.

The sub-band energy, the total density-of-states function and the 2D electron statistics in QWTs of IV–VI materials in accordance with the Lax model can, respectively, be expressed as

$$E_{n_{z8}}(1+\alpha E_{n_{z8}}) = \frac{\hbar^2}{2m_3}(n_z\pi/d_z)^2 \tag{3.52}$$

$$N_{2DT}(E) = \frac{g_v\sqrt{m_1m_2}}{\pi\hbar^2}\sum_{n_z=1}^{n_{z\max}}(1+2\alpha E)H(E-E_{n_{z8}}) \tag{3.53}$$

$$n_{2D} = \frac{g_v\sqrt{m_1m_2}k_BT}{\pi\hbar^2}\sum_{n_z=1}^{n_{z\max}}[(1+2\alpha E_{n_{z8}})F_0(\eta_{y2})+2\alpha k_BTF_1(\eta_{y2})] \tag{3.54}$$

where

$$\eta_{y_2} = \frac{E_{Fs} - E_{n_{z8}}}{k_B T}$$

Using (1.4) and (3.54) we can study the QC in this case.

(e) The Dimmock model

The DR of the conduction electrons in IV–VI semiconductors can be expressed in accordance with Dimmock [9] as

$$\left[\bar{\varepsilon} - \frac{E_{g_0}}{2} - \frac{\hbar^2 k_s^2}{2m_t^-} - \frac{\hbar^2 k_z^2}{2m_l^-}\right]\left[\bar{\varepsilon} + \frac{E_{g_0}}{2} + \frac{\hbar^2 k_s^2}{2m_t^-} + \frac{\hbar^2 k_z^2}{2m_l^-}\right] = P_\perp^2 k_s^2 + P_\parallel^2 k_z^2 \tag{3.55}$$

where $\bar{\varepsilon}$ is the energy as measured from the center of the band gap E_{g_0}, $m_t^\pm$ and $m_l^\pm$ represent the contributions to the transverse and longitudinal effective masses of the external L_6^+ and L_6^- bands arising from the $\vec{k} \cdot \vec{p}$ perturbations with the other bands taken to the second order.

Using $\varepsilon = E + (E_{g_0}/2)$, $P_\perp^2 = \frac{\hbar^2 E_{g_0}}{2m_t^*}$, $P_\parallel^2 = \frac{\hbar^2 E_{g_0}}{2m_l^*}$ (m_t^* and m_l^* are the transverse and longitudinal effective electron masses at $k = 0$) in (3.55), we can write

$$\left[E - \frac{\hbar^2 k_s^2}{2m_t^-} - \frac{\hbar^2 k_z^2}{2m_l^-}\right]\left[1 + \alpha E + \alpha\frac{\hbar^2 k_s^2}{2m_t^+} + \alpha\frac{\hbar^2 k_z^2}{2m_l^+}\right] = \frac{\hbar^2 k_s^2}{2m_t^*} + \frac{\hbar^2 k_z^2}{2m_l^*} \tag{3.56}$$

The 2D DR of the conduction electrons in QWTs of IV–VI materials in accordance with Dimmock model for the dimensional quantization along z direction can be expressed as

$$\begin{aligned}
&E(1+\alpha E) + \alpha E\left(\frac{\hbar^2 k_x^2}{2x_4} + \frac{\hbar^2 k_y^2}{2x_5}\right) + \alpha E\frac{\hbar^2}{2x_6}\left(\frac{n_z\pi}{d_z}\right)^2 \\
&\quad - (1+\alpha E)\left(\frac{\hbar^2 k_x^2}{2x_1} + \frac{\hbar^2 k_y^2}{2x_2}\right) - \alpha\left(\frac{\hbar^2 k_x^2}{2x_1} + \frac{\hbar^2 k_y^2}{2x_2}\right)\left(\frac{\hbar^2 k_x^2}{2x_4} + \frac{\hbar^2 k_y^2}{2x_5}\right) \\
&\quad - \alpha\left(\frac{\hbar^2 k_x^2}{2x_1} + \frac{\hbar^2 k_y^2}{2x_2}\right)\frac{\hbar^2}{2x_6}\left(\frac{n_z\pi}{d_z}\right)^2 - (1+\alpha E)\frac{\hbar^2}{2x_3}\left(\frac{n_z\pi}{d_z}\right)^2 \\
&\quad - \alpha\frac{\hbar^2}{2x_3}\left(\frac{n_z\pi}{d_z}\right)^2\left(\frac{\hbar^2 k_x^2}{2x_4} + \frac{\hbar^2 k_y^2}{2x_5}\right) - \alpha\frac{\hbar^2}{2x_3}\left(\frac{n_z\pi}{d_z}\right)^2\frac{\hbar^2}{2x_6}\left(\frac{n_z\pi}{d_z}\right)^2 \\
&= \frac{\hbar^2 k_x^2}{2m_1} + \frac{\hbar^2 k_y^2}{2m_2} + \frac{\hbar^2}{2m_3}\left(\frac{n_z\pi}{d_z}\right)^2
\end{aligned} \tag{3.57}$$

where

$$x_4 = m_t^+,\ x_5 = \frac{m_t^+ + 2m_l^+}{3},\ x_6 = \frac{3m_t^+ m_l^+}{2m_l^+ + m_t^+},\ x_1 = m_t^-,$$

$$x_2 = \frac{m_t^- + 2m_l^-}{3},\ x_3 = \frac{3m_t^- m_l^-}{2m_l^- + m_t^-},\ m_1 = m_t^*,\ m_2 = \frac{m_t^* + 2m_l^*}{3}$$

and $\quad m_3 = \dfrac{3m_l^* m_t^*}{m_t^* + 2m_l^*}$

Substituting $k_x = r\mathrm{Cos}\theta$ and $k_y = r\mathrm{Sin}\theta$ (where r and θ are 2D polar coordinates in 2D wave vector space) in (3.57), we can write

$$\begin{aligned}
&r^4\left[\alpha\frac{1}{4}\left(\frac{\hbar^2\mathrm{Cos}^2\theta}{x_1}+\frac{\hbar^2\mathrm{Sin}^2\theta}{x_2}\right)\left(\frac{\hbar^2\mathrm{Cos}^2\theta}{x_4}+\frac{\hbar^2\mathrm{Sin}^2\theta}{x_5}\right)\right]\\
&+r^2\frac{1}{2}\left[\left(\frac{\hbar^2\mathrm{Cos}^2\theta}{m_1}+\frac{\hbar^2\mathrm{Sin}^2\theta}{m_2}\right)+\alpha\frac{\hbar^2}{2x_3}\left(\frac{n_z\pi}{d_z}\right)^2\right.\\
&\times\left(\frac{\hbar^2\mathrm{Cos}^2\theta}{x_4}+\frac{\hbar^2\mathrm{Sin}^2\theta}{x_5}\right)+\alpha\left(\frac{\hbar^2\mathrm{Cos}^2\theta}{x_1}+\frac{\hbar^2\mathrm{Sin}^2\theta}{x_2}\right)\frac{\hbar^2}{2x_6}\left(\frac{n_z\pi}{d_z}\right)^2\\
&\left.+\hbar^2(1+\alpha E)\left(\frac{\mathrm{Cos}^2\theta}{x_1}+\frac{\mathrm{Sin}^2\theta}{x_2}\right)-\hbar^2\alpha E\left(\frac{\mathrm{Cos}^2\theta}{x_4}+\frac{\mathrm{Sin}^2\theta}{x_5}\right)\right]\\
&-\left[E(1+\alpha E+\alpha E\frac{\hbar^2}{2x_6}\left(\frac{n_z\pi}{d_z}\right)^2-(1+\alpha E)\frac{\hbar^2}{2x_3}\left(\frac{n_z\pi}{d_z}\right)^2\right.\\
&\left.-\alpha\left(\frac{\hbar^4}{4x_3x_6}\left(\frac{n_z\pi}{d_z}\right)^4\right)\right]=0
\end{aligned}\tag{3.58}$$

The area $A(E, n_z)$ of the 2D wave vector space can be expressed as

$$A(E,n_z) = \bar{J}_1 - \bar{J}_2 \tag{3.59}$$

where

$$\bar{J}_1 \equiv 2\int_0^{\pi/2}\frac{c}{b}d\theta \tag{3.60}$$

and

$$\bar{J}_2 \equiv 2\int_0^{\pi/2}\frac{ac^2}{b^3}d\theta \tag{3.61}$$

in which

$$a \equiv \left[\alpha\left(\frac{\hbar^4}{4}\right)\left(\frac{\mathrm{Cos}^2\theta}{x_1}+\frac{\mathrm{Sin}^2\theta}{x_2}\right)\left(\frac{\mathrm{Cos}^2\theta}{x_4}+\frac{\mathrm{Sin}^2\theta}{x_5}\right)\right],$$

$$b \equiv \left(\frac{\hbar^2}{2}\right)\left[\left(\frac{\mathrm{Cos}^2\theta}{m_1}+\frac{\mathrm{Sin}^2\theta}{m_2}\right)+\alpha\left(\frac{\hbar^2}{2x_3}\right)\left(\frac{n_z\pi}{d_z}\right)^2\left(\frac{\mathrm{Cos}^2\theta}{x_4}+\frac{\mathrm{Sin}^2\theta}{x_5}\right)\right.$$
$$+\alpha\left(\frac{\hbar^2}{2x_6}\right)\left(\frac{n_z\pi}{d_z}\right)^2\left(\frac{\mathrm{Cos}^2\theta}{m_1}+\frac{\mathrm{Sin}^2\theta}{m_2}\right)+(1+\alpha E)\left(\frac{\mathrm{Cos}^2\theta}{x_1}+\frac{\mathrm{Sin}^2\theta}{x_2}\right)$$
$$\left.-\alpha E\left(\frac{\mathrm{Cos}^2\theta}{x_4}+\frac{\mathrm{Sin}^2\theta}{x_5}\right)\right] \quad \text{and}$$

$$c \equiv \left[E(1+\alpha E)+\alpha E\left(\frac{\hbar^2}{2x_6}\right)\left(\frac{n_z\pi}{d_z}\right)^2-(1+\alpha E)\left(\frac{\hbar^2}{2x_3}\right)\left(\frac{n_z\pi}{d_z}\right)^2\right.$$
$$\left.-\alpha\left(\frac{\hbar^4}{4x_3x_6}\right)\left(\frac{n_z\pi}{d_z}\right)^4\right]$$

The (3.60) can be expressed as

$$\bar{J}_1 = 2\int_0^{\pi/2}\frac{t_3(E,n_z)d\theta}{A_1(E,n_z)\mathrm{Cos}^2\theta+B_1(E,n_z)\mathrm{Sin}^2\theta},$$

where

$$t_3(E,n_z)\equiv c,\ A_1(E,n_z)\equiv\frac{\hbar^2}{2m_1}t_1(E,n_z),$$

$$t_1(E,n_z)\equiv\left[1+m_1\left[\frac{1}{x_4}\frac{\alpha\hbar^2}{2x_3}\left(\frac{n_z\pi}{d_z}\right)^2+\frac{\alpha\hbar^2}{2x_1x_6}\left(\frac{n_z\pi}{d_z}\right)^2+\frac{1+\alpha E}{x_1}-\frac{\alpha E}{x_4}\right]\right],$$

$$B_1(E,n_z)\equiv\frac{\hbar^2}{2m_2}t_2(E,n_z) \quad \text{and}$$

$$t_2(E,n_z)\equiv\left[1+m_2\left[\frac{\alpha\hbar^2}{2x_3x_5}\left(\frac{n_z\pi}{d_z}\right)^2+\frac{\alpha\hbar^2}{2x_2x_6}\left(\frac{n_z\pi}{d_z}\right)^2+\frac{1+\alpha E}{x_2}-\frac{\alpha E}{x_5}\right]\right]$$

Performing the integration, we get

$$\bar{J}_1 = \pi t_3(E,n_z)[A_1(E,n_z)B_1(E,n_z)]^{-1/2} \tag{3.62}$$

From (3.61) we can write

$$\bar{J}_2 = \frac{\alpha t_3^2(E,n_z)\hbar^4}{2B_1^3(E,n_z)}I \tag{3.63}$$

where

$$I \equiv \int_0^\infty \frac{(a_1 + a_2 z^2)(a_3 + a_4 z^2)dz}{[(\bar{a})^2 + z^2]^3} \tag{3.64}$$

in which

$$a_1 \equiv \frac{1}{x_1},\ a_2 \equiv \frac{1}{x_2},\ z = \tan\theta,$$

θ is a new variable

$$a_3 \equiv \frac{1}{x_4},\ a_4 \equiv \frac{1}{x_5} \quad \text{and} \quad (\bar{a})^2 \equiv \left(\frac{A_1(E,n_z)}{B_1(E,n_z)}\right)$$

The use of the Residue theorem of Complex Analysis leads to the evaluation of the integral in (3.64) as

$$I = \frac{\pi}{4\bar{a}}[a_1 a_4 + 3a_2 a_4] \tag{3.65}$$

Therefore, the 2D area of the 2D wave vector space can be written as

$$A(E,n_z) = \frac{\pi t_3(E,n_z)}{\sqrt{A_1(E,n_z)B_1(E,n_z)}}\left[1 - \frac{1}{x_5}\left(\frac{1}{x_1} + \frac{3}{x_2}\right)\frac{\alpha t_3(E,n_z)\hbar^4}{8B_1^2(E,n_z)}\right] \tag{3.66}$$

The EEM for the QWTs of IV–VI materials can thus be written as

$$m^*(E,n_z) = \frac{\hbar^2}{2}[\theta_5(E,n_z)]\Big|_{E=E_{Fs}} \tag{3.67}$$

where

$$\begin{aligned}
\theta_5(E,n_z) \equiv{}& \left[1 - \frac{1}{x_5}\left(\frac{1}{x_1} + \frac{3}{x_2}\right)\frac{\alpha t_3(E,n_z)\hbar^4}{8[B_1(E,n_z)]^2}\right][A_1(E,n_z)B_1(E,n_z)]^{-1} \\
&\times\left[\sqrt{A_1(E,n_z)B_1(E,n_z)}\{t_3(E,n_z)\}' - t_3(E,n_z)\right. \\
&\times\left\{\frac{1}{2}\{A_1(E,n_z)\}'\left[\frac{B_1(E,n_z)}{A_1(E,n_z)}\right]^{1/2} + \frac{1}{2}\{B_1(E,n_z)\}'\right. \\
&\left.\left.\times\left[\frac{A_1(E,n_z)}{B_1(E,n_z)}\right]^{1/2}\right\}\right] - \frac{1}{8}\frac{t_3(E,n_z)\alpha\hbar^4}{\sqrt{A_1(E,n_z)B_1(E,n_z)}}\frac{1}{x_5}\left(\frac{1}{x_1} + \frac{3}{x_2}\right) \\
&\times[B_1(E,n_z)]^{-4}[\{B_1(E,n_z)\}^2\{t_3(E,n_z)\}' \\
&- 2B_1(E,n_z)\{B_1(E,n_z)\}' t_3(E,n_z)]
\end{aligned}$$

Thus, the EEM is a function of Fermi energy and the quantum number due to the band non-parabolicity.

The total density-of-states function can be written as

$$N_{2DT}(E) = \left(\frac{g_v}{2\pi}\right) \sum_{n_z=1}^{n_{z\max}} \theta_5(E, n_z) HE - E_{n_{z10}} \tag{3.68}$$

where the sub-band energy $(E_{n_{z10}})$ in this case can be written as

$$E_{n_{z10}}(1+\alpha E_{n_{z10}}) + \alpha E_{n_{z10}} \frac{\hbar^2}{2x_6}\left(\frac{n_z\pi}{d_z}\right)^2 - (1+\alpha E_{n_{z10}})\frac{\hbar^2}{2x_3}\left(\frac{n_z\pi}{d_z}\right)^2$$
$$-\alpha\frac{\hbar^2}{2x_3}\left(\frac{n_z\pi}{d_z}\right)^2 \frac{\hbar^2}{2x_6}\left(\frac{n_z\pi}{d_z}\right)^2 - \left[\frac{\hbar^2}{2m_3}\left(\frac{n_z\pi}{d_z}\right)^2\right] = 0 \tag{3.69}$$

The use (3.68) leads to the expression of 2D electron statistics in QWTs of IV–VI materials at $T \to 0$ as

$$n_{2D} = \frac{g_v}{2\pi} \sum_{n_z=1}^{n_{z\max}} [T_{59}(E_{Fs}, n_z)] \tag{3.70}$$

where

$$T_{59}(E_{Fs}, n_z) \equiv \frac{A(E_{Fs}, n_z)}{\pi}$$

Using (1.4) and (3.70) we can study the QC in this case.

(f) The ellipsoidal parabolic model

The 2D DR, the EEM, the sub-band energy $(\bar{E}_{n_{z9}})$, the total density-of states and the 2D electron statistics for this model can respectively be written as

$$E = \left(\frac{\hbar^2 k_x^2}{2m_1}\right) + \left(\frac{\hbar^2 k_y^2}{2m_2}\right) + \left(\frac{\hbar^2}{2m_3}\right)\left(\frac{n_z\pi}{d_z}\right)^2 \tag{3.71}$$

$$m^*(E_{Fs}) = (\sqrt{m_1 m_2}) \tag{3.72}$$

$$N_{2DT}(E) = \frac{g_v\sqrt{m_1 m_2}}{\pi\hbar^2} \sum_{n_z=1}^{n_{z\max}} H(E - \bar{E}_{n_{z9}}) \tag{3.73}$$

$$E_{n_{z9}} = \left(\frac{\hbar^2}{2m_3}\right)\left(\frac{n_z\pi}{d_z}\right)^2 \tag{3.74}$$

$$n_{2D} = \left[\frac{k_B T g_v\sqrt{m_1 m_2}}{\pi\hbar^2}\right] \sum_{n_z=1}^{n_{z\max}} F_0(\eta_{y3}) \tag{3.75}$$

where

$$\eta_{y3} \equiv (k_B T)^{-1}[E_{Fs} - \bar{E}_{n_{z9}}].$$

Using (1.4) and (3.75) we can study the QC in this case.

3.2.5 *The QC in QWTs of stressed Kane type materials*

The 2D electron energy spectrum in QWTs of stressed Kane type materials can be expressed following (2.21) as

$$\frac{k_x^2}{[\bar{a}_0(E)]^2} + \frac{k_y^2}{[\bar{b}_0(E)]^2} + \frac{1}{[\bar{c}_0(E)]^2}(n_z\pi/d_z)^2 = 1 \tag{3.76}$$

The area of 2D wave vector space enclosed by (3.76) can be written as

$$A(E, n_z) = \pi P^2(E, n_z)\bar{a}_0(E)\bar{b}_0(E) \tag{3.77}$$

where $P^2(E, n_z) = [1 - [n_z\pi/d_z\bar{c}_0(E)]^2]$.

The expression of the surface EEM in this case can be written as

$$m^*(E_{Fs}, n_z) = \frac{\hbar^2}{2}[\theta_6(E, n_z)]\Big|_{E=E_{Fs}} \tag{3.78}$$

in which $\theta_6(E, n_z) = [2P(E, n_z)\{P(E, n_z)\}'\bar{a}_0(E)\bar{b}_0(E) + \{P(E, n_z)\}^2 \{\bar{a}_0(E)\}'\bar{b}_0(E) + \{P(E, n_z)\}^2\{\bar{b}_0(E)\}'\bar{a}_0(E)]$

The EEM in this case is the function of Fermi energy and the size quantization number due to the presence of stress only.

Thus, the total 2D density-of-states function can be expressed as

$$N_{2DT}(E) = \left(\frac{g_v}{2\pi}\right)\sum_{n_z=1}^{n_{z\max}} \theta_6(E, n_z)H(E - E_{n_{z11}}) \tag{3.79}$$

The sub-band energies $(E_{n_{z11}})$ are given by

$$\bar{c}_0(E_{n_{z11}}) = n_z\pi/d_z \tag{3.80}$$

The 2D surface electron concentration per unit area for QWTs of stressed Kane type compounds can be written at $T \to 0$ as

$$n_{2D} = \frac{g_v}{2\pi}\sum_{n_z=1}^{n_{z\max}} [T_{61}(E_{Fs}, n_z)] \tag{3.81}$$

where $T_{61}(E_{Fs}, n_z) \equiv [P^2(E_{Fs}, n_z)\bar{a}_0(E_{Fs})\bar{b}_0(E_{Fs})]$.

Thus, by using (1.4) and (3.81), we can study the QC in this case.

3.2.6 The QC in QWTs of tellurium

Using (2.133), the 2D electron energy spectrum in QWTs of Te can be written as

$$k_s^2 = \psi_5(E) - \psi_6\left(\frac{\pi n_z}{d_z}\right)^2 \pm \psi_7\left[\psi_8^2(E) - \left(\frac{\pi n_z}{d_z}\right)^2\right]^{1/2} \tag{3.82}$$

where

$$\psi_5(E) = \left[\frac{E}{\psi_2} + \frac{\psi_4^2}{2\psi_2^2}\right], \psi_6 = \frac{\psi_1}{\psi_2}, \psi_7 = \frac{\psi_4\sqrt{\psi_1}}{\psi_2^{3/2}},$$

$$\psi_8^2(E) = \frac{\psi_4^4 + 4E\psi_2\psi_4^2 + 4\psi_2^2\psi_3^2}{4\psi_1\psi_2\psi_4^2}$$

The EEM in this case is given by

$$m^*(E_{Fs}, n_z) = \frac{\hbar^2}{2}[t'_{40}(E, n_z)]\Big|_{E=E_{Fs}} \tag{3.83}$$

where

$$t_{40}(E, n_z) = \left[\psi_5(E) - \psi_6\left(\frac{\pi n_z}{d_z}\right)^2 \pm \psi_7\left[\psi_8^2(E) - \left(\frac{\pi n_z}{d_z}\right)^2\right]^{1/2}\right]^{1/2}$$

It appears that the EEM in QWTs of Te is a function of Fermi energy and size quantum number which is the characteristics of such systems.

Thus, the total 2D density-of-states function can be expressed as

$$N_{2DT}(E) = \left(\frac{g_v}{\pi}\right)\sum_{n_z=1}^{n_{z\max}} t'_{40}(E, n_z)H(E - E_{n_{z12}}) \tag{3.84}$$

The sub-band energies ($E_{n_{z12}}$) are given by

$$E_{n_{z12}} = \psi_1(n_z\pi/d_z)^2 \pm \psi_3(n_z\pi/d_z) \tag{3.85}$$

The 2D surface electron concentration per unit area for QWTs of Te can be written at $T \to 0$ as

$$n_{2D} = \frac{g_v}{\pi}\sum_{n_z=1}^{n_{z\max}}[t_{40}(E_{Fs}, n_z)] \tag{3.86}$$

where

$$t_{41}(E_{Fs}, n_z) \equiv \sum_{r=1}^{s} L(r)t_{40}(E_{Fs}, n_z)$$

Thus, using (1.4) and (3.86) we can study the QC in this case.

3.2.7 The QC in QWTs of gallium phosphide

The 2D electron DR in QWTs GaP can be expressed using (2.72) as

$$E = ak_s^2 + C(n_z\pi/d_z)^2 + |V_G| - [Dk_s^2 + |V_G|^2 + D(n_z\pi/d_z)^2]^{1/2} \quad (3.87)$$

in which

$$a \equiv \frac{\hbar^2}{2m_\perp^*} + \frac{\hbar^2}{2m_\parallel^*}, C \equiv \frac{\hbar^2}{2m_\parallel^*} \quad \text{and} \quad D \equiv (\hbar^2 k_0/m_\parallel^*)^2$$

The sub-band energy ($E_{n_{z13}}$) are given by

$$E_{n_{z13}} = C(\pi n_z/d_z)^2 + |V_G| - [|V_G|^2 + D(\pi n_z/d_z)^2]^{1/2} \quad (3.88)$$

The (3.88) can be expressed as

$$k_s^2 = t_{42}(E, n_z) \quad (3.89)$$

in which, $t_{42}(E, n_z) \equiv [\{2a(E - t_1) + D\} - \{[2a(E - t_1) + D]^2 - 4a^2[(E - t_1)^2 - t_2]\}^{1/2}]$, $t_1 \equiv |V_G| + C(\pi n_z/d_z)^2$ and $t_2 \equiv |V_G|^2 + D(\pi n_z/d_z)^2$.

The EEM can be expressed from (3.89) as

$$m^*(E_{Fs}, n_z) = \frac{\hbar^2}{2} t'_{42}(E_{Fs}, n_z) \quad (3.90)$$

It appears that the EEM in QWTs of GaP is a function of Fermi energy and size quantum number due to the presence of the system constant k_0.

The total DOS function is given by

$$N_{2DT}(E) = \frac{g_v}{4\pi a^2} \sum_{n_z=1}^{n_{z\max}} [t'_{42}(E, n_z)] H(E - E_{n_{z13}}) \quad (3.91)$$

The electron statistics in QWTs in GaP at $T \to 0$ assumes the form

$$n_{2D} = \frac{g_v}{4\pi a^2} \sum_{n_z=1}^{n_{z\max}} [t_{42}(E_{F2D}, n_z)] \quad (3.92)$$

Thus, using (1.4) and (3.92) we can study the QC in this case.

3.2.8 *The QC in QWTs of platinum antimonide*

The DR for the n-type $PtSb_2$ can be written as [11]

$$\left(\bar{E}+\lambda_0\frac{(\bar{\bar{a}})^2}{4}k^2-lk_S^2\frac{(\bar{\bar{a}})^2}{4}\right)\left(\bar{E}+\delta_0-\upsilon\frac{(\bar{\bar{a}})^2}{4}k^2-\bar{\bar{n}}\frac{(\bar{\bar{a}})^2}{4}k_S^2\right)=I\frac{(\bar{\bar{a}})^4}{16}k^4 \tag{3.93}$$

where

$$\omega_1\equiv\left(\lambda_0\frac{(\bar{\bar{a}})^2}{4}-l\frac{(\bar{\bar{a}})^2}{4}\right),\omega_2\equiv\lambda_0\frac{(\bar{\bar{a}})^2}{4},\omega_3\equiv\left(\bar{\bar{n}}\frac{(\bar{\bar{a}})^2}{4}+\upsilon\frac{(\bar{\bar{a}})^2}{4}\right),$$

$$\omega_4\equiv\upsilon\frac{(\bar{\bar{a}})^2}{4},$$

$I_1\equiv I\left(\frac{(\bar{\bar{a}})^2}{4}\right)^2$, $\lambda_0, l, \delta_0, \nu$ and $\bar{\bar{n}}$ are the band constants and $\bar{\bar{a}}$ is the lattice constant.

The (3.93) can be expressed as

$$[E+\omega_1k_s^2+\omega_2k_z^2][E+\delta_0-\omega_3k_s^2-\omega_4k_z^2]=I_1(k_z^2+k_s^2)^2 \tag{3.94}$$

The use of (3.94) leads to the expression of the 2D dispersion law in QWTs of $PtSb_2$ as

$$k_s^2=t_{44}(E,n_z) \tag{3.95}$$

$$t_{44}(E,n_z)=[2A_9]^{-1}\left[-A_{10}(E,n_z)+\sqrt{A_{10}^2(E,n_z)+4A_9A_{11}(E,n_z)}\right] \tag{3.96}$$

where

$$A_9\equiv[I_1+\omega_1\omega_3],A_{10}(E,n_z)\equiv\left[\omega_3E+\omega_1\left\{E+\delta_0-\omega_4\left(\frac{\pi n_z}{d_z}\right)^2\right\}\right.$$
$$\left.+\ \omega_2\omega_3\left(\frac{\pi n_z}{d_z}\right)^2+2I_1\left(\frac{\pi n_z}{d_z}\right)^2\right]$$

and

$$A_{11}(E,n_z)\equiv\left[E\left[E+\delta_0-\omega_4\left(\frac{\pi n_z}{d_z}\right)^2\right]\right.$$
$$\left.+\ \omega_2\left(\frac{\pi n_z}{d_z}\right)^2\left[E+\delta_0-\omega_4\left(\frac{\pi n_z}{d_z}\right)^2\right]-I_1\left(\frac{\pi n_z}{d_z}\right)^4\right]$$

The area of k_s space can be expressed as

$$A(E, n_z) = \frac{\pi}{2A_9} t_{44}(E, n_z) \tag{3.97}$$

The EEM can be written as

$$m^*(E_{Fs}, n_z) = \frac{\hbar^2}{4A_9} t'_{44}(E, n_z) \tag{3.98}$$

It appears that the EEM in QWTs of $PtSb_2$ is a function of Fermi energy and size quantum number which is the characteristics of such systems.

The total DOS function assumes the form

$$N_{2DT}(E) = \frac{g_v}{4\pi A_9} \sum_{n_z=1}^{n_{z\max}} [t'_{44}(E, n_z)] H(E - E_{n_{z14}}) \tag{3.99}$$

where the quantized levels $E_{n_{z14}}$ can be expressed through the equation

$$\begin{aligned} E_{n_{z14}} = (2)^{-1} \Bigg[- \left[\omega_2 \left(\frac{\pi n_z}{d_z} \right)^2 + \delta_0 - \omega_4 \left(\frac{\pi n_z}{d_z} \right)^2 \right] \\ + \Bigg\{ \left[\omega_2 \left(\frac{\pi n_z}{d_z} \right)^2 + \delta_0 - \omega_4 \left(\frac{\pi n_z}{d_z} \right)^2 \right]^2 \\ + 4 \left[I_1 \left(\frac{\pi n_z}{d_z} \right)^4 + \omega_2 \omega_4 \left(\frac{\pi n_z}{d_z} \right)^4 - \omega_2 \delta_0 \left(\frac{\pi n_z}{d_z} \right)^2 \right] \Bigg\}^{1/2} \Bigg] \end{aligned} \tag{3.100}$$

The electron statistics in QWTs of $PtSb_2$ at $T \to 0$ can be written as

$$n_{2D} = \frac{g_v}{4\pi A_9} \sum_{n_z=1}^{n_{z\max}} [t_{44}(E_{F2D}, n_z)] \tag{3.101}$$

Thus, using (1.4) and (3.101) we can study the QC in this case.

3.2.9 *The QC in QWTs of bismuth telluride*

The 2D electron dispersion law in QWTs of Bi_2Te_3 can be expressed as

$$E(1 + \alpha E) = \mathrm{I}_1 \left(\frac{n_x \pi}{d_x} \right)^2 + \mathrm{I}_2 k_y^2 + \mathrm{I}_3 k_z^2 + 2\bar{\mathrm{I}}_4 k_z k_y \tag{3.102}$$

where

$$I_1 = \frac{\hbar^2}{2m_0}\alpha_{11}, I_2 = \frac{\hbar^2}{2m_0}\alpha_{22}, I_3 = \frac{\hbar^2}{2m_0}\alpha_{33} \quad \text{and} \quad \bar{\mathrm{I}}_4 = \frac{\hbar^2}{2m_0}\alpha_{23}$$

The area of the ellipse is given by

$$A_n(E, n_x) = \frac{\pi}{\sqrt{\alpha_{22}\alpha_{33} - 4\alpha_{23}^2}}\left[\frac{2m_0E(1+\alpha E)}{\hbar^2} - \alpha_{11}\left(\frac{n_x\pi}{d_x}\right)^2\right] \tag{3.103}$$

The EEM can be expressed as

$$m^*(E_{Fs}) = \frac{m_0(1+2\alpha E_{Fs})}{\sqrt{\alpha_{22}\alpha_{33} - 4\alpha_{23}^2}} \tag{3.104}$$

It appears that the EEM in QWTs of Bi_2Te_3 is a function of Fermi energy due to the presence of the band non-parabolicity.

The total DOS function assumes the form

$$N_{2DT}(E) = \frac{g_v m_0}{\pi\hbar^2\sqrt{\alpha_{22}\alpha_{33} - 4\alpha_{23}^2}}\sum_{n_x=1}^{n_{x\max}}(1+2\alpha E)H(E - E_{n_{x15}}) \tag{3.105}$$

where $(E_{n_{x15}})$ can be expressed through the equation

$$E_{n_{x15}}(1+\alpha E_{n_{x15}}) = \mathrm{I}_1\left(\frac{n_x\pi}{d_x}\right)^2 \tag{3.106}$$

The electron concentration in QWTs of Bi_2Te_3 can be written as

$$n_{2D} = \frac{k_BTg_v}{\pi\hbar^2}\left(\frac{m_0}{\sqrt{\alpha_{22}\alpha_{33} - 4\alpha_{23}^2}}\right)\sum_{n_x=1}^{n_{x\max}}[(1+2\alpha E_{n_{x15}})F_0(\eta_{n_{15}}) + 2\alpha k_BTF_1(\eta_{n_{15}})] \tag{3.107}$$

where

$$\eta_{n_{15}} = \frac{E_{Fs} - E_{n_{x15}}}{k_BT}$$

Thus, using (1.4) and (3.107) we can study the QC in this case.

3.2.10 *The QC in QWTs of germanium*

(a) Cardona *et al.*

In the presence of size quantization along k_z direction, the 2D DR of the conduction relations in QWTs of Ge can be written in accordance with the model of Cardona *et al.* [12] and using (2.47) as

$$\frac{\hbar^2k_x^2}{2m_1^*} + \frac{\hbar^2k_y^2}{2m_2^*} = \gamma(E, n_z) \tag{3.108}$$

where

$$m_1^* \equiv m_\perp^*,\ m_2^* = \frac{m_\perp^* + 2m_\parallel^*}{3},$$

$$\gamma(E, n_z) \equiv \left[E(1+\alpha E) - (1+2\alpha E)\frac{\hbar^2}{2m_3^*}\left(\frac{n_z\pi}{d_z}\right)^2 + \alpha\left[\frac{\hbar^2}{2m_3^*}\left(\frac{n_z\pi}{d_z}\right)^2\right]^2\right]$$

and

$$m_3^* \equiv \frac{3m_\parallel^* m_\perp^*}{2m_\parallel^* + m_\perp^*}$$

The area of ellipse of the 2D surface as given by (3.108) can be written as

$$A(E, n_z) = \frac{2\pi\sqrt{m_1^* m_2^*}}{\hbar^2}\gamma(E, n_z) \tag{3.109}$$

The EEM can be expressed using (3.109) as

$$m^*(E_{Fs}, n_z) \equiv \sqrt{m_1^* m_2^*}\left[(1+2\alpha E_{Fs}) - (2\alpha)\frac{\hbar^2}{2m_3^*}\left(\frac{n_z\pi}{d_z}\right)^2\right] \tag{3.110}$$

Therefore, the EEM is a function of Fermi energy and size quantum number due to the presence of band non-parabolicity only.

The DOS function per sub band can be expressed as

$$N_{2D}(E) = \frac{4\sqrt{m_1^* m_2^*}}{\pi\hbar^2}\left[1 + 2\alpha E - 2\alpha\left(\frac{\hbar^2}{2m_3^*}\left(\frac{\pi n_z}{d_z}\right)^2\right)\right] \tag{3.111}$$

The total DOS function is given by

$$N_{2DT}(E) = \frac{4}{\pi\hbar^2}\sqrt{m_1^* m_2^*}\sum_{n_z=1}^{n_{z\max}}\left[1 + 2\alpha E - 2\alpha\left(\frac{\hbar^2}{2m_3^*}\left(\frac{\pi n_z}{d_z}\right)^2\right)\right] \times H(E - E_{n_{z16}}) \tag{3.112}$$

where $E_{n_{z16}}$ is the positive root of the following equation

$$E_{n_{z16}}(1 + \alpha E_{n_{z16}}) - (1 + 2\alpha E_{n_{z16}})\left(\frac{\hbar^2}{2m_3^*}\left(\frac{\pi n_z}{d_z}\right)^2\right) + \alpha\left(\frac{\hbar^2}{2m_3^*}\left(\frac{\pi n_z}{d_z}\right)^2\right)^2 = 0 \tag{3.113}$$

Thus, combining (3.112) with the Fermi Dirac occupation probability factor, the electron statistics in QWTs of Ge can be written as

$$n_{2D} = \frac{4\sqrt{m_1^* m_2^*} k_B T}{\hbar^2} \sum_{n_z=1}^{n_{z\max}} [(A_1(n_z) + 2\alpha E_{n_{z16}}) F_0(\eta_{n_{z16}}) + 2\alpha k_B T F_1(\eta_{n_{z16}})] \quad (3.114)$$

where, $A_1(n_z) \equiv [1 + 2\alpha(\hbar^2/2m_3^*)(\pi n_z/d_z)^2]$ and $\eta_{n_{z16}} \equiv \frac{1}{k_B T}[E_{F2D} - E_{n_{z16}}]$

Thus, using (1.4) and (3.114) we can study the QC in this case.

(b) Wang and Ressler [13] model

The 2D dispersion law can be expressed for this model using (2.53) as

$$E = A_5(n_z) + A_6(n_z)\beta - \bar{c}_1\beta^2 \quad (3.115)$$

where

$$A_5(n_z) \equiv \frac{\hbar^2}{2m_3^*}\left(\frac{\pi n_z}{d_z}\right)^2 \left[1 - \bar{e}_1\left(\frac{\hbar^2}{2m_3^*}\right)\left(\frac{\pi n_z}{d_z}\right)^2\right]$$

$$A_6(n_z) \equiv \left[1 - \bar{d}_1\left(\frac{\hbar^2}{2m_3^*}\right)\left(\frac{\pi n_z}{d_z}\right)^2\right] \quad \text{and} \quad \beta \equiv \frac{\hbar^2 k_x^2}{2m_1^*} + \frac{\hbar^2 k_y^2}{2m_2^*}$$

The (3.115) can be written as

$$\frac{\hbar^2 k_x^2}{2m_1^*} + \frac{\hbar^2 k_y^2}{2m_2^*} = I_1(E, n_z) \quad (3.116)$$

where $I_1(E, n_z) \equiv (2\bar{c}_1)^{-1}[A_6(n_z) - [A_6^2(n_z) - 4\bar{c}_1 E + 4\bar{c}_1 A_5(n_z)]^{1/2}]$.

From (3.116), the area of the 2D k_s-space is given by

$$A(E, n_z) = \frac{2\pi\sqrt{m_1^* m_2^*}}{\hbar^2} I_1(E, n_z) \quad (3.117)$$

The EEM can be expressed using (3.117) as

$$m^*(E_{Fs}, n_z) \equiv \sqrt{m_1^* m_2^*}[I_1(E, n_z)]' \quad (3.118)$$

where

$$\{I_1(E, n_z)\}' \equiv \frac{\partial}{\partial E}[I_1(E, n_z)]$$

Therefore, the EEM according to this model is a function of Fermi energy and size quantum number due to the presence of band non-parabolicity only.

The DOS function per sub-band can be written as

$$N_{2D}(E) = \frac{4}{\pi}\frac{\sqrt{m_1^* m_2^*}}{\hbar^2}\{I_1(E, n_z)\}' \tag{3.119}$$

The total DOS function assumes the form

$$N_{2DT}(E) = \frac{4\sqrt{m_1^* m_2^*}}{\pi\hbar^2}\sum_{n_z=1}^{n_{z\max}}\{I_1(E, n_z)\}' H(E - E_{n_{z17}}) \tag{3.120}$$

where the sub-band energy ($E_{n_{z17}}$) are given by

$$E_{n_{z17}} = \frac{\hbar^2}{2m_3^*}\left(\frac{\pi n_z}{d_z}\right)^2 - \bar{e}_1\left[\frac{\hbar^2}{2m_3^*}\left(\frac{\pi n_z}{d_z}\right)^2\right]^2 \tag{3.121}$$

The electron statistics in QWTs of Ge can be written as

$$n_{2D} = \frac{4\sqrt{m_1^* m_2^*}}{\pi\hbar^2}\sum_{n_z=1}^{n_{z\max}}[t_{46}(E_{Fs}, n_z)] \tag{3.122}$$

where $t_{46}(E_{Fs}, n_z) \equiv I_1(E_{Fs}, n_z)$.

Thus, using (1.4) and (3.122) we can study the QC in this case.

3.2.11 *The QC in QWTs of gallium arsenide*

(a) Rossler model [14]

The 2D DR for this model can be expressed using (2.57) as

$$k_s^2 = \gamma_{4\pm}(E) - \left(\frac{n_z\pi}{d_z}\right)^2 \tag{3.123}$$

The EEM can be written as

$$m^*(E_{Fs}) = \frac{\hbar^2}{2}\gamma_{4\pm}(E_{Fs}) \tag{3.124}$$

The density-of-states is given by

$$N_{2DT}(E) = \frac{g_v}{4\pi}\sum_{n_z=1}^{n_{z\max}}\{\gamma_{4\pm}(E)\}' H(E - E_{n_{z18}}) \tag{3.125}$$

where the sub-band energy $E_{n_{z18}}$ is given by

$$\gamma_{4\pm}(E_{n_{z18}}) - \left(\frac{n_z\pi}{d_z}\right)^2 = 0 \tag{3.126}$$

The electron concentration in QWTs of GaAs at $T \to 0$ assumes the form

$$N_S = \frac{g_v}{4\pi}\sum_{n_z=1}^{n_{z\max}}[T_{47\pm}(E_{Fs}, n_z)] \tag{3.127}$$

where

$$T_{47\pm}(E_{Fs}, n_z) = \left[\gamma_{4\pm}(E_{Fs}) - \left(\frac{n_z\pi}{d_z}\right)^2\right]$$

Using (1.4) and (3.127) we can study the QC in this case.

(b) Model of Stillman *et al.* [15]

The 2D DR for this model can be expressed using (2.65) as

$$k_s^2 = \gamma_5(E) - \left(\frac{n_z\pi}{d_z}\right)^2 \tag{3.128}$$

The EEM can be written as

$$m^*(E_{Fs}) = \frac{\hbar^2}{2}\gamma_5(E_{Fs}) \tag{3.129}$$

The density-of-states is given by

$$N_{2DT}(E) = \frac{g_v}{2\pi}\sum_{n_z=1}^{n_{z\max}}\{\gamma_5(E)\}' H(E - E_{n_{z19}}) \tag{3.130}$$

where the sub-band energy $E_{n_{z19}}$ is given by

$$\gamma_5(E_{n_{z19}}) - \left(\frac{n_z\pi}{d_z}\right)^2 = 0 \tag{3.131}$$

The electron concentration in QWTs of GaAs at $T \to 0$ assumes the form

$$N_S = \frac{g_v}{2\pi}\sum_{n_z=1}^{n_{z\max}}[T_{48}(E_{Fs}, n_z)] \tag{3.132}$$

where

$$[T_{48}(E_{Fs}, n_z)] = \gamma_5(E_{Fs}) - \left(\frac{n_z\pi}{d_z}\right)^2$$

Using (1.4) and (3.132) we can study the QC in this case.

3.2.12 *The QC in QWTs of gallium antimonide*

(a) Model of Seiler *et al.* [16]

The 2D DR of the conduction electrons in GaSb can be written following (2.77) as

$$k_s^2 = \gamma_6(E) - \left(\frac{n_z\pi}{d_z}\right)^2 \tag{3.133}$$

The EEM can be written as

$$m^*(E_{Fs}) = \frac{\hbar^2}{2}\gamma_6(E_{Fs}) \tag{3.134}$$

The density-of-states is given by

$$N_{2DT}(E) = \frac{g_v}{2\pi}\sum_{n_z=1}^{n_{z\max}} \{\gamma_6(E)\}' H(E - E_{n_{z20}}) \tag{3.135}$$

where the sub-band energy $E_{n_{z20}}$ is given by

$$\gamma_6(E_{n_{z20}}) - \left(\frac{n_z \pi}{d_z}\right)^2 = 0 \tag{3.136}$$

The electron concentration in QWTs of GaSb at $T \to 0$ assumes the form

$$N_S = \frac{g_v}{2\pi}\sum_{n_z=1}^{n_{z\max}} [T_{49}(E_{Fs}, n_z)] \tag{3.137}$$

where

$$[T_{49}(E_{Fs}, n_z)] = \gamma_6(E_{Fs}) - \left(\frac{n_z \pi}{d_z}\right)^2$$

Using (1.4) and (3.137) we can study the QC in this case.

(b) Model of Mathur *et al.* [17]

The 2D DR of the conduction electrons in QWTs of GaSb can be written following (2.86) as

$$k_s^2 = \gamma_7(E) - \left(\frac{n_z \pi}{d_z}\right)^2 \tag{3.138}$$

The EEM can be written as

$$m^*(E_{Fs}) = \frac{\hbar^2}{2}\gamma_7(E_{Fs}) \tag{3.139}$$

The density-of-states is given by

$$N_{2DT}(E) = \frac{g_v}{2\pi}\sum_{n_z=1}^{n_{z\max}} \{\gamma_7(E)\}' H(E - E_{n_{z21}}) \tag{3.140}$$

where the sub-band energy $E_{n_{z21}}$ is given by

$$\gamma_7(E_{n_{z21}}) - \left(\frac{n_z \pi}{d_z}\right)^2 = 0 \tag{3.141}$$

The electron concentration in QWTs of GaSb at $T \to 0$ assumes the form

$$N_S = \frac{g_v}{2\pi} \sum_{n_z=1}^{n_{z\max}} [T_{50}(E_{Fs}, n_z)] \tag{3.142}$$

where

$$[T_{50}(E_{Fs}, n_z)] = \gamma_7(E_{Fs}) - \left(\frac{n_z \pi}{d_z}\right)^2$$

Using (1.4) and (3.142) we can study the QC in this case.

(c) Model of Zhang [18]

The 2D dispersion relation of the conduction electrons in GaSb can be written following (2.95) as

$$k_s^2 = \gamma_8(E) - \left(\frac{n_z \pi}{d_z}\right)^2 \tag{3.143}$$

The EEM can be written as

$$m^*(E_{Fs}) = \frac{\hbar^2}{2}\gamma_8(E_{Fs}) \tag{3.144}$$

The density-of-states is given by

$$N_{2DT}(E) = \frac{g_v}{2\pi} \sum_{n_z=1}^{n_{z\max}} \{\gamma_8(E)\}' H(E - \overline{E}_{n_{z22}}) \tag{3.145}$$

where the sub-band energy $\overline{E}_{n_{z22}}$ is given by

$$\gamma_8(\overline{E}_{n_{z22}}) - \left(\frac{n_z \pi}{d_z}\right)^2 = 0 \tag{3.146}$$

The electron concentration in QWTs of GaSb at $T \to 0$ assumes the form

$$N_S = \frac{g_v}{2\pi} \sum_{n_z=1}^{n_{z\max}} [T_{51}(E_{Fs}, n_z)] \tag{3.147}$$

where

$$[T_{51}(E_{Fs}, n_z)] = \gamma_8(E_{Fs}) - \left(\frac{n_z \pi}{d_z}\right)^2$$

Using (1.4) and (3.147) we can study the QC in this case.

3.2.13 *The QC in QWTs of indium antimonide*

(a) Model of Palik *et al.* [19]

The 2D DR of the conduction electrons in QWTs of InSb can be written following (2.104) as

$$k_s^2 = \gamma_9(E) - \left(\frac{n_z \pi}{d_z}\right)^2 \tag{3.148}$$

The EEM can be written as

$$m^*(E_{Fs}) = \frac{\hbar^2}{2}\gamma_9(E_{Fs}) \tag{3.149}$$

The density-of-states function is given by

$$N_{2DT}(E) = \frac{g_v}{2\pi}\sum_{n_z=1}^{n_{z\max}} \{\gamma_9(E)\}' H(E - E_{n_{z23}}) \tag{3.150}$$

where the sub-band energy $E_{n_{z23}}$ is given by

$$\gamma_9(E_{n_{z23}}) - \left(\frac{n_z \pi}{d_z}\right)^2 = 0 \tag{3.151}$$

The electron concentration in QWTs of InSb at $T \to 0$ assumes the form

$$N_S = \frac{g_v}{2\pi}\sum_{n_z=1}^{n_{z\max}} [T_{52}(E_{Fs}, n_z)] \tag{3.152}$$

where

$$[T_{52}(E_{Fs}, n_z)] = \gamma_9(E_{Fs}) - \left(\frac{n_z \pi}{d_z}\right)^2$$

Using (1.4) and (3.152) we can study the QC in this case.

(b) Model of Johnson *et al.* [20]

The 2D DR of the conduction electrons in QWTs of InSb can be written following (2.114) as

$$k_s^2 = \gamma_{10}(E) - \left(\frac{n_z \pi}{d_z}\right)^2 \tag{3.153}$$

The EEM can be written as

$$m^*(E_{Fs}) = \frac{\hbar^2}{2}\gamma_{10}(E_{Fs}) \tag{3.154}$$

The density-of-states is given by

$$N_{2DT}(E) = \frac{g_v}{2\pi} \sum_{n_z=1}^{n_{z\max}} \{\gamma_{10}(E)\}' H(E - E_{n_{z24}}) \tag{3.155}$$

where the sub-band energy $E_{n_{z24}}$ is given by

$$\gamma_{10}(E_{n_{z24}}) - \left(\frac{n_z \pi}{d_z}\right)^2 = 0 \tag{3.156}$$

The electron concentration in QWTs of InSb at $T \to 0$ assumes the form

$$N_S = \frac{g_v}{2\pi} \sum_{n_z=1}^{n_{z\max}} [T_{53}(E_{Fs}, n_z)] \tag{3.157}$$

where

$$[T_{53}(E_{Fs}, n_z)] = \gamma_{10}(E_{Fs}) - \left(\frac{n_z \pi}{d_z}\right)^2$$

Using (1.4) and (3.157) we can study the QC in this case.

(c) Model of Agafonov. *et al.* [21]

The 2D DR of the conduction electrons in InSb can be written following (2.124) as

$$k_s^2 = \gamma_{11}(E) - \left(\frac{n_z \pi}{d_z}\right)^2 \tag{3.158}$$

The EEM can be written as

$$m^*(E_{Fs}) = \frac{\hbar^2}{2} \gamma_{11}(E_{Fs}) \tag{3.159}$$

The density-of-states function is given by

$$N_{2DT}(E) = \frac{g_v}{2\pi} \sum_{n_z=1}^{n_{z\max}} \{\gamma_{11}(E)\}' H(E - E_{n_{z25}}) \tag{3.160}$$

where the sub-band energy $E_{n_{z25}}$ is given by

$$\gamma_{11}(E_{n_{z25}}) - \left(\frac{n_z \pi}{d_z}\right)^2 = 0 \tag{3.161}$$

The electron concentration in QWTs of InSb at $T \to 0$ assumes the form

$$N_S = \frac{g_v}{2\pi} \sum_{n_z=1}^{n_{z\max}} [T_{54}(E_{Fs}, n_z)] \tag{3.162}$$

where

$$[T_{54}(E_{Fs}, n_z)] = \gamma_{11}(E_{Fs}) - \left(\frac{n_z\pi}{d_z}\right)^2$$

Using (1.4) and (3.162) we can study the QC in this case.

3.3 Summary and Conclusions

In this chapter, in Section 3.2.1, we have formulated the electron statistics as given by (3.7) and by combining the same with (1.4) we can investigate the QC in QWTs of tetragonal and nonlinear optical materials. The expressions of 2D electron concentration for QWT of ternary and quaternary materials whose energy band structures are defined by three- and two-band models of Kane together with parabolic energy bands are given by (3.12), (3.17) and (3.21) in Section 3.2.2, respectively, and by combining each of them with (1.4) we can study the QC of the QWTs of the said materials in accordance the said types of band models. In Section 3.2.3, we have formulated the N_s for QWTs of II–VI materials as given by (3.26) and by combining the same with (1.4) we can study the QC in the present case.

The expressions of 2D N_s in QWT of IV–VI materials in accordance with the models of McClure and Choi, Hybrid, Cohen, Lax, Dimmock and the ellipsoidal parabolic are given by (3.35), (3.42), (3.48), (3.54), (3.70), and (3.75) in Section 3.2.4, respectively, and by combining each of them with (1.4) we can study the QC of the QWT of IV–VI materials in accordance with the said types of band models. In Section 3.2.5, we have formulated the N_s for QWTs of stressed Kane type materials as given by (3.81) and by combining the same with (1.4) we can study the QC in the present case. In Section 3.2.6, we have formulated the N_s for QWTs of Tellurium as given by (3.86) and by combining the same with (1.4) we can study the QC in the present case. In Section 3.2.7, we have formulated the N_s for QWTs of GaP as given by (3.92) and by combining the same with (1.4) we can study the QC in the present case. In Section 3.2.8, we have formulated the N_s for QWTs of $PtSb_2$ as given by (3.101) and by combining the same with (1.4) we can study the QC in the present case. In Section 3.2.9, we have formulated the N_s for QWTs of Bi_2Te_3 as given by (3.107) and by combining the same with (1.4) we can study the QC in the present case.

The expressions of 2D electron concentration for QWT of Ge whose energy band structures are defined by the models of Cardona *et al.* and

Wang *et al.* are given by (3.114) and (3.122) in Section 3.2.10, respectively, and by combining each of them with (1.4) we can study the QC of the QWTs of Ge in accordance the said types of band models. The expressions of 2D electron concentration for QWT of GaAs whose energy band structures are defined by the models of Rossler *et al.* and Stillman *et al.* are given by (3.127) and (3.132) in Section 3.2.11, respectively, and by combining each of them with (1.4) we can study the QC of the QWTs of GaAs in accordance with the said types of band models.

The expressions of 2D electron concentration for QWT of GaSb whose energy band structures are defined by the models of Seiler *et al.*, Mathur *et al.* and Zhang are given by (3.137), (3.142) and (3.147) in Section 3.2.12, respectively, and by combining each of them with (1.4) we can study the QC of the QWTs of GaSb in accordance with the said types of band models. The expressions of 2D electron concentration for QWT of InSb whose energy band structures are defined by the models of Palik *et al.*, Johnson *et al.* and Agafonov *et al.* are given by (3.152), (3.157) and (3.162) in Section 3.2.13, respectively, and by combining each of them with (1.4) we can study the QC of the QWTs of InSb in accordance with the said types of band models.

It may be noted that if the direction of application of the surface electric field applied perpendicular to the surface is taken as either k_x or k_y and not as k_z as assumed in the present work, the QCs would be different analytically for both the limits. Nevertheless, the arbitrary choice of the direction normal to the surface would not result in a change of the basic qualitative feature of the QCs in QWTs of the said materials. We wish to note that the manybody effects, the hot electron effects, the formation of band tails, arbitrary orientation of the direction of the electric quantization and the effects of surface of states have been neglected in our simplified theoretical formalism due to the lack of availability in the literature of the proper analytical techniques for including them for the generalized systems as considered in this chapter. Our simplified approach will be useful for the purpose of comparison, when, the methods of tackling the aforementioned formidable problems for the present generalized system appear. The inclusion of the said effects would certainly increase the accuracy of our results, and the qualitative features of the QCs as discussed in this chapter would not change in the presence of the aforementioned influences.

We have not given the graphs of the QCs for QWTs made of the materials as discussed here since we feel that our readers will perform the intricate numerical computation to obtain the oscillatory plots for better assessment and joy of understanding. Incidentally, one can easily infer how little is presented and how much more there is yet to be investigated in the research field of QCs in this case for different QWTs having different band structures in general which is the signature of the co-existence of new physics and advanced mathematics. Finally, we again realize that it is the band structure that changes in a fundamental way and consequently, all the physical properties of all the electronic materials change radically leading to new physical concepts.

3.4 Open Research Problems

R.3.1 Investigate the QCs in the presence of an arbitrarily oriented electric quantization for QWTs of all the materials as studied in this chapter. Draw the corresponding graphs and explain the results.

R.3.2 Investigate the QCs in QWTs of IV–VI, II–VI and stressed Kane type compounds in the presence of an arbitrarily oriented quantizing non-uniform electric field. Draw the corresponding graphs and explain the results.

R.3.3 Investigate the QCs in QWTs of all the materials as stated in R.1.1 of Chapter 1 in the presence of strain. Draw the corresponding graphs and explain the results.

R.3.4 Investigate the QCs in the presence of an arbitrarily oriented non-quantizing magnetic field in QWTs of all the materials as studied in this chapter by including the electron spin. Draw the corresponding graphs and explain the results.

R.3.5 Investigate the QCs in QWTs of IV–VI, II–VI and stressed Kane type compounds in the presence of an arbitrarily oriented non-quantizing magnetic field by including the electron spin. Draw the corresponding graphs and explain the results.

R.3.6 Investigate the QCs in QWTs of all the materials as stated in R.1.1 of Chapter 1 in the presence of an arbitrarily oriented non-quantizing magnetic field by including electron spin. Draw the corresponding graphs and explain the results.

R.3.7 Investigate the QCs in inversion ion layers for all the problems from R.2.1 to R.2.6 in the presence of an additional arbitrarily oriented electric field. Draw the corresponding graphs and explain the results.

R.3.8 Investigate the QCs in QWTs of all the problems from R.2.1 to R.2.8 in the presence of surface states. Draw the corresponding graphs and explain the results.

R.3.9 Investigate the QCs in QWTs of all the problems from R.2.1 to R.2.8 in the presence of hot electron effects. Draw the corresponding graphs and explain the results.

References

1. M. A. Hines and G. D. Scholes, *Adv. Mater.* **15**, 1844 (2003).
2. C. A. Wang, R. K. Huang, D. A. Shiau, M. K. Connors, P. G. Murphy, P. W. O'Brien, A. C. Anderson, D. M. DePoy, G. Nichols and M. N. Palmisiano, *Appl. Phys. Lett.* **83**, 1286 (2003).
3. C. W. Hitchcock, R. J. Gutmann, J. M. Borrego, I. B. Bhat and G. W. Charache, *IEEE Trans. Electron. Dev.* **46**, 2154 (1999).
4. J. J. Hopfield, *J. Appl. Phys.* **32**, 2277 (1961).
5. J. W. Mcclure and K. H. Choi, *Solid. State Commun.* **21**, 10 (1977).
6. S. Takoka, H. Kawamura, K. Murase and S. Takano, *Phys. Rev. B* **13**, 1428 (1976).
7. M. H. Cohen, *Phys. Rev.* **121**, 387 (1961).
8. B. Lax, J. G. Mavroides, H. J. Jdiger and R. J. Keys, *Phys. Rev. Lett.* **5**, 241 (1960).
9. J. O. Dimmock, *The Physics of Semimetals and Narrowgap Semiconductors*, by D. L. Carter, R. T. Bates (eds.) (Pergamon Press, Oxford, 1971).
10. M. Abramowitz and Iastegun, *Handbook of Mathematical Functions* (Dover Publication, New York, 1965).
11. P. R. Emtage, *Phys. Rev.* **138**, A246 (1965).
12. M. Cardona, W. Paul and H. brooks, *Helv. Acta. Phys.* **33**, 329 (1960).
13. C. C. Wang and N. W. Ressler, *Phys. Rev.* **2**, 1827 (1970).
14. U. Rossler, *Solid State Commun.* **49**, 943 (1984).
15. G. E. Stillman, C. M. Wolfe and J. C. Dinmockin, *Semiconductors and Semimetals*, R. K. Willardson and A. C. Beer (eds.) **12**, 169 (1977).
16. D. G. Seiler, W. M. Beeker and L. M. Roth, *Phys. Rev.* **1**, 764 (1970).
17. P. C. Mathur and S. Jain, *Phys. Rev.* **19**, 3159 (1979).
18. H. I. Zhang, *Phys. Rev.* **1B**, 3450 (1970).
19. E. D. Palik, G. S. Picus, S. Teither and R. F. Wallis, *Phys. Rev.* **122**, 475 (1961).
20. E. J. Johnson and D. H. Dicley, *Phys. Rev.* **1**, 2676 (1970).
21. V. G. Agafonov, P. M. Valov, B. S. Ryvkin and I. D. Yaroshetskii, *Sov. Phys. Semicond.* **12**, 1182 (1978).

Chapter 4

The QC in Quantized Transistors (QTs) Under Magnetic Quantization

I must try to convert the love for power to the power of love

4.1 Introduction

The importance of quantizing magnetic field in studying the physical properties of semiconducting materials has been discussed in the preface. In this chapter, from Sections 4.2.1–4.2.11, we shall study the QC in magneto MOSFETs of nonlinear optical, III–V, II–VI, IV–VI, stressed Kane type, Bi_2Te_3, Ge, GaAs, GaSb, InSb and Te, respectively. Besides, from Section 4.2.12–4.2.23, we shall further study in magneto QWFETs of nonlinear optical, III–V, II–VI, IV–VI, stressed Kane type, Bi_2Te_3, Ge, GaAs, GaSb, InSb, Te, $PtSb_2$ and GaP, respectively, by using various band models. Section 4.3 contains results and discussions. In Section 4.4, we present few open research problems.

4.2 Theoretical Background

4.2.1 *The QC in magneto MOSFETs of tetragonal and nonlinear optical materials*

(i) Weak electric field limit

Thus, the electron dispersion law in QMOSFETs operated under inversion mode and made of tetragonal and nonlinear optical materials under the weak electric field limit in the presence of a quantizing magnetic field B along z-direction can approximately be written following the area quantization rule of Landau [1] as

$$\psi_1(E_{41}) = P_7(E_{41}, i)\left(\frac{2eB}{\hbar}\right)\left(n + \frac{1}{2}\right) + Q_7(E_{41}, i) \quad (4.1)$$

where $n(= 0, 1, 2, \ldots)$ is the Landau quantum number, E_{41} is the totally quantized energy in this case and the other notations have already been defined in Chapter 1.

The density-of-states function assumes the form

$$N(E) = \left(\frac{g_v eB}{\pi\hbar}\right)\sum_{i=0}^{i_{\max}}\sum_{n=0}^{n_{\max}} \delta'(E - E_{41}) \quad (4.2)$$

where δ' is the Dirac's delta function.

Combining (4.2) with the Fermi–Dirac occupation probability factor, the surface electron concentration can be written as

$$N_s = \left(\frac{g_v eB}{\pi\hbar}\right)\sum_{i=0}^{i_{\max}}\sum_{n=0}^{n_{\max}} F_{-1}(\eta_1) \quad (4.3)$$

where,

$$\eta_1 = (k_B T)^{-1}[E_{F1} - E_{41}],\ E_{F1}\left(= eV_g - \frac{e^2 N_s}{C_{oX}} - \bar{E}_{fB}\right),$$

E_{F1} is the Fermi energy as measured from the edge of the conduction band at the surface in the presence of magnetic quantization and $\bar{E}_{fB}$ is the energy separation between the Fermi level and the conduction band edge in the bulk of the substrate material under present condition.

Combining (1.4) and (4.3) we can study the magneto QC in this case.

(ii) Strong electric field limit

The electron dispersion law in QMOSFETs operated under inversion mode and made of tetragonal and nonlinear optical materials under the strong electric field limit in the presence of a quantizing magnetic field B along z-direction can approximately be written following the area quantization rule of Landau [1] as

$$\left(\frac{2eB}{\hbar}\right)\left(n + \frac{1}{2}\right) = P_2(E_{42}, i) \quad (4.4)$$

where E_{42} is the totally quantized energy in this case.

The density-of-states function assumes the form

$$N(E) = \left(\frac{g_v eB}{\pi\hbar}\right) \sum_{i=0}^{i_{\max}} \sum_{n=0}^{n_{\max}} \delta'(E - E_{42}) \tag{4.5}$$

Combining (4.5) with the Fermi–Dirac occupation probability factor, the surface electron concentration can be written as

$$N_s = \left(\frac{g_v eB}{\pi\hbar}\right) \sum_{i=0}^{i_{\max}} \sum_{n=0}^{n_{\max}} F_{-1}(\eta_2) \tag{4.6}$$

where $\eta_2 = (k_B T)^{-1}[E_{F1} - E_{42}]$.

Combining (1.4) and (4.6) we can study the magneto QC in this case.

4.2.2 *The QC in magneto MOSFETs of III–V, ternary and quaternary materials*

(a) Weak electric field limit

The 2D magneto electron dispersion law in this case can be written following (1.24) as

$$I_{11}(E_{43}) = \left(n + \frac{1}{2}\right)\hbar\omega_0 + S_i\left[\frac{\hbar|e|F_s[I_{11}(E_{43})]'}{\sqrt{2m_c}}\right]^{2/3} \tag{4.7}$$

where $\omega_0 = \frac{eB}{m_c}$ and E_{43} is the totally quantized energy in this case.

The (4.7) represents the magneto dispersion relation of the 2D electrons in QMOSFETs of III–V, ternary and quaternary materials under the weak electric field limit in the presence of magnetic quantization whose bulk electrons obey the **three-band model of Kane**.

The density-of-states function assumes the form

$$N(E) = \left(\frac{g_v eB}{\pi\hbar}\right) \sum_{i=0}^{i_{\max}} \sum_{n=0}^{n_{\max}} \delta'(E - E_{43}) \tag{4.8}$$

Combining (4.8) with the Fermi–Dirac occupation probability factor, the surface electron concentration can be written as

$$N_s = \left(\frac{g_v eB}{\pi\hbar}\right) \sum_{i=0}^{i_{\max}} \sum_{n=0}^{n_{\max}} F_{-1}(\eta_3) \tag{4.9}$$

where, $\eta_3 = (k_B T)^{-1}[E_{F1} - E_{43}]$.

Combining (1.4) and (4.9) we can study the magneto QC in this case.

Using the constraints $\Delta \gg E_g$ or $\Delta \ll E_g$, from the (1.33b) the 2D magneto dispersion law can be written as

$$E_{44}(1+\alpha E_{44}) = \left(n+\frac{1}{2}\right)\hbar\omega_0 + S_i\left[\frac{\hbar|e|F_s(1+2\alpha E_{44})}{\sqrt{2m_c}}\right]^{2/3} \tag{4.10}$$

where E_{44} is the totally quantized energy in this case.

It may be noted that (4.10) is valid for the well-known **two-band model of Kane**.

The density-of-states function assumes the form

$$N(E) = \left(\frac{g_v eB}{\pi\hbar}\right)\sum_{i=0}^{i_{\max}}\sum_{n=0}^{n_{\max}}\delta'(E-E_{44}) \tag{4.11}$$

Combining (4.11) with the Fermi–Dirac occupation probability factor, the surface electron concentration can be written as

$$N_s = \left(\frac{g_v eB}{\pi\hbar}\right)\sum_{i=0}^{i_{\max}}\sum_{n=0}^{n_{\max}}F_{-1}(\bar{\eta}_4) \tag{4.12}$$

where $\bar{\eta}_4 = (k_BT)^{-1}[E_{F1}-E_{44}]$.

Combining (1.4) and (4.12) we can study the magneto QC in this case.

For $\alpha \to 0$, as for inversion layers, whose bulk electrons are defined by the parabolic energy bands, the magneto 2D dispersion relation can be expressed from (1.46) as

$$E_{45} = \left(n+\frac{1}{2}\right)\hbar\omega_0 + S_i\left[\frac{\hbar|e|F_s}{\sqrt{2m_c}}\right]^{2/3} \tag{4.13}$$

where E_{45} is the totally quantized energy in this case.

Equation (4.13) is valid for all values of the surface electric field [1].

The density-of-states function assumes the form

$$N(E) = \left(\frac{g_v eB}{\pi\hbar}\right)\sum_{i=0}^{i_{\max}}\sum_{n=0}^{n_{\max}}\delta'(E-E_{45}) \tag{4.14}$$

Combining (4.14) with the Fermi–Dirac occupation probability factor, the surface electron concentration can be written as

$$N_s = \left(\frac{g_v eB}{\pi\hbar}\right)\sum_{i=0}^{i_{\max}}\sum_{n=0}^{n_{\max}}F_{-1}(\bar{\eta}_4) \tag{4.15}$$

where $\bar{\eta}_4 = (k_BT)^{-1}[E_{F1}-E_{45}]$.

Combining (1.4) and (4.15) we can study the magneto QC in this case.

(b) Strong electric field limit

(i) The three-band Kane model

Using (1.63), the 2D magneto dispersion relation in this case can be written as

$$\left[I_{11}(E_{47}) - \left\{\frac{|e|F_s\hbar}{\sqrt{2m_c}}\left(\frac{2\sqrt{2}(S_i)^{3/2}}{3}\right)\sqrt{[I_{11}(E_{47})]''}\right\}\right] = \left(n+\frac{1}{2}\right)\hbar\omega_0 \tag{4.16}$$

where E_{47} is the totally quantized energy in this case.

The density-of-states function assumes the form

$$N(E) = \left(\frac{g_v eB}{\pi\hbar}\right)\sum_{i=0}^{i_{\max}}\sum_{n=0}^{n_{\max}}\delta'(E-E_{47}) \tag{4.17}$$

Combining (4.17) with the Fermi–Dirac occupation probability factor, the surface electron concentration can be written as

$$N_s = \left(\frac{g_v eB}{\pi\hbar}\right)\sum_{i=0}^{i_{\max}}\sum_{n=0}^{n_{\max}}F_{-1}(\eta_6) \tag{4.18}$$

where $\eta_6 = (k_BT)^{-1}[E_{F1} - E_{47}]$.

Combining (1.4) and (4.18) we can study the magneto QC in this case.

(ii) Two-band model of Kane

Using (1.72), the 2D magneto dispersion relation in this case can be written as

$$E_{48}(1+\alpha E_{48}) = \left(n+\frac{1}{2}\right)\hbar\omega_0 + \frac{2\sqrt{2}}{3}(S_i)^{3/2}\left[\frac{\hbar|e|F_s}{\sqrt{2m_cE_g}}\right] \tag{4.19}$$

where E_{48} is the totally quantized energy in this case.

The density-of-states function assumes the form

$$N(E) = \left(\frac{g_v eB}{\pi\hbar}\right)\sum_{i=0}^{i_{\max}}\sum_{n=0}^{n_{\max}}\delta'(E-E_{48}) \tag{4.20}$$

Combining (4.20) with the Fermi–Dirac occupation probability factor, the surface electron concentration can be written as

$$N_s = \left(\frac{g_v eB}{\pi\hbar}\right)\sum_{i=0}^{i_{\max}}\sum_{n=0}^{n_{\max}}F_{-1}(\eta_7) \tag{4.21}$$

where $\eta_7 = (k_BT)^{-1}[E_{F1} - E_{48}]$.

Combining (1.4) and (4.21) we can study the magneto QC in this case.

4.2.3 *The QC in magneto MOSFETs of II–VI materials*

The magneto dispersion relation in this case can be written using (2.3a) as

$$\left(n+\frac{1}{2}\right)\left(\frac{\hbar eB}{m_{\perp}^{*}}\right)\pm C_0\left(\left(\frac{2eB}{\hbar}\right)\left(n+\frac{1}{2}\right)\right)^{1/2}+S_i\left(\frac{\hbar eF_S}{\sqrt{2m_{\|}^{*}}}\right)^{2/3}-E_{49}=0 \tag{4.22}$$

where E_{49} is the totally quantized energy in this case.

The density-of-states function assumes the form

$$N(E)=\left(\frac{g_v eB}{2\pi\hbar}\right)\sum_{i=0}^{i_{\max}}\sum_{n=0}^{n_{\max}}\delta'(E-E_{49}) \tag{4.23}$$

Combining (4.23) with the Fermi–Dirac occupation probability factor, the surface electron concentration can be written as

$$N_s=\left(\frac{g_v eB}{2\pi\hbar}\right)\sum_{i=0}^{i_{\max}}\sum_{n=0}^{n_{\max}}F_{-1}(\eta_8) \tag{4.24}$$

where $\eta_8=(k_BT)^{-1}[E_{F1}-E_{49}]$.

Combining (1.4) and (4.24) we can study the magneto QC in this case.

4.2.4 *The QC in magneto MOSFETs of IV–VI materials*

(a) Weak electric field limit

The 2D magneto dispersion relation in this case can be expressed following (2.10) as

$$\begin{aligned}\left(\frac{2\pi eB}{\hbar}\right)\left(n+\frac{1}{2}\right)=\frac{4}{3}\left(\frac{r_{71}}{p_{71}}\right)^{1/2}&[\{a_{71w}(E_{50},i)\}^2+\{b_{71w}(E_{50},i)\}^2]^{1/2}\\&\times\left[\{a_{71w}(E_{50},i)\}^2F\left[\frac{\pi}{2},\ell_{71w}(E_{50},i)\right]\right.\\&-(\{a_{71w}(E_{50},i)\}^2+\{b_{71w}(E_{50},i)\}^2)\\&\left.\times\ E\left[\frac{\pi}{2},\ell_{71w}(E_{50},i)\right]\right]\end{aligned} \tag{4.25}$$

where E_{50} is the totally quantized energy in this case.

The density-of-states function assumes the form

$$N(E)=\left(\frac{g_v eB}{\pi\hbar}\right)\sum_{i=0}^{i_{\max}}\sum_{n=0}^{n_{\max}}\delta'(E-E_{50}) \tag{4.26}$$

Combining (4.26) with the Fermi–Dirac occupation probability factor, the surface electron concentration can be written as

$$N_s = \left(\frac{g_v eB}{\pi\hbar}\right)\sum_{i=0}^{i_{\max}}\sum_{n=0}^{n_{\max}} F_{-1}(\eta_9) \tag{4.27}$$

where $\eta_9 = (k_B T)^{-1}[E_{F1} - E_{50}]$.

Combining (1.4) and (4.27) we can study the magneto QC in this case.

(b) Strong electric field limit

The 2D magneto dispersion relation in this case can be written as

$$\begin{aligned}\left(\frac{2\pi eB}{\hbar}\right)\left(n+\frac{1}{2}\right) = \frac{4}{3}\left(\frac{r_{72}}{p_{72}}\right)^{1/2} & [\{a_{72s}(E_{51},i)\}^2 + \{b_{72s}(E_{51},i)\}^2]^{1/2} \\ & \times \left[\{a_{72s}(E_{51},i)\}^2 F\left[\frac{\pi}{2}, \ell_{72s}(E_{51},i)\right]\right. \\ & - (\{a_{72s}(E_{51},i)\}^2 - \{b_{72s}(E_{51},i)\}^2) \\ & \left.\times\; E\left[\frac{\pi}{2}, \ell_{72s}(E_{51},i)\right]\right]\end{aligned} \tag{4.28}$$

in which,

$$\{a_{72s}(E_{51},i)\}^2 \equiv \left[\frac{q_{72}(E_{51},i)}{2r_{72}} + \frac{1}{2}\left[\frac{\{q_{72}(E_{51},i)\}^2}{(r_{72})^2} + \frac{4\gamma_{72}(E_{51},i)}{r_{72}}\right]^{1/2}\right]$$

$$\{b_{72s}(E_{51},i)\}^2 \equiv \left[\frac{1}{2}\left[\frac{\{q_{72}(E_{51},i)\}^2}{(r_{72})^2} + \frac{4\gamma_{72}(E_{51},i)}{r_{72}}\right]^{1/2} - \left(\frac{q_{72}(E_{51},i)}{2r_{72}}\right)\right]$$

$$\ell_{72s}(E_{51},i) \equiv \frac{b_{72s}(E_{51},i)}{\sqrt{\{a_{72s}(E_{51},i)\}^2 + \{b_{72s}(E_{51},i)\}^2}}, \quad F\left[\frac{\pi}{2}, \ell_{72s}q_{72}(E_{51},i)\right]$$

$$\text{and} \quad E\left[\frac{\pi}{2}, \ell_{72s}q_{72}(E_{51},i)\right]$$

are the complete elliptic integral of the first and second kinds respectively and E_{51} is the totally quantized energy in this case.

The density-of-states function assumes the form

$$N(E) = \left(\frac{g_v eB}{\pi\hbar}\right)\sum_{i=0}^{i_{\max}}\sum_{n=0}^{n_{\max}} \delta'(E - E_{51}) \tag{4.29}$$

Combining (4.29) with the Fermi–Dirac occupation probability factor, the surface electron concentration can be written as

$$N_s = \left(\frac{g_v eB}{\pi\hbar}\right)\sum_{i=0}^{i_{\max}}\sum_{n=0}^{n_{\max}} F_{-1}(\eta_{10}) \tag{4.30}$$

where $\eta_{10} = (k_B T)^{-1}[E_{F1} - E_{51}]$.

Combining (1.4) and (4.30) we can study the magneto QC in this case.

4.2.5 *The QC in magneto MOSFETs of stressed Kane type materials*

(a) Weak electric field limit

The 2D magneto dispersion relation in this case can be written following (2.24) as

$$\left(\frac{2eB}{\hbar}\right)\left(n+\frac{1}{2}\right) = \frac{T_{77}(E_{52}, i)}{\sqrt{T_{57}(E_{52}, i)T_{67}(E_{52}, i)}} \tag{4.31}$$

where E_{52} is the totally quantized energy in this case.

The density-of-states function assumes the form

$$N(E) = \left(\frac{g_v eB}{\pi\hbar}\right)\sum_{i=0}^{i_{\max}}\sum_{n=0}^{n_{\max}} \delta'(E - E_{52}) \tag{4.32}$$

Combining (4.32) with the Fermi–Dirac occupation probability factor, the surface electron concentration can be written as

$$N_s = \left(\frac{g_v eB}{\pi\hbar}\right)\sum_{i=0}^{i_{\max}}\sum_{n=0}^{n_{\max}} F_{-1}(\eta_{11}) \tag{4.33}$$

where $\eta_{11} = (k_B T)^{-1}[E_{F1} - E_{52}]$.

Combining (1.4) and (4.33) we can study the magneto QC in this case.

(b) Strong electric field limit

The 2D magneto dispersion relation in this case can be written following (2.32) as

$$\left(\frac{2eB}{\hbar}\right)\left(n+\frac{1}{2}\right) = \frac{T_{137}(E_{53}, i)}{\sqrt{T_{117}(E_{53}, i)T_{127}(E_{53}, i)}} \tag{4.34}$$

where E_{53} is the totally quantized energy in this case.

The density-of-states function assumes the form

$$N(E) = \left(\frac{g_v eB}{\pi\hbar}\right) \sum_{i=0}^{i_{\max}} \sum_{n=0}^{n_{\max}} \delta'(E - E_{53}) \tag{4.35}$$

Combining (4.35) with the Fermi–Dirac occupation probability factor, the surface electron concentration can be written as

$$N_s = \left(\frac{g_v eB}{\pi\hbar}\right) \sum_{i=0}^{i_{\max}} \sum_{n=0}^{n_{\max}} F_{-1}(\eta_{12}) \tag{4.36}$$

where $\eta_{12} = (k_B T)^{-1}[E_{F1} - E_{53}]$.

Combining (1.4) and (4.36) we can study the magneto QC in this case.

4.2.6 *The QC in magneto MOSFETs of bismuth telluride*

(a) Weak electric field limit

The 2D magneto dispersion relation in this case can be written following (2.41) as

$$\left(\frac{2eB}{\hbar}\right)\left(n + \frac{1}{2}\right) = (I_2 I_3 - I_4^2)^{-1/2} T_{46}(E_{54}, i) \tag{4.37}$$

where E_{54} is the totally quantized energy in this case.

The density-of-states function assumes the form

$$N(E) = \left(\frac{g_v eB}{\pi\hbar}\right) \sum_{i=0}^{i_{\max}} \sum_{n=0}^{n_{\max}} \delta'(E - E_{54}) \tag{4.38}$$

Combining (4.38) with the Fermi–Dirac occupation probability factor, the surface electron concentration can be written as

$$N_s = \left(\frac{g_v eB}{\pi\hbar}\right) \sum_{i=0}^{i_{\max}} \sum_{n=0}^{n_{\max}} F_{-1}(\eta_{13}) \tag{4.39}$$

where $\eta_{13} = (k_B T)^{-1}[E_{F1} - E_{54}]$.

Combining (1.4) and (4.39) we can study the magneto QC in this case.

(b) Strong electric field limit

The 2D magneto dispersion relation in this case can be written following (2.44) as

$$\left(\frac{2eB}{\hbar}\right)\left(n + \frac{1}{2}\right) = (I_2 I_3 - I_4^2)^{-1/2} T_{47}(E_{55}, i) \tag{4.40}$$

where E_{55} is the totally quantized energy in this case.

The density-of-states function assumes the form

$$N(E) = \left(\frac{g_v eB}{\pi\hbar}\right) \sum_{i=0}^{i_{\max}} \sum_{n=0}^{n_{\max}} \delta'(E - E_{55}) \tag{4.41}$$

Combining (4.41) with the Fermi–Dirac occupation probability factor, the surface electron concentration can be written as

$$N_s = \left(\frac{g_v eB}{\pi\hbar}\right) \sum_{i=0}^{i_{\max}} \sum_{n=0}^{n_{\max}} F_{-1}(\eta_{14}) \tag{4.42}$$

where $\eta_{14} = (k_B T)^{-1}[E_{F1} - E_{55}]$.

Combining (1.4) and (4.42) we can study the magneto QC in this case.

4.2.7 *The QC in magneto MOSFETs of germanium*

(a) Model of Cardona *et al.* [2]

The 2D magneto dispersion relation in this case can be written following (2.49) as

$$\left(\frac{eB}{\hbar}\right)\left(n + \frac{1}{2}\right) = \frac{\sqrt{m_1 m_2}}{\hbar^2}[E_{57}(1 + \alpha E_{57}) + \alpha E_{i2}^2 - E_{i2}(1 + 2\alpha E_{57})] \tag{4.43}$$

where E_{56} is the totally quantized energy in this case.

The density-of-states function assumes the form

$$N(E) = \left(\frac{g_v eB}{\pi\hbar}\right) \sum_{i=0}^{i_{\max}} \sum_{n=0}^{n_{\max}} \delta'(E - E_{57}) \tag{4.44}$$

Combining (4.44) with the Fermi–Dirac occupation probability factor, the surface electron concentration can be written as

$$N_s = \left(\frac{g_v eB}{\pi\hbar}\right) \sum_{i=0}^{i_{\max}} \sum_{n=0}^{n_{\max}} F_{-1}(\eta_{15}) \tag{4.45}$$

where $\eta_{15} = (k_B T)^{-1}[E_{F1} - E_{57}]$.

Combining (1.4) and (4.45) we can study the magneto QC in this case.

(b) Model of Wang *et al.* [3]

The 2D magneto dispersion relation in this case can be written following (2.54) as

$$
\begin{aligned}
&\left[\left[1 - d_1\left(\left(n+\frac{1}{2}\right)\hbar\omega_{01}\right)\right]3^{-1}\right] \times \left[\left(1 - d_1\left(\left(n+\frac{1}{2}\right)\hbar\omega_{01}\right)\right)\right.\\
&\quad \times \left\{\left(1 - d_1\left(\left(n+\frac{1}{2}\right)\hbar\omega_{01}\right)\right)^2 - 4e_1\left(E_{59} - \left(\left(n+\frac{1}{2}\right)\hbar\omega_{01}\right)\right)\right.\\
&\quad \left.\left. + 4\left(\left(n+\frac{1}{2}\right)\hbar\omega_{01}\right) + 4\left(\left(n+\frac{1}{2}\right)\hbar\omega_{01}\right)^2\right\}^{1/2}\right]^{3/2}\\
&\quad - \frac{1}{5}\left[\left(1 - d_1\left(\left(n+\frac{1}{2}\right)\hbar\omega_{01}\right)\right) - \left[\left(1 - d_1\left(\left(n+\frac{1}{2}\right)\hbar\omega_{01}\right)\right)^2\right.\right.\\
&\quad \left.\left. - 4e_1\left(E_{59} - \left(\left(n+\frac{1}{2}\right)\hbar\omega_{01}\right)\right) + C_1\left(\left(n+\frac{1}{2}\right)\hbar\omega_{01}\right)^2\right]^{1/2}\right]^{5/2}\\
&= \frac{2}{3}\hbar e F_S (e_1 S_i)^{3/2}(m_3)^{-1/2}
\end{aligned}
\tag{4.46}
$$

where E_{59} is the totally quantized energy in this case.

The density-of-states function assumes the form

$$N(E) = \left(\frac{g_v eB}{\pi\hbar}\right)\sum_{i=0}^{i_{\max}}\sum_{n=0}^{n_{\max}}\delta'(E - E_{59}) \tag{4.47}$$

Combining (4.47) with the Fermi–Dirac occupation probability factor, the surface electron concentration can be written as

$$N_s = \left(\frac{g_v eB}{\pi\hbar}\right)\sum_{i=0}^{i_{\max}}\sum_{n=0}^{n_{\max}} F_{-1}(\eta_{16}) \tag{4.48}$$

where $\eta_{16} = (k_B T)^{-1}[E_{F1} - E_{59}]$.

Combining (1.4) and (4.48) we can study the magneto QC in this case.

4.2.8 *The QC in magneto MOSFETs of gallium arsenide*

(i) Rossler's Model [4]

(a) Weak electric field limit

The 2D magneto dispersion relation in this case can be written following (2.58) as

$$\gamma_{4\pm}(E_{60}) = \left(\frac{2eB}{\hbar}\right)\left(n+\frac{1}{2}\right) + S_i[eF_s\gamma'_{4,\pm}(E_{60})]^{2/3} \tag{4.49}$$

where E_{60} is the totally quantized energy in this case.

The density-of-states function assumes the form

$$N(E) = \left(\frac{g_v eB}{2\pi\hbar}\right)\sum_{i=0}^{i_{\max}}\sum_{n=0}^{n_{\max}}\delta'(E-E_{60}) \tag{4.50}$$

Combining (4.50) with the Fermi–Dirac occupation probability factor, the surface electron concentration can be written as

$$N_s = \left(\frac{g_v eB}{2\pi\hbar}\right)\sum_{i=0}^{i_{\max}}\sum_{n=0}^{n_{\max}}F_{-1}(\eta_{17}) \tag{4.51}$$

where $\eta_{17} = (k_BT)^{-1}[E_{F1} - E_{60}]$.

Combining (1.4) and (4.51) we can study the magneto QC in this case.

(b) Strong electric field limit

The 2D magneto dispersion relation in this case can be written following (2.61) as

$$\gamma_{4\pm}(E_{61}) = \left(\frac{2eB}{\hbar}\right)\left(n+\frac{1}{2}\right) + \frac{2\sqrt{2}eF_s}{3}S_i^{3/2}[\gamma'_{4,\pm}(E_{61})]^{1/2} \tag{4.52}$$

where E_{61} is the totally quantized energy in this case.

The density-of-states function assumes the form

$$N(E) = \left(\frac{g_v eB}{2\pi\hbar}\right)\sum_{i=0}^{i_{\max}}\sum_{n=0}^{n_{\max}}\delta'(E-E_{61}) \tag{4.53}$$

Combining (4.53) with the Fermi–Dirac occupation probability factor, the surface electron concentration can be written as

$$N_s = \left(\frac{g_v eB}{2\pi\hbar}\right)\sum_{i=0}^{i_{\max}}\sum_{n=0}^{n_{\max}}F_{-1}(\eta_{18}) \tag{4.54}$$

where $\eta_{18} = (k_BT)^{-1}[E_{F1} - E_{61}]$.

Combining (1.4) and (4.54) we can study the magneto QC in this case.

(ii) Model of Stillman *et al.* [5]

(a) Weak electric field limit

The 2D magneto dispersion relation in this case can be written following (2.66) as

$$\gamma_5(E_{62}) = \left(\frac{2eB}{\hbar}\right)\left(n + \frac{1}{2}\right) + S_i[eF_s\gamma_5'(E_{62})]^{2/3} \tag{4.55}$$

where E_{62} is the totally quantized energy in this case.

The density-of-states function assumes the form

$$N(E) = \left(\frac{g_v eB}{\pi\hbar}\right)\sum_{i=0}^{i_{\max}}\sum_{n=0}^{n_{\max}}\delta'(E - E_{62}) \tag{4.56}$$

Combining (4.56) with the Fermi–Dirac occupation probability factor, the surface electron concentration can be written as

$$N_s = \left(\frac{g_v eB}{\pi\hbar}\right)\sum_{i=0}^{i_{\max}}\sum_{n=0}^{n_{\max}}F_{-1}(\eta_{19}) \tag{4.57}$$

where $\eta_{19} = (k_BT)^{-1}[E_{F1} - E_{62}]$.

Combining (1.4) and (4.57) we can study the magneto QC in this case.

(b) Strong electric field limit

The 2D magneto dispersion relation in this case can be written following (2.69) as

$$\gamma_5(E_{63}) = \left(\frac{2eB}{\hbar}\right)\left(n + \frac{1}{2}\right) + \frac{2\sqrt{2}eF_s}{3}S_i^{3/2}[\gamma_5'(E_{63})]^{1/2} \tag{4.58}$$

where E_{63} is the totally quantized energy in this case.

The density-of-states function assumes the form

$$N(E) = \left(\frac{g_v eB}{\pi\hbar}\right)\sum_{i=0}^{i_{\max}}\sum_{n=0}^{n_{\max}}\delta'(E - E_{63}) \tag{4.59}$$

Combining (4.59) with the Fermi–Dirac occupation probability factor, the surface electron concentration can be written as

$$N_s = \left(\frac{g_v eB}{\pi\hbar}\right)\sum_{i=0}^{i_{\max}}\sum_{n=0}^{n_{\max}}F_{-1}(\eta_{20}) \tag{4.60}$$

where $\eta_{20} = (k_BT)^{-1}[E_{F1} - E_{63}]$.

Combining (1.4) and (4.60) we can study the magneto QC in this case.

4.2.9 The QC in magneto MOSFETs of gallium antimonide

(i) The model of Sailer *et al.* [6]

(a) Weak electric filed limit

The 2D magneto dispersion relation in this case can be written following (2.78) a

$$\left(\frac{2eB}{\hbar}\right)\left(n+\frac{1}{2}\right)=\gamma_6(E_{64})-S_i[eF_s\gamma_6'(E_{64})]^{2/3} \tag{4.61}$$

where E_{64} is the totally quantized energy in this case.

The density-of-states function assumes the form

$$N(E)=\left(\frac{g_veB}{2\pi\hbar}\right)\sum_{i=0}^{i_{\max}}\sum_{n=0}^{n_{\max}}\delta'(E-E_{64}) \tag{4.62}$$

Combining (4.62) with the Fermi–Dirac occupation probability factor, the surface electron concentration can be written as

$$N_s=\left(\frac{g_veB}{2\pi\hbar}\right)\sum_{i=0}^{i_{\max}}\sum_{n=0}^{n_{\max}}F_{-1}(\eta_{21}) \tag{4.63}$$

where $\eta_{21}=(k_BT)^{-1}[E_{F1}-E_{64}]$.

Combining (1.4) and (4.63) we can study the magneto QC in this case.

(b) Strong electric field limit

The 2D magneto dispersion relation in this case can be written following (2.81) as

$$\left(\frac{2eB}{\hbar}\right)\left(n+\frac{1}{2}\right)=\gamma_6(E_{65})-\frac{2\sqrt{2}}{3}eF_sS_i^{3/2}[\gamma''_6(E_{65})]^{1/2} \tag{4.64}$$

where E_{65} is the totally quantized energy in this case.

The density-of-states function assumes the form

$$N(E)=\left(\frac{g_veB}{2\pi\hbar}\right)\sum_{i=0}^{i_{\max}}\sum_{n=0}^{n_{\max}}\delta'(E-E_{65}) \tag{4.65}$$

Combining (4.65) with the Fermi–Dirac occupation probability factor, the surface electron concentration can be written as

$$N_s=\left(\frac{g_veB}{2\pi\hbar}\right)\sum_{i=0}^{i_{\max}}\sum_{n=0}^{n_{\max}}F_{-1}(\eta_{22}) \tag{4.66}$$

where $\eta_{22}=(k_BT)^{-1}[E_{F1}-E_{65}]$.

Combining (1.4) and (4.66) we can study the magneto QC in this case.

(ii) The model of Mathur *et al.* [7]

(a) Weak electric field limit

The 2D magneto dispersion relation in this case can be written following (2.87) as

$$\left(\frac{2eB}{\hbar}\right)\left(n+\frac{1}{2}\right) = \gamma_7(E_{66}) - S_i[eF_s\gamma_7'(E_{66})]^{2/3} \tag{4.67}$$

where E_{66} is the totally quantized energy in this case.

The density-of-states function assumes the form

$$N(E) = \left(\frac{g_v eB}{\pi\hbar}\right)\sum_{i=0}^{i_{\max}}\sum_{n=0}^{n_{\max}} \delta'(E - E_{66}) \tag{4.68}$$

Combining (4.68) with the Fermi–Dirac occupation probability factor, the surface electron concentration can be written as

$$N_s = \left(\frac{g_v eB}{\pi\hbar}\right)\sum_{i=0}^{i_{\max}}\sum_{n=0}^{n_{\max}} F_{-1}(\eta_{23}) \tag{4.69}$$

where $\eta_{23} = (k_B T)^{-1}[E_{F1} - E_{66}]$.

Combining (1.4) and (4.69) we can study the magneto QC in this case.

(b) Strong electric field limit

The 2D magneto dispersion relation in this case can be written following (2.90) as

$$\left(\frac{2eB}{\hbar}\right)\left(n+\frac{1}{2}\right) = \gamma_7(E_{67}) - \frac{2\sqrt{2}}{3}eF_s S_i^{3/2}[\gamma''{}_7(E_{67})]^{1/2} \tag{4.70}$$

where E_{67} is the totally quantized energy in this case.

The density-of-states function assumes the form

$$N(E) = \left(\frac{g_v eB}{\pi\hbar}\right)\sum_{i=0}^{i_{\max}}\sum_{n=0}^{n_{\max}} \delta'(E - E_{67}) \tag{4.71}$$

Combining (4.71) with the Fermi–Dirac occupation probability factor, the surface electron concentration can be written as

$$N_s = \left(\frac{g_v eB}{\pi\hbar}\right)\sum_{i=0}^{i_{\max}}\sum_{n=0}^{n_{\max}} F_{-1}(\eta_{24}) \tag{4.72}$$

where $\eta_{24} = (k_B T)^{-1}[E_{F1} - E_{67}]$.

Combining (1.4) and (4.72) we can study the magneto QC in this case.

(iii) The model of Zhang *et al.* [8]

(a) Weak electric field limit

The 2D magneto dispersion relation in this case can be written following (2.96) as

$$\left(\frac{2eB}{\hbar}\right)\left(n+\frac{1}{2}\right)=\gamma_8(E_{68})-S_i[eF_s\gamma_8'(E_{68})]^{2/3} \tag{4.73}$$

where E_{68} is the totally quantized energy in this case.

The density-of-states function assumes the form

$$N(E)=\left(\frac{g_v eB}{\pi\hbar}\right)\sum_{i=0}^{i_{\max}}\sum_{n=0}^{n_{\max}}\delta'(E-E_{68}) \tag{4.74}$$

Combining (4.74) with the Fermi–Dirac occupation probability factor, the surface electron concentration can be written as

$$N_s=\left(\frac{g_v eB}{\pi\hbar}\right)\sum_{i=0}^{i_{\max}}\sum_{n=0}^{n_{\max}}F_{-1}(\eta_{25}) \tag{4.75}$$

where $\eta_{25}=(k_BT)^{-1}[E_{F1}-E_{68}]$.

Combining (1.4) and (4.75) we can study the magneto QC in this case.

(b) Strong electric field limit

The 2D magneto dispersion relation in this case can be written following (2.99) as

$$\left(\frac{2eB}{\hbar}\right)\left(n+\frac{1}{2}\right)=\gamma_8(E_{69})-\frac{2\sqrt{2}}{3}eF_sS_i^{3/2}[\gamma''_8(E_{69})]^{1/2} \tag{4.76}$$

where E_{69} is the totally quantized energy in this case.

The density-of-states function assumes the form

$$N(E)=\left(\frac{g_v eB}{\pi\hbar}\right)\sum_{i=0}^{i_{\max}}\sum_{n=0}^{n_{\max}}\delta'(E-E_{69}) \tag{4.77}$$

Combining (4.77) with the Fermi–Dirac occupation probability factor, the surface electron concentration can be written as

$$N_s=\left(\frac{g_v eB}{\pi\hbar}\right)\sum_{i=0}^{i_{\max}}\sum_{n=0}^{n_{\max}}F_{-1}(\eta_{26}) \tag{4.78}$$

where $\eta_{26}=(k_BT)^{-1}[E_{F1}-E_{69}]$.

Combining (1.4) and (4.78) we can study the magneto QC in this case.

4.2.10 *The QC in magneto MOSFETs of indium antimonide*

(i) The model of Palik *et al.* [9]

(a) Weak electric field limit

The 2D magneto dispersion relation in this case can be written following (2.105) as

$$\left(\frac{2eB}{\hbar}\right)\left(n+\frac{1}{2}\right)=\gamma_9(E_{70})-S_i[eF_s\gamma_9'(E_{70})]^{2/3} \tag{4.79}$$

where E_{70} is the totally quantized energy in this case.

The density-of-states function assumes the form

$$N(E)=\left(\frac{g_v eB}{\pi\hbar}\right)\sum_{i=0}^{i_{\max}}\sum_{n=0}^{n_{\max}}\delta'(E-E_{70}) \tag{4.80}$$

Combining (4.80) with the Fermi–Dirac occupation probability factor, the surface electron concentration can be written as

$$N_s=\left(\frac{g_v eB}{\pi\hbar}\right)\sum_{i=0}^{i_{\max}}\sum_{n=0}^{n_{\max}}F_{-1}(\eta_{27}) \tag{4.81}$$

where $\eta_{27}=(k_BT)^{-1}[E_{F1}-E_{70}]$.

Combining (1.4) and (4.81) we can study the magneto QC in this case.

(b) Strong electric field limit

The 2D magneto dispersion relation in this case can be written following (2.109) as

$$\left(\frac{2eB}{\hbar}\right)\left(n+\frac{1}{2}\right)=\gamma_9(E_{71})-\frac{2\sqrt{2}}{3}eF_sS_i^{3/2}[\gamma''_9(E_{71})]^{1/2} \tag{4.82}$$

where E_{71} is the totally quantized energy in this case.

The density-of-states function assumes the form

$$N(E)=\left(\frac{g_v eB}{\pi\hbar}\right)\sum_{i=0}^{i_{\max}}\sum_{n=0}^{n_{\max}}\delta'(E-E_{71}) \tag{4.83}$$

Combining (4.83) with the Fermi–Dirac occupation probability factor, the surface electron concentration can be written as

$$N_s=\left(\frac{g_v eB}{\pi\hbar}\right)\sum_{i=0}^{i_{\max}}\sum_{n=0}^{n_{\max}}F_{-1}(\eta_{28}) \tag{4.84}$$

where $\eta_{28}=(k_BT)^{-1}[E_{F1}-E_{71}]$.

Combining (1.4) and (4.84) we can study the magneto QC in this case.

(ii) The model of Johnson *et al.* [10]

(a) Weak electric field limit

The 2D magneto dispersion relation in this case can be written following (2.115) as

$$\left(\frac{2eB}{\hbar}\right)\left(n+\frac{1}{2}\right)=\gamma_{10}(E_{72})-S_i[eF_s\gamma'_{10}(E_{72})]^{2/3} \tag{4.85}$$

where E_{72} is the totally quantized energy in this case.

The density-of-states function assumes the form

$$N(E)=\left(\frac{g_veB}{\pi\hbar}\right)\sum_{i=0}^{i_{\max}}\sum_{n=0}^{n_{\max}}\delta'(E-E_{72}) \tag{4.86}$$

Combining (4.86) with the Fermi–Dirac occupation probability factor, the surface electron concentration can be written as

$$N_s=\left(\frac{g_veB}{\pi\hbar}\right)\sum_{i=0}^{i_{\max}}\sum_{n=0}^{n_{\max}}F_{-1}(\eta_{29}) \tag{4.87}$$

where $\eta_{29}=(k_BT)^{-1}[E_{F1}-E_{72}]$.

Combining (1.4) and (4.87) we can study the magneto QC in this case.

(b) Strong electric field limit

The 2D magneto dispersion relation in this case can be written following (2.115) as

$$\left(\frac{2eB}{\hbar}\right)\left(n+\frac{1}{2}\right)=\gamma_{10}(E_{73})-\frac{2\sqrt{2}}{3}eF_sS_i^{3/2}[\gamma''_{10}(E_{73})]^{1/2} \tag{4.88}$$

where E_{73} is the totally quantized energy in this case.

The density-of-states function assumes the form

$$N(E)=\left(\frac{g_veB}{\pi\hbar}\right)\sum_{i=0}^{i_{\max}}\sum_{n=0}^{n_{\max}}\delta'(E-E_{73}) \tag{4.89}$$

Combining (4.89) with the Fermi–Dirac occupation probability factor, the surface electron concentration can be written as

$$N_s=\left(\frac{g_veB}{\pi\hbar}\right)\sum_{i=0}^{i_{\max}}\sum_{n=0}^{n_{\max}}F_{-1}(\eta_{30}) \tag{4.90}$$

where $\eta_{30}=(k_BT)^{-1}[E_{F1}-E_{73}]$.

Combining (1.4) and (4.90) we can study the magneto QC in this case.

(iii) The model of Agafonov *et al.* [11]

(a) Weak electric field limit

The 2D magneto dispersion relation in this case can be written following (2.125) as

$$\left(\frac{2eB}{\hbar}\right)\left(n+\frac{1}{2}\right) = \gamma_{11}(E_{74}) - S_i[eF_s\gamma'_{11}(E_{74})]^{2/3} \tag{4.91}$$

where E_{74} is the totally quantized energy in this case.

The density-of-states function assumes the form

$$N(E) = \left(\frac{g_v eB}{\pi\hbar}\right)\sum_{i=0}^{i_{\max}}\sum_{n=0}^{n_{\max}}\delta'(E-E_{74}) \tag{4.92}$$

Combining (4.92) with the Fermi–Dirac occupation probability factor, the surface electron concentration can be written as

$$N_s = \left(\frac{g_v eB}{\pi\hbar}\right)\sum_{i=0}^{i_{\max}}\sum_{n=0}^{n_{\max}}F_{-1}(\eta_{31}) \tag{4.93}$$

where $\eta_{31} = (k_B T)^{-1}[E_{F1} - E_{74}]$.

Combining (1.4) and (4.93) we can study the magneto QC in this case.

(b) Strong electric field limit

The 2D magneto dispersion relation in this case can be written following (2.129) as

$$\left(\frac{2eB}{\hbar}\right)\left(n+\frac{1}{2}\right) = \gamma_{11}(E_{75}) - \frac{2\sqrt{2}}{3}eF_s S_i^{3/2}[\gamma''_{11}(E_{75})]^{1/2} \tag{4.94}$$

where E_{75} is the totally quantized energy in this case.

The density-of-states function assumes the form

$$N(E) = \left(\frac{g_v eB}{\pi\hbar}\right)\sum_{i=0}^{i_{\max}}\sum_{n=0}^{n_{\max}}\delta'(E-E_{75}) \tag{4.95}$$

Combining (4.95) with the Fermi–Dirac occupation probability factor, the surface electron concentration can be written as

$$N_s = \left(\frac{g_v eB}{\pi\hbar}\right)\sum_{i=0}^{i_{\max}}\sum_{n=0}^{n_{\max}}F_{-1}(\eta_{32}) \tag{4.96}$$

where $\eta_{32} = (k_B T)^{-1}[E_{F1} - E_{75}]$.

Combining (1.4) and (4.96) we can study the magneto QC in this case.

4.2.11 *The QC in magneto MOSFETs of tellurium*

The magneto 2D dispersion relation in this case can be written following (2.135) as

$$\left[\frac{a_1(E_{76})D_1}{b_1} \pm \frac{D_1\sqrt{C_1(E_{76})}}{b_1}\right]^{2/3} - \left[\frac{a_1(E_{76})D_1}{b_1} - D_1\bar{z}_{t,\mp} \pm \left(\frac{D_1}{b_1}\right)\right.$$

$$\left.\times\sqrt{C_1(E_{76}) - D_1\bar{z}_{t,\mp}}\right]^{2/3} \mp \frac{3}{4}\left(\frac{D_1}{b_1}\right)\left[\beta_1(E_{76})\sqrt{\beta_1^2(E_{76}) - e_1^2(E_{76})}\right.$$

$$- \beta_2(E_{76})\sqrt{\beta_2^2(E_{76}) - e_1^2(E_{76})}$$

$$\left. + e_1^2(E_{76})l_n\left|\frac{\beta_2(E_{76}) + \sqrt{\beta_2^2(E_{76}) - e_1^2(E_{76})}}{\beta_1(E_{76}) + \sqrt{\beta_1^2(E_{76}) - e_1^2(E_{76})}}\right|\right] = \frac{(S_0D_1)^{3/2}}{\sqrt{b_1}} \quad (4.97)$$

where,

$$a_1(E_{76}) = (2\psi_1^2)^{-1}\left[E_{76}\psi_1 + \psi_3^2 - 2\psi_1\psi_2\left(\frac{2eB}{\hbar}\right)\left(n + \frac{1}{2}\right)\right],$$

$$C_1(E_{76}) = (4\psi_1^4)^{-1}\left[\psi_3^4 + 4\psi_1\psi_3^2E_{76} - 4\psi_1\psi_2\psi_3^2k_s^2\right.$$

$$\left. + 4\psi_1^2\psi_4^2\left(\frac{2eB}{\hbar}\right)\left(n + \frac{1}{2}\right)\right]^{1/2}$$

$$e_1^2(E_{76}) = \left[\frac{D_1^2}{4b_1} + C_1(E_{76}) - \frac{a_1(E_{76})D_1}{b_1}\right], \quad \beta_1(E_{76}) = \sqrt{C_1(E_{76})} \pm \frac{D_1}{2b_1},$$

$$\beta_2(E_{76}) = \sqrt{C_1(E_{76}) - D_1\bar{z}_{t,\mp}} \pm \frac{D_1}{2b_1},$$

$$\bar{z}_{t,\mp} = (eF_s)^{-1}\left[E_{76} - \psi_2\left(\frac{2eB}{\hbar}\right)\left(n + \frac{1}{2}\right) \mp \psi_4\sqrt{\left(\frac{2eB}{\hbar}\right)\left(n + \frac{1}{2}\right)}\right]$$

where E_{76} is the totally quantized energy in this case.

The density-of-states function assumes the form

$$N(E) = \left(\frac{g_v eB}{2\pi\hbar}\right)\sum_{i=0}^{i_{\max}}\sum_{n=0}^{n_{\max}}\delta'(E - E_{76}) \quad (4.98)$$

Combining (4.98) with the Fermi–Dirac occupation probability factor, the surface electron concentration can be written as

$$N_s = \left(\frac{g_v eB}{2\pi\hbar}\right)\sum_{i=0}^{i_{\max}}\sum_{n=0}^{n_{\max}}F_{-1}(\eta_{33}) \quad (4.99)$$

where $\eta_{33} = (k_BT)^{-1}[E_{F1} - E_{76}]$.

Combining (1.4) and (4.99) we can study the magneto QC in this case.

4.2.12 *The QC in magneto QWFETs of nonlinear optical materials*

The magneto 2D dispersion relation in this case can be written following (3.1) as

$$\psi_1(E_{77}) = \psi_2(E_{77})\left(\frac{2eB}{\hbar}\right)\left(n+\frac{1}{2}\right) + \psi_3(E_{77})(n_z\pi/d_z)^2 \tag{4.100}$$

where E_{77} is the totally quantized energy in this case.

The density-of-states function assumes the form

$$N(E) = \left(\frac{g_v eB}{\pi\hbar}\right)\sum_{n=0}^{n_{\max}}\sum_{n_z=1}^{n_{z\max}}\delta'(E-E_{77}) \tag{4.101}$$

Combining (4.101) with the Fermi–Dirac occupation probability factor, the surface electron concentration can be written as

$$N_s = \left(\frac{g_v eB}{\pi\hbar}\right)\sum_{n=0}^{n_{\max}}\sum_{n_z=1}^{n_{z\max}}F_{-1}(\eta_{34}) \tag{4.102}$$

where $\eta_{34} = (k_BT)^{-1}[E_{F1} - E_{77}]$.

Combining (1.4) and (4.102) we can study the magneto QC in this case

4.2.13 *The QC in magneto QWFETs of III–V, ternary and quaternary materials*

(i) Three-band model of Kane

The magneto 2D dispersion relation in this case can be written following (3.8) as

$$\left(n+\frac{1}{2}\right)\hbar\omega_0 + \frac{\hbar^2}{2m_c}(n_z\pi/d_z)^2 = I_{11}(E_{78}) \tag{4.103}$$

where E_{78} is the totally quantized energy in this case.

The density-of-states function assumes the form

$$N(E) = \left(\frac{g_v eB}{\pi\hbar}\right)\sum_{n=0}^{n_{\max}}\sum_{n_z=1}^{n_{z\max}}\delta'(E-E_{78}) \tag{4.104}$$

Combining (4.104) with the Fermi–Dirac occupation probability factor, the surface electron concentration can be written as

$$N_s = \left(\frac{g_v eB}{\pi\hbar}\right)\sum_{n=0}^{n_{\max}}\sum_{n_z=1}^{n_{z\max}}F_{-1}(\eta_{35}) \tag{4.105}$$

where $\eta_{35} = (k_BT)^{-1}[E_{F1} - E_{78}]$.

Combining (1.4) and (4.105) we can study the magneto QC in this case.

(ii) Two-band model of Kane

The magneto 2D dispersion relation in this case can be written following (3.13) as

$$E_{79}(1+\alpha E_{79}) = \left(n+\frac{1}{2}\right)\hbar\omega_0 + \frac{\hbar^2}{2m_c}\left(\frac{n_z\pi}{d_z}\right)^2 \tag{4.106}$$

where E_{79} is the totally quantized energy in this case.

The density-of-states function assumes the form

$$N(E) = \left(\frac{g_v eB}{\pi\hbar}\right)\sum_{n=0}^{n_{\max}}\sum_{n_z=1}^{n_{z\max}}\delta'(E-E_{79}) \tag{4.107}$$

Combining (4.107) with the Fermi–Dirac occupation probability factor, the surface electron concentration can be written as

$$N_s = \left(\frac{g_v eB}{\pi\hbar}\right)\sum_{n=0}^{n_{\max}}\sum_{n_z=1}^{n_{z\max}}F_{-1}(\eta_{36}) \tag{4.108}$$

where $\eta_{36} = (k_BT)^{-1}[E_{F1}-E_{79}]$.

Combining (1.4) and (4.108) we can study the magneto QC in this case.

(iii) Parabolic band model

The magneto 2D dispersion relation in this case can be written following (3.18a) as

$$E_{80} = \left(n+\frac{1}{2}\right)\hbar\omega_0 + \frac{\hbar^2}{2m_c}\left(\frac{n_z\pi}{d_z}\right)^2 \tag{4.109}$$

where E_{80} is the totally quantized energy in this case.

The density-of-states function assumes the form

$$N(E) = \left(\frac{g_v eB}{\pi\hbar}\right)\sum_{n=0}^{n_{\max}}\sum_{n_z=1}^{n_{z\max}}\delta'(E-E_{80}) \tag{4.110}$$

Combining (4.110) with the Fermi–Dirac occupation probability factor, the surface electron concentration can be written as

$$N_s = \left(\frac{g_v eB}{\pi\hbar}\right)\sum_{n=0}^{n_{\max}}\sum_{n_z=1}^{n_{z\max}}F_{-1}(\eta_{37}) \tag{4.111}$$

where $\eta_{37} = (k_BT)^{-1}[E_{F1}-E_{80}]$.

Combining (1.4) and (4.111) we can study the magneto QC in this case.

4.2.14 *The QC in magneto QWFETs of II–VI materials*

The magneto 2D dispersion relation in this case can be written following (3.22) as

$$E_{81} = a_0'\left(\frac{2eB}{\hbar}\right)\left(n+\frac{1}{2}\right) + b_0'\left(\frac{n_z\pi}{d_z}\right)^2 \pm C_0\sqrt{\left(\frac{2eB}{\hbar}\right)\left(n+\frac{1}{2}\right)} \tag{4.112}$$

where E_{81} is the totally quantized energy in this case.

The density-of-states function assumes the form

$$N(E) = \left(\frac{g_v eB}{2\pi\hbar}\right)\sum_{n=0}^{n_{\max}}\sum_{n_z=1}^{n_{z\max}}\delta'(E - E_{81}) \tag{4.113}$$

Combining (4.113) with the Fermi–Dirac occupation probability factor, the surface electron concentration can be written as

$$N_s = \left(\frac{g_v eB}{2\pi\hbar}\right)\sum_{n=0}^{n_{\max}}\sum_{n_z=1}^{n_{z\max}}F_{-1}(\eta_{38}) \tag{4.114}$$

where $\eta_{38} = (k_B T)^{-1}[E_{F1} - E_{81}]$.

Combining (1.4) and (4.114) we can study the magneto QC in this case.

4.2.15 *The QC in magneto QWFETs of IV–VI materials*

(a) The McClure and Choi model

The magneto 2D dispersion relation in this case can be written following (3.31) as

$$\left(\frac{2eB}{\hbar}\right)\left(n+\frac{1}{2}\right) = \frac{4}{3}\left[\frac{R_1(E_{82},n_z)}{p_1}\right]^{1/2}[a^2(E_{82},n_z)+b^2(E_{82},n_z)]^{1/2}$$
$$\left[a^2(E_{82},n_z)F\left[\frac{\pi}{2},\ell(E_{82},n_z)\right] - [a^2(E_{82},n_z) - b^2(E_{82},n_z)]\right.$$
$$\left.\times E\left[\frac{\pi}{2},\ell(E_{82},n_z)\right]\right] \tag{4.115}$$

where E_{82} is the totally quantized energy in this case.

The density-of-states function assumes the form

$$N(E) = \left(\frac{g_v eB}{\pi\hbar}\right)\sum_{n=0}^{n_{\max}}\sum_{n_z=1}^{n_{z\max}}\delta'(E - E_{82}) \tag{4.116}$$

Combining (4.116) with the Fermi–Dirac occupation probability factor, the surface electron concentration can be written as

$$N_s = \left(\frac{g_v eB}{\pi\hbar}\right) \sum_{n=0}^{n_{\max}} \sum_{n_z=1}^{n_{z\max}} F_{-1}(\eta_{39}) \tag{4.117}$$

where $\eta_{39} = (k_B T)^{-1}[E_{F1} - E_{82}]$.

Combining (1.4) and (4.117) we can study the magneto QC in this case.

(b) The hybrid model

The magneto 2D dispersion relation in this case can be written following (3.38) as

$$\left(\frac{2eB}{\hbar}\right)\left(n + \frac{1}{2}\right) = \frac{2\pi\sqrt{m_1 m_3}}{\hbar^2} t_{29}(E_{83}, n_y) \tag{4.118}$$

where E_{83} is the totally quantized energy in this case.

The density-of-states function assumes the form

$$N(E) = \left(\frac{g_v eB}{\pi\hbar}\right) \sum_{n=0}^{n_{\max}} \sum_{n_z=1}^{n_{z\max}} \delta'(E - E_{83}) \tag{4.119}$$

Combining (4.119) with the Fermi–Dirac occupation probability factor, the surface electron concentration can be written as

$$N_s = \left(\frac{g_v eB}{\pi\hbar}\right) \sum_{n=0}^{n_{\max}} \sum_{n_z=1}^{n_{z\max}} F_{-1}(\eta_{40}) \tag{4.120}$$

where $\eta_{40} = (k_B T)^{-1}[E_{F1} - E_{83}]$.

Combining (1.4) and (4.120) we can study the magneto QC in this case.

(c) The Cohen model

The magneto 2D dispersion relation in this case can be written following (5.29) as

$$\left(\frac{2eB}{\hbar}\right)\left(n + \frac{1}{2}\right) = \frac{4}{3}\left[\frac{R_{1c}(E_{84}, n_z)}{p_1}\right]^{1/2} [a_c^2(E_{84}, n_z) + b_c^2(E_{84}, n_z)]^{1/2}$$
$$\left[a_c^2(E_{84}, n_z) F\left[\frac{\pi}{2}, \ell(E_{84}, n_z)\right] - [a_c^2(E_{84}, n_z) - b_c^2(E_{84}, n_z)]\right.$$
$$\left. \times E\left[\frac{\pi}{2}, \ell(E_{84}, n_z)\right]\right] \tag{4.121}$$

where

$$a_c^2(E_{84}, n_z) \equiv \frac{q_2(E_{84})}{2R_2} + \frac{1}{2}\left[\frac{q_2^2(E_{84})}{R_2^2} + \frac{4\gamma_1(E_{84}, n_z)}{R_2}\right]^{1/2}$$

$$b_c^2(E_{84}, n_z) \equiv \frac{1}{2}\left[\frac{q_2^2(E_{84})}{R_2^2} + \frac{4\gamma_1(E_{84}, n_z)}{R_2}\right]^{1/2} - \left(\frac{q_2(E_{84})}{2R_2}\right)$$

$$\ell(E_{84}, n_z) \equiv \frac{b_c(E_{84}, n_z)}{\sqrt{a_c^2(E_{84}, n_z) + b_c^2(E_{84}, n_z)}}, \quad F\left[\frac{\pi}{2}, \ell(E_{84}, n_z)\right],$$

$$\text{and} \quad E\left[\frac{\pi}{2}, \ell(E_{84}, n_z)\right]$$

are the complete elliptic integral of the first and second kinds respectively in which E_{84} is the totally quantized energy in this case.

The density-of-states function assumes the form

$$N(E) = \left(\frac{g_v eB}{\pi\hbar}\right)\sum_{n=0}^{n_{\max}}\sum_{n_z=1}^{n_{z\max}} \delta'(E - E_{84}) \tag{4.122}$$

Combining (4.122) with the Fermi–Dirac occupation probability factor, the surface electron concentration can be written as

$$N_s = \left(\frac{g_v eB}{\pi\hbar}\right)\sum_{n=0}^{n_{\max}}\sum_{n_z=1}^{n_{z\max}} F_{-1}(\eta_{41}) \tag{4.123}$$

where $\eta_{41} = (k_B T)^{-1}[E_{F1} - E_{84}]$.

Combining (1.4) and (4.123) we can study the magneto QC in this case.

(d) The Lax model

The magneto 2D dispersion relation in this case can be written following (3.50) as

$$E_{85}(1 + \alpha E_{85}) = \left(n + \frac{1}{2}\right)\hbar\frac{eB}{\sqrt{m_1 m_2}} + \frac{\hbar^2}{2m_3}\left(\frac{n_z\pi}{d_z}\right)^2 \tag{4.124}$$

where E_{85} is the totally quantized energy in this case.

The density-of-states function assumes the form

$$N(E) = \left(\frac{g_v eB}{\pi\hbar}\right)\sum_{n=0}^{n_{\max}}\sum_{n_z=1}^{n_{z\max}} \delta'(E - E_{85}) \tag{4.125}$$

Combining (4.125) with the Fermi–Dirac occupation probability factor, the surface electron concentration can be written as

$$N_s = \left(\frac{g_v eB}{\pi\hbar}\right) \sum_{n=0}^{n_{\max}} \sum_{n_z=1}^{n_{z\max}} F_{-1}(\eta_{42}) \tag{4.126}$$

where $\eta_{42} = (k_B T)^{-1}[E_{F1} - E_{85}]$.

Combining (1.4) and (4.126) we can study the magneto QC in this case.

(e) The Dimmock model

The magneto 2D dispersion relation in this case can be written following (3.66) as

$$\left(\frac{2eB}{\hbar}\right)\left(n+\frac{1}{2}\right) = \frac{t_3(E_{86}, n_z)}{\sqrt{A_1(E_{86}, n_z)B_1(E_{86}, n_z)}} \times \left[1 - \frac{1}{x_5}\left(\frac{1}{x_1} + \frac{3}{x_2}\right)\frac{\alpha t_3(E_{86}, n_z)\hbar^4}{8B_1^2(E_{86}, n_z)}\right] \tag{4.127}$$

where E_{86} is the totally quantized energy in this case.

The density-of-states function assumes the form

$$N(E) = \left(\frac{g_v eB}{\pi\hbar}\right) \sum_{n=0}^{n_{\max}} \sum_{n_z=1}^{n_{z\max}} \delta'(E - E_{86}) \tag{4.128}$$

Combining (4.128) with the Fermi–Dirac occupation probability factor, the surface electron concentration can be written as

$$N_s = \left(\frac{g_v eB}{\pi\hbar}\right) \sum_{n=0}^{n_{\max}} \sum_{n_z=1}^{n_{z\max}} F_{-1}(\eta_{43}) \tag{4.129}$$

where $\eta_{43} = (k_B T)^{-1}[E_{F1} - E_{86}]$.

Combining (1.4) and (4.129) we can study the magneto QC in this case.

(f) The ellipsoidal parabolic model

The magneto 2D dispersion relation in this case can be written following (3.71) as

$$E_{87} = \left(n+\frac{1}{2}\right)\hbar\frac{eB}{\sqrt{m_1 m_2}} + \left(\frac{\hbar^2}{2m_3}\right)\left(\frac{n_z\pi}{d_z}\right)^2 \tag{4.130}$$

where E_{87} is the totally quantized energy in this case.

The density-of-states function assumes the form

$$N(E) = \left(\frac{g_v eB}{\pi\hbar}\right) \sum_{n=0}^{n_{\max}} \sum_{n_z=1}^{n_{z\max}} \delta'(E - E_{87}) \tag{4.131}$$

Combining (4.131) with the Fermi–Dirac occupation probability factor, the surface electron concentration can be written as

$$N_s = \left(\frac{g_v eB}{\pi\hbar}\right)\sum_{n=0}^{n_{\max}}\sum_{n_z=1}^{n_{z\max}} F_{-1}(\eta_{44}) \tag{4.132}$$

where $\eta_{44} = (k_B T)^{-1}[E_{F1} - E_{87}]$.

Combining (1.4) and (4.132) we can study the magneto QC in this case.

4.2.16 *The QC in magneto QWFETs of stressed Kane type materials*

The magneto 2D dispersion relation in this case can be written following (3.77) as

$$\left(\frac{2eB}{\hbar}\right)\left(n+\frac{1}{2}\right) = P^2(E_{88}, n_z)\bar{a}_0(E_{88})\bar{b}_0(E_{88}) \tag{4.133}$$

where E_{88} is the totally quantized energy in this case.

The density-of-states function assumes the form

$$N(E) = \left(\frac{g_v eB}{\pi\hbar}\right)\sum_{n=0}^{n_{\max}}\sum_{n_z=1}^{n_{z\max}} \delta'(E - E_{88}) \tag{4.134}$$

Combining (4.134) with the Fermi–Dirac occupation probability factor, the surface electron concentration can be written as

$$N_s = \left(\frac{g_v eB}{\pi\hbar}\right)\sum_{n=0}^{n_{\max}}\sum_{n_z=1}^{n_{z\max}} F_{-1}(\eta_{45}) \tag{4.135}$$

where $\eta_{45} = (k_B T)^{-1}[E_{F1} - E_{88}]$.

Combining (1.4) and (4.135) we can study the magneto QC in this case.

4.2.17 *The QC in magneto QWFETs of bismuth telluride*

The magneto 2D dispersion relation in this case can be written following (3.103) as

$$\left(\frac{2eB}{\hbar}\right)\left(n+\frac{1}{2}\right) = \frac{1}{\sqrt{\alpha_{22}\alpha_{33} - 4\alpha_{23}^2}} \times \left[\frac{2m_0 E_{89}(1+\alpha E_{89})}{\hbar^2} - \alpha_{11}\left(\frac{n_x \pi}{d_x}\right)^2\right] \tag{4.136}$$

where E_{89} is the totally quantized energy in this case.

The density-of-states function assumes the form

$$N(E) = \left(\frac{g_v eB}{\pi\hbar}\right) \sum_{n=0}^{n_{\max}} \sum_{n_z=1}^{n_{z\max}} \delta'(E - E_{89}) \tag{4.137}$$

Combining (4.137) with the Fermi–Dirac occupation probability factor, the surface electron concentration can be written as

$$N_s = \left(\frac{g_v eB}{\pi\hbar}\right) \sum_{n=0}^{n_{\max}} \sum_{n_z=1}^{n_{z\max}} F_{-1}(\eta_{46}) \tag{4.138}$$

where $\eta_{46} = (k_B T)^{-1}[E_{F1} - E_{89}]$.

Combining (1.4) and (4.138) we can study the magneto QC in this case.

4.2.18 *The QC in magneto QWFETs of germanium*

(a) Model of Cardona *et al.*

The magneto 2D dispersion relation in this case can be written following (3.109) as

$$\left(\frac{eB}{\hbar}\right)\left(n + \frac{1}{2}\right) = \frac{\sqrt{m_1^* m_2^*}}{\hbar^2}\gamma\left(E_{90}, n_z\right) \tag{4.139}$$

where E_{90} is the totally quantized energy in this case.

The density-of-states function assumes the form

$$N(E) = \left(\frac{g_v eB}{\pi\hbar}\right) \sum_{n=0}^{n_{\max}} \sum_{n_z=1}^{n_{z\max}} \delta'(E - E_{90}) \tag{4.140}$$

Combining (4.140) with the Fermi–Dirac occupation probability factor, the surface electron concentration can be written as

$$N_s = \left(\frac{g_v eB}{\pi\hbar}\right) \sum_{n=0}^{n_{\max}} \sum_{n_z=1}^{n_{z\max}} F_{-1}(\eta_{47}) \tag{4.141}$$

where $\eta_{47} = (k_B T)^{-1}[E_{F1} - E_{90}]$.

Combining (1.4) and (4.141) we can study the magneto QC in this case.

(b) The model of Wang *et al.*

The magneto 2D dispersion relation in this case can be written following (3.117) as

$$(eB)\left(n + \frac{1}{2}\right) = \frac{\sqrt{m_1^* m_2^*}}{\hbar} I_1(E_{91}, n_z) \tag{4.142}$$

where E_{91} is the totally quantized energy in this case.

The density-of-states function assumes the form

$$N(E) = \left(\frac{g_v eB}{\pi\hbar}\right)\sum_{n=0}^{n_{\max}}\sum_{n_z=1}^{n_{z\max}}\delta'(E - E_{91}) \tag{4.143}$$

Combining (4.143) with the Fermi–Dirac occupation probability factor, the surface electron concentration can be written as

$$N_s = \left(\frac{g_v eB}{\pi\hbar}\right)\sum_{n=0}^{n_{\max}}\sum_{n_z=1}^{n_{z\max}}F_{-1}(\eta_{48}) \tag{4.144}$$

where $\eta_{48} = (k_B T)^{-1}[E_{F1} - E_{91}]$.

Combining (1.4) and (4.144) we can study the magneto QC in this case.

4.2.19 *The QC in magneto QWFETs of gallium arsenide*

(a) Rossler model

The magneto 2D dispersion relation in this case can be written following (3.123) as

$$\left(\frac{2eB}{\hbar}\right)\left(n + \frac{1}{2}\right) = \gamma_{4\pm}(E_{92}) - \left(\frac{n_z\pi}{d_z}\right)^2 \tag{4.145}$$

where E_{92} is the totally quantized energy in this case.

The density-of-states function assumes the form

$$N(E) = \left(\frac{g_v eB}{2\pi\hbar}\right)\sum_{n=0}^{n_{\max}}\sum_{n_z=1}^{n_{z\max}}\delta'(E - E_{92}) \tag{4.146}$$

Combining (4.146) with the Fermi–Dirac occupation probability factor, the surface electron concentration can be written as

$$N_s = \left(\frac{g_v eB}{2\pi\hbar}\right)\sum_{n=0}^{n_{\max}}\sum_{n_z=1}^{n_{z\max}}F_{-1}(\eta_{49}) \tag{4.147}$$

where $\eta_{49} = (k_B T)^{-1}[E_{F1} - E_{92}]$.

Combining (1.4) and (4.147) we can study the magneto QC in this case.

(b) Model of Stillman *et al.*

The magneto 2D dispersion relation in this case can be written following (3.128) as

$$\left(\frac{2eB}{\hbar}\right)\left(n + \frac{1}{2}\right) = \gamma_5(E_{93}) - \left(\frac{n_z\pi}{d_z}\right)^2 \tag{4.148}$$

where E_{93} is the totally quantized energy in this case.

The density-of-states function assumes the form

$$N(E) = \left(\frac{g_v eB}{\pi\hbar}\right)\sum_{n=0}^{n_{\max}}\sum_{n_z=1}^{n_{z\max}}\delta'(E - E_{93}) \tag{4.149}$$

Combining (4.149) with the Fermi–Dirac occupation probability factor, the surface electron concentration can be written as

$$N_s = \left(\frac{g_v eB}{\pi\hbar}\right)\sum_{n=0}^{n_{\max}}\sum_{n_z=1}^{n_{z\max}} F_{-1}(\eta_{50}) \tag{4.150}$$

where $\eta_{50} = (k_B T)^{-1}[E_{F1} - E_{93}]$.

Combining (1.4) and (4.150) we can study the magneto QC in this case.

4.2.20 *The QC in magneto QWFETs of gallium antimonide*

(a) Model of Seiler *et al.*

The magneto 2D dispersion relation in this case can be written following (3.133) as

$$\left(\frac{2eB}{\hbar}\right)\left(n + \frac{1}{2}\right) = \gamma_6(E_{94}) - \left(\frac{n_z\pi}{d_z}\right)^2 \tag{4.151}$$

where E_{94} is the totally quantized energy in this case.

The density-of-states function assumes the form

$$N(E) = \left(\frac{g_v eB}{2\pi\hbar}\right)\sum_{n=0}^{n_{\max}}\sum_{n_z=1}^{n_{z\max}}\delta'(E - E_{94}) \tag{4.152}$$

Combining (4.152) with the Fermi–Dirac occupation probability factor, the surface electron concentration can be written as

$$N_s = \left(\frac{g_v eB}{2\pi\hbar}\right)\sum_{n=0}^{n_{\max}}\sum_{n_z=1}^{n_{z\max}} F_{-1}(\eta_{51}) \tag{4.153}$$

where $\eta_{51} = (k_B T)^{-1}[E_{F1} - E_{94}]$.

Combining (1.4) and (4.153) we can study the magneto QC in this case.

(b) Model of Mathur *et al.*

The magneto 2D dispersion relation in this case can be written following (3.138) as

$$\left(\frac{2eB}{\hbar}\right)\left(n + \frac{1}{2}\right) = \gamma_7(E_{95}) - \left(\frac{n_z\pi}{d_z}\right)^2 \tag{4.154}$$

where E_{95} is the totally quantized energy in this case.

The density-of-states function assumes the form

$$N(E) = \left(\frac{g_v eB}{\pi\hbar}\right) \sum_{n=0}^{n_{\max}} \sum_{n_z=1}^{n_{z\max}} \delta'(E - E_{95}) \tag{4.155}$$

Combining (4.155) with the Fermi–Dirac occupation probability factor, the surface electron concentration can be written as

$$N_s = \left(\frac{g_v eB}{\pi\hbar}\right) \sum_{n=0}^{n_{\max}} \sum_{n_z=1}^{n_{z\max}} F_{-1}(\eta_{52}) \tag{4.156}$$

where $\eta_{52} = (k_B T)^{-1}[E_{F1} - E_{95}]$.

Combining (1.4) and (4.156) we can study the magneto QC in this case.

(c) Model of Zhang

The magneto 2D dispersion relation in this case can be written following (3.143) as

$$\left(\frac{2eB}{\hbar}\right)\left(n + \frac{1}{2}\right) = \gamma_8(E_{96}) - \left(\frac{n_z \pi}{d_z}\right)^2 \tag{4.157}$$

where E_{96} is the totally quantized energy in this case.

The density-of-states function assumes the form

$$N(E) = \left(\frac{g_v eB}{\pi\hbar}\right) \sum_{n=0}^{n_{\max}} \sum_{n_z=1}^{n_{z\max}} \delta'(E - E_{96}) \tag{4.158}$$

Combining (4.158) with the Fermi–Dirac occupation probability factor, the surface electron concentration can be written as

$$N_s = \left(\frac{g_v eB}{\pi\hbar}\right) \sum_{n=0}^{n_{\max}} \sum_{n_z=1}^{n_{z\max}} F_{-1}(\eta_{53}) \tag{4.159}$$

where $\eta_{53} = (k_B T)^{-1}[E_{F1} - E_{96}]$.

Combining (1.4) and (4.159) we can study the magneto QC in this case.

4.2.21 *The QC in magneto QWFETs of indium antimonide*

(a) Model of Palik *et al.*

The magneto 2D dispersion relation in this case can be written following (3.148) as

$$\left(\frac{2eB}{\hbar}\right)\left(n + \frac{1}{2}\right) = \gamma_9(E_{97}) - \left(\frac{n_z \pi}{d_z}\right)^2 \tag{4.160}$$

where E_{97} is the totally quantized energy in this case.

The density-of-states function assumes the form

$$N(E) = \left(\frac{g_v eB}{\pi\hbar}\right) \sum_{n=0}^{n_{\max}} \sum_{n_z=1}^{n_{z\max}} \delta'(E - E_{97}) \tag{4.161}$$

Combining (4.161) with the Fermi–Dirac occupation probability factor, the surface electron concentration can be written as

$$N_s = \left(\frac{g_v eB}{\pi\hbar}\right) \sum_{n=0}^{n_{\max}} \sum_{n_z=1}^{n_{z\max}} F_{-1}(\eta_{54}) \tag{4.162}$$

where $\eta_{54} = (k_B T)^{-1}[E_{F1} - E_{97}]$.

Combining (1.4) and (4.162) we can study the magneto QC in this case.

(b) Model of Johnson *et al.*

The magneto 2D dispersion relation in this case can be written following (3.153) as

$$\left(\frac{2eB}{\hbar}\right)\left(n + \frac{1}{2}\right) = \gamma_{10}(E_{98}) - \left(\frac{n_z \pi}{d_z}\right)^2 \tag{4.163}$$

where E_{98} is the totally quantized energy in this case.

The density-of-states function assumes the form

$$N(E) = \left(\frac{g_v eB}{\pi\hbar}\right) \sum_{n=0}^{n_{\max}} \sum_{n_z=1}^{n_{z\max}} \delta'(E - E_{98}) \tag{4.164}$$

Combining (4.164) with the Fermi–Dirac occupation probability factor, the surface electron concentration can be written as

$$N_s = \left(\frac{g_v eB}{\pi\hbar}\right) \sum_{n=0}^{n_{\max}} \sum_{n_z=1}^{n_{z\max}} F_{-1}(\eta_{55}) \tag{4.165}$$

where $\eta_{55} = (k_B T)^{-1}[E_{F1} - E_{98}]$.

Combining (1.4) and (4.165) we can study the magneto QC in this case.

(c) Model of Agafonov *et al.*

The magneto 2D dispersion relation in this case can be written following (3.158) as

$$\left(\frac{2eB}{\hbar}\right)\left(n + \frac{1}{2}\right) = \gamma_{11}(E_{99}) - \left(\frac{n_z \pi}{d_z}\right)^2 \tag{4.166}$$

where E_{99} is the totally quantized energy in this case.

The density-of-states function assumes the form

$$N(E) = \left(\frac{g_v eB}{\pi\hbar}\right)\sum_{n=0}^{n_{\max}}\sum_{n_z=1}^{n_{z\max}}\delta'(E - E_{99}) \tag{4.167}$$

Combining (4.167) with the Fermi–Dirac occupation probability factor, the surface electron concentration can be written as

$$N_s = \left(\frac{g_v eB}{\pi\hbar}\right)\sum_{n=0}^{n_{\max}}\sum_{n_z=1}^{n_{z\max}}F_{-1}(\eta_{56}) \tag{4.168}$$

where $\eta_{56} = (k_B T)^{-1}[E_{F1} - E_{99}]$.

Combining (1.4) and (4.168) we can study the magneto QC in this case.

4.2.22 *The QC in magneto QWFETs of tellurium*

The magneto 2D dispersion relation in this case can be written following (3.82) as

$$\left(\frac{2eB}{\hbar}\right)\left(n + \frac{1}{2}\right) = \psi_5(E_{100}) - \psi_6\left(\frac{\pi n_z}{d_z}\right)^2 \pm \psi_7\left[\psi_8^2(E_{100}) - \left(\frac{\pi n_z}{d_z}\right)^2\right]^{1/2} \tag{4.169}$$

where E_{100} is the totally quantized energy in this case.

The density-of-states function assumes the form

$$N(E) = \left(\frac{g_v eB}{2\pi\hbar}\right)\sum_{n=0}^{n_{\max}}\sum_{n_z=1}^{n_{z\max}}\delta'(E - E_{100}) \tag{4.170}$$

Combining (4.170) with the Fermi–Dirac occupation probability factor, the surface electron concentration can be written as

$$N_s = \left(\frac{g_v eB}{2\pi\hbar}\right)\sum_{n=0}^{n_{\max}}\sum_{n_z=1}^{n_{z\max}}F_{-1}(\eta_{57}) \tag{4.171}$$

where $\eta_{57} = (k_B T)^{-1}[E_{F1} - E_{100}]$.

Combining (1.4) and (4.171) we can study the magneto QC in this case.

4.2.23 *The QC in magneto QWFETs of platinum antimonide*

The magneto 2D dispersion relation in this case can be written following (3.95) as

$$\left(\frac{2eB}{\hbar}\right)\left(n+\frac{1}{2}\right)=t_{44}(E_{101},n_z) \tag{4.172}$$

where E_{101} is the totally quantized energy in this case.

The density-of-states function assumes the form

$$N(E)=\left(\frac{g_v eB}{2\pi\hbar}\right)\sum_{n=0}^{n_{\max}}\sum_{n_z=1}^{n_{z\max}}\delta'(E-E_{101}) \tag{4.173}$$

Combining (4.173) with the Fermi–Dirac occupation probability factor, the surface electron concentration can be written as

$$N_s=\left(\frac{g_v eB}{2\pi\hbar}\right)\sum_{n=0}^{n_{\max}}\sum_{n_z=1}^{n_{z\max}}F_{-1}(\eta_{58}) \tag{4.174}$$

where $\eta_{58}=(k_BT)^{-1}[E_{F1}-E_{101}]$.

Combining (1.4) and (4.174) we can study the magneto QC in this case.

4.2.24 *The QC in magneto MOSFETs of gallium phosphide*

The magneto 2D dispersion relation in this case can be written following (3.89) as

$$\left(\frac{2eB}{\hbar}\right)\left(n+\frac{1}{2}\right)=t_{42}(E_{102},n_z) \tag{4.175}$$

where E_{102} is the totally quantized energy in this case.

The density-of-states function assumes the form

$$N(E)=\left(\frac{g_v eB}{2\pi\hbar}\right)\sum_{n=0}^{n_{\max}}\sum_{n_z=1}^{n_{z\max}}\delta'(E-E_{102}) \tag{4.176}$$

Combining (4.176) with the Fermi–Dirac occupation probability factor, the surface electron concentration can be written as

$$N_s=\left(\frac{g_v eB}{2\pi\hbar}\right)\sum_{n=0}^{n_{\max}}\sum_{n_z=1}^{n_{z\max}}F_{-1}(\eta_{59}) \tag{4.177}$$

where $\eta_{59}=(k_BT)^{-1}[E_{F1}-E_{102}]$.

Combining (1.4) and (4.177) we can study the magneto QC in this case.

4.3 Result and Discussions

Using the appropriate equations and the parameters as given in Chapter 18, we have determined the dependence of the QC taking the MOSFETs of $CdGeAs_2$, InSb and Hg_{1-x} Cd_x Te as examples operated under inversion layers modes on a quantizing magnetic field at 4.2K in the weak electric field limit, as shown in Figs. 4.1 and 4.2, respectively. The surface electric field is chosen for the weak field limit. The period of oscillations can easily expressed as low temperatures by $(1/B) = g_s g_v e/(2\pi\hbar n_s)$ and is found graphically to be 0.9 and 0.897T^{-1}, as calculated from the relation for the period corresponding to a surface concentration $5.4 \times 10^{14}\text{m}^{-2}$ induced at the surface by the surface field considered. The depths of the spikes increase with increasing thickness of the insulting layer. The oscillatory nature of the magnetic field dependence of the surface and QCs is found to be somewhat similar to that observed experimentally in Si-MOS structures having n-channel inversion layers, with decreasing magnetic field, i.e., with increasing 1/B, the MOS capacitance should approach its value corresponding to zero magnetic field.

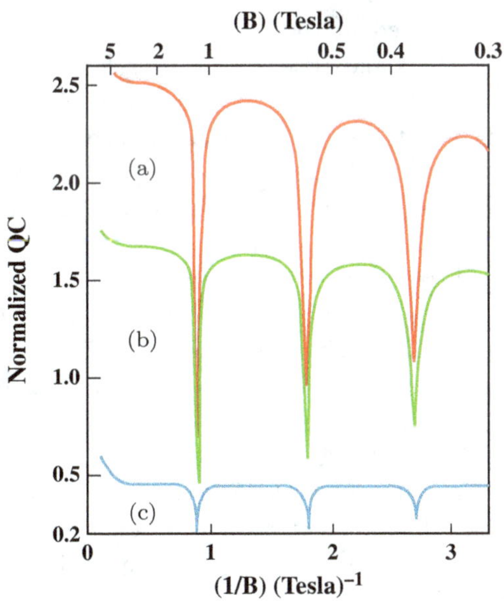

Fig. 4.1. Plot showing the oscillatory magnetic field dependence of the normalized QC in magneto MOSFETs of (a)$CdGeAs_2$ (b)$InSb$ and (c) $Hg_{1-x}Cd_xTe$ operated under inversion layer mode.

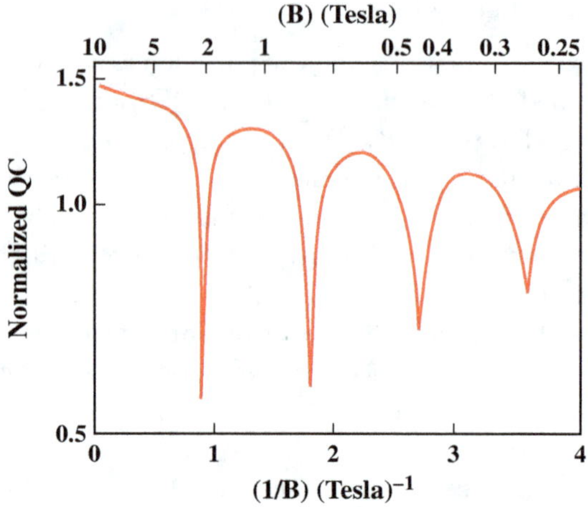

Fig. 4.2. Plot showing the oscillatory magnetic field dependence of the normalized QC in magneto MOSFETs of $In_{1-x}Ga_xAs_yP_{1-y}$ lattice matched to InP operated under inversion layer mode.

With varying magnetic field, each time a Landau level crosses the Fermi level, a change is reflected in the capacitance through redistribution of the carriers among the Landau levels. It may be noted that the three-dimensional quantization list to discrete energy levels in the real sense of the term somewhat like atomic energy levels and, consequently, the crossing of the Fermi level by the Landau levels in steps produce very sharp changes. This is what is reflected in both the figures by the increasing sharpness of the spikes with increasing magnetic field.

Incidentally, however, the collision broadening effects that have been neglected here would, of course, reduce the sharpness of the spikes to some extent. Even then, the sharpness would be much greater than that observed for oscillatory phenomena based on the same physics in 3D electron gases of semiconductors. This follows from the inherent feature of the 3D quantization of the 2D electron gas dealt with here. Under such quantization three remains no free electron state between any two successive Landau levels unlike that found for 3D electron gases of semiconductors under 2D quantization in the wave vector space in the presence of a quantizing magnetic field. Consequently, the crossing of the Fermi level by the Landau levels under 3D quantization would have much greater impact on the redistribution of the electron amongst the available states, as compared to that found for 2D quantization. It is basically this impact that results in the increased sharpness of the oscillatory spikes. The neglect of collision broadening does it, therefore, not render meaningless to stress this point.

It may be noted that though the spin splitting of the Landau levels are usually large in small gap semiconductor these have not been taken into consideration in the present calculation. The spin effects would simply increase the number of spikes splitting each of the spikes into two and reduce their amplitudes. In a calculation of the present type, the amplitudes are really arbitrary and depend strictly on the resolution of computation. Thus, for simplicity, the spin effects have been neglected without losing much of the information on the oscillatory features of the magneto- quantization.

In Section 4.2.3, we have formulated the N_s for magneto MOSFETs of II–VI materials as given by (4.24) and by combining the same with (1.4) we can study the QC in the present case. The expressions of 2D N_s for magneto MOSFET of IV–VI materials for both weak and strong electric field limits are given by (4.27) and (4.30) of Section 4.2.4 and combining each of them with (1.4) we can study the QC for both the limits, respectively. The expressions of 2D N_s for magneto MOSFET of stressed materials for both weak and strong electric field limits are given by (4.33) and (4.36) of Section 4.2.5 and combining each of them with (1.4) we can study the QC for both the limits respectively. The expressions of 2D N_s for magneto MOSFET of Bismuth Telluride for both weak and strong electric field limits are given by (4.39) and (4.42) of Section 4.2.6 and combining each of them with (1.4) we can study the QC for both the limits, respectively.

The expressions of 2D N_s for magneto MOSFET of Germanium for the models of Cardona *et al.* and Wang *et al.* are given by (4.45) and (4.48) of Section 4.2.7 and combining each of them with (1.4) we can study the QC respectively in this case.

In Section 4.2.8, we have formulated the expressions of 2D N_s for magneto MOSFET of gallium arsenide for the models of Rossler and Stillman *et al.* and they are given by (4.51) and (4.57) for weak electric field limit and (4.54) and (4.60) for strong electric field limit and combining each of them with (1.4) we can study the QC for both the limits, respectively. In Section 4.2.9, we have formulated the expressions of 2D N_s for magneto MOSFET of Gallium Antimonide in accordance with the models of Seiler *et al.*, Mathur *et al.* and Zhang and they are given by (4.63) (4.69) and (4.75) for weak electric field limit and (4.66), (4.72) and (4.78) for strong electric field limit and combining each of them with (1.4) we can study the QC for both the limits, respectively.

In Section 4.2.10, we have formulated the expressions of 2D N_s for magneto MOSFET of indium antimonide in accordance with the models of Palik *et al.*, Johnson *et al.*, and Agafonov *et al.* and they are given by (4.81) (4.87) and (4.93) for weak electric field limit and (4.84), (4.93) and (4.96) for strong electric field limit and combining each of them with (1.4) we can study the QC for both the limits, respectively. In Section 4.2.11, we have formulated the N_s for magneto MOSFETs of Te as given by (4.99) and by combining the same with (1.4) we can study the QC in the present case.

In Section 4.2.12, we have formulated the N_s for magneto QWFETs of nonlinear optical materials as given by (4.102) and by combining the same with (1.4) we can study the QC in the present case. In Section 4.2.13, we have formulated the N_s for magneto QWFETs of III–V ternary and quaternary materials in accordance with the models of three- and two-band models of Kane together with parabolic energy bands as given by (4.105), (4.108) and (4.111) and by combining each of them with (1.4) we can study the QC in each case, respectively. In Section 4.2.14, we have formulated the N_s for magneto QWFETs of II–VI materials as given by (4.114) and by combining the same with (1.4) we can study the QC in this case. The expressions of 2D N_s in magneto QWFETs of IV–Vi materials in accordance with the models of the McClure and Choi, Hybrid, Cohen, Lax, Dimmock and the ellipsoidal parabolic are given by (4.117), (4.120), (4.123), (4.126), (4.129), and (4.132) in Section 4.2.15, respectively and by combining each of them with (1.4) we can study the QC of the magneto of IV–VI materials in accordance with the said types of band models.

In Section 4.2.16, we have formulated the N_s for QWFETs of stressed Kane type materials as given by (4.135) and by combining the same with (1.4) we can study the QC in the present case. In Section 4.2.17, we have formulated the N_s for QWFETs of bismuth telluride as given by (4.138) and by combining the same with (1.4) we can study the QC in the present case. The expressions of 2D N_s for magneto QWFETs of Germanium for the models of Cardona *et al.* and Wang *et al.* are given by (4.145) and (4.148) of Section 4.2.18 and combining each of them with (1.4) we can study the QC respectively in this case. In Section 4.2.19, we have formulated the expressions of 2D N_s for magneto QWFETs of gallium arsenide for the models of Rossler. and Stillman *et al.* and they are given by (4.147) and (4.150) and combining each of them with (1.4) we can study the QC respectively in this case. In Section 4.2.20, we have formulated the expressions of 2D N_s for magneto QWFETs of gallium antimonide in accordance with the models of Seiler *et al.*, Mathur *et al.* and Zhang and they are given by (4.153) (4.156) and (4.159) and combining each of them with (1.4) we can study the QC in this case, respectively. In Section 4.2.21, we have formulated the expressions of 2D N_s for magneto QWFETs of Indium Antimonide in accordance with the models of Palik *et al.*, Johnson *et al.*, and Agafonov *et al.* and they are given by (4.162) (4.165) and (4.168) and combining each of them with (1.4) we can study the QC for all the said models, respectively. In Section 4.2.22, we have formulated the N_s for magneto QWFETs of Te as given by (4.171) and by combining the same with (1.4) we can study the QC in the present case. In Section 4.2.23, we have formulated the N_s for magneto QWFETs of platinum antimonide as given by (4.174) and by combining the same with (1.4) we can study the QC in the present case. In Section 4.2.24, we have formulated the N_s for magneto QWFETs of GaP as given by (4.177) and by combining the same with (1.4) we can study the QC in the present case.

We have not given the graphs of the QCs' for MOSFETs and QWFETs made of many materials as discussed here since we feel that our readers will perform the intricate numerical computation to obtain the oscillatory plots for better assessment and joy of understanding. Incidentally, one can easily infer how little is presented and how much more there is yet to be investigated in the research field of QCs' in this case for different 2D DRs having different band structures in general which is the signature of the co-existence of new physics and advanced mathematics. Finally, we again realize that it is the band structure that changes in a fundamental way and consequently, all the physical properties of all the electronic materials change radically leading to new physical concepts.

4.4 Open Research Problems

R.4.1 Investigate the QCs in the presence of an arbitrarily oriented electric quantization for QMOSFET and QWFET of nonlinear optical materials as studied in this chapter. Draw the corresponding graphs and explain the results.

R.4.2 Investigate the QCs in QMOSFET and QWFET of IV–VI, II–VI and stressed Kane type compounds in the presence of an arbitrarily oriented quantizing non-uniform electric field. Draw the corresponding graphs and explain the results.

R.4.3 Investigate the QCs in QMOSFET and QWFET of all the materials as stated in R.1.1 of Chapter 1 in the presence of an arbitrarily oriented quantizing electric field. Draw the corresponding graphs and explain the results.

R.4.4 Investigate the QCs in the presence of an arbitrarily oriented non-quantizing magnetic field in QMOSFET and QWFET of all the materials as studied in this chapter by including the electron spin. Draw the corresponding graphs and explain the results.

R.4.5 Investigate the QCs in QMOSFET and QWFET of IV–VI, II–VI and stressed Kane type compounds in the presence of an arbitrarily oriented non-quantizing magnetic field by including the electron spin. Draw the corresponding graphs and explain the results.

R.4.6 Investigate the QCs in QMOSFET and QWFET of all the materials as stated in R.1.1 of Chapter 1 in the presence of an arbitrarily oriented non-quantizing magnetic field by including electron spin. Draw the corresponding graphs and explain the results.

R.4.7 Investigate the QCs in QMOSFET and QWFET of all the problems from R.2.1 to R.2.6 in the presence of an additional arbitrarily oriented electric field. Draw the corresponding graphs and explain the results.

R.4.8 Investigate the QCs in QMOSFET and QWFET of all the problems from R.2.1 to R.2.3 in the presence of arbitrarily oriented crossed electric and magnetic fields. Draw the corresponding graphs and explain the results.

R.4.9 Investigate the QCs in QMOSFET and QWFET of all the problems from R.2.1 to R.2.8 in the presence of surface states. Draw the corresponding graphs and explain the results.

R.4.10 Investigate the QCs in QMOSFET and QWFET of all the problems from R.2.1 to R.2.8 in the presence of hot electron effects. Draw the corresponding graphs and explain the results.

R.4.11 Investigate the QCs in QMOSFET and QWFET of all the problems from R.2.1 to R.2.6 by including the occupancy of the electrons in various electric sub-bands. Draw the corresponding graphs and explain the results.

References

1. L. D. Landau and E. M. Lifshitz, *Statistical Physics*, (Pergamon Press, London, 1970).
2. M. Cardona, W. Paul and H. Brooks Helv, *Acta Phys.* **33**, 329 (1960); A. F. Gibson in *Proceeding of International School of Physics "ENRICO FERMI"* course XIII, R. A Smith (ed.) (Academic Press, 1963), p. 171.

3. C. C. Wang and N. W. Ressler, *Phys. Rev.* **2**, 1827 (1970).
4. U. Rossler, *Solid State Commun.* **49**, 943 (1984).
5. G. E. Stillman, C. M. Wolfe and J. C. Dinmockin, in Semiconductors and Semimetals, Vol. 12 R. K. Willardson and A. C. Beer (eds.), (1977), p. 169.
6. D. G. Seiler, W. M. Beeker and L. M. Roth, *Phys. Rev.* **1**, 764 (1970).
7. P. C. Mathur and S. Jain, *Phys. Rev.* **19**, 3159 (1979).
8. P. R. Zhang, *Phys. Rev.* **1B**, 3450 (1970).
9. E. D. Palik, G. S. Picus, S. Teither and R. F. Wallis, *Phys. Rev.* **122**, 475 (1961).
10. E. J. Johnson and D. H. Dicley, *Phys. Rev.* **1**, 2676 (1970).
11. V. G. Agafonov, P. M. Valov, B. S. Ryvkin and I. D. Yaroshetskii, *Sov. Phys. Semicond.* **12**, 1182 (1978).

Chapter 5

The QC in 2D QWTs of Important Materials Under Cross-Fields Configuration

The difference between ordinary and extraordinary is that little extra.

5.1 Introduction

The importance of studying the effect of crossed electric and quantizing magnetic field on the various physical properties of different technological materials has already been discussed in the preface and extensive works in this particular aspect have also been presented in the literature [1–14]. In this chapter, we shall study the QC in 2D QWT's of tetragonal and nonlinear optical, III–V, ternary, quaternary, II–VI, IV–VI and stressed Kane type materials under cross-fields configurations in Sections 5.2.1–5.2.5, respectively. Section 5.3 contains the summary and conclusion and in Section 5.4, we present few open research problems in this context.

5.2 Theoretical Background

5.2.1 *The QC in QWFETs of nonlinear optical materials under cross-fields configuration*

From Chapter 1, the (1.6) can be written as

$$U(E) = \frac{p_s^2}{2M_\perp} + \frac{p_z^2}{2M_{||}}V(E) \tag{5.1}$$

where

$$U(E) \equiv \left[E(1+\alpha E)\left[(E+E_g)(E+E_g+\Delta_{\parallel}) + \delta\left(E+E_g+\frac{2}{3}\Delta_{\parallel}\right)+\frac{2}{9}(\Delta_{\parallel}^2-\Delta_{\perp}^2)\right]\right] \times \left[(E+E_g)+\left(E+E_g+\frac{2}{3}\Delta_{\parallel}\right)+\delta\left(E+E_g+\frac{1}{3}\Delta_{\parallel}\right) + \frac{1}{9}(\Delta_{\parallel}^2-\Delta_{\perp}^2)\right]^{-1}$$

$$p_s=\hbar k_s,\quad M_{\parallel}=\frac{m_{\parallel}^*\left(E_g+\frac{2}{3}\Delta_{\parallel}\right)}{(E_g+\Delta_{\parallel})},\quad M_{\perp}=\frac{m_{\perp}^*\left(E_g+\frac{2}{3}\Delta_{\perp}\right)}{(E_g+\Delta_{\perp})},\quad p_z=\hbar k_z$$

$$V(E) \equiv \left[(E+E_g)\left(E+E_g+\frac{2}{3}\Delta_{\parallel}\right)\right]\left[(E+E_g)\left(E+E_g+\frac{2}{3}\Delta_{\parallel}\right) + \delta\left(E+E_g+\frac{1}{3}\Delta_{\parallel}\right)+\frac{1}{9}(\Delta_{\parallel}^2-\Delta_{\perp}^2)\right]^{-1}$$

and for this chapter $\bar{E}$ is replaced by E.

We know that from electromagnetic theory that,

$$\vec{B} = \nabla \times \vec{A} \tag{5.2}$$

where $\vec{A}$ is the vector potential. In the presence of quantizing magnetic field B along z-direction, the (5.2) assumes the form

$$0\hat{i}+0\hat{j}+B\hat{k} = \begin{vmatrix} \hat{i} & \hat{j} & \hat{k} \\ \frac{\partial}{\partial x} & \frac{\partial}{\partial y} & \frac{\partial}{\partial z} \\ A_x & A_y & A_z \end{vmatrix}$$

where $\hat{i}, \hat{j}$ and $\hat{k}$ are orthogonal triads. Thus, we can write

$$\begin{aligned} \frac{\partial A_z}{\partial y} - \frac{\partial A_y}{\partial z} &= 0 \\ \frac{\partial A_x}{\partial z} - \frac{\partial A_z}{\partial x} &= 0 \\ \frac{\partial A_y}{\partial x} - \frac{\partial A_x}{\partial y} &= B \end{aligned} \tag{5.3}$$

This particular set of equations is being satisfied for $A_x = 0, A_y = Bx$ and $A_z = 0$.

Therefore, in the presence of the electric field E_o along x-axis and the quantizing magnetic field B along z-axis for the present case following [5.1] one can approximately write,

$$U(E) + |e|E_o\widehat{x}\rho(E) = \frac{\widehat{p}_x^2}{2M_\perp} + \frac{(\widehat{p}_y - |e|B\widehat{x})^2}{2M_\perp} + \frac{\widehat{p}_z^2}{2a(E)} \tag{5.4}$$

where $\rho(E) \equiv \frac{\partial}{\partial E}[U(E)]$ and $a(E) \equiv M_\parallel[V(E)]^{-1}$.

Let us define the operator $\widehat{\theta}$ as

$$\widehat{\theta} = -\widehat{p}_y + |e|B\widehat{x} - \frac{M_\perp E_o \rho(E)}{B} \tag{5.5}$$

Eliminating the operator $\widehat{x}$, between Equations (5.4) and (5.5), the DR of the conduction electron in QWTs of tetragonal and nonlinear optical materials in the presence of cross-fields configuration is given by

$$U(E) = \left[\left(\left(n+\frac{1}{2}\right)\hbar\omega_{01}\right) + \left(\frac{[\hbar k_z(E)]^2}{2a(E)}\right) - \left(\frac{E_o\hbar k_y \rho(E)}{B}\right) - \left(\frac{M_\perp \rho^2(E)E_o^2}{2B^2}\right)\right] \tag{5.6}$$

where $\omega_{01} \equiv \frac{|e|B}{M_\perp}$.

The E–k_y electron energy spectrum in this case can be written as

$$U(E) = \left[\left(\left(n+\frac{1}{2}\right)\hbar\omega_{01}\right) + \left(\frac{\hbar^2}{2a(E)}\right)\left(\frac{n_z\pi}{d_z}\right)^2 - \left(\frac{E_o\hbar k_y \rho(E)}{B}\right) - \left(\frac{M_\perp \rho^2(E)E_o^2}{2B^2}\right)\right] \tag{5.7}$$

The Landau sub-band energy (e_{A1}) and the EEM $m_y^*(E, n, n_z, E_0)$ can respectively be expressed as

$$U(e_{A1}) = \left[\left(\left(n+\frac{1}{2}\right)\hbar\omega_{01}\right) + \left(\frac{\hbar^2}{2a(e_{A1})}\right)\left(\frac{n_z\pi}{d_z}\right)^2 - \left(\frac{M_\perp \rho^2(e_{A1})E_o^2}{2B^2}\right)\right] \tag{5.8}$$

$$m_y^2(E, n, n_z, E_0) = \left(\frac{B}{E_0}\right)^2 \frac{1}{\rho^2(E)}\left[U(E) - \left(n+\frac{1}{2}\right)\hbar\omega_{01} - \frac{\hbar^2}{2a(E)}\left(\frac{n_z\pi}{d_z}\right)^2 + \frac{M_\perp^2 \rho^2(E)E_0^2}{2B^2}\right]$$

$$\times \left[U'(E) + \frac{\hbar^2}{2a^2(E)} a'(E) \left(\frac{n_z \pi}{d_z} \right)^2 + \frac{M_\perp^2 \rho(E)\rho'(E) E_0^2}{B^2} - \frac{\rho'(E)}{\rho(E)} \right.$$
$$\left. \times \left[U(E) - \left(n + \frac{1}{2} \right) \hbar\omega_{01} - \frac{\hbar^2}{2a(E)} \left(\frac{n_z \pi}{d_z} \right)^2 + \frac{M_\perp^2 \rho^2(E) E_0^2}{2B^2} \right] \right] \tag{5.9}$$

The surface electron concentration can be evaluated from the formula as follows:

$$N_s = \frac{-g_v}{\pi L_x} \sum_{n=0}^{n_{\max}} \sum_{n_z=1}^{n_{z\max}} \int_{e_{A1}}^{\infty} I_A \frac{\partial f_0(E)}{\partial(E)} dE \tag{5.10}$$

where $f_0(E)$ is the Fermi-Dirac distribution function and

$$I_A = \int_{x_l}^{x_h} dk_y = x_h - x_l \tag{5.11}$$

in which $x_h = x_l + \frac{eBL_x}{\hbar}$.

Thus, $I_A = \frac{eBL_x}{\hbar}$ and it is E–k relation independent

$$N_s = \frac{g_v eB}{\pi\hbar} \sum_{n=0}^{n_{\max}} \sum_{n_z=1}^{n_{z\max}} F_{-1}(\eta_{A1}) \tag{5.12}$$

where

$$\eta_{A1} = \frac{E_{F1} - e_{A1}}{k_B T}$$

Using (1.4) and (5.12) we can study the QC in this case.

5.2.2 *The QC in QWFETs of III–V, ternary and quaternary materials under cross-fields configuration*

(a) Three-band model of Kane

Under the conditions $\delta = 0$, $\Delta_\parallel = \Delta_\perp = \Delta$ and $m_\parallel^* = m_\perp^* = m_c$, the (5.6) assumes the form

$$I_{11}(E) = \left(n + \frac{1}{2} \right) \hbar\omega_0 + \frac{[\hbar k_z(E)]^2}{2m_c} - \frac{E_0}{B} \hbar k_y \{I_{11}(E)\}' - \frac{m_c E_0^2 [\{I_{11}(E)\}']^2}{2B^2} \tag{5.13}$$

The $E{-}k_y$ electron energy spectrum in this case can be written as

$$I_{11}(E) = \left(n+\frac{1}{2}\right)\hbar\omega_0 + \left(\frac{\hbar^2}{2m_c}\right)\left(\frac{n_z\pi}{d_z}\right)^2 - \frac{E_0}{B}\hbar k_y\{I_{11}(E)\}' - \frac{m_c E_0^2[\{I_{11}(E)\}']^2}{2B^2} \tag{5.14}$$

From (5.14) the Landau sub-band energy (e_{A2}) and the EEM $m_y^*(E, n, n_z, E_0)$ can respectively be expressed as

$$I_{11}(e_{A2}) = \left(n+\frac{1}{2}\right)\hbar\omega_0 + \left(\frac{\hbar^2}{2m_c}\right)\left(\frac{n_z\pi}{d_z}\right)^2 - \frac{m_c E_0^2[\{I_{11}(e_{A2})\}']^2}{2B^2} \tag{5.15}$$

$$\begin{aligned} m_y^2(E,n,n_z,E_0) = {} & \left(\frac{B}{E_0}\right)^2 [I_{11}'(E)]^{-2}\Bigg[I_{11}(E) - \left(n+\frac{1}{2}\right)\hbar\omega_0 \\ & - \frac{\hbar^2}{2m_c}\left(\frac{n_z\pi}{d_z}\right)^2 + \frac{m_c E_0^2}{2B^2}[I_{11}'(E)]^2\Bigg]\Bigg[I_{11}'(E)\left\{1+\frac{m_c E_0^2}{B^2}I_1''(E)\right\} \\ & - \frac{I_{11}''(E)}{I_1'(E)}\Bigg[I_{11}(E) - \left(n+\frac{1}{2}\right)\hbar\omega_0 - \frac{\hbar^2}{2m_c}\left(\frac{n_z\pi}{d_z}\right)^2 \\ & + \frac{m_c E_0^2}{2B^2}[I_{11}'(E)]^2\Bigg]\Bigg] \end{aligned} \tag{5.16}$$

The N_s can be written as

$$N_s = \frac{g_v eB}{\pi\hbar}\sum_{n=0}^{n_{\max}}\sum_{n_z=1}^{n_{z\max}} F_{-1}(\eta_{A2}) \tag{5.17}$$

where

$$\eta_{A2} = \frac{E_{F1} - e_{A2}}{k_B T}$$

Using (1.4) and (5.17) we can study the QC in this case.

(b) The two-band model of Kane

Under the condition $\Delta \gg E_g$, the (5.13) assumes the well-known form [1]

$$E(1+\alpha E) = \left(n+\frac{1}{2}\right)\hbar\omega_0 - \frac{E_0}{B}\hbar k_y(1+2\alpha E) - \frac{m_c E_0^2}{2B^2}(1+2\alpha E)^2 + \frac{[\hbar k_z(E)]^2}{2m_c} \quad (5.18)$$

The $E{-}k_y$ electron energy spectrum in this case can be written as

$$E(1+\alpha E) = \left(n+\frac{1}{2}\right)\hbar\omega_0 - \frac{E_0}{B}\hbar k_y(1+2\alpha E) - \frac{m_c E_0^2}{2B^2}(1+2\alpha E)^2 + \frac{\hbar^2}{2m_c}\left(\frac{n_z\pi}{d_z}\right)^2 \quad (5.19)$$

From (5.19) the Landau sub-band energy (e_{A3}) and the EEM $m_y^*(E, n, n_z, E_0)$ can respectively be expressed as

$$e_{A3}(1+\alpha e_{A3}) = \left(n+\frac{1}{2}\right)\hbar\omega_0 - \frac{m_c E_0^2}{2B^2}(1+2\alpha e_{A3})^2 + \frac{\hbar^2}{2m_c}\left(\frac{n_z\pi}{d_z}\right)^2 \quad (5.20)$$

$$m_y^*(E, n, n_z, E_0) = \left[\left(\frac{B}{E_0}\right)^2 (1+2\alpha E)^{-2} \times E(1+\alpha E) - \left(n+\frac{1}{2}\right)\hbar\omega_0 - \frac{\hbar^2}{2m_c}\left(\frac{n_z\pi}{d_z}\right)^2 + \frac{m_c E_0^2}{2B^2}(1+2\alpha E)^2\right]\left[(1+2\alpha E)\left(1+\frac{2m_c\alpha E_0^2}{B^2}\right) - \frac{2\alpha}{(1+2\alpha E)}\left\{E(1+\alpha E) - \left(n+\frac{1}{2}\right)\hbar\omega_0 - \frac{\hbar^2}{2m_c}\left(\frac{n_z\pi}{d_z}\right)^2 + \frac{m_c E_0^2}{2B^2}(1+2\alpha E)^2\right\}\right] \quad (5.21)$$

The N_s can be written as

$$N_s = \frac{g_v eB}{\pi\hbar}\sum_{n=0}^{n_{\max}}\sum_{n_z=1}^{n_{z\max}} F_{-1}(\eta_{A3}) \quad (5.22)$$

where

$$\eta_{A3} = \frac{E_{F1} - e_{A3}}{k_B T}$$

Using (1.4) and (5.22) we can study the QC in this case.

(c) Parabolic band models

For parabolic energy bands, $\alpha \to 0$ and we can write from (5.18) as

$$E = \left(n+\frac{1}{2}\right)\hbar\omega_0 + \frac{[\hbar k_z(E)]^2}{2m_c} - \frac{1}{2}m_c\left(\frac{E_0}{B}\right)^2 - \frac{E_0}{B}\hbar k_y \tag{5.23}$$

The $E{-}k_y$ electron energy spectrum in this case can be written as

$$E = \left(n+\frac{1}{2}\right)\hbar\omega_0 + \frac{\hbar^2}{2m_c}\left(\frac{n_z\pi}{d_z}\right)^2 - \frac{1}{2}m_c\left(\frac{E_0}{B}\right)^2 - \frac{E_0}{B}\hbar k_y \tag{5.24a}$$

From (5.24b) the Landau sub-band energy (e_{A4}) and the EEM $m_y^*(E, n, n_z, E_0)$ can respectively be expressed as

$$e_{A4} = \left(n+\frac{1}{2}\right)\hbar\omega_0 + \frac{\hbar^2}{2m_c}\left(\frac{n_z\pi}{d_z}\right)^2 - \frac{1}{2}m_c\left(\frac{E_0}{B}\right)^2 \tag{5.24b}$$

$$m_y^*(E, n, n_z, E_0) = \left(\frac{B}{E_0}\right)^2\left[E - \left(n+\frac{1}{2}\right)\hbar\omega_0 - \frac{\hbar^2}{2m_c}\left(\frac{n_z\pi}{d_z}\right)^2 + \frac{m_c E_0^2}{2B^2}\right] \tag{5.25}$$

The N_s can be written as

$$N_s = \frac{g_v eB}{\pi\hbar}\sum_{n=0}^{n_{\max}}\sum_{n_z=1}^{n_{z\max}} F_{-1}(\eta_{A4}) \tag{5.26}$$

where

$$\eta_{A4} = \frac{E_{F1} - e_{A4}}{k_B T}$$

Using (1.4) and (5.26) we can study the QC in this case.

5.2.3 *The QC in QWFETs of II–VI materials under cross-fields configuration*

In the presence of electric field along x-axis and the quantizing magnetic field B along z-axis, from equation (2.1) of Chapter 2, we can write

$$\hat{E} + |e|E_0\hat{x} = \frac{\hat{p}_x^2}{2m_\perp^*} + \frac{(\hat{p}_y - |e|B\hat{x})^2}{2m_\perp^*} + \frac{\hat{p}_z^2}{2m_\parallel^*} + D\left[\frac{\hat{p}_x^2}{2m_\perp^*} + \frac{(\hat{p}_y - |e|B\hat{x})^2}{2m_\perp^*}\right]^{1/2} \tag{5.27}$$

where

$$D \equiv \pm\frac{C_0\sqrt{2m_\perp^*}}{\hbar}$$

Let us define the operator $\widehat{\theta}$ as

$$\widehat{\theta} = -\hat{p}_y + |e|B\hat{x} - \frac{E_0 m_\perp^*}{B} \tag{5.28}$$

Eliminating $\hat{x}$ between (5.27) and (5.28), one obtains

$$\hat{E} + \frac{E_0}{B}\widehat{\theta} + \frac{E_0}{B}\hat{p}_y + \left(\frac{E_0}{B}\right)^2 m_\perp^* = \frac{\hat{p}_x^2}{2m_\perp^*} + \frac{\hat{\theta}^2}{2m_\perp^*} + \frac{E_0^2 m_\perp^*}{2B^2} + \frac{\hat{\theta}E_0}{B}$$
$$+ D\left[\frac{\hat{p}_x^2}{2m_\perp^*} + \frac{\hat{\theta}^2}{2m_\perp^*} + \frac{E_0^2 m_\perp^*}{2B^2}\right]^{1/2} + \frac{\hat{p}_z^2}{2m_\parallel^*} \tag{5.29}$$

Thus, the electron energy spectrum in this case can be expressed a

$$E = (\beta_1(n, E_0)) + \left(\frac{[\hbar k_z(E)]^2}{2m_\parallel^*}\right) - \left(\frac{E_0}{B}\hbar k_y\right) \tag{5.30}$$

where

$$\beta_1(n, E_0) \equiv \left[\left(n + \frac{1}{2}\right)\hbar\omega_{02} - \left(\frac{E_0^2 m_\perp^*}{2B^2}\right)\right.$$
$$\left. + D\left\{\left(n + \frac{1}{2}\right)\hbar\omega_{02} + \left(\frac{E_0^2 m_\perp^*}{2B^2}\right)\right\}^{1/2}\right] \quad \text{and} \quad \omega_{02} \equiv \frac{|e|B}{m_\perp^*}$$

The E–k_y electron energy spectrum in quantum wells in this case can be written as

$$E = (\beta_1(n, E_0)) + \left(\frac{\hbar^2}{2m_\parallel^*}\right)\left(\frac{n_z\pi}{d_z}\right)^2 - \left(\frac{E_0}{B}\hbar k_y\right) \tag{5.31a}$$

From (5.31) the Landau sub-band energy (e_{A5}) and the EEM $m_y^*(E, n, n_z, E_0)$ can respectively be expressed as

$$e_{A5} = (\beta_1(n, E_0)) + \left(\frac{\hbar^2}{2m_\parallel^*}\right)\left(\frac{n_z\pi}{d_z}\right)^2 \tag{5.31b}$$

$$m_y^*(E, E_0, n, n_z) = \left(\frac{B}{E_0}\right)^2 \left[E - \beta_1(n, E_0) - \frac{\hbar^2}{2m_\parallel}\left(\frac{n_z\pi}{d_z}\right)^2\right] \tag{5.32}$$

The N_s can be written as

$$N_s = \frac{g_v eB}{\pi\hbar} \sum_{n=0}^{n_{\max}} \sum_{n_z=1}^{n_{z\max}} F_{-1}(\eta_{A5}) \tag{5.33}$$

where

$$\eta_{A5} = \frac{E_{F1} - e_{A5}}{k_B T}$$

Using (1.4) and (5.33) we can study the QC in this case.

5.2.4 *The QC in QWFETs of IV–VI materials under cross-fields configuration*

(a) The McClure and Choi model

In the presence of an electric field E_0 along trigonal- axis (z-direction) and the quantizing magnetic field B along bisectrix axis (y-direction) from (3.27), we can write

$$\begin{aligned} &E(1+\alpha E) + |e|E_0\widehat{z}(1+2\alpha E) \\ &\quad = \frac{(\widehat{p}_x - |e|B\widehat{z})^2}{2m_1} + \frac{\hat{p}_y^2}{2m_2} + \frac{\hat{p}_z^2}{2m_3} + \frac{\hat{p}_y^2}{2m_2}\alpha E\left(1 - \frac{m_2}{m_2'}\right) \\ &\qquad - \alpha\frac{\hat{p}_y^2\hat{p}_z^2}{4m_2m_3} + \alpha\frac{\hat{p}_y^4}{4m_2m_2'} - \alpha\frac{(\hat{p}_x^2 - |e|B\widehat{z})^2\hat{p}_y^2}{4m_1m_2} \end{aligned} \tag{5.34}$$

Let us define the operator $\widehat{\theta}$ as

$$\widehat{\theta} = |e|B\widehat{z} - \hat{p}_x - \frac{m_1E_0}{B}(1+2\alpha E) \tag{5.35}$$

Eliminating $\hat{z}$ between Equations (5.34) and (5.35) one obtains

$$\begin{aligned} &E\,(1+\alpha E) + \frac{E_0}{B}(1+2\alpha E)\widehat{\theta} + \frac{E_0}{B}(1+2\alpha E)\widehat{p}_x + m_1\left(\frac{E_0}{B}\right)^2(1+2\alpha E)^2 \\ &= \left(\frac{\widehat{\theta}^2}{2m_1} + \frac{\hat{p}_z^2}{2m_3}\right) + \frac{\hat{p}_y^2}{2m_2} + \frac{\hat{p}_y^2}{2m_2}\alpha E\left(1 \quad \frac{m_2}{m_2'}\right) + \frac{\alpha\hat{p}_y^4}{4m_2m_2'} \\ &\quad - \frac{\alpha\hat{p}_y^2}{2m_2}\left[\frac{\widehat{\theta}^2}{2m_1} + \frac{\hat{p}_z^2}{2m_3}\right] - \frac{\alpha\hat{p}_y^2m_1E_0^2}{4m_2B^2}(1+2\alpha E)^2 \\ &\quad + \frac{E_0}{B}\widehat{\theta}(1+2\alpha E) + \frac{1}{2}m_1\left(\frac{E_0}{B}\right)^2(1+2\alpha E)^2 \end{aligned} \tag{5.36}$$

Therefore, the required DR is given by

$$E(1+\alpha E) = \left(n+\frac{1}{2}\right)\hbar\omega_{03} + \frac{\hbar^2}{2m_2}\left(\frac{n_y\pi}{d_y}\right)^2 - \frac{E_0}{B}(1+2\alpha E)\hbar k_x - \frac{1}{2}m_1\left(\frac{E_0}{B}\right)^2(1+2\alpha E)^2 + \frac{\hbar^2}{2m_2}\left(\frac{n_y\pi}{d_y}\right)^2\alpha E\left(1-\frac{m_2}{m_2'}\right) + \frac{\alpha\hbar^4}{4m_2m_2'}\left(\frac{n_y\pi}{d_y}\right)^4 - \frac{\alpha\hbar^2}{2m_2}\left(\frac{n_y\pi}{d_y}\right)^2\left(n+\frac{1}{2}\right)\hbar\omega_{03} - \frac{\alpha\hbar^2 m_1 E_0^2}{4m_2B^2}(1+2\alpha E)^2\left(\frac{n_y\pi}{d_y}\right)^2 \quad (5.37)$$

where

$$\omega_{03} \equiv \frac{|e|B}{\sqrt{m_1m_3}}$$

From (5.37) the Landau sub-band energy (e_{A6}) and the EEM $m_x^*(E,n, n_y, E_0)$ can respectively be expressed as

$$e_{A6}(1+\alpha e_{A6}) = \left(n+\frac{1}{2}\right)\hbar\omega_{03} + \frac{\hbar^2}{2m_2}\left(\frac{n_y\pi}{d_y}\right)^2 - \frac{1}{2}m_1\left(\frac{E_0}{B}\right)^2 \times (1+2\alpha e_{A6})^2 + \frac{\hbar^2}{2m_2}\left(\frac{n_y\pi}{d_y}\right)^2\alpha e_{A6}\left(1-\frac{m_2}{m_2'}\right) + \frac{\alpha\hbar^4}{4m_2m_2'}\left(\frac{n_y\pi}{d_y}\right)^4 - \frac{\alpha\hbar^2}{2m_2}\left(\frac{n_y\pi}{d_y}\right)^2\left(n+\frac{1}{2}\right)\hbar\omega_{03} - \frac{\alpha\hbar^2 m_1 E_0^2}{4m_2B^2}(1+2\alpha e_{A6})^2\left(\frac{n_y\pi}{d_y}\right)^2 \quad (5.38)$$

$$m_x^*(E,n,n_y,E_0) = \left(\frac{B}{E_0}\right)^2(1+2\alpha E)^{-1}\left[E(1+\alpha E) - \left(n+\frac{1}{2}\right)\hbar\omega_{03} - \frac{\hbar^2}{2m_2}\left(\frac{n_y\pi}{d_y}\right)^2 + \frac{1}{2}m_1\left(\frac{E_0}{B}\right)^2 \times (1+2\alpha E)^2 - \frac{\hbar^2}{2m_2}\alpha E\left(1-\frac{m_2}{m_2'}\right)\left(\frac{n_y\pi}{d_y}\right)^2 - \frac{\alpha\hbar^4}{4m_2m_2'}\left(\frac{n_y\pi}{d_y}\right)^4 + \frac{\alpha\hbar^2}{2m_2}\left(\frac{n_y\pi}{d_y}\right)^2\left(n+\frac{1}{2}\right)\hbar\omega_{03} + \frac{\alpha\hbar^2}{4m_2B^2}\left(\frac{n_y\pi}{d_y}\right)^2 m_1\right.$$

$$
\begin{aligned}
&\times E_0^2(1+2\alpha E)^2\Big]\left[-\frac{2\alpha}{(1+2\alpha E)}\left[E(1+\alpha E)-\left(n+\frac{1}{2}\right)\hbar\omega_{03}\right.\right.\\
&-\frac{\hbar^2}{2m_2}\left(\frac{n_y\pi}{d_y}\right)^2+\frac{1}{2}m_1\left(\frac{E_0}{B}\right)^2(1+2\alpha E)^2\\
&-\frac{\hbar^2}{2m_2}\alpha E\left(1-\frac{m_2}{m_2'}\right)\left(\frac{n_y\pi}{d_y}\right)^2-\frac{\alpha\hbar^4}{4m_2m_2'}\left(\frac{n_y\pi}{d_y}\right)^4\\
&\left.+\frac{\alpha\hbar^2}{2m_2}\left(\frac{n_y\pi}{d_y}\right)^2\left(n+\frac{1}{2}\right)\hbar\omega_{03}+\frac{\alpha\hbar^2}{4m_2B^2}\left(\frac{n_y\pi}{d_y}\right)^2m_1E_0^2(1+2\alpha E)^2\right]\\
&+1+2\alpha m_1\left(\frac{E_0}{B}\right)^2+\frac{\alpha^2\hbar^2}{m_2B^2}\left(\frac{n_y\pi}{d_y}\right)^2m_1E_0^2\\
&\left.-\frac{\hbar^2}{2m_2}\alpha\left(1-\frac{m_2}{m_2'}\right)\left(\frac{n_y\pi}{d_y}\right)^2(1+2\alpha E)^{-1}\right]
\end{aligned}
\tag{5.39}
$$

The surface electron concentration can be evaluated from the formula as follows:

$$
N_s=\frac{-g_v}{\pi L_z}\sum_{n=0}^{n_{\max}}\sum_{n_z=1}^{n_{z\max}}\int_{e_{A6}}^{\infty}J_A\frac{\partial f_0(E)}{\partial(E)}dE \tag{5.40}
$$

where, J_A is the Fermi-Dirac distribution function and

$$
J_A=\int_{\bar{x}_l}^{\bar{x}_h}dk_x=\bar{x}_h-\bar{x}_l \tag{5.41}
$$

in which $\bar{x}_h=\bar{x}_l+\frac{eBL_z}{\hbar}$

Thus, $J_A=\frac{eBL_z}{\hbar}$ and it is E–k relation independent.

$$
N_s=\frac{g_veB}{\pi\hbar}\sum_{n=0}^{n_{\max}}\sum_{n_z=1}^{n_{z\max}}F_{-1}(\eta_{A6}) \tag{5.42}
$$

where

$$
\eta_{A6}=\frac{E_{F1}-e_{A6}}{k_BT}
$$

and we have used the relation

$$
-\int_{e_{A6}}^{\infty}\frac{\partial f_0}{\partial E}dE=F_{-1}(\eta_{A6}) \tag{5.43}
$$

Using (1.4) and (5.43) we can study the QC in this case.

(b) The Cohen model

In the presence of an electric field E_o along trigonal axis and the quantizing magnetic field B along bisectrix axis for this case, the (2.7) assumes the form

$$E(1+\alpha E)+|e|E_o\widehat{z}(1+2\alpha E)=\frac{(\widehat{p}_x-|e|B\widehat{z})^2}{2m_1}+\frac{\widehat{p}_z^2}{2m_3}-\frac{\alpha E\widehat{p}_y^2}{2m_2'}+\frac{\widehat{p}_y^2}{2m_2}(1+\alpha E)+\frac{\alpha\widehat{p}_y^4}{4m_2m_2'} \quad (5.44)$$

Using the same operator $\widehat{\theta}$ as defined by (5.35) and eliminating $\widehat{z}$ between (5.35) and (5.44) one can write

$$E(1+\alpha E)=\left(\frac{\widehat{\theta}^2}{2m_1}+\frac{\widehat{p}_z^2}{2m_3}\right)-\frac{E_o}{B}\widehat{p}_x(1+2\alpha E)-\frac{E_0^2m_1}{2B^2}(1+2\alpha E)^2-\frac{\alpha E\widehat{p}_y^2}{2m_2'}+\frac{\widehat{p}_y^2}{2m_2}(1+\alpha E)+\frac{\alpha\widehat{p}_y^4}{4m_2m_2'} \quad (5.45)$$

Thus, the electron energy spectrum can be expressed as

$$E(1+\alpha E)=\left[\left(n+\frac{1}{2}\right)\hbar\omega_{03}-\frac{E_0}{B}\hbar k_x(1+2\alpha E)-\frac{1}{2}m_1\left(\frac{E_0}{B}\right)^2(1+2\alpha E)^2-\left(\frac{\alpha E[\hbar k_y(E)]^2}{2m_2'}\right)+\frac{[\hbar k_y(E)]^2}{2m_2}(1+\alpha E)+\left[\frac{\alpha[\hbar k_y(E)]^4}{4m_2m_2'}\right]\right] \quad (5.46)$$

The $E{-}k_x$ dispersion relation in QWTs in this case can be written as

$$E(1+\alpha E)=\left[\left(n+\frac{1}{2}\right)\hbar\omega_{03}-\frac{E_0}{B}\hbar k_x(1+2\alpha E)-\frac{1}{2}m_1\left(\frac{E_0}{B}\right)^2(1+2\alpha E)^2-\left(\frac{\alpha E\hbar^2}{2m_2'}\right)\left(\frac{n_y\pi}{d_y}\right)^2+\left(\frac{\hbar^2}{2m_2}\right)\left(\frac{n_y\pi}{d_y}\right)^2(1+\alpha E)+\left[\frac{\alpha\hbar^4}{4m_2m_2'}\right]\left(\frac{n_y\pi}{d_y}\right)^4\right] \quad (5.47)$$

From (5.47) the Landau sub-band energy (e_{A7}) and the EEM $m_x^*(E, n, n_y, E_0)$ can respectively be expressed as

$$e_{A7}(1+\alpha e_{A7}) = \left[\left(n+\frac{1}{2}\right)\hbar\omega_{03} - \frac{1}{2}m_1\left(\frac{E_0}{B}\right)^2(1+2\alpha e_{A7})^2 \right.$$
$$- \left(\frac{\alpha e_{A7}\hbar^2}{2m_2'}\right)\left(\frac{n_y\pi}{d_y}\right)^2 + \left(\frac{\hbar^2}{2m_2}\right)\left(\frac{n_y\pi}{d_y}\right)^2(1+\alpha e_{A7})$$
$$\left. + \left[\frac{\alpha\hbar^4}{4m_2m_2'}\right]\left(\frac{n_y\pi}{d_y}\right)^4\right] \tag{5.48}$$

$$m_x^*(E, n, n_y, E_0) = \left(\frac{B}{E_0}\right)^2\left(\frac{1}{(1+2\alpha E)^2}\right)\left[E(1+\alpha E) - \left(n+\frac{1}{2}\right)\hbar\omega_{03}\right.$$
$$+ \frac{1}{2}m_1\left(\frac{E_0}{B}\right)^2(1+2\alpha E)^2 + \frac{\alpha E\hbar^2}{2m_2'}\left(\frac{n_y\pi}{d_y}\right)^2 - \frac{\hbar^2}{2m_2}\left(\frac{n_y\pi}{d_y}\right)^2(1+\alpha E)$$
$$\left. - \frac{\alpha\hbar^4}{4m_2m_2'}\left(\frac{n_y\pi}{d_y}\right)^4\right]\left[\left[(1+2\alpha E) + \frac{\alpha\hbar^2}{2m_2'}\left(\frac{n_y\pi}{d_y}\right)^2 - \frac{\alpha\hbar^2}{2m_2}\left(\frac{n_y\pi}{d_y}\right)^2\right]\right.$$
$$- \frac{2\alpha}{(1+2\alpha E)}\left[E(1+\alpha E) - \left(n+\frac{1}{2}\right)\hbar\omega_{03} + \frac{1}{2}m_1\left(\frac{E_0}{B}\right)^2(1+2\alpha E)^2\right.$$
$$\left.\left. + \frac{\alpha E\hbar^2}{2m_2'}\left(\frac{n_y\pi}{d_y}\right)^2 - \frac{\hbar^2}{2m_2}\left(\frac{n_y\pi}{d_y}\right)^2(1+\alpha E) - \frac{\alpha\hbar^4}{4m_2m_2'}\left(\frac{n_y\pi}{d_y}\right)^4\right]\right] \tag{5.49}$$

The surface concentration N_s in this case assumes the form

$$N_s = \frac{g_v eB}{\pi\hbar}\sum_{n=0}^{n_{\max}}\sum_{n_z=1}^{n_{z\max}} F_{-1}(\eta_{A7}) \tag{5.50}$$

where

$$\eta_{A7} = \frac{E_{F1} - e_{A7}}{k_B T}$$

Using (1.4) and (5.50) we can study the QC in this case.

(c) The Lax model

Under cross-field configuration from the (3.49) of Chapter 3, one can write

$$E(1+\alpha E)+|e|E_o\hat{z}(1+2\alpha E)=\frac{(p_x-|e|B\hat{z})^2}{2m_1}+\frac{\hat{p}_y^2}{2m_2}+\frac{\hat{p}_z^2}{2m_3} \tag{5.51}$$

Using the same operator $\widehat{\theta}$ as used for McClure and Choi model we get

$$E(1+\alpha E)+\frac{E_0}{B}(1+2\alpha E)\widehat{\theta}+\frac{E_0}{B}(1+2\alpha E)\widehat{p}_x+m_1\left(\frac{E_0}{B}\right)^2(1+2\alpha E)^2$$
$$=\frac{\widehat{\theta}^2}{2m_1}+\frac{\widehat{p}_z^2}{2m_3}+\frac{m_1}{2}\left(\frac{E_0}{B}\right)^2(1+2\alpha E)^2+\frac{\widehat{\theta}E_0}{B}(1+2\alpha E)+\frac{\widehat{p}_y^2}{2m_2} \tag{5.52}$$

Therefore, the electron DR assumes the form

$$E(1+\alpha E)=\left(n+\frac{1}{2}\right)\hbar\omega_{03}-\frac{E_0}{B}(1+2\alpha E)\hbar k_x$$
$$+\frac{[\hbar k_y]^2}{2m_2}-\frac{m_1}{2}\left(\frac{E_0}{B}\right)^2(1+2\alpha E)^2 \tag{5.53}$$

The E–k_x dispersion relation in this case can be written as

$$E(1+\alpha E)=\left(n+\frac{1}{2}\right)\hbar\omega_{03}-\frac{E_0}{B}(1+2\alpha E)\hbar k_x$$
$$+\frac{\hbar^2}{2m_2}\left(\frac{n_y\pi}{d_y}\right)^2-\frac{m_1}{2}\left(\frac{E_0}{B}\right)^2(1+2\alpha E)^2 \tag{5.54}$$

From (5.54) the Landau sub-band energy (e_{A7}) and the EEM $m_x^*(E,n,n_y,E_0)$ can respectively be expressed as

$$e_{A8}(1+\alpha e_{A8})=\left(n+\frac{1}{2}\right)\hbar\omega_{03}+\frac{\hbar^2}{2m_2}\left(\frac{n_y\pi}{d_y}\right)^2$$
$$-\frac{m_1}{2}\left(\frac{E_0}{B}\right)^2(1+2\alpha e_{A8})^2 \tag{5.55}$$

$$m_x^*(E,n,n_y,E_0)=\left(\frac{B}{E_0}\right)^2(1+2\alpha E)^{-2}\left[E(1+\alpha E)-\left(n+\frac{1}{2}\right)\hbar\omega_{03}\right.$$
$$\left.-\frac{\hbar^2}{2m_2}\left(\frac{n_y\pi}{d_y}\right)^2+\frac{m_1}{2}\left(\frac{E_0}{B}\right)^2(1+2\alpha E)^2\right]$$

$$\times\left[1+2\alpha E+2\alpha m_1\left(\frac{E_0}{B}\right)^2(1+2\alpha E)-\frac{2\alpha}{(1+2\alpha E)}\left[E(1+\alpha E)\right.\right.$$

$$\left.\left.-\left(n+\frac{1}{2}\right)\hbar\omega_{03}-\frac{\hbar^2}{2m_2}\left(\frac{n_y\pi}{d_y}\right)^2+\frac{m_1}{2}\left(\frac{E_0}{B}\right)^2(1+2\alpha E)^2\right]\right] \quad (5.56)$$

The surface concentration N_s in this case assumes the form

$$N_s=\frac{g_veB}{\pi\hbar}\sum_{n=0}^{n_{\max}}\sum_{n_z=1}^{n_{z\max}}F_{-1}(\eta_{A8}) \quad (5.57)$$

where

$$\eta_{A8}=\frac{E_{F1}-e_{A8}}{k_BT}$$

Using (1.4) and (5.56) we can study the QC in this case.

(c) Parabolic ellipsoidal model

For this model, the electron DR for the present case is given by

$$E=\left(n+\frac{1}{2}\right)\hbar\omega_{03}-\frac{E_0}{B}\hbar k_x+\frac{[\hbar k_y]^2}{2m_2}-\frac{m_1}{2}\left(\frac{E_0}{B}\right)^2 \quad (5.58)$$

The E–k_x dispersion relation in this case can be written as

$$E=\left(n+\frac{1}{2}\right)\hbar\omega_{03}-\frac{E_0}{B}\hbar k_x+\frac{\hbar^2}{2m_2}\left(\frac{n_y\pi}{d_y}\right)^2-\frac{m_1}{2}\left(\frac{E_0}{B}\right)^2 \quad (5.59)$$

From (5.59) the Landau sub-band energy (e_{A8}) and the EEM $m_x^*(E,n,n_y,E_0)$ can respectively be expressed as

$$e_{A8}=\left(n+\frac{1}{2}\right)\hbar\omega_{03}+\frac{\hbar^2}{2m_2}\left(\frac{n_y\pi}{d_y}\right)^2-\frac{m_1}{2}\left(\frac{E_0}{B}\right)^2 \quad (5.60)$$

$$m_x^*(E,n,n_y,E_0)=\left(\frac{B}{E_0}\right)^2\left[E-\left(n+\frac{1}{2}\right)\hbar\omega_{03}-\frac{\hbar^2}{2m_2}\left(\frac{n_y\pi}{d_y}\right)^2\right.$$

$$\left.+\frac{m_1}{2}\left(\frac{E_0}{B}\right)^2\right] \quad (5.61)$$

The surface concentration N_s in this case assumes the form

$$N_s=\frac{g_veB}{\pi\hbar}\sum_{n=0}^{n_{\max}}\sum_{n_z=1}^{n_{z\max}}F_{-1}(\eta_{A9}) \quad (5.62)$$

where

$$\eta_{A9} = \frac{E_{F1} - e_{A9}}{k_B T}$$

Using (1.4) and (5.61) we can study the QC in this case.

5.2.5 *The QC in QWFETs of stressed Kane type materials under cross-fields configuration*

In the presence of quantizing magnetic field B along z-direction and the electric field along x-axis, from (2.22) one obtains

$$\frac{\hat{p}_x^2}{2M_{\parallel}(E)} + \frac{(\hat{p}_y - |e|\, B\hat{x})^2}{2M_{\perp}(E)} + R(E)\hat{p}_z^2 = \rho_5(E) + |e|E_0\hat{x}[\rho_5(E)]' \tag{5.63}$$

where

$$M_{\parallel}(E) \equiv \frac{1}{2P(E)}, \quad P(E) \equiv \frac{1}{\hbar^2}(E - \alpha_1), \quad M_{\perp}(E) \equiv \frac{1}{2Q(E)},$$

$$Q(E) \equiv \frac{1}{\hbar^2}(E - \alpha_2), \quad R(E) \equiv \frac{1}{\hbar^2}(E - \alpha_3) \quad \text{and}$$

$$\rho_5(E) \equiv [t_1 E^3 - t_2 E^2 + t_3 E + t_4]$$

Let us define the operator $\hat{\theta}$ as

$$\hat{\theta} = -\hat{p}_y + |e|B\hat{x} - \frac{M_{\perp}(E)E_0[\rho_5(E)]'}{B} \tag{5.64}$$

Combining Equations (5.62) and (5.63), we can write

$$\frac{\hat{p}_x^2}{2M_{\parallel}(E)} + \frac{\hat{\theta}^2}{2M_{\perp}(E)} + \frac{M_{\perp}(E)E_0^2\{[\rho_5(E)]'\}^2}{2B^2} + \hat{\theta}\frac{E_0[\rho_5(E)]'}{B} + R(E)\hat{p}_z^2$$
$$= \rho_5(E) + \frac{E_0[\rho_5(E)]'\hat{\theta}}{B} + \frac{E_0}{B}[\rho_5(E)]'\hat{p}_y + \frac{E_0^2}{B^2}M_{\perp}(E)\{[\rho_5(E)]'\}^2 \tag{5.65}$$

Therefore, the electron DR in stressed Kane type semiconductors in the presence of cross-field configuration can be expressed as

$$\rho_5(E) = \left(n + \frac{1}{2}\right)\hbar\bar{\omega}(E) + R(E)[\hbar k_z(E)]^2 - \frac{E_0}{B}[\rho_5(E)]'\hbar k_y(E)$$
$$- \left\{\frac{M_{\perp}(E)E_0^2\{[\rho_5(E)]'\}^2}{2B^2}\right\} \tag{5.66}$$

where

$$\bar{\omega}(E) \equiv \frac{|e|B}{\sqrt{M_{\|}(E)M_{\perp}(E)}}$$

The $E{-}k_x$ DR in this case can be written as

$$\rho_5(E) = \left(n+\frac{1}{2}\right)\hbar\bar{\omega}(E) + R(E)\left[\hbar\frac{n_z\pi}{d_z}\right]^2 - \frac{E_0}{B}[\rho_5(E)]'\hbar k_y(E) - \left\{\frac{M_{\perp}(E)E_0^2\{[\rho_5(E)]'\}^2}{2B^2}\right\} \quad (5.67)$$

From (5.67) the Landau sub-band energy (e_{A8}) and the EEM $m_y^*(E,n,n_y,E_0)$ can respectively be expressed as

$$\rho_5(e_{A10}) = \left(n+\frac{1}{2}\right)\hbar\bar{\omega}(e_{A10}) + R(e_{A10})\hbar^2\left(\frac{n_z\pi}{d_z}\right)^2 - \left\{\frac{M_{\perp}(e_{A10})E_0^2\{[\rho_5(e_{A10})]'\}^2}{2B^2}\right\} \quad (5.68)$$

$$m_y^*(E,n,n_y,E_0) = \left(\frac{B}{E_0}\right)^2\left[\frac{1}{\rho_5'(E)}\right]^2\left[\rho_{51} - \left(n+\frac{1}{2}\right)\hbar[\bar{\omega}(E)] - R(E)\left(\frac{n_z\pi\hbar}{d_z}\right)^2\right]\left[\rho_{51}'(E) - \left(n+\frac{1}{2}\right)\hbar[\bar{\omega}(E)]' - R(E)\left(\frac{n_z\pi\hbar}{d_z}\right)^2\right] - \left[\frac{\rho_5''(E)}{\rho_5'(E)}\right]\left[\rho_{51}(E) - \left(n+\frac{1}{2}\right)\hbar[\bar{\omega}(E)] - R(E)\left(\frac{n_z\pi\hbar}{d_z}\right)^2\right] \quad (5.69)$$

where $\rho_{51}(E) = \rho_5(E) + \phi_7(E)$ and $\phi_7(E) = [M_{\perp}^2(E)E_0^2[\rho_5'(E)]^2/2B^2]$.

The surface concentration N_s in this case assumes the form

$$N_s = \frac{g_v eB}{\pi\hbar}\sum_{n=0}^{n_{\max}}\sum_{n_z=1}^{n_{z\max}} F_{-1}(\eta_{A10}) \quad (5.70)$$

where

$$\eta_{A10} = \frac{E_{F1} - e_{A10}}{k_B T}$$

Using (1.4) and (5.70) we can study the QC in this case.

5.3 Summary and Conclusions

In Section 5.2.1, we have formulated the expressions of Landau sub-band energy, the EEM and the electron statistics in QWTs of nonlinear optical materials under cross-fields configurations as given by (5.8), (5.9) and (5.12), respectively. By combining (5.12) and (1.4) we can study QC in this case. In Section 5.2.2, we have formulated the expressions of Landau sub-band energy, the EEM and the electron statistics in QWTs of III–V, ternary and quaternary compounds under cross-fields configurations as given by (5.15), (5.16) and (5.17), respectively on the basis of three-band model of Kane. By combining (5.17) and (1.4) we can study QC in this case. In the same section, we have also formulated the expressions of Landau sub-band energy, the EEM and the electron statistics in QWTs of III–V, ternary and quaternary compounds under cross-fields configurations as given by (5.20), (5.21) and (5.22), respectively on the basis of two-band model of Kane and by combining (5.22) and (1.4) we can study QC in this case. In the same section, we have also studied the results for isotropic parabolic energy bands under the condition $\alpha \to 0$. In Section 5.2.3, we have formulated the expressions of Landau sub-band energy, the EEM and the electron statistics in QWTs of II–VI compounds under cross-fields configurations as given by (5.31), (5.32) and (5.33), respectively on the basis of three-band model of Kane. By combining (5.33) and (1.4) we can study QC in this case.

In Section 5.2.4, we have formulated the expressions of Landau sub-band energy, the EEM and the electron statistics in QWTs of IV–VI compounds under cross-fields configurations as given by (5.38), (5.39) and (5.42), respectively on the basis of McClure and Choi model, (5.48), (5.49) and (5.50) in accordance with the model of Cohen, (5.50) (5.51) and (5.52) in accordance with the model of Lax and (5.60), (5.61) and (5.62) on the basis of parabolic Ellipsoidal model, respectively. Combining (5.42), (5.50) and (5.62) with (1.4) we can study QC in each case of the aforementioned models. In Section 5.2.5, we have formulated the expressions of Landau sub-band energy, the EEM and the electron statistics in QWTs of stressed Kane type materials under cross-fields configurations as given by (5.68), (5.69) and (5.70), respectively. By combining (5.70) and (1.4) we can study QC in this case.

We have not given the graphs of the QCs' for 2D QWT's under cross-field configurations made of many materials as discussed here since we feel that our readers will perform the intricate numerical computation to obtain the oscillatory plots for better assessment and joy of understanding. Incidentally, one can easily infer that how little is presented and how much more there is yet to be investigated in the research field of QCs' in this case for different 2D QWTs having different band structures in general which is the signature of the co-existence of new physics, advanced mathematics. Finally, we again realize that it is the band structure that changes in a fundamental way and consequently, all the physical properties of all the electronic materials change radically leading to new physical concepts.

5.4 Open Research Problems

R.5.1 Investigate the QCs in the presence of an arbitrarily oriented crossed electric and quantizing magnetic fields in 2DQWTs of nonlinear optical materials as studied in this chapter. Draw the corresponding graphs and explain the results.

R.5.2 Investigate the QCs in 2D QWTs of IV–VI, II–VI and stressed Kane type compounds in the presence of an arbitrarily oriented crossed electric and quantizing magnetic fields. Draw the corresponding graphs and explain the results.

R.5.3 Investigate the QCs in 2D QWTs of all the materials as stated in R.1.1 of Chapter 1 in the presence of an arbitrarily oriented crossed electric and quantizing magnetic fields. Draw the corresponding graphs and explain the results.

R.5.4 Investigate the QCs in 2D QWTs of all the materials as studied in this chapter by including the electron spin in the presence of an arbitrarily oriented crossed non-uniform electric and oscillatory quantizing magnetic fields. Draw the corresponding graphs and explain the results.

R.5.5 Investigate the QCs in 2D QWTs of IV–VI, II–VI and stressed Kane type compounds in the presence of an arbitrarily oriented crossed non-uniform electric and oscillatory quantizing magnetic fields by including the electron spin. Draw the corresponding graphs and explain the results.

R.5.6 Investigate the QCs in 2D QWTs of all the materials as stated in R.1.1 of Chapter 1 in the presence of an arbitrarily oriented crossed non-uniform electric and oscillatory quantizing magnetic fields by including electron spin. Draw the corresponding graphs and explain the results.

R.5.7 Investigate the QCs in 2D QWTs of all the problems from R.2.1 to R.2.6 in the presence of an arbitrarily oriented crossed non-uniform electric and oscillatory quantizing magnetic fields. Draw the corresponding graphs and explain the results.

R.5.8 Investigate the QCs in 2D QWTs of all the problems from R.2.1 to R.2.3 in the presence of arbitrarily oriented crossed electric and magnetic fields in the presence of non-uniform strain. Draw the corresponding graphs and explain the results.

R.5.9 Investigate the QCs in 2D QWTs of all the problems from R.2.1 to R.2.8 in the presence of surface states. Draw the corresponding graphs and explain the results.

R.5.10 Investigate the QCs in 2D QWTs of all the problems from R.2.1 to R.2.8 in the presence of hot electron effects. Draw the corresponding graphs and explain the results.

References

1. W. Zawadzki and B. Lax, *Phys. Rev. Lett.* **16**, 1001 (1966).
2. M. J. Harrison, *Phys. Rev.* A **29**, 2272 (1984); J. Zak and W. Zawadzki, *Phys. Rev.* **145**, 536 (1966).
3. W. Zawadzki, Q. H. Vrehen and B. Lax, *Phys. Rev.* **148**, 849 (1966); Q. H. Vrehen, W. Zawadzki and M. Reine, *Phys. Rev.* **158**, 702 (1967); M. H. Weiler, W. Zawadzki and B. Lax, *Phys. Rev.* **163**, 733 (1967).
4. W. Zawadzki and J. Kowalski, *Phys. Rev. Lett.* **27**, 1713 (1971); C. Chu, M.-S Chu, and T. Ohkawa, *Phys. Rev. Lett.* **41**, 653 (1978); P. Hu and C. S. Ting, *Phys. Rev. B* **36**, 9671 (1987).
5. E. I. Butikov, A. S. Kondratev and A. E. Kuchma, *Sov. Phys. Sol. State.* **13**, 2594 (1972).
6. K. P. Ghatak, J. P. Banerjee, B. Goswami and B. Nag, Nonlinear Optics and Quantum Optics, **16**, 241 (1996); M. Mondal and K. P. Ghatak, *Physica Status Solidi (b)* **133**, K67 (1986).
7. M. Mondal, N. Chattopadhyay and K. P. Ghatak, Journal of Low temperature Physics, **66**, 131 (1987); K. P. Ghatak and M. Mondal, *Zeitschrift fur Physik B* **69**, 471 (1988).
8. M. Mondal and K.P. Ghatak, in Physics Letters A, **131A**, 529 (1988); M. Mondal and K. P. Ghatak, in *Physica Status Solidi (b) Germany* **147**, K179 (1988); B. Mitra and K. P. Ghatak, *Phys. Lett.* **137A**, 413 (1989).
9. B. Mitra, A. Ghoshal and K. P. Ghatak, *Physica Status Solidi (b)* **154**, K147 (1989).
10. B. Mitra and K. P. Ghatak, *Physica Status Solidi (b)* **164**, K13 (1991); K. P. Ghatak and B. Mitra, *Int. J. Electron.* **70**, 345 (1991); K. P. Ghatak, B. Goswami, M. Mitra and B. Nag, *Nonlinear Opt.* **16**, 9 (1996).

11. K. P. Ghatak, M. Mitra, B. Goswami and B. Nag, *Nonlin. Opt.* **16**, 167, (1996); K. P. Ghatak, J. P. Banerjee, B. Goswami and B. Nag, in *Nonlin. Opt. Quant. Opt.* **16**, 241 (1996); K. P. Ghatak, D. K. Basu and B. Nag, *J. Phys. Chem. Sol.* **58**, 133 (1997).
12. K. P. Ghatak, N. Chattopadhyay and S. N. Biswas, Optoelectronic materials, devices, packaging and interconnects, *Proc. Society of Photo-optical and Instrumentation Engineers (SPIE)* **836**, 203 (1988); K. P. Ghatak, M. Mondal and S. Bhattacharyya, *SPIE* **1284**, 113 (1990); K. P. Ghatak, Photonic materials and optical bistability, *SPIE* **1280**, 53 (1990).
13. K. P. Ghatak and S.N. Biswas, Growth and characterization of materials for infrared detectors and nonlinear optical switches, *SPIE* **1484**, 149 (1991).
14. K. P. Ghatak, SPIE, *Fiber Optic and Laser Sensors* IX **1584**, 435 (1992).

Chapter 6

The QC in 2D MOSFETs of Important Materials Under Cross-Fields Configuration

Courage is not the absence of fear; it the tremendous effort for positive action in spite of real fear

6.1 Introduction

In this chapter, we shall study the QC in 2D MOSFETs of tetragonal and nonlinear optical, III–V, ternary, quaternary, II–VI, IV–VI and stressed Kane type materials under cross-fields configurations in Sections 6.2.1–6.2.5, respectively. Section 6.3 contains the summary and conclusion and in Section 6.4, we present few open research problems in this context.

6.2 Theoretical Background

6.2.1 *The QC in MOSFETs of tetragonal and nonlinear optical materials undcr cross-fields configuration*

In the presence of a surface electric field along z-direction, the k_z component from (5.6) assumes the form

$$k_z = \frac{\sqrt{2}}{\hbar}[\alpha_{51}(E, k_y) - eF_s Z \beta_{51}(E, k_y)]^{1/2} \tag{6.1}$$

where

$$\alpha_{51}(E, k_y) = \left[\left[a(E)U(E) - \left(n + \frac{1}{2}\right)\hbar\omega_{01}a(E) + \frac{E_0 \hbar k_y}{B}\rho(E)a(E)\right.\right.$$
$$\left. + \frac{M_\perp E_0^2}{2B^2}[\rho^2(E)a(E)]\right]$$

and

$$\beta_{51}(E,k_y) = [a'(E)U(E) + U'(E)a(E) - \left(n+\frac{1}{2}\right)\hbar\omega_{01}a'(E) + \frac{M_\perp E_0^2}{2B^2}[2\rho(E)\rho'(E)a(E) + a'(E)\rho^2(E)] + \frac{E_0\hbar k_y}{B}[a'(E)\rho(E) + \rho'(E)a(E)]$$

The use of (1.7) and (6.1) leads to the 2D electron DR in MOSFETs of nonlinear optical materials under cross-fields configurations as

$$\alpha_{51}(E,k_y) = \left(\frac{\hbar e F_s}{\sqrt{2}}\right)^{2/3} \times S_i[\beta_{51}(E,k_y)]^{2/3} \tag{6.2}$$

The 2D surface electron concentration can be written as

$$N_s = \frac{g_v e B}{\pi\hbar}\sum_{n=0}^{n_{\max}}\sum_{n_z=1}^{n_{z\max}} F_{-1}(\eta_{51}) \tag{6.3}$$

where

$$\eta_1 = (k_B T)^{-1}[\bar{E}_{F1} - e_{51}], \quad \bar{E}_{F1} = \left[eV_g - \frac{e^2 N_s}{C_{oX}} - \bar{E}_{fCL}\right], \quad \bar{E}_{fCL}$$

being the separation of the Fermi level and the conduction band edge in the substrate away from the surface under cross-fields configuration, and e_{51} can be expressed through the equation

$$\alpha_{51}(e_{51},0) = \left(\frac{\hbar e F_s}{\sqrt{2}}\right)^{2/3} \times S_i[\beta_{51}(e_{51},0)]^{2/3} \tag{6.4}$$

Using (1.4) and (6.3) we can study the QC in this case

6.2.2 *The QC in MOSFETs of III–V, ternary and quaternary materials under cross-fields configuration*

(a) Three-band model of Kane

In the presence of a surface electric field along z-direction, the k_z component from (5.13) assumes the form

$$k_z = \frac{\sqrt{2m_c}}{\hbar}[\alpha_{52}(E,k_y) - eF_s Z\beta_{52}(E,k_y)]^{1/2} \tag{6.5}$$

where

$$\alpha_{52}(E, k_y) = [I_{11}(E) - \left(n + \frac{1}{2}\right)\hbar\omega_0 + \frac{E_0}{B}\hbar k_y I'_{11}(E) + \frac{m_c E_0^2}{2B^2}[I'_{11}(E)]^2]$$

and

$$\beta_{52}(E, k_y) = \left[I'_{11}(E) + \frac{E_0}{B}\hbar k_y I'_{11}(E) + \frac{m_c E_0^2}{2B^2}[I'_{11}(E) I''_{11}(E)]\right]$$

The use of (1.7) and (6.5) leads to the 2D electron DR in MOSFETs of nonlinear optical materials under cross-fields configurations as

$$\alpha_{52}(E, k_y) = \left(\frac{\hbar e F_s}{\sqrt{2m_c}}\right)^{2/3} \times S_i[\beta_{52}(E, k_y)]^{2/3} \tag{6.6}$$

The 2D surface electron concentration can be written as

$$N_s = \frac{g_v eB}{\pi\hbar} \sum_{n=0}^{n_{\max}} \sum_{n_z=1}^{n_{z\max}} F_{-1}(\eta_{52}) \tag{6.7}$$

where

$$\eta_{52} = \frac{\bar{E}_{F1} - e_{52}}{k_B T}$$

and e_{52} can be expressed through the equation

$$\alpha_{52}(e_{52}, 0) = \left(\frac{\hbar e F_s}{\sqrt{2m_c}}\right)^{2/3} \times S_i[\beta_{52}(e_{52}, 0)]^{2/3} \tag{6.8}$$

Using (1.4) and (6.7) we can study the QC in this case

(b) Two-band model of Kane

In the presence of a surface electric field along z-direction, the k_z component from (5.18) assumes the form

$$k_z = \frac{\sqrt{2m_c}}{\hbar}[\alpha_{53}(E, k_y) - eF_s Z\beta_{53}(E, k_y)]^{1/2} \tag{6.9}$$

where

$$\alpha_{53}(E, k_y) = \left[E(1 + \alpha E) - \left(n + \frac{1}{2}\right)\hbar\omega_0 + \frac{E_0}{B}\hbar k_y[1 + 2\alpha E] - \frac{m_c E_0^2}{2B^2}[1 + 2\alpha E]^2\right]$$

and

$$\beta_{53}(E,k_y)=\left[1+2\alpha E+\frac{E_0}{B}\hbar k_y 2\alpha+\frac{m_c E_0^2}{2B^2}[(1+2\alpha E)(2\alpha)]\right]$$

The use of (1.7) and (6.9) leads to the 2D electron DR in this case as

$$\alpha_{53}(E,k_y)=\left(\frac{\hbar eF_s}{\sqrt{2m_c}}\right)^{2/3}\times S_i[\beta_{53}(E,k_y)]^{2/3} \tag{6.10}$$

The 2D surface electron concentration can be written as

$$N_s=\frac{g_v eB}{\pi\hbar}\sum_{n=0}^{n_{\max}}\sum_{n_z=1}^{n_{z\max}}F_{-1}(\eta_{53}) \tag{6.11}$$

where

$$\eta_{53}=\frac{\bar{E}_{F1}-e_{53}}{k_B T}$$

and e_{53} can be expressed through the equation

$$\alpha_{53}(e_{53},0)=\left(\frac{\hbar eF_s}{\sqrt{2m_c}}\right)^{2/3}\times S_i[\beta_{53}(e_{53},0)]^{2/3} \tag{6.12}$$

Using (1.4) and (6.11) we can study the QC in this case.

(c) Parabolic band model

In the presence of a surface electric field along z-direction, the k_z component from (5.23) assumes the form

$$k_z=\frac{\sqrt{2m_c}}{\hbar}[\alpha_{54}(E,k_y)-eF_s Z]^{1/2} \tag{6.13}$$

where

$$\alpha_{54}(E,k_y)=\left[E-\left(n+\frac{1}{2}\right)\hbar\omega_0+\frac{E_0}{B}\hbar k_y+\frac{m_c E_0^2}{2B^2}\right]$$

The use of (1.7) and (6.13) leads to the 2D electron DR in this case as

$$\alpha_{54}(E,k_y)=\left(\frac{\hbar eF_s}{\sqrt{2m_c}}\right)^{2/3}\times S_i \tag{6.14}$$

The 2D surface electron concentration can be written as

$$N_s=\frac{g_v eB}{\pi\hbar}\sum_{n=0}^{n_{\max}}\sum_{n_z=1}^{n_{z\max}}F_{-1}(\eta_{54}) \tag{6.15}$$

where

$$\eta_{54} = \frac{\bar{E}_{F1} - e_{54}}{k_B T}$$

and e_{54} can be expressed through the equation

$$\alpha_{54}(e_{54}, 0) = \left(\frac{\hbar e F_s}{\sqrt{2m_c}}\right)^{2/3} \times S_i \tag{6.16}$$

Using (1.4) and (6.15) we can study the QC in this case.

6.2.3 *The QC in MOSFETs of II–VI materials under cross-fields configuration*

In the presence of a surface electric field along z-direction, the k_z component from (5.30) assumes the form

$$k_z = \frac{\sqrt{2m_\parallel^*}}{\hbar}[\alpha_{55}(E, k_y) - eF_s Z]^{1/2} \tag{6.17}$$

where

$$\alpha_{55}(E, k_y) = [E - \beta_1(n, E_0) + \left(\frac{E_0}{B}\hbar k_y\right)$$

The use of (1.7) and (6.17) leads to the 2D electron DR in this case as

$$\alpha_{55}(E, k_y) = \left(\frac{\hbar e F_s}{\sqrt{2m_\parallel^*}}\right)^{2/3} \times S_i \tag{6.18}$$

The 2D surface electron concentration can be written as

$$N_s = \frac{g_v e B}{\pi\hbar} \sum_{n=0}^{n_{\max}} \sum_{n_z=1}^{n_{z\max}} F_{-1}(\eta_{55}) \tag{6.19}$$

where

$$\eta_{55} = \frac{\bar{E}_{F1} - e_{55}}{k_B T}$$

and e_{55} can be expressed through the equation

$$\alpha_{55}(e_{55}, 0) = \left(\frac{\hbar e F_s}{\sqrt{2m_\parallel^*}}\right)^{2/3} \times S_i \tag{6.20}$$

Using (1.4) and (6.19) we can study the QC in this case.

6.2.4 The QC in MOSFETs of IV–VI materials under cross-fields configuration

In accordance with Bangert *et al.* [1] the dispersion relation is given by

$$g(E) = \bar{F}_1(E)k_s^2 + \bar{F}_2(E)k_z^2 \tag{6.21}$$

where

$$g(E) \equiv 2E, \quad \bar{F}_1(E) \equiv \frac{R_1^2}{E+E_g} + \frac{S_1^2}{E+\Delta_c'} + \frac{Q_1^2}{E+E_g},$$

$$\bar{F}_2(E) \equiv \frac{2C_5^2}{E+E_g} + \frac{(S_1+Q_1)^2}{E+\Delta_c''}$$

$$R_1^2 = 2.3\times 10^{-19}(eVm)^2, \quad C_5^2 = 0.83\times 10^{-19}(eVm)^2, \quad Q_1^2 = 1.3R_1^2,$$

$$S_1^2 = 4.6R_1^2, \quad \Delta_c' = 3.07eV, \quad \Delta_c'' = 3.28eV$$

The (6.21) can be written as

$$\frac{p_s^2}{2\bar{M}_1(E)} + \frac{p_z^2}{2\bar{M}_2(E)} = g(E) \tag{6.22}$$

where

$$\bar{M}_1(E) = \frac{\hbar^2}{2\bar{F}_1(E)} \quad \text{and} \quad \bar{M}_2(E) = \frac{\hbar^2}{2\bar{F}_2(E)}$$

In the presence of a magnetic field B along z-direction and electric field E_0 along x-direction, from (6.22) we can write that

$$\frac{\hat{p}_x^2}{2\bar{M}_1(E)} + \frac{(\hat{p}_y - eB\hat{x})^2}{2\bar{M}_1(E)} + \frac{\hat{p}_z^2}{2\bar{M}_2(E)} = 2E + 2E_0\hat{x} \tag{6.23}$$

where the hats denote the corresponding operators.

Let us define the operator $\hat{\theta}$ as

$$\hat{\theta} = -\hat{p}_y + eB\hat{x} - \frac{2E_0\bar{M}_1(E)}{B} \tag{6.24}$$

Using (2.23) and (2.24) we get the DR in bulk specimens of IV–VI materials under cross-fields configuration as

$$2E = \left(n+\frac{1}{2}\right)\hbar\bar{\omega}_4(E) + \frac{\hbar^2 k_z^2}{2\bar{M}_2(E)} - \frac{2E_0}{B}\hbar k_y - \frac{2E_0^2}{B^2}\bar{M}_1(E) \tag{6.25}$$

where

$$\bar{\omega}_4(E) = \frac{eB}{\bar{M}_1(E)}$$

In the presence of a surface electric field F_s along z-direction from (6.25) we can write that

$$k_z = \frac{\sqrt{2m_2(E)}}{\hbar}[\alpha_{56}(E,k_y) - eF_sZ\beta_{56}(E)]^{1/2} \tag{6.26}$$

where

$$\alpha_{56}(E,k_y) = \left[2E - \left(n+\frac{1}{2}\right)\frac{\hbar eB}{\bar{M}_1(E)} + \frac{2E_0}{B}\hbar k_y + \frac{2E_0^2}{B^2}\bar{M}_1(E)\right],$$

$$\beta_{56}(E) = \left[2 + \frac{\left(n+\frac{1}{2}\right)\hbar eB}{\bar{M}_1^2(E)}\bar{M}'_1(E) + \frac{2E_0^2}{B^2}\bar{M}_1(E)\bar{M}'_1(E)\right]$$

Using (6.26) and (1.7), the DR in this case is given by

$$\alpha_{56}(E,k_y) = S_i\left[\left(\frac{\hbar eF_s\beta_{56}(E)}{\sqrt{2\bar{M}_2(E)}}\right]^{2/3} \tag{6.27}$$

The 2D surface electron concentration can be written as

$$N_s = \frac{g_v eB}{\pi\hbar}\sum_{n=0}^{n_{\max}}\sum_{n_z=1}^{n_{z\max}} F_{-1}(\eta_{56}) \tag{6.28}$$

where

$$\eta_{56} = \frac{\bar{E}_{F1} - e_{56}}{k_BT}$$

and e_{56} can be expressed through the equation

$$\alpha_{56}(e_{56},0) = S_i\left[\left(\frac{\hbar eF_s\beta_{56}(e_{56})}{\sqrt{2\bar{M}_2(e_{56})}}\right]^{2/3} \tag{6.29}$$

Using (1.4) and (6.28) we can study the QC in this case.

6.2.5 *The QC in MOSFETs of stressed Kane type materials under cross-fields configuration*

Using (5.65) we can write

$$k_z = \frac{1}{\hbar}[\alpha_{57}(E,k_y) - eF_sZ\beta_{57}(E,k_y)]^{1/2} \tag{6.30}$$

where

$$\alpha_{57}(E,k_y)=\left[\rho_5(E)-\left(n+\frac{1}{2}\right)\hbar\bar{\omega}(E)+\frac{E_0}{B}\rho_5'(E)\hbar k_y+\frac{M_\perp E^2}{2B^2}[\rho_5'(E)]^2\right],$$

$$\beta_{57}(E,k_y)=\left[\rho_5'(E)-\left(n+\frac{1}{2}\right)\hbar[\bar{\omega}(E)]'+\frac{E_0}{B}\rho_5''(E)\hbar k_y\right.$$
$$\left.+\frac{M_\perp^2 E^2}{B^2}\rho_5'(E)\rho_5''(E)\right]$$

Using (6.30) and (1.7), the DR in this case assumes the form

$$\alpha_{57}(E,k_y)=S_i[\hbar eF_s\beta_{57}(E,k_y)]^{2/3} \tag{6.31}$$

The 2D surface electron concentration can be written as

$$N_s=\frac{g_v eB}{\pi\hbar}\sum_{n=0}^{n_{\max}}\sum_{n_z=1}^{n_{z\max}}F_{-1}(\eta_{57}) \tag{6.32}$$

where

$$\eta_{57}=\frac{\bar{E}_{F1}-e_{57}}{k_B T}$$

and e_{57} can be expressed through the equation

$$\alpha_{57}(e_{57},0)=S_i[\hbar eF_s\beta_{57}(e_{57},0)]^{2/3} \tag{6.33}$$

Using (1.4) and (6.32) we can study the QC in this case.

6.3 Summary and Conclusions

In Section 6.2.1, we have formulated the expressions of Landau sub-band energy and the electron statistics in 2D MOSFETs of nonlinear optical materials under cross-fields configurations as given by (6.3) and (6.4), respectively. By combining (6.3) and (1.4), we can study QC in this case. In Section 6.2.2, we have formulated the expressions of Landau sub-band energy and the electron statistics in 2D MOSFETs of III–V, ternary and quaternary compounds under cross-fields configurations as given by (6.7) and (6.8), respectively, on the basis of three-band model of Kane. By combining (6.7) and (1.4), we can study QC in this case. In the same section, we have also formulated the expressions of Landau sub-band energy and the electron statistics in 2D MOSFETs of III–V, ternary and quaternary compounds under cross-fields configurations as given by (6.11) and (6.12), respectively, on the basis of two-band model of Kane and by combining

(6.11) and (1.4) we can study QC in this case. In the same section, we have also studied the results for isotropic parabolic energy bands under the condition $\alpha \to 0$. In Section 6.2.3, we have formulated the expressions of Landau sub-band energy and the electron statistics in 2D MOSFETs of II–VI compounds under cross-fields configurations as given by (6.19) and (6.20), respectively, on the basis of three-band model of Kane. By combining (6.19) and (1.4) we can study QC in this case.

In Section 6.2.4, we have formulated the expressions of Landau sub-band energy, and the electron statistics in 2D MOSFETs of IV–VI compounds under cross-fields configurations as given by (6.28) and (6.29), respectively. By combining (6.28) and (1.4), we can study QC in this case. In Section 6.2.5, we have formulated the expressions of Landau sub-band energy and the electron statistics in 2D MOSFETs of stressed Kane type materials under cross-fields configurations as given by (6.32) and (6.33), respectively. By combining (6.32) and (1.4), we can study QC in this case.

We have not given the graphs of the QCs for 2D MOSFETs under cross-field configurations made of many materials as discussed here since we feel that our readers will perform the intricate numerical computation to obtain the oscillatory plots for better assessment and joy of understanding.

6.4 Open Research Problems

R.6.1 Investigate the QCs in the presence of an arbitrarily oriented crossed electric and quantizing magnetic fields in 2D MOSFETs of nonlinear optical materials as studied in this chapter. Draw the corresponding graphs and explain the results.

R.6.2 Investigate the QCs in 2D MOSFETs of IV–VI, II–VI and stressed Kane type compounds in the presence of an arbitrarily oriented crossed electric and quantizing magnetic fields. Draw the corresponding graphs and explain the results.

R.6.3 Investigate the QCs in 2D MOSFETs of all the materials as stated in R.1.1 of chapter 1 in the presence of an arbitrarily oriented crossed electric and quantizing magnetic fields. Draw the corresponding graphs and explain the results.

R.6.4 Investigate the QCs in 2D MOSFETs of all the materials as studied in this chapter by including the electron spin in the presence of an arbitrarily oriented crossed non-uniform electric and oscillatory quantizing magnetic fields. Draw the corresponding graphs and explain the results.

R.6.5 Investigate the QCs in 2D MOSFETs of IV–VI, II–VI and stressed Kane type compounds in the presence of an arbitrarily oriented crossed non-uniform electric and oscillatory quantizing magnetic fields by including the electron spin. Draw the corresponding graphs and explain the results.

R.6.6 Investigate the QCs in 2D MOSFETs of all the materials as stated in R.1.1 of chapter 1 in the presence of an arbitrarily oriented crossed non-uniform electric and oscillatory quantizing magnetic fields by including electron spin. Draw the corresponding graphs and explain the results.

R.6.7 Investigate the QCs in 2D QWTs' of all the problems from R.2.1 to R.2.6 in the presence of an arbitrarily oriented crossed non-uniform electric and oscillatory quantizing magnetic fields. Draw the corresponding graphs and explain the results.

R.6.8 Investigate the QCs in 2D QWTs' of all the problems from R.2.1 to R.2.3 in the presence of arbitrarily oriented crossed electric and magnetic fields in the presence of non-uniform strain. Draw the corresponding graphs and explain the results.

R.6.9 Investigate the QCs in 2D QWTs' of all the problems from R.2.1 to R.2.8 in the presence of surface states. Draw the corresponding graphs and explain the results.

R.6.10 Investigate the QCs in 2D QWTs' of all the problems from R.2.1 to R.2.8 in the presence of hot electron effects. Draw the corresponding graphs and explain the results.

Reference

1. E. Bangert and P. Kastner, *Phys. Stat. Sol. (b)* **61**, 503 (1974).

Chapter 7

The QC in Nanowire (NW) FETs of Important Materials

My greatest glory is not in never falling, but in rising every time I fall

7.1 Introduction

A carrier in semiconducting samples, moves on the closed 3D wave vector (k) space wherever the overall electron energy (E) is constant. The system spatiality will manufacture 1D electron motion once; these carriers are forced to move in 1 direction where as other two wave vector components assume quantized values. Hence, the 1D energy wave vector DRs become straight lines whose length decreases with the increase of the size quantum numbers leading to quantization. The physical properties of such 1D systems have been investigated in the recent literature extensively [1–41]. Although many researchers had reported their investigations in various forms, even so, it seems that the study of other new aspects of such 1D structure is becoming more and more necessary. In this chapter, we shall study the QC in NWFETs. From Sections 7.2.1 up to 7.2.11 of theoretical background 7.2, we have investigated the QC in NWFETs of nonlinear optical, III–V, ternary and quaternary, II–VI, IV–VI, stressed Kane type, bismuth telluride, II–V, GaP and $PtSb_2$ materials, respectively.

In this context, it may be noted that with the discovery of carbon nanotubes (CNs) in 1991 by Iijima [42], the CNs have been recognized as fascinating materials with nanometer dimensions uncovering new phenomena in the sphere of low-dimensional science and technology. The significant physical properties of these quantum materials make them ideal candidates to reveal new phenomena in nanoelectronics. The CNs find wide applications in conductive [43, 44] and high-strength composites [45], chemical sensors [46], field emission displays [47, 48], hydrogen storage media [49, 50], nanotweezeres [51], nanogears [52], nanocantilever devices [53], nanomotors [54] and nanoelectronic devices [55, 56]. Single-walled carbon nanotubes (SWCNs) emerge to be excellent materials for single-molecule electronics [57–61] such as nanotube-based diodes [62], single electron transistors [63], random access memory cells [64], logic circuits [65], gigahertz oscillators [66], data storage nanodevices [67], nanorelay [68] and in other low-dimensional devices. The CNs can be bespoke into a metal or a semiconductor based on the diameter and the chiral index numbers (m, n), where the integers m and n denote the number of unit vectors along two directions in the honeycomb crystal lattice of graphene [69]. For armchair and zigzag nanotubes, the chiral indices are given as $m = n$ and $m = 0$, respectively [70]. Another class of CN called as chiral CN has distinct integers m and n. Besides, a CN can be a metallic if $m - n = 3q$; where $q = 1, 2, 3, \ldots$ otherwise it is a semiconductor. Metallic SWCNs have received substantial attention as potential substitutions for traditional interconnect materials like Cu due to their excellent inherent electrical and thermal properties. Since the carriers are confined, in a metallic SWCN, the inclusion of the sub-band energy owing to Born–Von Karman (BVK) boundary conditions [72] for their unique band structure becomes prominent. The quantization of the motion of the carriers in such structures leads to the discontinuity in the density-of-state (DOS) function due to van Hove singularity (VHS) of the wave vectors. In the Section 7.2.12, we shall explore the QC in CN.

Besides, we have also studied three important electronic quantities, namely EEM, ER, field emission and photoemission in this context. Section 7.3 contains the results and discussion and Section 7.4 presents few open research problems in this context.

7.2 Theoretical Background

7.2.1 *The QC in NWFETs of tetragonal and nonlinear optical materials*

The 1D dispersion relations from (3.1) can be written as

$$k_z^2 = A_{11}(E, n_x, n_y) \tag{7.1}$$

where

$$A_{11}(E, n_x, n_y) = [f_2(E)]^{-1}[\gamma(E) - \phi_1(n_x, n_y) f_1(E)],$$

$$\phi_1(n_x, n_y) = \left(\frac{n_x \pi}{d_x}\right)^2 + \left(\frac{n_y \pi}{d_y}\right)^2, \; n_x = (1, 2, 3, \ldots), \; n_y = (1, 2, 3, \ldots)$$

are the size quantum numbers along x and y directions, respectively, d_x and d_y are the nano-thickness along x and y directions, respectively.

The quantized sub-band energy (E_{11}) is given by

$$\gamma(E_{11}) = f_1(E_{11})\phi_1(n_x, n_y) \tag{7.2}$$

The EEM can be expressed as

$$m^*(E, n_x, n_y) = \frac{\hbar^2}{2} A'_{11}(E, n_x, n_y) \tag{7.3}$$

The electron concentration per unit length under the condition of extreme carrier degeneracy can be expressed as

$$n_s = \frac{2g_v}{\pi} S[B_{11}(E'_{F1D}, n_x, n_y)] \tag{7.4}$$

where

$$S = \sum_{n_x=1}^{n_{x\max}} \sum_{n_y=1}^{n_{y\max}}, \quad B_{11}(E'_{F1D}, n_x, n_y) = [A_{11}(E'_{F1D}, n_x, n_y)]^{1/2},$$

$$E'_{F1D} = \left[e\bar{V}_g - E_{fb} - \frac{e^2 n_0}{C_{ins}}\right],$$

E_{fb} is Fermi energy of the bulk material,

$$C_{ins} = \frac{2\pi\varepsilon_{ins}}{\ln\left(\frac{2t_{ins} + t_{wire}}{t_{wire}}\right)}$$

ε_{ins} is the permittivity of the insulator layer, t_{ins} is the thickness of the insulator layer and t_{wire} is the diameter of the wire and the other notations are given in [72].

Using (7.4) and (1.4) the C_g can be expressed as

$$C_g = \left[\frac{2g_v e^2}{\pi} S(\theta_1)\right]\left[1 + \frac{2g_v e^2}{\pi C_{ins}} S(\theta_1)\right]^{-1} \tag{7.5}$$

where

$$\theta_1 = \frac{\partial}{\partial E'_{F1D}}(B_{11}(E'_{F1D}, n_x, n_y))$$

The ER in this case can be expressed as

$$\frac{D}{\mu} = \frac{1}{e}[S[B_{11}(E'_{F1D}, n_x, n_y)]][S[\theta_1]]^{-1} \tag{7.6}$$

The current due to Fowler–Nordheim (FN) field emission (I_F) can be written as

$$I_F = \frac{1}{2}\left[S\int_{E_{11}}^{\infty}[e.vn_1t_{11}]\right] \tag{7.7}$$

where v is the velocity of the electron and given by

$$v = \frac{1}{\hbar}\frac{\partial E}{\partial k_z}, \quad n_1 = N_{1D}(E).f(E)dE,$$

$N_{1D}(E)$ is the 1D DOS function per sub-bands and can be written as $N_{1D}(E) = \frac{2g_v}{\pi}\frac{\partial k_z}{\partial E}$, $f(E)$ is the Fermi–Dirac occupation probability factor. It appears that I is a function of the product of the carrier velocity, concentration and transmission coefficient. These three quantities in turn depend totally on the dispersion relation of the material. From (7.7) we can write

$$I_F = \frac{1}{2}\left[S\left[\int_{E_{11}}^{\infty}\left[\frac{e}{\hbar}\frac{\partial E}{\partial k_z}.\frac{2g_v}{\pi}\frac{\partial k_z}{\partial E}f(E)dE\right]\bar{t}_{11}\right]\right] \tag{7.8}$$

The term "$\frac{\partial E}{\partial k_z}$" from velocity and the term $\frac{\partial k_z}{\partial E}$ from the DOS function per sub-bands cancel each other leaving the constant pre-factor.

Therefore, (1.6b) assumes the form

$$I_F = \frac{e.g_v}{\hbar\pi}\left[S\left[\int_{E_{11}}^{\infty}[f(E)dE].\bar{t}_{11}\right]\right] \tag{7.9}$$

The term "t_{11}" in this case can be expressed as

$$\bar{t}_{11} = \exp(-\beta_{11}) \tag{7.10}$$

in which

$$\beta_{11} = \frac{4[A_{11}(V_0, n_x, n_y)]^{3/2}}{3eF_{sz}[A'_{11}(V_0, n_x, n_y)]} \tag{7.11}$$

F_{sz} is the surface electric field along z-axis,

$$A_{11}(V_0, n_x, n_y) = [f_2(V_0)]^{-1}[\gamma(V_0) - \phi_1(n_x, n_y)f_1(V_0)],$$

V_0 is equal to the addition of the Fermi energy in the corresponding case and the work function ϕ_w of the material,

$$f_2(V_0) \equiv \frac{\hbar^2 E_{go}(E_{go} + \Delta_{||})}{\left[2m_{||}^*\left(E_{go} + \frac{2}{3}\Delta_{||}\right)\right]}\left[(V_0 + E_{go})\left(V_0 + E_{go} + \frac{2}{3}\Delta_{||}\right)\right]$$

$$\gamma(V_0) \equiv V_0(V_0 + E_{go})[(V_0 + E_{go})(V_0 + E_{go} + \Delta_{||}) + \delta\left(V_0 + E_{go} + \frac{2}{3}\Delta_{||}\right) + \frac{2}{9}(\Delta_{||}^2 - \Delta_{\perp}^2)]$$

$$f_1(V_0) \equiv \frac{\hbar^2 E_{go}(E_{go} + \Delta_{\perp})}{\left[2m_{\perp}^*\left(E_{go} + \frac{2}{3}\Delta_{\perp}\right)\right]}\left[\delta\left(V_0 + E_{go} + \frac{1}{3}\Delta_{||}\right) + (V_0 + E_{go})\left(V_0 + E_{go} + \frac{2}{3}\Delta_{||}\right) + \frac{1}{9}(\Delta_{||}^2 - \Delta_{||}^2)\right]$$

$$A'_{11}(V_0, n_x, n_y) = \left[\frac{-A_{11}(V_0, n_x, n_y) f'_2(V_0)}{f_2(V_0)} + [f_2(V_0)]^{-1}[\gamma'(V_0) - f'_1(V_0)\phi_1(n_x, n_y)]\right],$$

$$f'_2(V_0) = \left[[\hbar^2 E_{g0}(E_{g0} + \Delta_{||})]\left[2m_{||}^*\left(E_{g0} + \frac{2}{3}\Delta_{||}\right)\right]^{-1}\left[2V_0 + 2E_{g0} + \frac{2}{3}\Delta_{||}\right]\right]$$

$$f'_1(V_0) = \left[[\hbar^2 E_{g0}(E_{g0} + \Delta_{\perp})]\left[2m_{\perp}^*\left(E_{g0} + \frac{2}{3}\Delta_{\perp}\right)\right]^{-1} \times \left[2V_0 + 2E_{g0} + \frac{2}{3}\Delta_{||} + \delta\right]\right]$$

and

$$\gamma'(V_0) = \left[\frac{\gamma(V_0)(2V_0 + E_{g0})}{V_0(V_0 + E_{g0})} + (V_0(V_0 + E_{g0}))[2V_0 + 2E_{g0} + \Delta_{||} + \delta]\right]$$

Thus, we can write

$$I_F = \frac{2g_v e k_B T}{h} \sum_{n_x=1}^{n_{x\max}} \sum_{n_y=1}^{n_{y\max}} [F_0(\eta_{11})\,\bar{t}_{11}] \tag{7.12}$$

where

$$\eta_{11} = \frac{E'_{F1D} - E_{11}}{k_B T}$$

Therefore, the field emitted current is given by

$$I_F = \frac{2g_v e k_B T}{h}[SF_0(\eta_{11})\exp(-\beta_{11})] \tag{7.13}$$

In this connection, the photocurrent can be expressed as

$$I_p = \frac{\alpha_o e g_v}{2}\left[S\int_{\Delta_1}^{\infty}\left(\frac{2}{\pi}\frac{\partial k_z}{\partial E}\right)\left(\frac{1}{\hbar}\frac{\partial E}{\partial k_z}\right)f(E)dE\right] \tag{7.14}$$

where α_0 is the probability of photoemission $\Delta_1 \equiv E_{11} + W - h\nu$, W is the electron affinity, $h\nu$ is the energy of the incident photon along z axis.

Using (7.14), one can write,

$$I_p = \frac{\alpha_o e g_v k_B T}{\pi\hbar}[SF_0(\eta_6)] \tag{7.15}$$

where

$$\eta_6 \equiv \left[\frac{E'_{F1D} - (E_{11} + W - h\nu)}{k_B T}\right]$$

7.2.2 *The QC in NWFETs of III–V, ternary and quaternary materials*

(i) The Three-band model of Kane

Under the conditions, $\delta = 0$, $\Delta_{\parallel} = \Delta_{\perp} = \Delta$ (isotropic spin-orbit splitting constant) and $m^*_{\parallel} = m^*_{\perp} = m_c$ (isotropic effective electron mass at the edge of the conduction band), (7.1) to (7.5) get simplified as

$$k_z^2 = A_{12}(E, n_x, n_y) \tag{7.16}$$

$$I_{11}(E_{12}) = \frac{\hbar^2}{2m_c}\phi_1(n_x, n_y) \tag{7.17}$$

$$m^*(E) = m_c I'_{11}(E) \tag{7.18}$$

$$n_s = \frac{2g_v}{\pi}S\left[B_{13}(E'_{F1D}, n_x, n_y)\right] \tag{7.19}$$

$$C_g = \left[\frac{2g_v e^2}{\pi}S(\theta_2)\right]\left[1 + \frac{2g_v e^2}{\pi C_{ins}}S(\theta_2)\right]^{-1} \tag{7.20}$$

where

$$A_{12}(E, n_x, n_y) = \left[\frac{2m_c}{\hbar^2}I_{11}(E) - \phi_1(n_x, n_y)\right],$$

E_{12} is the sub-band energy in this case,

$$B_{13}(E'_{F1D}, n_x, n_y) = [A_{12}(E'_{F1D}, n_x, n_y)]^{1/2},$$

$$\theta_2 = \frac{\partial}{\partial E'_{F1D}}(B_{13}(E'_{F1D}, n_x, n_y))$$

The ER in this case can be expressed as

$$\frac{D}{\mu} = \frac{1}{e}[S[B_{13}(E'_{F1D}, n_x, n_y)]][S[\theta_2]]^{-1} \tag{7.21}$$

The field emitted current (I_F) can be written as

$$I_F = \frac{2g_v e k_B T}{h}[SF_0(\eta_{12})\,\bar{t}_{12}] \tag{7.22}$$

where

$$\eta_{12} = \frac{E'_{F1D} - E_{12}}{k_B T}, \quad \bar{t}_{12} = \exp(-\beta_{12}),$$

$$\beta_{12} = \frac{4}{3}[A_{12}(V_0, n_x, n_y)]^{3/2} \cdot [eF_{sz}A'_{12}(V_0)]^{-1},$$

$$A_{12}(V_0, n_x, n_y) = \left[\frac{2m_c}{\hbar^2}I_{11}(V_0) - \phi_1(n_x, n_y)\right],$$

$$I_{11}(V_0) \equiv \frac{V_0(V_0 + E_{g_0})(V_0 + E_{g_0} + \Delta)(E_{g_0} + \frac{1}{3}\Delta)}{E_{g_0}(E_{g_0} + \Delta)(V_0 + E_{g_0} + \frac{2}{3}\Delta)},$$

and

$$A'_{12}(V_0) = \left[\frac{2m_c}{\hbar^2}I_{11}(V_0)\left[\frac{1}{V_0} + \frac{1}{V_0 + E_{g0}} + \frac{1}{V_0 + E_{g0} + \Delta} - \frac{1}{V_0 + E_{g0} + (2/3)\Delta}\right]\right]$$

Therefore, the field emitted current is given by

$$I_F = \frac{2g_v e k_B T}{h}[SF_0(\eta_{12})\exp(-\beta_{12})] \tag{7.23}$$

The photo-emitted current in this case can be written as

$$I_p = \frac{\alpha_o e g_v k_B T}{\pi\hbar}[SF_0(\eta_{63})] \tag{7.24}$$

$$\eta_{63} \equiv \left[\frac{E'_{F1D} - (E_{12} + W - h\nu)}{k_B T}\right]$$

(ii) The Two-band model of Kane

Under the conditions, $\Delta \gg E_g$ or $\Delta \ll E_g, \Delta_{\|} = \Delta_{\perp} = \Delta$ (isotropic spin-orbit splitting constant) and $m_{\|}^* = m_{\perp}^* = m_c$ (isotropic effective electron mass at the edge of the conduction band), (7.16) to (7.20) get simplified as

$$k_z^2 = A_{13}(E, n_x, n_y) \tag{7.25}$$

$$E_{13}(1 + \alpha E_{13}) = \frac{\hbar^2}{2m_c}\phi_1(n_x, n_y) \tag{7.26}$$

$$m^*(E) = m_c[1 + 2\alpha E] \tag{7.27}$$

$$n_s = \frac{2g_v}{\pi}S\left[B_{15}(E'_{F1D}, n_x, n_y)\right] \tag{7.28}$$

$$C_g = \left[\frac{2g_v e^2}{\pi}S(\theta_3)\right]\left[1 + \frac{2g_v e^2}{\pi C_{ins}}S(\theta_3)\right]^{-1} \tag{7.29}$$

where

$$A_{13}(E, n_x, n_y) = \left[\frac{2m_c}{\hbar^2}E(1 + \alpha E) - \phi_1(n_x, n_y)\right],$$

E_{13} is the sub-band energy in this case,

$$B_{15}(E'_{F1D}, n_x, n_y) = [A_{13}(E'_{F1D}, n_x, n_y)]^{1/2},$$

$$\theta_3 = \frac{\partial}{\partial E'_{F1D}}(B_{15}(E'_{F1D}, n_x, n_y))$$

The ER in this case can be expressed as

$$\frac{D}{\mu} = \frac{1}{e}[S[B_{15}(E'_{F1D}, n_x, n_y)]][S[\theta_3]]^{-1} \tag{7.30}$$

The field emitted current can be expressed as

$$I_F = \frac{2g_v e k_B T}{h}[F_0(\eta_{13})F_0(\eta_{13})\,\bar{t}_{13}] \tag{7.31}$$

where

$$\eta_{13} = \frac{E'_{F1D} - E_{13}}{k_B T}, \quad \bar{t}_{13} = \exp(-\beta_{13}),$$

$$\beta_{13} = \frac{4}{3}[A_{13}(V_0, n_x, n_y)]^{3/2} \cdot [eF_{sz}A'_{13}(V_0)]^{-1},$$

$$A'_{13}(V_0) = \left[\frac{2m_c}{\hbar^2}(1 + 2\alpha V_0)\right]$$

Therefore, the field emitted current is given by

$$I_F = \frac{2g_v e k_B T}{h}[SF_0(\eta_{13})\exp(-\beta_{13})] \tag{7.32}$$

The photo-emitted current in this case can be written as

$$I_p = \frac{\alpha_o e g_v k_B T}{\pi\hbar}[SF_0(\eta_{62})] \tag{7.33}$$

where

$$\eta_{62} \equiv \left[\frac{E'_{F1D} - (E_{13} + W - h\nu)}{k_B T}\right]$$

(iii) The isotropic parabolic energy bands

Under the conditions, $\alpha \to 0$, (7.25) to (7.29) get simplified as

$$k_z^2 = \bar{A}_{13}(E, n_x, n_y) \tag{7.34}$$

$$\bar{E}_{13} = \frac{\hbar^2}{2m_c}\phi_1(n_x, n_y) \tag{7.35}$$

$$m^*(E) = m_c \tag{7.36}$$

$$n_s = \frac{2g_v}{\pi} S\,[\bar{B}_{15}(E'_{F1D}, n_x, n_y)] \tag{7.37}$$

$$C_g = \left[\frac{2g_v e^2}{\pi} S(\bar{\theta}_3)\right]\left[1 + \frac{2g_v e^2}{\pi C_{ins}} S(\bar{\theta}_3)\right]^{-1} \tag{7.38}$$

where

$$\bar{A}_{13}(E, n_x, n_y) = \left[\frac{2m_c}{\hbar^2}E - \phi_1(n_x, n_y)\right],$$

$\bar{E}_{13}$ is the sub-band energy in this case,

$$\bar{B}_{15}(E'_{F1D}, n_x, n_y) = [\bar{A}_{13}(E'_{F1D}, n_x, n_y)]^{1/2},$$

$$\bar{\theta}_3 = \frac{\partial}{\partial E'_{F1D}}(\bar{B}_{15}(E'_{F1D}, n_x, n_y))$$

The ER in this case can be expressed as

$$\frac{D}{\mu} = \frac{1}{e}[S[\bar{B}_{15}(E'_{F1D}, n_x, n_y)]][S[\bar{\theta}_3]]^{-1} \tag{7.39}$$

The field emitted current can be expressed as

$$I_F = \frac{2g_v e k_B T}{h}[SF_0(\bar{\eta}_{13})\,\bar{t}_{14}] \tag{7.40}$$

where

$$\bar{\eta}_{13} = \frac{E'_{F1D} - \bar{E}_{13}}{k_B T}, \quad \bar{t}_{14} = \exp(-\bar{\beta}_{13}),$$

$$\bar{\beta}_{13} = \frac{4}{3}[\bar{A}_{13}(V_0, n_x, n_y)]^{3/2} \cdot [eF_{sz}(\bar{A}_{13})']^{-1} \quad \text{and} \quad (\bar{A}_{13})' = \left[\frac{2m_c}{\hbar^2}\right]$$

Therefore, the field emitted current is given by

$$I_F = \frac{2g_v e k_B T}{h}[F_0(\bar{\eta}_{13}) \exp(-\bar{\beta}_{13})] \tag{7.41}$$

The photo-emitted current in this case can be written as

$$I_p = \frac{\alpha_o e g_v k_B T}{\pi \hbar}[SF_0(\bar{\eta}_{62})] \tag{7.42}$$

where

$$\bar{\eta}_{62} \equiv \left[\frac{E'_{F1D} - (\bar{E}_{13} + W - h\nu)}{k_B T}\right]$$

(iv) The model of Stillman *et al.*

The 1D DRs, sub-band energy, the EEM, the concentration per unit length and the QC for this model can respectively be expressed as

$$k_z^2 = A_{14}(E, n_x, n_y) \tag{7.43}$$

$$\gamma_5(E_{14}) = \frac{\hbar^2}{2m_c}\phi_1(n_x, n_y) \tag{7.44}$$

$$m^*(E) = m_c \gamma'_5(E) \tag{7.45}$$

$$n_s = \frac{2g_v}{\pi} S\,[B_{17}(E'_{F1D}, n_x, n_y)] \tag{7.46}$$

$$C_g = \left[\frac{2g_v e^2}{\pi} S(\theta_4)\right]\left[1 + \frac{2g_v e^2}{\pi C_{ins}} S(\theta_4)\right]^{-1} \tag{7.47}$$

where

$$A_{14}(E, n_x, n_y) = \left[\frac{2m_c}{\hbar^2}\gamma_5(E) - \phi_1(n_x, n_y)\right],$$

E_{14} is the sub-band energy in this case,

$$B_{17}(E'_{F1D}, n_x, n_y) = [A_{14}(E'_{F1D}, n_x, n_y)]^{1/2},$$

$$\theta_4 = \frac{\partial}{\partial E'_{F1D}}(B_{17}(E'_{F1D}, n_x, n_y))$$

The ER in this case can be expressed as

$$\frac{D}{\mu} = \frac{1}{e}[S[B_{17}(E'_{F1D}, n_x, n_y)]][S[\theta_4]]^{-1} \tag{7.48}$$

The field emitted current can be expressed as

$$I_F = \frac{2g_v e k_B T}{h}[SF_0(\eta_{15})\bar{t}_{15}] \tag{7.49}$$

where

$$\eta_{15} = \frac{E'_{F1D} - E_{14}}{k_B T}, \quad \bar{t}_{15} = \exp(-\beta_{15}),$$

$$\beta_{15} = \frac{4}{3}[A_{14}(V_0, n_x, n_y)]^{3/2} \cdot [eF_{sz}A'_{14}(V_0)]^{-1} \quad \text{and}$$

$$A'_{14}(V_0) = \left[\frac{2m_c}{\hbar^2}\gamma'_5(V_0)\right]$$

Therefore, the field emitted current is given by

$$I_F = \frac{2g_v e k_B T}{h}[SF_0(\eta_{15})\exp(-\beta_{15})] \tag{7.50}$$

The photo-emitted current in this case can be written as

$$I_p = \frac{\alpha_o e g_v k_B T}{\pi\hbar}[SF_0(\eta_{65})] \tag{7.51}$$

where

$$\eta_{65} \equiv \left[\frac{E'_{F1D} - (E_{14} + W - h\nu)}{k_B T}\right]$$

(v) The model of Newson and Kurobe

From (R1.13) of Chapter 1, the 1-D E-k_z relation in accordance with the model of Newson and Kurobe, assumes the form

$$k_z^2 = A_{15}(E, n_x, n_y) \tag{7.52}$$

where

$$A_{15}(E, n_x, n_y) = (2a_{13})^{-1}[-\overline{L_1}(n_x, n_y) + [\{\overline{L_1}(n_x, n_y)\}^2 - 4a_{13}[\overline{L_2}(n_x, n_y) - E)]^{1/2}]]$$

$$\overline{L_1}(n_x, n_y) = \frac{\hbar^2}{2m_c} + a_{14}\left[\left(\frac{n_x\pi}{d_x}\right)^2 + \left(\frac{n_y\pi}{d_y}\right)^2\right]$$

and

$$\overline{L_2}(n_x, n_y) = \left[\frac{\hbar^2}{2m_c}\phi_1(n_x, n_y) + a_{14}\left(\frac{n_x\pi}{d_x}\cdot\frac{n_y\pi}{d_y}\right)^2 + a_{13}\left[\left(\frac{n_x\pi}{d_x}\right)^4 + \left(\frac{n_y\pi}{d_y}\right)^4\right]\right]$$

The quantized sub-band energy (E_{16}) is given b

$$E_{16} = \overline{L_2}(n_x, n_y) \tag{7.53}$$

The electron concentration per unit length can be written as

$$n_s = \frac{2g_v}{\pi} S\left[B_{19}(E'_{F1D}, n_x, n_y)\right] \tag{7.54}$$

where $B_{19}(E'_{F1D}, n_x, n_y) = [A_{15}(E'_{F1D}, n_x, n_y)]^{1/2}$.

Using (1.4) and (7.54) the C_g can be expressed as

$$C_g = \left[\frac{2g_v e^2}{\pi} S(\theta_5)\right]\left[1 + \frac{2g_v e^2}{\pi C_{ins}} S(\theta_5)\right]^{-1} \tag{7.55}$$

where

$$\theta_5 = \frac{\partial}{\partial E'_F}(B_{19}(E'_F, n_x, n_y))$$

The ER in this case can be expressed as

$$\frac{D}{\mu} = \frac{1}{e}[S[B_{19}(E'_{F1D}, n_x, n_y)]][S[\theta_5]]^{-1} \tag{7.56}$$

The field emitted current can be expressed as

$$I_F = \frac{2g_v e k_B T}{h}[S F_0(\eta_{16})\,\bar{t}_{16}] \tag{7.57}$$

where

$$\eta_{16} = \frac{E'_{F1D} - E_{16}}{k_B T}, \quad \bar{t}_{16} = \exp(-\beta_{16}),$$

$$\beta_{16} = \frac{4}{3}[A_{15}(V_0, n_x, n_y)]^{3/2} \cdot [eF_{sz}A'_{15}(V_0)]^{-1} \quad \text{and}$$

$$A'_{15}(V_0) = \left[\frac{2m_c}{\hbar^2}\gamma'_5(V_0)\right]$$

Therefore, the field emitted current is given by

$$I_F = \frac{2g_v e k_B T}{h}[S[F_0(\eta_{15})\exp(-\beta_{15})]] \tag{7.58}$$

The photo-emitted current in this case can be written as

$$I_p = \frac{\alpha_o e g_v k_B T}{\pi \hbar}[S[F_0(\eta_{67})]] \tag{7.59}$$

where

$$\eta_{67} \equiv \left[\frac{E'_{F1D} - (E_{16} + W - h\nu)}{k_B T}\right]$$

(vi) The model of Palik *et al.*

The 1D DRs, sub-band energy, the EEM, the concentration per unit length and the QC for this model can respectively be expressed as

$$k_z^2 = \bar{A}_{15}(E, n_x, n_y) \tag{7.60}$$

where

$$\bar{A}_{15}(E, n_x, n_y) = \left[\frac{\hbar^2}{2m_c}\gamma_9(E) - \phi_1(n_x, n_y)\right]$$

The quantized sub-band energy ($\bar{E}_{16}$) is given by

$$\gamma_9(\bar{E}_{16}) = 0 \tag{7.61}$$

The electron concentration per unit length can be written as

$$n_s = \frac{2g_v}{\pi} S\left[B_{21}(E'_{F1D}, n_x, n_y)\right] \tag{7.62}$$

where $B_{21}(E'_{F1D}, n_x, n_y) = [\bar{A}_{15}(E'_{F1D}, n_x, n_y)]^{1/2}$.

Using (1.4) and (7.54) the C_g can be expressed as

$$C_g = \left[\frac{2g_v e^2}{\pi} S(\theta_6)\right]\left[1 + \frac{2g_v e^2}{\pi C_{ins}} S(\theta_6)\right]^{-1} \tag{7.63}$$

where

$$\theta_6 - \frac{\partial}{\partial E'_{F1D}}(B_{21}(E'_{F1D}, n_x, n_y))$$

The ER in this case can be expressed as

$$\frac{D}{\mu} = \frac{1}{e}[S[B_{21}(E'_{F1D}, n_x, n_y)]][S[\theta_6]]^{-1} \tag{7.64}$$

The field emitted current can be expressed as

$$I_F = \frac{2g_v e k_B T}{h}[F_0(\bar{\eta}_{16})\, \bar{t}_{17}] \tag{7.65}$$

where

$$\bar{\eta}_{16} = \frac{E'_{F1D} - \bar{E}_{16}}{k_B T}, \quad \bar{t}_{17} = \exp(-\bar{\beta}_{16}),$$

$$\bar{\beta}_{16} = \frac{4}{3}[\bar{A}_{15}(V_0, n_x, n_y)]^{3/2} \cdot [eF_{sz}(\bar{A}_{15}(V_0))']^{-1}$$

and

$$(\bar{A}_{15}(V_0))' = \left[\frac{2m_c}{\hbar^2}\gamma'_9(V_0)\right]$$

Therefore, the field emitted current is given by

$$I_F = \frac{2g_v e k_B T}{h}[SF_0(\bar{\eta}_{16})\exp(-\bar{\beta}_{16})] \tag{7.66}$$

The photo-emitted current in this case can be written as

$$I_p = \frac{\alpha_o e g_v k_B T}{\pi\hbar}[S[F_0(\eta_{68})]] \tag{7.67}$$

where

$$\eta_{68} \equiv \left[\frac{E'_{F1D} - (\bar{E}_{16} + W - h\nu)}{k_B T}\right]$$

7.2.3 *The QC in NWFETs of II–VI materials*

The (2.1) can be written as

$$E = A_o k_s^2 + B_o k_z^2 \pm C_o k_s \tag{7.68}$$

where

$$A_0 \equiv \hbar^2/2m_\perp^* \quad \text{and} \quad B_0 \equiv \hbar^2/2m_\parallel^*.$$

The 1D DRs for quantum wires of II–VI semiconductors can be expressed as

$$E = B_0 k_z^2 + G_{3,\pm}(n_x, n_y) \tag{7.69}$$

where

$$G_{3,\pm}(n_x, n_y) \equiv \left[A_0\left\{\left(\frac{\pi n_x}{d_x}\right)^2 + \left(\frac{\pi n_y}{d_y}\right)^2\right\} \pm C_0\left\{\left(\frac{\pi n_x}{d_x}\right)^2 + \left(\frac{\pi n_y}{d_y}\right)^2\right\}^{1/2}\right]$$

The 1D electron statistics under the condition of unity Fermi function at low temperatures can be written as

$$n_s = \frac{g_v}{\pi\sqrt{B_0}} S[t_7(E'_{F1D}, n_x, n_y)] \tag{7.70}$$

where

$$t_7(E'_{F1D}, n_x, n_y) \equiv [E'_{F1D} - [G_{3,+}(n_x, n_y)]]^{1/2} + [E'_{F1D} - [G_{3,-}(n_x, n_y)]]^{1/2}.$$

Using (1.4) and (7.70) the C_g can be expressed as

$$C_g = \left[\frac{g_v e^2}{\pi\sqrt{B_0}} S(\theta_7)\right]\left[1 + \frac{g_v e^2}{C_{ins}\pi\sqrt{B_0}} S(\theta_7)\right]^{-1} \tag{7.71}$$

where

$$\theta_7 = \frac{\partial}{\partial E'_{F1D}}(t_7(E'_{F1D}, n_x, n_y))$$

The ER in this case can be expressed as

$$\frac{D}{\mu} = \frac{1}{e}[S[t_7(E'_{F1D}, n_x, n_y)]][S[\theta_7]]^{-1} \tag{7.72}$$

The field emitted current can be expressed as

$$\begin{aligned} I_F = \frac{e g_v k_B T}{h} S[\{F_0\{(k_B T)^{-1}[E'_{F1D} - [G_{3,+}(n_x, n_y)]]\} \exp(-\overline{\beta}_{016,+})\} \\ + F_0\{(k_B T)^{-1}[E'_{F1D} - [G_{3,-}(n_x, n_y)]]\} \exp(-\overline{\beta}_{016,-})] \end{aligned} \tag{7.73}$$

and

$$\overline{\beta}_{016,\pm} = \frac{4}{3}[(V_0 - G_{3,\pm}(n_x, n_y))]^{3/2} \cdot [eF_{sz}\hbar]^{-1}\sqrt{2m^*_\parallel}$$

The photo-emitted current in this case can be written as

$$I_p = \frac{\alpha_o e g_v k_B T}{2\pi\hbar}[SF_0(\eta_{69})] \tag{7.74}$$

where

$$\eta_{69} \equiv \left[\frac{E'_{F1D} - (\bar{E}_{161} + W - h\nu)}{k_B T}\right]$$

and $(\bar{E}_{161})$ is given by

$$\bar{E}_{161} = G_{3,\pm}(n_x, n_y) \tag{7.75}$$

7.2.4 The QC in NWFETs of IV–VI materials

The 1D DRs in this case can be written from (3.56) as

$$k_z^2 = A_{23}(E, n_x, n_y) \tag{7.76}$$

where

$$A_{23}(E, n_x, n_y) = (2h_4)^{-1}[h_6(E, n_x, n_y) - [h_6^2(E, n_x, n_y) + 4h_4 h_7(E, n_x, n_y)]^{1/2}], \quad h_4 = \left[\frac{\alpha\hbar^4}{4x_3 x_6}\right]$$

$$x_3 = \frac{3m_t^- m_l^-}{2m_l^- + m_t^-}, \quad x_6 = \frac{3m_t^+ m_l^+}{2m_l^+ + m_t^+},$$

$$h_6(E, n_x, n_y) = \left[\frac{\alpha E\hbar^2}{2x_6} - \frac{\alpha\hbar^2}{2x_6}\left[\left(\frac{\pi n_x}{d_x}\right)^2 \frac{\hbar^2}{2x_1} + \left(\frac{\pi n_y}{d_y}\right)^2 \frac{\hbar^2}{2x_2}\right] -,\right.$$

$$\left.\frac{\alpha\hbar^2}{2x_3}\left[\left(\frac{\pi n_x}{d_x}\right)^2 \frac{\hbar^2}{2x_4} + \left(\frac{\pi n_y}{d_y}\right)^2 \frac{\hbar^2}{2x_5}\right] - \frac{\hbar^2}{2m_3} - \frac{(1+\alpha E)\hbar^2}{2x_3}\right],$$

$$x_1 = m_t^-, \quad x_2 = \frac{m_t^- + 2m_l^-}{3}, \quad x_4 = m_t^+, \quad x_5 = \frac{m_t^+ + 2m_l^+}{3},$$

$$m_3 = \frac{3m_t^* m_l^*}{m_t^* + 2m_l^*},$$

$$\begin{aligned} h_7(E, n_x, n_y) = \Bigg[& E(1+\alpha E) + \alpha E\left[\left(\frac{\pi n_x}{d_x}\right)^2 \frac{\hbar^2}{2x_4} + \left(\frac{\pi n_y}{d_y}\right)^2 \frac{\hbar^2}{2x_5}\right] \\ & -(1+\alpha E)\left[\left(\frac{\pi n_x}{d_x}\right)^2 \frac{\hbar^2}{2x_1} + \left(\frac{\pi n_y}{d_y}\right)^2 \frac{\hbar^2}{2x_2}\right] \\ & -\alpha\left[\left(\frac{\pi n_x}{d_x}\right)^2 \frac{\hbar^2}{2x_1} + \left(\frac{\pi n_y}{d_y}\right)^2 \frac{\hbar^2}{2x_2}\right] \\ & \times\left[\left(\frac{\pi n_x}{d_x}\right)^2 \frac{\hbar^2}{2x_4} + \left(\frac{\pi n_y}{d_y}\right)^2 \frac{\hbar^2}{2x_5}\right] \\ & -\left[\left(\frac{\pi n_x}{d_x}\right)^2 \frac{\hbar^2}{2m_1} + \left(\frac{\pi n_y}{d_y}\right)^2 \frac{\hbar^2}{2m_2}\right]\Bigg], \quad m_1 = m_t^* \end{aligned}$$

and

$$m_2 = \frac{m_t^* + 2m_l^*}{3}$$

From (7.76), the EEM can be written as

$$m^*(E, n_x, n_y) = \frac{\hbar^2}{2} A'_{23}(E, n_x, n_y) \tag{7.77}$$

The sub-band energy ($\bar{E}_{22}$) can be written as

$$A_{23}(\bar{E}_{22}, n_x, n_y) = 0 \tag{7.78}$$

The electron statistics under the condition of unity Fermi function at low temperatures can be written as

$$n_s = \frac{2g_v}{\pi} S\left[B_{32}(E'_{F1D}, n_x, n_y)\right] \tag{7.79}$$

where

$$B_{32}(E'_{F1D}, n_x, n_y) = [A_{32}(E'_{F1D}, n_x, n_y)]^{1/2}.$$

Using (1.4) and (7.79) the C_g can be expressed as

$$C_g = \left[\frac{2g_v e^2}{\pi} S(\theta_8)\right]\left[1 + \frac{2g_v e^2}{\pi C_{ins}} S(\theta_8)\right]^{-1} \tag{7.80}$$

where

$$\theta_8 = \frac{\partial}{\partial E'_F}(B_{32}(E'_{F1D}, n_x, n_y))$$

The ER in this case can be expressed as

$$\frac{D}{\mu} = \frac{1}{e}[S[B_{32}(E'_{F1D}, n_x, n_y)]][S[\theta_8]]^{-1} \tag{7.81}$$

The field emitted current can be written as

$$I_F = \frac{2g_v e k_B T}{h}[SF_0(\eta_{22})\exp(-\beta_{22})] \tag{7.82}$$

where

$$\eta_{22} = \frac{E_{F1D} - \bar{E}_{22}}{k_B T}$$

The photo-emitted current in this case can be written as

$$I_p = \frac{\alpha_o e g_v k_B T}{2\pi\hbar}[SF_0(\eta_{70})] \tag{7.83}$$

where

$$\eta_{70} \equiv \left[\frac{E'_{F1D} - (\bar{E}_{22} + W - h\nu)}{k_B T}\right]$$

7.2.5 The QC in NWFETs of stressed Kane type materials

The 1D DRs in this case can be written following (2.21) as

$$k_z^2 = A_{24}(E, n_x, n_y) \tag{7.84}$$

where

$$A_{24}(E, n_x, n_y) = [\bar{c}_0(E)]^2 \left[1 - \left(\frac{\pi n_x}{d_x}\right)^2 [\bar{a}_0(E)]^{-2} - \left(\frac{\pi n_y}{d_y}\right)^2 [\bar{b}_0(E)]^{-2}\right]$$

The sub-band energy $\bar{E}_{23}$ assumes the form

$$\left[\left(\frac{\pi n_x}{d_x}\right)^2 [\bar{a}_0(\bar{E}_{23})]^{-2} + \left(\frac{\pi n_y}{d_y}\right)^2 [\bar{b}_0(\bar{E}_{23})]^{-2}\right] = 1 \tag{7.85}$$

The 1D carrier statistics under the condition of extreme carrier degeneracy can be expressed as

$$n_s = \frac{2g_v}{\pi} S\left[B_{34}(E'_{F1D}, n_x, n_y)\right] \tag{7.86}$$

where

$$B_{34}(E'_{F1D}, n_x, n_y) = \sqrt{A_{24}(E'_{F1D}, n_x, n_y)}$$

Using (1.4) and (7.86) the C_g can be expressed as

$$C_g = \left[\frac{2g_v e^2}{\pi} S(\theta_9)\right]\left[1 + \frac{2g_v e^2}{\pi C_{ins}} S(\theta_9)\right]^{-1} \tag{7.87}$$

where

$$\theta_9 = \frac{\partial}{\partial E'_{F1D}}(B_{34}(E'_{F1D}, n_x, n_y))$$

The ER in this case can be expressed as

$$\frac{D}{\mu} = \frac{1}{e}[S[B_{34}(E'_{F1D}, n_x, n_y)]][S[\theta_9]]^{-1} \tag{7.88}$$

The EEM can be written as

$$m^*(E, n_x, n_y) = \frac{\hbar^2}{2} A'_{24}(E, n_x, n_y) \tag{7.89}$$

The field emitted current is given by

$$I_F = \frac{2g_v e k_B T}{h} \sum_{n_x=1}^{n_{x\max}} \sum_{n_y=1}^{n_{y\max}} F_0(\eta_{23}) \exp(-\beta_{23}) \tag{7.90}$$

where

$$\eta_{23} = \frac{E_{F1D} - E_{23}}{k_B T}, \quad \beta_{23} = \frac{4}{3}[A_{24}(V_0, n_x, n_y)]^{3/2} \cdot [eF_{sz}A'_{24}(V_0, n_x, n_y)]^{-1},$$

The photo-emitted current is given by

$$I_p = \frac{\alpha_o e g_v k_B T}{2\pi\hbar}[SF_0(\eta_{71})] \tag{7.91}$$

where

$$\eta_{71} \equiv \left[\frac{E'_{F1D} - (\bar{E}_{23} + W - h\nu)}{k_B T}\right]$$

7.2.6 *The QC in NWFETs of bismuth telluride*

The 1D DR from (2.39) in this case can be written as

$$k_x^2 = A_{28}(E, n_y, n_z) \tag{7.92}$$

where

$$A_{28}(E, n_y, n_z) = \left[E(1+\alpha E) - \mathrm{I}_2\left(\frac{n_y\pi}{d_y}\right)^2 - \mathrm{I}_3\left(\frac{n_z\pi}{d_z}\right)^2 - \mathrm{I}_4\left(\frac{n_y\pi}{d_y}\right)\left(\frac{n_z\pi}{d_z}\right)\right](\mathrm{I}_1)^{-1}$$

The sub-band energy (E_{30}) can be expressed as

$$E_{30} = (2\alpha)^{-1}[-1 + \sqrt{1 + 4\alpha\theta_{32}(n_y, n_z)}] \tag{7.93}$$

where

$$\theta_{32}(n_y, n_z) = \left[\mathrm{I}_2\left(\frac{n_y\pi}{d_y}\right)^2 + \mathrm{I}_3\left(\frac{n_z\pi}{d_z}\right)^2 + \mathrm{I}_4\left(\frac{n_y\pi}{d_y}\right)\left(\frac{n_z\pi}{d_z}\right)\right]$$

The electron concentration per unit length is given by

$$n_0 = \frac{2g_v}{\pi} S_2[B_{42}(E'_{F1D}, n_z, n_y)] \tag{7.94}$$

where

$$S_2 = \sum_{n_z=1}^{n_{z\max}} \sum_{n_y=1}^{n_{y\max}} \quad \text{and} \quad B_{42}(E'_{F1D}, n_z, n_y) = \sqrt{A_{28}(E'_{FID}, n_y, n_z)}$$

Using (1.4) and (7.94) the C_g can be expressed as

$$C_g = \left[\frac{2g_v e^2}{\pi} S(\bar{\theta}_9)\right]\left[1 + \frac{2g_v e^2}{\pi C_{ins}} S_2(\bar{\theta}_9)\right]^{-1} \tag{7.95}$$

where

$$\bar{\theta}_9 = \frac{\partial}{\partial E'_{F1D}}(B_{42}(E'_{F1D}, n_x, n_y))$$

The EEM can be written as

$$m^*(E) = \frac{\hbar^2}{2I_1}(1 + 2\alpha E) \tag{7.96}$$

The ER can be written as

$$\frac{D}{\mu} = S_2[B_{42}(E'_{F1D}, n_z, n_y)][S_2[B'_{42}(E'_{F1D}, n_z, n_y)]]^{-1} \tag{7.97}$$

The field emitted current in this case can be written as

$$I_F = \frac{2g_v e k_B T}{h} S_2[F_0(\eta_{28}) \exp(-\beta_{28})] \tag{7.98}$$

where

$$\eta_{28} = \frac{E'_{F1D} - E_{30}}{k_B T}, \quad \beta_{28} = \frac{4}{3}[A_{28}(V_0, n_y, n_z)]^{3/2} \cdot [eF_{sz}A'_{28}(V_0)]^{-1} \quad \text{and}$$

$$A'_{28}(V_0) = (\bar{\omega}_1)^{-1}[1 + 2\alpha V_0]$$

The photo-emitted current is given by

$$I_p = \frac{\alpha_o e g_v k_B T}{2\pi\hbar} S_2 F_0(\eta_{72}) \tag{7.99}$$

where

$$\eta_{72} \equiv \left[\frac{E'_{F1D} - (E_{30} + W - h\nu)}{k_B T}\right]$$

7.2.7 *The QC in NWFETs of Germanium*

(i) Model of Cardona *et al.*

The 1D DRs in this case can be written following (2.46) as

$$k_y^2 = A_{29}(E, n_x, n_z) \tag{7.100}$$

where

$$A_{29}(E, n_x, n_z) = \left[\left[\gamma_{15}(E, n_z) - \left(\frac{\hbar^2}{2m_1^*}\right)\left(\frac{n_x \pi}{d_x}\right)^2\right](2m_2^*/\hbar^2)\right]$$

$$\gamma_{15}(E,n_z) = \left[E(1+\alpha E) - (1+2\alpha E)\left(\frac{\hbar^2}{2m_3^*}\right)\left(\frac{n_z\pi}{d_z}\right)^2 + \alpha\left[\left(\frac{\hbar^2}{2m_3^*}\right)\left(\frac{n_z\pi}{d_z}\right)^2\right]^2\right],$$

$$m_1^* = m_\perp^*, \quad m_2^* = \frac{m_\perp^* + 2m_\parallel^*}{3} \quad \text{and} \quad m_3^* = \frac{3m_\perp^* m_\parallel^*}{m_\perp^* + 2m_\parallel^*}$$

The quantized energy levels (E_{31}) can be expressed through the equation

$$E_{31} = (2\alpha)^{-1}[-\rho_{91}(n_z) + \sqrt{\rho_{91}(n_z)^2 - 4\alpha\rho_{92}(n_z)}] \tag{7.101}$$

where

$$\rho_{91}(n_z) = \left[1 - 2\alpha\frac{\hbar^2}{2m_3^*}\left(\frac{n_z\pi}{d_z}\right)^2\right] \quad \text{and}$$

$$\rho_{92}(n_z) = \left[\frac{\hbar^2}{2m_3^*}\left(\frac{n_z\pi}{d_z}\right)^2 - \alpha\left[\frac{\hbar^2}{2m_3^*}\left(\frac{n_z\pi}{d_z}\right)^2\right]^2\right]$$

The electron concentration per unit length can be written as

$$n_s = \frac{2g_v}{\pi}S_3\left[B_{44}(E'_{F1D}, n_x, n_z)\right] \tag{7.102}$$

where

$$S_3 = \sum_{n_x=1}^{n_{x\max}}\sum_{n_z=1}^{n_{z\max}} \quad \text{and} \quad B_{44}(E'_{F1D}, n_x, n_z) = \sqrt{A_{29}(E'_{F1D}, n_x, n_z)}$$

Using (1.4) and (7.102) the C_g can be expressed as

$$C_g = \left[\frac{2g_v e^2}{\pi}S_3(\theta_{13})\right]\left[1 + \frac{2g_v e^2}{\pi C_{ins}}S(\theta_{13})\right]^{-1} \tag{7.103}$$

where

$$\theta_{13} = \frac{\partial}{\partial E'_F}(B_{44}(E'_{F1D}, n_x, n_z))$$

The ER can be written as

$$\frac{D}{\mu} = S_3[B_{44}(E'_{F1D}, n_x, n_z)][S_3[B'_{44}(E'_{F1D}, n_x, n_z)]]^{-1} \tag{7.104}$$

The EEM can be written as

$$m^*(E,n_x,n_z)=\frac{\hbar^2}{2}A'_{29}(E,n_x,n_z) \tag{7.105}$$

The field emitted current assumes the form

$$I_F=\frac{2g_v e k_B T}{h}[S_3[F_0(\eta_{29})\exp(-\beta_{29})]] \tag{7.106}$$

where

$$\eta_{29}=\frac{E'_{F1D}-E_{31}}{k_BT},\quad \beta_{29}=\frac{4}{3}[A_{29}(V_0,n_x,n_z)]^{3/2}\cdot[eF_{sz}A'_{29}(V_0,n_x,n_z)]^{-1}$$

and

$$A'_{29}(V_0,n_x,n_z)=\left[\left[1+2\alpha V_0-\left(\alpha\frac{\hbar^2}{m_3^*}\right)\left(\frac{n_z\pi}{d_z}\right)^2\right](2m_2^*/\hbar^2)\right]$$

The photo-emitted current is given by

$$I_p=\frac{\alpha_o e g_v k_B T}{2\pi\hbar}[S_3F_0(\eta_{73})] \tag{7.107}$$

where

$$\eta_{73}\equiv\left[\frac{E'_{F1D}-(E_{31}+W-h\nu)}{k_BT}\right]$$

(ii) Model of Wang and Ressler

The 1D DRs in this case can be written (2.53) as

$$k_y^2=A_{30}(E,n_x,n_z) \tag{7.108}$$

where

$$A_{30}(E,n_x,n_z)=\left[\left[I_{29}(E,n_z)-\left(\frac{\hbar^2}{2m_1^*}\right)\left(\frac{n_x\pi}{d_x}\right)^2\right](2m_2^*/\hbar^2)\right],$$

$$I_{29}(E,n_z)=[2\overline{C_1}]^{-1}[\overline{A_6}(n_z)+[\overline{A_6}^2(n_z)-4\bar{C}_1E+4(\overline{C_1})\overline{A_5}(n_z)]^{\frac{1}{2}}]$$

$$\overline{A_5}(n_z)=\left(\frac{\hbar^2}{2m_3^*}\right)\left(\frac{n_z\pi}{d_z}\right)^2\left[1-\bar{e}_1\left(\frac{\hbar^2}{2m_3^*}\right)\left(\frac{n_z\pi}{d_z}\right)^2\right]\sum_{i=1}^{n}X_i^2$$

and

$$\overline{A_6}(n_z)=\left[1-\bar{d}_1\left(\frac{\hbar^2}{2m_3^*}\right)\left(\frac{n_z\pi}{d_z}\right)^2\right] \tag{7.109}$$

The quantized energy levels (E_{32}) can be expressed through the equation

$$E_{32} = \bar{A}_5(n_z) + \left(\frac{1}{4\bar{C}_1}\right)\left[\left[\frac{\bar{C}_1\hbar^2}{m_1^*}\left(\frac{n_x\pi}{d_x}\right)^2\right]^2 - 2\bar{A}_6(n_z)\frac{\bar{C}_1\hbar^2}{m_1^*}\left(\frac{n_x\pi}{d_x}\right)^2\right] \tag{7.110}$$

The 1D electron statistics under the condition of unity Fermi function at low temperatures can be written following [41] as

$$n_s = \frac{2g_v}{\pi}S_3\left[B_{46}(E'_{F1D}, n_x, n_z)\right] \tag{7.111}$$

where $B_{46}(E'_{F1D}, n_x, n_z) = \sqrt{A_{30}(E'_{F1D}, n_x, n_z)}$.

Using (1.4) and (7.111) the C_g can be expressed as

$$C_g = \left[\frac{2g_v e^2}{\pi}S(\theta_{14})\right]\left[1 + \frac{2g_v e^2}{\pi C_{ins}}S(\theta_{14})\right]^{-1} \tag{7.112}$$

where

$$\theta_{14} = \frac{\partial}{\partial E'_{F1D}}(B_{46}(E'_{F1D}, n_x, n_z))$$

The ER can be written as

$$\frac{D}{\mu} = S_3[B_{46}(E'_{F1D}, n_z, n_x)][S_3[B'_{46}(E'_{F1D}, n_z, n_x)]]^{-1} \tag{7.113}$$

The EEM can be written as

$$m^*(E, n_x, n_z) = \frac{\hbar^2}{2}A'_{30}(E, n_x, n_z) \tag{7.114}$$

The field emitted current assumes the form

$$I_F = \frac{2g_v e k_B T}{h}S_3[F_0(\eta_{30})\exp(-\beta_{30})] \tag{7.115}$$

where

$$\eta_{30} = \frac{E'_{F1D} - E_{32}}{k_B T},\ \beta_{30} = \frac{4}{3}[A_{30}(V_0, n_x, n_z)]^{3/2} \cdot [eF_{sz}A'_{30}(V_0, n_z)]^{-1},$$

$$A'_{30}(V_0, n_z) = (2m_2^*/\hbar^2)I'_{29}(V_0, n_z)$$

and

$$I'_{29}(V_0, n_z) = [[\overline{A_6}^2(n_z) - 4\bar{C}_1 V_0 + 4(\bar{C}_1)\overline{A_5}(n_z)]^{-1/2}]$$

The photo-emitted current is given by

$$I_p = \frac{\alpha_o e g_v k_B T}{2\pi\hbar}[S_2 F_0(\eta_{74})] \tag{7.116}$$

where

$$\eta_{74} \equiv \left[\frac{E'_{F1D} - (E_{32} + W - h\nu)}{k_B T}\right]$$

7.2.8 *The QC in NWFETs of gallium antimonide*

(i) Model of Seiler *et al.*

The 1D DRs in this case can be written by using (2.77) as

$$k_z^2 = \bar{\gamma}_6(E, n_x, n_y) \tag{7.117}$$

where

$$\bar{\gamma}_6(E, n_x, n_y) = \left[\gamma_6(E) - \left(\frac{n_x\pi}{d_x}\right)^2 - \left(\frac{n_y\pi}{d_y}\right)^2\right]$$

The 1D electron concentration can be written as

$$n_s = \frac{2g_v}{\pi} S\left[B_{47}(E'_{F1D}, n_x, n_y)\right] \tag{7.118}$$

where $B_{47}(E'_{F1D}, n_x, n_y) = \sqrt{\bar{\gamma}_6(E'_{F1D}, n_x, n_y)}$.

Using (1.4) and (7.117) the C_g can be expressed as

$$C_g = \left[\frac{2g_v e^2}{\pi} S(\bar{\theta}_{15})\right]\left[1 + \frac{2g_v e^2}{\pi C_{ins}} S(\bar{\theta}_{15})\right]^{-1} \tag{7.119}$$

where

$$\bar{\theta}_{15} = \frac{\partial}{\partial E'_{F1D}}(B_{47}(E'_{F1D}, n_x, n_y))$$

The ER can be written as

$$\frac{D}{\mu} = S[B_{47}(E'_{F1D}, n_x, n_y)][S[B'_{47}(E'_{F1D}, n_x, n_y)]]^{-1} \tag{7.120}$$

The EEM can be written as

$$m^*(E) = \frac{\hbar^2}{2}\gamma'_6(E) \tag{7.121}$$

The field emitted current assumes the form

$$I_F = \frac{2g_v e k_B T}{h}[S[F_0(\eta_{31})\exp(-\beta_{31})]] \tag{7.122}$$

where

$$\eta_{31} = \frac{E'_{F1D} - \bar{E}_{32}}{k_B T}, \quad \beta_{31} = \frac{4}{3}[\bar{\gamma}_6(V_0, n_x, n_y)]^{3/2} \cdot [eF_{sz}\gamma'_6(V_0)]^{-1}$$

and $\bar{E}_{32}$ can be expressed through the equation

$$\bar{\gamma}_6(\bar{E}_{32}, n_x, n_y) = 0 \tag{7.123}$$

The photo-emitted current is given by

$$I_p = \frac{\alpha_o e g_v k_B T}{2\pi\hbar}[S[F_0(\eta_{75})]] \tag{7.124}$$

where

$$\eta_{75} \equiv \left[\frac{E'_{F1D} - (\bar{E}_{32} + W - h\nu)}{k_B T}\right]$$

(ii) Model of Mathur and Jain

Following (2.86) the 1D DRs in this case can be written as

$$k_z^2 = \bar{\gamma}_7(E, n_x, n_y) \tag{7.125}$$

where

$$\bar{\gamma}_7(E, n_x, n_y) = \left[\gamma_7(E, n_x, n_y) - \left(\frac{n_x \pi}{d_x}\right)^2 - \left(\frac{n_y \pi}{d_y}\right)^2\right]$$

The 1D electron concentration can be written as

$$n_s = \frac{2g_v}{\pi} S\left[B_{471}(E'_{F1D}, n_x, n_y)\right] \tag{7.126}$$

where $B_{471}(E'_{F1D}, n_x, n_y) = \sqrt{\bar{\gamma}_7(E'_{F1D}, n_x, n_y)}$.

Using (1.4) and (7.126) the C_g can be expressed as

$$C_g = \left[\frac{2g_v e^2}{\pi} S(\bar{\theta}_{151})\right]\left[1 + \frac{2g_v e^2}{\pi C_{ins}} S(\bar{\theta}_{151})\right]^{-1} \tag{7.127}$$

where

$$\bar{\theta}_{151} = \frac{\partial}{\partial E'_{F1D}}(B_{471}(E'_{F1D}, n_x, n_y))$$

The ER can be written as

$$\frac{D}{\mu} = s[B_{471}(E'_{F1D}, n_x, n_y)][S[B'_{471}(E'_{F1D}, n_x, n_y)]]^{-1} \tag{7.128}$$

The EEM can be written as

$$m^*(E) = \frac{\hbar^2}{2}\gamma'_7(E) \tag{7.129}$$

The field emitted current assumes the form

$$I_F = \frac{2g_v e k_B T}{h}[s[\exp(-\beta_{311})]] \tag{7.130}$$

where

$$\eta_{311} = \frac{E'_{F1D} - \bar{E}_{321}}{k_B T}, \quad \beta_{311} = \frac{4}{3}[\bar{\gamma}_7(V_0, n_x, n_y)]^{3/2} \cdot [eF_{sz}\gamma'_7(V_0)]^{-1},$$

and $\bar{E}_{321}$ can be expressed through the equation

$$\bar{\gamma}_7(\bar{E}_{321}, n_x, n_y) = 0 \tag{7.131}$$

The photo-emitted current is given by

$$I_p = \frac{\alpha_o e g_v k_B T}{2\pi\hbar}[S[F_0(\eta_{76})] \tag{7.132}$$

where

$$\eta_{76} \equiv \left[\frac{E'_{F1D} - (\bar{E}_{321} + W - h\nu)}{k_B T}\right]$$

(iii) Model of Zhang

Following (2.97) the 1D DRs in this case can be written as

$$k_z^2 = \bar{\gamma}_8(E, n_x, n_y) \tag{7.133}$$

where

$$\bar{\gamma}_8(E, n_x, n_y) = \left[\gamma_8(E, n_x, n_y) - \left(\frac{n_x \pi}{d_x}\right)^2 - \left(\frac{n_y \pi}{d_y}\right)^2\right]$$

The 1D electron concentration can be written as

$$n_s = \frac{2g_v}{\pi} S\left[B_{472}(E'_{F1D}, n_x, n_y)\right] \tag{7.134}$$

where

$$B_{472}(E'_{F1D}, n_x, n_y) = \sqrt{\bar{\gamma}_8(E'_{F1D}, n_x, n_y)}$$

Using (1.4) and (7.134) the C_g can be expressed as

$$C_g = \left[\frac{2g_v e^2}{\pi} S(\bar{\theta}_{152})\right]\left[1 + \frac{2g_v e^2}{\pi C_{ins}} S(\bar{\theta}_{152})\right]^{-1} \tag{7.135}$$

where

$$\bar{\theta}_{152} = \frac{\partial}{\partial E'_{F1D}}(B_{472}(E'_{F1D}, n_x, n_y)) \tag{7.136}$$

The ER can be written as

$$\frac{D}{\mu} = S[B_{472}(E'_{F1D}, n_x, n_y)][S[B'_{472}(E'_{F1D}, n_x, n_y)]]^{-1} \quad (7.137)$$

The EEM can be written as

$$m^*(E) = \frac{\hbar^2}{2}\gamma'_8(E) \quad (7.138)$$

The field emitted current assumes the form

$$I_F = \frac{2g_v e k_B T}{h} \sum_{n_x=1}^{n_{x\max}} \sum_{n_y=1}^{n_{y\max}} [S[F_0(\eta_{312})\exp(-\beta_{312})]] \quad (7.139)$$

where

$$\eta_{312} = \frac{E'_{F1D} - \bar{E}_{322}}{k_B T}, \quad (7.140)$$

$$\beta_{312} = \frac{4}{3}[\bar{\gamma}_8(V_0, n_x, n_y)]^{3/2} \cdot [eF_{sz}\gamma'_8(V_0)]^{-1} \quad (7.141)$$

and $\bar{E}_{322}$ can be expressed through the equation

$$\bar{\gamma}_8(\bar{E}_{322}, n_x, n_y) = 0 \quad (7.142)$$

The photo-emitted current is given by

$$I_p = \frac{\alpha_o e g_v k_B T}{2\pi\hbar} SF_0(\eta_{77}) \quad (7.143)$$

where

$$\eta_{77} \equiv \left[\frac{E'_{F1D} - (\bar{E}_{322} + W - h\nu)}{k_B T}\right] \quad (7.144)$$

7.2.9 *The QC in NWFETs of GaP*

The 1D DRs in this case can be written following (2.72) as

$$k_z^2 = A_{26}(E, n_x, n_y) \quad (7.145)$$

where

$$A_{26}(E, n_x, n_y) = [t_1 E + t_2 - t_3\phi_1(n_x, n_y) - [t_4\phi_1(n_x, n_y) + t_6 E + t_5]^{1/2}],$$

$$t_1 = (b_0)^{-1}, \quad b_0 = (\hbar^2/2m^*_{\parallel}), \quad t_2 = [-2b_0 D_0 + C][2b_0^2]^{-1}, \quad D_0 = |V_g|,$$

$$C = (\hbar^2 k_0/m^*_{\parallel})^2, \quad t_3 = (a_0/b_0), \quad a_0 = [(\hbar^2/2m^*_{\perp}) + (A\hbar^2/2m^*_{\parallel})],$$

$$t_4 = (\bar{a}_1/(4b_0^4)), \quad \bar{a} = 4b_0 C(b_0 - C), \quad t_6 = (\bar{a}_3/(4b_0^4)), \quad \bar{a}_3 = 4Cb_0,$$

$$t_5 = (\bar{a}_2/(4b_0^4)) \quad \text{and} \quad \bar{a}_2 = [C^2 + 4b_0^2 D_0^2 - 4b_0 C D_0].$$

The sub-band energy $\bar{E}_{27}$ can be written as

$$\bar{E}_{27} = a_0\phi_1(n_x, n_y) - (C\phi_1(n_x, n_y) + D_0^2)^{1/2} + D_0 \tag{7.146}$$

The electron concentration per unit length can be expressed as

$$n_s = \frac{2g_v}{\pi} S\left[B_{38}(E'_{F1D}, n_x, n_y)\right] \tag{7.147}$$

where $B_{38}(E'_{F1D}, n_x, n_y) = \sqrt{A_{26}(E'_{F1D}, n_x, n_y)}$.

Using (1.4) and (7.147) the C_g can be expressed as

$$C_g = \left[\frac{2g_v e^2}{\pi} S(\theta_{101})\right]\left[1 + \frac{2g_v e^2}{\pi C_{ins}} S(\theta_{101})\right]^{-1} \tag{7.148}$$

where

$$\theta_{101} = \frac{\partial}{\partial E'_{F1D}}(B_{38}(E'_{F1D}, n_x, n_y))$$

The ER can be written as

$$\frac{D}{\mu} = S[B_{38}(E'_{F1D}, n_x, n_y)][S[B'_{38}(E'_{F1D}, n_x, n_y)]]^{-1} \tag{7.149}$$

The EEM can be written as

$$m^*(E, n_x, n_y) = \frac{\hbar^2}{2} A'_{26}(E, n_x, n_y) \tag{7.150}$$

The field emitted current assumes the form

$$I_F = \frac{2g_v e k_B T}{h} S[F_0(\eta_{261}) \exp(-\beta_{261})] \tag{7.151}$$

where

$$\eta_{261} = \frac{E'_{F1D} - \bar{E}_{27}}{k_B T}, \ \beta_{261} = \frac{4}{3}[A_{26}(V_0, n_x, n_y)]^{3/2} \cdot [eF_{sz} A'_{26}(V_0, n_x, n_y)]^{-1}$$

and

$$A'_{26}(V_0, n_x, n_y) = \left[t_1 - \frac{1}{2} t_6 [t_4\phi_1(n_x, n_y) + t_6 V_0 + t_5]^{-1/2}\right]$$

The photo-emitted current is given by

$$I_p = \frac{\alpha_o e g_v k_B T}{2\pi\hbar} S F_0(\eta_{78}) \tag{7.152}$$

where

$$\eta_{78} \equiv \left[\frac{E'_{F1D} - (\bar{E}_{27} + W - h\nu)}{k_B T}\right]$$

7.2.10 *The QC in NWFETs of PtSb₂*

The 1D DRs from (R1.11) can be written as

$$k_x^2 = [2\overline{A_9}]^{-1}[-\overline{A_{10}}(E, n_z) + [\overline{A_{10}}^2(E, n_z) + 4(\bar{A}_9)\overline{A_{11}}(E, n_z)]]^{1/2} - \left(\frac{n_y \pi}{d_Y}\right)^2 \quad (7.153)$$

where

$$\overline{A_9} = (I_1 + \omega_1\omega_3), \quad I_1 = I\frac{(\bar{a})^4}{16}$$

$$\overline{A_{10}}(E, n_z) = \left[\omega_3 E + \omega_1\left[E + \overline{\delta_0} - \omega_4\left(\frac{n_z\pi}{d_z}\right)^2\right] + \omega_2\omega_3\left(\frac{n_z\pi}{d_z}\right)^2 + 2I_1\left(\frac{n_z\pi}{d_z}\right)^2\right],$$

$$\omega_1 = \frac{(\bar{a})^2}{4}[(\bar{\lambda}_0) - \bar{l}], \quad \omega_2 = (\overline{\lambda_0}\frac{(\bar{a})^2}{4}), \quad \omega_3 = \frac{(\bar{a})^2}{4}(\bar{n} + \bar{\nu}), \quad \omega_4 = \frac{(\bar{a})^2}{4}\bar{\nu}$$

and

$$\overline{A_{11}}(E, n_z) = \left[E\left[E + \overline{\delta_0} - \omega_4\left(\frac{n_z\pi}{d_z}\right)^2\right] + \omega_2\left(\frac{n_z\pi}{d_z}\right)^2\left[E + \overline{\delta_0} - \omega_4\left(\frac{n_z\pi}{d_z}\right)^2\right] - I_1\left(\frac{n_z\pi}{d_z}\right)^4\right]$$

The (7.153) can be expressed as

$$k_x^2 = A_{27}(E, n_y, n_z) \quad (7.154)$$

where

$$A_{27}(E, n_y, n_z) = [2\overline{A_9}]^{-1}[-\overline{A_{10}}(E, n_z) + [\overline{A_{10}}^2(E, n_z) + 4(\bar{A}_9)\overline{A_{11}}(E, n_z)]]^{1/2} - \left(\frac{n_y\pi}{d_y}\right)^2$$

The 1D electron statistics under the condition of unity Fermi function at low temperatures can be written following (7.154) as

$$n_s = \frac{2g_v}{\pi} S_2\left[B_{40}(E'_{F1D}, n_y, n_z)\right] \tag{7.155}$$

where $B_{40}(E'_{F1D}, n_y, n_z) = \sqrt{A_{27}(E'_{F1D}, n_y, n_z)}$.

Using (1.4) and (7.155) the C_g can be expressed as

$$C_g = \left[\frac{2g_v e^2}{\pi} S_2(\theta_{11})\right]\left[1 + \frac{2g_v e^2}{\pi C_{ins}} S_2(\theta_{11})\right]^{-1} \tag{7.156}$$

where

$$\theta_{11} = \frac{\partial}{\partial E'_{F1D}}(B_{40}(E'_{F1D}, n_y, n_z))$$

The ER can be written as

$$\frac{D}{\mu} = [S_2[B_{40}(E'_{F1D}, n_x, n_y)]][S_2[B'_{40}(E'_{F1D}, n_x, n_y)]]^{-1} \tag{7.157}$$

The EEM can be written as

$$m^*(E, n_y, n_z) = \frac{\hbar^2}{2} A'_{27}(E, n_y, n_z) \tag{7.158}$$

The field emitted current assumes the form

$$I = \frac{2g_v e k_B T}{h} S_2[F_0(\bar{\eta}_{27}) \exp(-\bar{\beta}_{27})] \tag{7.159}$$

where

$$\bar{\eta}_{27} = \frac{E'_{F1D} - \bar{E}_{28}}{k_B T}$$

E_{28} is the root of the equation

$$[2\overline{A_9}]^{-1}[-\overline{A_{10}}(E_{28}, n_z) + [\overline{A_{10}}^2(E_{28}, n_z) + 4(\bar{A}_9)\overline{A_{11}}(E_{28}, n_z)]]^{1/2} - \left(\frac{n_y \pi}{d_y}\right)^2 = 0 \tag{7.160}$$

where

$$\beta_{27} = \frac{4}{3}[A_{27}(V_0, n_x, n_y)]^{3/2} \cdot [eF_{sz} A'_{27}(V_0, n_x, n_y)]^{-1}$$

$$A'_{27}(E, n_y, n_z) = [2\overline{A_9}]^{-1}[-(\overline{A_{10}})' + [\overline{A_{10}}^2(V_0, n_z) + 4(\bar{A}_9)\overline{A_{11}}(V_0, n_z)]^{-1/2} \cdot [\overline{A_{10}}(V_0, n_z)(\overline{A_{10}})' + 2(\bar{A}_9)(\overline{A_{11}}(V_0, n_z))']],$$

$$A'_{10} = (\omega_1 + \omega_3)$$

and

$$(\overline{A_{11}}(V_0, n_z))' = \left[2V_0 + \overline{\delta_0} - \omega_4\left(\frac{n_z\pi}{d_z}\right)^2 + \omega_2\left(\frac{n_z\pi}{d_z}\right)^2\right]$$

The photo-emitted current is given by

$$I = \frac{\alpha_o e g_v k_B T}{2\pi\hbar} S_2 F_0(\eta_{79}) \tag{7.161}$$

where

$$\eta_{79} \equiv \left[\frac{E'_{F1D} - (\bar{E}_{28} + W - h\nu)}{k_B T}\right]$$

7.2.11 *The QC in NWFETs of II–V materials*

Using (R1.19), the 1D DRs in this case can be written as

$$k_z^2 = A_{32,\pm}(E, n_x, n_y) \tag{7.162}$$

where

$$\begin{aligned}
A_{32,\pm}(E, n_x, n_y) &= \alpha_{4,\pm}(n_x, n_y) + \beta_4 E \pm [\beta_5 E^2 + E\beta_{6,\pm}(n_x, n_y) \\
&\quad + \beta_{7,\pm}(n_x, n_y)]^{\frac{1}{2}} \\
\alpha_{4,\pm}(n_x, n_y) &= [2(\theta_3^2 - \theta_7^2)]^{-1}[2\theta_7\alpha_2(n_x, n_y) - 2\alpha_{1,\mp}(n_x, n_y)\theta_3], \\
\alpha_{1,\mp}(n_x, n_y) &= \theta_1\left(\frac{n_x\pi}{d_x}\right)^2 + \theta_2\left(\frac{n_y\pi}{d_y}\right)^2 + \delta_4\left(\frac{n_x\pi}{d_x}\right) \mp \Delta_3, \\
\alpha_2(n_x, n_y) &= \left[\theta_5\left(\frac{n_x\pi}{d_x}\right)^2 + \theta_6\left(\frac{n_y\pi}{d_y}\right)^2 + \delta_5\left(\frac{n_x\pi}{d_x}\right)\right], \\
\beta_4 &= 2\theta_3[2(\theta_3^2 - \theta_7^2)]^{-1}, \quad \beta_5 = [2(\theta_3^2 - \theta_7^2)]^{-2}[4\theta_7^2], \\
\beta_{6,\pm}(n_x, n_y) &= [8\theta_3\theta_7\alpha_2(n_x, n_y) - 8\theta_7^2\alpha_{1,\mp}(n_x, n_y)][2(\theta_3^2 - \theta_7^2)]^{-1}, \\
\beta_{7,\pm}(n_x, n_y) &= [2(\theta_3^2 - \theta_7^2)]^{-2}[4\theta_7^2\alpha_2^2(n_x, n_y) \\
&\quad - 8\theta_7\theta_3\alpha_2(n_x, n_y)\alpha_{1,\mp}(n_x, n_y) + 4\theta_3^2\alpha_3(n_y) \\
&\quad + 4\theta_7^2\alpha_{1,\mp}(n_x, n_y) - 4\alpha_3(n_y)\theta_7^2] \quad \text{and} \\
\alpha_3(n_y) &= G_3^2\left(\frac{n_y\pi}{d_y}\right)^2 + \Delta_3^2.
\end{aligned}$$

The quantized energy levels $(E_{34,\pm})$ can be expressed through the equation

$$E_{34,\pm} = \alpha_{1,\mp}(n_x, n_y) \pm [\alpha_2^2(n_x, n_y) + \alpha_3(n_y)]^{\frac{1}{2}} \tag{7.163}$$

The 1D electron statistics under the condition of unity Fermi function at low temperatures can be written following [41] as

$$n_s = \frac{g_v}{\pi} S\left[B_{49}(E'_{F1D}, n_x, n_y)\right] \tag{7.164}$$

where

$$B_{49}(E'_{F1D}, n_x, n_y) = [\sqrt{A_{32,+}(E'_{F1D}, n_x, n_y)} + \sqrt{A_{32,-}(E'_{F1D}, n_x, n_y)}].$$

Using (1.4) and (7.164) the C_g can be expressed as

$$C_g = \left[\frac{2g_v e^2}{\pi} S(\bar{\theta}_{166})\right]\left[1 + \frac{2g_v e^2}{\pi C_{ins}} S(\bar{\theta}_{166})\right]^{-1} \tag{7.165}$$

where

$$\theta_{166} = \frac{\partial}{\partial E'_{F1D}}(B_{49}(E'_{F1D}, n_x, n_y))$$

The ER can be written as

$$\frac{D}{\mu} = [S[B_{49}(E'_{F1D}, n_x, n_y)]][S[B'_{49}(E'_{F1D}, n_x, n_y)]]^{-1} \tag{7.166}$$

The EEM can be expressed as

$$m^*_{,\pm}(E, n_x, n_y) = \frac{\hbar^2}{2} A'_{32,\pm}(E, n_x, n_y) \tag{7.167}$$

The field emitted current assumes the form

$$I = \frac{g_v e k_B T}{h} S[\exp(-\beta_{32,+}) + F_0(\eta_{32,-})\exp(-\beta_{32,-})] \tag{7.168}$$

where

$$\eta_{32,\pm} = \frac{E_{F1D} - E_{34,\pm}}{k_B T},$$

$$\beta_{32,\pm} = \frac{4}{3}[A_{32,\pm}(V_0, n_x, n_y)]^{3/2} \cdot [eF_{sz} A'_{32,\pm}(V_0, n_x, n_y)]^{-1}$$

and

$$A'_{32,\pm}(V_0, n_x, n_y) = \left[\beta_4 \pm \frac{1}{2}[2\beta_5 V_0 + \beta_{6,\pm}(n_x, n_y)] \cdot [\beta_5 V_0^2 + \beta_{6,\pm}(n_x, n_y)V_0 + \beta_{7,\pm}(n_x, n_y)]^{-\frac{1}{2}}\right]$$

The photo emitted current is given by

$$I = \frac{\alpha_o e g_v k_B T}{\pi\hbar}[SF_0(\eta_{80})] \tag{7.169}$$

where

$$\eta_{80} \equiv \left[\frac{E'_{F1D} - (\bar{E}_{34,\pm} + W - h\nu)}{k_B T}\right]$$

7.2.12 *The QC in CNTFETs*

For (n, n) and $(n, 0)$ carbon nanotubes, the energy DRs are given by [72,73]

$$E = \pm t_c\sqrt{1 + 4\cos\frac{m\pi}{n}\cos\frac{k_y a_c}{2} + 4\cos^2\frac{k_y a_c}{2}},$$
$$-\frac{\pi}{a_c} < k_y < \frac{\pi}{a_c} \quad \text{and} \quad m(= 1, 2, \ldots, 2n) \tag{7.170}$$

$$E = \pm t_c\sqrt{1 + 4\cos\frac{\sqrt{3}k_x a_c}{n}\cos\frac{m\pi}{n} + 4\cos^2\frac{m\pi}{n}},$$
$$-\frac{\pi}{\sqrt{3}a_c} < k_x < \frac{\pi}{\sqrt{3}a_c} \quad \text{and} \quad m\,(= 1, 2, \ldots, 2n) \tag{7.171}$$

where $t_c(= -0.9eV)$ [50] is the tight binding parameter and m and n are the chiral indices [53].

Using (7.170) and (7.171) the E-k_y relation can approximately written as [83]

$$E^2 - E_i^2 = Ak_y^2 \tag{7.172}$$

where

$$E_i = \frac{|3i - n_1 + n_2|}{2}|t_c| \cdot \left(\frac{d}{r}\right), \quad i = 1, 2, 3, \ldots, i_{\max}$$

a_c is the nearest neighbor C–C bonding distance, r is the nanotube radius and $A = (9t_c^2 d^2/4\pi^2)$. From (7.172), it appears that when $k_y \to 0$, $E \to E_i$.

Using this idea from (7.169), we can write

$$4\cos\left(\frac{m\pi}{n}\right) = \frac{E_i^2}{t_c^2} - 5 \tag{7.173}$$

By Eliminating $\cos\left(\frac{m\pi}{n}\right)$ between (7.172) and (7.173), we can write

$$4z^2 + \left(\frac{E_i^2}{t_c^2} - 5\right)z + 1 - \frac{E_i^2}{t_c^2} = 0 \tag{7.174}$$

where

$$z = \cos\left(\frac{a_c k_y \sqrt{3}}{2}\right)$$

From (7.174), we can write that the E–k_y relation for arm chair nanotube is given by

$$k_y = \frac{2}{a_c\sqrt{3}} f_1(E, m, n) \tag{7.175}$$

where $f_1(E, m, n) = \cos^{-1}[\theta_1(E, m, n)]$ and

$$\theta_1(E, m, n) = \frac{1}{8}\left[-\left(\frac{E_i^2}{t_c^2} - 5\right) + \left[\left(\frac{E_i^2}{t_c^2} - 5\right)^2 + 16\left(\frac{E_i^2}{t_c^2} - 1\right)\right]^{1/2}\right]$$

The electron statistics can be written as

$$n_{1D} = \frac{8}{a_c \pi\sqrt{3}} \sum_{i=0}^{i_{\max}} [f_1(E_{F1}, m, n)] \tag{7.176}$$

where E_{F_1} is the Fermi energy.

Using (7.176) and (1.4) we can study the QC in this case

The EMM in this case is given by

$$m^*(E_{F_1}, m, n) = 4(\hbar/a_c)^2 f_1(E_{F_1}, m, n) \frac{\theta_1'(E_{F_1}, m, n)}{\sqrt{1 - \theta_1^2(E_{F_1}, m, n)}} \tag{7.177}$$

where

$$\theta_1'(E, m, n) = \frac{4E_{F1}}{t_c^2}\left[\left(\frac{E_i^2}{t_c^2} - 5\right)^2 + 16\left(\frac{E_{F1}^2}{t_c^2} - 1\right)\right]^{-1/2}$$

The ER can be written as

$$\frac{D}{\mu} = \frac{1}{e}\left[\sum_{i=0}^{i_{\max}} [f_1(E_{F1}, m, n)]\right]\left[\sum_{i=0}^{i_{\max}} [f_1'(E_{F1}, m, n)]\right]^{-1} \tag{7.178}$$

Similarly, the E-k_y relation of a zigzag nanotube is given by

$$k_y = \frac{2}{3a_c} f_2(E, m, n) \tag{7.179}$$

where $f_2(E, m, n) = \cos^{-1}[\theta_2(E, m, n)]$ and

$$\theta_2(E, m, n) = \left[\left[\left(\frac{E_i^2}{t_c^2} - 1\right) - \left(\frac{E_i}{t_c} - 1\right)^2\right]\left[\frac{2E_i}{t_c} - 1\right]^{-1}\right]$$

The electron concentration can be written as

$$n_{1D} = \frac{8}{3a_c\pi}\sum_{i=0}^{i_{\max}}[f_2(E_{F1},m,n)] \quad (7.180)$$

Using (7.180) and (1.4) we can study the QC in this case

Using (7.180) the EEM can be written as

$$m^*(E_{F_1},m,n) = (4\hbar^2/9a_c^2)f_2(E_{F_1},m,n)\frac{\theta_2'(E_{F_1},m,n)}{\sqrt{1-\theta_2^2(E_{F_1},m,n)}} \quad (7.181)$$

where

$$\theta_2'(E_{F_1},m,n) = \left[\left(\frac{2E_{F1}}{t_c^2}\right)\right]\left[\frac{2E_i}{t_c}-1\right]^{-1}$$

The ER can be written as

$$\frac{D}{\mu} = \frac{1}{e}\left[\sum_{i=0}^{i_{\max}}[f_2(E_{F1},m,n)]\right]\left[\sum_{i=0}^{i_{\max}}[f_2'(E_{F1},m,n)]\right]^{-1} \quad (7.182)$$

7.3 Results and Discussion

The normalized QC in NWFETs made of $CdGeAs_2$ with film thickness as the independent variable has been premeditated in Fig. 7.1 by taking

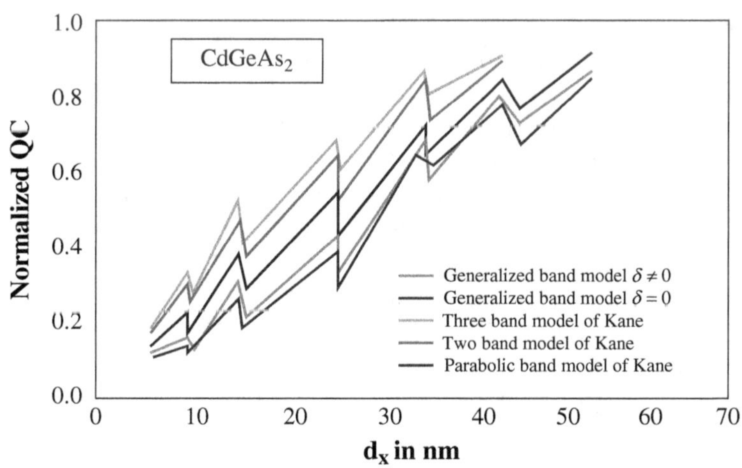

Fig. 7.1. The normalized QC in NWFETs made of $CdGeAs_2$ with film thickness as the independent variable has been premeditated using (a) (7.5) (orange), (b) (7.5) with $\delta = 0$ (brown), (c) (7.20) (green), (d) (7.29) (red) and (e) (7.38) (blue).

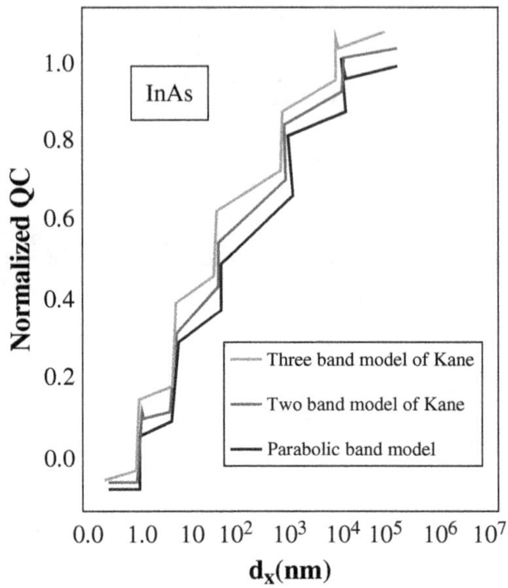

Fig. 7.2. The normalized QC in NWFETs made of InAs with film thickness as the independent variable has been premeditated using (a) (7.20) (green), (d) (7.29) (red) and (e) (7.38) (blue).

the different values of the energy band parameters as given in Chapter 18 using (a) (7.5) (orange), (b) (7.5) with $\delta = 0$ (brown), (c) (7.20) (green) (d) (7.29) (red) and (e) (7.38) (blue). From Fig. 7.1, it appears that the normalized QC enhances in a quantized saw-tooth manner with increasing film thickness. The normalized QC in NWFETs made of InAs with film thickness as the independent variable has been premeditated in Fig. 7.2 by taking the different values of the energy band parameters as given in Chapter 18 for all cases of Fig. 7.1. From Fig. 7.2 it appears that the normalized QC enhances in a quantized distorted saw-tooth manner with increasing film thickness. The normalized QC in NWFETs made of InSb with film thickness as the independent variable has been premeditated in Fig. 7.3 by taking the different values of the energy band parameters as given in Chapter 18 using (a) (7.20) (green), (d) (7.29) (red), and (e) (7.38) (blue). From Fig. 7.3, it appears that the normalized QC in this case enhances in a quantized saw-tooth manner with increasing film thickness. The normalized QC in NWFETs made of $Hg_{1-x}Cd_xTe$ with film thickness as the independent variable has been premeditated in Fig. 7.4 by taking the

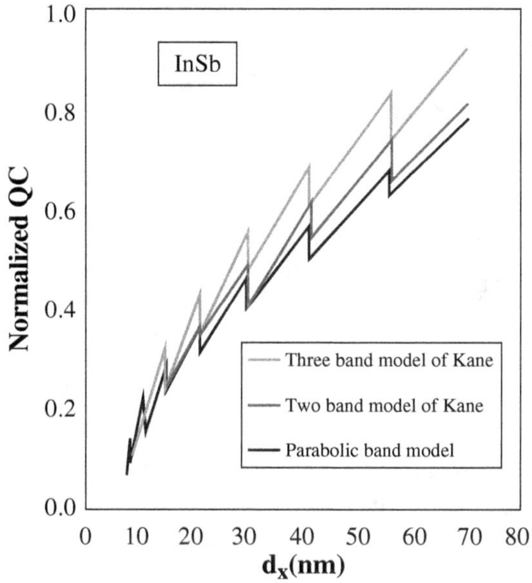

Fig. 7.3. The normalized QC in NWFETs made of InSb with film thickness as the independent variable has been premeditated using (a) (7.20) (green), (d) (7.29) (red) and (e) (7.38) (blue).

different values of the energy band parameters as given in Chapter 18 using (a) (7.20) (green), (d) (7.29) (red) and (e) (7.38) (blue). From Fig. 7.4 it appears that the normalized QC in this case enhances in a quantized saw-tooth manner with increasing film thickness. The normalized QC in NWFETs made of $Ga_{1-x}Al_xAs$ with film thickness as the independent variable has been premeditated in Fig. 7.5 by taking the different values of the energy band parameters as given in Chapter 18 using (a) (7.20) (green), (d) (7.29) (red) and (e) (7.38) (blue). From Fig. 7.5, it appears that the normalized QC in this case enhances in a quantized saw-tooth manner with increasing film thickness.

The normalized QC in NWFETs made of $In_{1-x}As_xGa_yP_{1-y}$ with film thickness as the independent variable has been premeditated in Fig. 7.6 by taking the different values of the energy band parameters as given in Chapter 18 using (a) (7.20) (green), (d) (7.29) (red) and (e) (7.38) (blue). From Fig. 7.6, it appears that the normalized QC enhances in a quantized saw-tooth manner with increasing film thickness. The normalized QC in NWFETs made of Cd_3As_2 with 1D electron statistics as the independent

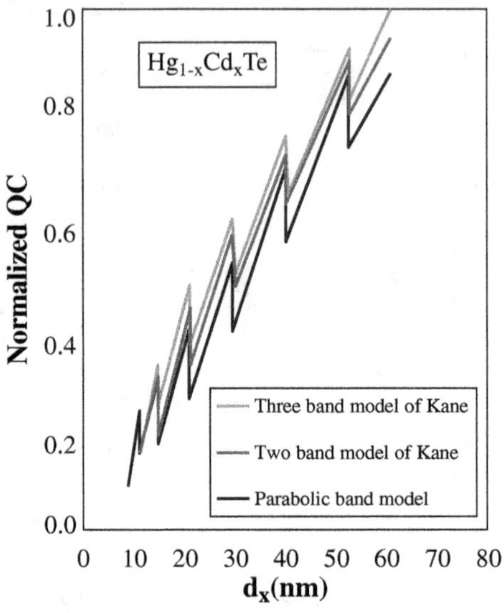

Fig. 7.4. The normalized QC in NWFETs made of $Hg_{1-x}Cd_xTe$ with film thickness as the independent variable has been premeditated using (a) (7.20) (green), (d) (7.29) (red) and (e) (7.38) (blue).

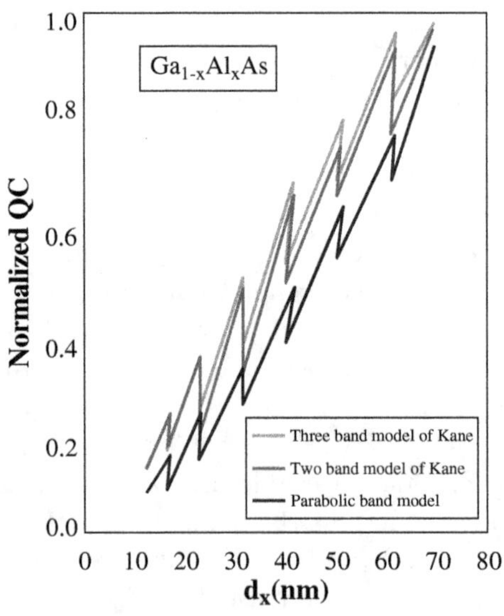

Fig. 7.5. The normalized QC in NWFETs made of $Ga_{1-x}Al_xAs$ with film thickness as the independent variable has been premeditated using (a) (7.20) (green), (d) (7.29) (red) and (e) (7.38) (blue).

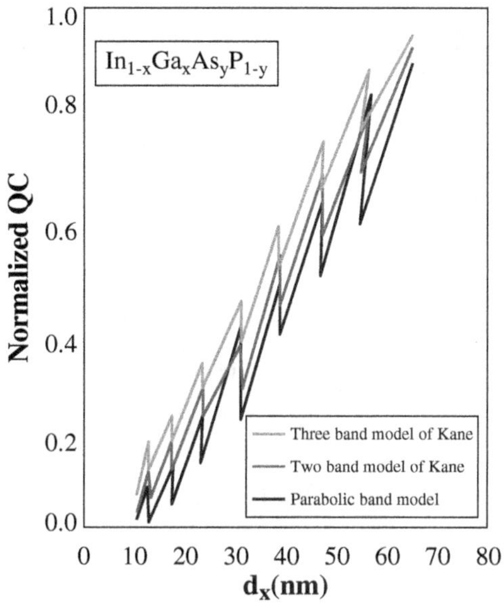

Fig. 7.6. The normalized QC in NWFETs made of $In_{1-x}Ga_xAs_yP_{1-y}$ with film thickness as the independent variable has been premeditated using (a) (7.20) (green), (d) (7.29) (red) and (e) (7.38) (blue).

variable has been premeditated in Fig. 7.7 by taking the different values of the energy band parameters as given in Chapter 18 using (a) (7.20) (green), (d) (7.29) (red) and (e) (7.38) (blue). From Fig. 7.7, it appears that the normalized QC enhances in a quantized step-like manner with increasing 1D electron statistics. The normalized QC in NWFETs made of $CdGeAs_2$ with 1D electron statistics as the independent variable has been premeditated in Fig. 7.8 by taking the different values of the energy band parameters as given in Chapter 18 using (a) (7.5) (brown), (b) (7.5) with $\delta = 0$ (orange), (c) (7.20) (green), (d) (7.29) (red) and (e) (7.38) (blue). From Fig. 7.8, it appears that the normalized QC in this case enhances in a quantized step-like manner with increasing 1D electron statistics. The normalized QC in NWFETs made of InAs with 1D electron statistics as the independent variable has been premeditated in Fig. 7.9 by taking the different values of the energy band parameters as given in Chapter 18 using (a) (7.20) (green), (d) (7.29) (red) and (e) (7.38) (blue). From Fig. 7.9, it appears that the normalized QC enhances in quantized step like manner with increasing 1D electron statistics. The normalized QC in NWFETs made of InSb with 1D electron statistics as the independent variable has

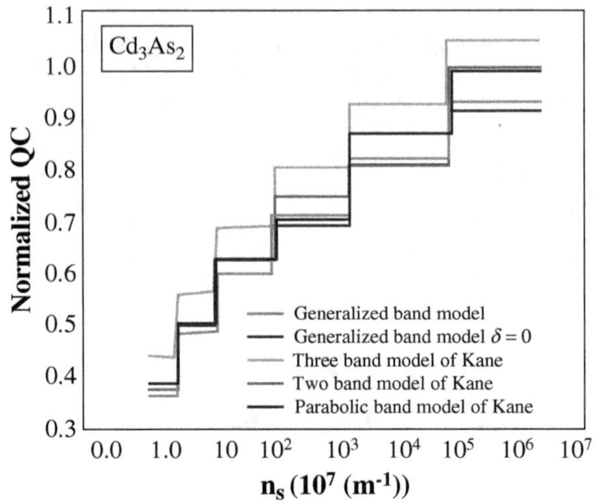

Fig. 7.7. The normalized QC in NWFETs made of Cd_3As_2 with 1D electron statistics as the independent variable has been premeditated using (a) (7.5) (orange), (b) (7.5) with $\delta = 0$ (brown), (c) (7.20) (green), (d) (7.29) (red) and (e) (7.38) (blue).

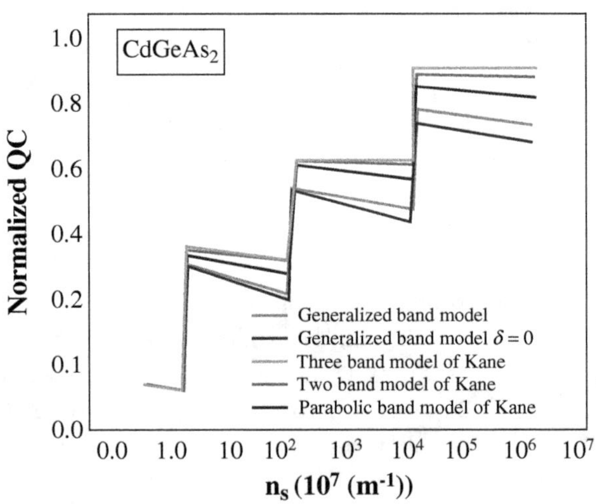

Fig. 7.8. The normalized QC in NWFETs made of $CdGeAs_2$ with 1D electron statistics as the independent variable has been premeditated using (a) (7.5) (orange), (b) (7.5) with $\delta = 0$ (brown), (c) (7.20) (green), (d) (7.29) (red) and (e) (7.38) (blue).

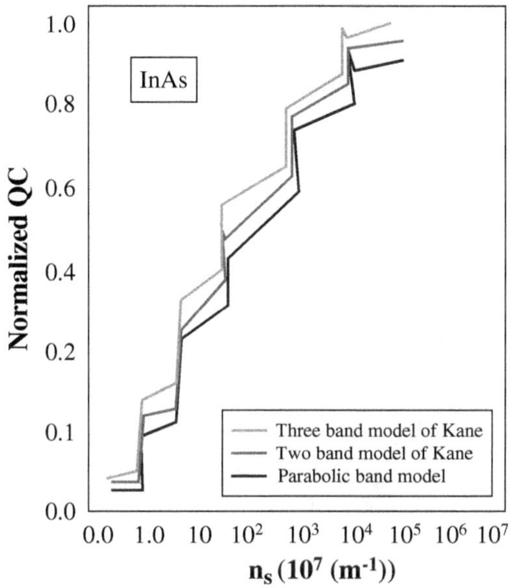

Fig. 7.9. The normalized QC in NWFETs made of InAs with 1D electron statistics as the independent variable has been premeditated using (a) (7.20) (green), (d) (7.29) (red) and (e) (7.38) (blue).

been premeditated in Fig. 7.10 by taking the different values of the energy band parameters as given in Chapter 18 using (a) (7.20) (green), (d) (7.29) (red) and (e) (7.38) (blue). From Fig. 7.10, it appears that the normalized QC enhances in a quantized step-like manner with increasing 1D electron statistics.

The normalized QC in NWFETs made of $Hg_{1-x}Cd_xTe$ with 1D electron statistics as the independent variable has been premeditated in Fig. 7.11 by taking the different values of the energy band parameters as given in Chapter 18 using (a) (7.20) (green), (d) (7.29) (red) and (e) (7.38) (blue). From Fig. 7.11, it appears that the normalized QC enhances in a quantized step-like manner with increasing 1D electron statistics. The normalized QC in NWFETs made of $Ga_{1-x}Al_xAs$ with 1D electron statistics as the independent variable has been premeditated in Fig. 7.12 by taking the different values of the energy band parameters as given in Chapter 18 using (a) (7.20) (green), (d) (7.29) (red) and (e) (7.38) (blue). From Fig. 7.12, it appears that the normalized QC enhances in quantized step like manner with increasing 1D electron statistics. The normalized QC in NWFETs

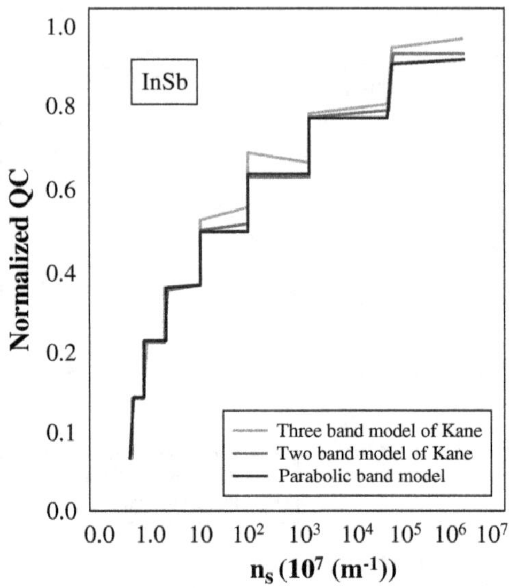

Fig. 7.10. The normalized QC in NWFETs made of InSb with 1D electron statistics as the independent variable has been premeditated using (a) (7.20) (green), (d) (7.29) (red) and (e) (7.38) (blue).

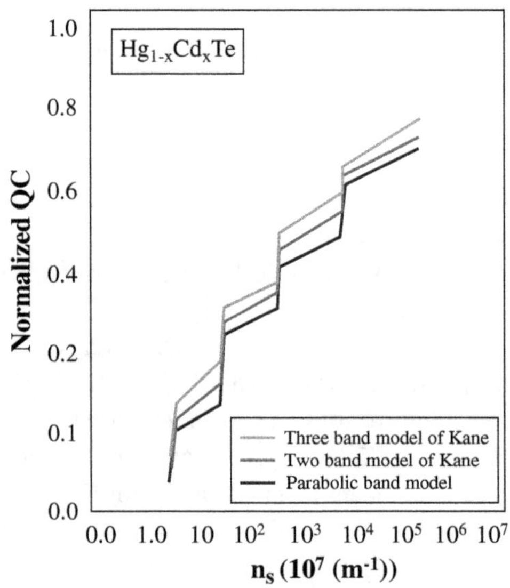

Fig. 7.11. The normalized QC in NWFETs made of $Hg_{1-x}Cd_xTe$ with 1D electron statistics as the independent variable has been premeditated using (a) (7.20) (green), (d) (7.29) (red) and (e) (7.38) (blue).

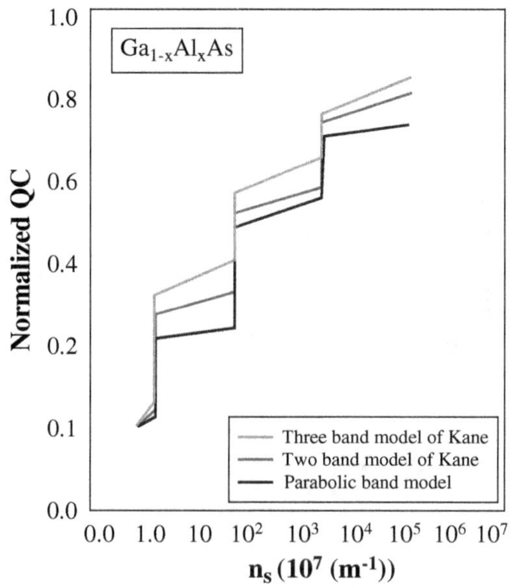

Fig. 7.12. The normalized QC in NWFETs made of $Ga_{1-x}Al_xAs$ with 1D electron statistics as the independent variable has been premeditated using (a) (7.20) (green), (d) (7.29) (red) and (e) (7.38) (blue).

made of Cd_3As_2 with V_g as the independent variable has been premeditated in Fig. 7.13 by taking the different values of the energy band parameters as given in Chapter 18 using (a) (7.5) (violet), (b) (7.5) with $\delta = 0$ (orange), (c) (7.20) (green), (d) (7.29) (red) and (e) (7.38) (blue). From Fig. 7.13, it appears that the normalized QC decreases with increasing gate voltage in a quantized oscillatory fashion for all cases as considered in this case. The rate of oscillatory decrease changes from one type of band model to another type.

The normalized QC in NWFETs made of $CdGeAs_2$ with V_g as the independent variable has been premeditated in Fig. 7.14 by taking the different values of the energy band parameters as given in Chapter 18 using (a) (7.5) (brown), (b) (7.5) with $\delta = 0$ (orange), (c) (7.20) (green), (d) (7.29) (red) and (e) (7.38) (blue). From Fig. 7.14, it appears that the normalized QC decreases with increasing gate voltage in a quantized oscillatory fashion for all cases as considered in this case. The rate of oscillatory decrease changes from one type of band model to another type. The normalized QC in NWFETs made of InAs with V_g as the independent variable has been premeditated in Fig. 7.15 by taking the different values of the energy band

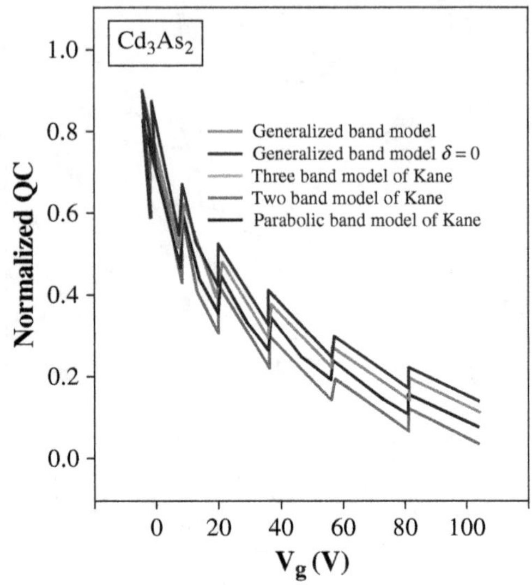

Fig. 7.13. The normalized QC in NWFETs made of Cd_3As_2 with V_g as the independent variable has been premeditated using (a) (7.5) (orange), (b) (7.5) with $\delta = 0$ (brown), (c) (7.20) (green), (d) (7.29) (red) and (e) (7.38) (blue).

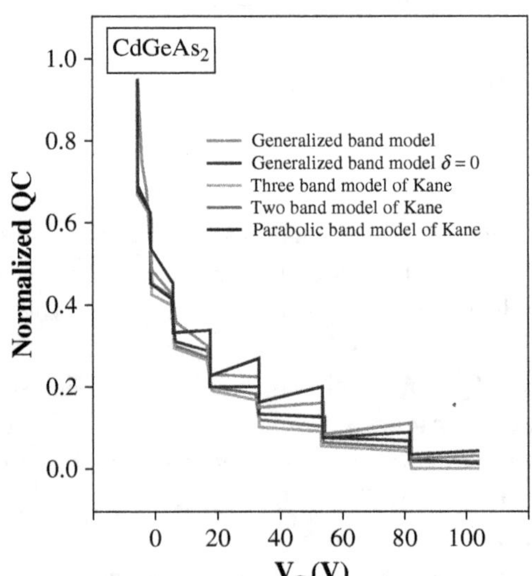

Fig. 7.14. The normalized QC in NWFETs made of $CdGeAs_2$ with V_g as the independent variable has been premeditated using (a) (7.5) (orange), (b) (7.5) with $\delta = 0$ (brown), (c) (7.20) (green), (d) (7.29) (red) and (e) (7.38) (blue).

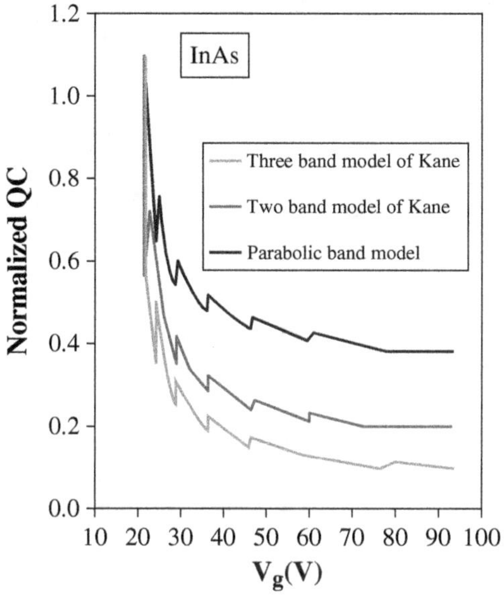

Fig. 7.15. The normalized QC in NWFETs made of InAs with V_g as the independent variable has been premeditated using (a) (7.20) (green), (d) (7.29) (red) and (e) (7.38) (blue).

parameters as given in Chapter 18 using (a) (7.20) (green), (d) (7.29) (red) and (e) (7.38) (blue). From Fig. 7.15, it appears that the normalized QC decreases with increasing gate voltage in quantized oscillatory fashion for all cases as consider in this case. The rate of oscillatory decrease changes from one type of band model to another type. The normalized QC in NWFETs made of InSb with V_g as the independent variable has been premeditated in Fig. 7.16 by taking the different values of the energy band parameters as given in Chapter 18 using (a) (7.20) (green), (d) (7.29) (red) and (e) (7.38) (blue).

From Fig. 7.16, it appears that the normalized QC decreases with increasing gate voltage in a quantized oscillatory fashion for all cases as considered in this case. The rate of oscillatory decrease changes from one type of band model to another type. The normalized QC in NWFETs made of $Hg_{1-x}Cd_xTe$ with V_g as the independent variable has been premeditated in Fig. 7.17 by taking the different values of the energy band parameters as given in Chapter 18 using (a) (7.20) (green), (d) (7.29) (red) and (e) (7.38) (blue). From Fig. 7.17, it appears that the normalized QC decreases with increasing gate voltage in a quantized oscillatory fashion for all cases as

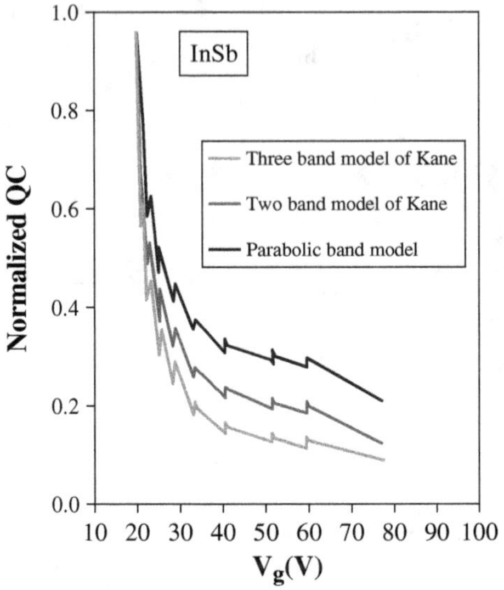

Fig. 7.16. The normalized QC in NWFETs made of InSb with V_g as the independent variable has been premeditated using (a) (7.20) (green), (d) (7.29) (red) and (e) (7.38) (blue).

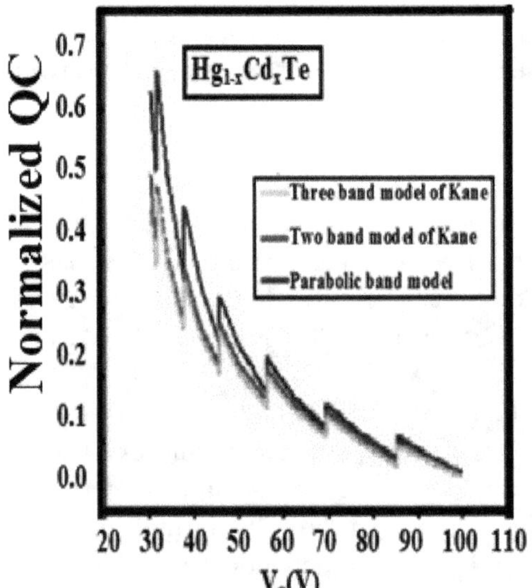

Fig. 7.17. The normalized QC in NWFETs made of $Hg_{1-x}Cd_xTe$ with V_g as the independent variable has been premeditated using (a) (7.20) (green), (d) (7.29) (red) and (e) (7.38) (blue).

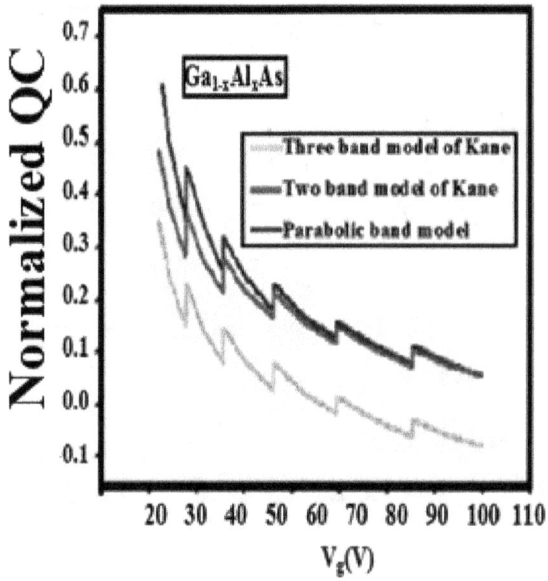

Fig. 7.18. The normalized QC in NWFETs made of $Ga_{1-x}Al_xAs$ with V_g as the independent variable has been premeditated using (a) (7.20) (green), (d) (7.29) (red) and (e) (7.38) (blue).

considered in this case. The rate of oscillatory decrease changes from one type of band model to another type. The normalized QC in NWFETs made of $Ga_{1-x}Al_xAs$ with V_g as the independent variable has been premeditated in Fig. 7.18 by taking the different values of the energy band parameters as given in Chapter 18 using (a) (7.20) (green), (d) (7.29) (red) and (e) (7.38) (blue).

From Fig. 7.18, it appears that the normalized QC decreases with increasing gate voltage in quantized oscillatory fashion for all cases as consider in this case. The rate of oscillatory decrease changes from one type of band model to another type. The normalized QC in NWFETs made of $In_{1-x}As_xGa_yP_{1-y}$ with V_g as the independent variable has been premeditated in Fig. 7.19 by taking the different values of the energy band parameters as given in Chapter 18 using (a) (7.20) (green), (d) (7.29) (red) and (e) (7.38) (blue). From Fig. 7.19 it appears that the normalized QC decreases with increasing gate voltage in a quantized oscillatory fashion for all cases as considered in this case. The rate of oscillatory decrease changes from one type of band model to another type.

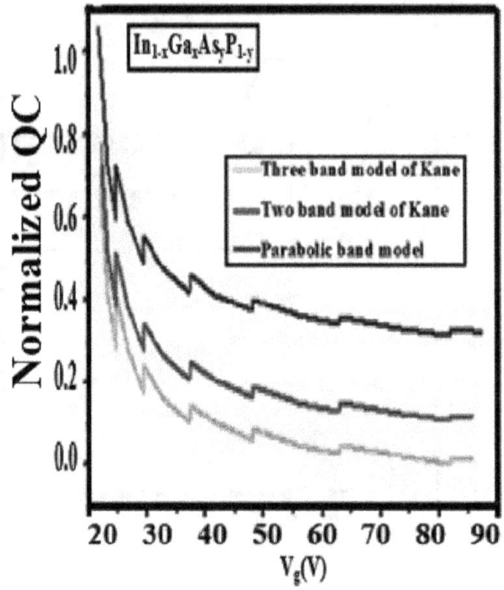

Fig. 7.19. The normalized QC in NWFETs made of $In_{1-x}Ga_xAs_yP_{1-y}$ with V_g as the independent variable has been premeditated using (a) (7.20) (green), (d) (7.29) (red) and (e) (7.38) (blue).

Using the appropriate equations and taking the energy band constants as given in the Chapter 18, we have plotted the field emitted current from NWs of $CdGeAs_2$ as a function of d_y as shown by the dotted plot of Fig. 7.20, in which the plot corresponds to the solid line represents the same for the two-band model of Kane. Figure 7.21 exhibits the plot field emitted current from NWs of n-InSb as a function of film thickness in accordance with the three- and two-band models of Kane together with the models of Stillman *et al.*, Newson *et al.*, and Palik *et al.*, respectively for the purpose of assessing the influence of dispersion relations on the field emitted current from NWs of III–V semiconductors. Figure 7.22 exhibits the variation of the field emitted current with electric field which is normal to the two directions of quantization's in the wave-vector space of the material or the longitudinal direction which is along the axis of the wire for all cases Fig. 7.21. Figure 7.23 presents the field emitted current as a function of film thickness for two different values of alloy compositions for NWs of n-$Hg_{1-x}Cd_xTe$. Figure 7.24 shows the carrier concentration dependence of field emitted current from NWs of $Hg_{1-x}Cd_xTe$, InSb, InAs and GaAs,

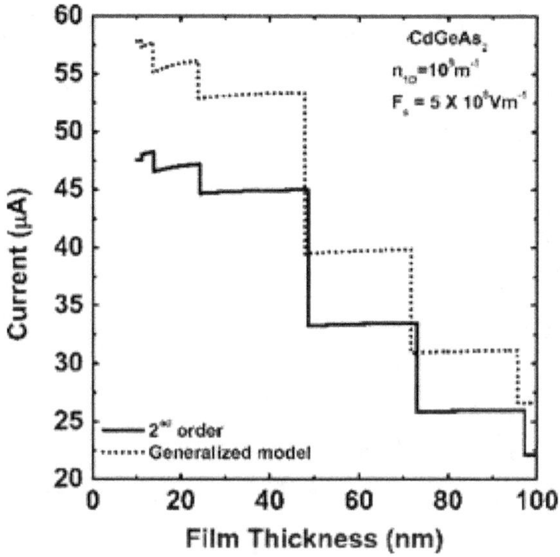

Fig. 7.20. Plot of the FN field emission current as function of film thickness d_y for NWs of $CdGeAs_2$. The dotted and solid curves correspond to the generalized and the two-band models of Kane respectively, where $d_z = 30\,\text{nm}$.

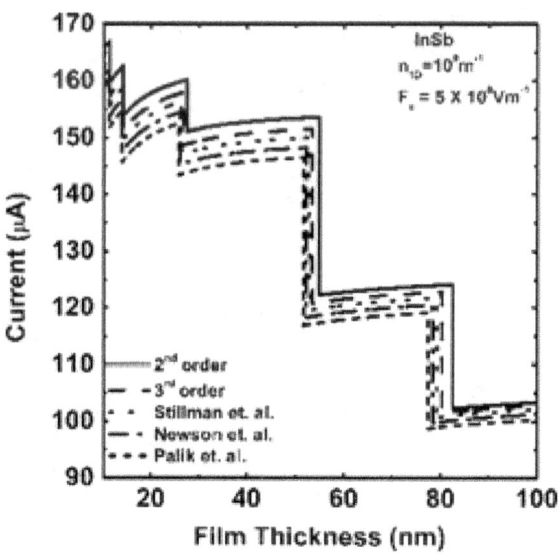

Fig. 7.21. Plot of the FN field emission current as a function of film thickness d_y for NWs of InSb in accordance with the three and two-band models of Kane together with the models of Stillman *et al.*, Newson *et al.* and Palik *et al.*, respectively, where $d_z = 30\,\text{nm}$.

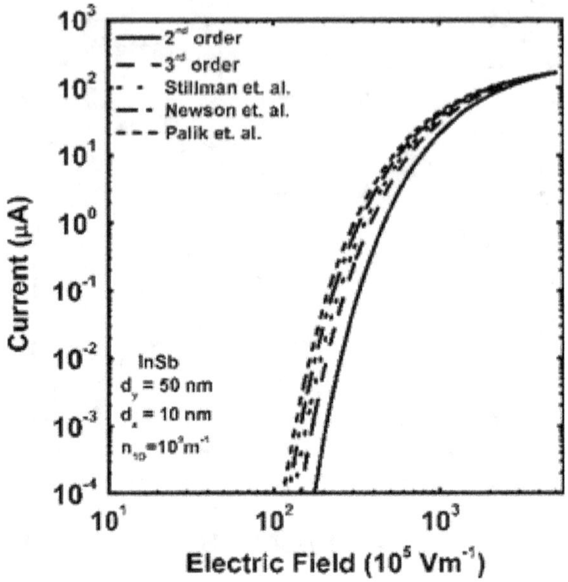

Fig. 7.22. Plot of the FN field emission current as a function of electric field for NWs of InSb in accordance with all the cases of Fig. 1.2.

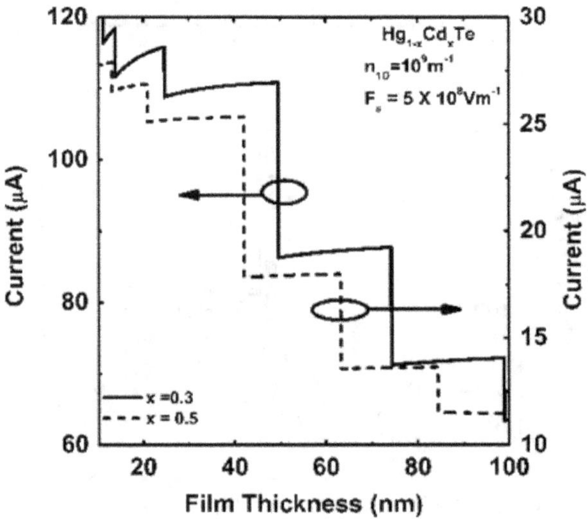

Fig. 7.23. Plot of the FN field emission current as a function of film thickness d_y for NWs of n-$Hg_{1-x}Cd_xTe$ in accordance with the two-band model of Kane for two different values of alloy composition.

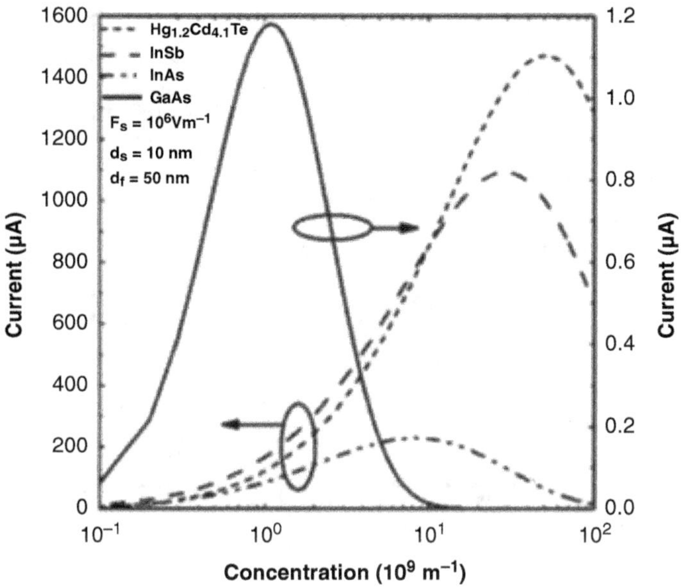

Fig. 7.24. Plot of the FN field emission current as a function of carrier concentration for NWs of $Hg_{0.7}Cd_{0.3}Te$, InSb, InAs and GaAs in accordance with the two-band model of Kane at the lowest sub-band.

respectively, for the purpose of assessing the influence of different energy band constants on the field emitted current from NWs of III–V materials. In Fig. 7.25, exhibits the film thickness dependence of field emitted current from NWs of $In_{1-x}Ga_xAs_yP_{1-y}$ in accordance with the three- and two-band models of Kane together with the models of Stillman *et al.*, Newson *et al.* and Palik *et al.*, respectively. Figure 7.26 shows the dependence of field emitted current on alloy composition from NWs of ternary and quaternary materials in accordance with the two-band models of Kane.

Figure 7.27 exhibits the film thickness dependence of field emitted current from NWs of II–VI materials taking p-CdS as an example. Figure 7.28 shows the field emitted current as a function of carrier concentration for the case of Fig. 7.8. Figure 7.29, exhibits the variation of the field emitted current as a function of film thickness for NWs of stressed materials taking stressed InSb as an example. Figures 7.30 and 7.31 explore the stress dependence of the field emitted current from NWs of stressed InSb for different values of doping for the purpose of assessing the influence of carrier concentration on the field emitted current in this case. In Fig. 7.32, the field

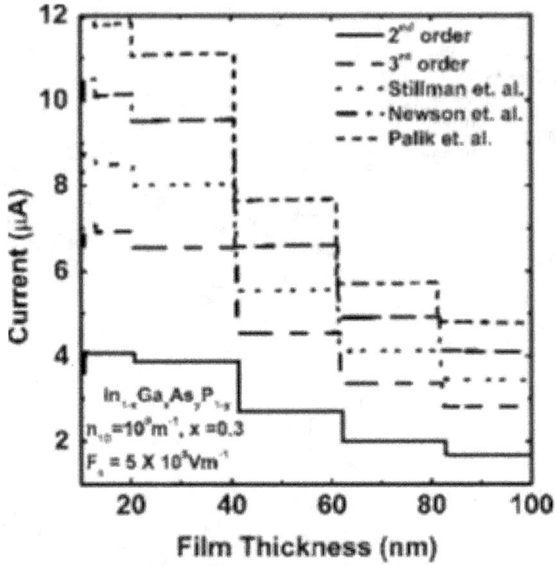

Fig. 7.25. Plot of the field current as a function of film thickness d_y for nanowires of $In_{1-x}Ga_xAs_yP_{1-y}$ in accordance with the three and two band models of Kane together with the models of Stillman, *et al.*, Newson *et al.* and Palik *et al.* respectively, where $d_z = 30\,\text{nm}$.

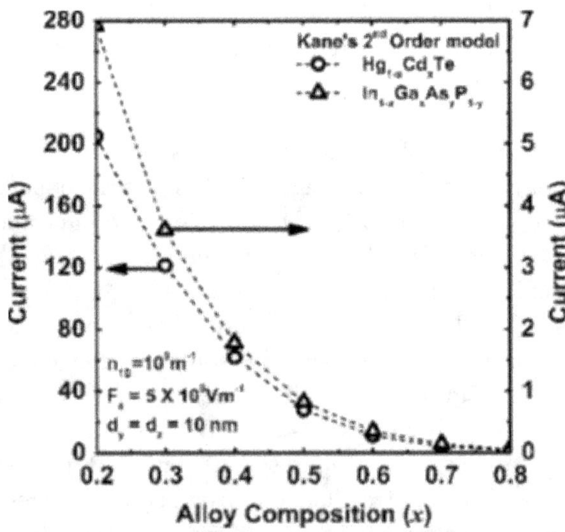

Fig. 7.26. Plot of the FNFE current as a function of alloy composition for NWs of $Hg_{1-x}Cd_xTe$ and $In_{1-x}Ga_xAs_yP_{1-y}$ in accordance with the two-band model of Kane.

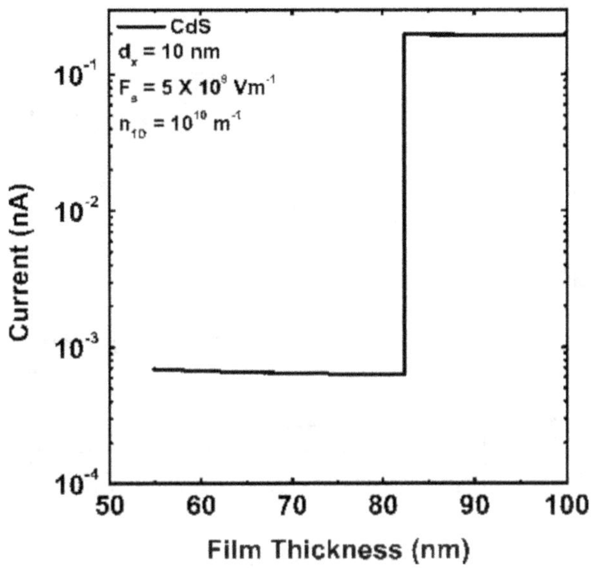

Fig. 7.27. Plot of the field emitted current as a function of film thickness d_y for NWs of CdS.

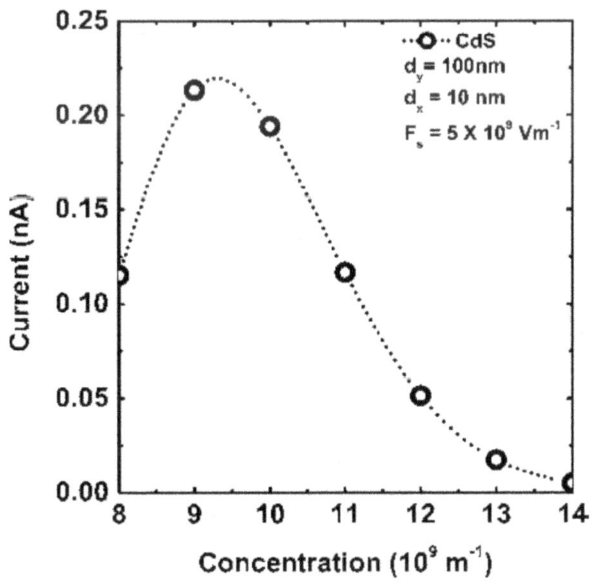

Fig. 7.28. Plot of the field emitted current as a function of concentration for NWs of CdS.

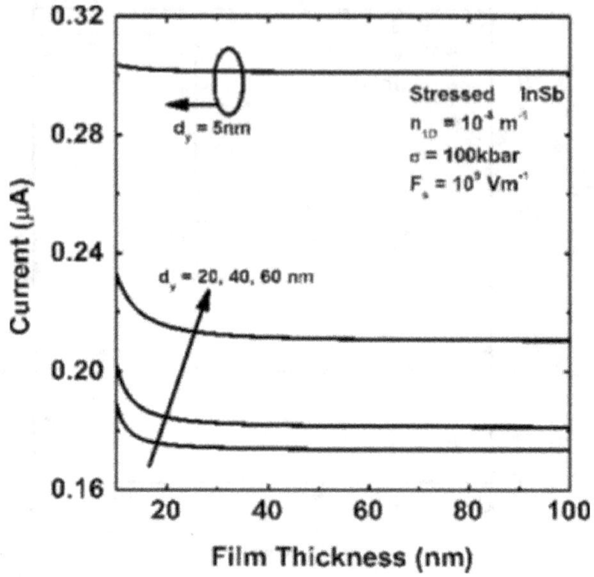

Fig. 7.29. Plot of the field emitted current as a function of film thickness for NWs of stressed InSb for different values of d_y.

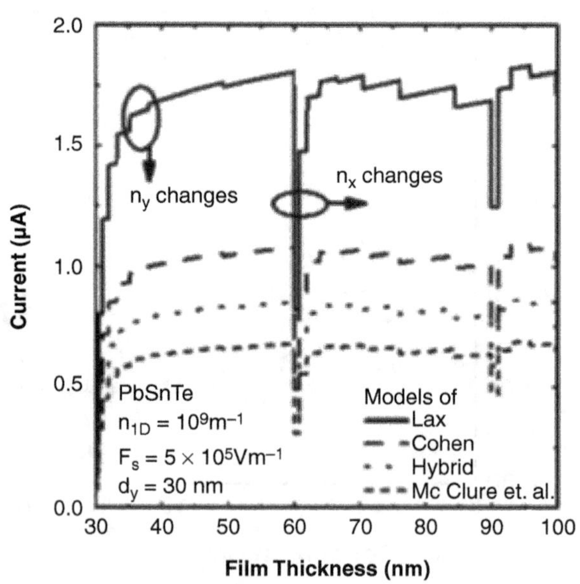

Fig. 7.30. Plot of the field emitted current as a function of stress for NWs of stressed PbSnTe for two different values of carrier concentration as shown.

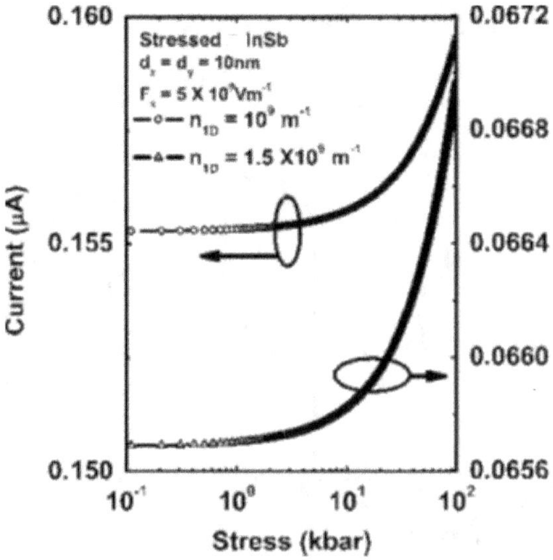

Fig. 7.31. Plot of the field emitted current as a function of stress for NWs of stressed InSb for two different values of carrier concentration as shown.

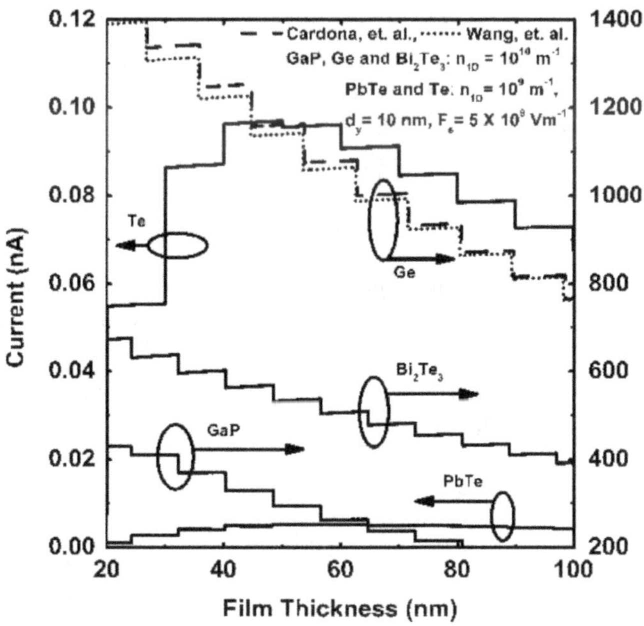

Fig. 7.32. Plot of the field emitted current as a function of film thickness for NWs of Ge (in accordance with both types of band models), GaP, Bi_2Te_3, PbTe and Te, respectively.

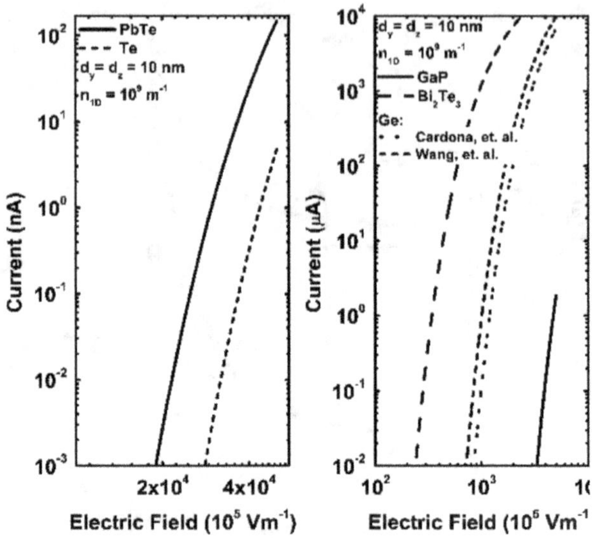

(a) (b)

Fig. 7.33. Plot of the field current as a function of electric field for nanowires of (a) PbTe and Te and (b) GaP, Bi_2Te_3 and Ge.

emitted current as a function of film thickness has been plotted for NWs of Ge (in accordance with both types of band models of Ge), GaP, PbTe and p-Bi_2Te_3, respectively. The plot (a) of Fig. 7.33 shows the variation of field emitted current on the electric field for NWs of PbTe and Te, while (b) exhibits the same for NWs of Ge, GaP and Bi_2Te_3, respectively.

The salient features of the above figures are described as follows:

From Fig. 7.20, we observe that the field emission current exhibits a step-functional decreasing dependence with an increase in film thickness for NWs of n-$CdGeAs_2$. The combined influence of the anisotropies of the energy band constants and the crystal field splitting is to enhance the field emitted current as compared with the corresponding which is based on two-band model of Kane in the whole range of thicknesses as considered in Fig. 7.20. The periodicity with respect to the film thickness is same in both the cases and is invariant of the energy band constants.

It should be noted from the field-emitted current in general, is a product of two quantities inside the summation signs. One of them is $F_0(\eta_{ij})$, ($i = 1, 2, 3, \ldots$ and $j = 1, 2, 3, \ldots$) and the other one is $\exp(-\beta_{ij})$ which is a function of Fermi energy, electric field, effective mass and various parameters of the energy bands in a complex way. Thus, we observe that field-emitted current depends totally on the product of these two functions within the summation signs and a pre-factor outside the summation sign. The product of these two functions ultimately determines the behavior of the field emitted current. Although we know that the Fermi energy of low-dimensional systems decreases with increasing size, one cannot be always certain that I will decrease with increasing film thickness due to the particular form of field-emitted current in case of NWs. If the rate of increase of $F_0(\eta_{ij})$ overcomes the rate of change of $\exp(-\beta_{ij})$, I will increase, whereas, for the opposite case, I will decrease. This important physical fact determines the magnitude of the field emitted current and its dependence with respect to any other physical variable.

From Fig. 7.21, we observe that I decrease with increasing film thickness for the three- and two-band models of Kane together with the models of Stillman *et al.*, Newson *et al.* and Palik *et al.*, respectively. The I exhibits a different magnitude which is the direct signature of the dispersion relation on the field emitted current. Numerical computations reflect the fact that the field emitted current for a $30 \times 10\,\text{nm}^2$ size quantum wire can reach nearly $160\,\mu$A for n-InSb at low temperatures and at a field strength of $5 \times 10^8\,\text{Vm}^{-1}$ with a linear doping density of $10^9\,\text{m}^{-1}$. From Fig. 7.22, it can be stated that field strength of nearly $10^7\,\text{Vm}^{-1}$ is sufficient to produce a tenth of microampers. Incidentally, due to the velocity saturation phenomena, we observe that beyond $10^8\,\text{Vm}^{-1}$, I saturates converging to a unique value and become invariant of DRs. The field emitted current from NWs of $\text{Hg}_{1-x}\text{Cd}_x\text{Te}$ as a function of film thickness has been exhibited in Fig. 7.23 and can be compared with the corresponding cases of earlier figures. It appears that for lower values of film thickness, I increases for a particular sub-band. As thickness increases, generation of different sub-bands occurs which leads to the overall decrease in I. The variation of I over a large range of carrier concentration has been plotted in Fig. 7.24 for the quantum limit case. It appears that I initially increase due to low value of Fermi energy and as the concentration increases, the magnitude of the current decreases sharply exhibiting a peak. The amount of broadening is highly sensitive to the spectrum parameters of a particular semiconductor. It appears that for GaAs, the broadening is more as compared with others as shown in the figure. Figure 7.25 exhibits the variation of I as a function of film thickness for NWs of n-$\text{In}_{1-x}\text{As}_x\text{Ga}_y\text{P}_{1-y}$ which shows almost a constant step functional dependence of the current. This implies that the difference in the Fermi energy and the quantized sub-band energy is almost invariant of film thickness. From Fig. 7.26, one observes that as alloy composition increases, the field emitted current decreases for NWs of both ternary and quaternary compounds. For lower values of alloy composition, I from NWs of quaternary semiconductor is more than that of the corresponding ternary one, whereas for higher values of alloy composition, it converges. Step functional dependence of field emitted current with film thickness for NWs of CdS has been observed in Fig. 7.27. We see a nearly constant field-emitted current per sub-band, until a new sub-band is being occupied. Comparing with the earlier figures on the thickness dependence, we observe in this case an opposite trend because of the fact of overriding of the term $F_0(\eta_{ij})$ by $\exp(-\beta_{ij})$. Figure 7.28 exhibits the fact that for p-CdS, the field current is of the order of a few nanoamperes even at field strength of $5 \times 10^9\,\mathrm{Vm}^{-1}$. It should be noted that the behavior of composite oscillations is a general feature in thickness dependence of field emission from quantum wires. The energy band parameters determine the composite periodicity. The effect of film thickness on the field current in stressed InSb has been plotted in Fig. 7.29 in the quantum limit case, where the current decreases different values of d_y. For $d_y = 5\,\mathrm{nm}$, the field emitted current is highest although is approximately a constant quantity with respect to d_z. We observe here that the product of the two terms namely $F_0(\eta_{ij})$ and $\exp(-\beta_{ij})$ becomes independent of d_z for large values.

The effect of stress on the field-emitted current has been exhibited in Figs. 7.30 and 7.31 for different values of carrier concentration. From Fig. 7.30, it appears that with the increase in stress, the current increases having a magnitude of few tenths of microamperes. With the increase in carrier degeneracy, the current almost reduces to about 10 times. From Fig. 7.32, we observe that the field current from NWs of Ge (in accordance with the models of Cardona *et al.* and Wang *et al.*, respectively), Bi_2Te_3, GaP decreases as film thickness increases because of the fact that the term $\exp(-\beta_{ij})$ dominates over the term $F_0(\eta_{ij})$ in the whole range of thickness as considered here. For NWs of PbTe and Te, the current initially increases and then decreases exhibiting the fact that of the dominancy of the two terms $F_0(\eta_{ij})$ and $\exp(-\beta_{ij})$. For NWs of Ge and Bi_2Te_3, the field current has been observed to be in the order of hundreds of microamperes, where as for other 1D systems as considered in this figure the currents are in the order of few nanoamperes exhibiting the influence of the carrier dispersion relation of a particular semiconductor. In Fig. 7.33(a), the cut-in electric field for the field currents is in the order of $10^9\,\mathrm{Vm}^{-1}$ for NWs of PbTe and

Te, while about 10^7–$10^8\,\mathrm{Vm}^{-1}$ field strength is sufficient for field current to reach few microamperes.

The influence of quantum confinement is immediately apparent from Figs. 7.20, 7.21, 7.24, 7.26, 7.28, 7.29 and 7.33 since the field-emitted current depends strongly on the thickness of the quantum-confined materials in contrast with the corresponding bulk specimens. The current changes with increasing film thickness in an oscillatory way with different numerical magnitudes. It appears from the aforementioned figures that the FNFE exhibits spikes for particular values of film thickness which, in turn, depends on the particular band structure of the specific semiconductor. Moreover, the FNFE from NWs of different compounds can be smaller than of bulk specimens of the same materials, which is also a direct signature of quantum confinement. This oscillatory dependence will be less and less prominent with increasing film thickness. For bulk specimens of the same material, the FNFE will be found to increase continuously with increasing electron degeneracy in a non-oscillatory manner. The appearance of the discrete jumps in the respective figures are due to the redistribution of the electrons among the quantized energy levels when the size quantum number corresponding to the highest occupied level changes from one fixed value to the others.

With varying electron degeneracy, a change is reflected in the FNFE through the redistribution of the electrons among the size-quantized levels. It may be noted that at the transition zone from one sub-band to another, the height of the peaks between any two sub-bands decreases with the increasing in the degree of quantum confinement and is clearly shown in all the curves. It should be noted that although, the FNFE changes in various manners with all the variables as evident from all the figures, the rates of variations are totally band-structure dependent.

By taking the different values of energy band constants from Chapter 18, we have plotted the normalized EEM in nanowires of various materials for different physical variables from Figs. 7.34 to 7.53, respectively. Throughout our formalism, we have assumed the NWs to be of rectangular cross-sectional dimensions so that the usual "particle-in-a-box" concept can be applied along the quantized directions. The NWs are assumed to be degenerately doped with a carrier density starting from $10^8\,\mathrm{m}^{-1}$. While deriving the closed-form analytical solutions of normalized EEM in all the materials, we have also assumed that the constants of the energy band structures of the materials are independent of thickness in the range

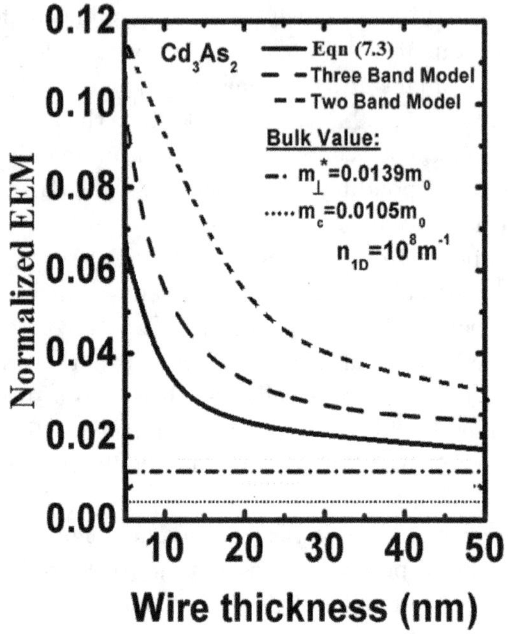

Fig. 7.34. Plot of the normalized EEM as a function of wire thickness for NWs of Cd_3As_2 considering (7.3). The plots for three- and two-band models of Kane have also been exhibited with their corresponding anisotropic bulk values as presented in the image.

beyond 5nm. We have assumed the invariant property of the material energy spectrum constants and evaluated the normalized EEM in NWs of Cd_3As_2 and $CdGeAs_2$ as a function of wire thickness along their respective transport directions in Figs. 7.34 and 7.35.

It appears from Figs. 7.34 and 7.35 that the normalized EEM at the lowest sub-band in both the cases is strong cross-sectional functions of the dimensions which converge to their corresponding bulk values at larger dimensions. The effect of crystal field splitting in case of dispersion relation of Cd_3As_2 lets the asymptotic fall to be closer to the bulk value. It should be noted that all the curves have been evaluated at $T = 4\,\text{K}$ where the average thermal energy, i.e., $E_F + k_B T$ is very less than that of the difference of the adjacent sub-band energies. This leads the carrier to reside in the lowest sub-bands only. Figures 7.37–7.38 exhibit the variation of the normalized EEM with the wire thickness for III–V materials namely InAs, InSb and GaAs in accordance with the well-known standard non-parabolic dispersion relation of Kane.

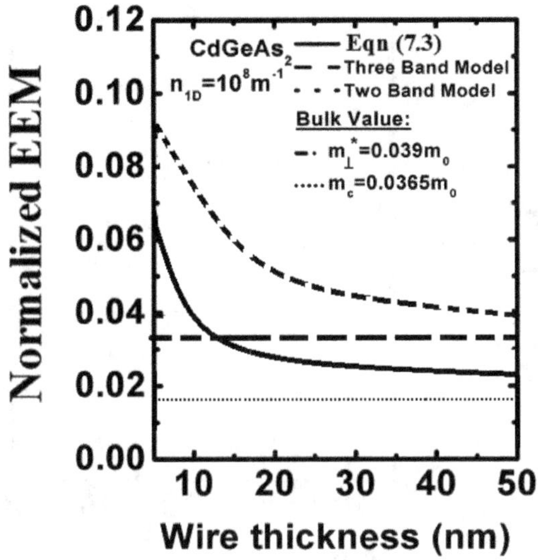

Fig. 7.35. Plot of the normalized EEM as a function of wire thickness for NWs of $CdGeAs_2$ for all the cases of Fig. 7.34.

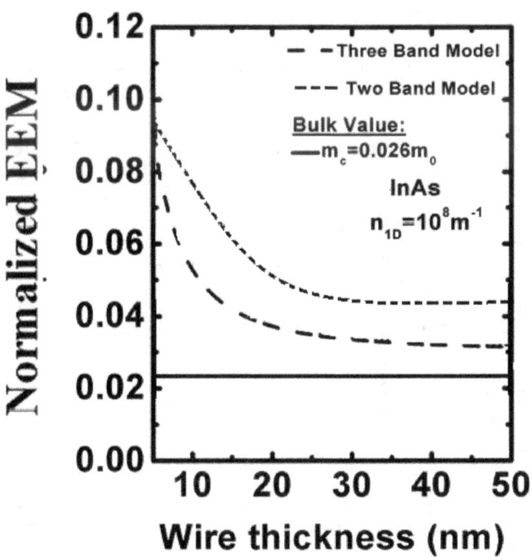

Fig. 7.36. Plot of the normalized EEM as a function of wire thickness for NWs of InAs considering the three- and two-band models of Kane with the corresponding isotropic bulk value.

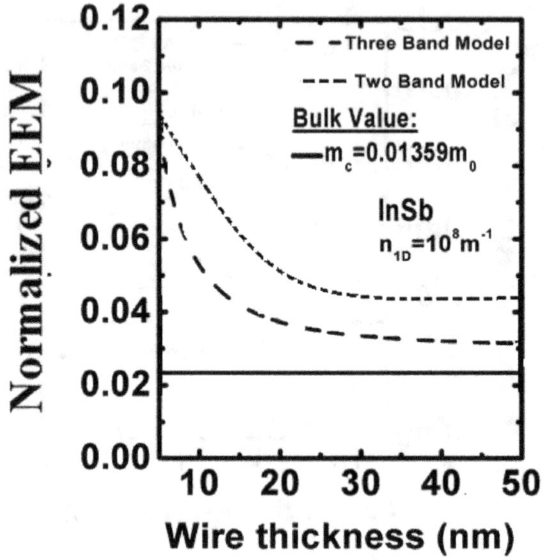

Fig. 7.37. Plot of the normalized EEM as a function of wire thickness for NWs of InSb considering the three- and two-band models of Kane with the corresponding isotropic bulk value.

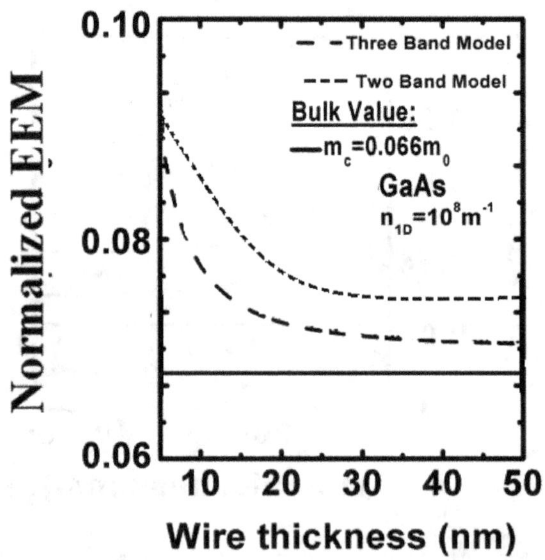

Fig. 7.38. Plot of the normalized EEM as a function of wire thickness for NWs of GaAs considering the three- and two-band models of Kane with the corresponding isotropic bulk value.

It appears from these figures that the difference in the energy band models in predicting the normalized EEM is almost insignificant. Hence for all practical purposes for the determination of normalized EEM, the second-order model of Kane can fit well. It should be noted that the normalized EEM for these materials as presented here can be compared with that of the normalized EEM in 2D systems as given in Chapter 1.

A quick view can lead to interpret that the normalized EEM for both the case is same. However, it should also be kept in mind about the difference in the wire thickness and carrier concentration. All the curves in this chapter have been evaluated at those concentrations for which the normalized EEM stand close to that of their corresponding 2D systems. Further in deriving the results, we have assumed that the conduction band valley does not split along the channel transport direction, which is a usual case with Silicon nanowire along [110] and [111] valleys (Chapter 8).

Figures 7.39 and 7.40 exhibit the variation of the normalized EEM for $Hg_{1-x}Cd_xTe$ and $In_{1-x}As_xGa_yP_{1-y}$ considering all the aforementioned cases at $x = 0.3$. In Figs. 7.41–7.47, we have exhibited the variation of the normalized EEM as a function of carrier degeneracy. It appears from

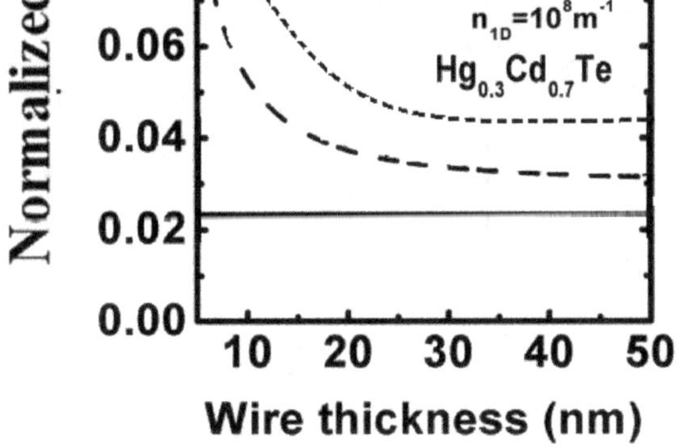

Fig. 7.39. Plot of the normalized EEM as a function of wire thickness for NWs of $Hg_{1-x}Cd_xTe$ considering the three- and two-band models of Kane with the corresponding isotropic bulk value.

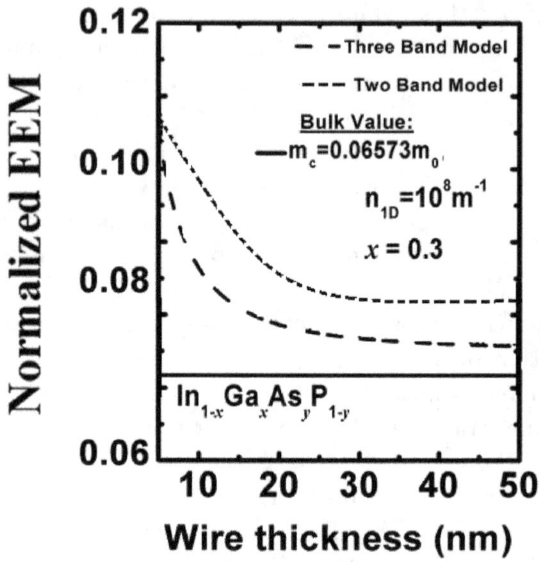

Fig. 7.40. Plot of the normalized EEM as a function of wire thickness for NWs of $In_{1-x}As_xGa_yP_{1-y}$ considering the three- and two-band models of Kane with the corresponding isotropic bulk value at $x = 0.3$.

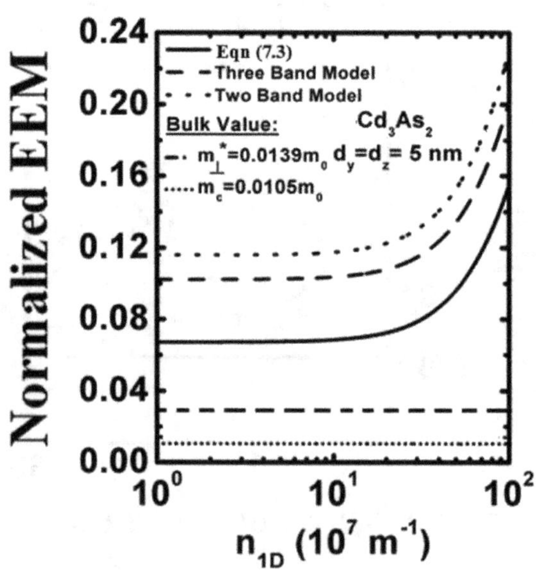

Fig. 7.41. Plot of the normalized EEM as a function of carrier degeneracy for Cd_3As_2 nanowire for all cases of Fig. 7.34.

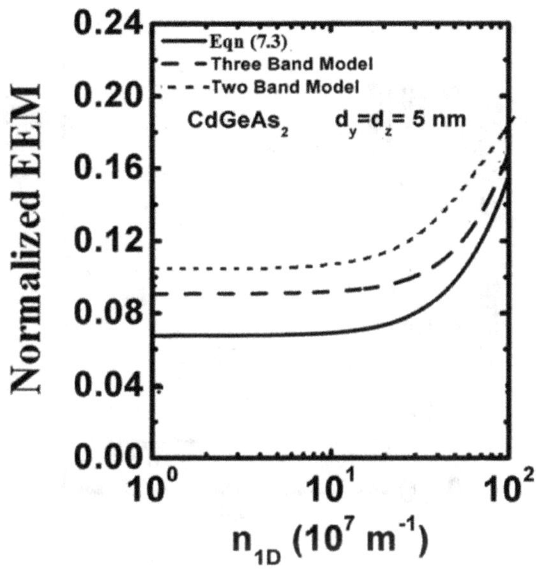

Fig. 7.42. Plot of the normalized EEM as a function of carrier degeneracy for NWs of $CdGeAs_2$ for all cases of Fig. 7.35.

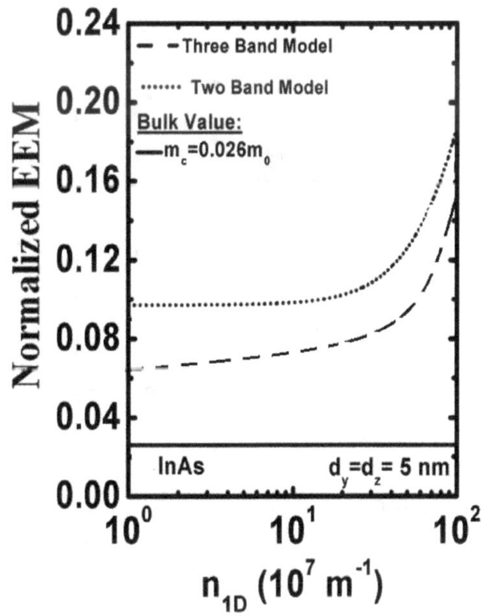

Fig. 7.43. Plot of the normalized EEM as a function of carrier degeneracy for InAs nanowire for all cases of Fig. 7.36.

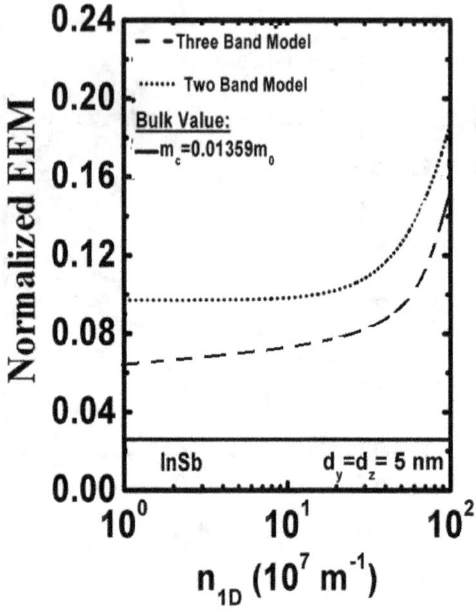

Fig. 7.44. Plot of the normalized EEM as a function of carrier degeneracy for InSb nanowire for all cases of Fig. 7.37.

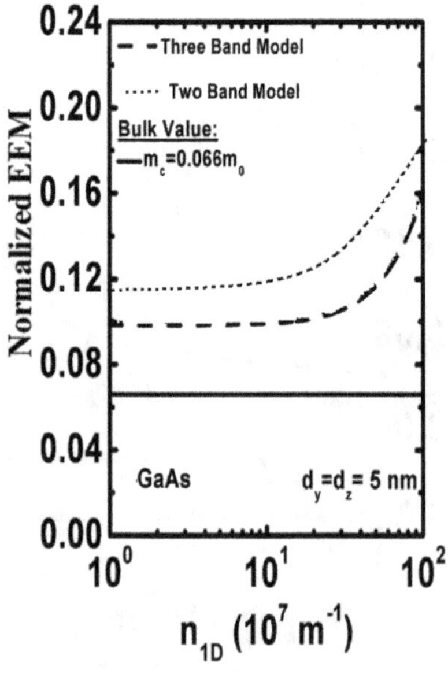

Fig. 7.45. Plot of the normalized EEM as a function of carrier degeneracy for GaAs nanowire for all cases of Fig. 7.38.

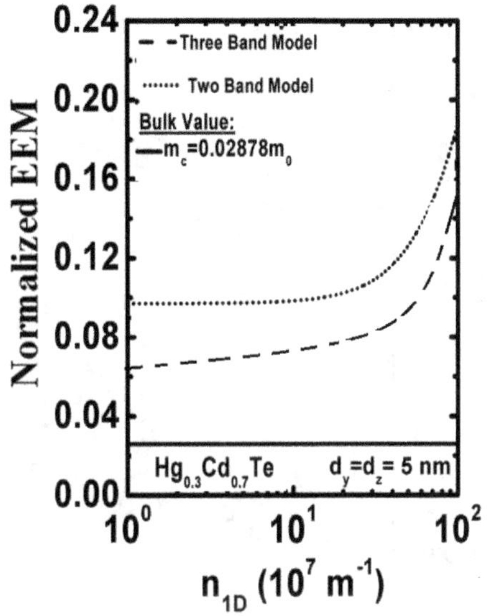

Fig. 7.46. Plot of the normalized EEM as a function of carrier degeneracy for $Hg_{1-x}Cd_xTe$ nanowire for all cases of Fig. 7.39.

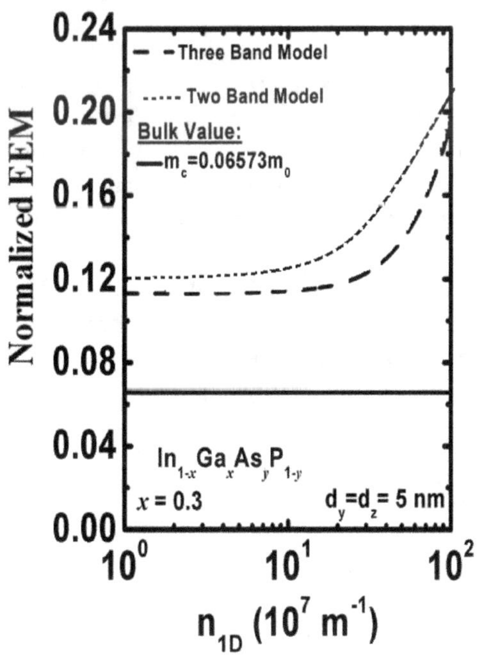

Fig. 7.47. Plot of the normalized EEM as a function of carrier degeneracy for $In_{1-x}As_xGa_yP_{1-y}$ nanowire for all cases of Fig. 7.41.

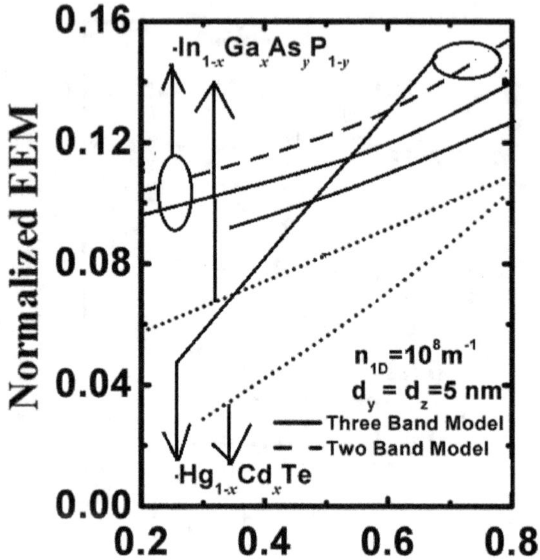

Fig. 7.48. Plot of the normalized EEM as a function of alloy composition for $Hg_{1-x}Cd_xTe$ and $In_{1-x}As_xGa_yP_{1-y}$ nanowires.

the said figures that the normalized EEM increases with the increase in degeneracy. The normalized EEM rises sharply above $10^7\,m^{-1}$ for all the materials in an exponential way due to the presence of the Fermi–Dirac probability factor in the respective carrier concentration equation.

Figure 7.48 exhibits the variation of the normalized EEM as a function of the alloy composition in ternary and quaternary systems. It appears that as in the previous cases of quantum confinements, the normalized EEM in this case also exhibits an increasing variation with x. We have also plotted the variation of the bulk effective mass as x varies to present a comparative view. The influence of band non-parabolicity on the normalized EEM in these two materials can easily be seen. In both the cases, we see that the normalized EEM is a slow variation function of x due to the change in band gap.

The effect of dimensionality on nanowire of PbSnTe has been exhibited in Figs. 7.49 and 7.50 for the energy band models of Mc Clure and Choi, Hybrid, Cohen and Lax. It appears that as the dimension reduces in Fig. 7.49, the normalized EEM increases which is generally accepted. However, using Mc Clure and Choi model the normalized EEM tends to decrease in the sub-5 nm regime there by unfolding the validity of the model

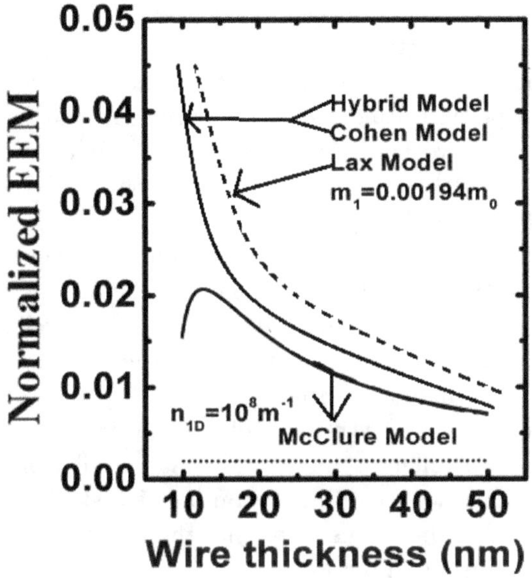

Fig. 7.49. Plot of the normalized EEM as a function of wire thickness for PbSnTe nanowire.

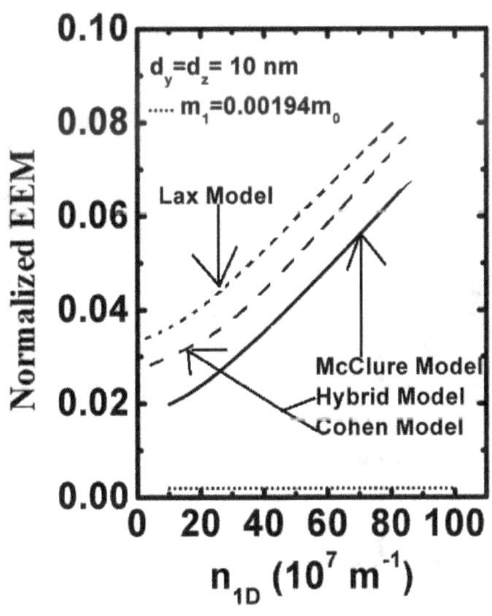

Fig. 7.50. Plot of the normalized EEM as a function of carrier degeneracy for PbSnTe nanowire.

in this zone. However, at large thicknesses, the entire model tends to their corresponding bulk value $0.00194\,m_0$.

The variation of the normalized EEM as a function of carrier degeneracy has further been plotted using the aforementioned band structure models. It appears from the two figures that the effect of different band structure models has significantly less deviation from one another except for the model of Mc Clure and Choi.

The normalized EEM is found to increase almost linear with degeneracy $2 \times 10^8\,m^{-1}$ and beyond using all the models. In case of PbTe, the normalized EEM rises sharply with decreasing wire thickness below sub-15 nm from the bulk value $0.098\,m_0$ at carrier degeneracy of $10^9\,m^{-1}$, which can affect the carrier mobility strongly.

The effect of strain on stressed InSb nanowires has been exhibited in Fig. 7.52 for two different momentum matrix elements to signify its importance as dimension reduces. A compressive strain of 3% has been applied along x and z directions to predict the variation of the normalized EEM along the transport direction.

Figure 7.53 exhibits the effect of cross-sectional dimension on the normalized EEM in Ge, GaSb and topological insulators like Bi_2Te_3.

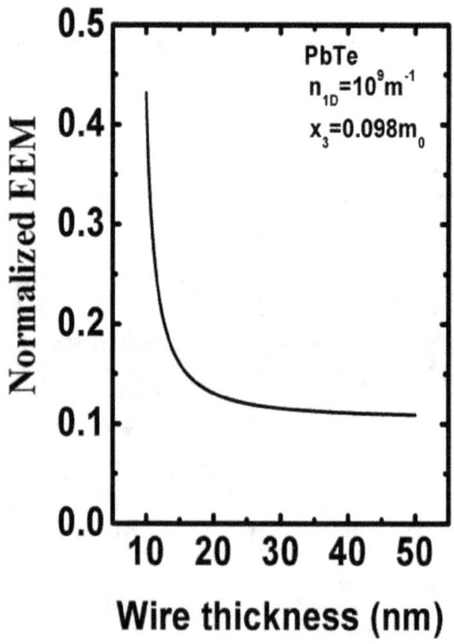

Fig. 7.51. Plot of the normalized EEM as a function of carrier degeneracy for PbTe nanowire.

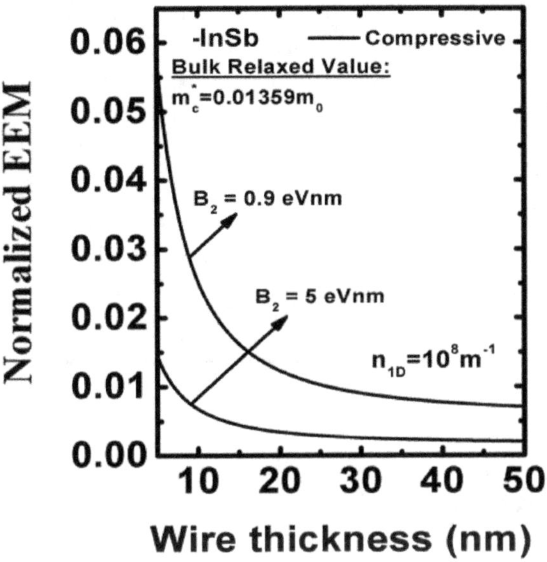

Fig. 7.52. Plot of the normalized EEM as a function of wire thickness for stressed InSb nanowire.

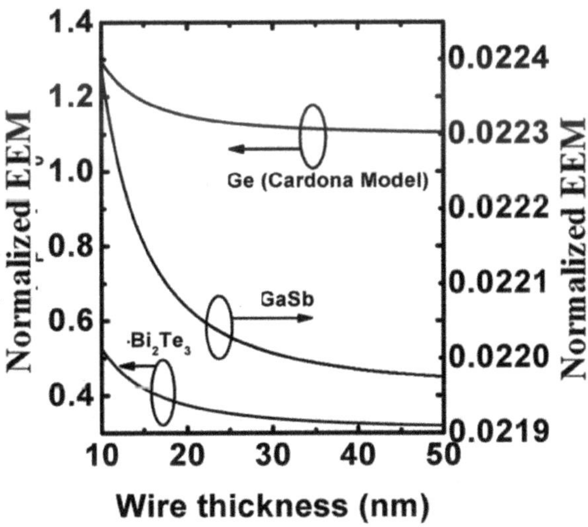

Fig. 7.53. Plot of the normalized EEM as a function of wire thickness for Ge, GaSb and Bi_2Te_3 nanowires.

We have used the Cardona model to evaluate the normalized EEM in Ge nanowires. The reader is left with the corresponding evaluation of the normalized EEM for Wang *et al.* model of the same. It appears that the normalized EEM in Ge and Bi_2Te_3 evolve from its corresponding bulk value significantly from below 15 nm. However in case of GaSb, the EEM is almost invariant.

By taking the different values of energy band constants from Chapter 18, we have plotted the normalized ER in nanowires of various materials for different physical variables from Figs. 7.54 to 7.75, respectively. In Figs. 7.54

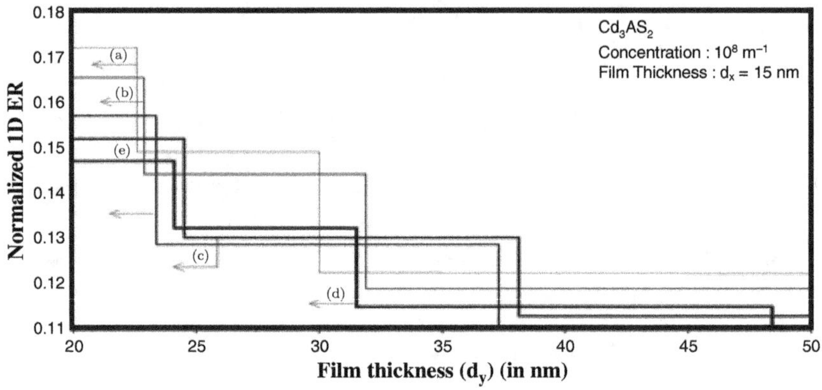

Fig. 7.54. Plot of the normalized 1D ER as a function of film thickness for the NWs of Cd_3As_2 in accordance with (a) the generalized band model with $\delta \neq 0$, (b) the generalized band model with $\delta = 0$, (c) the simplified three-band model of Kane, (d) the two-band model of Kane and (e) the parabolic energy bands.

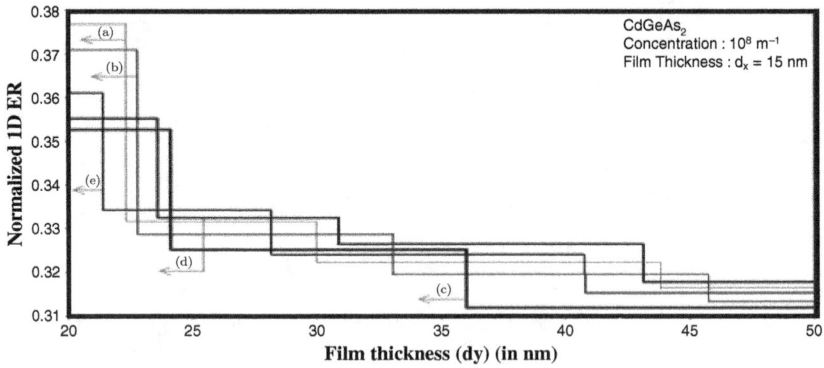

Fig. 7.55. Plot of the normalized 1D ER as a function of film thickness for the NWs of $CdGeAs_2$ for all cases of Fig. 7.54.

and 7.55, plotted the 1D ER at low temperatures, where the quantum effects become prominent for the NWs of Cd_3As_2 and $CdGeAs_2$ as a function of nanothickness. The curves (a) and (b) corresponds to $\delta \neq 0$ and $\delta = 0$, respectively for the purpose of assessing the influence of crystal field splitting on the 1D ER in NWs of tetragonal materials. The crystal field splitting enhances the numerical values of the 1D ER in both the cases.

Using (7.21), we have plotted the curve (c) which reflects the influence of the dispersion relation of the tetragonal materials in accordance with the three band Kane model on the 1D ER in NWs. The curves (d) and (e) have been plotted using (7.30) and (7.39) in accordance with the two-band model of Kane and that of the parabolic energy bands respectively.

The influence of quantum confinement is immediately apparent from all the curves of Figs. 7.54 and 7.55, since, the 1D ER depends strongly on the nanothickness, which is in direct contrast with the corresponding bulk specimens.

It appears from the said figures that the 1D ER decreases with the increasing film thickness in a step-like manner as considered here although the numerical values vary widely and determined by the constants of the energy spectra. The oscillatory dependence is due to the crossing over of the Fermi level by the quantized levels. For each coincidence of a quantized level with the Fermi level, there would be a discontinuity in the DOS function resulting in a peak of oscillations. With large values of film thickness, the height of the steps decreases and the ER decreases with increasing film thickness in non-oscillatory manner and exhibits monotonic decreasing dependence. The height of step size and the rate of decrement are totally dependent on the band structure. The numerical values of the 1D ER in accordance with the three-band model of Kane is greater than that of the corresponding two-band model, which reflects the fact that the presence of the spin-orbit splitting constant enhances the magnitude of the 1D ER. It may be noted that the presence of the band non-parabolicity in accordance with the two-band model of Kane enhances the peaks of the oscillatory 1D ER for all cases of quantum confinements. The appearance of the humps of the respective curves is due to the redistribution of the electrons among the quantized energy levels when the quantum numbers corresponding to the highest occupied level changes from one fixed value to the others. With varying electron concentrations, a change is reflected in the 1D ER through the redistribution of the electrons among the quantized levels. Although the 1D ER varies in various manners with all the variables in all the limiting cases as evident from all the curves of figures of 1D ERs, the rates of variations are totally band-structure dependent.

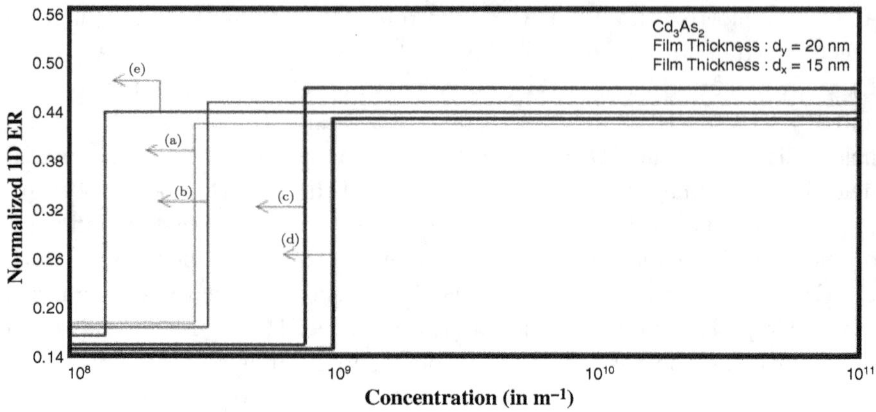

Fig. 7.56. Plot of the normalized 1D ER as a function of electron concentration per unit length for the NWs of Cd_3As_2 in accordance with (a) the generalized band model with $\delta \neq 0$, (b) the generalized band model with $\delta = 0$, (c) the simplified three-band model of Kane, (d) the two-band model of Kane and (e) the parabolic energy bands.

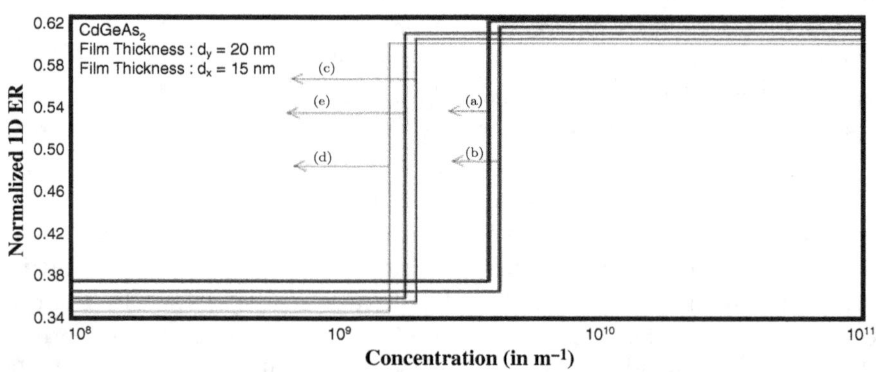

Fig. 7.57. Plot of the normalized 1D ER as a function of electron concentration per unit length for the NWs of $CdGeAs_2$ for all cases of Fig. 7.54.

In Figs. 7.56 and 7.57, we have plotted the 1D ER as a function of surface electron concentration per unit length for all cases of Figs. 7.54 and 7.55, respectively. It appears that the 1D ER increases with increasing carrier degeneracy and also reflects the signature of the 1D confinement through the step-like functional dependence with the 1D electron statistics. This oscillatory dependence will be less and less prominent with increasing

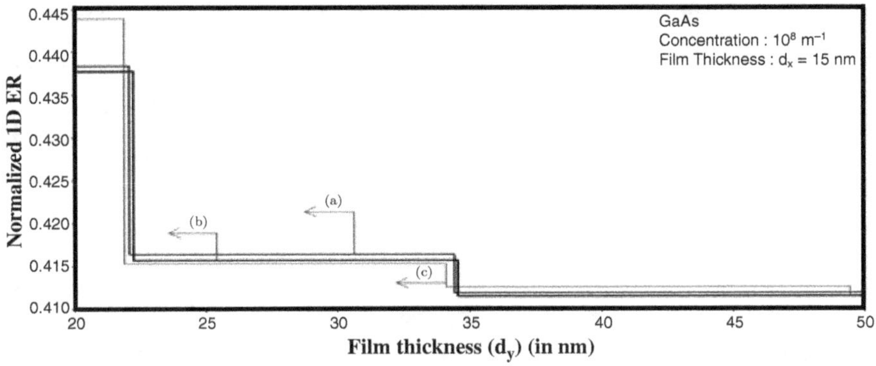

Fig. 7.58. Plot of the normalized 1D ER as a function of film thickness for the NWs of GaAs in accordance with (a) the simplified three-band model of Kane, (b) the two-band model of Kane and (c) the parabolic energy bands.

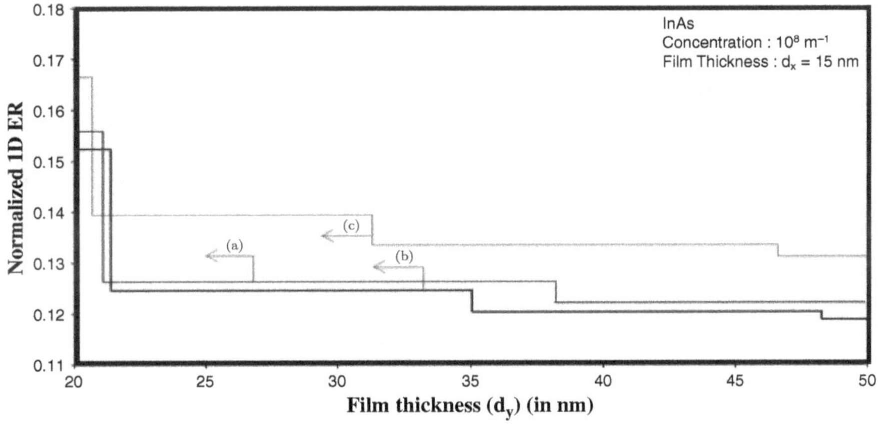

Fig. 7.59. Plot of the normalized 1D ER as a function of film thickness for the NWs of InAs for all cases of Fig. 7.36.

carrier concentration and ultimately, for bulk specimens of the same material, the ER will be found to increase continuously with increasing electron concentration in a non-oscillatory manner. Using Equations (7.21), (7.30) and (7.39), we have plotted the 1D ER as functions of nanothickness for the NWs of GaAs, InAs and InSb in Figs. 7.58, 7.59 and 7.60, where curves (a), (b) and (c) correspond to the three- and the two-band models of Kane together with the parabolic energy band, respectively.

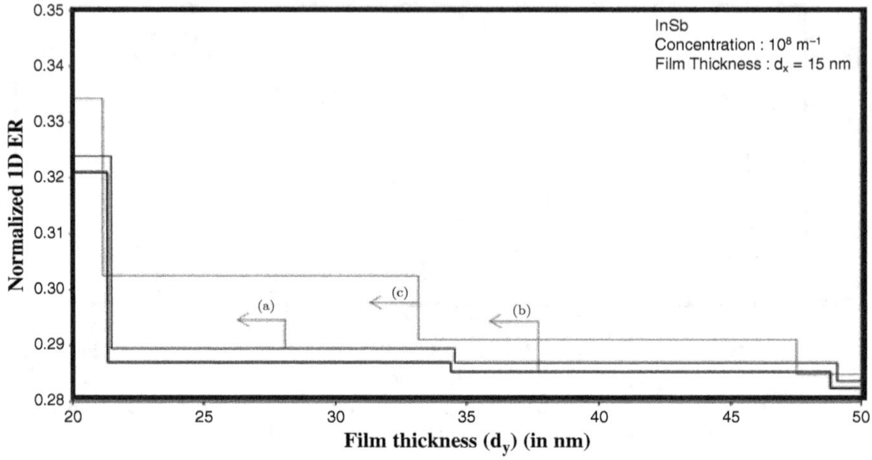

Fig. 7.60. Plot of the normalized 1D ER as a function of film thickness for the NWs of InAs for all cases of Fig. 7.36.

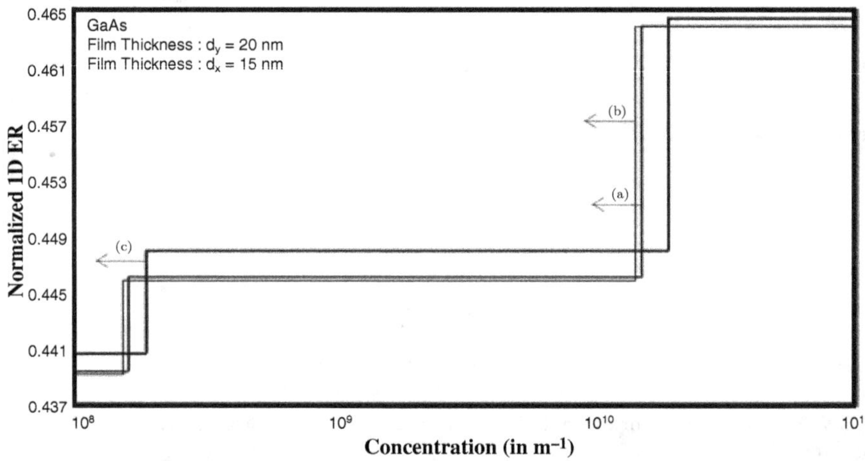

Fig. 7.61. Plot of the normalized 1D ER as a function of electron concentration per unit length for the NWs of GaAs in accordance with (a) the simplified three-band model of Kane, (b) the two-band model of Kane and (c) the parabolic energy bands.

The dependence of the 1D ER on the electron concentration per unit length for NWs of all the said compounds is shown in Figs. 7.61, 7.62 and 7.63, respectively. Using the same set of equations as for III–V materials, we have plotted the 1D ER for $Hg_{1-x}Cd_xTe$ and $In_{1-x}As_xGa_yP_{1-y}$ as functions of

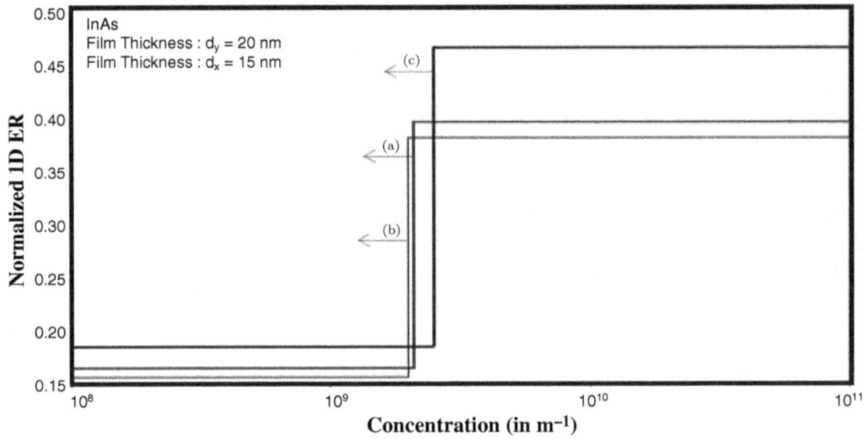

Fig. 7.62. Plot of the normalized 1D ER as a function of electron concentration per unit length for the NWs of InAs for all cases of Fig. 7.61.

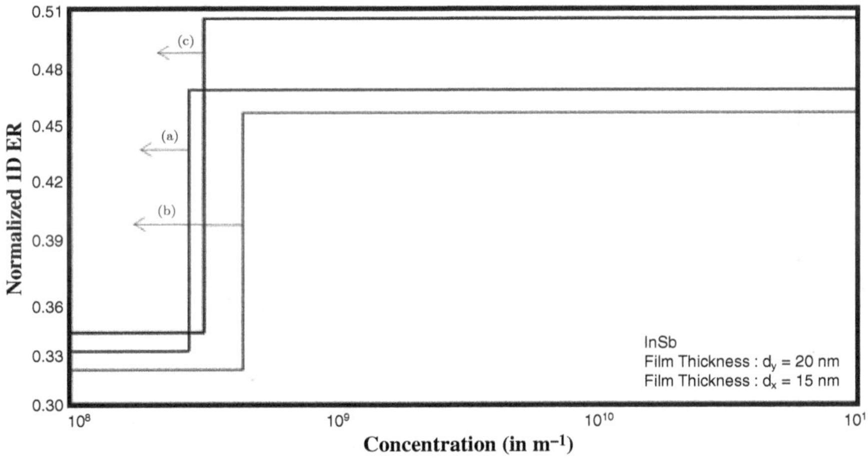

Fig. 7.63. Plot of the normalized 1D ER as a function of electron concentration per unit length for the NWs of InSb for all cases of Fig. 7.37.

nanothickness and 1D carrier degeneracy as shown in Figs. 7.64, 7.65 and 7.66, 7.67, respectively in which, curves (a), (b) and (c) corresponds to the three- and the two-band models of Kane together with the parabolic energy band, respectively. The numerical values of the 1D ER depend on the energy band constants of different materials.

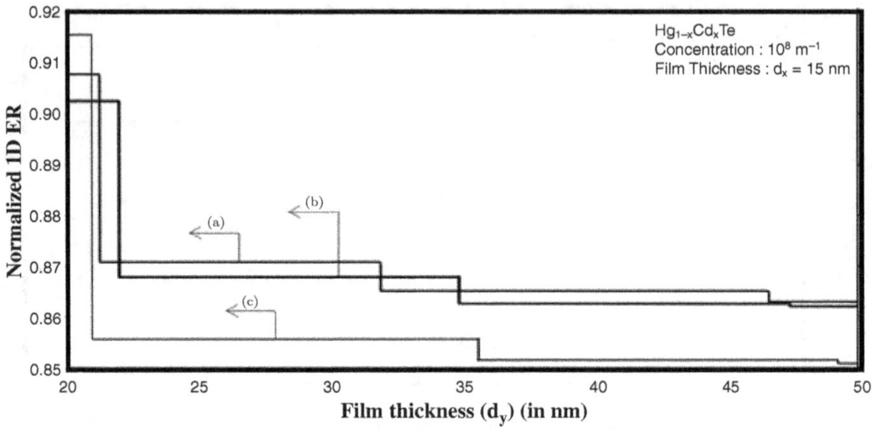

Fig. 7.64. Plot of the normalized 1D ER as a function of film thickness for the NWs of $Hg_{1-x}Cd_xTe$ in accordance with (a) the simplified three-band model of Kane, (b) the two-band model of Kane and (c) the parabolic energy bands.

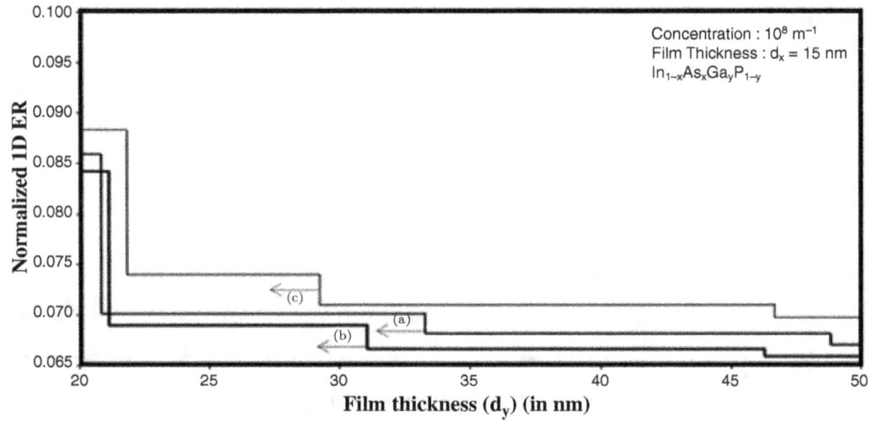

Fig. 7.65. Plot of the normalized 1D ER as a function of film thickness for the NWs of $In_{1-x}As_xGa_yP_{1-y}$ for all cases of Fig. 7.48.

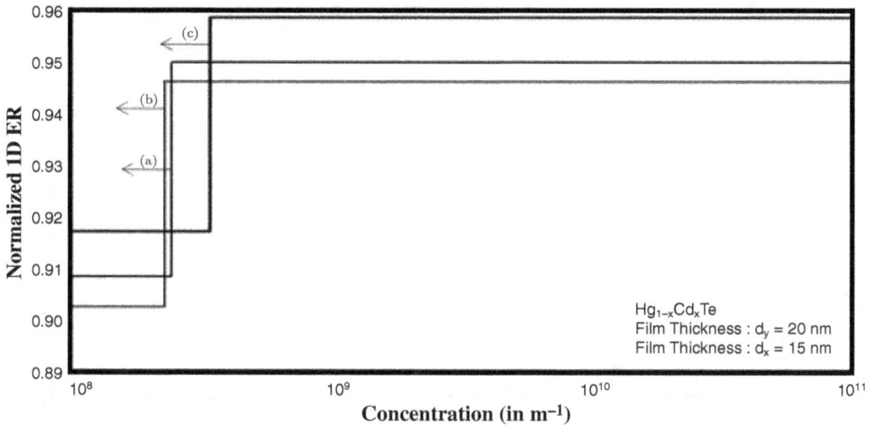

Fig. 7.66. Plot of the normalized 1D ER as a function of electron concentration per unit length for the NWs of $Hg_{1-x}Cd_xTe$ in accordance with (a) the simplified three-band model of Kane, (b) the two-band model of Kane and (c) the parabolic energy bands.

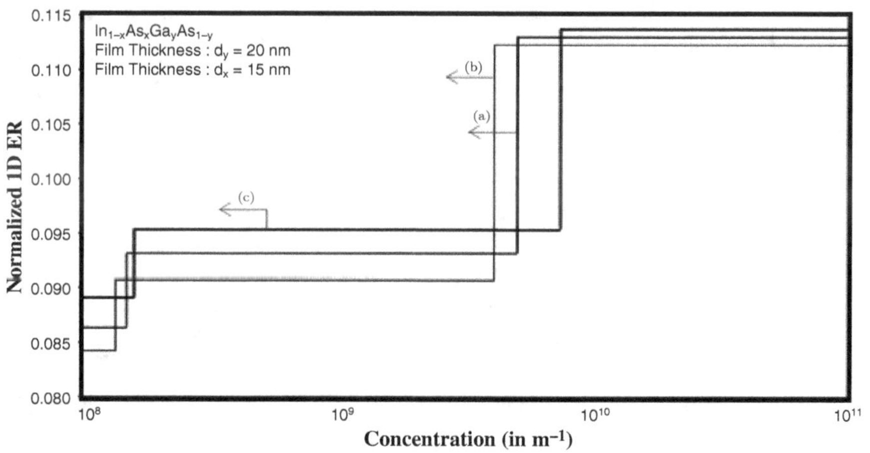

Fig. 7.67. Plot of the normalized 1D ER as a function of electron concentration per unit length for the NWs of $In_{1-x}As_xGa_yP_{1-y}$ for all cases of Fig. 7.48.

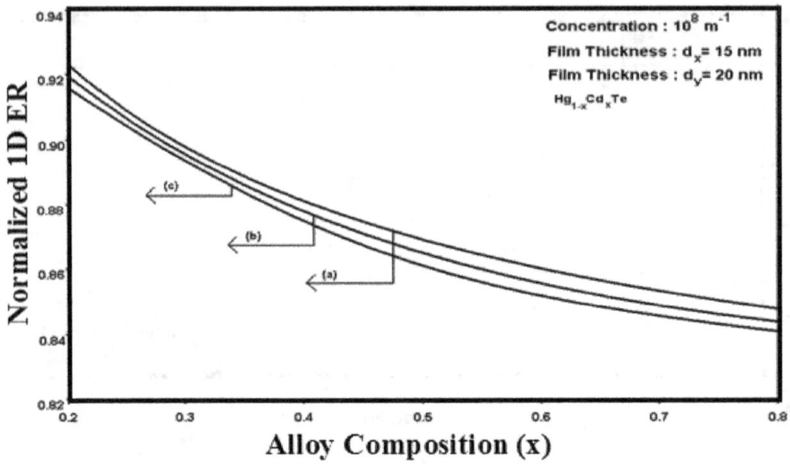

Fig. 7.68. Plot of the normalized 1D ER as a function of alloy composition (x) for the NWs of $Hg_{1-x}Cd_xTe$ in accordance with (a) the simplified three-band model of Kane, (b) the two-band model of Kane and (c) the parabolic energy bands.

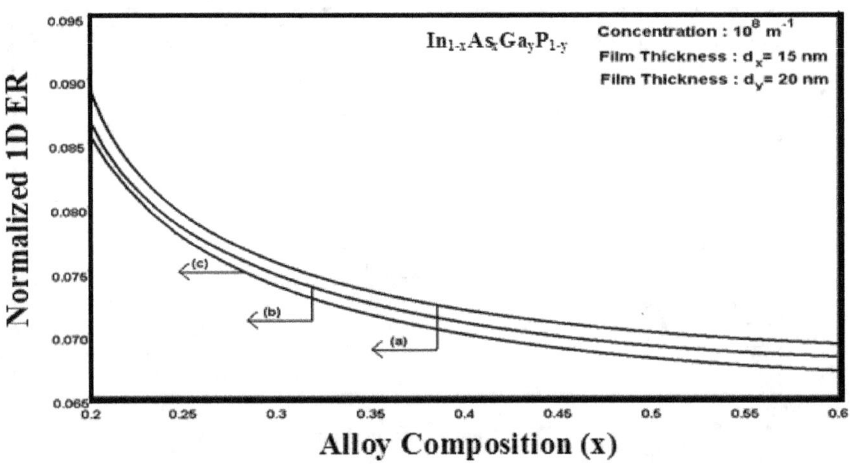

Fig. 7.69. Plot of the normalized 1D ER as a function of alloy composition (x) for the NWs of $In_{1-x}As_xGa_yP_{1-y}$ for all cases of Fig. 7.48.

Figures 7.68 and 7.69 exhibit the same dependence as a function of alloy composition for all the cases of the ternary and quaternary materials as considered above. From both the figures, it is apparent that the ER decreases with increasing alloy composition.

Using (7.71), we have plotted the 1D ER, for the NWs of p-CdS materials as functions of nanothickness and electron concentration per unit length in Figs. 7.70 and 7.71, respectively, where the curve (a) refers to $C_0 \neq 0\, eVm$ and the curve (b) refers to $C_0 = 0\, eVm$. Both the curves have

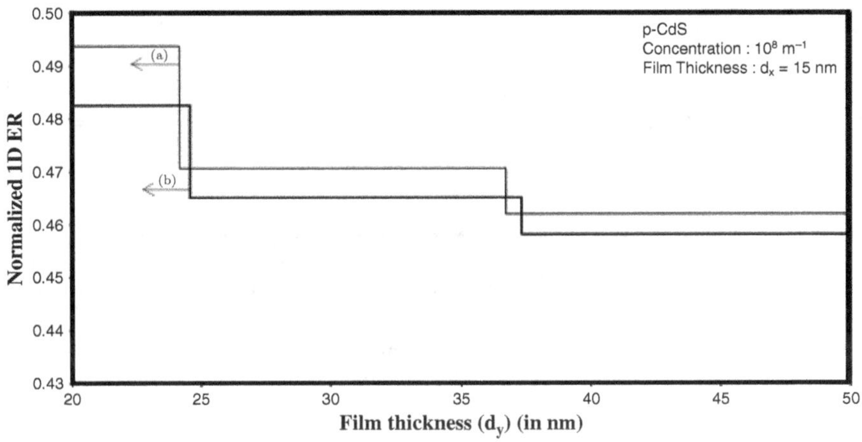

Fig. 7.70. Plot of the normalized 1D ER as a function of film thickness for NWs of p-CdS in accordance with (a) Hopfield model with $C_0 \neq 0$ and (b) Hopfield model with $C_0 = 0$.

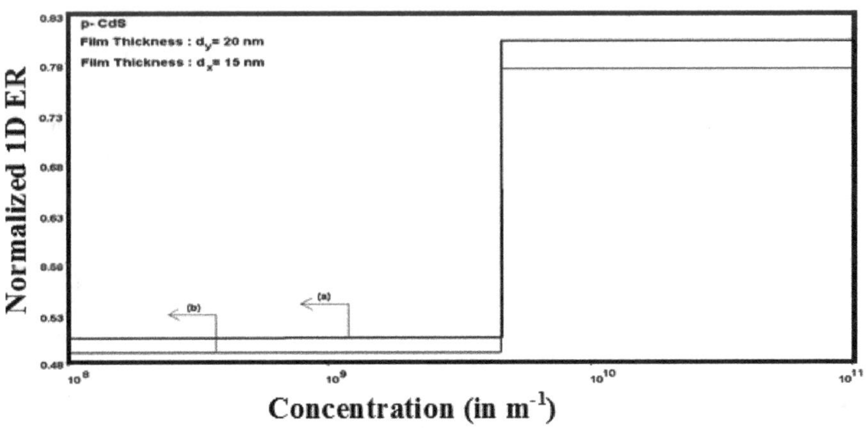

Fig. 7.71. Plot of the normalized 1D ER as a function of electron concentration per unit length for NWs of p-CdS in accordance with (a) Hopfield model with $C_0 \neq 0$ and (b) Hopfield model with $C_0 = 0$.

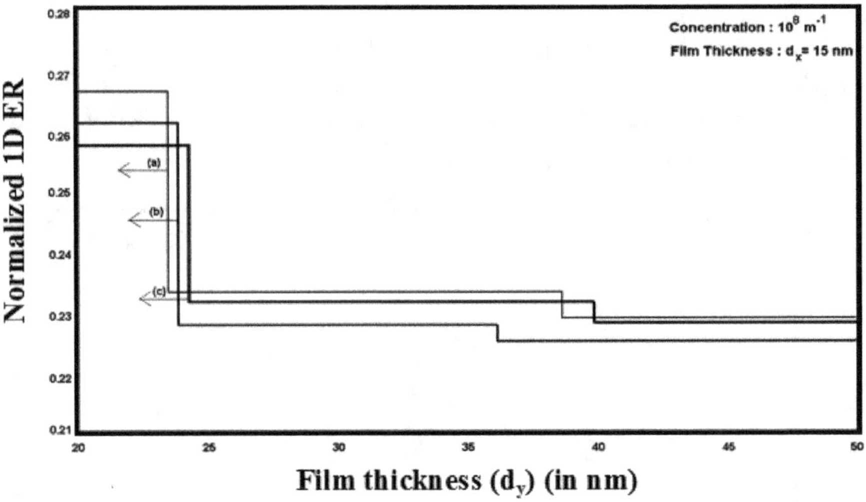

Fig. 7.72. Plot of the normalized 1D ER as a function of film thickness for the NWs of (a) PbTe, (b) PbSnTe and (c) $Pb_{1-x}Sn_xSe$.

been drawn for the purpose of assessing the influence of the splitting of the two-spin states by the spin-orbit coupling and the crystalline field in this case. It appears from Figs. 7.70 and 7.71, that the influence of the term C_0 is the reduction of the quantum jumps of the oscillatory 1D ER in NWs of II–VI materials for both the said variables.

Using equations (7.81), Figs. 7.72 and 7.73 exhibit the 1D ER for the NWs of IV–VI materials as functions of nanothickness and electron concentration per unit length, respectively. Curves (a), (b) and (c) correspond to PbTe, PbSnTe and $Pb_{1-x}Sn_xSe$, respectively. The influence of the energy band constants on the ER in both the cases is apparent for all the three different materials as considered here. The 1D ER for stressed Kane type InSb has been plotted in Figs. 7.74 and 7.75 as functions of nanothickness and electron concentration per unit length respectively using (7.88) as shown in plot (a) in the presence of stress while the plot (b) exhibits the same in the absence of stress for the purpose of assessing the influence of stress on the 1D ER in NWs of stressed InSb. In the presence of stress, the magnitude of the 1D ER is increased as compared with the same under stress-free condition. The basic aim of this chapter is not only to derive and plot the QC and few electronic properties of NWFETs but also to realize

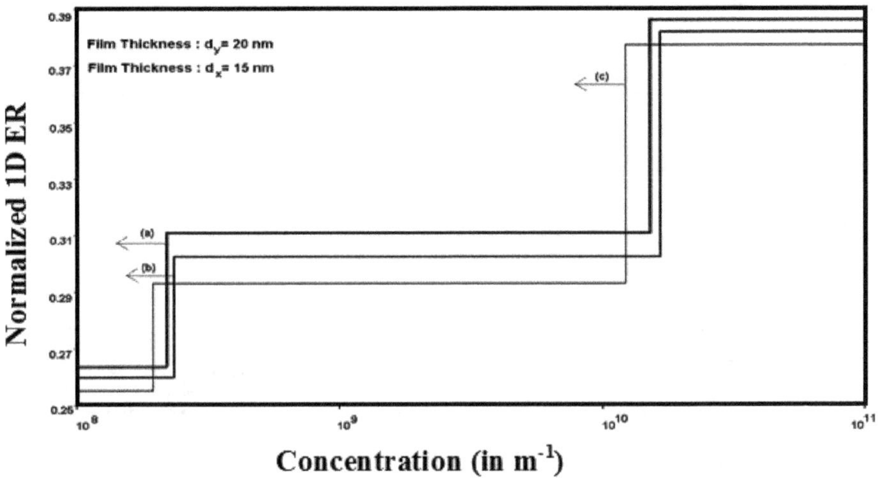

Fig. 7.73. Plot of the normalized 1D ER as a function of electron concentration per unit length for the NWs of (a) PbTe, (b) PbSnTe and (c) $Pb_{1-x}Sn_xSe$.

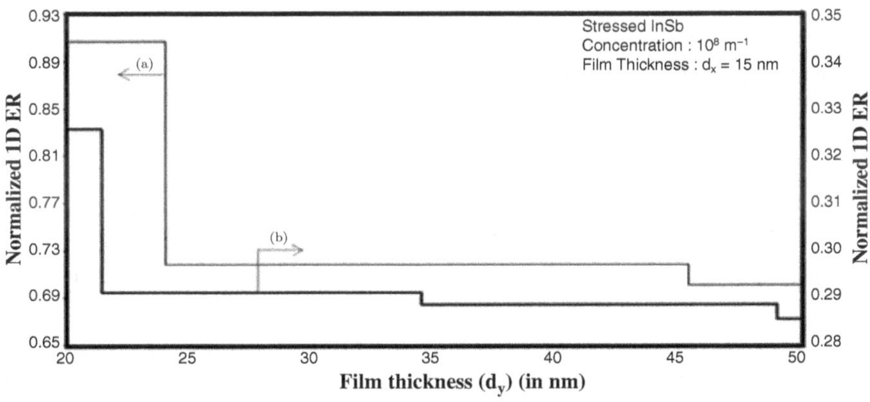

Fig. 7.74. Plot of the normalized 1D ER as a function of film thickness for the NWs of stressed InSb in which the curve (a) shows the 1D ER in the presence of stress while the curve and (b) is applicable in the absence of the stress.

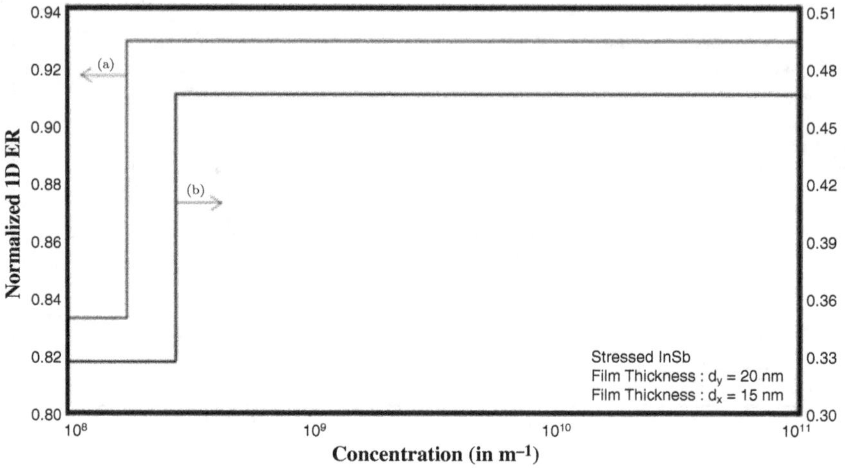

Fig. 7.75. Plot of the normalized 1D ER as a function of electron concentration per unit length for the NWs of stressed InSb in which the curve (a) shows the 1D ER in the presence of stress while the curve and (b) is applicable in the absence of the stress.

the role of band structure in different cases which generates, in turn, new concepts that are truly amazing.

7.4 Open Research Problems

R.7.1 Investigate the QCs in the presence of an arbitrarily oriented electric field in NWFETs of nonlinear optical materials as stated in this chapter. Draw the corresponding graphs and explain the results.

R.7.2 Investigate the QCs in the presence of an arbitrarily oriented electric field in NWFETs of IV–VI, II–VI and stressed Kane type compounds as stated in this chapter. Draw the corresponding graphs and explain the results.

R.7.3 Investigate the QCs in the presence of an arbitrarily oriented electric field in NWFETs of all the materials as stated in R.1.1 of Chapter 1. Draw the corresponding graphs and explain the results.

R.7.4 Investigate the QCs in the presence of an arbitrarily oriented electric field in NWFETs of all the materials as studied in this chapter in the presence of an arbitrarily oriented non-uniform electric field. Draw the corresponding graphs and explain the results.

R.7.5 Investigate the QCs in the presence of an arbitrarily oriented electric field in NWFETs of IV–VI, II–VI and stressed Kane type compounds in the presence of an arbitrarily oriented non-uniform electric field. Draw the corresponding graphs and explain the results.

R.7.6 Investigate the QCs in the presence of an arbitrarily oriented electric field in NWFETs of all the materials as stated in R.1.1 of Chapter 1 in the presence of an arbitrarily oriented non-uniform electric field. Draw the corresponding graphs and explain the results.

R.7.7 Investigate the QCs in NWFETs of all the problems from R.2.1 to R.2.6 in the presence of an arbitrarily oriented non-uniform electric field. Draw the corresponding graphs and explain the results.

References

1. K. P. Ghatak and M. Mondal, *J. Appl. Phys.* **70**, 299 (1991).
2. K. P. Ghatak and S. N. Biswas, *J. Vac. Sci. Technol.* **7B**, 104 (1989).
3. B. Mitra and K. P. Ghatak, *J. Solid-State Electron.* **32**, 177 (1989).
4. K. P. Ghatak and M. Mondal, *J. Appl. Phys.* **62**, 922 (1987).
5. M. Mondal and K. P. Ghatak, *J. Magn. Magn. Mater.* **62**, 115 (1986).
6. M. Mondal and K. P. Ghatak, *J. Phys. Scr.* **31**, 613 (1985).
7. K. P. Ghatak and M. Mondal, *J. Zeitschrift für Physik B Condens Matter.* **64**, 223 (1986).
8. K. P. Ghatak and S. N. Biswas, *J. Solid-State Electron.* **37**, 1437 (1994).
9. K. P. Ghatak, N. Chattopadhyay and M. Mondal, *J. Appl. Phys. A* **48**, 365 (1989).
10. K. P. Ghatak and M. Mondal, *Z. fur Physik B* **64**, 223 (1986).
11. K. P. Ghatak and S. N. Biswas, *Sol. State Electron.* **37**, 1437 (1994).
12. D. R. Choudhury, A. K. Chowdhury, K. P. Ghatak and A. N. Chakravarti, *Phys. Stat. Sol. (b)* **98**, K141 (1980).
13. A. N. Chakravarti, A. K. Chowdhury and K. P. Ghatak, *Phys. Stat. Sol. (a)* **63**, K97 (1981).
14. M. Mondal and K. P. Ghatak, *Acta Phys. Polon. A* **67**, 983 (1985); M. Mondal and K. P. Ghatak, *Phys. Stat. Sol. (b)* **128**, K21 (1985).
15. M. Mondal and K. P. Ghatak, *Phys. Stat. Sol. (a)* **93**, 377 (1986).
16. K. P. Ghatak and M. Mondal, *Phys. Stat. Sol. (b)* **135**, 819 (1986).
17. M. Mondal and K. P. Ghatak, *Phys. Stat. Sol. (b)* **139**, 185 (1987).
18. K. P. Ghatak, N. Chattopadhyay and S. N. Biswas, *OE/Fibers'* **87**, 203 (1987).
19. K. P. Ghatak, N. Chatterjee and M. Mondal, *Phys. Stat. Sol. (b)* **139**, K25 (1987).
20. K. P. Ghatak and M. Mondal, *Phys. Stat. Sol. (b)* **148**, 645 (1988); K. P. Ghatak and A. Ghosal, *Phys. Stat. Sol. (b)* **151**, K135 (1989); K P. Ghatak, N. Chattopadhyay and M. Mondal, *Phys. A* **48**, 365 (1989).
21. P. K. Das and K. P. Ghatak and *J. Nanosci, Nanotechnol.* **19**, 2909 (2019); K. P. Ghatak, S. Chakrabarti and B. Chatterjee, *Mater. Focus* **7**, 361 (2018); K. P. Ghatak, S. Chakrabarti, B. Chatterjee, P. K. Das, P. Dutta

and A. Halder, *Mater. Focus* **7**, 390 (2018); B. Chatterjee, N. Debbarma, M. Mitra, T. Datta and K. P. Ghatak, *J. Nanosci. Nanotechnol.* **17**, 3352 (2017).

22. K. P. Ghatak and D. De, *Mater. Focus* **6**, 114 (2017); P. K. Das, P. Dutta, A. Halder, J. Pal, N. Debbarma, S. Debbarma and K. P. Ghatak, *Mater. Focus* **6**, 167 (2017); P. K. Das, P. Dutta, A. Halder, R. Bhattacharjee and K. P. Ghatak, *Mater. Focus* **6**, 133 (2017); R. Bhattacharjee and K. P. Ghatak, *J. Nanosci. Nanotechnol.* **17**, 640 (2017).
23. M.Mitra, T. N. Sen, T. Datta, R. Bhattacharjee, L. S. Singh and K. P. Ghatak, *J. Nanosci. Nanotechnol.* **17**, 256 (2017); T. N. Sen and K. P. Ghatak, *Quant. Matter* **5**, 732 (2016); T. N. Sen and K. P. Ghatak, *Quant. Matter* **5**, 721 (2016); K. P. Ghatak, K. Sarkar, S. Chakrabarti, M. Kumar, M. Debbarma, T. N. Sen and M. Chakraborty, *Rev. Theor. Sci.* **4**, 199 (2016).
24. R. Bhattacharya, K. Sarkar, M. Kumar, B. Chatterjee and K. P. Ghatak, *Quant. Matter* **5**, 557 (2016); K. P. Ghatak, K. Sarkar, N. Debbarma and L. Suraj Singh, *Quant. Matter* **5**, 427 (2016); K. P. Ghatak and D. De, *J. Nanoeng. Nanomanuf.* **6**, 1 (2016); S. K. Biswas, M. Mitra and K. P. Ghatak, *J. Nanoeng. Nanomanuf.* **6**, 63 (2016).
25. N. Paitya and K. P. Ghatak, *Quant. Matter* **5**, 191 (2016); B.Chatterjee, S. Chakrabarti, S. K. Sen, M. Mitra and K. P. Ghatak, *Quant. Matter* **5**, 85 (2016); K. P. Ghatak, in *Magneto Thermoelectric Power in Heavily Doped Quantized Structures*, Series on the Foundations of Natural Science and Technology (World Scientific Publishing Company, 2016), vol. 7, pp. 758; K. P. Ghatak, in *Dispersion Relations in Heavily-Doped Nanostructures* (Springer International Publishing, 2016).
26. B. Chatterjee, K. Sarkar and K. P. Ghatak, in *Advances in Optical Science and Engineering* (Springer, India, 2015), p. 621; K. P. Ghatak, in *Einstein's Photoemission* (Springer International Publishing, 2015).
27. S. M. Adhikari and K. P. Ghatak, *J. Adv. Phys.* **2**, 130 (2013); S. Bhattacharya, D. De, S. Ghosh and K. P. Ghatak, *J. Comput. Theor. Nanosci.* **10**, 664 (2013); N. Paitya, S. Bhattacharya, D. De, S. Ghosh and K. P. Ghatak, *J. Nanoeng. Nanomanuf.* **2**, 211 (2012).
28. S. Pahari, S. Bhattacharya and K. P. Ghatak, *J. Comput. Theor. Nanosci.* **6**, 2088 (2009); L. J.Singh, S. Choudhury, S. Singha Roy and K. P. Ghatak, *Electr. Eng.* **87**, 19 (2005); P. K. Chakraborty, G. C. Datta and K. P. Ghatak, *Phys. Scr.* **68**, 368 (2003).
29. K. P. Ghatak, J. Mukhopadhyay and J. P.Banerjee, *SPIE Proc. Ser.* **4746**, 1292 (2002); K. P. Ghatak, J. Mukhopadhyay and J. P. Banerjee, *SPIE Proc. Ser.* **4746**, 1296 (2002); K. P. Ghatak, J. Mukhopadhyay and J. P. Banerjee, *Proceedings-SPIE Int. Soc. Opt. Eng.* **2**, 1296 (2002); K. P. Ghatak, J. Mukhopadhyay and J. P. Banerjee, *SPIE Proc. Ser.* **4746**, 347 (2002).
30. K. P. Ghatak, P. K. Bose and J. P. Banerjee, *SPIE Proc. Ser.* **4746**, 351 (2002); K. P. Ghatak, *SPIE Proc. Ser.* **4746**, 292 (2002); K. P. Ghatak, *J. Wave Mater. Interact.* **14**, 157 (1999); K. P. Ghatak and P. K. Bose, *J. Wave Mater. Interact.* **12**, 53 (1997); K. P. Ghatak, B. Nag, *J. Wave Mater. Interact.* **12**, 85 (1997); B. Mitra, D. K. Basu, B. Nag and K. P.

Ghatak, Nonlinear Optics-Read. **17**, 171 (1997); K. P. Ghatak and P. K. Bose, *J. Wave Mater. Interact.* **12**, 60 (1997); P. K. Bose and K. P. Ghatak, *J. Wave Mater. Interact.* **12**, 67 (1997).

31. K. P. Ghatak, P. K. Bose, G. Majumder, *MRS Proc.* **494**, 157 (1997); K. P. Ghatak, D. K. Basu, D. Basu and B. Nag, *Il Nuovo Cimento D* **18**, 947 (1996); K. P. Ghatak, S. Dutta, A. Ali, S. Banerjee and B. Nag, *J. Wave Mater. Interact.* **11**, 127 (1996); P. K. Chakrabarty, B. Nag, S. Dutta and K. P. Ghatak, *J. Wave Mater. Interact.* **11**, 55 (1996).
32. B. Nag, P. K. Chakrabarty, K. P. Ghatak, *J. Wave Mater. Interact.* **11**, 211 (1996); K. P. Ghatak, P. K. Chakrabarty and B. Nag, *J. Wave Mater. Interact.* **11**, 159 (1996); K. P. Ghatak, S. N. Banik, *FIZIKA A-ZAGREB* **5**, 31 (1996).
33. K. P. Ghatak, J. P. Banerjee, P. K. Chakrabarty and B. Nag, *J. Wave Mater. Interact.* **11**, 166 (1996); P. K. Chakraborty, B. Nag, S. Dutta, K. P. Ghatak, *J. Wave Mater. Interact.* **11**, 111 (1996).
34. K.P Ghatak, S. Bera, A. Ali and B. Nag, *FIZIKA A-ZAGREB* **5**, 111 (1996); K. P. Ghatak, J. P. Banerjee, B. Goswami and B. Nag, *Nonlin. Opt. Read.* **16**, 241 (1996); K. P. Ghatak, M. Mitra, B. Goswami and B. Nag, *Nonlinear Optics-Read.* **16**, 167 (1996); K. P. Ghatak, D. Bhattacharya, D. Basu and B. Nag, *Phys. Status Sol. (b)* **191**, 141 (1995).
35. K. P. Ghatak, B. Nag, M. Mitra and J. P. Bannerjee, *J. Wave Mater. Interact.* **10**, 11 (1995); K. P. Ghatak, D. K. Basu and B. Nag, *J. Wave Mater. Interact.* **10**, 29 (1995); K. P. Ghatak, S. Dutta, D. Basu and B. Nag, *J. Wave Mater. Interact.* **10**, 1 (1995).
36. K. P. Ghatak, B. Nag and G. Mazumder, *MRS Proc.* **379**, 85 (1995); K. P. Ghatak, B. Nag, G. Mazumder, *MRS Proc.* **379**, 109 (1995); K. P. Ghatak, S. N Banik, *FIZIKA A* **3**, 155 (1994); S. N. Banik, K. P. Ghatak and S. N. Biswas, *FIZIKA A* **3**, 77 (1994); K. P. Ghatak, *Fizika A* **2**, 133 (1993); K. P. Ghatak and S. N. Biswas, *MRS Proc.* **313**, 375 (1993); K. P. Ghatak and S. N. Biswas, *MRS Proc.* **308**, 445 (1993).
37. K. P. Ghatak and D. Bhattacharyya, *J. Wave Mater. Interact.* **8**, 233 (1993); K. P. Ghatak, S. N. Biswas, *Acta Phys. Slovaca* **43**, 425 (1993); K. P. Ghatak and S. N. Biswas, *Nanostruct. Mater.* **2**, 91 (1993).
38. K. P. Ghatak and M. Mondal, *Fizika A* **1**, 113 (1992); K. P. Ghatak and B. De, *MRS Proc.* **262**, 911 (1992); K. P. Ghatak and B. De, *MRS Proc.* **242**, 373 (1992); K. P. Ghatak and B. De, *MRS Online Proc. Libr. Archive* **228**, (1992); K. P. Ghatak, *Fizika A* **1**, 197 (2006).
39. K. P. Ghatak, S. Bhattacharyya and S. N. Biswas, *Acta Phys. Hung.* **70**, 83 (1991); K. P. Ghatak, *Acta Phys. Hung.* **69**, 121 (1991); K. P. Ghatak and B. De, *MRS Proc.* **234**, 59 (1991); K. P. Ghatak and B. De, *MRS Proc.* **234**, 55 (1991); K. P. Ghatak, *Acta Phys. Hung.* **68**, 253 (1990); K. P. Ghatak, *Acta Phys. Hung.* **67**, 407 (1990); K. P. Ghatak, *MRS Proc.* **216**, 469 (1990); K. P. Ghatak, *MRS Proc.* **216**, 465 (1990).
40. K. P. Ghatak, B. De, M. Mondal and S. N. Biswas, *MRS Proc.* **184**, 261 (1990); K. P. Ghatak, B. De, M. Mondal and S. N. Biswas, *MRS Proc.* **198**, 333 (1990); B. Mitra, K. P. Ghatak, *Solid-State Electron.* **32**, 810 (1989);

K. P. Ghatak and M. Mondal, *J. Appl. Phys.* **62**, 922 (1987); S. N. Biswas and K. P. Ghatak, Megagauss Technol. *Pulsed Power Appl.* **8**, 219 (1987); M. Mondal and K. P. Ghatak, *Czechoslovak J. Phys.* **36**, 1396 (1986); A. Arti, K. P. Ghatak, K. K. Ghosh, S. Ghosh and A. Dhar, *Physica Status Solidi B-Basic Res.* **103**, K55 (1981).
41. S. Bhattacharya and K. P. Ghatak, *Fowler-Nordheim Field Emission*, Springer Series in Solid-State Sciences, vol. 170 (Springer-Verlag, Germany, 2012).
42. S. Iijima, *Nature* **354**, 56 (1991).
43. V. N. Popov, *Mater. Sci. Eng.* ***R*** **43**, 61 (2004).
44. J. Sandler, M. S. P. Shaffer, T. Prasse, W. Bauhofer, K. Schulte and A. H. Windle, *Polymer* **40**, 5967 (1999).
45. D. Qian, E. C. Dickey, R. Andrews and T. Rantell, *Appl. Phys. Lett.* **76**, 2868 (2000).
46. J. Kong, N. R. Franklin, C. W. Zhou, M. G. Chapline, S. Peng, K. J. Cho and H. J. Dai, *Science*, **287**, 622 (2000).
47. W. A. Deheer, A. Chatelain, D. Ugarte, *Science* **270**, 1179 (1995).
48. A. G. Rinzler, J. H. Hafner, P. Nikolaev, L. Lou, S. G. Kim, D. Tomanek and P. Nordlander.
49. D. T. olbert and R. E. Smalley, *Science* **269**, 1550 (1995).
50. A. C. Dillon, K. M. Jones, T. A. Bekkedahl, C. H. Kiang, D. S. Bethune and M. J. Heben, *Nature* **386**, 377 (1997).
51. C. Liu, Y. Y. Fan, M. Liu, H. T. Cong, H. M. Cheng and M. S. Dresselhaus, *Science* **286**, 1127 (1999).
52. P. Kim and C. M. Lieber, *Science* **286**, 2148 (1999).
53. D. Srivastava, *Nanotechnology* **8**, 186 (1997).
54. C. Ke and H. D. Espinosa, *Appl. Phys. Lett.* **85**, 681 (2004).
55. J. W. Kang and H. J. Hwang, *Nanotechnology* **15**, 1633 (2004); J. Cumings and A. Zettl, *Science* **289**, 602 (2000).
56. S. J. Tans, M. H. Devoret, H. J. Dai, A. Thess, R. E. Smalley, L J. Geerligs and C. Dekker, *Nature*, **386**, 474 (1997).
57. S. J. Tans, A. R. M. Verschueren and C. Dekker, *Nature* **393**, 49 (1998).
58. P. Avouris, *Acc. Chem. Res.* **35**, 1026 (2002).
59. P. G. Collins, A. Zettl, H. Bando, A. Thess and R. E. Smalley, *Science* **278**, 100 (1997).
60. S. Saito, *Science* **278**, 77 (1997); J. C. Charlier, *Acc. Chem. Res.* **35**, 1063 (2002); R. H. Baughman, A. A. Zakhidov and W. A. de Heer, *Science* **297**, 787 (2002).
61. Z. Yao, H. W. C. Postma, L. Balents and C. Dekker, *Nature* **402**, 273 (1999); C. W. Zhou, J. Kong, E. Yenilmez and H. Dai, *Science* **290**, 1552 (2000).
62. S. J. tans, A. R. M. Verschueren and C. Dekker, *Nature* **393**, 49 (1998); H. W. Ch. Postma, T. Teepen, Z. Yao, M. Grifoni and C. Dekker, *Science* **293**, 76 (2001).
63. T. Rueckes, K. Kim, E. Joselevich, G. Y. Tseng, C. L. Cheung and C. M. Lieber, *Science* **289**, 94 (2000).

64. A. Bacthtold, P. Hadley, T. Nakanish and C. Dekker, *Science* **294**, 1317 (2001).
65. Q. Zheng and Q. Jiang, *Phys. Rev. Lett.* **88**, 045503 (2002); Q. Zheng, J. S. Liu and Q. Jiang, *Phys. Rev B.* **65**, 245409 (2002); Y. Zhao, C.-C. Ma, G. H. Chen and Q. Jiang, *Phys. Rev. Lett.* **91**, 175504 (2003); S. B. Legoas, V. R. Coluci, S. F. Braga, P. Z. Coura, S. O. Dantus and D. S. Galvao, *Phys. Rev. Lett.* **90**, 055504 (2003); S. B. Legaos, V. R. Coluci, S. F. Braga, P. Z. Coura, S. O. Dantus and D. S. Galvao, *Nanotechnology* **15**, S184 (2004); J. W. Kang and H. W. Hwang, *J. Appl. Phys.* **96**, 3900 (2004).
66. W. Y. Choi, J. W. Kang and H. W. Hwang, *Physica E* **23**, 125 (2004); H. J. Hwang and K. R. Byun.
67. J. W. Kang, *Physica E* **23**, 208 (2004); J. W. Kang and H. J. Hwang, *Physica E* **23**, 36 (2004).
68. J. W. Kang and H. J. Hwang, *J. Appl. Phys.* **73**, 4447 (2004); J. W. Kang and H. J. Hwang, *J. Phys. Soc. Jpn.* **73**, 1077 (2004); J. W. Kang, Y. W. Choi and H. J. Hwang, *J. Comp. Theor. Nanosc.* **1**, 199 (2004).
69. M. Dequesnes, S. V. Rotkin and N. R. Aluru, *Nanotechnology* **13**, 120 (2002); J. M. Kinaret.
70. T. Nord and S. Viefers, *Appl. Phys. Lett.* **82**, 1287 (2003); C. Ke and H. D. Espinosa, *Appl. Phys. Lett.* **85**, 681 (2004); L. M. Jonsson, T. Nord, J. M. Kinaret and S. Viefers, *J. Appl. Phys.* **96**, 629 (2004); L. M. Jonsson, S. Axelsson, T. Nord, S. Viefers and J. M. Kinaret, *Nanotechnology* **15**, 1497 (2004); S. W. Lee, D. S. Lee, R. E. Morjan, S. H. Jhang and M. Sveningsson.
71. O. A. Nerushev, Y. W. Park and E. E. B. Campbell, *Nano. Lett.* **4**, 2027 (2004).
72. M. Lundstrom and J. Guo, *Nanoscale Transistors: Device Physics, Modeling and simulation* (Springer, New York, 2006).
73. R. Saito, G. Dresselhaus and M. S. Dresselhaus, *Physical Properties of Carbon Nanotubes* (Imperial College Press, London, 1998).
74. R. Heyd, A. Charlier, E. McRae, *Phys. Rev. B.* **55**, 6820 (1997).

Chapter 8

The QC in 2D QMOSFETs of Important Materials Under Accumulation Mode of Operation

I must bow down my head to the dust of all persons feet in order to enjoy God's Laugh from every heart.

8.1 Introduction

It may be noted that the QC in 2D QMOSFETs of important materials operated under accumulation mode has relatively been less investigated. In this chapter, we study the same in 2D QMOSFETs of tetragonal and nonlinear optical materials in Section 8.2.1 of theoretical background 8.2, III–V, ternary and quaternary materials in Section 8.2.2 respectively. From Sections 8.2.3 up to 8.2.5, we shall investigate the QC in 2D QMOSFETs of II–VI, IV–VI, and stressed materials, respectively. Section 8.3 contains summary and conclusion and the last Section 8.4 contains open research problems.

8.2 Theoretical Background

8.2.1 *The QC in 2D QMOSFETs of tetragonal and nonlinear optical materials under accumulation mode of operation*

From (1.6) we can write

$$\frac{\hbar^2 k_z^2}{2m_{\parallel}^*} + \left(\frac{b_{\parallel}}{b_{\perp}}\frac{c_{\perp}}{c_{\parallel}}\right)\frac{\hbar^2 k_s^2}{2m_{\perp}^*} = \left\{\frac{E(\alpha E+1)(b_{\parallel}E+1)}{(c_{\parallel}E+1)} + \frac{\alpha b_{\parallel}}{c_{\parallel}}\left[\delta E + \frac{2}{9}(\Delta_{\parallel}^2 - \Delta_{\perp}^2)\right] - \left(\frac{2}{9}\right)\frac{\alpha b_{\parallel}}{c_{\parallel}}\frac{(\Delta_{\parallel}^2 - \Delta_{\perp}^2)}{(c_{\parallel}E+1)}\right\}$$

$$-\left(\frac{\hbar^2 k_s^2}{2m_\perp^*}\right)\left\{\left(\frac{b_\parallel}{b_\perp}\frac{c_\perp}{c_\parallel}\right)\left[\left(\frac{\delta}{2}+\frac{\Delta_\parallel^2-\Delta_\perp^2}{6\Delta_\parallel}\right)\frac{\alpha_\parallel}{\alpha_\parallel E+1}\right.\right.$$
$$\left.\left.+\left(\frac{\delta}{2}-\left\{\frac{\Delta_\parallel^2-\Delta_\perp^2}{6\Delta_\parallel}\right\}\right)\frac{c_\parallel}{c_\parallel E+1}\right]\right\} \tag{8.1}$$

where

$$b_\parallel \equiv (E_g+\Delta_\parallel)^{-1},\ c_\perp \equiv \left(E_g+\frac{2}{3}\Delta_\perp\right)^{-1},\ b_\perp \equiv (E_g+\Delta_\perp)^{-1},$$

$$c_\parallel = \left(E_g+\frac{2}{3}\Delta_\parallel\right)^{-1} \quad \text{and} \quad \alpha = (E_g)^{-1}.$$

It is well known that the band tails are being formed in the forbidden zone of heavily doped semiconductors (HDSs) and can be explained by the overlapping of the impurity band with the conduction and valence bands [1]. Kane [2] and Bonch Bruevich [3] have independently derived the theory of band tailing for materials having unperturbed parabolic energy bands. Kane's model [2] was used to explain the experimental results on tunneling [4] and the optical absorption edges [5, 6] in this context. Halperin and Lax [7] developed a model for band tailing applicable only to the deep tailing states. Although Kane's concept is often used in the literature for the investigation of band tailing [8, 9], it may be noted that this model [2, 10] suffers from serious assumptions in the sense that the local impurity potential is assumed to be small and slowly varying in space coordinates [9]. In this respect, the local impurity potential may be assumed to be a constant. In order to avoid these approximations, we have developed in this book, the electron energy spectra for HDS for studying the QC based on the concept of the variation of the kinetic energy [1, 9] of the electron with the local point in space coordinates. This kinetic energy is then averaged over the entire region of variation using a Gaussian-type potential energy. It may be noted that, a more general treatment of many-body theory for the density-of-states (DOS) function of HDS merges with one-electron theory under macroscopic conditions [1]. Also, the experimental results for the Fermi energy and others are the average effect of this macroscopic case. So, the present treatment of the one-electron system is more applicable to the experimental point of view and it is also easy to understand the overall effect in such a case [11]. In a HDS, each impurity atom is surrounded by electrons, assuming a regular distribution of atoms, and it is screened independently [7, 10, 12]. The interaction energy between electrons and impurities is known as the impurity screening potential. This energy is determined by the inter-impurity distance and the screening radius, which is known as the screening length. The screening radius changes with the electron concentration and the effective mass. Furthermore, these entities are important for HDS in characterizing the different properties of the materials [13, 14] and the modern electronic devices [8, 15]. The works on Fermi energy and the screening length in an n-type GaAs have already been initiated in the literature [16, 17], based on Kane's model. Incidentally, the limitations of Kane's model [2, 9], as mentioned above, are also present in their studies.

The Gaussian distribution F(V) of the impurity potential is given by [10]

$$F(V) = (\pi\eta_g^2)^{-1/2} \exp(-V^2/\eta_g^2) \tag{8.2}$$

where η_g is the impurity screening potential. It appears from (3.4) that the variance parameter η_g is not equal to zero, but the mean value is zero. Further, the impurities are assumed to be uncorrelated and the band mixing effect has been neglected in this simplified theoretical formalism.

We have to average the kinetic energy in the order to obtain the dispersion relation in nonlinear optical materials in the presence of band tails. Using (8.1) and (8.2), we get

$$\begin{aligned}
&\left[\frac{\hbar^2 k_z^2}{2m_\parallel^*}\int_{-\infty}^{E} F(V)dV\right] + \left[\left(\frac{b_\parallel}{b_\perp}\frac{c_1}{c_\parallel}\right)\frac{\hbar^2 k_s^2}{2m_\perp^*}\int_{-\infty}^{E} F(V)dV\right] \\
&\quad = \left\{\int_{-\infty}^{E} \frac{(E-V)[\alpha(E-V)+1][b_\parallel(E-V)+1}{[c_\parallel(E-V)+1]} F(V)dV\right. \\
&\qquad + \frac{\alpha b_\parallel}{c_\parallel}\left[\delta\int_{-\infty}^{E}(E-V)F(V)dV + \frac{2}{9}(\Delta_\parallel^2-\Delta_\perp^2)\int_{-\infty}^{E} F(V)dV\right] \\
&\qquad \left. - \left(\frac{2}{9}\right)\frac{\alpha b_\parallel}{c_\parallel}(\Delta_\parallel^2-\Delta_\perp^2)\int_{-\infty}^{E}\frac{F(V)dV}{[c_\parallel(E-V)+1]}\right\} - \left(\frac{\hbar^2 k_s^2}{2m_\perp^*}\right) \\
&\qquad \times \left\{\left(\frac{b_\parallel}{b_\perp}\frac{c_\perp}{c_\parallel}\right)\left[\left(\frac{\delta}{2}+\frac{\Delta_\parallel^2-\Delta_\perp^2}{6\Delta_\parallel}\right)\alpha\int_{-\infty}^{E}\frac{F(V)dV}{[\alpha(E-V)+1]}\right.\right. \\
&\qquad \left.\left. + \left(\frac{\delta}{2}-\frac{\Delta_\parallel^2-\Delta_\perp^2}{6\Delta_\parallel}\right)c_\parallel\int_{-\infty}^{E}\frac{F(V)dV}{[c_\parallel(E-V)+1]}\right]\right\}
\end{aligned} \tag{8.3}$$

(8.3) can be rewritten as

$$\frac{\hbar^2 k_z^2}{2m_{\|}^*} I(1) + \left(\frac{b_{\|}}{b_{\perp}}\frac{c_1}{c_{\|}}\right)\frac{\hbar^2 k_s^2}{2m_{\perp}^*} I(1) = \left\{ I_3(c_{\|}) + \frac{\alpha b_{\|}}{c_{\|}}\left[\delta I(4) + \frac{2}{9}(\Delta_{\|}^2 - \Delta_{\perp}^2) I(1)\right]\right.$$

$$\left. - \left(\frac{2}{9}\right)\frac{\alpha b_{\|}}{c_{\|}}(\Delta_{\|}^2 - \Delta_{\perp}^2) I_6(c_{\|})\right\} - \left(\frac{\hbar^2 k_s^2}{2m_{\perp}^*}\right)\left\{\left(\frac{b_{\|}}{b_{\perp}}\frac{c_{\perp}}{c_{\|}}\right)\right.$$

$$\left.\times\left[\left(\frac{\delta}{2} + \frac{\Delta_{\|}^2 - \Delta_{\perp}^2}{6\Delta_{\|}}\right)\alpha I(\alpha) + \left(\frac{\delta}{2}\left\{\frac{\Delta_{\|}^2 - \Delta_{\perp}^2}{6\Delta_{\|}}\right\}\right) c_{\|} I(c_{\|})\right]\right\} \tag{8.4}$$

where

$$I(1) = \int_{-\infty}^{E} F(V)dV \tag{8.5}$$

$$I_3(c_{\|}) = \int_{-\infty}^{E} \frac{(E-V)[\alpha(E-V)+1][b_{\|}(E-V)+1]}{[c_{\|}(E-V)+1]} F(V)dV \tag{8.6}$$

$$I(4) = \int_{-\infty}^{E} (E-V)F(V)dV \tag{8.7}$$

$$I(\alpha) = \int_{-\infty}^{E} \frac{F(V)dV}{[\alpha(E-V)+1]} \tag{8.8}$$

Substituting $E - V \equiv x$ and, $\frac{x}{\eta_g} \equiv t_0$ we get from (8.5)

$$I(1) = \left(\exp\left(-\frac{E^2}{\eta_g^2}\right)\Big/\sqrt{\pi}\right)\int_0^{\infty} \exp\left[-t_0^2 + \left(\frac{2Et_0}{\eta_g}\right)\right] dt_0$$

Thus,

$$I(1) = \left[\frac{1 + Erf(E/\eta_g)}{2}\right] \tag{8.9}$$

where $Erf(E/\eta_g)$ is the error function of (E/η_g).

From (8.7), one can write

$$I(4) = (1/\eta_g\sqrt{\pi})\int_{-\infty}^{E} (E-V)\exp(-V^2/\eta_g^2)dV = \frac{E}{2}[1 + Erf(E/\eta_g)]$$

$$- \left\{\frac{1}{\sqrt{\pi\eta_g^2}}\int_{-\infty}^{E} V\exp(-V^2/\eta_g^2)dV\right\} \tag{8.10}$$

After computing this simple integration, we can write

$$I(4) = \eta_g \exp(-E^2/\eta_g^2)(2\sqrt{\pi}) + \frac{E}{2}(1 + Erf(E/\eta_g)) = \gamma_0(E, \eta_g) \tag{8.11}$$

From (8.8), we can write

$$I(\alpha) = \frac{1}{\sqrt{\pi\eta_g^2}} \int_{-\infty}^{E} \frac{\exp(-V^2/\eta_g^2)dV}{[\alpha(E-V)+1]} \tag{8.12}$$

When

$$V \to \pm\infty, \frac{1}{[\alpha(E-V)+1]} \to 0 \quad \text{and;} \quad \exp(-V^2/\eta_g^2) \to 0$$

Thus, (8.12) can be expressed as

$$I(\alpha) = (1/\alpha\eta_g\sqrt{\pi}) \int_{-\infty}^{\infty} \exp(-t^2)(u-t)^{-1}dt \tag{8.13}$$

where

$$\frac{V}{\eta_g} = t \quad \text{and} \quad u \equiv \left(\frac{1+\alpha E}{\alpha\eta}\right)$$

It is well known that [18, 19]

$$W(Z) = (i/\pi) \int_{-\infty}^{\infty} (Z-t)^{-1} \exp(-t^2)dt \tag{8.14}$$

In which $i = \sqrt{-1}$ and Z, in general, is a complex number.

We also know [18, 19],

$$W(Z) = \exp(-Z^2)Erfc(-iZ) \tag{8.15}$$

where, $Erfc(Z) \equiv 1 - Erf(Z)$.

Thus, $Erfc(-iu) = 1 - Erf(-iu)$

Since, $Erf(-iu) = -Erf(iu)$

Therefore, we get,

$$I(\alpha) = [-i\sqrt{\pi}/\alpha\eta_g] \exp(-u^2)[1 + Erf(iu)] \tag{8.16}$$

It is well known that [18, 19]

$$Erf(x+iy) = Erf(x) + \left(\frac{e^{-x^2}}{2\pi x}\right)\Bigg[(1-\cos(2xy)) + i\sin(2xy) + \frac{2}{\pi}e^{-x^2}\sum_{p=1}^{\infty}\frac{\exp(-p^2/4)}{(p^2+4x^2)}\Bigg][f_p(x,y) + ig_p(x,y) + \varepsilon(x,y)] \tag{8.17}$$

where $f_p(x,y) = [2x - 2x\cosh(py)\cos(2xy) + p\sinh(py)\sin(2xy)]$

$$g_p(x,y) \equiv [2x\cosh(py)\sin(2xy) + p\sinh(py)\cos(2xy)],$$

$$|\varepsilon(x,y)| \approx 10^{-15}|Erf(x+iy)|$$

Substituting $x = 0$ and $y = u$ in (8.17), one obtains,

$$Erf(iu) = \left(\frac{2i}{\pi}\right)\sum_{p=1}^{\infty}\left\{\frac{\exp(-p^2/4)}{p}\sinh(pu)\right\} \tag{8.18}$$

Therefore, one can write

$$I(\alpha) = C_{21}(\alpha, E, \eta_g) - iD_{21}(\alpha, E, \eta_g) \tag{8.19}$$

where

$$C_{21}(\alpha, E, \eta_g) = \left[\frac{2}{\alpha\eta_g\sqrt{\pi}}\right]\exp(-u^2)\left[\sum_{p=1}^{\infty}\left\{\frac{\exp(-p^2/4)}{p}\sinh(pu)\right\}\right]$$

and $D_{21}(\alpha, E, \eta_g) = \left[\frac{\sqrt{\pi}}{\alpha\eta_g}\exp(-u^2)\right]$

(8.19) consists of both real and imaginary parts and therefore, $I(\alpha)$ is complex, which can also be proved by using the method of analytic continuation of complex analysis.

The integral $I_3(c_{\|})$ in (8.6) can be written as

$$I_3(c_{\|}) = \left(\frac{\alpha b_{\|}}{c_{\|}}\right)I(5) + \left(\frac{\alpha c_{\|} + b_{\|}c_{\|} - \alpha b_{\|}}{c_{\|}^2}\right)I(4)$$
$$+\frac{1}{c_{\|}}\left(1-\frac{\alpha}{c_{\|}}\right)\left(1-\frac{b_{\|}}{c_{\|}}\right)I(1) - \left\{\frac{1}{c_{\|}}\left(1-\frac{\alpha}{c_{\|}}\right)\left(1-\frac{b_{\|}}{c_{\|}}\right)I(c_{\|})\right\} \tag{8.20}$$

where

$$I(5) = \int_{-\infty}^{E}(E-V)^2F(V)dV. \tag{8.21}$$

From (8.21) one can write

$$I(5) = \frac{1}{\sqrt{\pi\eta_g^2}}\left[E^2\int_{-\infty}^{E}\exp\left(\frac{-V^2}{\eta_g^2}\right)dV - 2E\int_{-\infty}^{E}V\exp\left(\frac{-V^2}{\eta_g^2}\right)dV\right.$$
$$\left.+\int_{-\infty}^{E}V^2\exp\left(\frac{-V^2}{\eta_g^2}\right)dV\right] \tag{8.22}$$

The evaluations of the component integrals lead us to write

$$I(5) = \frac{\eta_g E}{2\sqrt{\pi}} \exp\left(\frac{-E^2}{\eta_g^2}\right) + \frac{1}{4}(\eta_g^2 + 2E^2)\left[1 + Erf\left(\frac{E}{\eta_g}\right)\right] = \theta_0(E, \eta_g) \tag{8.23}$$

Thus, combining the aforementioned equations, $I_3(c_{\|})$ can be expressed as

$$I_3(c_{\|}) = A_{21}(E, \eta_g) + iB_{21}(E, \eta_g) \tag{8.24}$$

where

$$\begin{aligned} A_{21}(E,\eta) = &\left[\frac{\alpha b_{\|}}{c_{\|}}\left[\frac{\eta_g E}{2\sqrt{\pi}} \exp\left(\frac{-E^2}{\eta_g^2}\right) - \frac{1}{4}(\eta_g^2 + 2E^2)\left\{1 + Erf\left(\frac{E}{\eta_g}\right)\right\}\right]\right. \\ &+\left[\frac{\alpha c_{\|} + b_{\|}c_{\|} - \alpha b_{\|}}{c_{\|}^2}\right]\left\{\frac{E}{2}\left[1 + Erf\left(\frac{E}{\eta_g}\right)\right] + \frac{\eta \exp(-E^2/\eta_g^2)}{2\sqrt{\pi}}\right\} \\ &+\frac{1}{c_{\|}}\left(1 - \frac{\alpha}{c_{\|}}\right)\left(1 - \frac{b_{\|}}{c_{\|}}\right)\frac{1}{2}\left[1 + Erf\left(\frac{E}{\eta_g}\right)\right] \\ &-\left\{\frac{2}{c_{\|}^2 \eta_g \sqrt{\pi}}\left(1 - \frac{\alpha}{c_{\|}}\right)\left(1 - \frac{b_{\|}}{c_{\|}}\right)\exp(-u_1^2)\right\} \\ &\times\left[\sum_{p=1}^{\infty}\left\{\frac{\exp(-p^2/4)}{p}\sinh(pu_1)\right\}\right], \quad u_1 \equiv \left[\frac{1 + c_{\|}E}{c_{\|}\eta_g}\right] \quad \text{and} \end{aligned}$$

$$B_{21}(E, \eta_g) \equiv \frac{\sqrt{\pi}}{c_{\|}^2 \eta_g}\left(1 - \frac{\alpha}{c_{\|}}\right)\left(1 - \frac{b_{\|}}{c_{\|}}\right)\exp(-u_1^2)$$

Therefore, the combination of all the appropriate integrals together with algebraic manipulations leads to the expression of the dispersion relation of the conduction electrons of HD tetragonal and nonlinear optical materials forming Gaussian band tails as

$$\frac{\hbar^2 k_z^2}{2m_{\|}^* T_{21}(E, \eta_g)} + \frac{\hbar^2 k_s^2}{2m_{\perp}^* T_{22}(E, \eta_g)} = 1 \tag{8.25}$$

where $T_{21}(E, \eta_g)$ and $T_{22}(E, \eta_g)$ have both real and complex parts and are given by

$$T_{21}(E, \eta_g) = [T_{27}(E, \eta_g) + iT_{28}(E, \eta_g)], \; T_{27}(E, \eta_g) = \left[\frac{T_{23}(E, \eta_g)}{T_5(E, \eta_g)}\right]$$

$$T_{23}(E, \eta_g) \equiv \left[A_{21}(E, \eta_g) + \frac{\alpha b_{\|}}{c_{\|}}\left[\delta\gamma_0(E, \eta_g) + \frac{1}{9}(\Delta_{\|}^2 - \Delta_{\perp}^2)\right.\right.$$

$$\times[1+Erf(E/\eta_g)]\Big]-\left\{\frac{2}{9}\left(\frac{\alpha b_\parallel}{c_\parallel}\right)(\Delta_\parallel^2-\Delta_\perp^2)G_{21}(c_\parallel,E,\eta_g)\right\}\Big],$$

$$G_{21}(E,\eta_g)\equiv\frac{2}{c_\parallel\eta_g\sqrt{\pi}}\exp(-u_1^2)\sum_{p=1}^{\infty}\left\{\frac{\exp(-p^2/4)}{p}\sinh(pu_1)\right\},$$

$$T_5(E,\eta_g)\equiv\frac{1}{2}[1+Erf(E/\eta_g)]$$

$$T_{28}(E,\eta_g)\equiv\left[\frac{T_{24}(E,\eta_g)}{T_5(E,\eta_g)}\right],$$

$$T_{24}(E,\eta_g)\equiv\left[B_{21}(E,\eta_g)+\frac{2}{9}\frac{\alpha b_\parallel}{c_\parallel}(\Delta_\parallel^2-\Delta_\perp^2)H_{21}(c_\parallel,E,\eta_g)\right]$$

$$H_{21}(c_\parallel,E,\eta_g)\equiv\left[\frac{\sqrt{\pi}}{\eta_g c_\parallel}\exp(-u_1^2)\right],\ T_{22}(E,\eta_g)\equiv[T_{29}(E,\eta_g)+iT_{30}(E,\eta_g)],$$

$$T_{29}(E,\eta_g)\equiv\frac{T_3(E,\eta_g)T_{25}(E,\eta_g)-T_{24}(E,\eta_g)T_{26}(E,\eta_g)}{[(T_{25}(E,\eta_g))^2+(T_{26}(E,\eta_g))^2]}$$

$$T_{25}(E,\eta_g)\equiv\left[\left(\frac{b_\parallel}{b_\perp}\frac{c_\perp}{c_\parallel}\right)\frac{1}{2}\left[1+Erf\left(\frac{E}{\eta_g}\right)\right]\right.$$
$$+\left(\frac{b_\parallel}{b_\perp}\frac{c_\parallel}{c_\perp}\right)\left(\frac{\delta}{2}+\left[\frac{\Delta_\parallel^2-\Delta_\perp^2}{6\Delta_\parallel}\right]\right)\alpha_\parallel C_{21}(\alpha_\parallel,E,\eta_g)$$
$$\left.+\left(\frac{b_\parallel c_\perp}{b_\perp}\right)\left(\frac{\delta}{2}-\left[\frac{\Delta_\parallel^2-\Delta_\perp^2}{6\Delta_\parallel}\right]\right)G_{21}(\alpha_\parallel,E,\eta_g)\right],$$

$$C_{21}(\alpha,E,\eta_g)\equiv\left[\frac{2}{\alpha\sqrt{\pi}\eta_g}\exp(-u^2)\left[\sum_{p=1}^{\infty}\frac{\exp(-p^2/4)}{p}\sinh(pu)\right]\right],$$

$$T_{26}(E,\eta_g)\equiv\left(\frac{b_\parallel}{b_\perp}\frac{c_\perp}{c_\parallel}\right)\left(\frac{\delta}{2}-\frac{\Delta_\parallel^2-\Delta_\perp^2}{6\Delta_\parallel}\right)\alpha D_{21}(\alpha,E,\eta_g)$$
$$+\frac{b_\parallel c_\perp}{b_\perp}\left(\frac{\delta}{2}-\frac{\Delta_\parallel^2-\Delta_\perp^2}{6\Delta_\parallel}\right)H_{21}(c_\parallel,E,\eta_g)\quad\text{and}$$

$$T_{30}(E,\eta_g)\equiv\frac{T_{24}(E,\eta_g)T_{25}(E,\eta_g)+T_{23}(E,\eta_g)T_{26}(E,\eta_g)}{[(T_{25}(E,\eta_g))^2+(T_{26}(E,\eta_g))^2]}$$

From (8.25), it appears that the energy spectrum in HD nonlinear optical materials is complex. The complex nature of the electron dispersion law in HD materials occurs from the existence of the essential poles in the corresponding electron energy spectrum in the absence of band tails. It may be noted that the complex band structures have already been studied for bulk materials and super lattices without heavy doping and bears no relationship with the complex electron dispersion law as indicated by (3.26). The physical picture behind the formulation of the complex energy spectrum in HDS is the interaction of the impurity atoms in the tails with the splitting constants of the valance bands. More is the interaction; more is the prominence of the complex part than the other case. In the absence of band tails, $\eta_g \to 0$, and there is no interaction of the impurity atoms in the tails with the spin-orbit constants. As a result, there exists no complex energy spectrum and (8.25) gets converted into (1.6) when $\eta_g \to 0$. Besides, the complex spectra are not related to the same evanescent modes in the band tails and the conduction bands. The transverse and the longitudinal EEMs at the Fermi energy E_{F_h} of HD nonlinear optical materials can, respectively, be expressed as

$$m_{\perp}^{*}(E,\eta_g) = m_{\perp}^{*}\{T_{29}(E,\eta_g)\}' \tag{8.26}$$

and

$$m_{\parallel}^{*}(E,\eta_g) = m_{\parallel}^{*}\{T_{27}(E,\eta_g)\}' \tag{8.27}$$

In the absence of band tails $\eta_g \to 0$ and we get

$$m_{\perp}^{*}(\bar{E},O) = \frac{\hbar^2}{2}\left[\frac{\psi_2(\bar{E})\{\psi_1(\bar{E})\}' - \psi_1(\bar{E})\{\psi_2(\bar{E})\}'}{\{\psi_2(\bar{E})\}}\right] \tag{8.28}$$

and

$$m_{\parallel}^{*}(\bar{E},O) = \frac{\hbar^2}{2}\left[\frac{\psi_3(\bar{E})\{\psi_1(\bar{E})\}' - \{\psi_1(\bar{E})\}\{\psi_3(\bar{E})\}'}{\{\psi_3(\bar{E})\}}\right] \tag{8.29}$$

Comparing the aforementioned equations, one can infer that the effective masses exist in the forbidden zone, which is impossible without the effect of band tailing. For semiconductors, in the absence of band tails, the effective mass in the band gap is infinity.

The DOS function is given by

$$N_{HD}(E,\eta_g) = \frac{2g_\nu m_{\perp}^{*}\sqrt{2m_{\parallel}^{*}}}{3\pi^2\hbar^3} R_{11}(E,\eta_g)\cos[\psi_{11}(E,\eta_g)] \tag{8.30}$$

where

$$R_{11}(E,\eta_g) \equiv \left[\left[\{T_{29}(E,\eta_g)\}'\sqrt{x(E,\eta_g)} + \frac{T_{29}(E,\eta_g)\{x(E,\eta_g)\}'}{2\sqrt{x(E,\eta_g)}}\right.\right.$$

$$\left. -\{T_{30}(E,\eta_g)\}'\sqrt{y(E,\eta_g)}\frac{T_{30}(E,\eta_g)\{y(E,\eta_g)\}'}{2\sqrt{y(E,\eta_g)}}\right]^2$$

$$+\left[\{T_{29}(E,\eta_g)\}'\sqrt{y(E,\eta_g)} + \frac{T_{29}(E,\eta_g)\{y(E,\eta_g)\}'}{2\sqrt{y(E,\eta_g)}}\right.$$

$$\left.\left. + \{T_{30}(E,\eta_g)\}'\sqrt{x(E,\eta_g)}\frac{T_{30}(E,\eta_g)\{x(E,\eta_g)\}'}{2\sqrt{x(E,\eta_g)}}\right]^2\right]^{1/2},$$

$$x(E,\eta_g) \equiv \frac{1}{2}\left[T_{27}(E,\eta_g) + \sqrt{\{T_{27}(E,\eta_g)\}^2 + \{T_{28}(E,\eta_g)\}^2}\right]$$

$$y(E,\eta_g) \equiv \frac{1}{2}\left[\sqrt{\{T_{27}(E,\eta_g)\}^2 + \{T_{28}(E,\eta_g)\}^2} - T_{27}(E,\eta_g)\right]$$

and

$$\psi_{11}(E,\eta_g) \equiv \tan^{-1}\left[\left[\{T_{29}(E,\eta_g)\}'\sqrt{y(E,\eta_g)} + \frac{T_{29}(E,\eta_g)}{2\sqrt{y(E,\eta_g)}}\right.\right.$$

$$\left. + \{T_{30}(E,\eta_g)\}'\sqrt{x(E,\eta_g)} + \frac{T_{30}\{x(E,\eta_g)\}'}{2\sqrt{x(E,\eta_g)}}\right]$$

$$\times\left[\{T_{29}(E,\eta_g)\}'\sqrt{x(E,\eta_g)} + \frac{T_{29}(E,\eta_g)\{x(E,\eta_g)\}'}{2\sqrt{x(E,\eta_g)}}\right.$$

$$\left.\left. - \{T_{30}(E,\eta_g)\}'\sqrt{y(E,\eta_g)} + \frac{T_{30}\{T(E,\eta_g)\}'}{2\sqrt{y(E,\eta_g)}}\right]^{-1}\right]$$

The oscillatory nature of the DOS for HD nonlinear optical materials is apparent from (8.30). For, $\psi_{11}(E,\eta_g) \geq \pi$, the cosine function becomes negative leading to the negative values of the DOS. The electrons cannot exist for the negative values of the DOS and therefore, this region is forbidden for electrons, which indicates that in the band tail, there appears a new forbidden zone in addition to the normal band gap of the semiconductor.

The use of (8.30) leads to the expression of the electron concentration as

$$n_0 = \text{Real part of } \bar{n}_0 \tag{8.31}$$

where

$$\bar{n}_0 = \frac{2g_v m_{\parallel}^* \sqrt{2m_{\perp}^*}}{3\pi^2\hbar^3}[\bar{I}_{11}(E_{FB}, \eta_g)] \tag{8.32}$$

in which E_{FB} is the Fermi energy in the present case and $\bar{I}_{11}(E_{FB}, \eta_g) = [T_{21}(E_{FB}, \eta_g)\sqrt{T_{22}(E_{FB}, \eta_g)}]$

In the presence of a surface electric field F_s along z-direction and perpendicular to the surface, (8.25) assumes the form

$$\frac{\hbar^2 k_z^2}{2m_{\parallel}^*} + \frac{\hbar^2 k_s^2}{2m_{\perp}^*}\frac{T_{21}(E - |e|F_s z, \eta_g)}{T_{22}(E - |e|F_s z, \eta_g)} = T_{21}(E - |e|F_s z, \eta_g) \tag{8.33}$$

where for this chapter, E represents the electron energy as measured from the edge of the conduction band at the surface in the vertically upward direction.

The use of (1.7) and (8.33) leads to the DR of the 2D electrons in QMOSFETs of tetragonal and nonlinear optical materials operated under accumulation mode can be written under weak electric field limit as

$$\frac{\hbar^2 k_s^2}{2m_{\parallel}^*} = L_6(E, i, \eta_g) \tag{8.34}$$

where

$$L_6(E, i, \eta_g) = \frac{T_{21}(E, \eta_g) - L_3(E, i, \eta_g)}{L_4(E, i, \eta_g)},$$

$$L_3(E, i, \eta_g) = S_i[T_{21}'(E, \eta_g)]^{2/3}\left[\frac{\hbar|e|F_s}{\sqrt{2m_{\parallel}^*}}\right]^{2/3}$$

and

$$L_4(E, i, \eta_g) = \left[\frac{T_{21}(E, \eta_g)}{T_{22}(E, \eta_g)} + L_3(E, i, \eta_g)\frac{T_{21}(E, \eta_g)}{T_{21}'(E, \eta_g)T_{22}(E, \eta_g)}\right.$$
$$\left. \cdot \frac{2}{3}\left\{\frac{T_{21}'(E, \eta_g)}{T_{21}(E, \eta_g)} - \frac{T_{22}'(E, \eta_g)}{T_{22}(E, \eta_g)}\right\}\right]$$

The EEM in this case can be written as

$$m^*(E_f', i, \eta_g) = m_{\parallel}^* \text{ Real part of } [L_6'(E_f', i, \eta_g)] \tag{8.35}$$

where

$$E_f' = eV_g - \frac{e^2 n_s}{C_{ox}},$$

n_s is the surface electron concentration.

The sub-band energy E_i can be determined from the equation

$$0 = \text{ Real part of } L_6(E_i, i, \eta_g) \tag{8.36}$$

The total surface electron concentration in the regime of very low temperatures where the quantum effects become prominent can be written in this case as

$$n_s = \text{ Real Part of } \bar{n}_s \tag{8.37}$$

where

$$\bar{n}_s = g_v \sum_{i=0}^{i_{\max}} \left[\left[\frac{m_\perp^*}{\pi\hbar^2} L_6(E'_f, i, \eta_g) \right] + \frac{2m_\parallel^* \sqrt{2m_\perp^*}}{3\pi^2\hbar^3} t_{i1} [\bar{I}_{11}(E_{FB}, \eta_g)] \right],$$

$$t_{i1} = \frac{E_{i\max}}{eF_s(1 + i_{\max})},$$

$E_{i\max}$ is the root of the real part of the equation.

$$T_{21}(E_{i\max}, \eta_g) - L_3(E_{i\max}, i_{\max}, \eta_g) = 0 \tag{8.38}$$

Thus, combining the appropriate equations with (1.4) we can study the QC in this case.

8.2.2 *The QC in 2D QMOSFETs of III–V, ternary and quaternary materials under accumulation mode of operation*

(a) Three-band model of Kane

Under the conditions $\delta = 0, \Delta_\parallel = \Delta_\perp = \Delta_0$, (the isotropic spin-orbit splitting), $m_\parallel^* = m_\perp^* = m_c$ (the isotropic effective electron mass at the edge of the conduction band), the generalized dispersion law as given by (8.25) gets reduced into the form as

$$\frac{\hbar^2 k^2}{2m_c} = T_{31}(E, \eta_g) + iT_{32}(E, \eta_g) \tag{8.39}$$

where

$$\begin{aligned} T_{31}(E, \eta_g) \equiv & \left(\frac{2}{1 + Erf(E/\eta_g)} \right) \left[\frac{ab}{c} \theta_0(E, \eta_g) + \left[\frac{\alpha c + bc - \alpha b}{c^2} \right] \gamma_0(E, \eta_g) \right. \\ & + \frac{1}{c}\left(1 - \frac{\alpha}{c}\right)\left(1 - \frac{b}{c}\right)\frac{1}{2}\left[1 + Erf\left(\frac{E}{\eta_g}\right)\right] \\ & - \frac{1}{c}\left(1 - \frac{\alpha}{c}\right)\left(1 - \frac{b}{c}\right)\frac{2}{c\eta_g\sqrt{\pi}} \exp(-u_2^2) \\ & \left. \times \left[\sum_{p=1}^{\infty} \frac{\exp(-p^2/4)}{p} \sinh(pu_2) \right] \right], \end{aligned}$$

$$b \equiv (E_g + \Delta)^{-1},\ c \equiv \left(E_g + \frac{2}{3}\Delta\right)^{-1},\ a = \frac{1}{E_g} = \alpha,$$

$$u_2 \equiv \frac{1+cE}{c\eta_g} \quad \text{and} \quad T_{32}(E,\eta_g) \equiv \left(\frac{2}{1+Erf(E/\eta_g)}\right)\frac{1}{c}\left(1-\frac{\alpha}{c}\right)$$
$$\times \left(1-\frac{b}{c}\right)\frac{\sqrt{\pi}}{c\eta_g}\frac{\sqrt{\pi}}{c\eta_g}\exp(-u_2^2)$$

Thus, the complex energy spectrum occurs due to the term $T_{32}(E,\eta_g)$ and this imaginary band is quite different from the forbidden energy band. The EEM at the Fermi level is given by

$$m^*(E_{F_s},\eta_g) = m_c\{T_{31}(E,\eta_g)\}'\Big| \tag{8.40}$$

Thus, the EEM in HD III–V, ternary and quaternary materials exists in the band gap, which is the new attribute of the theory of band tailing.

In the absence of band tails, $\eta_g \to 0$ and the EEM assumes the form

$$m^*(E_F) = m_c\{I_{11}(E_F,\eta_g)\}'\Big| \tag{8.41}$$

The DOS function in this case can be written as

$$N_{HD}(E,\eta_g) = \frac{g_v}{3\pi^2}\left(\frac{2m_c}{\hbar^2}\right)^{3/2} R_{21}(E,\eta_g)\cos[\vartheta_{21}(E,\eta_g)] \tag{8.42}$$

where

$$R_{21}(E,\eta_g) \equiv \left[\frac{[\{\alpha_{11}(E,\eta_g)\}']^2}{4\alpha_{11}(E,\eta_g)} + \frac{[\{\beta_{11}(E,\eta_q)\}']^2}{4\beta_{11}(E,\eta_g)}\right]^{1/2}$$

$$\alpha_{11}(E,\eta_g) \equiv \frac{1}{2}\left[T_{33}(E,\eta_g) + \sqrt{\{T_{33}(E,\eta_g)\}^2 + \{T_{34}(E,\eta_g)\}^2}\right],$$

$$T_{33}(E,\eta_g) \equiv [\{T_{31}(E,\eta_g)\}^3 - 3T_{31}(E,\eta_g)\{T_{32}(E,\eta_g)\}^2],$$

$$T_{34}(E,\eta_g) \equiv [3T_{32}(E,\eta_g)\{T_{31}(E,\eta_g)\}^2 - \{T_{32}(E,\eta_g)\}^3],$$

$$\beta_{11}(E,\eta_g) \equiv \frac{1}{2}\left[\sqrt{\{T_{33}(E,\eta_g)\}^2 + \{T_{34}(E,\eta_g)\}^2} - T_{33}(E,\eta_g)\right]$$

and

$$\vartheta_{21}(E,\eta_g) \equiv \tan^{-1}\left[\frac{\{\beta_{11}(E,\eta_g)\}'}{\{\alpha_{11}(E,\eta_g)\}'}\sqrt{\frac{\alpha_{11}(E,\eta_g)}{\beta_{11}(E,\eta_g)}}\right]$$

Thus, the oscillatory DOS function becomes negative for $\vartheta_{21}(E, \eta_g) \geq \pi$ and a new forbidden zone will appear in addition to the normal band gap.

The electron concentration can be expressed as

$$n_0 = \text{Real part of } \bar{n}_0 \tag{8.43}$$

where

$$n_0 = \frac{g_v}{3\pi^2}\left(\frac{2m_c}{\hbar^2}\right)^{3/2} [\bar{I}_{111e1}(E_{FB}, \eta_g)] \tag{8.44}$$

in which

$$\bar{I}_{111e1}(E_{FB}, \eta_g) = \{T_{31}(E_{FB}, \eta_g) + iT_{31}(E_{FB}, \eta_g)\}^{3/2}$$

Under the said approximations, the (8.34) gets simplified as

$$T_{90}(E, \eta_g) = \frac{\hbar^2 k_s^2}{2m_c} + S_i \left[\frac{\hbar|e|F_s[T_{90}(E, \eta_g)]'}{\sqrt{2m_c}}\right]^{2/3} \tag{8.45}$$

where $T_{90}(E, \eta_g) = T_{31}(E, \eta_g) + iT_{32}(E, \eta_g)$.

Equation (8.45) represents the DR of the 2D electrons in accumulation layers of HD MOSFETs of III–V, ternary and quaternary materials under the weak electric field limit whose bulk electrons obey the HD three-band model of Kane. Since the electron energy spectrum in accordance with the HD three-band model of Kane is complex in nature, the (8.45) will also be complex. Both complexities occur due to the presence of poles in the finite complex plane of the dispersion relation of the materials in the absence of band tails.

The EEM can be expressed as

$$m^*(E'_f, i, \eta_g) = m_c \,\text{Real part of } P'_{3HD}(E'_f, i, \eta_g) \tag{8.46}$$

where

$$P_{3HD}(E'_f, i, \eta_g) = \left[T_{90}(E'_f, i, \eta_g) - S_i \left[\frac{\hbar|e|F_s[T_{90}(E'_f, i, \eta_g)]'}{\sqrt{2m_c}}\right]^{2/3}\right]$$

Thus, one can observe that the EEM is a function of the sub-band index, surface electric field, the Fermi energy and the other spectrum constants due to the combined influence of E_g and Δ.

The sub-band energy E_{i1} is given by

$$0 = \text{Real part of } [T_{90}(E_{i1}, \eta_g) - S_i[\hbar|e|F_s[T_{90}(E_{i1}, \eta_g)]' \cdot (2m_c)^{-1/2}]^{2/3}] \tag{8.47}$$

The DOS function can be written as

$$N_{2D_i}(E) = \frac{m_c g_v}{\pi\hbar^2} \sum_{i=0}^{i_{\max}} [P_{3HD}(E, i, \eta_g) H(E - E_{i1})] \tag{8.48}$$

Thus, the DOS function is complex in nature.

The total surface electron concentration is given by

$$n_s = \text{Real Part of } \bar{n}_s \tag{8.49a}$$

where

$$\bar{n}_s = g_v \sum_{i=0}^{i_{\max}} \left[\left[\frac{m_c}{\pi\hbar^2} P_{3HD}(E_f', i, \eta_g) \right] + \frac{t_{i2}}{3\pi^2} \left(\frac{2m_c}{\hbar^2} \right)^{3/2} [\bar{I}_{111e1}(E_{FB}, \eta_g)] \right],$$

$$t_{i2} = \frac{E_{i2\max}}{eF_s(1 + i_{\max})},$$

$E_{i2\max}$ is the root of the real part of the equation.

$$\begin{aligned} 0 = \; & \text{Real part of } [T_{90}(E_{i2\max}, \eta_g) \\ & - S_i[\hbar|e|F_s[T_{90}(E_{i2\max}, \eta_g)]' \cdot (2m_c)^{-1/2}]^{2/3}] \end{aligned} \tag{8.49b}$$

Thus, combining the appropriate equations with (1.4) we can study the QC in this case.

(b) Two-band model of Kane

The dispersion relation in HD III–V, ternary and quaternary materials whose energy spectrum in the absence of band tails obeys the two-band model of Kane as defined by (1.33a), can be written as

$$\frac{\hbar^2 k^2}{2m_c} = \gamma_2(E, \eta_g) \tag{8.50}$$

where

$$\gamma_2(E, \eta_g) \equiv \left[\frac{2}{1 + Erf(E/\eta_g)} \right] [\gamma_0(E, \eta_g) + \alpha\theta_0(E, \eta_g)]$$

The EEM in this case can be written as

$$m^*(E, \eta_g) = m_c\{\gamma_2(E, \eta_g)\}' \tag{8.51}$$

Thus, one again observes that the EEM in this case exists in the band gap.

In the absence of band tails, $\eta_g \to 0$ and the EEM assumes the well-known form

$$m^*(E_F) = m_c\{1 + 2\alpha E\}|_{E=E_F} \tag{8.52}$$

The DOS function in this case can be written as

$$N_{HD}(E, \eta_g) = \frac{g_v}{2\pi^2}\left(\frac{2m_c}{\hbar^2}\right)^{3/2} \sqrt{\gamma_2(E, \eta_g)}\{\gamma_2(E, \eta_g)\}' \tag{8.53}$$

Since, the poles of the original two-band Kane model are at infinity and no finite poles with respect to energy, therefore the HD counterpart will be totally real and the complex band vanishes.

The electron concentration is given by

$$n_0 = \frac{g_v}{3\pi^2}\left(\frac{2m_c}{\hbar^2}\right)^{3/2} [\bar{I}_{114}(E_{FB}, \eta_g)] \tag{8.54}$$

where $\bar{I}_{114}(E_{FB}, \eta_g) = \{\gamma_2(E_{FB}, \eta_g)\}^{3/2}$.

Equation (8.45) under the low electric field limit for two-band model of Kane assumes the form

$$\gamma_2(E, \eta_g) = \frac{\hbar^2 k_s^2}{2m_c} + S_i\left[\frac{\hbar|e|F_s[\gamma_2(E, \eta_g)]'}{\sqrt{2m_c}}\right]^{2/3} \tag{8.55}$$

Equation (8.45) represents the dispersion relation of the 2D electrons in accumulation layers of HD MOSFETs of III–V, ternary and quaternary materials under the weak electric field limit whose bulk electrons obey the HD two-band model of Kane.

The EEM can be expressed as

$$m^*(E'_f, i, \eta_g) = m_c P'_{3HD1}(E'_f, i, \eta_g) \tag{8.56}$$

where

$$P_{3HD1}(E'_f, i, \eta_g) = \left[\gamma_2(E'_f, \eta_g) - S_i\left[\frac{\hbar|e|F_s[\gamma_2(E'_f, \eta_g)]'}{\sqrt{2m_c}}\right]^{2/3}\right]$$

Thus, one can observe that the EEM is a function of the sub-band index, surface electric field, the Fermi energy and the other spectrum constants due to the combined influence of E_g and Δ.

The sub-band energy E_{i2} is given by

$$0 = [\gamma_2(E_{i2}, \eta_g) - S_i[\hbar|e|F_s[\gamma_2(E_{i2}, \eta_g)]' \cdot (2m_c)^{-1/2}]^{2/3}] \tag{8.57}$$

The DOS function can be written as

$$N_{2D_i}(E) = \frac{m_c g_v}{\pi\hbar^2}\sum_{i=0}^{i_{\max}}[P_{3HD1}(E,i,\eta_g)H(E-E_{i2})] \tag{8.58}$$

The surface electron concentration is given by

$$n_S = g_v\sum_{i=0}^{i_{\max}}\left[\left[\frac{m_c}{\pi\hbar^2}P_{3HD1}(E'_f,i,\eta_g)\right]+\frac{t_{i3}}{3\pi^2}\left(\frac{2m_c}{\hbar^2}\right)^{3/2}[\bar{I}_{114}(E_{FB},\eta_g)]\right]$$

$$t_{i3} = \frac{E_{i3\max}}{eF_s(1+i_{\max})}, \tag{8.59a}$$

$E_{i3\max}$, can be determine from the equation.

$$0 = [\gamma_2(E_{i3\max},\eta_g) - S_i[\hbar|e|F_s[\gamma_2(E_{i3\max},\eta_g)]'\cdot(2m_c)^{-1/2}]^{2/3}] \tag{8.59b}$$

Thus, combining the appropriate equations with (1.4) we can study the QC in this case.

(c) Parabolic energy bands

The dispersion relation in HDS whose energy spectrum in the absence of band tails obeys the parabolic energy bands is given by

$$\frac{\hbar^2k^2}{2m_c} = \gamma_3(E,\eta_g) \tag{8.60}$$

where

$$\gamma_3(E,\eta_g) \equiv \left[\frac{2}{(1+Erf(E/\eta_g))}\right]\gamma_0(E,\eta_g)$$

Since the dispersion relation in accordance with the said model is all-zero function with no pole in the finite complex plane, therefore the HD counterpart will be totally real, which is also apparent form the expression (8.60).

The EEM in this case can be written as

$$m^*(E,\eta_g) = m_c\{\gamma_3(E,\eta_g)\}' \tag{8.61}$$

In the absence of band tails, $n_g \to 0$ and the EEM assumes the form

$$m^*(E) = m_c \tag{8.62}$$

It is well-known that the EEM in unperturbed parabolic energy bands is a constant quantity in general excluding cross-fields configuration. However, the same mass in the corresponding HD bulk counterpart becomes a complicated function of Fermi energy and the impurity potential together with the fact that the EEM also exists in the band gap solely due to the presence of finite η_g.

The DOS function in this case can be written as

$$N_{HD}(E,\eta_g) = \frac{g_v}{2\pi^2}\left(\frac{2m_c}{\hbar^2}\right)^{3/2}\sqrt{\gamma_3(E,\eta_g)}\{\gamma_3(E,\eta_g)\}' \tag{8.63}$$

The electron concentration is given by

$$n_0 = \frac{g_v}{3\pi^2}\left(\frac{2m_c}{\hbar^2}\right)^{3/2}[\bar{I}_{115}(E_{FB},\eta_g)] \tag{8.64}$$

where $\bar{I}_{115}(E_{F_h},\eta_g) = \{\gamma_3(E_{FB},\eta_g)\}^{3/2}$.

Equation (8.55) under the low electric field limit in this case assumes the form

$$\gamma_3(E,\eta_g) = \frac{\hbar^2 k_s^2}{2m_c} + S_i\left[\frac{\hbar|e|F_s[\gamma_3(E,\eta_g)]'}{\sqrt{2m_c}}\right]^{2/3} \tag{8.65}$$

(8.65) represents the dispersion relation of the 2D electrons in accumulation layers of HD MOSFETs of III–V, ternary and quaternary materials under the weak electric field limit whose bulk electrons obey the HD parabolic band model.

The EEM can be expressed as

$$m^*(E'_f,i,\eta_g) = m_c P'_{3HD2}(E'_f,i,\eta_g) \tag{8.66}$$

where

$$P_{3HD2}(E'_f,i,\eta_g) = \left[\gamma_3(E'_f,\eta_g) - S_i\left[\frac{\hbar|e|F_s[\gamma_3(E'_f,\eta_g)]'}{\sqrt{2m_c}}\right]^{2/3}\right]$$

Thus, one can observe that the EM is a function of the sub-band index, surface electric field, the Fermi energy and the other spectrum constants due to the combined influence of E_g and Δ.

The sub-band energy E_{i3} is given by

$$0 = [\gamma_3(E_{i3},\eta_g) - S_i[\hbar|e|F_s[\gamma_3(E_{i3},\eta_g)]' \cdot (2m_c)^{-1/2}]^{2/3}] \tag{8.67}$$

The DOS function can be written as

$$N_{2D_i}(E) = \frac{m_c g_v}{\pi\hbar^2}\sum_{i=0}^{i_{\max}}[P_{3HD2}(E,i,\eta_g)H(E-E_{i3})] \tag{8.68}$$

The total surface electron concentration is given by

$$n_s = g_v \sum_{i=0}^{i_{\max}} \left[\left[\frac{m_c}{\pi \hbar^2} P_{3HD2}(E'_f, i, \eta_g) \right] + \frac{t_{i4}}{3\pi^2} \left(\frac{2m_c}{\hbar^2} \right)^{3/2} [\bar{I}_{115}(E_{FB}, \eta_g)] \right]$$

$$t_{i4} = \frac{E_{i4\max}}{eF_s(1 + i_{\max})}, \tag{8.69a}$$

$E_{i4\max}$ can be determine from the equation.

$$0 = [\gamma_3(E_{i4\max}, \eta_g) - S_i[\hbar|e|F_s[\gamma_3(E_{i4\max}, \eta_g)]' \cdot (2m_c)^{-1/2}]^{2/3}] \tag{8.69b}$$

Thus, combining the appropriate equations with (1.4) we can study the QC in this case.

8.2.3 *The QC in 2D QMOSFETs of II–VI materials under accumulation mode of operation*

Using (2.1) and (8.2), in this case we can write

$$\gamma_3(E, \eta_g) = \frac{\hbar^2 k_z^2}{2m_\parallel^*} + a_0' k_s^2 \pm \bar{\lambda}_0 k_s \tag{8.70}$$

where

$$a_0' = \frac{\hbar^2}{2m_\parallel^*} \quad \text{and} \quad C_0 = \bar{\lambda}_0$$

Thus, the energy spectrum in this case is real since the corresponding E–k relation in the absence of band tails as given by (8.70) is a no-pole function in the finite complex plane.

The transverse and the longitudinal EEMs masses are, respectively, given by

$$m_\perp^*(E, \eta_g) = m_\perp^* \{\gamma_3(E, \eta_g)\}' \left[1 + \left(\frac{\bar{\lambda}_0}{\sqrt{(\bar{\lambda}_0)^2 + 4a_0' \gamma_3(E, \eta_g)}} \right) \right] \tag{8.71}$$

and

$$m_\parallel^*(E, \eta_g) = m_\parallel^* \{\gamma_3(E, \eta_g)\}' \Big|_{E - E_{Fn}} \tag{8.72}$$

Thus, the transverse EEM in HD II–VI semiconductors is a function of electron energy and is double valued due to the presence of $\bar{\lambda}_0$ and due to heavy doping the same mass exists in the band gap.

In the absence of band tails, $\eta_g \to 0$, we get

$$m_\perp^*(E_F) = m_\perp^*\left[1 + \left(\frac{\bar{\lambda}_0}{\sqrt{(\bar{\lambda}_0)^2 + 4a_0'E)}}\right)\right] \tag{8.73}$$

and

$$m_\parallel^*(E_F) = m_\parallel^* \tag{8.74}$$

The volume in k-space as enclosed in (8.70) can be expressed as

$$V(E,\eta_g) = \frac{4\pi}{3a_0'\sqrt{b_0'}}\left[\{\gamma_3(E,\eta_g)\}^{3/2} + \frac{3}{8}\frac{(\bar{\lambda}_0)^2\sqrt{\gamma_3(E,\eta_g)}}{a_0'}\right.$$
$$\left.\pm\left(\frac{3}{4}\frac{\bar{\lambda}_0}{\sqrt{a_0'}}\right)\left(\gamma_3(E,\eta_g) + \frac{(\bar{\lambda}_0)^2}{4a_0'}\right)\sin^{-1}\left[\frac{\sqrt{\gamma_3(E,\eta_g)}}{\sqrt{\gamma_3(E,\eta_g) + \frac{(\bar{\lambda}_0)^2}{4d_0}}}\right]\right] \tag{8.75}$$

Therefore, the electron concentration can be written as

$$n_0 = \frac{g_v}{3\pi^2 a_0'\sqrt{b_0'}}[\bar{I}_{127}(E_{FB},\eta_g)] \tag{8.76}$$

where

$$\bar{I}_{127}(E_{FB},\eta_g) = \left[\{\gamma_3(E_{F_h},\eta_g)\}^{3/2} + \frac{3}{8}\frac{(\bar{\lambda}_0)^2\sqrt{\gamma_3(E,\eta_g)}}{a_0'}\right]$$

The use of (1.7) and (8.70) leads to the expression of the quantization integral as

$$\frac{\sqrt{2m_\parallel^*}}{\hbar}\int_0^{z_t}[\gamma_3(E,\eta_g) - |e|F_s z\gamma_3'(E,\eta_g) - a_0'k_s^2 \mp (\bar{\lambda}_0)k_s] \tag{8.77}$$

where $z_t \equiv (|e|F_s\gamma_3'(E,\eta_g))^{-1}[\gamma_3(E,\eta_g) - a_0'k_s^2 \mp (\bar{\lambda}_0)k_s]$.

Therefore, the 2D electron dispersion law in QMOSFETs of II–VI materials operated under accumulation mode can be expressed as

$$\gamma_3(E,\eta_g) = a_0'k_s^2 \pm (\bar{\lambda}_0)k_s + S_i\left[\frac{\hbar|e|F_s\gamma_3'(E,\eta_g)}{\sqrt{2m_\parallel^*}}\right]^{2/3} \tag{8.78}$$

The area of the 2D surface as enclosed by (8.78) can be expressed as

$$A(E,\eta_g,i) = \frac{\pi}{a_0'^2}\Delta_{10}(E,\eta_g,i) \tag{8.79}$$

where

$$\Delta_{10}(E,\eta_g,i) = \left[(\bar{\lambda}_0)^2 - 2a_0'\left[-\gamma_3(E,\eta_g) + S_i\left(\frac{\hbar|e|F_s\gamma_3'(E,\eta_g)}{\sqrt{2m_\parallel^*}}\right)^{2/3}\right]\right]$$

The EEM in this case assumed the form

$$m^*(E_f',\eta_g,i) = m_\perp^*\Delta_{10}(E_f',\eta_g,i) \tag{8.80}$$

The sub-band energy E_{i5} can be written as

$$\gamma_3(E_{i5},\eta_g) = S_i\left(\frac{\hbar|e|F_s\gamma_3'(E_{i5},\eta_g)}{\sqrt{2m_\parallel^*}}\right)^{2/3} \tag{8.81}$$

The total surface electron concentration can be written as

$$n_s = g_v\sum_{i=0}^{i_{\max}}\left[\left(\frac{m_\perp^*}{\pi\hbar^2}\right)[\Delta_{10}(E_f',i,\eta_g)] + \frac{t_{i5}}{3\pi^2 a_0'\sqrt{b_0'}}[\bar{I}_{127}(E_{FB},\eta_g)]\right]$$

$$t_{i5} = \frac{E_{i5\max}}{eF_s(1+i_{\max})}, \tag{8.82a}$$

$E_{i5\max}$ can be determined from the equation

$$\gamma_3(E_{i5\max},\eta_g) = S_i\left(\frac{\hbar|e|F_s\gamma_3'(E_{i5\max},\eta_g)}{\sqrt{2m_\parallel^*}}\right)^{2/3} \tag{8.82b}$$

Thus, combining the appropriate equations with (1.4) we can study the QC in this case.

8.2.4 *The QC in 2D QMOSFETs of IV–VI materials under accumulation mode of operation*

The (R1.3) can be rewritten as

$$\omega_1(E)k_s^2 + \omega_2(E)k_z^2 = 1 \tag{8.83}$$

where

$$\omega_1(E) = (2E)^{-1}\left[\frac{(\bar{R})^2}{E_{g_0}(1+\alpha_1 E)} + \frac{(\bar{S})^2}{\Delta_c'(1+\alpha_3 E)} + \frac{(\bar{Q})^2}{\Delta_c''(1+\alpha_3 E)}\right] \quad \text{and}$$

$$\omega_2(E) = (2E)^{-1}\left[\frac{(\bar{A})^2}{E_{g0}(1+\alpha_1 E)} + \frac{(\bar{S}+\bar{Q})^2}{\Delta_c''(1+\alpha_3 E)}\right],$$

$$(\overline{R})^2 = 2.3 \times 10^{-10}(eVm)^2$$

$$(\bar{S})^2 = 4.6(\bar{R})^2, \quad \alpha_1 = \frac{1}{E_{g0}}, \quad \alpha_2 = \frac{1}{\Delta_c'}, \quad \alpha_3 = \frac{1}{\Delta_c''},$$

$$\Delta_c' = 3.07eV, \quad \Delta_c'' = 3.28eV \quad \text{and} \quad (\bar{A})^2 = 0.166 \times 10^{-19}(eVm)^2,$$

Using (8.83) and (8.2) in this case we can write

$$\begin{aligned} 2I(4) = k_s^2 \Bigg[& \{c_1(\alpha_1 E, E_g) - iD_1(\alpha_1, E, E_g)\}\frac{(\bar{R})^2}{E_{g_0}} + \{c_2(\alpha_2 E, E_g) \\ & -iD_2(\alpha_2, E, E_g)\}\frac{(\bar{S})^2}{\Delta_c'} + \{c_3(\alpha_3 E, E_g) - iD_3(\alpha_3 E, E_g)\}\frac{(\bar{Q})^2}{\Delta_c''}\Bigg] \\ & + k_z^2\Bigg[\frac{2(\bar{A})^2}{E_{g_0}}\{c_1(\alpha_1 E, E_g) - iD_1(\alpha_1 E, E_g)\} \\ & + \frac{(\bar{S}+\bar{Q})^2}{\Delta_c''}\{c_3(\alpha_3 E, E_g) - iD_3(\alpha_3 E, E_g)\}\Bigg] \end{aligned} \tag{8.84}$$

where

$$G_t = \frac{1+\alpha_i E}{\eta_g \alpha_i}, c_i(\alpha_i E, \eta_g) = \left[\frac{2}{\alpha_i \eta_g \sqrt{\pi}}\right]\exp(-u_i^2)$$
$$\times \left[\sum_{p=1}^{\infty}\{\exp(-p^2/4)(\sinh(pu_i))\}p^{-1}\right],$$
$$i = 1, 2, 3, 4, \ldots \quad \text{and} \quad D_i(\alpha_i, E, \eta_g) = \left[\frac{\sqrt{\pi}}{\alpha_i \eta_g}\right]\exp(-u_i^2)$$

Therefore, (8.84) can be written as,

$$\bar{F}_1(E, \eta_g)k_s^2 + \bar{F}_2(E, \eta_g)k_z^2 = 1 \tag{8.85}$$

where

$$\begin{aligned} \bar{F}_1(E, \eta_g) = [2\gamma_0(E, \eta_g)]^{-1}\Bigg[& \frac{(\bar{R})^2}{E_g}\{C_1(\alpha_1 E, E_g) - iD_1(\alpha_1, E, E_g)\} \\ & + \frac{(\bar{S})^2}{\Delta_c'}\{C_2(\alpha_2, E, E_g) - iD_2(\alpha_2, E, E_g)\} \\ & + \frac{(\bar{Q})^2}{\Delta_c''}\{C_3(\alpha_3, E, E_g) - iD_3(\alpha_3, E, E_g)\}\Bigg] \end{aligned}$$

and

$$\bar{F}_2(E,\eta_g) = [2\gamma_0(E,\eta_g)]^{-1}\left[\frac{2(\bar{A})^2}{E_g}\{C_1(\alpha_1,E,\eta_g) - iD_1(\alpha_1,E,\eta_g)\} + \frac{(\bar{S}+\bar{Q})^2}{\Delta_c''}\{C_3(\alpha_3,E,\eta_g) - iD_3(\alpha_3,E,\eta_g)\}\right]$$

Since $\bar{F}_1(E,\eta_g)$ and $\bar{F}_2(E,\eta_g)$ are complex, the energy spectrum is also complex in the presence of Gaussian band tails.

The EEMs can be written as

$$m_{\perp}^{*}(E,\eta_g) = \left(\frac{\hbar^2}{2}\right) \text{ Real part of } \left[\frac{\{\bar{F}_1(E,\eta_g)\}'}{\{\bar{F}_1(E,\eta_g)\}^2}\right] \tag{8.86}$$

$$m_{\parallel}^{*}(E,\eta_g) = \left(\frac{\hbar^2}{2}\right) \text{ Real part of } \left(\frac{\{\bar{F}_2(E,\eta_g)\}'}{\{\bar{F}_2(E,\eta_g)\}^2}\right) \tag{8.87}$$

It appears then that, the evolution of the masses needs an expression of the carrier concentration, which in turn is determined by the DOS function.

The DOS function in this case can be expressed as

$$N_{HD}(\mathrm{E},\eta_\gamma) = \frac{g_v}{3\pi^2}F_3'(\mathrm{E},\eta_g), \quad F_3(\mathrm{E},\eta_g) = \left[F_1(\mathrm{E},\eta_g)\sqrt{F_2(\mathrm{E},\eta_g)}\right]^{-1} \tag{8.88}$$

The electron concentration is given by

$$n_0 = \text{ Real part of } \bar{n}_0 \tag{8.89}$$

where

$$\bar{n}_0 - \frac{g_v}{3\pi^2}[F_3(E_{FB},\eta_g)]. \tag{8.90}$$

The 2D electron DR in accumulation layers of IV–VI semiconductors can be written as

$$\theta_1(E,i,\eta_g)k_x^2 + \theta_2(E,i,\eta_g)k_y^2 = \theta_3(E,i,\eta_g) \tag{8.91}$$

where $\theta_1(E,i,\eta_g) = [F_1(E,\eta_g)S_i(eF_s a_1(E,\eta_g))^{2/3}F_2(E,\eta_g)]$

$$a_1(E,\eta_g) = \frac{1}{F_2(E,\eta_g)}\left[\frac{F_2'(E,\eta_g)}{F_2(E,\eta_g)}F_1(E,\eta_g) - F_1'(E,\eta_g)\right]$$

$$\theta_2(E,i,\eta_g) = \left[\left[F_1(E,\eta_g) + \frac{2a_2(E,\eta_g)}{3a_1(E,\eta_g)}(eF_s a_1(E,\eta_g))^{2/3}S_i F_1(E,\eta_g)\right]\right]$$

$$a_2(E,\eta_g) = \frac{1}{F_2(E,\eta_g)}\left[\frac{F_2'(E,\eta_g)}{F_2(E,\eta_g)}F_1(E,\eta_g) - F_1'(E,\eta_g)\right]$$

$$\theta_3(E,i,\eta_g) = \left[1 + \frac{2C(E,\eta_g)}{3a_1(E,\eta_g)} S_i(eF_s a_1(E,\eta_g))^{2/3} F_2(E,\eta_g)\right] \quad \text{and}$$

$$C(E,\eta_g) = \left[\frac{F_2'(E,\eta_g)}{F_2^2(E,\eta_g)}\right]$$

The EEM can be expressed as

$$m^*(E_f',i,\eta_g) = \frac{\hbar^2}{2}\theta_4'(E_f',i,\eta_g) \tag{8.92}$$

where

$$\theta_4(E_f',i,\eta_g) = \frac{\theta_3(E_f',i,\eta_g)}{\sqrt{\theta_1(E_f',i,\eta_g)\theta_2(E_f',i,\eta_g)}}$$

The sub-band energy E_{i8} is given by

$$\theta_3(E_{i8},i,\eta_g) = 0 \tag{8.93}$$

The 2D electron concentration in accumulation layer of MOSFETs of IV–VI materials under the condition of extreme degeneracy and weak electric field limit can be written as

$$n_s = \text{ Real Part of } \bar{n}_s \tag{8.94a}$$

$$\bar{n}_S = g_v \sum_{i=0}^{i_{\max}} \left[\theta_4(E_f',i,\eta_g) + \frac{t_{i6}}{3\pi^2}[F_3(E_{FB},\eta_g)]\right]$$

$$t_{i6} = \frac{E_{i6\max}}{eF_s(1+i_{\max})},$$

$E_{i6\max}$, can be determined from the real part of the equation.

Thus, combining the appropriate equations with (1.4) we can study the QC in this case.

8.2.5 *The QC in 2D QMOSFETs of stressed Kane type materials under accumulation mode of operation*

Equation (2.22) can be re-written as

$$Ek^2 - T_{17}k_x^2 - T_{27}k_y^2 - T_{37}k_z^2 = [q_{67}E^3 - R_{67}E^2 + V_{67}E + \rho_{67}] \tag{8.95}$$

where $T_{17} = \alpha_1$, $T_{27} = \alpha_2$, $T_{37} = \alpha_3$, $t_1 = q_{67}$, $t_2 = R_{67}$, $t_3 = V_{67}$ and $t_4 = \rho_{67}$.

Under the condition of heavy doping, using (8.2) and (8.95) we get

$$\begin{aligned} I(4)k^2 - T_{17}I(1)k_x^2 - T_{27}I(1)k_y^2 - T_{37}k_z^2 I(1) \\ = [q_{67}I(6) - R_{67}I(5) + V_{67}I(4) + \rho_{67}I(1)] \end{aligned} \tag{8.96}$$

where

$$I(6)=\int_{-\infty}^{E}(E-V)^3F(V)dV. \tag{8.97}$$

Equation (8.97) can be written as

$$I(6)=E^3I(1)-3E^2I(7)+3EI(8)-I(9) \tag{8.98}$$

In which,

$$I(7)=\int_{-\infty}^{E}VF(V)dV \tag{8.99}$$

$$I(8)=\int_{-\infty}^{E}V^2F(V)dV \tag{8.100}$$

$$I(9)=\int_{-\infty}^{E}V^3F(V)dV \tag{8.101}$$

Using (8.2), together with simple algebraic manipulations, one obtains

$$I(7)=\frac{-\eta_g}{2\sqrt{\pi}}\exp\left(\frac{-E^2}{\eta_g^2}\right) \tag{8.102}$$

$$I(8)=\frac{\eta_g^2}{4}\left[1+Erf\left(\frac{E}{\eta_g}\right)\right] \tag{8.103}$$

and

$$I(9)=\frac{-\eta_g^3}{2\sqrt{\pi}}\exp\left(\frac{-E^2}{\eta_g^2}\right)\left[1+\frac{E^2}{\eta_g^2}\right] \tag{8.104}$$

Thus, (8.98) can be written as

$$I(6)=\left[\frac{E}{2}\left[1+Erf\left(\frac{E}{\eta_g}\right)\right]\left[E^2+\frac{3}{2}\eta_g^2\right]+\frac{\eta_g}{2\sqrt{\pi}}\exp\left(\frac{-E^2}{\eta_g^2}\right)[4E^2+\eta_g^2]\right] \tag{8.105}$$

Thus, combining the appropriate equations, the DRs of the conduction electrons in HD stressed materials can be expressed as

$$P_{11}(E,\eta_g)k_x^2+Q_{11}(E,\eta_g)k_y^2+S_{11}(E,\eta_g)k_z^2=1 \tag{8.106}$$

where

$$P_{11}(E,\eta_g)\equiv\left[\frac{\gamma_0(E,\eta_g)-(T_{17}/2)[1+Erf(E/\eta_g)]}{\Delta_{14}(E,\eta_g)}\right]$$

$$\Delta_{14}(E,\eta_g)\equiv\left[q_{67}\left\{\frac{E}{2}[1+Erf(E/\eta_g)]\left[E^2+\frac{3}{2}\eta_g^2\right]\right.\right.$$

$$+\frac{\eta_g}{2\sqrt{\pi}}\exp(-E^2/\eta_g^2)[4E^2+\eta_g^2]\Big\},$$

$$-R_{67}\theta_0(E,\eta_g)+V_{67}\gamma_0(E,\eta_g)+\frac{\rho_{67}}{2}[1+Erf(E/\eta_g)]\Bigg],$$

$$Q_{11}(E,\eta_g)\equiv\left[\frac{\gamma_0(E,\eta_g)-(T_{27}/2)[1+Erf(E/\eta_g)]}{\Delta_{14}(E,\eta_g)}\right]$$

and

$$S_{11}(E,\eta_g)\equiv\left[\frac{\gamma_0(E,\eta_g)-(T_{37}/2)[1+Erf(E/\eta_g)]}{\Delta_{14}(E,\eta_g)}\right]$$

Thus, the energy spectrum in this case is real since the dispersion relation of the corresponding materials in the absence of band tails as given by (8.106) has no poles in the finite complex plane.

The EEMs along x, y and z-directions in this case can be written as

$$\begin{aligned}m_{xx}^*(E_{F_h},\eta_g)&=\frac{\hbar^2}{2}\Bigg[[\gamma_0(E_{FB},\eta_g)-(T_{17}/2)[1+Erf(E_{FB},\eta_g)]]^{-2}\\&\times[\{\Delta_{14}(E_{FB},\eta_g)\}'[\gamma_0(E_{FB},\eta_g)-(T_{27}/2)[1+Erf(E_{FB},\eta_g)]]]\\&-\Delta_{14}(E_{FB},\eta_g)\left[\frac{1}{2}[1+Erf(E_{FB}/\eta_g)]-\left\{\frac{T_{17}}{\eta_g\sqrt{\pi}}\exp(E_{F_h}^2/\eta_g^2)\right\}\right]\Bigg]\end{aligned} \tag{8.107}$$

$$\begin{aligned}m_{yy}^*(E_{FB},\eta_g)&=\frac{\hbar^2}{2}\Bigg[[\gamma_0(E_{FB},\eta_g)-(T_{27}/2)[1+Erf(E_{FB},\eta_g)]]^{-2}\\&\times[\Delta_{14}(E_{FB},\eta_g)\}'[\gamma_0(E_{FB},\eta_g)-(T_{27}/2)[1+Erf(E_{FB},\eta_g)]]]\\&-\Delta_{14}(E_{FB},\eta_g)\left[\frac{1}{2}[1+Erf(E_{FB},\eta_g)]-\left\{\frac{T_{27}}{\eta_g\sqrt{\pi}}\exp(E_{FB}^2/\eta_g^2)\right\}\right]\Bigg]\end{aligned} \tag{8.108}$$

$$\begin{aligned}m_{zz}^*(E_{FB},\eta_g)&=\frac{\hbar^2}{2}\Bigg[[\gamma_0(E_{FB},\eta_g)-(T_{37}/2)[1+Erf(E_{FB},\eta_g)]]^{-2}\\&\times[\Delta_{14}(E_{FB},\eta_g)\}'[\gamma_0(E_{FB},\eta_g)-(T_{37}/2)[1+Erf(E_{FB},\eta_g)]]]\\&-\Delta_{14}(E_{FB},\eta_g)\left[\frac{1}{2}[1+Erf(E_{FB}/\eta_g)]-\left\{\frac{T_{37}}{\eta_g\sqrt{\pi}}\exp(E_{FB}^2/\eta_g^2)\right\}\right]\Bigg]\end{aligned} \tag{8.109}$$

Thus, we can see that the EEMs in this case exist within the band gap.

The DOS function in this case can be written as

$$N_{HD}(E,\eta_g) = \frac{g_v}{3\pi^2}\Delta_{100}(E,\eta_g) \tag{8.110}$$

where

$$\Delta_{100}(E,\eta_g)\{\Delta_{15}(E,\eta_g)\}^{-2}\left[\frac{3}{2}\{\Delta_{15}(E,\eta_g)\}\sqrt{\Delta_{14}(E,\eta_g)}\{\Delta_{14}(E,\eta_g)\}'\right.$$
$$\left. - \{\Delta_{14}(E,\eta_g)\}^{3/2}\{\Delta_{15}(E,\eta_g)\}'\right]$$
$$\Delta_{15}(E/\eta_g) \equiv [[\gamma_0(E/\eta_g) - (T_{17}/2)[1 + Erf(E/\eta_g)]][\gamma_0(E/\eta_g)$$
$$-(T_{27}/2)[1 + Erf(E/\eta_g)]]$$
$$[\gamma_0(E,\eta_g) - (T_{37}/2)[1 + Erf(E/\eta_g)]]]^{1/2}$$

Using (8.110), the electron concentration can be written as

$$n_0 = \frac{g_v}{3\pi^2}[\bar{I}_{126}(E_{FB},\eta_g)] \tag{8.111}$$

where

$$\bar{I}_{126}(E_{FB},\eta_g) = \left[\frac{\{\Delta_{14}(E_{FB},\eta_g)^{3/2}}{\Delta_{15}(E_{FB},\eta_g)}\right]$$

The 2D electron DR in accumulation layers of stressed III–V semiconductors can be written as

$$\theta_{13}(E,i,\eta_g)k_x^2 + \theta_{23}(E,i,\eta_g)k_y^2 = \theta_{33}(E,i,\eta_g) \tag{8.112}$$

where $\theta_{13}(E,i,\eta_g) = [P_{11}(E,\eta_g) + S_i(eF_s a_{13}(E,\eta_g))^{2/3}S_{11}(E,\eta_g)]$

$$a_{13}(E,\eta_g) = \frac{1}{S_{11}(E,\eta_g)}\left[\frac{S'_{11}(E,\eta_g)}{S_{11}(E,\eta_g)}P_{11}(E,\eta_g) - P'_{11}(E,\eta_g)\right]$$

$$\theta_{23}(E,i,\eta_g) = \left[\left[Q_{11}(E,\eta_g) + \frac{2a_{23}(E,\eta_g)}{3a_{13}(E,\eta_g)}(eF_s a_{13}(E,\eta_g))^{2/3}S_i Q_{11}(E,\eta_g)\right]\right]$$

$$a_{23}(E,\eta_g) = \frac{1}{S_{11}(E,\eta_g)}\left[\frac{S'_{11}(E,\eta_g)}{S_{11}(E,\eta_g)}Q_{11}(E,\eta_g) - Q'_{11}(E,\eta_g)\right]$$

$$\theta_{33}(E,i,\eta_g) = \left[1 + \frac{2C_3(E,\eta_g)}{3a_{13}(E,\eta_g)}S_i(eF_s a_{13}(E,\eta_g))^{2/3}S_{11}(E,\eta_g)\right],$$

$$C_3(E,\eta_g) = \left[\frac{S'_{11}(E,\eta_g)}{S_{11}^2(E,\eta_g)}\right]$$

and $P_{11}(E,\eta_g)$, $Q_{11}(E,\eta_g)$ and $S_{11}(E,\eta_g)$ are defined in Chapter 2, respectively.

The EEM can be expressed as

$$m^*(E'_f, i, \eta_g) = \frac{\hbar^2}{2}\theta'_{43}(E'_f, i, \eta_g) \tag{8.113}$$

where

$$\theta_{43}(E'_f, i, \eta_g) = \frac{\theta_{33}(E'_f, i, \eta_g)}{\sqrt{\theta_{13}(E'_f, i, \eta_g)\theta_{23}(E'_f, i, \eta_g)}}$$

The sub-band energy E_{i33} is given by

$$\theta_{33}(E_{i33}, i, \eta_g) = 0 \tag{8.114a}$$

The 2D electron concentration in accumulation layers of stressed III–V materials under the condition of extreme degeneracy and low electric field limit can be written as

$$\theta_{33}(E_{i33}, i, \eta_g) = 0 \tag{8.114b}$$

The 2D electron concentration in accumulation layers of stressed III–V materials under the condition of extreme degeneracy and low electric field limit can be written as

$$n_s = \bar{n}_s$$

where

$$\bar{n}_S = g_v \sum_{i=0}^{i_{\max}} \left[\theta_{43}(E'_f, i, \eta_g) + \frac{t_{i7}}{3\pi^2}[\bar{I}_{126}(E_{FB}, \eta_g)]\right]$$

$$t_{i7} = \frac{E_{i7\max}}{eF_s(1 + i_{\max})}, \tag{8.115a}$$

$E_{i7\max}$, can be determined from the real part of the equation

$$\theta_{33}(E_{i7\max}, i_{\max}, \eta_g) = 0 \tag{8.115b}$$

Thus, combining the appropriate equations with (1.4) we can study the QC in this case.

8.3 Summary and Conclusion

From theoretical background, we note that the expression for surface electron concentration of 2D MOSFETs operated under accumulation mode of nonlinear optical, III–V (three- and two-band models of Kane together with parabolic energy bands), II–VI, IV–VI and stressed Kane type materials are given by (8.37), (8.49a), (8.59a), (8.69a), (8.82a), (8.94a) and (8.115a), respectively. By combining each of them with (1.4) we can study the QC in each case.

We have not given the graphs of the QCs for various 2D QMOSFETs made of many technologically important materials as discussed here operated under accumulation layer mode since we feel that our readers will perform the intricate numerical computation to obtain the oscillatory plots for better assessment and joy of deep understanding.

8.4 Open Research Problems

R.8.1 Investigate the QCs in the presence of an arbitrarily oriented electric field in 2DMOSFETs of nonlinear optical materials operated under accumulation mode as stated in this chapter. Draw the corresponding graphs and explain the results.

R.8.2 Investigate the QCs in the presence of an arbitrarily oriented electric field in 2D MOSFETs of IV–VI, II–VI and stressed Kane type compounds operated under accumulation mode as stated in this chapter. Draw the corresponding graphs and explain the results.

R.8.3 Investigate the QCs in the presence of an arbitrarily oriented electric field in 2D MOSFETs operated under accumulation mode of all the materials as stated in R.1.1 of chapter 1. Draw the corresponding graphs and explain the results.

R.8.4 Investigate the QCs in the presence of an arbitrarily oriented electric field in 2D MOSFETs operated under accumulation mode of all the materials as studied in this chapter in the presence of an arbitrarily oriented non-uniform electric field. Draw the corresponding graphs and explain the results.

R.8.5 Investigate the QCs in the presence of an arbitrarily oriented electric field in 2D MOSFETs operated under accumulation mode of IV–VI, II–VI and stressed Kane type compounds in the presence of an arbitrarily oriented non-uniform electric field. Draw the corresponding graphs and explain the results.

R.8.6 Investigate the QCs in the presence of an arbitrarily oriented electric field in 2D MOSFETs operated under accumulation mode of all the materials as stated in R.1.1 of chapter 1 in the presence of an arbitrarily oriented non-uniform electric field. Draw the corresponding graphs and explain the results.

R.8.7 Investigate the QCs in 2D MOSFETs operated under accumulation mode of all the problems from R.2.1 to R.2.6 in the presence of an arbitrarily oriented non-uniform electric field. Draw the corresponding graphs and explain the results

References

1. R. K. Willardson, in *Materials and Semimetals*, Vol. 1 A. C. Beer, (ed.) (Academic Press, New York, 1966) p. 102.
2. E. O. Kane, *Phys. Rev.* **131**, 79 (1963); *Phys. Rev. B.*, **139**, 343 (1965).
3. V. L. Bonch Bruevich, *Sov. Phys. Sol. Stat.* **4**, 53 (1963).
4. R. A. Logan, A. G. Chynoweth, *Phys. Rev.* **131**, 89 (1963).
5. C. J. Hwang, *J. Appl. Phys.* **40**, 3731 (1969).
6. J. I. Pankove, *Phys. Rev. A*, **130**, 2059 (1965).
7. B. I. Halperin, M. Lax, *Phys. Rev.* **148**, 722 (1966).
8. R. A. Abram, G. J. Rees and B. L. H. Wilson, *Adv. Phys.* **27**, 799 (1978).
9. B. I. Shklovskii and A. L. Efros, *Electronic Properties of Doped Materials*, Springer series in Solid State Sciences, vol. 45 (Springer, Berlin, 1984).
10. E. O. Kane, Solid State Electron, **28**, 3 (1985).
11. P. K. Chakraborty and J. C. Biswas, *J. Appl. Phys.* **82**, 3328 (1997).
12. B. R. Nag, *Electron Transport in Compound Materials*, Springer Series in Solid State Sciences, vol. 11, (Springer, Heidelberg, 1980).
13. P. E. Schmid, *Phys. Rev. B* **23**, 5531 (1981).
14. Jr. G. E. Jellison, F. A. Modine, C. W. White, R. F. Wood and R. T. Young, *Phys. Rev. Lett.* **46**, 1414 (1981).
15. V. I. Fistul, *Heavily Doped Materials* (Plenum, New York, 1969), ch 7.
16. C. J. Hwang, *J. Appl. Phys.* **41**, 2668 (1970); W. Sritrakool, H. R. Glyde, V. Sa Yakanit *Can. J. Phys.* **60**, 373 (1982); H. Ikoma, *J. Phys. Soc. Japan*, **27**, 514 (1969).
17. M. H. Chen, C. C. Wu, C. J. Lin, *J. Low Temp. Phys.* **55**, 127 (1984).
18. M. Abramowitz and I. A. Stegun, *Handbook of Mathematical Functions with Formulas, Graphs and Mathematical Tables* (New York: Wiley, 1964).
19. I. S. Gradshteyn and I. M. Ryzhik, *Tables of Integrals, Series and Products* (Academic, New York, 1965).

Chapter 9

The Influence of Magnetic Quantization on the QC in QMOSFETs Operated Under Accumulation Mode

The secret of a great scholar is to know something new that nobody else knows

9.1 Introduction

It appears from the recent literature that the magneto QC in 2D QMOSFETs of important materials operated under accumulation mode has yet to be investigated. In this chapter, we study the magneto QC in 2D QMOSFETs of tetragonal and nonlinear optical materials in Section 9.2.1 of theoretical background 9.2. The magneto QC in 2D QMOSFETs of III–V, ternary and quaternary materials under accumulation mode has been studied in Section 9.2.2. From Sections 9.2.3 up to 9.2.5, we shall investigate the same in 2D QMOSFETs of II–VI, IV–VI, and stressed Kane type materials under accumulation mode, respectively. Section 9.3 contains summary and conclusion and the last Section 9.4 contains few open research problems.

9.2 Theoretical Background

9.2.1 *The QC in magneto 2D QMOSFETs of tetragonal and nonlinear optical materials operated under accumulation mode*

The DR under magnetic quantization in nonlinear optical materials can be written as

$$\frac{\hbar^2 k_z^2}{2m_{\|}^*} = U_1(E, n, \eta_g) \tag{9.1}$$

where

$$U_1(E,n,\eta_g) = \left[[T_{21}(E,\eta_g)]\left[1 - \frac{\hbar eB\left(n+\frac{1}{2}\right)}{m_{\perp}^{*}T_{22}(E,\eta_g)}\right]\right]$$

The EEM at the Fermi Level can be written from (9.1) as

$$m^*(E_{FBHD},n,\eta_g) = m_{||}^{*} \text{ Real part of } U_1'(E_{FBHD},n,\eta_g) \tag{9.2}$$

where E_{FBHD} is the Fermi energy in this case.

Therefore, the EEM in this case is a function of Fermi energy, magnetic field, quantum number and the scattering potential together with the fact that the EEM exists in the band gap which is the general characteristics of HD materials.

The electron concentration can be expressed as

$$n_0 = \text{Real part of } \bar{n}_0 \tag{9.3}$$

where

$$\bar{n}_0 = \frac{eB}{\pi^2\hbar^2}\sqrt{2m_{||}^{*}}\sum_{n=0}^{n_{\max}}[[\theta_1(E_{FBHD},n,\eta_g)] + \sum_{r=1}^{r_s}L(r)[\theta_1(E_{FBHD},n,\eta_g)]] \tag{9.4}$$

in which,

$$\theta_1(E_{FBHD},n,\eta_g) = [U_1(E_{FBHD},n,\eta_g)]^{1/2},$$

$$L(r) \equiv [2(k_BT)^{2r}(1-2^{1-2r})\xi(2r)]\left[\frac{\partial^{2r}}{\partial E_F^{2r}}\right],$$

$\xi(2r)$ is the zeta function of order (2r) and in general E_F is Fermi energy which in this case is $E_F = E_{FBHD}$

The DR of the 2D electrons in QMOSFETs of HD tetragonal and nonlinear optical materials operated in accumulation mode under the condition of weak electric field limit under magnetic quantization can be written as

$$\frac{heB\left(n+\frac{1}{2}\right)}{m_{||}^{*}} = L_6(E_{14,1},i,\eta_g) \tag{9.5}$$

where $E_{14,1}$ is the totally quantized energy in this case.

The DOS function is given by

$$N_{Bi} = \frac{g_v eB}{\pi\hbar} \sum_{n=0}^{n_{\max}} \sum_{i=0}^{n_{i\max}} \delta'(E - E_{14,1}) \tag{9.6}$$

The electron concentration per unit area can be written as

$$n_s = \text{Real Part of } \bar{n}_s \tag{9.7}$$

where

$$\bar{n}_s = \frac{g_v eB}{\pi\hbar} \sum_{n=0}^{n_{\max}} \sum_{i=0}^{n_{i\max}} F_{-1}(\eta_{14,1}) \tag{9.8}$$

in which $\eta_{14,1} = (E_{FHD2D14} - E_{14,1})(k_B T)^{-1}$ and $E_{FHD2D14}$ is the Fermi energy in this case and is given by

$$E_{FHD2D14} \equiv eV_g - \frac{e^2 N_s}{C_{OX}}$$

The total surface concentration in QMOSFETs of tetragonal and nonlinear optical materials in this case can be written as

$$N_s = \text{Real Part of } g_v \left[\sum_{n=0}^{n_{\max}} \sum_{i=0}^{n_{i\max}} \left[\frac{eB}{\pi\hbar}[F_{-1}(\eta_{14,1})] + \frac{eBt_{i1B}}{\pi^2\hbar^2}\sqrt{2m_{||}^*} \times \left[[\theta_1(E_{FBHD}, n, \eta_g)] + \sum_{r=1}^{r_s} L(r)[\theta_1(E_{FBHD}, n, \eta_g)]\right]\right]\right] \tag{9.9}$$

$$t_{i1B} = \frac{E_{14,1\max}}{eF_s(1 + i_{\max})},$$

$E_{14,1\max}$, is the maximum value of the real part of the equation

$$\frac{heB\left(n + \frac{1}{2}\right)}{m_{||}^*} = L_6(E_{14,1\max}, i_{\max}, \eta_g) \tag{9.10}$$

Thus, combining the appropriate equations with (1.4) we can study the QC in this case.

9.2.2 The QC in magneto 2D QMOSFETs of III–V, ternary and quaternary materials operated under accumulation mode

Three-band model of Kane The electron energy spectrum in HD III–V materials under magnetic quantization is given by

$$\left(n+\frac{1}{2}\right)\hbar\omega_0+\frac{\hbar^2k_z^2}{2m_c}=T_{31}(E,\eta_g)+iT_{32}(E,\eta_g) \tag{9.11}$$

where

$$\omega_0=\frac{eB}{m_c}$$

Therefore, the DR is given by

$$\frac{\hbar^2k_z^2}{2m_c}=U_3(E,n,\eta_g) \tag{9.12a}$$

where

$$U_3(E,n,\eta_g)=[T_{31}(E,\eta_g)+iT_{32}(E,\eta_g)-\left(n+\frac{1}{2}\right)\hbar\omega_0]$$

The EEM at the Fermi Level can be written from (9.12a) as

$$m^*(E_{FBHD},n,\eta_g)=m_c\text{Real part of }U_3'(E_{FBHD},n,\eta_g) \tag{9.12b}$$

Thus, the EEM is a function of Fermi energy, Landau quantum number and scattering potential, respectively.

The electron concentration can be expressed as

$$n_0=\text{Real part of }\bar{n}_0 \tag{9.13}$$

where

$$\bar{n}_0=\frac{eBg_v}{\pi^2\hbar^2}\sqrt{2m_c}\left[\sum_{n=0}^{n_{\max}}[[\theta_3(E_{FBHD},n,\eta_g)]+\sum_{r=1}^{r_s}L(r)[\theta_3(E_{FBHD},n,\eta_g)]\right]\right] \tag{9.14}$$

in which $\theta_3(E_{FBHD},n,\eta_g)=[U_3(E_{FBHD},n,\eta_g)]^{1/2}$.

The DR of the 2D electrons in QMOSFETs of HD III–V, ternary and quaternary materials operated in accumulation mode under the condition of weak electric field limit under magnetic quantization can be written as

$$T_{90}(E_{14,3}, \eta_g) = \left(n + \frac{1}{2}\right)\hbar\omega_0 + S_i \left[\frac{\hbar|e|F_s[T_{90}(E_{14,3}, \eta_g)]'}{\sqrt{2m_c}}\right]^{2/3} \tag{9.15}$$

where $E_{14,3}$ is the totally quantized energy in this case.

The DOS function is given by

$$N_{Bi} = \frac{g_v eB}{\pi\hbar} \sum_{n=0}^{n_{\max}} \sum_{i=0}^{n_{i\max}} \delta'(E - E_{14,3}) \tag{9.16}$$

Equation (9.15) represents the DR of the 2D electrons in accumulation layers of HD QMOSFETs of III–V, ternary and quaternary materials under the weak electric field limit under magnetic quantization whose bulk electrons obey the HD three-band model of Kane. Since the electron energy spectrum in accordance with the HD three-band model of Kane is complex in nature, (9.15) will also be complex. Both complexities occur due to the presence of poles in the finite complex plane of the dispersion relation of the materials in the absence of band tails.

The electron concentration per unit area can be written as

$$n_s = \text{Real Part of } \bar{n}_s \tag{9.17}$$

where

$$\bar{n}_s = \frac{g_v eB}{\pi\hbar} \sum_{n=0}^{n_{\max}} \sum_{i=0}^{n_{i\max}} F_{-1}(\eta_{14,3}) \tag{9.18}$$

in which $\eta_{14,3} = (E_{FHD2D14} - E_{14,3})(k_B T)^{-1}$.

The total surface concentration in QMOSFETs in this case can be written as

$$N_s = g_v \text{ Real Part of } \left[\sum_{n=0}^{n_{\max}} \sum_{i=0}^{n_{i\max}} \left[\frac{eB}{\pi\hbar}[F_{-1}(\eta_{14,3})] + \frac{eBt_{i3B}}{\pi^2\hbar^2}\sqrt{2m_c} \right.\right.$$
$$\left.\left. \times \left[[\theta_3(E_{FBHD}, n, \eta_g)] + \sum_{r=1}^{r_s} L(r)[\theta_3(E_{FBHD}, n, \eta_g)]\right]\right]\right] \tag{9.19}$$

$$t_{i3B} = \frac{E_{14,3\max}}{eF_s(1 + i_{\max})},$$

$E_{14,3\max}$, is the maximum value of the real part of the equation.

$$T_{90}(E_{14,3\max}, \eta_g) = \left(n + \frac{1}{2}\right)\hbar\omega_0 + S_{i_{\max}} \left[\frac{\hbar|e|F_s[T_{90}(E_{14,3\max}, \eta_g)]'}{\sqrt{2m_c}}\right]^{2/3} \tag{9.20}$$

Thus, combining the appropriate equations with (1.4) we can study the QC in this case.

(b) Two-band model of Kane

The magneto-dispersion law in this case is given by

$$\frac{\hbar^2 k_z^2}{2m_c} = \gamma_2(E,\eta_g) - \left(n+\frac{1}{2}\right)\hbar\omega_0 = U_5(E,n,\eta_g) \tag{9.21}$$

The EEM at the Fermi Level can be written from (9.21) as

$$m^*(E_{FBHD},\eta_g) = m_c\gamma_2'(E_{FBHD},\eta_g) \tag{9.22}$$

Thus, EEM is independent of quantum number.

The electron concentration can be expressed as

$$n_0 = \frac{eB}{\pi^2\hbar^2}\sqrt{2m_c}\left[\sum_{n=0}^{n_{\max}}[\theta_5(E_{FHD2D14},n,\eta_g)] + \sum_{r=1}^{r_s}L(r)[\theta_5(E_{FBHD},n,\eta_g)]\right] \tag{9.23}$$

where $\theta_5(E_{FHD2D14},n,\eta_g) = [U_5(E_{FHD2D14},n,\eta_g)]^{1/2}$.

The DR of the 2D electrons in QMOSFETs of HD nonlinear optical materials whose energy band structures are defined by two-band model of Kane operated in accumulation mode under the condition of weak electric field limit under magnetic quantization can be written as

$$\gamma_2(E_{14,5},\eta_g) = \left(n+\frac{1}{2}\right)\hbar\omega_0 + S_i\left[\frac{\hbar|e|F_s[\gamma_2(E_{14,5},\eta_g)]'}{\sqrt{2m_c}}\right]^{2/3} \tag{9.24}$$

where $E_{14,5}$ is the totally quantized energy in this case.

The DOS function is given by

$$N_{Bi} = \frac{g_v eB}{\pi\hbar}\sum_{n=0}^{n_{\max}}\sum_{i=0}^{n_{i\max}}\delta'(E-E_{14,5}) \tag{9.25}$$

The electron concentration per unit area can be written as

$$n_s = \bar{n}_s \tag{9.26}$$

where

$$\bar{n}_s = \frac{g_v eB}{\pi\hbar}\sum_{n=0}^{n_{\max}}\sum_{i=0}^{n_{i\max}}F_{-1}(\eta_{14,5}) \tag{9.27}$$

in which $\eta_{14,5} = (E_{FHD2D14} - E_{14,5})(k_BT)^{-1}$.

The total surface concentration in QMOSFETs operated under accumulation layer mode of nonlinear optical material can be written as

$$N_s = g_v \left[\sum_{n=0}^{n_{\max}} \sum_{i=0}^{n_{i\max}} \left[\frac{eB}{\pi\hbar}[F_{-1}(\eta_{14,5})] + \frac{eBt_{i5B}}{\pi^2\hbar^2}\sqrt{2m_c} \left[[\theta_5(E_{FBHD}, n, \eta_g)] + \sum_{r=1}^{r_s} L(r)[\theta_5(E_{FBHD}, n, \eta_g)] \right] \right] \right] \tag{9.28}$$

$$t_{i5B} = \frac{E_{14,5\max}}{eF_s(1+i_{\max})},$$

$E_{14,5\max}$, is the maximum value of the real part of the equation.

$$\gamma_2(E_{14,5\max}, \eta_g) = \left(n + \frac{1}{2}\right)\hbar\omega_0 + S_i \left[\frac{\hbar|e|F_s[\gamma_2(E_{14,5\max}, \eta_g))]'}{\sqrt{2m_c}} \right]^{2/3} \tag{9.29}$$

Thus, combining the appropriate equations with (1.4) we can study the QC in this case.

(c) Parabolic energy bands

The magneto-dispersion law in this case is given by

$$\frac{\hbar^2 k_z^2}{2m_c} = \gamma_3(E, \eta_g) - \left(n + \frac{1}{2}\right)\hbar\omega_0 = U_6(E, n, \eta_g) \tag{9.30}$$

The EEM at the Fermi Level can be written from (9.30) as

$$m^*(E_{FBHD}, \eta_g) = m_c\gamma_3'(E_{FBHD}, \eta_g) \tag{9.31}$$

Thus, the EEM in heavily doped parabolic energy bands is a function of Fermi energy and scattering potential whereas in the absence of heavy doping the same mass is a constant quantity invariant of any variables.

The electron concentration can be expressed as

$$n_0 = \frac{eB}{\pi^2\hbar^2}\sqrt{2m_c} \left[\sum_{n=0}^{n_{\max}} [\theta_7(E_{FHD2D14}, n, \eta_g)] + \sum_{r=1}^{r_s} L(r)[\theta_7(E_{FBHD}, n, \eta_g)] \right] \tag{9.32}$$

where $\theta_7(E_{FHD2D14}, n, \eta_g) = [U_6(E_{FHD2D14}, n, \eta_g)]^{1/2}$.

The DR of the 2D electrons in QMOSFETs of HD nonlinear optical materials whose energy band structures are defined by parabolic energy bands operated in accumulation mode under the condition of weak electric field limit under magnetic quantization can be written as

$$\gamma_3(E_{14,7},\eta_g) = \left(n+\frac{1}{2}\right)\hbar\omega_0 + S_i\left[\frac{\hbar|e|F_s[\gamma_3(E_{14,7},\eta_g)]'}{\sqrt{2m_c}}\right]^{2/3} \tag{9.33}$$

where $E_{14,7}$ is the totally quantized energy in this case.

The DOS function is given by

$$N_{Bi} = \frac{g_v eB}{\pi\hbar}\sum_{n=0}^{n_{\max}}\sum_{i=0}^{n_{i\max}}\delta'(E-E_{14,7}) \tag{9.34}$$

The electron concentration per unit area can be written as

$$n_s = \bar{n}_s \tag{9.35}$$

where

$$\bar{n}_s = \frac{g_v eB}{\pi\hbar}\sum_{n=0}^{n_{\max}}\sum_{i=0}^{n_{i\max}}F_{-1}(\eta_{14,7}) \tag{9.36}$$

in which $\eta_{14,7} = (E_{FHD2D14} - E_{14,7})(k_BT)^{-1}$.

The total surface concentration in QMOSFETs of in this case can be written as

$$N_s = g_v\left[\sum_{n=0}^{n_{\max}}\sum_{i=0}^{n_{i\max}}\left[\frac{eB}{\pi\hbar}[F_{-1}(\eta_{14,7})] + \frac{eBt_{i7B}}{\pi^2\hbar^2}\sqrt{2m_c}\right.\right.$$
$$\left.\left.\times\left[[\theta_7(E_{FHD2D14},n,\eta_g)] + \sum_{r=1}^{r_s}L(r)[\theta_7(E_{FBHD},n,\eta_g)]\right]\right]\right] \tag{9.37}$$

$$t_{i7B} = \frac{E_{14,7\max}}{eF_s(1+i_{\max})},$$

$E_{14,7\max}$ is the maximum value of the equation

$$\gamma_3(E_{14,7\max},\eta_g) = \left(n+\frac{1}{2}\right)\hbar\omega_0 + S_{i_{\max}}\left[\frac{\hbar|e|F_s[\gamma_3(E_{14,7\max},\eta_g)]'}{\sqrt{2m_c}}\right]^{2/3} \tag{9.38}$$

Thus, combining the appropriate equations with (1.4) we can study the QC in this case.

9.2.3 The QC in magneto 2D QMOSFETs of II–VI materials operated under accumulation mode

The magneto-dispersion relation of the carriers in heavily doped II–VI semiconductors is given by

$$\gamma_3(E,\eta_g) = a'_0 \frac{2eB}{\hbar}\left(n+\frac{1}{2}\right) + b'_0 k_z^2 \pm \bar{\lambda}_0 \left[\frac{2eB}{\hbar}\left(n+\frac{1}{2}\right)\right]^{1/2} \tag{9.39}$$

where

$$a'_0 = \frac{\hbar^2}{2m_\perp^*},\ b'_0 = \frac{\hbar^2}{2m_\parallel^*} \text{ and } \bar{\lambda}_0 = C_0$$

Equation (9.39) can be written as

$$k_z^2 = U_{8,\pm}(E,n,\eta_g) \tag{9.40}$$

where

$$U_{8,\pm}(E,n,\eta_g) = (b'_0)^{-1}\left[\gamma_3(E,\eta_g) - \frac{2eBa'_0}{\hbar}\left(n+\frac{1}{2}\right) \mp \bar{\lambda}_0 \left[\frac{2eB}{\hbar}\left(n+\frac{1}{2}\right)\right]^{1/2}\right]$$

The EEM at the Fermi Level can be written from (9.40) as

$$m^*(E_{FBHD},\eta_g) = \frac{\hbar^2}{2} U'_{8,\pm}(E_{FBHD},n,\eta_g) \tag{9.41}$$

The electron concentration can be expressed as

$$n_0 = \frac{eB}{2\pi^2\hbar}\left[\sum_{n=0}^{n_{\max}} [\theta_{13,\pm}(E_{FBHD},n,\eta_g)] + \sum_{r=1}^{r_s} L(r)[\theta_{13,\pm}(E_{FBHD},n,\eta_g)]\right] \tag{9.42}$$

where $\theta_{13,\pm}(E_{FBHD},n,\eta_g) = [U_{8,\pm}(E_{FBHD},n,\eta_g)]^{1/2}$.

The DR of the 2D electrons in QMOSFETs of HD QMOSFETs of II–VI materials whose unperturbed energy band structures are defined by

Hopfield model operated in accumulation mode under the condition of weak electric field limit under magnetic quantization can be written as

$$\gamma_3(E_{14,9}, \eta_g) = a_0' \left[\frac{2eB}{\hbar}\left(n+\frac{1}{2}\right)\right] \pm (\bar{\lambda}_0)\left[\frac{2eB}{\hbar}\left(n+\frac{1}{2}\right)\right]^{1/2} + S_i \left[\frac{\hbar|e|F_s\gamma_3'(E_{14,9}, \eta_g)}{\sqrt{2m_{||}^*}}\right]^{2/3} \tag{9.43}$$

where $E_{14,9}$ is the totally quantized energy in this case.

The DOS function is given by

$$N_{Bi} = \frac{g_v eB}{2\pi\hbar} \sum_{n=0}^{n_{\max}} \sum_{i=0}^{n_{i_{\max}}} \delta'(E - E_{14,9}) \tag{9.44}$$

The electron concentration per unit area can be written as

$$n_s = \bar{n}_s \tag{9.45}$$

where

$$\bar{n}_s = \frac{g_v eB}{2\pi\hbar} \sum_{n=0}^{n_{\max}} \sum_{i=0}^{n_{i_{\max}}} F_{-1}(\eta_{14,9}) \tag{9.46}$$

in which $\eta_{14,9} = (E_{FBHD} - E_{14,9})(k_B T)^{-1}$.

The total surface concentration in QMOSFETs in this case can be written as

$$N_s = g_v \left[\sum_{n=0}^{n_{\max}} \sum_{i=0}^{n_{i_{\max}}} \left[\frac{eB}{2\pi\hbar}[F_{-1}(\eta_{14,9})] + \frac{eBt_{i9B}}{2\pi^2\hbar^2}\sqrt{2m_c} \times [\theta_{13,\pm}(E_{FBHD}, n, \eta_g)] + \sum_{r=1}^{r_s} L(r)[[\theta_{13,\pm}(E_{FBHD}, n, \eta_g)]]\right]\right] \tag{9.47}$$

$$t_{i9B} = \frac{E_{14,9\,\max}}{eF_s(1 + i_{\max})}$$

$E_{14,9\,\max}$ is the maximum value of the equation

$$\gamma_3(E_{14,9\,\max}, \eta_g) = a_0' \left[\frac{2eB}{\hbar}\left(n+\frac{1}{2}\right)\right] \pm (\bar{\lambda}_0)\left[\frac{2eB}{\hbar}\left(n+\frac{1}{2}\right)\right]^{1/2} + S_{i_{\max}} \left[\frac{\hbar|e|F_s\gamma_3'(E_{14,9\,\max}, \eta_g)}{\sqrt{2m_{||}^*}}\right]^{2/3} \tag{9.48}$$

Thus, combining the appropriate equations with (1.4) we can study the QC in this case.

9.2.4 *The QC in magneto 2D QMOSFETs of IV–VI materials operated under accumulation mode*

In accordance with the model of Bangert and Kastner [8], the carrier energy spectrum in HD IV–VI semiconductors can be written following (8.85) as

$$\frac{k_s^2}{\rho_{11}^2(E,\eta_g)}+\frac{k_y^2}{\rho_{12}^2(E,\eta_g)}=1 \tag{9.49}$$

where

$$\rho_{11}(E,\eta_g)=\frac{1}{\sqrt{S_1(E,\eta_g)}},\rho_{12}(E,\eta_g)=\frac{1}{\sqrt{S_2(E,\eta_g)}}$$

$$S_1(E,\eta_g)=[2\gamma_0(E,\eta_g)]^{=1}\left[\frac{(\bar{R})^2}{E_g}\{c_1(\alpha_1E,E_g)-iD_1(\alpha_1E,E_g)\}\right.$$
$$\times\frac{(\bar{S})^2}{\Delta_c'}\{c_2(\alpha_2E,E_g)-iD_2(\alpha_2E,E_g)\}$$
$$\left.+\frac{(\bar{Q})^2}{\Delta_c''}\{c_3(\alpha_3E,E_g)-iD_3(\alpha_3E,E_g)\}\right]$$

and

$$S_2(E,\eta_g)+[2\gamma_0(E,\eta_g)]^{-1}\left[\frac{(2\bar{A})^2}{E_g}\{c_1(\alpha_1E,E_g)-iD_1(\alpha_1E,E_g)\}\right.$$
$$\left.+\frac{(\bar{S}+\bar{Q})^2}{\Delta_c''}\{c_3(\alpha_3E,E_g)-iD_3(\alpha_3E,E_g)\}\right]$$

Since $S_1(E,\eta_g)$ and $S_2(E,\eta_g)$ are complex, the energy spectrum is also complex in the presence of Gaussian band tails.

Therefore, the magneto dispersion law in this case can be written as

$$k_z^2=U_{18}(E,n,\eta_g) \tag{9.50}$$

where

$$U_{18}(E,n,\eta_g)=[\rho_{12}^2(E,\eta_g)]-\left[\frac{2eB}{\hbar}\left(n+\frac{1}{2}\right)[(\rho_{11}^2(E,\eta_g)\right.$$
$$\left.\times\rho_{12}^2(E,\eta_g))^{-1}\{\rho_{12}^2(E,\eta_g)\}^{3/2}]\right]$$

The EEM at the Fermi Level can be written from (9.64) as

$$m^*(E_{FBHD}, n, \eta_g) = \frac{\hbar^2}{2} \text{ Real part of } [U_{18}(E_{FBHD}, n, \eta_g)]' \tag{9.51}$$

Thus, the EEM is a function of Fermi energy, Landau quantum number and the scattering potential and the orientation of the applied quantizing magnetic field.

The electron concentration can be expressed as

$$n_0 = \text{Real part of } \bar{n}_0 \tag{9.52}$$

where

$$\bar{n}_0 = \frac{eB}{\pi^2\hbar}\left[\sum_{n=0}^{n_{\max}}[\theta_{21}(E_{FBHD}, n, \eta_g)] + \sum_{r=1}^{r_s} L(r)[\theta_{21}(E_{FBHD}, n, \eta_g)]\right] \tag{9.53}$$

in which $\theta_{21}(E_{FBHD}, n, \eta_g) = [U_{18}(E_{FBHD}, n, \eta_g)]^{1/2}$.

The DR of the 2D electrons in QMOSFETs of HD IV–VI materials operated in accumulation mode under the condition of weak electric field limit under magnetic quantization can be written as

$$\left(n + \frac{1}{2}\right)\hbar \frac{eB}{M_{9.4}(E_{14,11}, i, \eta_g)} = \theta_3(E_{14,11}, i, \eta_g) \tag{9.54}$$

where

$$M_{9.4}(E_{14,11}, i, \eta_g) = \frac{\hbar^2}{2}[\theta_1(E_{14,11}, i, \eta_g)\theta_2(E_{14,11}, i, \eta_g)]^{-1/2}$$

where $E_{14,11}$ is the totally quantized energy in this case.

The DOS function is given by

$$N_{Bi} = \frac{g_v eB}{\pi\hbar}\sum_{n=0}^{n_{\max}}\sum_{i=0}^{n_{i\max}} \delta'(E - E_{14,11}) \tag{9.55}$$

The electron concentration per unit area can be written as

$$n_s = \text{Real part of } \bar{n}_s \tag{9.56}$$

where

$$\bar{n}_s = \frac{g_v eB}{\pi\hbar}\sum_{n=0}^{n_{\max}}\sum_{i=0}^{n_{i\max}} F_{-1}(\eta_{14,11}) \tag{9.57}$$

in which $\eta_{14,11} = (E_{FHD2D14} - E_{14,11})(k_B T)^{-1}$

The total surface concentration in QMOSFETs in this case can be written as

$$N_s = g_v \text{ Real Part of } \left[\sum_{n=0}^{n_{\max}} \sum_{i=0}^{n_{i\max}} \left[\frac{eB}{\pi\hbar} [F_{-1}(\eta_{14,11})] + \frac{eBt_{i11B}}{\pi^2\hbar} \right. \right.$$
$$\left. \left. \times \left[[\theta_{21}(E_{FBHD}, n, \eta_g)] + \sum_{r=1}^{r_s} L(r)[\theta_{21}(E_{FBHD}, n, \eta_g)] \right] \right] \right] \tag{9.58}$$

$$t_{i11B} = \frac{E_{14,11\max}}{eF_s(1 + i_{\max})},$$

$E_{14,11\max}$ is the maximum value of the real part of the equation

$$\left(n + \frac{1}{2}\right)\hbar \frac{eB}{M_{9.4}(E_{14,11\ \max}, i_{\max}, \eta_g)} = \theta_3(E_{14,11\ \max}, i_{\max}, \eta_g) \tag{9.59}$$

Thus, combining the appropriate equations with (1.4) we can study the QC in this case.

9.2.5 *The QC in magneto 2D QMOSFETs of stressed Kane type materials operated under accumulation mode*

The DR of the conduction electrons in heavily doped Kane type semiconductors can be written following (8.106) of Chapter 8 as

$$\frac{k_x^2}{a_{11}^2(E, \eta_g)} + \frac{k_y^2}{b_{11}^2(E, \eta_g)} + \frac{k_z^2}{c_{11}^2(E, \eta_g)} = 1 \tag{9.60}$$

where

$$a_{11}(E, \eta_g) = \frac{1}{\sqrt{P_{11}(E, \eta_g)}}, \quad b_{11}(E, \eta_g) = \frac{1}{\sqrt{Q_{11}(E, \eta_g)}} \text{ and}$$
$$c_{11}(E, \eta_g) = \frac{1}{\sqrt{S_{11}(E, \eta_g)}}$$

The electron energy spectrum in heavily doped Kane type semiconductors in the presence of an arbitrarily oriented quantizing magnetic field B along

z-axis can be written as

$$(k_z)^2 = U_{41}(E, n, \eta_g) \tag{9.61}$$

where

$$U_{41}(E, n, \eta_g) = I_2(E, \eta_g)[1 - I_3(E, n, \eta_g)], I_2(E, \eta_g) = [c_{11}(E, \eta_g)]_1^2$$

and

$$I_3(E, n, \eta_g) = \frac{2eB}{\hbar}\left(n + \frac{1}{2}\right)[a_{11}(E, \eta_g)b_{11}(E, \eta_g)c_{11}(E, \eta_g)]^{-1}[I_2(E, \eta_g)]^{1/2}$$

The EEM at the Fermi Level can be written from (9.61) as

$$m^*(E_{FBHD}, n, \eta_g) = \frac{\hbar^2}{2} \text{ Real part of } U'_{41}(E_{FBHD}, n, \eta_g) \tag{9.62}$$

From (9.62), we observe that the EEM is a function of Fermi energy, Landau quantum number, the scattering potential and the orientation of the applied quantizing magnetic field.

The electron concentration can be expressed as

$$n_0 = \frac{eB}{\pi^2\hbar}\left[\sum_{n=0}^{n_{\max}}\left[[\theta_{25}(E_{FBHD}, n, \eta_g)] + \sum_{r=1}^{r_s} L(r)[\theta_{25}(E_{FBHD}, n, \eta_g)]\right]\right] \tag{9.63}$$

where $\theta_{25}(E_{FBHD}, n, \eta_g) = [U_{41}(E_{FBHD}, n, \eta_g)]^{1/2}$.

The DR of the 2D electrons in QMOSFETs of HD stressed Kane type materials operated in accumulation mode under the condition of weak electric field limit under magnetic quantization can be written as

$$\left(n + \frac{1}{2}\right)\hbar\frac{eB}{M_{9.5}(E_{14,13}, i, \eta_g)} = \theta_3(E_{14,13}, i, \eta_g) \tag{9.64}$$

where

$$M_{9.5}(E_{14,13}, i, \eta_g) = \frac{\hbar^2}{2}[\theta_{13}(E_{14,13}, i, \eta_g)\theta_{23}(E_{14,13}, i, \eta_g)]^{-1/2}$$

where $E_{14,13}$ is the totally quantized energy in this case.

The DOS function is given by

$$N_{Bi} = \frac{g_v eB}{\pi\hbar} \sum_{n=0}^{n_{\max}} \sum_{i=0}^{n_{i\max}} \delta'(E - E_{14,13}) \tag{9.65}$$

The electron concentration per unit area can be written as

$$n_s = \bar{n}_s \tag{9.66}$$

where

$$\bar{n}_s = \frac{g_v eB}{\pi\hbar} \sum_{n=0}^{n_{\max}} \sum_{i=0}^{n_{i\max}} F_{-1}(\eta_{14,13}) \tag{9.67}$$

in which $\eta_{14,13} = (E_{FHD2D14} - E_{14,13})(k_B T)^{-1}$.

The total surface concentration in QMOSFETs in this case can be written as

$$N_s = g_v \left[\sum_{n=0}^{n_{\max}} \sum_{i=0}^{n_{i\max}} \left[\frac{eB}{\pi\hbar} [F_{-1}(\eta_{14,13})] + \frac{eBt_{i13B}}{\pi^2\hbar} \right.\right.$$
$$\left.\left. \times \left[[\theta_{25}(E_{FBHD}, n, \eta_g)] + \sum_{r=1}^{r_s} L(r) [\theta_{25}(E_{FBHD}, n, \eta_g)] \right] \right] \right] \tag{9.68}$$

$$t_{i13B} = \frac{E_{14,13\,\max}}{eF_s(1 + i_{\max})},$$

$E_{14,13\,\max}$, is the maximum value of the real part of the equation

$$\left(n + \frac{1}{2}\right) \hbar \frac{eB}{M_{9.5\,\max}(E_{14,13\,\max}, i_{\max}, \eta_g)} = \theta_3(E_{14,13\,\max}, i_{\max}, \eta_g) \tag{9.69}$$

where

$$M_{9.5\,\max}(E_{14,13\,\max}, i, \eta_g)$$
$$= \frac{\hbar^2}{2} [\theta_{13}(E_{14,13\,\max}, i_{\max}, \eta_g) \theta_{23}(E_{14,13\,\max}, i_{\max}, \eta_g)]^{-1/2}$$

Thus, combining the appropriate equations with (1.4) we can study the QC in this case.

9.3 Summary and Conclusions

From the theoretical background, we note that the expression for surface electron concentration under magnetic quantization of 2D QMOSFETs operated under accumulation mode of nonlinear optical, III–V (three- and two-band models of Kane together with parabolic energy bands), II–VI, IV–VI and stressed Kane type materials are given by (9.9), (9.19), (9.28), (9.37), (9.47), (9.58) and (9.68), respectively. By combining each of them with the (1.4) we can study the QC in each case.

We have not given the graphs of the magneto QCs' for various 2D QMOSFETs made of many materials as discussed here operated under accumulation layer mode since we feel that our readers will perform the intricate numerical computation to obtain the oscillatory plots for better assessment and joy of understanding.

9.4 Open Research Problems

(R.9.1) Investigate the magneto QCs in the presence of an arbitrarily oriented magnetic field in 2D QMOSFETs of nonlinear optical materials operated under accumulation mode as stated in this chapter. Draw the corresponding graphs and explain the results.

(R.9.2) Investigate the magneto QCs in the presence of an arbitrarily oriented magnetic field in 2D QMOSFETs of IV–VI, II–VI and stressed Kane type compounds operated under accumulation mode as stated in this chapter. Draw the corresponding graphs and explain the results.

(R.9.3) Investigate the magneto QCs in the presence of an arbitrarily oriented magnetic field in 2D QMOSFETs operated under accumulation mode of all the materials as stated in R.1.1 of chapter 1. Draw the corresponding graphs and explain the results.

(R.9.4) Investigate the QCs in the presence of an arbitrarily oriented magnetic field in 2D QMOSFETs operated under accumulation mode of all the materials as studied in this chapter in the presence of an arbitrarily oriented non-uniform magnetic field. Draw the corresponding graphs and explain the results.

(R.9.5) Investigate the QCs in the presence of an arbitrarily oriented magnetic field in 2D QMOSFETs operated under accumulation mode of IV–VI, II–VI and stressed Kane type compounds in the presence of an arbitrarily oriented non-uniform magnetic field. Draw the corresponding graphs and explain the results.

(R.9.6) Investigate the QCs in the presence of an arbitrarily oriented magnetic field in 2D QMOSFETs operated under accumulation mode of all the materials as stated in R.1.1 of chapter 1 in the presence of an arbitrarily oriented non-uniform magnetic field. Draw the corresponding graphs and explain the results.

(R.9.7) Investigate the QCs in 2D QMOSFETs operated under accumulation mode of all the problems from R.2.1 to R.2.6 in the presence of an arbitrarily oriented non-uniform magnetic field. Draw the corresponding graphs and explain the results.

Chapter 10

The QC in 2D MOSFETs of Important Materials Operated Under Accumulation Mode in the Presence of Cross-Fields Configuration

The best book is life and the best teacher is experience.

10.1 Introduction

It appears from the recent literature that the QC in 2D QMOSFETs of important materials operated under accumulation mode in the presence of cross-fields configuration has yet to be investigated. In this chapter, we study the QC under cross-fields configuration in 2D QMOSFETs of tetragonal and nonlinear optical materials in Section 10.2.1 of theoretical background 10.2. The QC in the presence of cross-fields configuration and under accumulation mode has been studied in 2D QMOSFETs of III–V, ternary and quaternary materials in Section 10.2.2. From Sections 10.2.3 up to 10.2.5, we shall investigate the same in 2D QMOSFETs of II–VI, IV–VI, and stressed Kane type materials, respectively. Section 10.3 contains summary and conclusion and Section 10.4 contains few open research problems.

10.2 Theoretical Background

10.2.1 *The QC in 2D MOSFETs of tetragonal and nonlinear optical materials operated under accumulation mode in the presence of cross-fields configuration*

Equation (8.25) of Chapter 8 can be expressed as

$$T_{22}(E,\eta_g) = \frac{p_s^2}{2m_\perp^*} + \frac{p_z^2}{2m_\parallel^*} T_{22}(E,\eta_g)[T_{21}(E,\eta_g)]^{-1} \tag{10.1}$$

where $p_s = \hbar k_s$ and $p_z = \hbar k_z$.

Therefore, in the presence of the electric field E_o along x-axis and the quantizing magnetic field B along z-axis for the present case, from (10.1) one can approximately write,

$$T_{22}(E,\eta_g) + |e|E_o\hat{x}\rho(e,\eta_g) = \frac{\hat{p}_x^2}{2m_\perp^*} + \frac{(\hat{p}_x - |e|B\hat{x})^2}{2m_\perp^*} + \frac{\hat{p}_z^2}{2a(E,\eta_g)} \tag{10.2}$$

where

$$\rho(E) \equiv \frac{\partial}{\partial E}[T_{22}(E,\eta_g)] \quad \text{and} \quad a(E,\eta_g) \equiv m_\parallel^*[T_{22}(E,\eta_g)]^{-1}[T_{21}(E,\eta_g)]$$

Let us define the operator $\hat{\theta}$ as

$$\hat{\theta} = -\hat{p}_y + |e|B\hat{x} - \frac{m_\perp^* E_o \rho(E,\eta_g)}{B} \tag{10.3}$$

Eliminating the operator $\hat{x}$, between (10.2) and (10.3) the dispersion relation of the conduction electron in tetragonal semiconductors in the presence of cross-fields configuration is given by

$$T_{22}(E,\eta_g) = \left[\left(\left(n+\frac{1}{2}\right)\hbar\omega_{01}\right)\right] + \left(\frac{[\hbar k_z(E)]^2}{2a(E,\eta_g)}\right) - \left(\frac{E_0\hbar k_y \rho(E,\eta_g)}{B}\right) - \left(\frac{m_\perp^* \rho^2(E,\eta_g) E_o^2}{2B^2}\right) \tag{10.4}$$

where

$$\omega_{01} = \frac{|e|B}{m_\perp^*}$$

The EEMs along z and y directions can, respectively be expressed from (10.4) as

$$m_z^*(\bar{E}_{FBDH}, \eta_g, n, E_0) = \text{ Real part of } \left[a'(\bar{E}_{FBDH}, \eta_g)\left[T_{22}(\bar{E}_{FBDH}, \eta_g) - \left(n+\frac{1}{2}\right)\hbar\omega_{01} + \frac{m_\perp^* \rho^2(\bar{E}_{FBDH}, \eta_g)E_0^2}{2B^2}\right] + \left[a(\bar{E}_{FBDH}, \eta_g) \times \left[T_{22}'(\bar{E}_{FBDH}, \eta_g) + \frac{m_\perp^* \rho(\bar{E}_{FBDH}, \eta_g)\rho'(\bar{E}_{FBDH}, \eta_g)E_0^2}{B^2}\right]\right] \quad (10.5)$$

and

$$m_y^*(\bar{E}_{FBDH}, \eta_g, n, E_0) = \left(\frac{B}{E_0}\right)^2 \text{ Real part of } \left[\rho(\bar{E}_{FBDH}, \eta_g)^{-3} \times \left[T_{22}(\bar{E}_{FBDH}, \eta_g) - \left(n+\frac{1}{2}\right)\hbar\omega_{01} + \frac{m_\perp^* \rho^2(\bar{E}_{FBDH}, \eta_g)E_0^2}{2B^2}\right]\right] \times \left[\rho(\bar{E}_{FBDH}, \eta_g)\left[T_{22}(\bar{E}_{FBDH}, \eta_g) + \frac{m_\perp^* \rho(\bar{E}_{FBDH}, \eta_g)\rho'(\bar{E}_{FBDH}, \eta_g)E_0^2}{B^2}\right] - \left[T_{22}(\bar{E}_{FBDH}, \eta_g) - \left(n+\frac{1}{2}\right)\hbar\omega_{01}\frac{m_\perp^* \rho^2(\bar{E}_{FBDH}, \eta_g)E_0^2}{2B^2}\right]\rho'(\bar{E}_{FBDH}, \eta_g)\right]\right] \quad (10.6)$$

where $\bar{E}_{FBHD}$ is the Fermi energy in the presence of cross-fields configuration and heavy doping as measured from the edge of the conduction band in the vertically upward direction in the absence of any quantization.

When $E_0 \to 0, m_z^*(\bar{E}_{FBHD}, \eta_g, n, E_0) \to \infty$, which is a physically justified result. The dependence of the EEM along y direction on the Fermi energy, electric field, magnetic field and the magnetic quantum number is an intrinsic property of cross-fields together with the fact in the present case of heavy doping, the EEM exists in the band gap.

Another characteristic feature of cross-field is that various transport coefficients will be sampled dimension dependent. These conclusions are valid for even isotropic parabolic energy bands and cross-fields introduce the index-dependent anisotropy in the effective mass.

The formulation of QC requires the expression of the electron concentration which can, in general, be written excluding the electron spin as

$$n_o = \frac{-g_v}{L_x \pi^2} \sum_{n=0}^{n_{\max}} \int_{\bar{E}_0}^{\infty} I(E,\eta_g) \frac{\partial f_o}{\partial E} dE \tag{10.7}$$

where L_x is the sample length along x direction, $\bar{E}_0$ is determined by the equation

$$I(E,\eta_g) = 0 \tag{10.8}$$

where

$$I(E,\eta_g) = \int_{x_l(E,\eta_g)}^{x_h(E,\eta_g)} k_z(E) dk_y$$

in which,

$$x_l(E,\eta_g) = \frac{-E_0 M_\perp \rho(E,\eta_g)}{\hbar B} \quad \text{and} \quad x_h(E,\eta_g) = \frac{|e|BL_x}{\hbar} + x_l(E,\eta_g)$$

Thus, we get

$$\begin{aligned} I(E,\eta_g) = \frac{2}{3}\left[\frac{B\sqrt{2a(E,\eta_g)}}{\hbar^2 E_0 \rho(E,\eta_g)}\right] & \left[\left[T_{22}(E,\eta_g) - \left(n+\frac{1}{2}\right)\frac{\hbar|e|B}{m_\perp^*}\right.\right. \\ & \left. + |e|E_0 L_x \rho(E,\eta_g) - \frac{m_\perp^* E_0^2[\rho(E,\eta_g)]^2}{2B^2}\right]^{\frac{3}{2}} \\ & \left.- \left[T_{22}(E,\eta_g) - \left(n+\frac{1}{2}\right)\frac{\hbar|e|B}{m_\perp^*} - \frac{m_\perp^* E_0^2[\rho(E,\eta_g)]^2}{2B^2}\right]^{\frac{3}{2}}\right] \end{aligned} \tag{10.9}$$

Therefore, the electron concentration is given by

$$n_0 = \text{Real part of } \bar{n}_0 \tag{10.10}$$

where

$$\begin{aligned} \bar{n}_0 = \left(\frac{2g_v B\sqrt{2}}{3L_x \pi^2 \hbar^2 E_0}\right) \sum_{n=0}^{n_{\max}} & \left[[T_{41FBHD}(n,\bar{E}_{FBHD},\eta_g)] \right. \\ & \left. + \sum_{r=1}^{s} L(r)[T_{41FBHD}(n,\bar{E}_{FBHD},\eta_g)] \right] \end{aligned} \tag{10.11}$$

in which

$$T_{41}(n,\bar{E}_{FBHD},\eta_g) \equiv \frac{\sqrt{a(\bar{E}_{FBHD},\eta_g)}}{\rho(\bar{E}_{FBHD},\eta_g)} \times \Bigg[\Bigg[T_{22}(\bar{E}_{FBHD},\eta_g) - \left(n+\frac{1}{2}\right)\frac{\hbar|e|B}{M_\perp} + |e|E_0 L_x \rho(\bar{E}_{FBHD},\eta_g) - \frac{m_\perp^*[\rho(\bar{E}_{FBHD},\eta_g)]^2}{2B^2}\Bigg]^{\frac{3}{2}} - \Bigg[T_{22}(\bar{E}_{FBHD},\eta_g) - \left(n+\frac{1}{2}\right)\frac{\hbar|e|B}{m_\perp^*} - \frac{m_\perp^* E_0^2[\rho(\bar{E}_{FBHD},\eta_g)]^2}{2B^2}\Bigg]^{\frac{3}{2}}\Bigg]$$

where $\bar{E}_{FBHD}$ is the Fermi energy in this case.

In the presence of a surface electric field F_s along z direction from (10.4) we can write

$$\alpha_{10,1}(E,\eta_g,n) - eF_S z\beta_{10,1}(E,\eta_g,n) = \frac{\hbar^2 k_z^2}{2} \tag{10.12}$$

where

$$\alpha_{10,1}(E,\eta_g,n) = \Bigg[T_{22}(E,\eta_g)a(E,\eta_g) - \left(n+\frac{1}{2}\right)\hbar\omega_{01}a(E,\eta_g) + \frac{E_0}{B}\hbar k_y\rho(E,\eta_g)a(E,\eta_g) + \frac{m_\perp^* E_0^2}{2B^2}[\rho^2(E,\eta_g)a(E,\eta_g)]$$

$$\beta_{10,1}(E,\eta_g,n) = \Bigg[T_{22}'(E,\eta_g)a(E,\eta_g) - \left(n+\frac{1}{2}\right)\hbar\omega_{01}a'(E,\eta_g) + T_{22}(E,\eta_g)a'(E,\eta_g) + \frac{E_0}{B}\hbar k_y[\rho'(E,\eta_g)a(E,\eta_g) + \rho(E,\eta_g)a'(E,\eta_g)] + \frac{m_\perp^* E_0^2}{2B^2}\,[a'(E,\eta_g)\rho^2(E,\eta_g) + 2\rho(E,\eta_g)\rho'(E,\eta_g)a(E,\eta_g)]\Bigg]$$

Using (1.7) of Chapter 1 and (10.12), the DR in this case is given by

$$\Bigg[T_{22}(E,\eta_g)a(E,\eta_g) - \left(n+\frac{1}{2}\right)\hbar\omega_{01}a(E,\eta_g) + \frac{E_0}{B}\hbar k_y\rho(E,\eta_g)a(E,\eta_g) + \frac{m_\perp^* E_0^2}{2B^2}[\rho^2(E,\eta_g)a(E,\eta_g)]\Bigg] = S_i\left(\frac{\hbar eF_s}{\sqrt{2}}\right)^{2/3}\Bigg[\Bigg[T_{22}'(E,\eta_g)a(E,\eta_g)$$

$$- \left(n+\frac{1}{2}\right)\hbar\omega_{01}a'(E,\eta_g) + T_{22}(E,\eta_g)a'(E,\eta_g)$$
$$+ \frac{E_0}{B}\hbar k_y[\rho'(E,\eta_g)a(E,\eta_g) + \rho(E,\eta_g)a'(E,\eta_g)]$$
$$+ \frac{m_\perp^* E_0^2}{2B^2}[a'(E,\eta_g)\rho^2(E,\eta_g) + 2\rho(E,\eta_g)\rho'(E,\eta_g)a(E,\eta_g)]\Big]\Big]^{2/3} \quad (10.13)$$

The sub-band energy $E_{10,1}$ in this case can be written as

$$\Big[\Big[T_{22}(E_{10,1},\eta_g)a(E_{10,1},\eta_g) - \left(n+\frac{1}{2}\right)\hbar\omega_{01}a(E_{10,1},\eta_g)$$
$$+ \frac{E_0}{B}\hbar k_y \rho(E_{10,1},\eta_g)a(E_{10,1},\eta_g) + \frac{m_\perp^* E_0^2}{2B^2}[\rho^2(E_{10,1},\eta_g)a(E_{10,1},\eta_g)]\Big]$$
$$= S_i\left(\frac{\hbar e F_s}{\sqrt{2}}\right)^{2/3}\Big[\Big[T'_{22}(E_{10,1},\eta_g)a(E_{10,1},\eta_g)$$
$$- \left(n+\frac{1}{2}\right)\hbar\omega_{01}a'(E_{10,1},\eta_g) + T_{22}(E_{10,1},\eta_g)a'(E_{10,1},\eta_g)$$
$$+ \frac{m_\perp^* E_0^2}{2B^2}[a'(E_{10,1},\eta_g)\rho^2(E_{10,1},\eta_g) + 2\rho(E_{10,1},\eta_g)\rho'(E_{10,1},\eta_g)$$
$$\times a(E_{10,1},\eta_g)]\Big]\Big]\Big]^{2/3} \quad (10.14)$$

Therefore, the total surface concentration per unit area in this case can be written as

$$N_s = g_v \text{ Real part of } \sum_{i=0}^{i_{\max}}\sum_{n=0}^{n_{\max}}\Big[t_{i10,1}\left(\frac{2B\sqrt{2}}{3L_x\pi^2\hbar^2 E_0}\right)\Big[T_{41FBHD}(n,\bar{E}_{FBHD},\eta_g)$$
$$+ \sum_{r=1}^{s} L(r)[T_{41FBHD}(n,\bar{E}_{FBHD},\eta_g)]\Big] + \frac{eB}{\pi\hbar}F_{-1}(\eta_{i10,1})\Big] \quad (10.15)$$

where

$$\eta_{i10,1} = \frac{E_{F10,1} - E_{10,1}}{k_B T},$$

$E_{F10,1}$ is the Fermi energy in this case.

$$t_{i10,1} = \frac{E_{10,1\max}}{eF_s(1+i_{\max})},$$

$E_{10,1\,\max}$ can be determined from the equation

$$\Bigg[\Bigg[T_{22}(E_{i10,1\,\max},\eta_g)a(E_{i10,1\,\max},\eta_g)-\left(n+\frac{1}{2}\right)\hbar\omega_{01}a(E_{i10,1\,\max},\eta_g)$$
$$+\frac{E_0}{B}\hbar k_y\rho(E_{i10,1\,\max},\eta_g)a(E_{i10,1\,\max},\eta_g)$$
$$+\frac{m_\perp^* E_0^2}{2B^2}[\rho^2(E_{i10,1\,\max},\eta_g)a(E_{i10,1\,\max},\eta_g)]\Bigg]=S_{i_{\max}}\left(\frac{\hbar eF_s}{\sqrt{2}}\right)^{2/3}$$
$$\times\Bigg[\Bigg[T_{22}'(E_{i10,1\,\max},\eta_g)a(E_{i10,1\,\max},\eta_g)-\left(n+\frac{1}{2}\right)\hbar\omega_{01}a'(E_{i10,1\,\max},\eta_g)$$
$$+T_{22}(E_{i10,1\,\max},\eta_g)a'(E_{i10,1\,\max},\eta_g)+\frac{m_\perp^* E_0^2}{2B^2}[a'(E_{i10,1\,\max},\eta_g)$$
$$\times\rho^2(E_{i10,1\,\max},\eta_g)+2\rho(E_{i10,1\,\max},\eta_g)\rho'(E_{i10,1\,\max},\eta_g)$$
$$\times a(E_{i10,1\,\max},\eta_g)]\Bigg]\Bigg]^{2/3} \quad (10.16)$$

Using the appropriate equations and (1.4) we can study the QC in this case.

10.2.2 *The QC in 2D MOSFETs of III–V, ternary and quaternary materials operated under accumulation mode in the presence of cross-fields configuration*

(a) Under the conditions $\delta=0$, $\Delta_\parallel=\Delta_\perp=\Delta$ and $m_\parallel^*=m_\perp^*=m_c$,, (10.4) assumes the form

$$T_{33}(E,\eta_g)=\left(n+\frac{1}{2}\right)\hbar\omega_0+\frac{[\hbar k_z(E)]^2}{2m_c}-\frac{E_0}{B}\hbar k_y\{T_{33}(E,\eta_g)\}'$$
$$-\frac{m_cE_0^2[\{T_{33}(E,\eta_g)'\}]^2}{2B^2} \quad (10.17)$$

where $T_{33}(E,\eta_g)=T_{31}(E,\eta_g)+iT_{32}(E,\eta_g)$.

The use of (10.17) leads to the expressions of the EEMs' along z and y directions as

$$m_z^*(\bar{E}_{FBHD},\eta_g,n,E_0)=m_c\text{ Real part of }\Bigg[\{T_{33}(\bar{E}_{FBHD},\eta_g)\}''$$
$$+\frac{m_cE_0^2\{T_{33}(\bar{E}_{FBHD},\eta_g)'\}\{T_{33}(\bar{E}_{FBHD},\eta_g)\}''}{B^2}\Bigg] \quad (10.18)$$

and

$$m_y^*(\bar{E}_{FBHD},\eta_g,n,E_0)=\left(\frac{B}{E_0}\right)^2 \text{ Real part of } [\{T_{33}(\bar{E}_{FBHD},\eta_g)\}']^{-1}$$

$$+\left[T_{33}(\bar{E}_{FBHD},\eta_g)-\left(n+\frac{1}{2}\right)\hbar\omega_0+\frac{m_cE_0^2[\{T_{33}(\bar{E}_{FBHD},\eta_g)\}'']^2}{2B^2}\right]$$

$$\times\left[\frac{-\{T_{33}(\bar{E}_{FBHD},\eta_g)\}''}{[\{T_{33}(\bar{E}_{FBHD},\eta_g)\}']^2}\left[T_{33}(\bar{E}_{FBHD},\eta_g)-\left(n+\frac{1}{2}\right)\hbar\omega_0\right.\right.$$

$$\left.\left.+\frac{m_cE_0^2[\{T_{33}(\bar{E}_{FBHD},\eta_g)\}']}{2B^2}\right]+1\frac{m_cE_0^2\{T_{33}(\bar{E}_{FBHD},\eta_g)\}''}{B^2}\right] \tag{10.19}$$

The Landau energy $(\bar{E}_{n_1})$ can be written as

$$T_{33}(\bar{E}_{n_1},\eta_g)=\left(n+\frac{1}{2}\right)\hbar\omega_0-\frac{m_cE_0^2[\{T_{33}(\bar{E}_{n_1},\eta_g)\}']^2}{2B^2} \tag{10.20}$$

The electron concentration in this case assumes the form

$$n_0 = \text{ Real part of } \bar{n}_0 \tag{10.21}$$

where

$$\bar{n}_0=\frac{2g_vB\sqrt{2m_c}}{3L_x\pi^2\hbar^2E_0}\sum_{n=0}^{n_{\max}}\left[[T_{43FBHD}(n,\bar{E}_{FBHD},\eta_g)]\right.$$

$$\left.+\sum_{r=1}^{s}L(r)[T_{43FBHD}(n,\bar{E}_{FBHD},\eta_g)]\right] \tag{10.22}$$

in which

$$T_{43FBHD}(n,\bar{E}_{FBHD},\eta_g)\equiv\left[\left[T_{33}(\bar{E}_{FBHD},\eta_g)-\left(n+\frac{1}{2}\right)\hbar\omega_0\right.\right.$$

$$\left.-\frac{m_cE_0^2}{2B^2}[\{T_{33}(\bar{E}_{FBHD},\eta_g)\}']^2+|e|E_0L_x[\{T_{33}(\bar{E}_{FBHD},\eta_g)\}']\right]^{\frac{3}{2}}$$

$$-\left[T_{33}(\bar{E}_{FBHD},\eta_g)-\left(n+\frac{1}{2}\right)\hbar\omega_0-\frac{m_cE_0^2}{2B^2}[\{T_{33}(\bar{E}_{FBHD},\eta_g)'\}]^2\right]^{\frac{3}{2}}$$

$$\left.\times\frac{1}{[\{T_{33}(\bar{E}_{FBHD},\eta_g)\}']}\right]$$

In the presence of a surface electric field F_s along z direction from (10.17) we can write

$$\alpha_{10,2}(E,\eta_g,n)-eF_Sz\beta_{10,2}(E,\eta_g,n)=\frac{\hbar^2k_z^2}{2m_c} \tag{10.23}$$

where

$$\alpha_{10,2}(E,\eta_g,n) = \left[T_{33}(E,\eta_g) - \left(n+\frac{1}{2}\right)\hbar\omega_0 + \frac{E_0}{B}\hbar k_y\{T_{33}(E,\eta_g)\}' + \frac{m_c E_0^2[\{T'_{33}(E,\eta_g)\}]^2}{2B^2}\right]$$

and

$$\beta_{10,2}(E,\eta_g,n) = \left[T'_{33}(E,\eta_g) + \frac{E_0}{B}\hbar k_y\{T_{33}(E,\eta_g)\}'' + \frac{m_c E_0^2[\{T'_{33}(E,\eta_g)\}][\{T''_{33}(E,\eta_g)\}]}{2B^2}\right]$$

Using (1.7) of Chapter 1 and (10.23), the dispersion relation in this case is given by

$$\left[T_{33}(E,\eta_g) - \left(n+\frac{1}{2}\right)\hbar\omega_0 + \frac{E_0}{B}\hbar k_y\{T_{33}(E,\eta_g)\}' + \frac{m_c E_0^2[\{T'_{33}(E,\eta_g)\}]^2}{2B^2}\right] = S_i\left(\frac{\hbar e F_s}{\sqrt{2m_c}}\right)^{2/3}\left[T'_{33}(E,\eta_g) + \frac{E_0}{B}\hbar k_y\{T_{33}(E,\eta_g)\}'' + \frac{m_c E_0^2[\{T'_{33}(E,\eta_g)\}][\{T''_{33}(E,\eta_g)\}]}{2B^2}\right]^{2/3} \quad (10.24)$$

The sub-band energy $E_{10,2}$ in this case can be written as

$$\left[T_{33}(E_{10,2},\eta_g) - \left(n+\frac{1}{2}\right)\hbar\omega_0 + \frac{m_c E_0^2[\{T'_{33}(E_{10,2},\eta_g)\}]^2}{2B^2}\right] = S_i\left(\frac{\hbar e F_s}{\sqrt{2m_c}}\right)^{2/3} \times \left[T'_{33}(E_{10,2},\eta_g) + \frac{m_c E_0^2[\{T'_{33}(E_{10,2},\eta_g)\}][\{T''_{33}(E_{10,2},\eta_g)\}]}{2B^2}\right]^{2/3} \quad (10.25)$$

Therefore, the surface concentration per unit area in this case can be written as

$$N_s = g_v \text{ Real part of } \sum_{i=0}^{i_{\max}}\sum_{n=0}^{n_{\max}}\left[t_{i10,2}\frac{2B\sqrt{2m_c}}{3L_x\pi^2\hbar^2 E_0}\left[T_{43FBHD}(n,\bar{E}_{FBHD},\eta_g) + \sum_{r=1}^{s} L(r)[T_{43FBHD}(n,\bar{E}_{FBHD},\eta_g)]\right] + \frac{eB}{\pi\hbar}F_{-1}(\eta_{i10,2})\right] \quad (10.26)$$

where

$$\eta_{i10,2} = \frac{E_{F10,1} - E_{10,2}}{k_B T}$$

$$t_{i10,2} = \frac{E_{10,2\,\max}}{eF_s(1 + i_{\max})},$$

$E_{10,2\,\max}$ can be determined from the equation

$$\left[T_{33}(E_{10,2\,\max}, \eta_g) - \left(n + \frac{1}{2}\right)\hbar\omega_0 + \frac{m_c E_0^2[\{T'_{33}(E_{10,2\,\max}, \eta_g)\}]^2}{2B^2}\right]$$
$$= S_{i_{\max}}\left(\frac{\hbar e F_s}{\sqrt{2m_c}}\right)^{2/3}\left[T'_{33}(E_{10,2\,\max}, \eta_g) + \frac{m_c E_0^2[\{T'_{33}(E_{10,2\,\max}\eta_g)\}][\{T''_{33}(E_{10,2\,\max}, \eta_g)\}]}{2B^2}\right]^{2/3} \quad (10.27)$$

Using the appropriate equations and (1.4) we can study the QC in this case.

(b) Under the condition $\Delta \gg E_g$, (10.17) assumes the form

$$\gamma_2(E, \eta_g) = \left(n + \frac{1}{2}\right)\hbar\omega_0 - \frac{E_0}{B}\hbar k_y \gamma'_2(E, \eta_g) - \frac{m_c E_0^2}{2B^2}(\gamma'_2(E, \eta_g))^2 + \frac{[\hbar k_z(E)]^2}{2m_c} \quad (10.28)$$

The use of (10.28) leads to the expressions of the EEMs' along z and y directions as

$$m_z^*(\bar{E}_{FBHD}, \eta_g, n, E_0) = m_c\left[\{\gamma_2(\bar{E}_{FBHD}, \eta_g)\}'' + \frac{m_c E_o^2\{\gamma_2(\bar{E}_{FBHD}, \eta_g)\}'\{\gamma_2(\bar{E}_{FBHD}, \eta_g)\}''}{B^2}\right] \quad (10.29)$$

$$m_y^*(\bar{E}_{FBHD}, \eta_g, n, E_0) = \left(\frac{B}{E_0}\right)^2 \frac{1}{[\{\gamma_2(\bar{E}_{FBHD}, \eta_g)\}']}\left[\gamma_2(\bar{E}_{FBHD}, \eta_g) - \left(n + \frac{1}{2}\right)\hbar\omega_0 + \frac{m_c E_o^2[\{\gamma_2(\bar{E}_{FBHD}, \eta_g)\}']^2}{2B^2}\right]$$
$$\times\left[\frac{-\{\gamma_2(\bar{E}_{FBHD}, \eta_g)\}''}{[\{\gamma_2(\bar{E}_{FBHD}, \eta_g)\}']^2}\left[\gamma_2(\bar{E}_{FBHD}, \eta_g) - \left(n + \frac{1}{2}\right)\hbar\omega_0 + \frac{m_c E_0^2[\{\gamma_2(\bar{E}_{FBHD}, \eta_g)\}']^2}{2B^2}\right] + 1 + \frac{m_c E_0^2\{\gamma_2(\bar{E}_{FBHD}, \eta_g)\}''}{B^2}\right] \quad (10.30)$$

The Landau energy ($\bar{E}_{n_2}$) can be written as

$$\gamma_2(\bar{E},\eta_g) = \left(n+\frac{1}{2}\right)\hbar\omega_0 - \frac{m_c E_0^2}{2B^2}(\gamma_2'(\bar{E}_{n_2},\eta_g))^2 \tag{10.31}$$

The expression for n_0 in this case assumes the form

$$n_0 = \frac{2g_v B\sqrt{2m_c}}{3L_x\pi^2\hbar^2 E_0}\sum_{n=0}^{n_{\max}}\left[T_{47FBHD}(n,\bar{E}_{FBHD},\eta_g) + \sum_{r=1}^{s} L(r)[T_{47FBHD}(n,\bar{E}_{FBHD},\eta_g)]\right] \tag{10.32}$$

where

$$\begin{aligned} T_{47FBHD}(n,\bar{E}_{FBHD},\eta_g) &\equiv \Bigg[\gamma_2(\bar{E}_{FBHD},\eta_g) - \left(n+\frac{1}{2}\right)\hbar\omega_0 \\ &+ |e|E_0L_x(\gamma_2'(\bar{E}_{FBHD},\eta_g)) - \frac{m_c E_0^2}{2B^2}(\gamma_2(\bar{E}_{FBHD},\eta_g))^2\Bigg]^{\frac{3}{2}} \\ &- \left[\left[(\gamma_2(\bar{E}_{FBHD},\eta_g)) - \left(n+\frac{1}{2}\right)\hbar\omega_0 - \frac{m_c E_0^2}{2B^2}(\gamma_2(\bar{E}_{FBHD},\eta_g))^2\right]^{\frac{3}{2}}\right] \\ &\times[\gamma_2'(\bar{E}_{FBHD},\eta_g)]^{-1} \end{aligned}$$

In the presence of a surface electric field F_s along z direction from (10.28) we can write

$$\alpha_{10,3}(E,\eta_g,n) - eF_S z\beta_{10,3}(E,\eta_g,n) = \frac{\hbar^2 k_z^2}{2m_c} \tag{10.33}$$

where

$$\alpha_{10,3}(E,\eta_g,n) = \left[\gamma_2(E,\eta_g) - \left(n+\frac{1}{2}\right)\hbar\omega_0 + \frac{E_0}{B}\hbar k_y\{\gamma_2(E,\eta_g)\}' + \frac{m_c E_0^2[\{\gamma_2'(E,\eta_g)\}]^2}{2B^2}\right]$$

and

$$\beta_{10,3}(E,\eta_g,n) = \left[\gamma_2'(E,\eta_g) + \frac{E_0}{B}\hbar k_y\{\gamma_2(E,\eta_g)\}'' + \frac{m_c E_0^2[\{\gamma_2'(E,\eta_g)\}][\{\gamma_2''(E,\eta_g)\}]}{2B^2}\right]$$

Using (1.7) of Chapter 1 and (10.33), the DR in this case is given by

$$\left[\gamma_2(E,\eta_g) - \left(n+\frac{1}{2}\right)\hbar\omega_0 + \frac{E_0}{B}\hbar k_y\{\gamma_2(E,\eta_g)\}' + \frac{m_c E_0^2[\{\gamma_2'(E,\eta_g)\}]^2}{2B^2}\right]$$

$$= S_i\left(\frac{\hbar e F_s}{\sqrt{2m_c}}\right)^{2/3}\left[\gamma_2'(E,\eta_g) + \frac{E_0}{B}\hbar k_y\{\gamma_2(E,\eta_g)\}'' + \frac{m_c E_0^2[\{\gamma_2'(E,\eta_g)\}][\{\gamma_2''(E,\eta_g)\}]}{2B^2}\right]^{2/3} \quad (10.34)$$

The sub-band energy $E_{10,3}$ in this case can be written as

$$\left[\gamma_2(E_{10,3},\eta_g) - \left(n+\frac{1}{2}\right)\hbar\omega_0 + \frac{m_c E_0^2[\{\gamma_2'(E_{10,3},\eta_g)\}]^2}{2B^2}\right] = S_i\left(\frac{\hbar e F_s}{\sqrt{2m_c}}\right)^{2/3}$$

$$\times\left[\gamma_2'(E_{10,3},\eta_g) + \frac{m_c E_0^2[\{\gamma_2'(E_{10,3},\eta_g)\}][\{\gamma_{233}''(E_{10,3},\eta_g)\}]}{2B^2}\right]^{2/3} \quad (10.35)$$

Therefore, the total surface concentration per unit area in this case can be written as

$$N_s = g_v\sum_{i=0}^{i_{\max}}\sum_{n=0}^{n_{\max}}\left[t_{i10,3}\frac{2B\sqrt{2m_c}}{3L_x\pi^2\hbar^2 E_0}\left[T_{47FBHD}(n,\bar{E}_{FB},\eta_g) + \sum_{r=1}^{s}L(r)[T_{47FBHD}(n,\bar{E}_{FBHD},\eta_g)]\right] + \frac{eB}{\pi\hbar}F_{-1}(\eta_{i10,3})\right] \quad (10.36)$$

where

$$\eta_{i10,3} = \frac{E_{F10,1} - E_{10,3}}{k_B T}$$

$$t_{i10,3} = \frac{E_{10,3\max}}{eF_s(1+i_{\max})},$$

$E_{10,3\max}$ can be determined from the equation

$$\left[\gamma_2(E_{10,3\max},\eta_g) - \left(n+\frac{1}{2}\right)\hbar\omega_0 + \frac{m_c E_0^2[\{\gamma_2'(E_{10,3\max},\eta_g)\}]^2}{2B^2}\right]$$

$$= S_{i_{\max}}\left(\frac{\hbar e F_s}{\sqrt{2m_c}}\right)^{2/3}\left[\gamma_2'(E_{10,3\max},\eta_g) + \frac{m_c E_0^2[\{\gamma_2'(E_{10,2\max}\eta_g)\}][\{\gamma_3''(E_{10,3\max},\eta_g)\}]}{2B^2}\right]^{2/3} \quad (10.37)$$

Using the appropriate equations and (1.4) we can study the QC in this case.

(c) For $a \to 0$ and we can write,

$$\gamma_3(E,\eta_g) = \left(n+\frac{1}{2}\right)\hbar\omega_0 - \frac{E_0}{B}\hbar k_y \gamma_3'(E,\eta_g) - \frac{m_c E_0^2}{2B^2}(\gamma_3'(E,\eta_g))^2 + \frac{[\hbar k_z(E)]^2}{2m_c} \tag{10.38}$$

The use of (10.38) leads to the expressions of the EEMs' along z and y directions as

$$m_y^*(\bar{E}_{FBHD},\eta_g,n,E_0) = \left(\frac{B}{E_0}\right)^2 \frac{1}{[\{\gamma_3(\bar{E}_{FBHD},\eta_g)\}']}\left[\gamma_3(\bar{E}_{FBHD},\eta_g) - \left(n+\frac{1}{2}\right)\hbar\omega_0 + \frac{m_c E_0^2[\{\gamma_3(\bar{E}_{FBHD},\eta_g)\}']^2}{2B^2}\right]\left[\frac{-\{\gamma_3(\bar{E}_{FBHD},\eta_g)\}''}{[\{\gamma_3(\bar{E}_{FBHD},\eta_g)\}']^2} \times \left[\gamma_3(\bar{E}_{FBHD},\eta_g) - \left(n+\frac{1}{2}\right)\hbar\omega_0 + \frac{m_c E_0^2[\{\gamma_3(\bar{E}_{FBHD},\eta_g)\}']^2}{2B^2}\right] + 1 + \frac{m_c E_0^2\{\gamma_3(\bar{E}_{FBHD},\eta_g)\}''}{B^2}\right] \tag{10.39}$$

The Landau energy ($\bar{E}_{n_3}$) can be written as

$$\gamma_3(\bar{E}_{n_3},\eta_g) = \left(n+\frac{1}{2}\right)\hbar\omega_0 - \frac{m_c E_0^2}{2B^2}(\gamma_3'(\bar{E}_{n_3},\eta_g))^2 \tag{10.40}$$

The expression for n_0 in this case assumes the form

$$n_0 = \frac{2g_v B\sqrt{2m_c}}{3L_x\pi^2\hbar^2 E_0}\sum_{n=0}^{n_{\max}}\left[T_{49FBHD}(n,\bar{E}_{FBHD},\eta_g) + \sum_{r=1}^{s} L(r)[T_{49FBHD}(n,\bar{E}_{FBHD},\eta_g)]\right] \tag{10.41}$$

$$T_{49FBHD}(n,\bar{E}_{FBHD},\eta_g) \equiv \left[\left[\gamma_3'(\bar{E}_{FBHD},\eta_g) - \left(n+\frac{1}{2}\right)\hbar\omega_0 + |e|E_0L_x(\gamma_3'(\bar{E}_{FBHD},\eta_g)) - \frac{m_c E_0^2}{2B^2}(\gamma_3'(\bar{E}_{FBHD},\eta_g))^2\right]^{3/2} - \left[(\gamma_3'(\bar{E}_{FBHD},\eta_g)) - \left(n+\frac{1}{2}\right)\hbar\omega_0 - \frac{m_c E_0^2}{2B^2}(\gamma_3'(\bar{E}_{FBHD},\eta_g))^2\right]^{3/2}\right] \times[\gamma_3'(\bar{E}_{FBHD},\eta_g)]^{-1}$$

In the presence of a surface electric field F_s along z direction from (10.38) we can write

$$\alpha_{10,4}(E,\eta_g,n) - eF_S z\beta_{10,4}(E,\eta_g,n) = \frac{\hbar^2 k_z^2}{2m_c} \tag{10.42}$$

where

$$\alpha_{10,4}(E,\eta_g,n) = \left[\gamma_3(E,\eta_g) - \left(n+\frac{1}{2}\right)\hbar\omega_0 + \frac{E_0}{B}\hbar k_y\{\gamma_3(E,\eta_g)\}' + \frac{m_c E_0^2[\{\gamma_3'(E,\eta_g)\}]^2}{2B^2}\right]$$

and

$$\beta_{10,4}(E,\eta_g,n) = \left[\gamma_3'(E,\eta_g) + \frac{E_0}{B}\hbar k_y\{\gamma_3(E,\eta_g)\}'' + \frac{m_c E_0^2[\{\gamma_3'(E,\eta_g)\}][\{\gamma_3''(E,\eta_g)\}]}{2B^2}\right]$$

Using (1.7) of Chapter 1 and (10.42), the DR in this case is given by

$$\begin{aligned}&\left[\gamma_3(E,\eta_g) - \left(n+\frac{1}{2}\right)\hbar\omega_0 + \frac{E_0}{B}\hbar k_y\{\gamma_3(E,\eta_g)\}' + \frac{m_c E_0^2[\{\gamma_3'(E,\eta_g)\}]^2}{2B^2}\right]\\ &= S_i\left(\frac{\hbar eF_s}{\sqrt{2m_c}}\right)^{2/3}\left[\gamma_3'(E,\eta_g) + \frac{E_0}{B}\hbar k_y\{\gamma_3(E,\eta_g)\}''\right.\\ &\left.+ \frac{m_c E_0^2[\{\gamma_3'(E,\eta_g)\}][\{\gamma_3''(E,\eta_g)\}]}{2B^2}\right]^{2/3}\end{aligned} \tag{10.43}$$

The sub-band energy $E_{10,4}$ in this case can be written as

$$\begin{aligned}&\left[\gamma_3(E_{10,4},\eta_g) - \left(n+\frac{1}{2}\right)\hbar\omega_0 + \frac{m_c E_0^2[\{\gamma_3'(E_{10,4},\eta_g)\}]^2}{2B^2}\right]\\ &= S_i\left(\frac{\hbar eF_s}{\sqrt{2m_c}}\right)^{2/3}\left[\gamma_3'(E_{10,4},\eta_g)\right.\\ &\left.+ \frac{m_c E_0^2[\{\gamma_3'(E_{10,3},\eta_g)\}][\{\gamma_3''(E_{10,3},\eta_g)\}]}{2B^2}\right]^{2/3}\end{aligned} \tag{10.44}$$

Therefore, the total surface concentration per unit area in this case can be written as

$$N_s = g_v \sum_{i=0}^{i_{\max}} \sum_{n=0}^{n_{\max}} \left[t_{i10,4} \frac{2B\sqrt{2m_c}}{3L_x \pi^2 \hbar^2 E_0} \left[T_{49FBHD}(n, \bar{E}_{FB}, \eta_g) + \sum_{r=1}^{s} L(r)[T_{49FBHD}(n, \bar{E}_{FBHD}, \eta_g)] \right] + \frac{eB}{\pi\hbar} F_{-1}(\eta_{i10,4}) \right] \quad (10.45)$$

where

$$\eta_{i10,4} = \frac{E_{F10,1} - E_{10,4}}{k_B T}$$

$$t_{i10,4} = \frac{E_{10,4\,\max}}{eF_s(1 + i_{\max})}$$

$E_{10,4\,\max}$ can be determined from the equation

$$\left[\gamma_3(E_{10,4\,\max}, \eta_g) - \left(n + \frac{1}{2}\right) \hbar\omega_0 + \frac{m_c E_0^2 [\{\gamma_3'(E_{10,4\,\max}, \eta_g)\}]^2}{2B^2} \right] = S_{i_{\max}} \left(\frac{\hbar e F_s}{\sqrt{2m_c}} \right)^{2/3} \left[\gamma_3'(E_{10,4\,\max}, \eta_g) + \frac{m_c E_0^2 [\{\gamma_3'(E_{10,3\,\max} \eta_g)\}][\{\gamma_3''(E_{10,4\,\max}, \eta_g)\}]}{2B^2} \right]^{2/3} \quad (10.46)$$

Using the appropriate equations and (1.4) we can study the QC in this case.

10.2.3 *The QC in 2D MOSFETs of II–VI materials operated under accumulation mode in the presence of cross-fields configuration*

The electron energy spectrum in HD II–VI semiconductors in the presence of electric field E_0 along x direction and quantizing magnetic field along z direction can approximately be written as

$$\gamma_3(E, \eta_g) = \beta_1(n, E_0) - \frac{E_0}{B} \hbar k_y \gamma_3'(E, \eta_g) - \frac{m_{\parallel}^* E_0^2}{2B^2} (\gamma_3'(E, \eta_g))^2 + \frac{[\hbar k_z(E)]^2}{2m_{\parallel}^*} \quad (10.47)$$

where

$$\beta_1(n,E_0) \equiv \left[\left(n+\frac{1}{2}\right)\hbar\omega_0 - \left(\frac{E_0^2 m_\perp^*}{2B^2}\right) + D\left\{\left(n+\frac{1}{2}\right)\hbar\omega_{02} - \left(\frac{E_0^2 m_\perp^*}{2B^2}\right)\right\}^{\frac{1}{2}}\right],$$

$$\omega_{02} \equiv \frac{|e|B}{m_\perp^*} \quad \text{and} \quad D \equiv \pm\frac{\bar{\lambda}_0\sqrt{2m_\perp^*}}{\hbar}$$

The use of (10.47) leads to the expressions of the EEMs' along z and y directions as

$$m_z^*(\bar{E}_{FBHD},\eta_g,n,E_0) = m_\parallel^*\left[\{\gamma_3(\bar{E}_{FBHD},\eta_g)\}'' + \frac{m_\parallel^* E_0^2\{\gamma_3(\bar{E}_{FBHD},\eta_g)\}'\{\gamma_3(\bar{E}_{FBHD},\eta_g)\}''}{B^2}\right] \tag{10.48}$$

$$\begin{aligned} m_y^*(\bar{E}_{FBHD},\eta_g,n,E_0) &= \left(\frac{B}{E_0}\right)^2 \frac{1}{[\{\gamma_3(\bar{E}_{FBHD},\eta_g)\}']} \\ &\times\left[\{\gamma_3(\bar{E}_{FBHD},\eta_g)\} - \beta_1(n,E_0) + \frac{m_\parallel^* E_0^2[\{\gamma_3(\bar{E}_{FBHD},\eta_g)\}']^2}{2B^2}\right] \\ &\times\left[\frac{-\{\gamma_3(\bar{E}_{FBHD},\eta_g)\}''}{[-\{\gamma_3(\bar{E}_{FBHD},\eta_g)\}']^2}\left[\gamma_3(\bar{E}_{FBHD},\eta_g) - \beta_1(n,E_0) + \frac{m_\parallel^* E_0^2[\{\gamma_3(\bar{E}_{FBHD},\eta_g)\}']^2}{2B^2}\right] + 1 + \frac{m_\parallel^* E_0^2[\{\gamma_3(\bar{E}_{FBHD},\eta_g)\}']}{B^2}\right] \end{aligned} \tag{10.49}$$

The Landau energy ($\bar{E}_{n_4}$) can be written as

The expression for n_0 in this case assumes the form

$$n_0 = \frac{g_v B\sqrt{2m_\parallel^*}}{3L_x\pi^2\hbar^2 E_0}\sum_{n=0}^{n_{\max}}\left[T_{53FBHD}(n,\bar{E}_{FBHD},\eta_g) + \sum_{r=1}^{s} L(r)[T_{53FBHD}(n,\bar{E}_{FBHD},\eta_g)]\right] \tag{10.50}$$

where

$$T_{53FBHD}(n, \bar{E}_{FBHD}, \eta_g) \equiv \Bigg[\Bigg[\gamma_3(\bar{E}_{FBHD}, \eta_g) - \beta_1(n, E_0)$$

$$+ |e|E_0 L_x(\gamma_3'(\bar{E}_{FBHD}, \eta_g)) - \frac{m_{\parallel}^* E_0^2}{2B^2}(\gamma_3'(\bar{E}_{FBHD}, \eta_g)^2\Bigg]^{3/2}$$

$$- \left[(\gamma_3(\bar{E}_{FBHD}, \eta_g)) - \beta_1(n, E_0) - \frac{m_{\parallel}^* E_0^2}{2B^2}(\gamma_3'(\bar{E}_{FBHD}, \eta_g))^2\right]^{3/2}\Bigg]$$

$$\times [\gamma_3'(\bar{E}_{FBHD}, \eta_g)]^{-1}$$

In the presence of a surface electric field F_s along z direction from (10.47) we can write

$$\alpha_{10,5}(E, \eta_g, n) - eF_S z\beta_{10,5}(E, \eta_g, n) = \frac{\hbar^2 k_z^2}{2} \tag{10.51}$$

where

$$\alpha_{10,5}(E, \eta_g, n) = \left[2m_{\parallel}^*\left[\gamma_3(E, \eta_g) - \beta_1(n, E_0) + \frac{E_0}{B}\hbar k_y \gamma_3'(E, \eta_g)\right]^2\right]$$

and

$$\beta_{10,5}(E, \eta_g, n) = \left[2m_{\parallel}^*\left[\gamma_3'(E, \eta_g) - \beta_1(n, E_0) + \frac{E_0}{B}\hbar k_y + \frac{m_{\parallel}^* E_0^2}{B^2}\gamma_3''(E, \eta_g)\right]\right]$$

Using (1.7) of Chapter 1 and (10.51), the DR in this case is given by

$$\left[2m_{\parallel}^*\left[\gamma_3(E, \eta_g) - \beta_1(n, E_0) + \frac{E_0}{B}\hbar k_y \gamma_3'(E, \eta_g)\right]^2\right] = S_i\left[\frac{\hbar e F_s}{\sqrt{2}}\right]^{2/3}$$
$$\times \left[2m_{\parallel}^*\left[\gamma_3'(E, \eta_g) - \beta_1(n, E_0) + \frac{E_0}{B}\hbar k_y + \frac{m_{\parallel}^* E_0^2}{B^2}\gamma_3''(E, \eta_g)\right]\right]^{2/3} \tag{10.52}$$

The sub-band energy $E_{10,5}$ in this case can be written as

$$[2m_{\|}^{*}[\gamma_3(E_{10,5},\eta_g)-\beta_1(n,E_0)]]=S_i\left[\frac{\hbar eF_s}{\sqrt{2}}\right]^{2/3}\times\left[2m_{\|}^{*}\left[\gamma_3'(E_{10,5},\eta_g)-\beta_1(n,E_0)+\frac{m_{\|}^{*}E_0^2}{B^2}\gamma_3''(E_{10,5},\eta_g)\right]\right]^{2/3} \tag{10.53}$$

Therefore, the total surface concentration per unit area in this case can be written as

$$N_s=g_v\sum_{i=0}^{i_{\max}}\sum_{n=0}^{n_{\max}}\left[t_{i10,5}\frac{B\sqrt{2m_{\|}^{*}}}{3L_x\pi^2\hbar^2E_0}\left[T_{53HD}(n,\bar{E}_{FB},\eta_g)+\sum_{r=1}^{s}L(r)[T_{53FBHD}(n,\bar{E}_{FBHD},\eta_g)]\right]+\frac{eB}{2\pi\hbar}F_{-1}(\eta_{i10,5})\right] \tag{10.54}$$

where

$$\eta_{i10,5}=\frac{E_{F10,1}-E_{10,5}}{k_BT}$$

$$t_{i10,5}=\frac{E_{10,5\,\max}}{eF_s(1+i_{\max})}$$

$E_{10,5\,\max}$ can be determine from the equation

$$[2m_{\|}^{*}[\gamma_3(E_{10,5\,\max},\eta_g)-\beta_1(n,E_0)]]=S_{i_{\max}}\left[\frac{\hbar eF_s}{\sqrt{2}}\right]^{2/3}\times\left[2m_{\|}^{*}\left[\gamma_3'(E_{10,5\,\max},\eta_g)-\beta_1(n,E_0)+\frac{m_{\|}^{*}E_0^2}{B^2}\gamma_3''(E_{10,5\,\max},\eta_g)\right]\right]^{2/3} \tag{10.55}$$

Using the appropriate equations and (1.4) we can study the QC in this case.

10.2.4 *The QC in 2D MOSFETs of IV–VI materials operated under accumulation mode in the presence of cross-fields configuration*

The (8.85) can be written as

$$\frac{p_s^2}{2M_1^*(E,\eta_g)}+\frac{p_z^2}{2M_3^*(E,\eta_g)}=g^*(E,\eta_g) \tag{10.56}$$

where

$$M_1^*(E,\eta_g) = \left[\frac{(\bar{R})}{E_g}\{c_1(\alpha_1,E,E_g) - iD_1(\alpha_1,E,E_g)\}\right.$$
$$+\frac{(\bar{S})}{\Delta_c}\{c_2(\alpha_2,E,E_g) - iD_2(\alpha_2,E,E_g)\}$$
$$\left.\times\frac{(\bar{Q})}{\Delta_c}\{c_3(\alpha_3,E,E_g) - iD_3(\alpha_3,E,E_g)\}\right]^{-1}$$

$$M_3^*(E,\eta_g) = \left[\frac{2(\bar{A})^2}{E_g}\{c_1(\alpha_1,E,E_g) - iD_1(\alpha_1,E,E_g)\}\right.$$
$$\left.+\frac{(\bar{S}+\bar{Q})^2}{\Delta_c}\{c_3(\alpha_3,E,E_g) - iD_3(\alpha_3,E,E_g)\}\right]^{-1}$$

and

$$g^*(E,\eta_g) = 2\hbar^2\gamma_0(E,\eta_g)$$

In the presence of quantizing magnetic field B along z direction and the electric field along x-axis, from the above equation one obtains

$$\frac{\hat{p}_x^2}{2M_1^*(E,\eta_g)} + \frac{(\hat{p}_y - |e|B\hat{x})^2}{2M_1^*(E,\eta_g)} + \frac{\hat{p}_z^2}{2M_3^*(E,\eta_g)} = g^*(E,\eta_g) + |e|E_0\hat{x}\rho_1^*(E,\eta_g) \tag{10.57}$$

where

$$\rho_1^*(E,\eta_g) = \frac{\partial}{\partial E}[g^*(E,\eta_g)]$$

Let us define the operator $\hat{\theta}$ as

$$\hat{\theta} = -\hat{p}_y + |e|B\hat{x} - \frac{\rho_1^*(E,\eta_g)E_0[M_1^*(E,\eta_g)]}{B} \tag{10.58}$$

Eliminating $\hat{x}$, between the above two equations, the dispersion relation of the conduction electrons in HD stressed Kane type semiconductors in the presence of cross-fields configuration can be expressed as

$$g^*(E,\eta_g) = \left(n+\frac{1}{2}\right)\hbar\bar{\omega}_{i1}(E,\eta_g) + \frac{\hbar^2k_z^2}{2M_3^*(E,\eta_g)} - \frac{E_0}{B}\rho_1^*(E,\eta_g)\hbar k_y$$
$$-\frac{E_0^2}{2B^2}[\rho_1^*(E,\eta_g)]^2M_1^*(E,\eta_g) \tag{10.59a}$$

where $\bar{\omega}_{i1}(E,\eta_g) = eB[M_1^*(E,\eta_g)]^{-1}$.

The use of (10.59a) leads to the expressions of the EEMs' along z and y directions as

$$m_z^*(\bar{E}_{FBHD}, \eta_g, n, E_0) = \text{Real part of } [M_3^*(\bar{E}_{FBHD}, \eta_g)]'$$
$$\times \left[g^*(\bar{E}_{FBHD}, \eta_g) - \left(n + \frac{1}{2}\right) \hbar\bar{\omega}_i(\bar{E}_{FBHD}, \eta_g) \right.$$
$$+ \frac{E_0^2}{2B^2}[\rho_1^*(\bar{E}_{FBHD}, \eta_g)^2 M_1^*(\bar{E}_{FBHD}, \eta_g)] + 2[M_3^*(\bar{E}_{FBHD}, \eta_g)]$$
$$\times \left[[g^*(\bar{E}_{FBHD}, \eta_g)]' - \left(n + \frac{1}{2}\right) \hbar[\bar{\omega}_i(\bar{E}_{FBHD}, \eta_g)]' \right] \tag{10.59b}$$

$$m_y^*(\bar{E}_{FBHD}, \eta_g, n, E_0) = (B/E_0)^2 \text{ Real part of } [\rho_1^*(\bar{E}_{FBHD}, \eta_g)]^{-3}$$
$$\times \left[g^*(\bar{E}_{FBHD}, \eta_g) - \left(n + \frac{1}{2}\right) \hbar\bar{\omega}_{i1}(\bar{E}_{FBHD}, \eta_g) \right.$$
$$+ \frac{E_0^2}{2B^2}[\rho_1^*(\bar{E}_{FBHD}, \eta_g)^2 M_1^*(\bar{E}_{FBHD}, \eta_g)][[\rho_1^*(\bar{E}_{FBHD}, \eta_g)]$$
$$\times [g^*(\bar{E}_{FBHD}, \eta_g)]' - \left(n + \frac{1}{2}\right) \hbar[\bar{\omega}_i(\bar{E}_{FBHD}, \eta_g)]'$$
$$+ \frac{E_0^2}{2B^2} \left[[\rho_1^*(\bar{E}_{FBHD}, \eta_g)^2 [M_1^*(\bar{E}_{FBHD}, \eta_g)]'] - [\rho_1^*(\bar{E}_{FBHD}, \eta_g)]' \right.$$
$$\times \left[g^*(\bar{E}_{FBHD}, \eta_g) - \left(n + \frac{1}{2}\right) \hbar\bar{\omega}_{i1}(\bar{E}_{FBHD}, \eta_g) \right.$$
$$\left.\left.\left. \times \frac{E_0^2}{2B^2}[\rho_1^*(\bar{E}_{FBHD}, \eta_g)]^2 M_1^*(\bar{E}_{FBHD}, \eta_g) \right]\right]\right] \tag{10.60}$$

The Landau level energy (E_{n9}) in this case can be expressed through the equation

$$g^*(E_{n_9}, \eta_g) = \left(n + \frac{1}{2}\right) \hbar\bar{\omega}_{i1}(E_{n_9}, \eta_g) - \frac{E_0^2}{2B^2}[\rho_1^*(E_{n_9}, \eta_g)]^2 M_1^*(E_{n_9}, \eta_g) \tag{10.61}$$

The electron concentration can be written as

$$n_0 = \text{Real part of } \bar{n}_0 \tag{10.62}$$

$$\bar{n}_0 = \frac{2Bg_v}{3L_x\pi^2\hbar^2 E_0} \sum_{n=0}^{n_{\max}} \left[T_{4131FBHD}(n, \bar{E}_{FBHD}, \eta_g) \right.$$
$$\left. + \sum_{r=1}^{s} L(r)[T_{4131FBHD}(n, \bar{E}_{FBHD}, \eta_g)] \right] \tag{10.63}$$

where

$$T_{4131FBHD}(n, \bar{E}_{FBHD}, \eta_g) = \left[\frac{\sqrt{2M_3^*(\bar{E}_{FBHD}, \eta_g)}}{\rho_1^*(\bar{E}_{FBHD}, \eta_g)}\right]$$

$$\times \left[\left[\left[T_{51}(n, \bar{E}_{FBHD}, \eta_g) + \frac{E_0}{B}\rho_1^*(\bar{E}_{FBHD}, \eta_g)\hbar x_{hHD1}(\bar{E}_{FBHD}, \eta_g)\right.\right.\right.$$

$$\left.\times\ \rho_1^*(\bar{E}_{FBHD}, \eta_g)\right]^{3/2} - \left[T_{51}(n, \bar{E}_{FBHD}, \eta_g) + \frac{E_0}{B}\rho_1^*(\bar{E}_{FBHD}, \eta_g)\right.$$

$$\left.\left.\times\ \hbar x_{hHD1}(\bar{E}_{FBHD}, \eta_g)\rho_1^*(\bar{E}_{FBHD}, \eta_g)\right]^{3/2}\right]$$

$$T_{51}(n, \bar{E}_{FBHD}, \eta_g) = \left[g^*(\bar{E}_{FBHD}, \eta_g) - \left(n + \frac{1}{2}\right)\hbar\bar{\omega}_{i1}(\bar{E}_{FBHD}, \eta_g)\right.$$

$$\left. + \frac{M_1^*(\bar{E}_{FBHD}, \eta_g)E_0^2}{2B^2}[\rho_1^*(\bar{E}_{FBHD}, \eta_g)]^2\right]$$

$$x_{1HD1}(\bar{E}_{FBHD}, \eta_g) = \frac{-M_1^*(\bar{E}_{FBHD}, \eta_g)E_0[\rho_1^*(\bar{E}_{FBHD}, \eta_g)]}{B},$$

$$x_{hHD1}(\bar{E}_{FBHD}, \eta_g) = \frac{|e|BL_x}{\hbar} + x_{1HD1}(\bar{E}_{FBHD}, \eta_g)$$

where in the presence of a surface electric field F_s along z direction from (10.59) we can write

$$\alpha_{10,6}(E, \eta_g, n) - eF_S z\beta_{10,6}(E, \eta_g, n) = \frac{\hbar^2 k_z^2}{2} \tag{10.64}$$

where

$$\alpha_{10,6}(E, \eta_g, n) = \left[\frac{M_2^*(E, \eta_g)}{M_1^*(E, \eta_g)}\left\{g^*(E, \eta_g) - \left(n + \frac{1}{2}\right)\hbar eB\right.\right.$$

$$\left.\left. + \frac{E_0}{B}\hbar k_y[\rho_1^*(E, \eta_g)M_1^*(E, \eta_g)]\right\} + \bar{A}\right]$$

$$\bar{A} = \left[\frac{E_0^2}{2B^2}[\{\rho_1^*(E, \eta_g)\}^2 M_1^*(E, \eta_g)M_2^*(E, \eta_g)]\right]$$

and

$$\beta_{10,6}(E, \eta_g, n) = \left[\frac{M_2^*(E, \eta_g)}{M_1^*(E, \eta_g)}\left[\{g^*(E, \eta_g)\}'\right.\right.$$

$$\left. + \frac{E_0}{B}\hbar k_y\{\rho_1^*(E, \eta_g)\}' M_1^*(E, \eta_g)\right] + \frac{E_0}{B}\hbar k_y[\{\rho_1^*(E, \eta_g)\}\{M_1^*(E, \eta_g)\}']$$

$$+\frac{E_0^2}{2B^2}[2\rho_1^*(E,\eta_g)M_1^*(E,\eta_g)M_2^*(E,\eta_g)+\{M_1^*(E,\eta_g)\}'M_2^*(E,\eta_g) \times\{\rho_1^*(E,\eta_g)\}^2+\{M_1^*(E,\eta_g)\}\{M_2^*(E,\eta_g)\}'\{\rho_1^*(E,\eta_g)\}^2]\Bigg]$$

Using (1.7) of Chapter 1 and (10.64), the DR in this case is given by

$$\left[\frac{M_2^*(E,\eta_g)}{M_1^*(E,\eta_g)}\left\{g^*(E,\eta_g)-\left(n+\frac{1}{2}\right)\hbar eB +\frac{E_0}{B}\hbar k_y[\rho_1^*(E,\eta_g)M_1^*(E,\eta_g)]\right\}+\bar{A}\right]=S_i\left(\frac{\hbar eF_s}{\sqrt{2}}\right)^{2/3}\Bigg[\frac{M_2^*(E,\eta_g)}{M_1^*(E,\eta_g)} \times\left[\{g^*(E,\eta_g)\}'+\frac{E_0}{B}\hbar k_y\{\rho_1^*(E,\eta_g)\}'M_1^*(E,\eta_g)\right]+\frac{E_0}{B}\hbar k_y[\{\rho_1^*(E,\eta_g)\} \times\{M_1^*(E,\eta_g)\}']+\frac{E_0^2}{2B^2}[2\rho_1^*(E,\eta_g)M_1^*(E,\eta_g)M_2^*(E,\eta_g)+\{M_1^*(E,\eta_g)\}' \times M_2^*(E,\eta_g)\{\rho_1^*(E,\eta_g)\}^2+\{M_1^*(E,\eta_g)\}\{M_2^*(E,\eta_g)\}'\{\rho_1^*(E,\eta_g)\}^2]\Bigg]^{2/3} \tag{10.65}$$

The sub-band energy $E_{10,6}$ in this case can be written as

$$\left[\frac{M_2^*(E_{10,6},\eta_g)}{M_1^*(E_{10,6},\eta_g)}\left\{g^*(E_{10,6},\eta_g)-\left(n+\frac{1}{2}\right)\hbar eB\right\}+\bar{A}\right]=S_i\left(\frac{\hbar eF_s}{\sqrt{2}}\right)^{2/3} \times\Bigg[\frac{M_2^*(E_{10,6},\eta_g)}{M_1^*(E_{10,6},\eta_g)}[\{g^*(E_{10,6},\eta_g)\}']+\frac{E_0^2}{2B^2}[2\rho_1^*(E_{10,6},\eta_g)M_1^*(E_{10,6},\eta_g) \times M_2^*(E_{10,6},\eta_g)+\{M_1^*(E_{10,6},\eta_g)\}'M_2^*(E_{10,6},\eta_g)\{\rho_1^*(E_{10,6},\eta_g)\}^2 +\{M_1^*(E_{10,6},\eta_g)\}\{M_2^*(E_{10,6},\eta_g)\}'\{\rho_1^*(E_{10,6},\eta_g)\}^2]\Bigg]^{2/3} \tag{10.66}$$

Therefore, the total surface concentration per unit area in this case can be written as

$$N_s=g_v\,\text{Real part of}\sum_{i=0}^{i_{\max}}\sum_{n=0}^{n_{\max}}\left[\frac{2Bt_{i10,6}}{3L_x\pi^2\hbar^2E_0}\left[T_{4131HD}(n,\bar{E}_{FBHD},\eta_g) +\sum_{r=1}^{s}L(r)[T_{4131FBHD}(n,\bar{E}_{FBHD},\eta_g)]\right]+\frac{eB}{\pi\hbar}F_{-1}(\eta_{i10,6})\right] \tag{10.67}$$

where

$$\eta_{i10,6} = \frac{E_{F10,1} - E_{10,6}}{k_B T},$$

$$t_{i10,6} = \frac{E_{10,6\,\max}}{eF_s(1 + i_{\max})}$$

$E_{10,6\,\max}$ can be determined from the equation

$$\begin{aligned}
&\left[\frac{M_2^*(E_{10,6\,\max},\eta_g)}{M_1^*(E_{10,6\,\max},\eta_g)}\left\{g^*(E_{10,6\,\max},\eta_g) - \left(n+\frac{1}{2}\right)\hbar eB\right\} + \bar{A}\right] \\
&\quad = S_{i_{\max}}\left(\frac{\hbar eF_s}{\sqrt{2}}\right)^{2/3}\left[\frac{M_2^*(E_{10,6\,\max},\eta_g)}{M_1^*(E_{10,6\,\max},\eta_g)}[\{g^*(E_{10,6\,\max},\eta_g)\}']\right. \\
&\qquad + \frac{E_0^2}{2B^2}[2\rho_1^*(E_{10,6\,\max},\eta_g)M_1^*(E_{10,6\,\max},\eta_g)M_2^*(E_{10,6\,\max},\eta_g) \\
&\qquad + \{M_1^*(E_{10,6\,\max},\eta_g)\}'M_2^*(E_{10,6\,\max},\eta_g)\{\rho_1^*(E_{10,6\,\max},\eta_g)\}^2 \\
&\qquad \left. + \{M_1^*(E_{10,6\,\max},\eta_g)\}\{M_2^*(E_{10,6\,\max},\eta_g)\}'\{\rho_1^*(E_{10,6\,\max},\eta_g)\}^2]\right]^{2/3}
\end{aligned} \tag{10.68}$$

Using the appropriate equations and (1.4) we can study the QC in this case.

10.2.5 *The QC in MOSFETs of stressed Kane type materials operated under accumulation mode in the presence of cross-fields configuration*

The use of (8.106) can be written as

$$\frac{p_x^2}{2m_1^*(E,\eta_g)} + \frac{p_y^2}{2m_2^*(E,\eta_g)} + \frac{p_z^2}{2m_3^*(E,\eta_g)} = G^*(E,\eta_g) \tag{10.69}$$

where

$$m_1^*(E,\eta_g) = \frac{\hbar^2}{2}[\gamma_0(E,\eta_g) - (T_{17}/2)[1 + Erf(E/\eta_g)]]^{-1}$$

$$m_2^*(E,\eta_g) = \frac{\hbar^2}{2}[\gamma_0(E,\eta_g) - (T_{27}/2)[1 + Erf(E/\eta_g)]]^{-1}$$

$$m_2^*(E,\eta_g) = \frac{\hbar^2}{2}[\gamma_0(E,\eta_g) - (T_{37}/2)[1 + Erf(E/\eta_g)]]^{-1},$$

$$G^*(E,\eta_g) = \Delta_{14}(E,\eta_g)$$

and the other symbols are written in Chapter 8.

In the presence of quantizing magnetic field B along z direction and the electric field along x-axis, from (10.69) one obtains

$$\frac{\hat{p}_x^2}{2m_1^*(E,\eta_g)}+\frac{(\hat{p}_y-|e|B\hat{x})^2}{2m_2^*(E,\eta_g)}+\frac{\hat{p}_z^2}{2m_3^*(E,\eta_g)}$$

$$=G^*(E,\eta_g)+|e|E_0\hat{x}\left[\frac{m_1^*(E,\eta_g)}{m_2^*(E,\eta_g)}\right]^{\frac{1}{2}}\rho^*(E,\eta_g) \tag{10.70}$$

where

$$\rho^*(E,\eta_g)=\frac{\partial}{\partial E}[G^*(E,\eta_g)]$$

Let us define the operator $\hat{\theta}$ as

$$\hat{\theta}=-\hat{p}_y+|e|B\hat{x}-\frac{\rho^*(E,\eta_g)E_0[m_1^*(E,\eta_g)m_2^*(E,\eta_g)]^{1/2}}{B} \tag{10.71}$$

Eliminating $\hat{x}$, between the above two equations, the DR of the conduction electrons in HD stressed Kane type semiconductors in the presence of cross-fields configuration can be expressed as

$$G^*(E,\eta_g)=\left(n+\frac{1}{2}\right)\hbar\bar{\omega}_i(E,\eta_g)+\frac{\hbar^2k_z^2}{2m_3^*(E,\eta_g)}-\frac{E_0}{B}\rho^*(E,\eta_g)$$

$$\times\left[\frac{m_1^*(E,\eta_g)}{m_2^*(E,\eta_g)}\right]^{1/2}\hbar k_y-\frac{E_0^2}{2B^2}[\rho^*(E,\eta_g)]^2m_1^*(E,\eta_g) \tag{10.72}$$

where $\bar{\omega}_i(E,\eta_g)=eB[m_1^*(E,\eta_g)m_2^*(E,\eta_g)]^{-\frac{1}{2}}$.

The use of (10.72) leads to the expressions of the EEMs' along z and y directions as

$$m_z^*(\bar{E}_{FBHD},\eta_g,n,E_0)=\Bigg[[m_3^*(\bar{E}_{FBHD},\eta_g)]'[G^*(\bar{E}_{FBHD},\eta_g)$$

$$-\left(n+\frac{1}{2}\right)\hbar\bar{\omega}_i(\bar{E}_{FBHD},\eta_g)+\frac{E_0^2}{2B^2}[\rho^*(\bar{E}_{FBHD},\eta_g)^2m_1^*(\bar{E}_{FBHD},\eta_g)]$$

$$+[m_3^*(\bar{E}_{FBHD},\eta_g)]\Bigg[[G^*(\bar{E}_{FBHD},\eta_g)]'-\left(n+\frac{1}{2}\right)\hbar[\bar{\omega}_i(\bar{E}_{FBHD},\eta_g)]'$$

$$+\frac{E_0^2}{2B^2}[2\,[\rho^*(\bar{E}_{FBHD},\eta_g)][\rho^*(\bar{E}_{FBHD},\eta_g)'][m_1^*(\bar{E}_{FBHD},\eta_g)]$$

$$+\,[m_1^*(\bar{E}_{FBHD},\eta_g)]'[\rho^*(\bar{E}_{FBHD},\eta_g)]^2]]\Bigg]\Bigg] \tag{10.73}$$

$$m_y^*(\bar{E}_{FBHD}, \eta_g, n, E_0) = (B/E_0)^2[m_4^*(\bar{E}_{FBHD}, \eta_g)]^{-3}\Bigg[G^*(\bar{E}_{FBHD}, \eta_g) - \left(n+\frac{1}{2}\right)\hbar\bar{\omega}_i(\bar{E}_{FBHD}, \eta_g) + \frac{E_0^2}{2B^2}[\rho^*(\bar{E}_{FBHD}, \eta_g)^2 m_1^*(\bar{E}_{FBHD}, \eta_g)] \times \Bigg[[m_4^*(\bar{E}_{FBHD}, \eta_g)][G^*(\bar{E}_{FBHD}, \eta_g)]' - \left(n+\frac{1}{2}\right)\hbar\,[\bar{\omega}_i(\bar{E}_{FBHD}, \eta_g)' + \frac{E_0^2}{2B^2}[[\rho^*(\bar{E}_{FBHD}, \eta_g)^2 m_1^*(\bar{E}_{FBHD}, \eta_g)]'][m_4^*(\bar{E}_{FBHD}, \eta_g)]' \times [G^*(\bar{E}_{FBHD}, \eta_g)] - \left(n+\frac{1}{2}\right)\hbar\bar{\omega}_i(\bar{E}_{FBHD}, \eta_g) + \frac{E_0^2}{2B^2}[\rho^*(\bar{E}_{FBHD}, \eta_g)^2 m_1^*(\bar{E}_{FBHD}, \eta_g)]\Bigg]\Bigg] \tag{10.74}$$

where

$$m_4^*(\bar{E}_{FBHD}, \eta_g) = \left[[\rho^*(\bar{E}_{FBHD}, \eta_g)]\left[\frac{m_1^*(\bar{E}_{FBHD}, \eta_g)}{m_2^*(\bar{E}_{FBHD}, \eta_g)}\right]^{\frac{1}{2}}\right]$$

The Landau level energy (E_{n_g}) in this case can be expressed through the equation

$$G^*(E_{n_9}, \eta_g) = \left(n+\frac{1}{2}\right)\hbar\bar{\omega}_i(E_{n_8}, \eta_g) - \frac{E_0^2}{2B^2}[\rho^*(E_{n_8}, \eta_g)^2 m_1^*(E_{n_8}, \eta_g) \tag{10.75}$$

The electron concentration can be written as

$$n_0 = \frac{2Bg_v}{3L_x\pi^2\hbar^2 E_0}\sum_{n=0}^{n_{\max}}\left[T_{413FBHD}(n, \overline{E}_{FBHD}, \eta_g) + \sum_{r=1}^{s} L(r)[T_{413FBHD}(n, \bar{E}_{FBHD}, \eta_g)]\right] \tag{10.76}$$

where

$$T_{413HD}(n, \overline{E}_{FBHD}, \eta_g) = \left[\frac{\sqrt{2m_3^*(\overline{E}_{FBHD}, \eta_g)}}{\rho^*(\overline{E}_{FBHD}, \eta_g)}\right]\Bigg[\Bigg[T_5(n, \overline{E}_{FBHD}, \eta_g) + \frac{E_0}{B}\rho^*(\overline{E}_{FBHD}, \eta_g)\hbar x_{hHD}(\overline{E}_{FBHD}, \eta_g)\rho^*(\overline{E}_{FBHD}, \eta_g)\Bigg]^{\frac{3}{2}}$$

$$- \left[T_5(n, \overline{E}_{FBHD}, \eta_g) + \frac{E_0}{B} \rho^*(\overline{E}_{FBHD}, \eta_g) \hbar x_{lHD}(\overline{E}_{FBHD}, \eta_g) \right.$$

$$\left. \times \left. \rho^*(\overline{E}_{FBHD}, \eta_g) \right]^{\frac{3}{2}} \right]$$

$$T_5(n, \overline{E}_{FBHD}, \eta_g) = \left[G^*(\overline{E}_{FBHD}, \eta_g) - \left(n + \frac{1}{2}\right) \hbar\overline{\omega_i}(\overline{E}_{FBHD}, \eta_g) \right.$$

$$\left. + \frac{m_1^*(\overline{E}_{FBHD}, \eta_g) E_0^2}{2B^2} [\rho^*(\overline{E}_{FBHD}, \eta_g)]^2 \right]$$

$$x_{lHD}(\overline{E}_{FBHD}, \eta_g) = \frac{-m_1^*(\overline{E}_{FBHD}, \eta_g) E_0 [\rho^*(\overline{E}_{FBHD}, \eta_g)]}{B}, \tag{10.77a}$$

and

$$x_{hHD}(\overline{E}_{FBHD}, \eta_g) = \frac{|e| B L_x}{\hbar} + x_{lHD}(\overline{E}_{FBHD}, \eta_g) \tag{10.77b}$$

In the presence of a surface electric field F_s along z direction from (10.72) we can write

$$\alpha_{10,7}(E, \eta_g, n) - eF_S z \beta_{10,7}(E, \eta_g, n) = \frac{\hbar^2 k_z^2}{2} \tag{10.78}$$

$$\alpha_{10,7}(E, \eta_g, n) = \left[J_1(E, \eta_g) - \left(n + \frac{1}{2}\right) J_2(E, \eta_g) + \hbar k_y J_3(E, \eta_g) \right.$$

$$\left. + \frac{E_0^2}{2B^2} J_4(E, \eta_g) \right]$$

$$J_1(E, \eta_g) = [G^*(E, \eta_g) m_3^*(E, \eta_g)], \quad J_2(E, \eta_g) = [\hbar\bar{\omega}_i(E, \eta_g) m_3^*(E, \eta_g)]$$

$$J_3(E, \eta_g) = \left[\frac{E_0}{B} \rho^*(E, \eta_g) \left[\frac{m_1^*(E, \eta_g)}{m_2^*(E, \eta_g)} \right]^{1/2} m_3^*(E, \eta_g) \right],$$

$$J_4(E, \eta_g) = [[\rho^*(E, \eta_g)]^2 m_3^*(E, \eta_g) m_2^*(E, \eta_g)]$$

$$\beta_{10,7}(E, \eta_g, n) = \left[J_1'(E, \eta_g) - \left(n + \frac{1}{2}\right) J_2'(E, \eta_g) + \hbar k_y J_3'(E, \eta_g) \right.$$

$$\left. + \frac{E_0^2}{2B^2} J_4'(E, \eta_g) \right]$$

Using (1.7) of Chapter 1 and (10.78), the DR in this case is given by

$$\left[J_1(E,\eta_g) - \left(n+\frac{1}{2}\right)J_2(E,\eta_g) + \hbar k_y J_3(E,\eta_g) + \frac{E_0^2}{2B^2}J_4(E,\eta_g)\right]$$
$$= S_i\left[\frac{\hbar e F_s}{\sqrt{2}}\right]^{2/3}\left[J_1'(E,\eta_g) - \left(n+\frac{1}{2}\right)J_2'(E,\eta_g) + \hbar k_y J_3'(E,\eta_g)\right.$$
$$\left. + \frac{E_0^2}{2B^2}J_4'(E,\eta_g)\right]^{2/3} \quad (10.79)$$

The sub-band energy $E_{10,7}$ in this case can be written as

$$\left[J_1(E_{10,7},\eta_g) - \left(n+\frac{1}{2}\right)J_2(E_{10,7},\eta_g) + \frac{E_0^2}{2B^2}J_4(E_{10,7},\eta_g)\right]$$
$$= S_i\left[\frac{\hbar e F_s}{\sqrt{2}}\right]^{2/3}\left[J_1'(E_{10,7},\eta_g) - \left(n+\frac{1}{2}\right)J_2'(E_{10,7},\eta_g)\right.$$
$$\left. + \frac{E_0^2}{2B^2}J_4'(E_{10,7},\eta_g)\right]^{2/3} \quad (10.80)$$

Therefore, the total surface concentration per unit area in this case can be written as

$$N_s = g_v\sum_{i=0}^{i_{\max}}\sum_{n=0}^{n_{\max}}\left[\frac{2Bt_{i\,10,7}}{3L_x\pi^2\hbar^2E_0}[T_{413FBHD}(n,\overline{E}_{FBHD},\eta_g)]\right.$$
$$\left. + \sum_{r=1}^{s}L(r)[T_{413FBHD}(n,\bar{E}_{FBHD},\eta_g)] + \frac{eB}{\pi\hbar}F_{-1}(\eta_{i10,7})\right] \quad (10.81)$$

where

$$\eta_{i10,7} = \frac{E_{F10,1} - E_{10,7}}{k_BT},$$

$$t_{i10,7} = \frac{E_{10,7\,\max}}{eF_s(1+i_{\max})}$$

$E_{10,7\,\max}$ can be determined from the equation

$$\left[J_1(E_{10,7\,\max},\eta_g) - \left(n+\frac{1}{2}\right)J_2(E_{10,7\,\max},\eta_g) + \frac{E_0^2}{2B^2}J_4(E_{10,7\,\max},\eta_g)\right]$$
$$= S_{i_{\max}}\left[\frac{\hbar eF_s}{\sqrt{2}}\right]^{2/3}\left[J_1'(E_{10,7\,\max},\eta_g) - \left(n+\frac{1}{2}\right)J_2'(E_{10,7\,\max},\eta_g)\right.$$
$$\left. + \frac{E_0^2}{2B^2}J_4'(E_{10,7\,\max},\eta_g)\right]^{2/3} \tag{10.82}$$

Using the appropriate equations and (1.4) we can study the QC in this case.

10.3 Summary and Conclusions

In Section 10.2.1, we have formulated the expressions of the EEM and the electron statistics in 2D MOSFETs of nonlinear optical materials under cross-fields configurations in the presence of accumulation mode of operation as given by (10.5), (10.6) and (10.15), respectively. By combining (10.15) and (1.4), we can study QC in this case. In Section 10.2.2, we have formulated the expressions of EEMs, sub-band energy and the electron statistics in 2D MOSFETs of III–V, ternary and quaternary compounds under cross-fields configurations in the presence of accumulation mode of operation as given by (10.18), (10.19), (10.25) and (10.26), respectively, on the basis of three-band model of Kane. By combining (10.26) and (1.4), we can study QC in this case. In the same section, we have also formulated the expressions of EEMs, sub-band energy and the electron statistics in 2D MOSFETs of III–V, ternary and quaternary compounds under cross-fields configurations in the presence of accumulation mode of operation as given by (10.29), (10.30), (10.35) and (10.36), respectively, on the basis of two-band model of Kane and by combining (10.36) and (1.4) we can study QC in this case. In the same section, we have also studied the results for isotropic parabolic energy bands under the condition $\alpha \to 0$. In Section 10.2.3, we have formulated the expressions of EEMs, sub-band energy and the electron statistics in 2D MOSFETs of II–VI compounds under cross-fields configurations in the presence of accumulation mode of operation as given by (10.48), (10.49), (10.53) and (10.54), respectively on the basis of three-band model of Kane. By combining (10.54) and (1.4), we can study QC in this case. In Section 10.2.4, we have formulated the expressions of sub-band energy, and the electron statistics in 2D MOSFETs of IV–VI compounds under cross-fields configurations in the presence of accumulation mode of operation as given by (10.59b), (10.60), (10.66) and (10.67), respectively. By combining (10.67) and (1.4), we can study QC in this case. In Section 10.2.5, we have formulated the expressions of EEMs, sub-band energy and the electron statistics in 2D MOSFETs of stressed Kane type materials under cross-field's configurations in the presence of accumulation mode of operation as given by (10.73), (10.74), (10.80) and (10.81), respectively. By combining (10.81) and (1.4) we can study QC in this case.

We have not given the graphs of the QCs' for 2D MOSFETs under cross-field configurations made of many materials as discussed here since we feel that our readers will perform the intricate numerical computation to obtain the oscillatory plots for better assessment and joy of understanding.

10.4 Open Research Problems

R.10.1 Investigate the QCs in the presence of an arbitrarily oriented crossed electric and quantizing magnetic fields in 2D MOSFETs of nonlinear optical materials under accumulation mode of operation as studied in this chapter. Draw the corresponding graphs and explain the results.

R.10.2 Investigate the QCs in 2D MOSFETs of IV–VI, II–VI and stressed Kane type compounds in the presence of an arbitrarily oriented crossed electric and quantizing magnetic fields under accumulation mode of operation. Draw the corresponding graphs and explain the results.

R.10.3 Investigate the QCs in 2D MOSFETs of all the materials as stated in R.1.1 of Chapter 1 in the presence of an arbitrarily oriented crossed electric and quantizing magnetic fields under accumulation mode of operation. Draw the corresponding graphs and explain the results.

R.10.4 Investigate the QCs in 2D MOSFETs of all the materials as studied in this chapter by including the electron spin in the presence of an arbitrarily oriented crossed non-uniform electric and oscillatory quantizing magnetic fields under accumulation mode of operation. Draw the corresponding graphs and explain the results.

R.10.5 Investigate the QCs in 2D MOSFETs of IV–VI, II–VI and stressed Kane type compounds in the presence of an arbitrarily oriented crossed non-uniform electric and oscillatory quantizing magnetic fields under accumulation mode of operation by including the electron spin. Draw the corresponding graphs and explain the results.

R.10.6 Investigate the QCs in 2D MOSFETs of all the materials as stated in R.1.1 of Chapter 1 in the presence of an arbitrarily oriented crossed non-uniform electric and oscillatory quantizing magnetic fields under accumulation mode of operation by including electron spin. Draw the corresponding graphs and explain the results.

R.10.7 Investigate the QCs in 2D MOSFETs of all the problems from R.2.1 to R.2.6 in the presence of an arbitrarily oriented crossed non-uniform electric and oscillatory quantizing magnetic fields under accumulation mode of operation. Draw the corresponding graphs and explain the results.

R.10.8 Investigate the QCs in 2D MOSFETs of all the problems from R.2.1 to R.2.3 in the presence of arbitrarily oriented crossed electric and magnetic fields under accumulation mode of operation in the presence of non-uniform strain. Draw the corresponding graphs and explain the results.

R.10.9 Investigate the QCs in 2D MOSFETs' of all the problems from R.2.1 to R.2.8 in the presence of surface states. Draw the corresponding graphs and explain the results.

R.10.10 Investigate the QCs in 2D MOSFETs of all the problems from R.2.1 to R.2.8 in the presence of hot electron effects. Draw the corresponding graphs and explain the results.

Chapter 11

The QC in Heavily Doped Quantum Well Transistors (HDQWT) of Important Materials

I must invite life long struggle to be desire free.

11.1 Introduction

It appears from the literature that the QC in HDQWTs of important materials has yet to be investigated. From Sections 11.2.1 up to 11.2.11 of theoretical background 11.2, we have investigated the QC in HDQWTs' made of tetragonal and nonlinear optical, III–V, ternary and quaternary, II–VI, IV–VI, stressed Kane type, Te, GaP, $PtSb_2$, Bi_2Te_3, Ge and GaSb, respectively, by formulating the electron statistics in each case. We have also derived the expressions of the EEMs. Section 11.3 contains the summary and conclusion and Section 11.4 presents few open research problems in this context.

11.2 Theoretical Background

11.2.1 *The QC in HDQWTs of tetragonal and nonlinear optical materials*

For dimensional quantization along z-direction, the DR of the 2D electrons in this case can be written following (8.25) as

$$\frac{\hbar^2(n_z\pi/d_z)^2}{2m_{||}^*T_{21}(E,\eta_g)}+\frac{\hbar^2k_s^2}{2m_\perp^*T_{22}(E,\eta_g)}=1 \tag{11.1}$$

where $n_z(=1,2,3,\ldots)$ and d_z are the size quantum number and the nanothickness along the z-direction, respectively.

It has already been written that the general expression of the total 2D DOS $N_{2DT}(E)$ can, in general, be expressed following (1.11) as

$$N_{2DT}(E) = \frac{2g_v}{(2\pi)^2} \sum_{n_z=1}^{n_{z\max}} \frac{\partial}{\partial E}[A(E, n_z)H(E - E_{n_z})] \qquad (11.2)$$

where $A(E, n_z)$ is the area of the constant energy 2D wave vector space, $H(E - E_{n_z})$ is the Heaviside step function and E_{n_z} is the corresponding sub-band energy.

Using (11.1) and (11.2), we get

$$N_{2DT}(E) = \frac{m_\perp^* g_v}{\pi\hbar^2} \sum_{n_z=1}^{n_{z\max}} T'_{1D}(E, \eta_g, n_z)H(E - E_{n_z D_1}) \qquad (11.3)$$

where

$$T_{1D}(E, \eta_g, n_z) = \left[1 - \frac{\hbar^2(n_z\pi/d_z)^2}{2m_{||}^* T_{21}(E, \eta_g)}\right] T_{22}(E, \eta_g)$$

and the sub-band energies $E_{n_z D_1}$ in this case is given by the following equation:

$$\frac{kh^2(n_z\pi/d_z)^2}{2m_{||}^* T_{21}(E_{n_z D_1}, \eta_g)} = 1 \qquad (11.4)$$

Thus, we observe that both the total DOS and sub-band energies in HDQWTs of tetragonal and nonlinear optical materials are complex due to the presence of the poles in energy axis of the corresponding materials in the absence of band tails.

The EEM in this case is given by

$$m^*(E_{F_1HD}, \eta_g, n_z) = m_\perp^*[\text{Real part of } T'_{1D}(E_{F_1HD}, \eta_g, n_z)] \qquad (11.5)$$

where E_{F_1HD}, is the Fermi energy in the presence of size quantization of the HDQWTs of tetragonal and nonlinear optical materials as measured from the edge of the conduction band in the vertically upward direction in the absence of any perturbation

Under the condition of extreme carrier degeneracy, the 2D carrier statistics in this case assumes the form

$$n_{2D} = \text{ Real part of } \bar{n}_{2D} \qquad (11.6)$$

where

$$\bar{n}_{2D} = \frac{m_\perp^* g_v}{\pi\hbar^2} \sum_{n_z=1}^{n_{\max}} [[T_{1D}(E_{F1HD}, \eta_g, n_z)]] \qquad (11.7a)$$

The QC can be written as

$$C_g = e\frac{\partial n_{2D}}{\partial V_g} \tag{11.7b}$$

Using (11.7b) and (11.7a) we can study the QC in this case.

11.2.2 *The QC in HDQWTs of III–V, ternary and quaternary materials*

(a) Three-band model of Kane

For dimensional quantization along z-direction, the DR of the 2D electrons in this case can be written using (8.39) as

$$\frac{\hbar^2(n_z\pi/d_z)^2}{2m_c} + \frac{\hbar^2(k_s)^2}{2m_c} = T_{31}(E,\eta_g) + iT_{32}(E,\eta_g) \tag{11.8}$$

The expression of the $N_{2DT}(E)$ in this case assumes the form

$$N_{2DT}(E) = \frac{m_c g_v}{\pi\hbar^2}\sum_{n_z=1}^{n_{z\max}} T'_{5D}(E,\eta_g,n_z)H(E-E_{n_z D5}) \tag{11.9}$$

where $T_{5D}(E,\eta_g,n_z) = [T_{31}(E,\eta_g) + iT_{32}(E,\eta_g) - \hbar^2(n_z\pi/d_z)^2(2m_c)^{-1}]$ and the sub-band energies $E_{n_z D5}$ in this case given by

$$\{\hbar^2(n_z/d_z)^2\}(2m_c)^{-1} = T_{31}(E_{n_z D5},\eta_g) \tag{11.10}$$

Thus we observe that both the total DOS in HDQWTs of III–V, ternary and quaternary compounds and the sub-band energies are complex due to the presence of the pole in energy axis of the corresponding materials in the absence of band tails.

The EEM in this case is given by

$$m^*(E_{F1HD},\eta_g,n_z) = m_c[T'_{31}(E_{F1HD},\eta_g,n_z)] \tag{11.11}$$

The 2D carrier statistics under the condition of extreme carrier degeneracy in this case can be expressed as

$$n_{2D} = \text{Real part of } \bar{n}_{2D} \tag{11.12}$$

where

$$\bar{n}_{2D} = \frac{m_c g_v}{\pi\hbar^2}\sum_{n_z-1}^{n_{\max}} [T_{5D}(E_{F1HD},\eta_g,n_z)] \tag{11.13}$$

Using (11.7b) and (11.13) we can study the QC in this case.

(b) Two-band model of Kane

For dimensional quantization along z-direction, the dispersion relation of the 2D electrons in this case can be written following (8.50) as

$$\frac{\hbar^2(n_z\pi/d_z)^2}{2m_c} + \frac{\hbar^2(k_s)^2}{2m_c} = \gamma_2(E, \eta_g) \tag{11.14}$$

The expression of the $N_{2DT}(E)$ in this case can be written as

$$N_{2DT}(E) = \frac{m_c g_v}{\pi\hbar^2} \sum_{n_z=1}^{n_{z\max}} [T_{7D}(E, \eta_g, n_z)]' H(E - E_{n_z D7}) \tag{11.15}$$

where $T_{7D}(E, \eta_g, n_z) = [\gamma_2(E, \eta_g) - \hbar^2(n_z\pi/d_z)^2(2m_c)^{-1}]$.

The sub-band energies $E_{n_z D7}$ in this case given by

$$\{\hbar^2(n_z\pi/d_z)^2\}(2m_c)^{-1} = \gamma_2(E_{n_z D7}, \eta_g) \tag{11.16}$$

Thus, we observe that both the total DOS and sub-band energies of HDQWTs of III–V ternary and quaternary compounds in accordance with two-band model of Kane are not at all complex since the DR in accordance with the said model has no pole in the finite complex plane.

The EEM in this case is given by

$$m^*(E_{F1HD}, \eta_g, n_z) = m_c[\gamma_2(E_{F1HD}, \eta_g, n_z)]' \tag{11.17}$$

The electron statistics under the condition of extreme carrier degeneracy in this case assumes the form

$$n_{2D} = \frac{m_c g_v}{\pi\hbar^2} \sum_{n_z=1}^{n_{z\max}} [T_{7D}(E_{FLHD}, \eta_g, n_z)] \tag{11.18}$$

Using (11.7b) and (11.18) we can study the QC in this case

(c) Parabolic band model

For dimensional quantization along z-direction, the DR of the 2D electrons in this case can be written following (8.60) as

$$\frac{\hbar^2(n_z z/d_z)^2}{2m_c} + \frac{\hbar^2(k_s)^2}{2m_c} = \gamma_3(E, \eta_g) \tag{11.19}$$

The expression of the $N_{2DT}(E)$ in this case can be written as

$$N_{2DT}(E) = \frac{m_c g_v}{\pi\hbar^2} \sum_{n_z=1}^{n_{z\max}} T'_{9D}(E, \eta_g, n_z) H(E - E_{n_z D9}) \tag{11.20}$$

where $T_{9D}(E, \eta_g, n_z) = [\gamma_3(E, \eta_g) - \hbar^2(n_z\pi/d_z)^2(2m_c)^{-1}]$.

The sub-band energies $E_{n_z D9}$ in this case given by

$$\{\hbar^2(n_z\pi/d_z)^2\}(2m_c)^{-1} = \gamma_3(E_{n_z D9}, \eta_g) \tag{11.21a}$$

The EEM in this case can be written as

$$m^*(E_{F1HD}, \eta_g, n_z) = m_c[\gamma_3'(E_{F1HD}, \eta_g)] \tag{11.21b}$$

The carrier statistics under the condition of extreme carrier degeneracy in this case assumes the form

$$n_{2D} = \frac{m_c g_v}{\pi\hbar^2} \sum_{n_z=1}^{n_{z\max}} [T_{9D}(E_{F1HD}, \eta_g, n_z)] \tag{11.22}$$

Using (11.7b) and (11.22) we can study the QC in this case.

(d) The model of Stillman *et al.*

In the presence of band tails, (2.64) gets transformed as

$$\frac{\hbar^2 k^2}{2m_c} = I_{12}(E, \eta_g) \tag{11.23}$$

where

$$I_{12}(E, \eta_g) = a_{11}[1 - (1 - a_{12}\gamma_3(E, \eta_g))^{1/2}] \quad \text{and}$$

$$a_{11} \equiv \left(\frac{\hbar^2 \alpha_5}{4m_c\alpha_6}\right) \quad \text{and} \quad a_{12} = \frac{4\alpha_6}{\alpha_5^2}$$

For dimensional quantization along z-direction, the DR of the 2D electrons in this case can be written following (11.23) as

$$\frac{\hbar^2(n_z\pi/d_z)^2}{2m_c} + \frac{\hbar^2(k_s)^2}{2m_c} = I_{12}(E, \eta_g) \tag{11.24}$$

The expression of the $N_{2DT}(E)$ in this case can be written as

$$N_{2DT}(E) = \frac{m_c g_v}{\pi\hbar^2} \sum_{n_z=1}^{n_{z\max}} T_{11D}'(E, \eta_g, n_z) H(E - E_{n_z D11}) \tag{11.25}$$

where $T_{11D}((E, \eta_g, n_z) = [I_{12}(E, \eta_g) - \hbar^2(n_z\pi/d_z)^2(2m_c)^{-1}]$.

The sub-band energies $E_{n_z D11}$ in this case given by

$$\{\hbar^2(n_z\pi/d_z)^2\}(2m_c)^{-1} = I_{12}(E_{n_z D11}, \eta_g) \tag{11.26}$$

The EEM in this case assumes the form

$$m^*(E_{F1HD}, \eta_g, n_z) = m_c\{I_{12}'(E_{F1HD}, \eta_g, n_z)] \tag{11.27}$$

The 2D electron statistics under the condition of extreme carrier degeneracy in this case can be written as

$$n_{2D} = \frac{m_c g_v}{\pi\hbar^2} \sum_{n_z=1}^{n_{z\max}} [T_{11D}(E_{F1HD}, \eta_g, n_z)] \tag{11.28}$$

Using (11.7b) and (11.28) we can study the QC in this case.

(e) The model of Palik *et al.*

The energy spectrum of the conduction electrons in III–V semiconductors up to the fourth-order in effective mass theory, taking into account the interactions of heavy hole, light hole and the split-off holes can be expressed in accordance with the model of Palik *et al.* [1] as

$$E = \frac{\hbar^2 k^2}{2m_c} - \bar{B}_{11} k^4 \tag{11.29}$$

where

$$\bar{B}_{11} = \left[\frac{\hbar^4}{4E_{g_0}(m_c)^2}\right]\left[\frac{1+\frac{x_{11}^2}{2}}{1+\frac{x_{11}}{2}}\right](1-y_{11})^2, \quad x_{11}\left[1+\left(\frac{\Delta}{E_{g_0}}\right)\right]^{-1} \text{ and}$$

$$y_{11} = \frac{m_c}{m_0}$$

The (11.29) gets simplified as

$$\frac{\hbar^2 k^2}{2m_c} = I_{13}(E) \tag{11.30}$$

where

$$I_{13}(E) = \bar{b}_{12}[\bar{a}_{12} - ((\bar{a}_{12})^2 - 4E\bar{B}_{11})^{1/2}], \ \bar{a}_{12} = \left(\frac{\hbar^2}{2m_c}\right) \text{ and } \bar{b}_{12} = \left[\frac{\bar{a}_{12}}{2\bar{B}_{11}}\right]$$

Under the condition of heavy doping forming Gaussian band tails, (11.30) assumes the form

$$\frac{\hbar^2 k^2}{2m_c} = I_{13}(E, \eta_g) \tag{11.31}$$

where $I_{13}(E, \eta_g) = \bar{b}_{12}[\bar{a}_{12} - ((\bar{a}_{12})^2 - 4\bar{B}_{11}\gamma_3(E, \eta_g))^{1/2}]$.

For dimensional quantization along z-direction, the DR of the 2D electrons in this case can be written following (11.31) as

$$\frac{h^2(n_z\pi/d_z)^2}{2m_c} + \frac{\hbar^2(k_s)^2}{2m_c} = I_{13}(E, \eta_g) \tag{11.32}$$

The expression of the $N_{2DT}(E)$ in this case can be written as

$$N_{2DT}(E) = \frac{m_c g_v}{\pi \hbar^2} \sum_{n_z=1}^{n_{z\max}} T'_{13D}(E, \eta_g, n_z) H(E - E_{n_z D13}) \tag{11.33}$$

where $T_{13D}(E, \eta_g, n_z) = [I_{13}(E, \eta_g) - \hbar^2 (n_z \pi / d_z)^2 (2m_c)^{-1}]$.

The sub-band energies $E_{n,D13}$ in this case given by

$$\{\hbar (n_z \pi / d_z)^2\} (2m_c)^{-1} = I_{13}(E_{n_z D13}, \eta_g) \tag{11.34}$$

The EEM in this case can be expressed as

$$m^*(E_{F1HD}, \eta_g, n_z) = m_c [I'_{13}(E_{F1HD}, \eta_g, n_z)] \tag{11.35}$$

The 2D electron statistics under the condition of extreme carrier degeneracy in this case can be written as

$$n_{2D} = \frac{m_c g_v}{\pi \hbar^2} \sum_{n_z=1}^{n_{\max}} [T_{13D}(E_{F1HD}, \eta_g, n_z)] \tag{11.36}$$

Using (11.7b) and (11.36) we can study the QC in this case.

11.2.3 *The QC in HDQWTs of II–VI materials*

The DR of the conduction electrons of HDQWTs of II–VI materials for dimensional quantization along z-direction can be written following (8.70) as

$$\gamma_3(E, \eta_g) = a'_0 k_s^2 + b'_0 \left(\frac{\pi n_z}{d_z} \right)^2 \pm \bar{\lambda}_0 k_s \tag{11.37}$$

The EEM can be expressed following (11.37) as

$$m^*(E_{F1HD}, n_z, \eta_g) = m_\perp^* \left[1 \mp \frac{(\bar{\lambda}_0)}{[(\bar{\lambda}_0)^2 - 4a'_0 b'_0 \left(\frac{n_z \pi}{d_z} \right)^2 + 4a'_0 \gamma_3(E_{F1HD}, \eta_g)]^{1/2}} \right] \times \gamma'_3(E_{F1HD}, \eta_g) \tag{11.38}$$

Thus, we observe that the doubled valued effective mass in 2D HDQWTs of II–VI materials is a function of Fermi energy, size quantum number and the screening potential respectively together with the fact that the same mass exists in the band gap due to the sole presence of the splitting of the two-spin states by the spin-orbit coupling and the crystalline field.

The sub-band energy in this case is given by

$$\gamma_3(E_{n_z D14}, \eta_g) = b_0' \left(\frac{\pi n_z}{d_z}\right)^2 \tag{11.39}$$

The surface electron concentration under the condition of extreme carrier degeneracy at low temperatures in this case assumes the form

$$n_{2D} = \frac{g_v m_\perp^*}{\pi \hbar^2} \sum_{nz=1}^{n_{z\max}} (\gamma_3(E_{F1HD}, \eta_g) - E_{nzD_{14}} + (\bar{\lambda})^2 m_\perp^* \hbar^{-2}) \tag{11.40}$$

Using (11.7b) and (11.40) we can study the QC in this case.

11.2.4 *The QC in HDQWTs of IV–VI materials*

(a) Dimmock model

Following (3.56), the HD 2D DR in this case assumes the form

$$\begin{aligned}
&\gamma_2(E, \eta_g) + \alpha\gamma_3(E, \eta_g)\left(\frac{\hbar^2 k_x^2}{2x_4} + \frac{\hbar^2 k_y^2}{2x_5}\right) + \alpha\gamma_3(E, \eta_g)\frac{\hbar^2}{2x_6}\left(\frac{n_z\pi}{d_z}\right)^2 \\
&\quad - (1 + \alpha\gamma_3(E, \eta_g))\left(\frac{\hbar^2 k_x^2}{2x_1} + \frac{\hbar^2 k_y^2}{2x_2}\right) - \alpha\left(\frac{\hbar^2 k_x^2}{2x_1} + \frac{\hbar^2 k_y^2}{2x_2}\right) \\
&\quad \times \left(\frac{\hbar^2 k_x^2}{2x_4} + \frac{\hbar^2 k_y^2}{2x_5}\right) - \alpha\left(\frac{\hbar^2 k_x^2}{2x_1} + \frac{\hbar^2 k_y^2}{2x_2}\right)\frac{\hbar^2}{2x_6}\left(\frac{n_z\pi}{d_z}\right)^2 \\
&\quad - (1 + \alpha\gamma_3(E, \eta_g))\frac{\hbar^2}{2x_3}\left(\frac{n_z\pi}{d_z}\right)^2 \\
&\quad - \alpha\frac{\hbar^2}{2x_3}\left(\frac{n_z\pi}{d_z}\right)^2\left(\frac{\hbar^2 k_x^2}{2x_4} + \frac{\hbar^2 k_y^2}{2x_5}\right) - \alpha\frac{\hbar^2}{2x_3}\left(\frac{n_z\pi}{d_z}\right)^2\frac{\hbar^2}{2x_6}\left(\frac{n_z\pi}{d_z}\right)^2 \\
&= \frac{\hbar^2 k_x^2}{2m_1} + \frac{\hbar^2 k_y^2}{2m_2} + \frac{\hbar^2}{2m_3}\left(\frac{n_z\pi}{d_z}\right)^2
\end{aligned} \tag{11.41}$$

Substituting, $k_x = r\cos\theta$ and $k_y = r\sin\theta$ (where r and θ are 2D polar coordinates in 2D wave vector space) in (11.41), we can write

$$\begin{aligned}
&r^4\left[\alpha\frac{1}{4}\left(\frac{\hbar^2\cos^2\theta}{x_1} + \frac{\hbar^2\sin^2\theta}{x_2}\right)\left(\frac{\hbar^2\cos^2\theta}{x_4} + \frac{\hbar^2\sin^2\theta}{x_5}\right)\right] \\
&\quad + r^2\frac{1}{2}\left[\left(\frac{\hbar^2\cos^2\theta}{m_1} + \frac{\hbar^2\sin^2\theta}{m_2}\right) + \alpha\frac{\hbar^2}{2x_3}\left(\frac{n_z\pi}{d_z}\right)^2\right.
\end{aligned}$$

$$
\begin{aligned}
&\times\left(\frac{\hbar^2\cos^2\theta}{x_4}+\frac{\hbar^2\sin^2\theta}{x_5}\right)+\alpha\left(\frac{\hbar^2\cos^2\theta}{x_1}+\frac{\hbar^2\sin^2\theta}{x_2}\right)\frac{\hbar^2}{2x_6}\left(\frac{n_z\pi}{d_z}\right)^2\\
&+\hbar^2(1+\alpha\gamma_3(E,\eta_g))\left(\frac{\cos^2\theta}{x_1}+\frac{\sin^2\theta}{x_2}\right)-\hbar^2\alpha\gamma_3(E,\eta_g)\\
&\times\left(\frac{\cos^2\theta}{x_4}+\frac{\sin^2\theta}{x_5}\right)-\left[\gamma_2(E,\eta_g)+\alpha\gamma_3(E,\eta_g)\frac{\hbar^2}{2x_6}\left(\frac{n_z\pi}{d_z}\right)^2\right.\\
&\left.-(1+\alpha\gamma_3(E,\eta_g))\frac{\hbar^2}{2x_3}\left(\frac{n_z\pi}{d_z}\right)^2-\alpha\left(\frac{\hbar^4}{4x_3x_6}\left(\frac{n_z\pi}{d_z}\right)^4\right)\right]=0
\end{aligned}
\tag{11.42}
$$

The area $A(E, n_z)$ of the 2D wave vector space can be expressed as

$$
A(E,\eta_g,n_z)=\bar{J}_{1HD}-\bar{J}_{2HD} \tag{11.43}
$$

where

$$
\bar{J}_{1HD}\equiv 2\int_0^{\pi/2}\frac{c_1}{b_1}d\theta \tag{11.44}
$$

and

$$
\bar{J}_{2HD}\equiv 2\int_0^{\pi/2}\frac{ac_1^2}{b_1^3}d\theta \tag{11.45}
$$

in which

$$
\begin{aligned}
b_1\equiv&\left(\frac{\hbar^2}{2}\right)\left[\left(\frac{\cos^2\theta}{m_1}+\frac{\sin^2\theta}{m_2}\right)+\alpha\left(\frac{\hbar^2}{2x_3}\right)\left(\frac{n_z\pi}{d_z}\right)^2\left[\left(\frac{\cos^2\theta}{x_4}+\frac{\sin^2\theta}{x_5}\right)\right.\right.\\
&+\alpha\left(\frac{\hbar^2}{2x_6}\right)\left(\frac{n_z\pi}{d_z}\right)^2\left[\left(\frac{\cos^2\theta}{m_1}+\frac{\sin^2\theta}{m_2}\right)\right.\\
&\left.\left.\left.+(1+\alpha\gamma_3(E,\eta_g))\left(\frac{\cos^2\theta}{x_1}+\frac{\sin^2\theta}{x_2}\right)-\alpha\gamma_3(E,\eta_g)\left(\frac{\cos^2\theta}{x_4}+\frac{\sin^2\theta}{x_5}\right)\right]\right]\right]
\end{aligned}
$$

and

$$
\begin{aligned}
c_1\equiv&\left[\gamma_2(E,\eta_g)+\alpha\gamma_3(E,\eta_g)\left(\frac{\hbar^2}{2x_6}\right)\left(\frac{n_z\pi}{d_z}\right)^2-(1+\alpha\gamma_3(E,\eta_g))\right.\\
&\left.\times\left(\frac{\hbar^2}{2x_3}\right)\left(\frac{n_z\pi}{d_z}\right)^2-\alpha\left(\frac{\hbar^4}{4x_3x_6}\right)\left(\frac{n_z\pi}{d_z}\right)^4\right]
\end{aligned}
$$

The (11.45) can be expressed as

$$\bar{J}_{1HD} = 2\int_0^{\pi/2} \frac{t_{31}(E,n_z)d\theta}{A_{11}(E,n_z)\cos^2\theta + B_{11}(E,n_z)\sin^2\theta} \tag{11.46}$$

where

$$t_{31}(E,n_z) \equiv c_1, A_{11}(E,n_z) \equiv \frac{\hbar^2}{2m_1}t_{11}(E,n_z)$$

$$t_{11}(E,n_z) \equiv \left[1 + m_1\left[\frac{1}{x_4}\frac{\alpha\hbar^2}{2x_3}\left(\frac{n_z\pi}{d_z}\right)^2 + \frac{\alpha\hbar^2}{2x_1x_6}\left(\frac{n_z\pi}{d_z}\right)^2 + \frac{1+\alpha\gamma_2(E,\eta_g)}{x_1} - \frac{\alpha\gamma_3(E,\eta_g)}{x_4}\right]\right]$$

$$B_{11}(E,n_z) \equiv \frac{\hbar^2}{2m_2}t_{21}(E,n_z)$$

and

$$t_{21}(E,n_z) \equiv \left[1 + m_2\left[\frac{\alpha\hbar^2}{2x_3x_5}\left(\frac{n_z\pi}{d_z}\right)^2 + \frac{\alpha\hbar^2}{2x_2x_6}\left(\frac{n_z\pi}{d_z}\right)^2 + \frac{1+\alpha\gamma_3(E,\eta_g)}{x_2} - \frac{\alpha\gamma_3(E,\eta_g)}{x_5}\right]\right]$$

Performing the integration, we get

$$\bar{J}_{1HD} = \pi t_{31}(E,n_z)[A_{11}(E,n_z)B_{11}(E,n_z)]^{-1/2} \tag{11.47}$$

From (3.148) we can write

$$\bar{J}_{2HD} = \frac{\alpha t_{31}^2(E,n_z)\hbar^4}{2B_{11}^3(E,n_z)}I_{HD} \tag{11.48}$$

where

$$I_{HD} \equiv \int_0^\infty \frac{(a_{1HD}+a_2z^2)(a_3+a_4z^2)dz}{[(\bar{a}_{HD})^2+z^2]^3}, \quad (\bar{a}_{HD})^2 = \left(\frac{A_{11}(E,n_z)}{B_{11}(E,n_z)}\right) \tag{11.49}$$

The use of the Residue theorem leads to the evaluation of the integral in (11.49) as

$$I_{HD} = \frac{\pi}{4\bar{a}_{HD}}[a_1 a_4 + 3a_2 a_4] \tag{11.50}$$

Therefore, the 2D area of the 2D wave vector space can be written as

$$A_{HD}(E, n_z) = \frac{\pi t_{31}(E, n_z)}{\sqrt{A_{11}(E, n_z) B_{11}(E, n_z)}} \left[1 - \frac{1}{x_5}\left(\frac{1}{x_1} + \frac{3}{x_2}\right) \times \frac{\alpha t_{31}(E, n_z)\hbar^4}{8B_{11}^2(E, n_z)}\right] \tag{11.51}$$

The EEM for the HD QWs of IV–VI materials can thus be written as

$$m^*(E, n_z) = \frac{\hbar^2}{2}[\theta_{5HD}(E, n_z)]\Big|_{E=E_{F1HD}} \tag{11.52}$$

where

$$\begin{aligned}
\theta_{5HD}(E, n_z) \equiv {} & \left[1 - \frac{1}{x_5}\left(\frac{1}{x_1} + \frac{3}{x_2}\right)\frac{\alpha t_{31}(E, n_z)\hbar^4}{8[B_{11}(E, n_z)]^2}\right] \\
& \times [A_{11}(E, n_z) B_{11}(E, n_z)]^{-1} \Bigg[\sqrt{A_{11}(E, n_z) B_{11}(E, n_z)} \\
& \times \{t_{31}(E, n_z)\}' - t_{31}(E, n_z)\left\{\frac{1}{2}\{A_{11}(E, n_z)\}'\right. \\
& \left.\times \left[\frac{B_{11}(E, n_z)}{A_{11}(E, n_z)}\right]^{1/2} + \frac{1}{2}\{B_{11}(E, n_z)\}'\left[\frac{A_{11}(E, n_z)}{B_{11}(E, n_z)}\right]^{1/2}\right\}\Bigg] \\
& - \frac{1}{8}\frac{t_{31}(E, n_z)\alpha\hbar^4}{\sqrt{A_{11}(E, n_z) B_{11}(E, n_z)}}\frac{1}{x_5}\left(\frac{1}{x_1} + \frac{3}{x_2}\right)[B_{11}(E, n_z)]^{-4} \\
& \times [\{B_{11}(E, n_z)\}^2\{t_{31}(E, n_z)\}' - 2B_{11}(E, n_z) \\
& \times \{B_{11}(E, n_z)\}' t_{31}(E, n_z)]
\end{aligned}$$

Thus, the EEM is a function of Fermi energy and the quantum number due to the band non-parabolicity.

The total DOS function can be written as

$$N_{2DT}(E) = \left(\frac{g_v}{2\pi}\right)\sum_{n_z=1}^{n_{z\max}} \theta_{5HD}(E, n_z) H(E - E_{n_{z7HD}}) \tag{11.53}$$

where the sub-band energy ($E_{n_{z7}}$) in this case can be written as

$$\gamma_2(E_{n_{27HD}},\eta_g)+\alpha\gamma_3(E_{n_{27HD}},\eta_g)\frac{\hbar^2}{2x_6}\left(\frac{n_z\pi}{d_z}\right)^2 - (1+\alpha\gamma_3(E_{n_{27HD}},\eta_g))\frac{\hbar^2}{2x_3}\left(\frac{n_z\pi}{d_z}\right)^2 - \alpha\frac{\hbar^{2=0}}{2x_3}\left(\frac{n_z\pi}{d_z}\right)^2\frac{\hbar^2}{2x_6}\left(\frac{n_z\pi}{d_z}\right)^2 - \left[\frac{\hbar^2}{2m_3}\left(\frac{n_z\pi}{d_z}\right)^2\right] \quad (11.54)$$

The use of (11.53) leads to the expression of 2D electron statistics under the condition of extreme carrier degeneracy in this case as

$$n_{2D}=\frac{g_v}{2\pi}\sum_{n_z=1}^{n_{\max}}[T_{55HD}(E_{F1HD},n_z)] \quad (11.55)$$

where

$$T_{55HD}(E_{F1HD},n_z)\equiv\frac{A_{HD}(E_{F1HD},n_z)}{\pi}$$

Using (11.7b) and (11.55) we can study the QC in this case.

(b) The model of Bangert and Kastner

The electron energy spectrum in heavily doped IV–VI materials in this case can be expressed using (6.21) as

$$2I(4)=k_s^2\left[\{c_1(\alpha_1E,E_g)-iD_1(\alpha_1,E,E_g)\}\frac{(\bar{R})^2}{E_{g0}}+\{c_2(\alpha_2E,E_g) - iD_2(\alpha_2,E,E_g)\}\frac{(\bar{S})^2}{\Delta_c'}+\{c_3(\alpha_3E,E_g)-iD_3(\alpha_3E,E_g)\} \times\frac{(\bar{Q})^2}{\Delta_c''}\right]+k_z^2\left[\frac{2(\bar{A})^2}{E_{g0}}\{c_1(\alpha_1E,E_g)-iD_1(\alpha_1E,E_g)\} + \frac{(\bar{S}+\bar{Q})^2}{\Delta_c''}\{c_3(\alpha_3E,E_g)-iD_3(\alpha_3E,E_g)\}\right] \quad (11.56)$$

where

$$\alpha_1=\frac{1}{E_g},\quad \alpha_2=\frac{1}{\Delta_c'},\quad \alpha_3=\frac{1}{\Delta_c''},\quad G_t=\frac{1+\alpha_iE}{\eta_g\alpha_i}$$

$$c_i(\alpha_iE,\eta_g)=\left[\frac{2}{\alpha_i\eta_g\sqrt{\pi}}\right]\exp(-u_i^2)\times\left[\sum_{p=1}^{\infty}\{\exp(-p^2/4)(\sinh(pu_i))\}p^{-1}\right],$$

$$i=1,2,3,4,\ldots$$

and

$$D_i(\alpha_i, E, \eta_g) = \left[\frac{\sqrt{\pi}}{\alpha_i \eta_g}\right] \exp(-u_i^2)$$

Therefore, (11.56) can be written as

$$F_{1HD}(E, \eta_g) k_s^2 + F_{2HD}(E, \eta_g) k_z^2 = 1 \tag{11.57}$$

where

$$\begin{aligned} F_{1HD}(E, \eta_g) = [2\gamma_0(E, \eta_g)]^{-1} \Bigg[& \frac{(\bar{R})^2}{E_g} \{C_1(\alpha_1 E, E_g) - iD_1(\alpha_1, E, E_g)\} \\ & + \frac{(\bar{S})^2}{\Delta_c'} \{C_2(\alpha_2, E, E_g) - iD_2(\alpha_2, E, E_g)\} \\ & + \frac{(\bar{Q})^2}{\Delta_c''} \{C_3(\alpha_3, E, E_g) - iD_3(\alpha_3, E, E_g)\} \Bigg] \end{aligned}$$

and

$$\begin{aligned} F_{2HD}(E, \eta_g) = [2\gamma_0(E, \eta_g)]^{-1} \Bigg[& \frac{2(\bar{A})^2}{E_g} \{C_1(\alpha_1, E, \eta_g) - iD_1(\alpha_1, E, \eta_g)\} \\ & + \frac{(\bar{S} + \bar{Q})^2}{\Delta_c''} \{C_3(\alpha_3, E, \eta_g) - iD_3(\alpha_3, E, \eta_g)\} \Bigg] \end{aligned}$$

Since $F_1(E, \eta_g)$ and $F_2(E, \eta_g)$ are complex, the energy spectrum is also complex in the presence of Gaussian band tails.

The EEMs can be written as

$$m_\perp^*(E, \eta_g) = \left(\frac{\hbar^2}{2}\right) \text{ Real part of } \left[\frac{-F_{1HD}'(E, \eta_g)}{F_{1HD}^2(E, \eta_g)}\right] \tag{11.58}$$

$$m_{11}^*(E, \eta_g) = \left(\frac{\hbar^2}{2}\right) \text{ Real part of } \left(\frac{-F_{2HD}'(E, \eta_g)}{F_{2HD}^2(E, \eta_g)}\right) \tag{11.59}$$

It appears then that, the evolution of the masses needs an expression of the carrier concentration, which in turn is determined by the DOS function.

The 2D DR in this case assumes the form

$$k_s^2 = F_6(E, \eta_g, n_z) \tag{11.60}$$

where

$$F_6(E, \eta_g, n_z) = \left[\frac{1 - F_2 F_6(E, \eta_g)(n_z \pi / d_z)}{F_1(E, \eta_g)}\right]$$

The EEM in this case is given by

$$m^*(E_{F1HD}, \eta_g, n_z) = \frac{\hbar^2}{2} \text{Real part of } [F_6'(E_{F1HD}, \eta_g, n_z)] \tag{11.61}$$

The total DOS function can be written as

$$N_{2DT}(E) = \frac{g_v}{2\pi} \sum_{n_z=1}^{n_{z\max}} F_6'(E, \eta_g, n_z) H(E - E_{n_z 71HD}) \tag{11.62}$$

where $E_{n_z 71HD}$ is the quantized energy in this case and is given by

$$1 = F_{26}(E_{n_z} 71HD, \eta_g)(\pi n_z/d_z)^2 \tag{11.63}$$

The surface electron concentration under the condition of extreme carrier degeneracy can be expressed as

$$n_0 = \text{Real part of } \bar{n}_{2D} \tag{11.64}$$

where

$$\bar{n}_{2D} = \frac{g_v}{2\pi} \left[\sum_{n_z=1}^{n_{z\max}} [F_6(E_{F1HD}, \eta_g, n_z)] \right] \tag{11.65}$$

Using (11.7b) and (11.64) we can study the QC in this case.

11.2.5 *The QC in HDQWTs of stressed Kane type materials*

The DR of the conduction electrons in HD QWs of Kane type semiconductors can be written using (8.106) as

$$P_{11}(E, \eta_g) k_x^2 + Q_{11}(E, \eta_g) k_y^2 + S_{11}(E, \eta_g) \left(\frac{\pi n_z}{d_z} \right)^2 \equiv 1 \tag{11.66}$$

The EEM can be expressed as

$$m^*(E_{F1HD}, \eta_g, n_z) = \frac{\hbar^2}{2} A_{56}'(E_{F1HD}, \eta_g, n_z) \tag{11.67}$$

where

$$A_{56}(E, \eta_g, n_z) = \frac{\pi[1 - S_{11}(E, \eta_g)(\eta_z \pi/d_z)^2]}{\sqrt{P_{11}(E, \eta_g) Q_{11}(E, \eta_g)}}$$

From (11.67), it appears that the EM is a function of Fermi energy, and size quantum number and the same mass exists in the band gap.

Thus, the total 2D DOS function can be expressed as

$$N_{2DT}(E) = \frac{g_v}{2\pi} \sum_{n_z=1}^{n_{z\max}} A_{56}'(E_{F1HD}, \eta_g, n_z) H(E - E_{n_{z8}HD}) \tag{11.68}$$

The sub-band energies ($E_{n_{z8HD}}$) are given by

$$S_{11}(E_{n_{z8HD}}, \eta_g)(\pi n_z/d_z)^2 = 1 \tag{11.69}$$

The 2D surface electron concentration per unit area under the condition of extreme carrier degeneracy in this case can be written as

$$n_{2D} = \frac{g_v}{2\pi}\sum_{n_z=1}^{n_{\max}}[T_{57HD}(E_{F1HD}, \eta_g, \eta_z)] \tag{11.70}$$

where $T_{57HD}(E_{F1HD}, \eta_g, \eta_z) \equiv A_{56HD}(E_{F1HD}, \eta_g, \eta_z)$.

Using (11.7b) and (11.70) we can study the QC in this case.

11.2.6 *The QC in HDQWTs of Te*

The carrier energy spectrum in HD Te can be written using (2.133) as

$$\gamma_3(E, \eta_g) = \psi_1 k_z^2 + \psi_2 k_s^2 \pm [\psi_3^2 k_z^2 + \psi_4^2 k_s^2]^{1/2} \tag{11.71}$$

The 2D electron energy spectrum in HDQWTs of Te can be written using (11.71) as

$$k_s^2 = \psi_{5HD}(E, \eta_g) - \psi_6\left(\frac{\pi n_z}{d_z}\right)^2 \pm \psi_7\left[\psi_{8HD}^2(E) - \left(\frac{\pi n_z}{d_z}\right)^2\right]^{1/2} \tag{11.72}$$

where

$$\psi_7 = \frac{\psi_4\sqrt{\psi_1}}{\psi_2^{3/2}} \text{ and } \psi_{8HD}^2(E, \eta_g) = \left[\frac{\psi_4 + 4\gamma_3(E, \eta_g)\psi_2\psi_4^2 + 4\psi_2^2\psi_3^2}{4\psi_1\psi_2\psi_4^2}\right]$$

The EEM in this case is given by

$$m^*(E_{F1HD}, \eta_g, n_z) = \frac{\hbar^2}{2}\left[\psi_{5HD}'(E_{F1HD}, \eta_g) \pm \frac{\psi_{8HD}(E_{F1HD}, \eta_g)\psi_{8HD}'(E_{F1HD}, \eta_g)}{\sqrt{\psi_{8HD}^2(E_{F1HD}, \eta_g) - (\pi n_z/d_z)^2}}\right] \tag{11.73}$$

The total DOS function in this case can be expressed as

$$N_{2DT}(E) = \frac{g_v}{\pi}\sum_{n_z=1}^{n_{z\max}}[\psi_{5HD}(E, \eta_g)]' H(E - E_{n_{z59HD}}) \tag{11.74}$$

where $E_{n_{z59HD}}$ is the lowest positive root of the equation

$$\psi_{5HD}(E_{n_{z59HD}},\eta_g) - \psi_6\left(\frac{\pi n_z}{d_z}\right)^2 \pm \psi_7\left[\psi_{8HD}^2(E_{n_{z59HD}},\eta_g) - \left(\frac{\pi n_z}{d_z}\right)^2\right]^{1/2} = 0 \tag{11.75}$$

The surface electron concentration under the condition of extreme carrier degeneracy in this case is given by

$$n_{2D} = \frac{g_v}{\pi}\sum_{n_z=1}^{n_{z\max}}[t_{1HDTE}(E_{F1HD},\eta_g,n_z)] \tag{11.76}$$

where

$$t_{1HDTE}(E_{F1HD},\eta_g,n_z) = \left[\psi_{5HD}(E_{F1HD},\eta_g,n_z) - \psi_6\left(\frac{\pi n_z}{d_z}\right)^2\right]$$

Using (11.7b) and (11.76) we can study the QC in this case.

11.2.7 *The QC in HDQWTs of GaP*

The dispersion relation of the conduction electrons in HD GaP can be expressed using (2.72) as

$$\gamma_3(E,\eta_g) = \frac{\hbar^2 k_s^2}{2m_\perp^*} + \frac{\hbar^2}{2m_{||}^*}[k_s^2 + k_z^2] - \left[\frac{\hbar^4 k_0^2}{m_{||}^{*2}}(k_s^2+k_z^2) + |V_G|^2\right]^{1/2} + |V_G| \tag{11.77}$$

The 2D DR in QW of HD GaP can be expressed following (11.77) as

$$k_s^2 = t_{11}\gamma_3(E,\eta_g) + t_{21} - t_{31}\left(\frac{\pi n_z}{d_z}\right)^2 - t_{41}\left[\left(\frac{\pi n_z}{d_z}\right)^2 + t_5^2(E,\eta_g)\right]^{1/2} \tag{11.78}$$

where

$t_{11} = \frac{1}{a}$, $a = \frac{\hbar^2}{2m_\perp^*} + b$, $t_{21} = \frac{g_1}{2a^2}$, $g_1 = -(C + 2aD)$,

$C = \left(\frac{\hbar^2 k_0}{m_{||}^*}\right)^2$, $D = |V_G|$, $t_{31} = \frac{b}{a}$, $t_{41} = \frac{\sqrt{g_3}}{2a^2}$,

$g_3 = (4abC + 4a^2C)$, $t_5^2(E_{F_h},\eta_g) = [g_2 - 4aC\gamma_3(E_{F_h},\eta_g)](g_3)^{-1}$ and $g_2 = (4a^2b^2 + C^2 + 4aCD)$

The EEM in this case can be written following (11.78) as

$$m^*(E_{F1HD}, \eta_g, n_z) = \frac{\hbar^2}{2}\left[t_{11}\gamma_3'(E_{F1HD}, \eta_g) - t_{41}t_5(E_{F1HD}, \eta_g) \times t_5'(E_{F1HD}, \eta_g)\left[\left(\frac{\pi n_z}{d_z}\right)^2 + t_5^2(E_{F1HD}, \eta_g)\right]^{-1/2}\right] \tag{11.79}$$

The total DOS function assumes the form

$$N_{2DT}(E, \eta_g) = \frac{g_v}{2\pi}\sum_{n_z=1}^{n_{z\max}}\left[t_{11}\gamma_3'(E_{F1HD}, \eta_g) - t_{41}t_5(E_{F1HD}, \eta_g) \times t_5'(E_{F1HD}, \eta_g)\left[\left(\frac{\pi n_z}{d_z}\right)^2 + t_5^2(E_{F1HD}, \eta_g)\right]^{-1/2}\right] \times H(E - E_{n_{z8THD}}) \tag{11.80}$$

where $E_{n_{z8THD}}$ is given by the equation

$$t_{11}y_3(E_{n_{z8THD}}) + t_{21} - t_{31}\left(\frac{\pi n_z}{d_z}\right)^2 - t_{41}\left[\left(\frac{\pi n_z}{d_z}\right)^2 + t_5^2(E_{n_{z8THD}}, \eta_g)\right]^{1/2} = 0 \tag{11.81}$$

The surface electron concentration under the condition of extreme carrier degeneracy in this case can be written as

$$n_{2D} = \frac{g_v}{\pi}\sum_{n_z=1}^{n_{\max}}[t_{3HDGaP}(E_{F1HD}, \eta_g, n_z)] \tag{11.82}$$

where

$$t_{3HDGap}(E_{F1HD}, \eta_g, n_z) = \left[t_{11}\gamma_3(E_{F1HD}, \eta_g) + t_{21}\left(\frac{\pi n_z}{d_z}\right)^2 - t_{41}\left[\left(\frac{\pi n_z}{d_z}\right)^2 + t_5^2(E_{F1HD}, \eta_g)\right]^{1/2}\right]$$

Using (11.7b) and (11.82) we can study the QC in this case.

11.2.8 The QC in HDQWTs of platinum antimonide

Equation (3.94) can be rewritten as

$$[E + \omega_1 k_s^2 + \omega_2 k_z^2][E + \delta_0 + \omega_3 k_s^2 - \omega_4 k_z^2] = I_1(k_s^2 + k_z^2)^2 \qquad (11.83)$$

where

$$\omega_1 = \left[\lambda_0 \frac{(\bar{\bar{a}})^2}{4} + l\frac{(\bar{\bar{a}})^2}{4}\right], \quad \omega_2 = \lambda_0 \frac{(\bar{\bar{a}})^2}{4}, \quad \omega_3 = \left[n\frac{(\bar{\bar{a}})^2}{4} - \nu\frac{(\bar{\bar{a}})^2}{4}\right],$$

$$\omega_4 = \nu\frac{(\bar{\bar{a}})^2}{4}, \quad I_1 = I\left(\frac{(\bar{\bar{a}})^2}{4}\right)^2,$$

$\lambda_0, l, \delta_0, v, n$ and a are the band constants.

The carrier dispersion law in HD $PtSb_2$ can be written as

$$T_{11}k_s^4 - k_s^2[T_{21}(E,\eta_g) - T_{31}k_z^2] + [T_{41}k_z^4 - T_{51}(E,\eta_g)k_z^2 - T_{61}(E,\eta_g)] = 0 \qquad (11.84)$$

where

$$T_{11} = (I_1 - \omega_2\omega_3), \;\; T_{21}(E,\eta_g) = [\omega_1\delta_0 + \omega_1\gamma_3(E,\eta_g) + \omega_3\gamma_3(E,\eta_g)]$$

$$T_{31} = [2I_1 + \omega_2\omega_4 - \omega_2\omega_3], \quad T_{41} = [2I_1 + \omega_2\omega_4],$$

$$T_{51}(E,\eta_g) = [\omega_2\gamma_0 - \omega_4\gamma_3(E,\eta_g) + \omega_2\gamma_3(E,\eta_g)],$$

$$T_{61}(E',\eta_g) = [\gamma_8(E,\eta_g) + \gamma_0\gamma_3(E,\eta_g)] \quad \text{and}$$

$$\gamma_8(E,\eta_g) = 2\theta_0(E,\eta_g)[1 + Erf(E/\eta_g)]^{-1}$$

The EEMs are given by

$$m_s^*(E,\eta_g) = \frac{\hbar^2}{2T_{11}}\left[T_{21}'(E,\eta_g) + \frac{(T_{21}(E,\eta_g)T_{21}'(E,\eta_g) + 2T_{11}T_{61}'(E,\eta_g))}{\sqrt{T_{21}^2(E,\eta_g) + 4T_{11}T_{61}'(E,\eta_g)}}\right] \qquad (11.85)$$

$$m_z^*(E,\eta_g) = \left(\frac{\hbar^2}{2T_{41}}\right)[T_{51}'(E,\eta_g) + [T_{51}'(E,\eta_g)T_{51}'(E,\eta_g) + 2T_{41}'T_{61}'(E,\eta_g)] \times [T_{51}^2(E,\eta_g) + 4T_{41}T_{61}(E,\eta_g)]^{-1/2}] \qquad (11.86)$$

The electron concentration under the condition of extreme carrier degeneracy assumes the form

$$n_0 = \frac{g_v}{3\pi^2}[\bar{I}_{128}(E_{F_h}, \eta_g)] \tag{11.87}$$

where E_{F_h} is the Fermi energy in this case, $\bar{I}_{128}(E_{F_h}, \eta_g) = [M_{6HD}(E_{F_h}, \eta_g)]$,

$$M_{6HD}(E_{F_h}, \eta_g) = \left[T_{91HD}(E_{F_h}, \eta_g)\rho_{2HD}(E_{F_h}, \eta_g) - T_{101}\frac{\rho_{2HD}^3(E_{F_h}, \eta_g)}{3} - T_{11}J_3(E_{F_h}, \eta_g)\right]$$

$$T_{91HD}(E_{F_h}, \eta_g) = \frac{T_{21}(E_{F_h}, \eta_g)}{2T_{11}}$$

$$\rho_{2HD}(E_{F_h}, \eta_g) = \left[(2T_{41})^{-1}\left[T_{51}(E_{F_h}, \eta_g) + \sqrt{T_{51}^2(E_{F_h}, \eta_g) + 4T_{41}T_{61}(E_{F_h}, \eta_g)}\right]\right]^{1/2},$$

$$T_{101} = [T_{31}/2T_{11}]$$

$$J_3(E_{F_h}, \eta_g) = \frac{\rho_{2HD}(E_{F_h}, \eta_g)}{3}[[A_{3HD}^2(E_{F_h}, \eta_g) + B_{3HD}^2(E_{F_h}, \eta_g)] \times E_0(\eta(E_{F_h}, \eta_g), \quad t(E_{F_h}, \eta_g)) - [A_{3HD}^2(E_{F_h}, \eta_g) - B_{3HD}^2(E_{F_h}, \eta_g)]F_0(\eta(E_{F_h}, \eta_g), t(E_{F_h}, \eta_g))] + \frac{\rho_{2HD}(E_{F_h}, \eta_g)}{3}[(A_{3HD}^2(E_{F_h}, \eta_g) - \rho_{2HD}^2(E_{F_h}, \eta_g)) \times (B_{3HD}^2(E_{F_h}, \eta_g) - \rho_{2HD}^2(E_{F_h}, \eta_g))]^{1/2}$$

$E_0(\eta(E_{F_h}, \eta_g)), t(E_{F_h}, \eta_g)$ and $F_0E_0(\eta(E_{F_h}, \eta_g)), t(E_{F_h}, \eta_g)$ are the incomplete elliptic integrals of second and first, respectively.

$$A_{3HD}^2(E_{F_h}, \eta_g) = \frac{1}{2}\left[\left[T_{12}(E_{F_h}, \eta_g) + \sqrt{T_{12}^2(E_{F_h}, \eta_g) - 4T_{13}(E_{F_h}, \eta_g)}\right]\right],$$

$$T_{12}(E_{F_h}, \eta_g) = [T_7(E_{F_h}, \eta_g)/\bar{T}_{61}]$$

$$\bar{T}_{61} = [T_{31}^2 - 4T_{11}T_{41}], \quad T_7(E_{F_h}, \eta_g) = [2T_{31}T_{21}(E_{F_h}, \eta_g) - 4T_{11}T_{51}(E_{F_h}, \eta_g)]$$

$$T_{13}(E_{F_h},\eta_g) = (T_8(E_{F_h},\eta_g)/\bar{T}_8), \quad T_8(E_{F_h},\eta_g) = [T_{21}^2(E_{F_h},\eta_g) + 4T_{11}T_{61}(E_{F_h},\eta_g)]$$

$$B_{3HD}^2(E_{F_h},\eta_g) = \frac{1}{2}\left[\left[T_{12}(E_{F_h},\eta_g) - \sqrt{T_{12}^2(E_{F_h},\eta_g) - 4T_{13}(E_{F_h},\eta_g)}\right]\right],$$

$$\bar{T}_{11} = [\sqrt{\bar{T}_{61}}/2T_{11}]$$

$$t(E_{F_h},\eta_g) = [B_3(E_{F_h},\eta_g)/A_3(E_{F_h},\eta_g)], \eta(E_{F_h},\eta_g) = \sin^{-1}\left[\frac{\rho_2(E_{F_h},\eta_g)}{B_3(E_{F_h},\eta_g)}\right]$$

From (11.84) the dispersion relation in HDQWTs of $PtSb_2$ can be expressed as

$$T_{11}k_s^4 - P_{1HD}(E,\eta_g,n_z)k_s^2 + P_{2HD}(E,\eta_g,n_z) = 0 \tag{11.88a}$$

where $P_{1HD}(E,\eta_g,n_z) = [T_{21}(E,\eta_g,n_z) - T_{31}(\pi n_z/d_z)]$

$$P_{2HD}(E,\eta_g,n_z) = [T_{41}(\pi n_z/d_z)^4 - T_{51}(E_{F_h},\eta_g)(\pi n_z/d_z)^2 - T_{61}(E_{F_h},\eta_g)]$$

From (11.88) the sub-band energy $(E_{n_{z9THD}})$ can be expressed as

$$P_{2HD}(E_{n_{z9THD}},\eta_g,n_z) = 0 \tag{11.88b}$$

(11.88) can be written as

$$k_s^2 = A_{60}(E_{F_h},\eta_g,n_z) \tag{11.89}$$

where

$$A_{60}(E_{F_h},\eta_g,n_z) = \left[P_{1HD}(E_{F_h},\eta_g,n_z) - \sqrt{P_{1HD}^2(E_{F_h},\eta_g,n_z) - 4T_{11}P_{2HD}(E_{F_h},\eta_g,n_z)}\right]$$

The EEM assumes the form

$$m^*(E_{F1HD},\eta_g,n_z) = \frac{\hbar^2}{2}A'_{60}(E_{F1HD},\eta_g,n_z) \tag{11.90}$$

The surface electron concentration under the condition of extreme carrier degeneracy in this case is given by

$$n_{2D} = \frac{g_v}{2\pi} \sum_{n_z=1}^{n_{z\max}} [A_{60}(E_{F1HD}, \eta_g, n_z)] \tag{11.91}$$

Using (11.7b) and (11.91) we can study the QC in this case

11.2.9 *The QC in HDQWTs of bismuth telluride*

Using (3.102) the DR in HD Bi_2Te_3 assumes the form

$$\gamma_2(E, \eta_g) = I_1 k_x^2 + I_2 k_y^2 + I_3 k_z^2 + 2\bar{I}_4 k_z k_y \tag{11.92}$$

The dispersion relation in HDQWTs of Bi_2Te_3 can be expressed as

$$\gamma_2(E, \eta_g) = I_1 \left(\frac{\pi n_x}{d_x}\right)^2 + I_2 k_y^2 + I_3 k_z^2 + 2\bar{I}_4 k_z k_y \tag{11.93}$$

The EEM can be expressed as

$$m^*(E_{F1HD}, \eta_g) = \frac{m_0}{\sqrt{\alpha_1 \alpha_{33} - 4\alpha_{23}^2}} \gamma_2'(E_{F1HD}, \eta_g) \tag{11.94a}$$

The sub-band energy ($E_{n_{z10THD}}$) can be written as

$$\gamma_2(E_{n_{z10THD}}, \eta_g) = I_1 \left(\frac{\pi n_x}{d_x}\right)^2 \tag{11.94b}$$

The DOS function can be written as

$$N_{2DT}(E, \eta_g) = \frac{g_v}{2\pi} \left[\sum_{n_x=1}^{n_{x\max}} R_{60}'(E_{F1HD}, \eta_g, n_x)\right] H(E - E_{n_{z10THD}}) \tag{11.94c}$$

where

$$R_{60}(E_{F1HD}, \eta_g, n_x) = \frac{1}{\sqrt{a_{11}\alpha_{33} - 4\alpha_{23}^2}} \times \left[\frac{2m_0 \gamma_2'(E_{F1HD}, \eta_g)}{\hbar^2} - \frac{2m_0}{\hbar^2}\left(\frac{\pi n_x}{d_x}\right)^2 \alpha_{11}\right]$$

The surface electron concentration under the condition of extreme carrier degeneracy in this case can be written as

$$n_{2D} = \frac{g_v}{2\pi} \left[\sum_{n_x=1}^{n_{x\max}} R_{60}(E_{F1HD}, \eta_g, n_x)\right] \tag{11.95}$$

Using (11.7b) and (11.95) we can study the QC in this case

11.2.10 *The QC in HDQWTs of germanium*

(a) Cardona *et al.*

In the presence of size quantization, the dispersion law in HDQWTs of Ge can be written using (2.47) as

$$\frac{\hbar^2 k_s^2}{2m_\perp^*} = \gamma_2(E,\eta_g) + \alpha\left(\frac{\hbar^2\left(\frac{n_z\pi}{d_z}\right)^2}{2m_\parallel^*}\right) - (1+2\alpha\gamma_3(E,\eta_g))\left(\frac{\hbar^2\left(\frac{n_z\pi}{d_z}\right)^2}{2m_\parallel^*}\right) \tag{11.96a}$$

The sub-band energy ($E_{n_{z11THD}}$) can be written from (11.96a) as

$$\gamma_2(E_{n_{z10THD}},\eta_g) + \alpha\left(\frac{\hbar^2\left(\frac{n_z\pi}{d_z}\right)}{2m_\parallel^*}\right) - (1+2\alpha\gamma_3(E_{n_{z11THD}},\eta_g)) \times\left(\frac{\hbar^2\left(\frac{n_z\pi}{d_z}\right)^2}{2m_\parallel^*}\right) = 0 \tag{11.96b}$$

The DOS function can be written as

$$N_{2DT}(E,\eta_g) = \frac{g_v m_\perp^*}{\pi\hbar^2}\left[\sum_{n_z=1}^{n_{z\max}} R_1'(E,\eta_g,n_z)\right] \tag{11.96c}$$

where

$$R_1(E,\eta_g,n_z) = \left[\gamma_2(E,\eta_g) + \alpha\left(\frac{\hbar^2\left(\frac{n_z\pi}{d_z}\right)^2}{2m_\parallel^*}\right) - (1+2\alpha\gamma_3(E,\eta_g)) \times\left(\frac{\hbar^2\left(\frac{n_z\pi}{d_z}\right)^2}{2m_\parallel^*}\right)\right]$$

The surface electron concentration under the condition of extreme carrier degeneracy in this case can be written as

$$n_{2D} = \frac{g_v m_\perp^*}{\pi\hbar^2}\left[\sum_{n_z=1}^{n_{z\max}} R_1(E_{F1HD},\eta_g,n_z)\right] \tag{11.97}$$

Using (11.7b) and (11.97) we can study the QC in this case.

(b) Wang and Ressler model

The DR of the conduction electron in bulk specimens of n-Ge can be expressed in accordance with the model of Wang and Ressler can be rewritten from (2.53) as

$$E = \frac{\hbar^2 k_z^2}{2m_{||}^*} + \frac{\hbar^2 k_s^2}{2m_\perp^*} - \bar{\alpha}_4 \left(\frac{\hbar^2 k_s^2}{2m_\perp^*}\right)^2 - \bar{\alpha}_5 \left(\frac{\hbar^2 k_s^2}{2m_\perp^*}\right)\left(\frac{\hbar^2 k_z^2}{2m_{||}^*}\right) - \bar{\alpha}_6 \left(\frac{\hbar^2 k_z^2}{2m_{||}^*}\right)^2 \tag{11.98}$$

where $\bar{\alpha}_4 = \beta_4\left(\frac{2m_\perp^*}{\hbar^2}\right)$, $\beta_4 = 1.4\beta_5$,

$$\beta_5 = \frac{\alpha\hbar^4}{4}[(m_\perp^*)^{-1} - (m_0)^{-1}]^2, \quad \bar{\alpha}_5 = \bar{\alpha}_7\left(\frac{4m_\perp^* m_{||}^*}{\hbar^4}\right), \quad \bar{\alpha}_7 = 0.8\beta_5 \text{ and}$$

$$\bar{\alpha}_6 = (0.005\beta_5)\left(\frac{2m_{||}^*}{\hbar^2}\right)^2$$

The energy spectrum under the condition of heavy doping can be written as

$$\gamma_3(E,\eta_g) = \frac{\hbar^2 k_z^2}{2m_{||}^*} + \frac{\hbar^2 k_s^2}{2m_\perp^*} - \bar{\alpha}_4 \left(\frac{\hbar^2 k_s^2}{2m_\perp^*}\right)^2 - \bar{\alpha}_5 \left(\frac{\hbar^2 k_s^2}{2m_\perp^*}\right)\left(\frac{\hbar^2 k_z^2}{2m_{||}^*}\right) - \bar{\alpha}_6 \left(\frac{\hbar^2 k_z^2}{2m_{||}^*}\right)^2 \tag{11.99}$$

(11.99) can be expressed as

$$\frac{\hbar^2 k_s^2}{2m_\perp^*} = \bar{\alpha}_8 - \bar{\alpha}_9 k_z^2 - \bar{\alpha}_{10}[k_z^4 + \bar{\alpha}_{11}k_z^2 + \bar{\alpha}_{12}(E,\eta_g)]^{1/2} \tag{11.100}$$

where

$$\bar{\alpha}_8 = \frac{1}{2\bar{\alpha}_4}, \quad \bar{\alpha}_9 = \frac{\bar{\alpha}_5}{2\alpha_4}\left(\frac{\hbar^2}{2m_{||}^*}\right), \quad \bar{\alpha}_{10} = \frac{1}{2\alpha_4}\left(\frac{\hbar^2}{2m_{||}^*}\right)\sqrt{\bar{\alpha}_5^2 - 4\bar{\alpha}_4\bar{\alpha}_6}$$

$$\bar{\alpha}_{11} = \frac{2m_{||}^*}{\hbar^2}\left[\frac{4\bar{\alpha}_4 - 2\bar{\alpha}_5}{\bar{\alpha}_5^2 - 4\bar{\alpha}_4\bar{\alpha}_6}\right]$$

The DR in HDQWTs of Ge can be written as

$$\frac{\hbar^2 k_s^2}{2m_\perp^*} = \bar{\alpha}_8 - \bar{\alpha}_9\left(\frac{n_z\pi}{d_z}\right)^2 - \bar{\alpha}_{10}\left[\left(\frac{n_z\pi}{d_z}\right)^4 + \bar{\alpha}_{11}\left(\frac{n_z\pi}{d_z}\right)^2 + \bar{\alpha}_{12}(E,\eta_g)\right]^{1/2} \tag{11.101}$$

The (11.101) can be expressed as

$$\frac{\hbar^2 k_s^2}{2m_\perp^*} = A_{75}(E, \eta_g, n_z) \tag{11.102}$$

where

$$A_{75}(E, \eta_g, n_z) = \left[\bar{\alpha}_8 - \bar{\alpha}_9\left(\frac{\pi n_z}{d_z}\right)^2 - \bar{\alpha}_{10}\left[\left(\frac{\pi n_z}{d_z}\right)^4 + \bar{\alpha}_{11}\left(\frac{\pi n_z}{d_z}\right)^2 + \bar{\alpha}_{12}(E, \eta_g)\right]^{1/2}\right]$$

The EEM is given by

$$m_s^2(E_{F1HD}, \eta_g, n_z) = m_\perp^* A'_{75}(E_{F1HD}, \eta_g, n_z) \tag{11.103}$$

The electron concentration per unit area under the condition of extreme carrier degeneracy in this case assumes the form as

$$n_{2D} = \frac{m_\perp^* g_v}{\pi\hbar^2}\sum_{n_z=1}^{n_{z\max}}[A_{75}(E_{F1HD}, \eta_g, n_z)] \tag{11.104}$$

Using (11.7b) and (11.104) we can study the QC in this case

11.2.11 *The QC in HDQWTs of gallium antimonide*

Under the condition of heavy doping, (2.86) assumes the form

$$\frac{\hbar^2 k^2}{2m_c} = I_{36}(E, \eta_g) \tag{11.105}$$

where

$$I_{36}(E, \eta_g) = \left[\gamma_3(E, \eta_g) + E'_g - \frac{m_c}{m_0}.\frac{E'_g}{2} - \left[\left(\frac{E'_g}{2}\right)^2 + \left[\frac{E'_g}{2}\left(1 - \frac{m_c}{m_0}\right)\right]^2 + \left(\frac{E'_g}{2}\right)^2\left(1 - \frac{m_c}{m_0}\right) + \gamma_3(E, \eta_g)E'_g\left(1 - \frac{m_c}{m_0}\right)\right]^{1/2}\right]$$

For dimensional quantization along z-direction, the dispersion relation of the 2D electrons in HDQWTs of GaSb can be written following (11.105) as

$$\frac{\hbar^2(n_z\pi/d_z)^2}{2m_c} + \frac{\hbar^2(k_s)^2}{2m_c} = I_{36}(E, \eta_g) \tag{11.106}$$

The expression of the $N_{2Dr}(E)$ in this case can be written as

$$N_{2DT}(E) = \frac{m_c g_v}{\pi \hbar^2} \sum_{n_z=1}^{n_{z\max}} T'_{119D}(E_{F_h}, \eta_g, n_z) H(E - E_{n_z D119}) \quad (11.107)$$

where $T_{119D}(E_{F_h}, \eta_g, n_z) = [I_{36}(E_{F_h}, \eta_g) - \hbar^2(n_z\pi/d_z)^2(2m_c)^{-1}]$.

The sub-band energies $E_{n_z D119}$ in this case given by

$$\{\hbar^2(n_z\pi/d_z)^2\}(2m_c)^{-1} = I_{36}(E_{n_z D119}, \eta_g) \quad (11.108)$$

The EEM in this case assumes the form

$$m^*(E_{F1HD}, \eta_g, n_z) = m_c[I'_{36}(E_{F1HD}, \eta_g, n_z)] \quad (11.109)$$

The 2D electron statistics under the condition of extreme carrier degeneracy in this case can be written as

$$n_{2D} = \left(\frac{m_c g_v}{\pi \hbar^2}\right) \sum_{n_z=1}^{n_{z\max}} [T_{119D}(E_{F1HD}, \eta_g, n_z)] \quad (11.110)$$

Using (11.7b) and (11.110) we can study the QC in this case

11.3 Summary and Conclusions

In Section 11.2.1, we have formulated the electron statistics in HDQWTs made of nonlinear optical materials as given by (11.6) and by combining it with (11.7b) we can study the QC in this case. In Section 11.2.2, we have formulated the electron statistics in HDQWTs of III–V, ternary and quaternary materials in accordance with the three-band model of Kane, two-band model of Kane, isotropic parabolic energy bands together with the models of Stillman *et al.* and Palik *et al.* as given by (11.12), (11.18), (11.22), (11.28) and (11.36) and by combining each of them with (11.7b) we can study the QC for each model in this case. In Section 11.2.3, we have formulated the electron statistics in HDQWTs of II–VI materials as given by (11.40) and by combining it with (11.7b) we can study the QC in this case.

In Section 11.2.4, we have formulated the electron statistics in HDQWTs of IV–VI materials in accordance with the models of Dimmock and Bangert and Kastner as given by (11.55) and (11.64) and by combining them with (11.7b) we can study the QC for each model in this case. In Section 11.2.5, we have formulated the electron statistics in HDQWTs of stressed Kane type as given by (11.70) and by combining it with (11.7b) we can study the QC in this case. In Section 11.2.6, we have formulated the electron statistics in HDQWTs of Te as given by (11.76) and by combining it with (11.7b) we can study the QC in this case. In Section 11.2.7, we have formulated the electron statistics in HDQWTs of GaP as given by (11.82) and by combining it with (11.7b) we can study the QC in this case. In Section 11.2.8, we have formulated the electron statistics in HDQWTs of $PtSb_2$ as given by (11.91) and by combining it with (11.7b) we can study the QC in this case. In Section 11.2.9, we have formulated the electron statistics in HDQWTs of Bi_2Te_3 as given by (11.95) and by combining it with (11.7b) we can study the QC in this case.

In Section 11.2.10, we have formulated the electron statistics in HDQWTs of Ge in accordance with the models of together with the models of Cardona *et al.* and Wang and Ressler as given by (11.97) and (11.104) and by combining each of them with (11.7b) we can study the QC for each model in this case. In Section 11.2.11, we have formulated the electron statistics in HDQWTs of GaSb as given by (11.110) and by combining it with (11.7b) we can study the QC in this case.

We have not given the graphs of the QCs' for 2D HD QWTs made of many materials as discussed here since we feel that our readers will perform the intricate numerical computation to obtain the oscillatory plots for better assessment and joy of understanding. It is the band structure which changes in a fundamental way and consequently all the physical properties of all the electronic materials changes radically leading to new physical concepts.

11.4 Open Research Problems

(R.11.1) Investigate the QC for all the HD QWTs' made by using the materials as given in problems in R.1.1 of chapter 1. Draw the corresponding graphs and explain the results.

(R.11.2) Investigate the QC for HD QWTs' made by using the IV–VI, II–VI and stressed Kane type compounds in the presence of exponential band tails. Draw the corresponding graphs and explain the results.

(R.11.3) Investigate the QC for all the HD QWTs' made by using the materials as stated in R.1.1 of chapter 1 in the presence of exponential band tails. Draw the corresponding graphs and explain the results.

(R.11.5) Investigate the QC in the presence of in the presence of exponential band tails in HD QWTs' made of the nonlinear optical materials Study all the special cases for HD III–V, ternary and quaternary materials in this context. Draw the corresponding graphs and explain the results.

Reference

1. E. D. Palik, G. S. Picus, S. Teither and R. F. Wallis, *Phys. Rev.* **122**, 475 (1961).

Chapter 12

The QC in Heavily Doped Quantum Well Transistors (HDQWT) Under Magnetic Quantization

I will never find a rainbow if I am looking down

12.1 Introduction

In recent years there has been considerable interest in studying the effect of magnetic quantization on the different electronic properties of quantized transistors [1–15]. In this chapter in Section 12.2.1, of the theoretical background, the QC has been investigated in HDQWTs of tetragonal and nonlinear optical semiconductors in the presence of a quantizing magnetic field. Section 12.2.2 contains the results for HDQWTs of III–V, ternary and quaternary compounds in accordance with the three- and the two-band models of Kane along with the parabolic model. In the same section the QC in accordance with the models of Stillman *et al.* [16] and Palik *et al.* [17] have also been studied for the purpose of relative comparison. Section 12.2.3 contains the study of the QC for HDQWTs of II–VI semiconductors under magnetic quantization. In Section 12.2.4, the QC in HDQWTs of IV–VI materials has been discussed in accordance with the models of Cohen, Lax, Dimmock, Bangert and Kastner and Foley and Landenberg respectively. In Section 12.2.5, the magneto-QC for the stressed HDQWTs of Kane type semiconductors has been investigated. In Section 12.2.6, the QC in HDQWTs of Te has been studied under magnetic quantization. In Section 12.2.7, the magneto-QC in HDQWTs of GaP has been studied. In Section 12.2.8, the QC in HDQWTs of $PtSb_2$ has been explored under magnetic quantization. In Section 12.2.9, the magneto-QC in HDQWTs of Bi_2Te_3 has been studied. In Section 12.2.10, the QC in HDQWTs of Ge has been studied under magnetic quantization in accordance with the models of Cardona *et al.* and Wang and Ressler respectively. In Sections 12.2.11 and 12.2.12, the magneto-QC in HDQWTs of GaSb and II–V compounds has respectively been studied. In Sections 12.2.13 the magneto QC in HDQWTs of $Pb_{1-x}Ge_xTe$ *has been discussed.* Section 12.3 explores the summary and conclusion and the last Section 12.4 contains 15 open research problems for this chapter.

12.2 Theoretical Background

12.2.1 *The QC in HDQWTs of tetragonal and nonlinear optical materials under magnetic quantization*

The DR under magnetic quantization in tetragonal and nonlinear optical materials can be written as

$$\gamma(E) = \frac{\hbar^2 k_s^2}{2m_\perp^*} f_3(E) + \frac{\hbar^2 k_z^2}{2m_{||}^*} f_4(E) \pm \frac{eB\hbar E_g}{6}\left[\frac{(E_{g_0}+\Delta_\perp)}{(E_{g_0}+\frac{2}{3}\Delta_\perp)}\right] \times \left[E + E_{g_0} + \delta + \frac{\Delta_{||}^2 - \Delta_\perp^2}{3\Delta_{||}}\right] \tag{12.1}$$

where

$$f_3(E) = \frac{E_g(E_g+\Delta_\perp)}{(E_g+\frac{2}{3}\Delta_\perp)}\left[(E+E_g)\left(E+E_g+\frac{2}{3}\Delta_{||}\right) + \delta\left(E+E_g+\frac{1}{3}\Delta_{||}\right) + \frac{1}{9}(\Delta_{||}^2-\Delta_\perp^2)\right]$$

and

$$f_4(E) = \frac{E_g(E_g+\Delta_{||})}{(E_g+\frac{2}{3}\Delta_{||})}\left[(E+E_g)\left(E+E_g+\frac{2}{3}\Delta_{||}\right)\right]$$

The (12.1) can be expressed as

$$\frac{\hbar^2 k_z^2}{2m_{||}^*} + \left(\frac{b_{||}c_\perp}{b_\perp c_{||}}\right)\left(\frac{\hbar^2 k_s^2}{2m_\perp^*}\right) = \left\{\left[\frac{ab_{||}}{c_{||}}E^2 + \frac{(ac_{||}+b_{||}-ab_{||})}{c_{||}^2}E + \frac{1}{c_{||}}\left(1-\frac{a}{c_{||}}\right)\left(1-\frac{b_{||}}{c_{||}}\right) - \frac{1}{c_{||}}\left(1-\frac{a}{c_{||}}\right)\left(1-\frac{b_{||}}{c_{||}}\right)\frac{1}{(c_{||}E+1)}\right]\right.$$

$$+\frac{ab_{||}}{c_{||}}\left[\delta E+\frac{2}{9}(\Delta_{||}^2-\Delta_{\perp}^2)\right]-\frac{2}{9}\frac{ab_{||}}{c_{||}}\frac{(\Delta_{||}^2-\Delta_{\perp}^2)}{(c_{||}E+1)}\Bigg\}$$

$$-\left(\frac{\hbar^2k_s^2}{2m_{\perp}^*}\right)\left\{\left(\frac{b_{||}c_{\perp}}{b_{\perp}c_{||}}\right)\left[\left(\frac{\delta}{2}+\frac{\Delta_{||}^2-\Delta_{\perp}^2}{6\Delta_{||}}\right)\frac{a}{(aE+1)}\right.\right.$$

$$\left.\left.+\left(\frac{\delta}{2}-\frac{\Delta_{||}^2-\Delta_{\perp}^2}{6\Delta_{||}}\right)\frac{c_{||}}{(c_{||}E+1)}\right]\right\}\pm e_1\left[\frac{\rho_1}{E+E_g}+\frac{\rho_2}{E+E_g+\frac{2}{3}\Delta_{||}}\right] \tag{12.2}$$

where

$$e_1=\left[\frac{eB\hbar(E_g+\frac{2}{3}\Delta_{||})}{6(E_g+\Delta_{\perp})}\frac{(E_g+\Delta_{\perp})}{(E_g+\frac{2}{3}\Delta_{||})}\right],$$

$$\rho_1=\frac{(-E_g+G_1)}{(\frac{2}{3}\Delta_{||})},\quad G_1=\left[E_{g_0}+\delta+\frac{\Delta_{||}^2-\Delta_{\perp}^2}{3\Delta_{||}}\right]$$

and

$$\rho_2=\frac{3}{2\Delta_{||}}\left[\frac{\Delta_{||}}{3}+\frac{\Delta_{\perp}^2}{3\Delta_{||}}-\delta\right]$$

Therefore, the DR of the conduction electrons in HDQWTs of tetragonal and nonlinear optical materials in the presence of a quantizing magnetic field B can be written following the methods as developed in Chapter 8 as

$$\frac{\hbar^2k_z^2}{2m_{||}^*}=U_{1,+}(E,n,\eta_g)+iU_{2,\pm}(E,n,\eta_g) \tag{12.3}$$

where

$$U_{1,\pm}(E,n,\eta_g)=\left[\frac{-eB}{m_{\perp}^*}\left(n+\frac{1}{2}\right)\left(\frac{b_{||}c_{\perp}}{c_{||}b_{\perp}}\right)+\left[\frac{1+Erf(E/\eta_g)}{2}\right]^{-1}\right.$$

$$\times\left[\left\{\left[\frac{ab_{||}}{c_{||}}\theta_0(E,\eta_g)+\frac{ac_{||}+b_{||}c_{||}-ab_{||}}{c_{||}^2}\gamma_0(E,\eta_g)\right.\right.\right.$$

$$+\frac{1}{c_{||}}\left(1-\frac{a}{c_{||}}\right)\left(1-\frac{b_{||}}{c_{||}}\right)\left[\frac{1+Erf(E/\eta_g)}{2}\right]$$

$$-\frac{1}{c_{||}}\left(1-\frac{a}{c_{||}}\right)\left(1-\frac{b_{||}}{c_{||}}\right)c(\beta_1,E,\eta_g)$$

$$+\frac{ab_{||}}{c_{||}}\left[\delta\gamma_0(E,\eta_g)+\frac{2}{9}\left(\Delta_{||}^2-\Delta_{\perp}^2\right)\left[\frac{1+Erf(E/\eta_g)}{2}\right]\right]$$

$$-\frac{2}{9}\frac{ab_{||}}{c_{||}}(\Delta_{||}^2-\Delta_{\perp}^2)c(\beta_1,E,\eta_g)\Big\}-\frac{\hbar eB}{m_{\perp}^*}\left(n+\frac{1}{2}\right)$$

$$\times\left\{\left(\frac{b_{||}c_{\perp}}{c_{||}b_{\perp}}\right)\left[\left(\frac{\delta}{2}+\frac{\Delta_{||}^2-\Delta_{\perp}^2}{6\Delta_{||}}\right)ac(\beta_2,E,\eta_g)\right.\right.$$

$$\left.\left.+\left(\frac{\delta}{2}+\frac{\Delta_{||}^2-\Delta_{\perp}^2}{6\Delta_{||}}\right)c_{||}c(\beta_1,E,\eta_g)\right]\right\}\pm\frac{\rho_1e_1}{E_g}(\beta_2,E,\eta_g)$$

$$\left.+\frac{\rho_2e_1}{(E_g+\frac{2}{3}\Delta_{||})}c(\beta_3,E,\eta_g)\right]$$

$$U_{2,\pm}(E,n,\eta_g)=\left[\left[\frac{1+Erf(E/\eta_g)}{2}\right]^{-1}\left[\frac{1}{c_{||}}\left(1-\frac{a}{c_{||}}\right)\left(1-\frac{b_{||}}{c_{||}}\right)\right.\right.$$

$$\times D(\beta_1,E,\eta_g)+\frac{2}{9}\frac{ab_{||}}{c_{||}}(\Delta_{||}^2-\Delta_{\perp}^2)D(\beta_1,E,\eta_g)$$

$$-\frac{\hbar eB}{m_{\perp}^*}\left(n+\frac{1}{2}\right)\left\{\left(\frac{b_{||}c_{\perp}}{c_{||}b_{\perp}}\right)\left[\left(\frac{\delta}{2}+\frac{\Delta_{||}^2-\Delta_{\perp}^2}{6\Delta_{||}}\right)\right.\right.$$

$$\left.\left.\times aD(\beta_2,E,\eta_g)+\left(\frac{\delta}{2}-\frac{\Delta_{||}^2-\Delta_{\perp}^2}{6\Delta_{||}}\right)c_{||}D(\beta_1,E,\eta_g)\right]\right\}$$

$$\left.\mp\frac{\rho_1e_1}{E_g}D(\beta_2,E,\eta_g)\pm\frac{\rho_2e_1}{(E_g+\frac{2}{3}\Delta_{||})}D(\beta_3,E,\eta_g)\right],$$

$$C(\beta_i,E,\eta_g)=\left[\frac{2}{\beta_i\eta_g\sqrt{\pi}}\right]\exp(-u_i^2)\times\left[\sum_{p=1}^{\infty}\left\{\frac{\exp\left(\frac{-p^2}{4}\right)}{p}\right\}\sinh(pu_i)\right],$$

$$u_i=\frac{1+\beta_iE}{\beta_i\eta_g}$$

and

$$D(\beta_i,E,\eta_g)=\left[\frac{\sqrt{\pi}}{\beta_i\eta_g}\exp(-u_i^2)\right]$$

The DR of the conduction electrons in HDQWTs of tetragonal and nonlinear optical semiconductors in the presence of a quantizing magnetic field B can be written following (12.3) as

$$\frac{\hbar^2\left(\frac{n_z\pi}{d_z}\right)^2}{2m_{||}^*} = U_{1,\pm}(e_{81}, n, \eta_g) + iU_{2,\pm}(e_{81}, n, \eta_g) \tag{12.4}$$

where e_{81} is the totally quantized energy in this case.

The DOS function is given by

$$N_{Bn_z} = \frac{g_v eB}{2\pi\hbar}\sum_{n=0}^{n_{\max}}\sum_{n_z=1}^{n_{z\max}} \delta'(E - e_{81}) \tag{12.5}$$

The electron concentration per unit area can be written as

$$N_s = \frac{g_v eB}{2\pi\hbar} \text{Real part of} \sum_{n=0}^{n_{\max}}\sum_{n_z=1}^{n_{zMax}} F_{-1}(\eta_{12,1}) \tag{12.6}$$

where $\eta_{12,1} = (k_B T)^{-1}[E_{F12HD} - e_{81}]$ and E_{F12HD} is the Fermi energy in the present case.

The QC can be expressed as

$$C_G = e\frac{\partial}{\partial V_g}(N_s) \tag{12.7}$$

Thus, by using (12.7) and other appropriate equations, we can study the QC in this case.

12.2.2 *The QC in HDQWTs of III–V, ternary and quaternary materials under magnetic quantization*

(a) Three-band model of Kane

Substituting $\delta = 0$, $\Delta_{||} - \Delta_{\perp} - \Delta$ and $m_{||}^* = m_{\perp}^* = m_c$ in (12.1), thc electron energy spectrum in III–V materials under magnetic quantization is given by

$$\frac{E(E+E_g)(E+E_g+\Delta)\left(E_g+\frac{2}{3}\Delta\right)}{E_g(E_g+\Delta)\left(E+E_g+\frac{2}{3}\Delta\right)} = \left(n+\frac{1}{2}\right)\hbar\omega_0 + \frac{\hbar^2 k_z^2}{2m_c} \pm \frac{eB\hbar\Delta}{6m_c\left(E+E_g+\frac{2}{3}\Delta\right)} \tag{12.8}$$

Under the condition of heavy doping, the (12.8) can be written as

$$\begin{aligned}&\left[\frac{ab}{c}E^2+\left(\frac{ac+bc-ab}{c^2}\right)E+\frac{1}{c}\left(1-\frac{a}{c}\right)\left(1-\frac{b}{c}\right)\right.\\&\quad\left.-\frac{1}{c}\left(1-\frac{a}{c}\right)\left(1-\frac{b}{c}\right)\frac{1}{(1+cE)}\right]\\&=\left(n+\frac{1}{2}\right)\hbar\omega_0+\frac{\hbar^2k_z^2}{2m_c}+\frac{eB\hbar\Delta}{6m_c(1+cE)\left(E_g+\frac{2}{3}\Delta\right)}\end{aligned}$$

Therefore,

$$\begin{aligned}&\frac{ab}{c}I(5)+\left(\frac{ac+bc-ab}{c^2}\right)I(4)+\left[\frac{1}{c}\left(1-\frac{a}{c}\right)\left(1-\frac{b}{c}\right)-\left(n+\frac{1}{2}\right)\hbar\omega_0\right]\\&\quad I(1)-g_{\pm}[G(C,E,\eta_g)-iH(C,E,\eta_g)]=\frac{\hbar^2k_z^2}{2m_c}I(1)\end{aligned}\tag{12.9}$$

where

$$g_{\pm}=\left[\frac{1}{c}\left(1-\frac{a}{c}\right)\left(1-\frac{b}{c}\right)\pm\frac{eB\hbar\Delta}{6m_c\left(E_g+\frac{2}{3}\Delta\right)}\right],$$

$$G(C,E,\eta_g)=\left[\frac{2}{C\eta_g\sqrt{\pi}}\right]\exp(-u^2)\times\left[\sum_{p=1}^{\infty}\left\{\frac{\exp\left(\frac{-p^2}{4}\right)}{p}\right\}\sinh(pu)\right],$$

$$u=\frac{1+cE}{C\eta_g},\ H(C,E,\eta_g)=\left[\frac{\sqrt{\pi}}{C\eta_g}\exp(-u^2)\right]$$

Therefore,

$$\begin{aligned}\frac{\hbar^2k_z^2}{2m_c}&=\left[\frac{ab}{c}\right]\theta_0(E,\eta_g)\left[\frac{1+Erf(E/\eta_g)}{2}\right]^{-1}\\&\quad+\left(\frac{ac+bc-ab}{c^2}\right)\gamma_0(E,\eta_g)\left[\frac{1+Erf(E/\eta_g)}{2}\right]^{-1}\\&\quad+\left[\frac{1}{c}\left(1-\frac{a}{c}\right)\left(1-\frac{b}{c}\right)-\left(n+\frac{1}{2}\right)\hbar\omega_0\right]\\&\quad-g_{\pm}\left[\frac{1+Erf(E/\eta_g)}{2}\right]^{-1}[G(C,E,\eta_g)-\mathrm{i}H(C,E,\eta_g)]\end{aligned}\tag{12.10}$$

Therefore, the DR is given by

$$\frac{\hbar^2 k_z^2}{2m_c} = U_{3,\pm}(E, n, \eta_g) + iU_{4,\pm}(E, n, \eta_g) \tag{12.11}$$

where

$$\begin{aligned} U_{3,\pm}(E, n, \eta_g) = \Bigg[& \frac{ab}{c}\theta_0(E, \eta_g) \times \left[\frac{1 + Erf(E/\eta_g)}{2}\right]^{-1} \\ & + \left(\frac{ac + bc - ab}{c^2}\right)\gamma_0(E, \eta_g) \times \left[\frac{1 + Erf(E/\eta_g)}{2}\right]^{-1} \\ & + \frac{1}{c}\left(1 - \frac{a}{c}\right)\left(1 - \frac{b}{c}\right) - \left(n + \frac{1}{2}\right)\hbar\omega_0 \\ & + g_{\pm}\left[\frac{1 + Erf(E/\eta_g)}{2}\right]^{-1} G(C, E, \eta_g)\Bigg] \end{aligned}$$

and

$$U_{4,\pm}(E, \eta_g) = g_{\pm}\left[\frac{1 + Erf(E/\eta_g)}{2}\right]^{-1} H(C, E, \eta_g)$$

In accordance with three-band model of Kane, the DR in the present case can be written following (12.11) as

$$\frac{\hbar^2 \left(\frac{n_z \pi}{d_z}\right)^2}{2m_{||}^*} = U_{3,\pm}(e_{82}, n, \eta_g) + iU_{4,\pm}(e_{82}, n, \eta_g) \tag{12.12a}$$

where e_{82} is the totally quantized energy in this case.

The DOS function is given by

$$N_{Bn_z} = \frac{g_v eB}{2\pi\hbar} \sum_{n=0}^{n_{\max}} \sum_{n_z=1}^{n_{z_{Max}}} \delta'(E - e_{82}) \tag{12.12b}$$

The electron concentration per unit area in this case can be written as

$$N_s = \frac{g_v eB}{2\pi\hbar} \text{Real part of} \sum_{n=0}^{n_{\max}} \sum_{n_z=1}^{n_{z_{Max}}} F_{-1}(\eta_{12,2}) \tag{12.13}$$

where $\eta_{12,2} = (k_B T)^{-1}[E_{F12HD} - e_{82}]$

Thus by using (12.7), (12.13) and other appropriate equations we can study the QC in this case.

(b) Two-band model of Kane

The DR in this case is given by

$$\frac{\hbar^2\left(\frac{n_z\pi}{d_z}\right)^2}{2m_{||}^*} = \gamma_2(e_{83},\eta_g) - \left(n+\frac{1}{2}\right)\hbar\omega_0 \mp \frac{1}{2}g^*\mu_0 B \tag{12.14}$$

where e_{83} is the totally quantized energy in this case.

The DOS function is given by

$$N_{Bn_z} = \frac{g_v eB}{2\pi\hbar}\sum_{n=0}^{n_{\max}}\sum_{n_z=1}^{n_{zMax}}\delta'(E-e_{83}) \tag{12.15}$$

The electron concentration per unit area in this case can be written as

$$N_s = \frac{g_v eB}{2\pi\hbar}\sum_{n=0}^{n_{\max}}\sum_{n_z=1}^{n_{zMax}}F_{-1}(\eta_{12,3}) \tag{12.16}$$

where $\eta_{12,3} = (k_BT)^{-1}[E_{F12HD} - e_{83}]$.

Thus, by using (12.7), (12.16) and other appropriate equations, we can study the QC in this case.

(c) Parabolic energy bands

The DR in this case is given by

$$\frac{\hbar^2\left(\frac{n_z\pi}{d_z}\right)^2}{2m_{||}^*} = \gamma_3(e_{84},\eta_g) - \left(n+\frac{1}{2}\right)\hbar\omega_0 \mp \frac{1}{2}g^*\mu_0 B \tag{12.17}$$

where e_{84} is the totally quantized energy in this case.

The DOS function is given by

$$N_{Bn_z} = \frac{g_v eB}{2\pi\hbar}\sum_{n=0}^{n_{\max}}\sum_{n_z=1}^{n_{zMax}}\delta'(E-e_{84}) \tag{12.18}$$

The electron concentration per unit area in this case can be written as

$$N_s = \frac{g_v eB}{2\pi\hbar}\sum_{n=0}^{n_{\max}}\sum_{n_z=1}^{n_{zMax}}F_{-1}(\eta_{12,4}) \tag{12.19}$$

where $\eta_{12,4} = (k_BT)^{-1}[E_{F12HD} - e_{84}]$.

Thus, by using (12.7), (12.19) and other appropriate equations, we can study the QC in this case.

(d) The model of Stillman *et al.* [16]

The (2.64) under the condition of band tailing assumes the form

$$k^2 = \frac{[\bar{t}_{11} - \sqrt{(\bar{t}_{11})^2 - 4\bar{t}_{12}\gamma_3(E,\eta_g)}]}{2\bar{t}_{12}} \tag{12.20}$$

where $\bar{t}_{11} = e_5$ and $\bar{t}_{12} = e_6$.

Therefore, the magneto DR is given by

$$k^2 = U_7(E, n, \eta_g) \tag{12.21}$$

where

$$U_7(E, n, \eta_g) = \left[\frac{\left[\bar{t}_{11} - \sqrt{(\bar{t}_{11})^2 - 4\bar{t}_{12}\gamma_3(E,\eta_g)}\right]}{2\bar{t}_{12}} - \frac{2eB}{\hbar}\left(n + \frac{1}{2}\right)\right]$$

Following (7.21) the DR in the present case can be written as

$$\left(\frac{n_z\pi}{d_z}\right)^2 = U_7(e_{85}, n, \eta_g) \tag{12.22}$$

where e_{85} is the totally quantized energy in this case.

The DOS function is given by

$$N_{Bn_z} = \frac{g_v eB}{\pi\hbar} \sum_{n=0}^{n_{\max}} \sum_{n_z=1}^{n_{zMax}} \delta'(E - e_{85}) \tag{12.23}$$

The electron concentration per unit area in this case can be written as

$$N_s = \frac{g_v eB}{\pi\hbar} \sum_{n=0}^{n_{\max}} \sum_{n_z=1}^{n_{zMax}} F_{-1}(\eta_{12,5}) \tag{12.24}$$

where $\eta_{12,5} = (k_B T)^{-1}[E_{F12HD} - e_{85}]$.

Thus, by using (12.7), (12.24) and other appropriate equations, we can study the QC in this case.

(e) The model of Palik *et al.* [17]

To the fourth-order in effective mass theory and taking into account the interactions of the conduction, light hole, heavy-hole and split-off hole bands, the electron energy spectrum in III–V semiconductors in the presence of a quantizing magnetic field $\vec{B}$ can be written [17] as

$$\begin{aligned}E = J_{31} + \left(n + \frac{1}{2}\right)\hbar\omega_0 + \frac{\hbar^2 k_z^2}{2m_c} \pm \frac{1}{4}\left(\frac{m_c}{m_0}\right)\hbar\omega_0 g_0^* \\ \pm k_{30}\alpha\left(n + \frac{1}{2}\right)(\hbar\omega_0)^2 \pm k_{31}\alpha\hbar\omega_0\frac{\hbar^2 k_z^2}{2m_c} \\ + k_{32}\alpha\left[\hbar\omega_0\left(n + \frac{1}{2}\right) + \frac{\hbar^2 k_z^2}{2m_c}\right]^2\end{aligned} \tag{12.25}$$

where

$$\begin{aligned}J_{31} &= -\frac{1}{2}\alpha\hbar\omega_0[(1-y_{11})/(2+x_{11})^2].J_{32},\\ J_{32} &= \left\{\left[\frac{1}{3}(1-x_{11})^2 - (2+x_{11})^2\right](2+x_{11}).y_{11} \right.\\ &\quad \left. + \frac{1}{2}(1-x_{11}^2)(1+x_{11})(1+y_{11})\right\}\\ g_0^* &= 2\left\{1 - \left[\frac{(1-x_{11})}{(2+x_{11})}\right]\left[\frac{(1-y_{11})}{y_{11}}\right]\right\},\\ k_{30} &= (1-y_{11})(1-x_{11})\left\{\left[\left(2+\frac{3}{2}x_{11}+x_{11}^2\right)\cdot\frac{(1-x_{11})}{(2+x_{11})^2}\right] - \frac{2}{3}y_{11}\right\}\\ k_{31} &= (1-y_{11})\left[\frac{(1-x_{11})}{(2+x_{11})}\right]\cdot\left\{\left[\left(2+\frac{3}{2}x_{11}+x_{11}^2\right)\cdot\frac{(1-y_{11})}{(2+x_{11})^2}\right] \right.\\ &\quad \left. - \frac{2}{3}(1-x_{11})y_{11}\right\}\\ k_{32} &= \left[\left(1+\frac{1}{2}x_{11}^2\right)\Big/\left(1+\frac{1}{2}x_{11}\right)(1-y_{11})^2\right.,\\ x_{11} &= \left[1+\left(\frac{\Delta}{E_g}\right)\right]^{-1} \quad \text{and} \quad y_{11} = \frac{m_c}{m_0}\end{aligned}$$

Under the condition of heavy doping, (12.25) assumes the form

$$J_{34}k_z^4 + J_{35,\pm}(n)k_z^2 + J_{36,\pm}(n) - \gamma_3(E,\eta_g) = 0 \tag{12.26}$$

where

$$J_{34} = \alpha k_{32} \left(\frac{\hbar^2}{2m_c}\right)^2,$$

$$J_{35,\pm}(n) = \left[\frac{\hbar^2}{2m_c} \pm \alpha k_{31}\hbar\omega_0 \cdot \frac{\hbar^2}{2m_c} + \alpha k_{32}\hbar\omega_0 \cdot \frac{\hbar^2}{2m_c}\left(n+\frac{1}{2}\right)\right]$$

$$J_{36,\pm}(n) = \left[J_{31} \pm \frac{1}{4}\left(\frac{m_c}{m_0}\right)\hbar\omega_0 g_0^* \pm k_{30}\alpha(\hbar\omega_0)^2\left(n+\frac{1}{2}\right) + k_{32}\alpha\left[(\hbar\omega_0)\left(n+\frac{1}{2}\right)\right]^2\right]$$

The (12.28) can be written as

$$k_z^2 = A_{35,\pm}(E, n, \eta_g) \tag{12.27}$$

where

$$A_{35,\pm}(E, n, \eta_g) = (2J_{34})^{-1}[-J_{35,\pm}(n) + \sqrt{(J_{35,\pm}(n))^2 - 4J_{34}[J_{36,\pm}(n) - \gamma_3(E, \eta_g)]}]$$

Following (12.27) the DR in the present case can be written as

$$\left(\frac{n_z\pi}{d_z}\right)^2 = A_{35,\pm}(e_{86}, n, \eta_g) \tag{12.28}$$

where e_{86} is the totally quantized energy in this case.

The DOS function is given by

$$N_{Bn_z} = \frac{g_v eB}{2\pi\hbar} \sum_{n=0}^{n_{\max}} \sum_{n_z=1}^{n_{z_{Max}}} \delta'(E - e_{86}) \tag{12.29}$$

The electron concentration per unit area in this case can be written as

$$N_s = \frac{g_v eB}{2\pi\hbar} \sum_{n=0}^{n_{\max}} \sum_{n_z=1}^{n_{z_{Max}}} F_{-1}(\eta_{12,6}) \tag{12.30}$$

where $\eta_{12,6} = (k_B T)^{-1}[E_{F12HD} - e_{86}]$.

Thus, by using (12.7), (12.31) and other appropriate equations, we can study the QC in this case.

12.2.3 The QC in HDQWTs of II–VI materials under magnetic quantization

Following (9.40), the DR in the present case can be written as

$$\left(\frac{n_z\pi}{d_z}\right)^2 = U_{8,\pm}(e_{87}, n, \eta_g) \tag{12.31}$$

where e_{87} is the totally quantized energy in this case.

The DOS function is given by

$$N_{Bn_z} = \frac{g_v eB}{2\pi\hbar}\sum_{n=0}^{n_{\max}}\sum_{n_z=1}^{n_{zMax}} \delta'(E - e_{87}) \tag{12.32}$$

The electron concentration per unit area in this case can be written as

$$N_s = \frac{g_v eB}{2\pi\hbar}\sum_{n=0}^{n_{\max}}\sum_{n_z=1}^{n_{zMax}} F_{-1}(\eta_{12,7}) \tag{12.33}$$

where $\eta_{12,7} = (k_BT)^{-1}[E_{F12HD} - e_{87}]$.

Thus, by using (12.7), (12.33) and other appropriate equations we can study the QC in this case.

12.2.4 The QC in HDQWTs of IV–VI materials under magnetic quantization

The electron energy spectrum in IV–VI semiconductors are defined by the models of Cohen, Lax, Dimmock and Bangert and Kastner, respectively. The magneto DR in HD IV–VI semiconductors is discussed in accordance with the said model for the purpose of relative comparison.

(a) Cohen model

Following (3.43), the magneto electron energy spectrum in IV–VI semiconductors in the presence of quantizing magnetic field B along z-direction can be written as

$$E(1+\alpha E) = \frac{p_x^2}{2m_1} + \frac{p_z^2}{2m_3} - \frac{\alpha E p_y^2}{2m_2'} + \frac{p_y^2(1+\alpha E)}{2m_2} + \frac{\alpha p_y^4}{4m_2 m_2'} \tag{12.34}$$

where

$$\omega(E) = \frac{|e|B}{\sqrt{m_1 m_2}}\left[1 + \alpha E\left(1 - \frac{m_2}{m_2'}\right)\right]^{1/2}$$

Therefore, the magneto dispersion law in HDQWTs of IV–VI materials can be expressed as

$$\frac{\hbar^2 k_z^2}{2m_3} = U_{16,\pm}(E, n, \eta_g) \tag{12.35}$$

where

$$\begin{aligned} U_{16,\pm}(E, n, \eta_g) = \Bigg[& \gamma_2(E, \eta_g) - \left(n + \frac{1}{2}\right) \frac{\hbar e B}{\sqrt{m_1 m_2}} \mp \frac{1}{2} g^* \mu_0 B \\ & - \frac{3\alpha}{8} \left(n^2 + n + \frac{1}{2}\right) \left(\frac{\hbar e B}{\sqrt{m_1 m_2}}\right)^2 \\ & - \gamma_3(E, \eta_g) \left[\frac{\alpha}{2} \left(n + \frac{1}{2}\right) \frac{\hbar e B}{\sqrt{m_1 m_2}} \left(1 - \frac{m_2}{m_2'}\right) \right. \\ & \left. + \frac{3\alpha^2}{8} \left(n^2 + n + \frac{1}{2}\right) \left(\frac{\hbar e B}{\sqrt{m_1 m_2}}\right)^2 \left(1 - \frac{m_2}{m_2'}\right) \right] \Bigg] \end{aligned}$$

The DR in the present case can be written as

$$\left(\frac{n_z \pi}{d_z}\right)^2 = U_{16,\pm}(e_{88}, n, \eta_g) \tag{12.36}$$

where e_{88} is the totally quantized energy in this case.

The DOS function is given by

$$N_{Bn_z} = \frac{g_v e B}{2\pi\hbar} \sum_{n=0}^{n_{\max}} \sum_{n_z=1}^{n_{z_{Max}}} \delta'(E - e_{88}) \tag{12.37}$$

The electron concentration per unit area in this case can be written as

$$N_s = \frac{g_v e B}{2\pi\hbar} \sum_{n=0}^{n_{\max}} \sum_{n_z=1}^{n_{z_{Max}}} F_{-1}(\eta_{12,8}) \tag{12.38}$$

where

$$\eta_{12,8} = (k_B T)^{-1} [E_{F12HD} - e_{88}]. \tag{12.39}$$

Thus, by using (12.7), (12.39) and other appropriate equations, we can study the QC in this case.

(b) Lax model

In accordance with this model, the magneto DR assumes the form

$$E(1+\alpha E) = \left(n+\frac{1}{2}\right)\hbar\omega_{03}(E) + \frac{\hbar^2 k_z^2}{2m_3} \pm \frac{1}{2}\mu_0 g^* B \tag{12.40}$$

where

$$\omega_{03}(E) \equiv \frac{eB}{\sqrt{m_1 m_2}}$$

The magneto DR in HDQWTs of IV–VI materials can be written following (12.40) as

$$\gamma_2(E,\eta_g) = (n+\frac{1}{2}) + \hbar\omega_{03}(E) + \frac{\hbar^2 k_z^2}{2m_3} \pm \frac{1}{2}g^*\mu_0 B \tag{12.41}$$

(12.42) can be written as

$$\frac{\hbar^2 k_z^2}{2m_3} = U_{17,\pm}(E,n,\eta_g) \tag{12.42}$$

where

$$U_{17,\pm}(E,n,\eta_g) = \gamma_2(E,\eta_g) - \left(n+\frac{1}{2}\right)\hbar\omega_{03}(E) \pm \frac{1}{2}g^*\mu_0 B$$

The DR in the present case can be written as

$$\left(\frac{n_z\pi}{d_z}\right)^2 = U_{17,\pm}(e_{89},n,\eta_g) \tag{12.43}$$

where e_{89} is the totally quantized energy in this case.

The DOS function is given by

$$N_{Bn_z} = \frac{g_v eB}{2\pi\hbar}\sum_{n=0}^{n_{\max}}\sum_{n_z=1}^{n_{zMax}} \delta'(E-e_{89}) \tag{12.44}$$

The electron concentration per unit area in this case can be written as

$$N_s = \frac{g_v eB}{2\pi\hbar}\sum_{n=0}^{n_{\max}}\sum_{n_z=1}^{n_{zMax}} F_{-1}(\eta_{12,9}) \tag{12.45}$$

where

$$\eta_{12,9} = (k_B T)^{-1}[E_{F12HD} - e_{89}]. \tag{12.46}$$

Thus, by using (12.7), (12.45) and other appropriate equations, we can study the QC in this case.

(c) Dimmock Model

Following (3.56) the dispersion relation under magnetic quantization in HDQWTs of IV–VI materials can be expressed in accordance with Dimmock model as

$$\begin{aligned}
&\gamma_2(E,\eta_g)+\alpha\gamma_3(E,\eta_g)\frac{2eB}{\hbar}\left(n+\frac{1}{2}\right)\frac{\hbar^2}{2}\left(\frac{1}{m_t^+}-\frac{1}{m_t^-}\right)\\
&\quad+\alpha\gamma_3(E,\eta_g)x\frac{\hbar^2}{2}\left(\frac{1}{m_l^+}-\frac{1}{m_l^-}\right)\\
&=\frac{\hbar^2k_s^2}{2m_t^*}+\frac{\hbar^2k_z^2}{2m_l^*}+\frac{\hbar^2k_s^2}{2m_t^-}+\frac{\hbar^2k_z^2}{2m_l^-}\\
&\quad+\alpha\left[\frac{\hbar^4k_s^4}{4m_t^-m_t^+}+\frac{\hbar^2k_s^2k_z^2}{4m_l^-m_l^+}+\frac{\hbar^2k_z^2k_s^2}{4m_t^+m_t^-}+\frac{\hbar^4k_z^4}{4m_l^-m_l^+}\right]\\
&=\frac{2eB}{\hbar}\left(n+\frac{1}{2}\right)\frac{\hbar^2}{2}\left(\frac{1}{m_t^*}-\frac{1}{m_t^-}\right)+x\frac{\hbar^2}{2}\left(\frac{1}{m_l^*}-\frac{1}{m_l^-}\right)\\
&\quad+\alpha\left[\frac{\hbar^4}{4m_t^-m_t^+}\left(\frac{2eB}{\hbar}\left(n+\frac{1}{2}\right)\right)^2\right.\\
&\quad\left.+x\left[\frac{\hbar^4eB}{2m_l^+m_t^-\hbar}+\frac{\hbar^4eB}{2m_t^+m_l^-\hbar}\right]\left(n+\frac{1}{2}\right)+\frac{\hbar^4}{4m_l^-m_l^+}x^2\right]
\end{aligned}\tag{12.47}$$

where $x = k_z^2$.

Therefore, the magneto dispersion relation in HDQWTs of IV–VI materials, whose unperturbed carriers obey the Dimmock model can be expressed as

$$k_z^2 = U_{170}(E,n,\eta_g)\tag{12.48}$$

where

$$\begin{aligned}
U_{170}(E,n,\eta_g)&=[2p_9]^{-1}[-q_9(E,n,\eta_g)\\
&\quad+[q_9^2(E,n,\eta_g)+4p_9R_9(E,n,\eta_g)]^{1/2}]
\end{aligned}$$

$$p_9=\frac{\alpha\hbar^4}{4m_l^-m_l^+},\quad q_9(E,n,\eta_g)=\left[\frac{\hbar^2}{2}\left(\frac{1}{m_l^*}+\frac{1}{m_l^-}\right)+\frac{\alpha\hbar^3eB}{2}\left(n+\frac{1}{2}\right)\right.$$
$$\left.\left(\frac{1}{m_t^-m_l^+}+\frac{1}{m_l^-m_t^+}\right)-\alpha\gamma_3(E,\eta_g)\left(\frac{1}{m_l^+}+\frac{1}{m_l^-}\right)\right]$$

and

$$R_9(E,n,\eta_g) = \gamma_3(E,\eta_g) + \alpha eB\gamma_3(E,\eta_g)\left(n+\frac{1}{2}\right)\hbar\left(\frac{1}{m_t^+}-\frac{1}{m_t^-}\right) - \frac{\alpha\hbar^2}{m_t^- m_t^+}\left[eB\left(n+\frac{1}{2}\right)^2\right]$$

The DR in the present case can be written as

$$\left(\frac{n_z\pi}{d_z}\right)^2 = U_{170}(e_{90},n,\eta_g)a \tag{12.49}$$

where e_{90} is the totally quantized energy in this case.

The DOS function is given by

$$N_{Bn_z} = \frac{g_v eB}{\pi\hbar}\sum_{n=0}^{n_{\max}}\sum_{n_z=1}^{n_{zMax}}\delta'(E-e_{90}) \tag{12.50}$$

The electron concentration per unit area in this case can be written as

$$N_s = \frac{g_v eB}{\pi\hbar}\sum_{n=0}^{n_{\max}}\sum_{n_z=1}^{n_{zMax}}F_{-1}(\eta_{12,10}) \tag{12.51}$$

where $\eta_{12,10} = (k_BT)^{-1}[E_{F12HD}-e_{90}]$.

Thus, by using (12.7), (12.51) and other appropriate equations, we can study the QC in this case.

(d) Model of Bangert and Kastner

Using (9.50) the DR in the present case can be written as

$$\left(\frac{n_z\pi}{d_z}\right)^2 = U_{18}(e_{91},n,\eta_g) \tag{12.52}$$

where e_{91} is the totally quantized energy in this case.

The DOS function is given by

$$N_{Bn_z} = \frac{g_v eB}{\pi\hbar}\sum_{n=0}^{n_{\max}}\sum_{n_z=1}^{n_{zMax}}\delta'(E-e_{91}) \tag{12.53}$$

The electron concentration per unit area in this case can be written as

$$N_s = \frac{g_v eB}{\pi\hbar} \text{ Real part of } \sum_{n=0}^{n_{\max}} \sum_{n_z=1}^{n_{zMax}} F_{-1}(\eta_{12,11}) \tag{12.54}$$

where $\eta_{12,11} = (k_B T)^{-1}[E_{F12HD} - e_{91}]$.

Thus, by using (12.7), (12.54) and other appropriate equations, we can study the QC in this case.

(e) Model of Foley and Langenberg

Using (R1.4), (8.2) and following the methods as given in Chapter 8, the DR in HDQWTs of IV–VI materials in the present case is given by

$$\left[\left[\gamma_3(E,\eta_g) + \frac{E_{g0}}{2}\right] - \left[\frac{\hbar^2 k_s^2}{2m_\perp^-} + \frac{\hbar^2 k_z^2}{2m_{||}^-}\right]\right]^2 = \left[\frac{\hbar^2 k_s^2}{2m_\perp^-} + \frac{\hbar^2 k_z^2}{2m_{||}^-}\right] + \frac{E_{g0}^2}{4} + E_{g0}\left[\frac{\hbar^2 k_s^2}{2m_\perp^-} + \frac{\hbar^2 k_z^2}{2m_{||}^-}\right] + P_{||}^2 k_z^2 + P_\perp^2 k_s^2 \tag{12.55}$$

Therefore, the magneto-DR in HDQWTs of IV–VI materials can be written as

$$\gamma_3^2(E,\eta_g) + \frac{E_g^2}{4} + E_g\gamma_3(E,\eta_g) + \left[\frac{\hbar eB}{m_\perp^-}\left(n+\frac{1}{2}\right) + \frac{\hbar^2 x}{2m_{||}^-}\right]^2 - 2\left[\gamma_3(E,\eta_g) + \frac{E_g}{2}\right]\left[\frac{\hbar eB\,(n+\frac{1}{2})}{m_\perp^-} + \frac{\hbar^2 x}{2m_{||}^-}\right] = \left[\frac{\hbar eB\,(n+\frac{1}{2})}{m_\perp^+} + \frac{\hbar^2 x}{2m_{||}^+}\right] + \left[\frac{\hbar eB}{m_\perp^+}\left(n+\frac{1}{2}\right) + \frac{\hbar^2 x}{2m_{||}^+}\right]^2 + P_{||}^2 x + P_\perp^2 \frac{2eB}{\hbar}\left(n+\frac{1}{2}\right) \tag{12.56}$$

where $k_z^2 = x$.

Therefore, the magneto DR in HDQWTs of IV–VI materials, where unperturbed carriers follow the model of Foley *et al.* can be expressed following (12.56)

$$k_z^2 = U_{19}(E, n, \eta_g) \tag{12.57}$$

where

$$U_{19}(E, n, \eta_g) = [[2p_{91}]^{-1}[-q_{91}(E, n, \eta_g) + \{q_{91}^2(E, n, \eta_g) + 4p_{91}R_{91}(E, n, \eta_g)\}^{1/2}]] \quad p_{91} = \frac{\hbar^4}{4}\left[\frac{1}{(m_{||}^2)^2} - \frac{1}{(m_{||}^-)^2}\right]$$

$$q_{91}(E, n, \eta_g) = \left[\frac{\hbar^2 eB}{m_\perp^- m_{||}^+}\left(n + \frac{1}{2}\right) + P_{||}^2 + \frac{\hbar^2 E_g}{2m_{||}^+} - \frac{\hbar^2 eB\left(n + \frac{1}{2}\right)}{m_\perp^- m_{||}^-} + \frac{\hbar^2}{m_{||}^-}(\gamma_3\left(E, \eta_g) + \frac{E_g}{2}\right)\right]$$

$$R_{91}(E, n, \eta_g) = \left[\gamma_3^2(E, \eta_g) + E_g\gamma_3(E, \eta_g) - \frac{2\hbar eB}{m_\perp^-}\left(\gamma_3(E, \eta_g) + \frac{E_g}{2}\right) \times \left(n + \frac{1}{2}\right) - E_g\frac{\hbar eB}{m_\perp^+}\left(n + \frac{1}{2}\right) - P_\perp^2 . \frac{2eB}{\hbar}\left(n + \frac{1}{2}\right)\right]$$

The DR in the present case can be written as

$$\left(\frac{n_z\pi}{d_z}\right)^2 = U_{19}(e_{91}, n, \eta_g) \tag{12.58}$$

where e_{91} is the totally quantized energy in this case.

The DOS function is given by

$$N_{Bn_z} = \frac{g_v eB}{\pi\hbar}\sum_{n=0}^{n_{\max}}\sum_{n_z=1}^{n_{zMax}}\delta'(E - e_{92}) \tag{12.59}$$

The electron concentration per unit area in this case can be written as

$$N_s = \frac{g_v eB}{\pi\hbar}\sum_{n=0}^{n_{\max}}\sum_{n_z=1}^{n_{zMax}}F_{-1}(\eta_{12,12}) \tag{12.60}$$

where $\eta_{12,12} = (k_B T)^{-1}[E_{F12HD} - e_{92}]$.

Thus, by using (12.7), (12.60) and other appropriate equations, we can study the QC in this case.

12.2.5 *The QC in HDQWTs of stressed Kane type materials under magnetic quantization*

Following (9.61), the DR in this case is given by

$$\left(\frac{n_z\pi}{d_z}\right)^2 = U_{41}(e_{93}, n, \eta_g) \tag{12.61}$$

where e_{93} is the totally quantized energy in this case.

The DOS function is given by

$$N_{Bn_z} = \frac{g_v eB}{\pi\hbar}\sum_{n=0}^{n_{\max}}\sum_{n_z=1}^{n_{z_{Max}}}\delta'(E - e_{93}) \tag{12.62}$$

The electron concentration per unit area in this case can be written as

$$N_s = \frac{g_v eB}{\pi\hbar}\sum_{n=0}^{n_{\max}}\sum_{n_z=1}^{n_{z_{Max}}}F_{-1}(\eta_{12,13}) \tag{12.63}$$

where $\eta_{12,13} = (k_BT)^{-1}[E_{F12HD} - e_{93}]$.

Thus, by using (12.7), (12.63) and other appropriate equations we can study the QC in this case.

12.2.6 *The QC in HDQWTs of Te under magnetic quantization*

Following (11.71) the DR in the present case can be written as

$$k_z^2 = U_{42,\pm}(E, n, \eta_g) \tag{12.64}$$

where

$$\begin{aligned}U_{42,\pm}(E, n, \eta_g) = (2\psi_1^2)^{-1}\Bigg[&\left\{2\gamma_3(E,\eta_g)\psi_1 + \psi_3^2 - 4\psi_1\psi_2\frac{eB}{\hbar}\left(n+\frac{1}{2}\right)\right\}\\&-\{\psi_3^4 + 4\psi_1\psi_3^2\gamma_3(E,\eta_g)\}\\&+\frac{8eB}{\hbar}\left(n+\frac{1}{2}\right)(\psi_1^2\psi_4^2 - \psi_1\psi_2\psi_3)^{-1/2}\Bigg]\end{aligned}$$

and $\bar{e}_{94}$ is the totally quantized energy in this case.

The DOS function is given by

$$N_{Bn_z} = \frac{g_v eB}{2\pi\hbar} \sum_{n=0}^{n_{\max}} \sum_{n_z=1}^{n_{z_{Max}}} \delta'(E - \bar{e}_{94}) \tag{12.65}$$

The electron concentration per unit area in this case can be written as

$$N_s = \frac{g_v eB}{2\pi\hbar} \sum_{n=0}^{n_{\max}} \sum_{n_z=1}^{n_{z_{Max}}} F_{-1}(\eta_{12,14}) \tag{12.66}$$

where $\eta_{12,14} = (k_B T)^{-1}[E_{F12HD} - \bar{e}_{94}]$.

Thus, by using (12.7), (12.66) and other appropriate equations, we can study the QC in this case.

12.2.7 *The QC in HDQWTs of gallium phosphide under magnetic quantization*

Using (11.77) and (8.2), the DR in the present case can be written (12.67)

$$\left(\frac{n_z \pi}{d_z}\right)^2 = U_{43}(\bar{e}_{941}, n, \eta_g) \tag{12.67}$$

where

$$U_{43}(\bar{e}_{941}, n, \eta_g) = (2b^2)^{-1}\left[\left\{2\gamma_3(\bar{e}_{941}, \eta_g)b + c - 2Db - 4ab\frac{eB}{\hbar}\left(n + \frac{1}{2}\right)\right\} + \left\{[c^2 + 4bc\gamma_3(\bar{e}_{941}, \eta_g) + 4D^2b^2 - 4cDb] - \frac{8eB}{\hbar}\left(n + \frac{1}{2}\right)(2ab^2D + 4\gamma_3(\bar{e}_{941}, \eta_g))b^2 a + abc - 2b^2 a\gamma_3(\bar{e}_{941}, \eta_g) - b^2 c\right\}^{-1/2}\right]$$

and $\bar{e}_{941}$ is the totally quantized energy in this case.

The DOS function is given by

$$N_{Bn_z} = \frac{g_v eB}{\pi\hbar} \sum_{n=0}^{n_{\max}} \sum_{n_z=1}^{n_{z_{Max}}} \delta'(E - \bar{e}_{941}) \tag{12.68}$$

The electron concentration per unit area in this case can be written as

$$N_s = \frac{g_v eB}{\pi\hbar} \sum_{n=0}^{n_{\max}} \sum_{n_z=1}^{n_{z_{Max}}} F_{-1}(\eta_{12,15}) \tag{12.69}$$

where $\eta_{12,15} = (k_B T)^{-1}[E_{F12HD} - \bar{e}_{941}]$.

Thus, by using (12.7), (12.69) and other appropriate equations, we can study the QC in this case.

12.2.8 *The QC in HDQWTs of platinum antimonide under magnetic quantization*

The magneto DR in HDQWTs of $PtSb_2$ can be written following (11.84) as

$$k_z^2 = U_{44}(E, n, \eta_g) \tag{12.70}$$

where

$$U_{44}(E, n, \eta_g) = \frac{1}{2T_{41}}[T_{71}(E, n, \eta_g) + \sqrt{T_{71}^2(E, n, \eta_g) + 4T_{41}T_{71}(E, n, \eta_g)}],$$

$$T_{71}(E, n, \eta_g) = \left[T_{51}(E, \eta_g) - T_{31}\frac{2eB}{\hbar}\left(n + \frac{1}{2}\right)\right] \quad \text{and}$$

$$T_{72}(E, n, \eta_g) = \left[T_{61}(E, \eta_g) + \frac{2eB}{\hbar}\left(n + \frac{1}{2}\right)T_{21}(E, \eta_g)\right]$$

The DR in the present case can be expressed as

$$\left(\frac{n_z\pi}{d_z}\right)^2 = U_{44}(e_{95}, n, \eta_g) \tag{12.71}$$

and e_{95} is the totally quantized energy in this case.

The DOS function is given by

$$N_{Bn_z} = \frac{g_v eB}{\pi\hbar}\sum_{n=0}^{n_{\max}}\sum_{n_z=1}^{n_{z_{Max}}}\delta'(E - e_{95}) \tag{12.72}$$

The electron concentration per unit area in this case can be written as

$$N_s = \frac{g_v eB}{\pi\hbar}\sum_{n=0}^{n_{\max}}\sum_{n_z=1}^{n_{z_{Max}}}F_{-1}(\eta_{12,16}) \tag{12.73}$$

where

$$\eta_{12,16} = (k_B T)^{-1}[E_{F12HD} - e_{95}] \tag{12.74}$$

Thus, by using (12.7), (12.73) and other appropriate equations, we can study the QC in this case.

12.2.9 *The QC in HDQWTs of bismuth telluride under magnetic quantization*

The magneto DR of the valance bands in bulk p-type HD bismuth telluride can be written expressed following (12.74) as

$$k_x^2 = U_{45}(E, \eta_g, n) \tag{12.75}$$

where

$$U_{45}(E, \eta_g, n) = \frac{\gamma_2(E, \eta_g) - \left(n + \frac{1}{2}\right) \frac{e\hbar B}{M_{31}}}{\bar{\omega}_1} \quad \text{and}$$

$$M_{31} = \frac{m_0}{\left(\bar{\alpha}_{22}\bar{\alpha}_{33} - \frac{(\bar{\alpha}_{23})^2}{4}\right)^{1/2}}$$

Therefore, in the present case the hole energy spectrum can be expressed as

$$\left(\frac{n_z \pi}{d_z}\right)^2 = U_{45}(e_{96}, n, \eta_g) \tag{12.76a}$$

where e_{96} is the totally quantized energy in this case.

The DOS function is given by

$$N_{Bn_z} = \frac{g_v eB}{\pi\hbar} \sum_{n=0}^{n_{\max}} \sum_{n_z=1}^{n_{z_{Max}}} \delta'(E - e_{96}) \tag{12.76b}$$

The electron concentration per unit area in this case can be written as

$$N_s = \frac{g_v eB}{\pi\hbar} \sum_{n=0}^{n_{\max}} \sum_{n_z=1}^{n_{z_{Max}}} F_{-1}(\eta_{12,17}) \tag{12.77}$$

where $\eta_{12,17} = (k_B T)^{-1}[E_{F12HD} - e_{96}]$.

Thus, by using (12.7), (12.77) and other appropriate equations we can study the QC in this case.

12.2.10 *The QC in HDQWTs of germanium under magnetic quantization*

(a) Model of Cardona *et al.*

The magneto DR under the condition of heavy doping can be written as

$$k_x^2 = U_{46}(E, \eta_g, n) \tag{12.78}$$

where

$$U_{46}(E, \eta_g, n) = \frac{2m_{||}^*}{\hbar^2}\left[\gamma_3(E, \eta_g) + \frac{E_{g_0}}{2} - \left[\frac{E_{g_0}^2}{4} + \frac{E_{g_0}\hbar^2}{m_\perp^*}\frac{2eB}{\hbar}\left(n + \frac{1}{2}\right)\right]\right]^{1/2} \tag{12.79}$$

The DR in the present case can be written as

$$\left(\frac{n_z\pi}{d_z}\right)^2 = U_{46}(e_{97}, n, \eta_g) \tag{12.72}$$

where e_{97} is the totally quantized energy in this case.

The DOS function is given by

$$N_{Bn_z} = \frac{g_v eB}{\pi\hbar}\sum_{n=0}^{n_{\max}}\sum_{n_z=1}^{n_{z_{Max}}}\delta'(E - e_{97}) \tag{12.80}$$

The electron concentration per unit area in this case can be written as

$$N_s = \frac{g_v eB}{\pi\hbar}\sum_{n=0}^{n_{\max}}\sum_{n_z=1}^{n_{z_{Max}}}F_{-1}(\eta_{12,18}) \tag{12.81}$$

$$\eta_{12,18} = (k_B T)^{-1}[E_{F12HD} - e_{97}] \tag{12.82}$$

Thus, by using (12.7), (12.82) and other appropriate equations, we can study the QC in this case.

(b) Model of Wang and Ressler

The magneto DR in HD Ge can be written expressed following (12.82) as

$$k_x^2 = U_{47}(E, \eta_g, n) \tag{12.83}$$

where

$$U_{47}(E, \eta_g, n) = \left(\frac{m_{||}^*}{\hbar^2 \bar{\alpha}_6}\right)\left[1 - \bar{\alpha}_5\left(n + \frac{1}{2}\right)\hbar\omega_\perp - \{\theta_7(n) - 4\bar{\alpha}_6\gamma_3(E, \eta_g)\}^{1/2}\right], \quad \omega_\perp = \frac{eB}{m_\perp^*}$$

and

$$\theta_7(n) = \left[1 + (\bar{\alpha}_5)^2\left\{\left(n + \frac{1}{2}\right)\hbar\omega_\perp\right\}^2 - 2\bar{\alpha}_5\left(n + \frac{1}{2}\right)\hbar\omega_\perp + 4\bar{\alpha}_6\left(n + \frac{1}{2}\right)\hbar\omega_\perp - 4\bar{\alpha}_6\bar{\alpha}_4\left\{\left(n + \frac{1}{2}\right)\hbar\omega_\perp\right\}^2\right]$$

The DR in the present case can be written as

$$\left(\frac{n_z\pi}{d_z}\right)^2 = U_{47}(e_{98}, n, \eta_g) \tag{12.84}$$

where e_{98} is the totally quantized energy in this case.

The DOS function is given by

$$N_{Bn_z} = \frac{g_v eB}{\pi\hbar}\sum_{n=0}^{n_{\max}}\sum_{n_z=1}^{n_{zMax}}\delta'(E - e_{98}) \tag{12.85}$$

The electron concentration per unit area in this case can be written as

$$N_s = \frac{g_v eB}{\pi\hbar}\sum_{n=0}^{n_{\max}}\sum_{n_z=1}^{n_{zMax}}F_{-1}(\eta_{12,19}) \tag{12.86}$$

$$\eta_{12,19} = (k_B T)^{-1}[E_{F12HD} - e_{98}] \tag{12.87}$$

Thus, by using (12.7), (12.86) and other appropriate, equations, we can study the QC in this case.

12.2.11 *The QC in HDQWTs of gallium antimonide under magnetic quantization*

The heavily doped magneto DR in the present case can be written as

$$k_z^2 = U_{48}(E, \eta_g, n)$$
$$U_{48}(e_{99}, n, \eta_g) = \left[I_{36}(e_{99}, \eta_g) - \frac{2eB}{\hbar}\left(n + \frac{1}{2}\right)\right] \tag{12.88}$$

and e_{99} is the totally quantized energy in this case.

The DOS function is given by

$$N_{Bn_z} = \frac{g_v eB}{\pi\hbar} \sum_{n=0}^{n_{\max}} \sum_{n_z=1}^{n_{zMax}} \delta'(E - e_{99}) \tag{12.89}$$

The electron concentration per unit area in this case can be written as

$$N_s = \frac{g_v eB}{\pi\hbar} \sum_{n=0}^{n_{\max}} \sum_{n_z=1}^{n_{zMax}} F_{-1}(\eta_{12,20}) \tag{12.90}$$

$$\eta_{12,20} = (k_B T)^{-1}[E_{F12HD} - e_{99}] \tag{12.91}$$

Thus, by using (12.7), (12.90) and other appropriate equations we can study the QC in this case.

12.2.12 *The QC in HDQWTs of II–V materials under magnetic quantization*

The hole energy spectrum in HDQWTs of II–V materials can be written following (R1.19) as

$$y_3(E, n_g) = \theta_1 k_x^2 + \theta_2 k_y^2 + \theta_3 k_z^2 + \delta_4 k_x \pm [\{\theta_5 k_x^2 + \theta_6 k_y^2 + \theta_7 k_z^2 + \delta_5 k_x\}^2$$
$$+ G_3^2 k_y^2 + \Delta_3^2]^{\frac{1}{2}} \pm \Delta_3 \tag{12.92}$$

the magneto-dispersion law in HDQWTs of II–V materials assumes the form

$$k_y^2 = U_{49,\pm}(E, \eta_g, n)$$
$$U_{49,\pm}(E, \eta_g, n) = [I_{35}\gamma_3(E, \eta_g) + I_{36,\pm}(n) \pm [\gamma_3^2(E, \eta_g)$$

$$+\gamma_3(E,\eta_g)I_{38,\pm}(n)+I_{39,\pm}(n)]^{1/2}]$$

$$I_{35}=\frac{\theta_2}{(\theta_2^2-\theta_5^2)},\quad I_{36,\pm}(n)=\frac{I_{33,\pm}(n)}{2(\theta_2^2-\theta_5^2)},$$

$$I_{38,\pm}(n)=(4\theta_5^2)^{-1}[4\theta_2 I_{33,\pm}(n)+8\theta_2^2 I_{31,\pm}(n)-\theta_5^2 I_{31,\pm}(n)]$$

$$I_{39,\pm}=(4\theta_5^2)^{-1}[I_{33,\pm}^2(n)+4\theta_2^3 I_{34,\pm}(n)-4\theta_5^3 I_{34,\pm}(n)],$$

$$I_{33,\pm}(n)=[G_3^2+2\theta_5 I_{32}(n)-2\theta_2 I_{31,\pm}(n)]$$

$$I_{34,\pm}(n)=[I_{32,\pm}^2(n)+\Delta_3^2-I_{31,\pm}(n)]$$

$$I_{31,\pm}(n)=\left[\left(n+\frac{1}{2}\right)\hbar\omega_{31}-\frac{\delta_4^2}{4\theta_1}\pm\Delta_3\right],$$

$$I_{32}(n)=\left[\left(n+\frac{1}{2}\right)\hbar\omega_{32}-\frac{\delta_5^2}{4\theta_5}\right]$$

$$\omega_{31}=\frac{eB}{\sqrt{M_{31}M_{32}}},\quad \omega_{32}=\frac{eB}{\sqrt{M_{33}M_{34}}},\quad M_{31}=\frac{\hbar^2}{2\theta_1},$$

$$M_{32}=\frac{\hbar^2}{2\theta_3},\quad M_{33}=\frac{\hbar^2}{2\theta_5}\quad\text{and}\quad M_{34}=\frac{\hbar^2}{2\theta_7}. \tag{12.93}$$

The DR in the present case can be written as

$$\left(\frac{n_z\pi}{d_z}\right)^2=U_{49,\pm}(e_{100},n,\eta_g) \tag{12.94}$$

where e_{100} is the totally quantized energy in this case.

The DOS function is given by

$$N_{Bn_z}=\frac{g_v eB}{2\pi\hbar}\sum_{n=0}^{n_{\max}}\sum_{n_z=1}^{n_{z_{Max}}}\delta'(E-e_{100}) \tag{12.95}$$

The electron concentration per unit area in this case can be written as

$$N_s=\frac{g_v eB}{2\pi\hbar}\sum_{n=0}^{n_{\max}}\sum_{n_z=1}^{n_{z_{Max}}}F_{-1}(\eta_{12,21}) \tag{12.96}$$

where $\eta_{12,21}=(k_BT)^{-1}[E_{F12HD}-e_{100}]$.

Thus, by using (12.7), (12.96) and other appropriate equations, we can study the QC in this case.

12.2.13 *The QC in HDQWTs of lead germanium telluride under magnetic quantization*

The DR of the carriers in n-type $Pb_{1-x}Ga_xTe$ with $x = 0.01$ can be written following Vassilev [12] as

$$[E - 0.606k_s^2 - 0.0722k_z^2][E + \bar{E}_g + 0.411k_s^2 + 0.0377k_z^2] = 0.23k_s^2 + 0.02k_z^2 \pm [0.06\bar{E}_g + 0.061k_s^2 + 0.0066k_z^2]k_s \quad (12.97)$$

where $\bar{E}_g (= 0.21\,\text{eV})$ is the energy gap for the transition point, the zero of the energy E is at the edge of the conduction band of the Γ point of the Brillouin zone and is measured positively upwards, k_x, k_y and k_z are in the units of $10^9\,\text{m}^{-1}$.

The magneto-dispersion law in HDQWTs of $Pb_{1-x}Ge_xTe$ can be expressed as

$$\begin{aligned}
&\left[\frac{2\theta_0(E,\eta_g)}{1 + Erf(E/\eta_g)}\right] + \gamma_3(E,\eta_g)\left[\overline{E_{g_0}} - 0.345x - 0.390\frac{eB}{\hbar}\left(n + \frac{1}{2}\right)\right] \\
&= \frac{0.46eB}{\hbar}\left(n + \frac{1}{2}\right) + 0.02x \pm \left[0.06\overline{E_{g_0}} + 0.122\frac{eB}{\hbar}\left(n + \frac{1}{2}\right)\right. \\
&\quad \left. + 0.0066x\right]\left(\frac{2eB}{\hbar}\left(n + \frac{1}{2}\right)\right)^{1/2} \\
&\quad + \left[\overline{E_{g_0}} + \frac{0.822eB}{\hbar}\left(n + \frac{1}{2}\right) + 0.377x\right] \\
&\quad \times \left[\frac{1.212cB}{\hbar}\left(n + \frac{1}{2}\right) + 0.722x\right]
\end{aligned} \quad (12.98)$$

The (12.98) assumes the form

$$k_z^2 = U_{50,\pm}(E,\eta_g,n) \quad (12.99)$$

where

$$\begin{aligned}
U_{50,\mp}(E,\eta_g,n) &= (2p_{10})^{-1}[q_{10}(E,\eta_g,n) - [q_{10}^2(E,\eta_g,n) \\
&\quad + 4p_{10}R_{10,\mp}(E,\eta_g,n)]^{1/2}]
\end{aligned}$$

$$p_{10}(0.377 \times 0.722), \quad q_{10}(E,\eta_g,n) = \Big[0.02 + 0.345\gamma_3(E,\eta_g)$$

$$\pm 0.0066\left(\frac{2eB}{\hbar}\left(n+\frac{1}{2}\right)\right)^{1/2}+0.377\times\frac{1.212eB}{\hbar}\left(n+\frac{1}{2}\right)$$

$$+0.722\left[\bar{E}_{g_0}+0.822\frac{eB}{\hbar}\left(n+\frac{1}{2}\right)\right]\Bigg]\quad\text{and}$$

$$R_{10,\mp}(E,\eta_g,n)=\left[\frac{2\theta_0(E,\eta_g)}{1+Eof(E/\eta_g)}+\gamma_3(E,\eta_g)\left[\bar{E}_{g_0}-0.390\frac{eB}{\hbar}\left(n+\frac{1}{2}\right)\right]\right.$$

$$\mp\left(0.06\bar{E}_{g_0}+0.122\frac{eB}{\hbar}\left(n+\frac{1}{2}\right)\right)\left(\frac{2eB}{\hbar}\left(n+\frac{1}{2}\right)\right)^{1/2}$$

$$\left.-\left(\bar{E}_{g_0}+0.822\frac{eB}{\hbar}\left(n+\frac{1}{2}\right)\right)\frac{1.212eB}{\hbar}\left(n+\frac{1}{2}\right)-\frac{0.46eB}{\hbar}\left(n+\frac{1}{2}\right)\right]$$

The DR in the present case can be written as

$$\left(\frac{n_z\pi}{d_z}\right)^2=U_{50\pm}(e_{101},n,\eta_g)\tag{12.100}$$

where e_{101} is the totally quantized energy in this case.

The DOS function is given by

$$N_{Bn_z}=\frac{g_v eB}{2\pi\hbar}\sum_{n=0}^{n_{\max}}\sum_{n_z=1}^{n_{z_{Max}}}\delta'(E-e_{101})\tag{12.101}$$

The electron concentration per unit area in this case can be written as

$$N_s=\frac{g_v eB}{2\pi\hbar}\sum_{n=0}^{n_{\max}}\sum_{n_z=1}^{n_{z_{Max}}}F_{-1}(\eta_{12,22})\tag{12.102}$$

where $\eta_{12,22}=(k_BT)^{-1}[E_{F12HD}-e_{101}]$.

Thus by, using (12.7), (12.102) and other appropriate equations, we can study the QC in this case.

12.3 Summary and Conclusion

The DRs in HDQWTs made of the said HD materials exhibit the fact that the total energy is quantized since the corresponding wave vector space is totally quantized by quantizing magnetic field and size quantization along z-direction. We note the following two points in this context.

1. The density-of-states (DOS) functions for all the materials in this case are series of non-uniformly distributed Dirac's Delta functions at specified quantized points in the respective energy axis. The spacing between

the consecutive Delta functions are functions of energy band constants and quantization of the wave vector space of a particular material. The DOS function needs two summations namely one summation over the Landau quantum number and the other one is due to size quantization.

2. It may be noted that the energy levels in magneto HD QWTs lead to discrete energy levels, somewhat like atomic energy levels, which produce very large changes. This follows from the inherent nature of the quantum confinement of the carrier gas dealt with here. In this case, there remain no free carrier states in between any two allowed sets of size-quantized levels unlike that found for QWs and NWs where the quantum confinements are 1D and 2D, respectively. Consequently, the crossing of the Fermi level by the size-quantized levels in this case would have much greater impact on the redistribution of the carriers among the allowed levels, as compared to that found for QWs and NWs, respectively.

In Section 12.2.1, we have formulated the electron statistics in HDQWTs of nonlinear optical materials under magnetic quantization as given by (12.6) and by combining it with (12.7) we can study the QC in this case. In Section 12.2.2, we have formulated the electron statistics in HDQWTs of III–V, ternary and quaternary materials under magnetic quantization in accordance with the three-band model of Kane, two-band model of Kane, isotropic parabolic energy bands together with the models of Stillman *et al.* [17] and Palik *et al.* [16] as given by (12.13), (12.16), (12.19), (12.24) and (12.31) and by combining each of them with (12.7) we can study the QC for each model in this case. In Section 12.2.3, we have formulated the electron statistics in HDQWTs of II–VI materials under magnetic quantization as given by (12.34) and by combining it with (12.7) we can study the QC in this case.

In Section 12.2.4, we have formulated the electron statistics in HDQWTs of IV–VI materials under magnetic quantization in accordance with the models of together with the models of Cohen, Lax, Dimmock, Bangert and Kastner and Foley and Langenberg as given by (12.39), (12.46), (12.51), (12.54) and (12.60) and by combining them with (12.7) we can study the QC for each model in this case. In Section 12.2.5, we have formulated the electron statistics in HDQWTs of stressed Kane type under magnetic quantization as given by (12.63) and by combining it with (12.7) we can study the QC in this case. In Section 12.2.6, we have formulated the electron statistics in HDQWTs of Te under magnetic quantization as given by (12.66) and by combining it with (12.7) we can study the QC in this case. In Section 12.2.7, we have formulated the electron statistics in HDQWTs of GaP under magnetic quantization as given by (12.68) and by combining it with (12.7) we can study the QC in this case. In Section 12.2.8, we have formulated the electron statistics in HDQWTs of $PtSb_2$ under magnetic quantization as given by (12.73) and by combining it with (12.7) we can study the QC in this case. In Section 12.2.9, we have formulated the electron statistics in HDQWTs of Bi_2Te_3 under magnetic quantization as given by (12.77) and by combining it with (12.7) we can study the QC in this case.

In Section 12.2.10, we have formulated the electron statistics in HDQWTs of Ge under magnetic quantization in accordance with the models of together with the models of Cardona *et al.* and Wang and Ressler as given by (12.80) and (12.86) and by combining them with (12.7) we can study the QC for each model in this case. In Section 12.2.11, we have formulated the electron statistics in HDQWTs of GaSb under magnetic quantization as given by (12.91) and by combining it with (12.7) we can study the QC in this case. In Section 12.2.12, we have formulated the electron statistics in HDQWTs of II–V materials under magnetic quantization as given by (12.96) and by combining it with (12.7) we can study the QC in this case. In Section 12.2.13, we have formulated the electron statistics in HDQWTs of lead germanium telluride under magnetic quantization as given by (12.102) and by combining it with (12.7) we can study the QC in this case.

We have not given the graphs of the QCs' for HDQWTs made of many materials in the present case as discussed here since we feel that our readers will perform the intricate numerical computation to obtain the oscillatory plots for better assessment and joy of understanding. It is the band structure that changes in a fundamental way and consequently all the physical properties of all the electronic materials changes radically leading to new physical concepts.

12.4 Open Research Problems

(R.12.1) Investigate the QC in the presence of an arbitrarily oriented quantizing magnetic field for all the HD QWTs' made by using the materials as given in problems in R. 1.1 of chapter 1. Draw the corresponding graphs and explain the results.

(R.12.2) Investigate the QC in the presence of an arbitrarily oriented quantizing magnetic field in HD QWTs' of the nonlinear optical semiconductors by including broadening and the electron spin. Study all the special cases for HD QWTs' made by using the III–V, ternary and quaternary materials in this context. Draw the corresponding graphs and explain the results.

(R.12.3) Investigate the QC for HD QWTs' made by using the IV–VI, II–VI and stressed Kane type compounds in the presence of an arbitrarily oriented quantizing magnetic field by including broadening and electron spin. Draw the corresponding graphs and explain the results.

(R.12.4) Investigate the QC for all the HD QWTs' made by using the materials as stated in R.1.1 of chapter 1 in the presence of an arbitrarily oriented quantizing magnetic field by including broadening and electron spin. Draw the corresponding graphs and explain the results.

(R.12.5) Investigate the QC in the presence of an arbitrarily oriented quantizing magnetic field and crossed electric fields in HD QWTs' made by using the nonlinear optical materials by including broadening and the electron spin. Study all the special cases for HD III–V, ternary and quaternary materials in this context. Draw the corresponding graphs and explain the results.

(R.12.6) Investigate the QC for HD QWTs' made by using the IV–VI, II–VI and stressed Kane type compounds in the presence of an arbitrarily oriented quantizing magnetic field and crossed electric field by including broadening and electron spin. Draw the corresponding graphs and explain the results.

(R.12.7) Investigate the QC for all the HD QWTs' made by using the materials as stated in R.1.1 of chapter 1 in the presence of an arbitrarily oriented quantizing magnetic field and crossed electric field by including broadening and electron spin. Draw the corresponding graphs and explain the results.

(R.12.8) Investigate the QC in HD QWTs' made by using the nonlinear optical, III–V, II–VI, IV–VI and stressed Kane type semiconductors. Draw the corresponding graphs and explain the results.

(R.12.9) Investigate the 2D QC for HD QWTs' made by using all the materials as considered in problems R.1.1. Draw the corresponding graphs and explain the results.

(R.12.10) Investigate the 2D QC in the presence of an arbitrarily oriented non-quantizing magnetic field for the HD QWTs' made by using the nonlinear optical semiconductors by including the electron spin. Study all the special cases for III–V, ternary and quaternary materials in this context. Draw the corresponding graphs and explain the results.

(R.12.11) Investigate the DRs in QWs HD IV–VI, II–VI and stressed Kane type compounds in the presence of an arbitrarily oriented non-quantizing magnetic field by including the electron spin. Draw the corresponding graphs and explain the results.

(R.12.12) Investigate the 2D QC for HD QWTs' made by using all the materials as stated in R.1.1 of Chapter 1 in the presence of an arbitrarily oriented magnetic field by including electron spin and broadening. Draw the corresponding graphs and explain the results.

(R.12.13) Investigate all the problems of this chapter by removing all the mathematical approximations and establishing the uniqueness conditions in each case. Draw the corresponding graphs and explain the results.

(R.12.14) a) Investigate the QC in all the HD QWTs' made by using the semiconductors as considered in this chapter in the presence of defects and magnetic quantization. Draw the corresponding graphs and explain the results.

(R.12.15) Investigate all the problems of this section by removing all the mathematical approximations and establishing the respective appropriate uniqueness conditions. Draw the corresponding graphs and explain the results.

References

1. N. Miura, *Physics of Semiconductors in High Magnetic Fields*, Series on Semiconductor Science and Technology (Oxford University Press, USA, 2007); K. H. J Buschow and F. R. de Boer, *Physics of Magnetism and Magnetic Materials* (Springer, New York, 2003); D. Sellmyer (Ed.) and R. Skomski (Ed.), *Advanced Magnetic Nanostructures* (Springer, New York, 2005); J. A. C. Bland (Ed.), B. Heinrich (Ed.), *Ultrathin Magnetic Structures III: Fundamentals of Nanomagnetism (Pt. 3)* (Springer-Verlag, Germany, 2005); B. K. Ridley, *Quantum Processes in semiconductors*, Fourth Edition (Oxford publications, Oxford, 1999); J. H. Davies, *Physics of low dimensional semiconductors* (Cambridge University Press, UK, 1998); S. Blundell, *Magnetism in Condensed Matter*, Oxford Master Series in Condensed Matter Physics (Oxford University Press, USA, 2001); C. Weisbuch and B. Vinter, *Quantum Semiconductor Structures: Fundamentals and Applications* (Academic Publishers, USA, 1991); D. Ferry, *Semiconductor Transport* (CRC,

USA, 2000); M. Reed (Ed.), *Semiconductors and Semimetals: Nanostructured Systems* (Academic Press, USA, 1992); T. Dittrich, *Quantum Transport and Dissipation* (Wiley-VCH Verlag GmbH, Germany, 1998); A. Y. Shik, *Quantum Wells: Physics & Electronics of Twodimensional Systems* (World Scientific, USA, 1997).

2. K. P. Ghatak and M. Mondal, *Zietschrift fur Naturforschung A* **41a**, 881 (1986); K. P. Ghatak and M. Mondal, *J. Appl. Phys.* **62**, 922 (1987); K. P. Ghatak and S. N. Biswas, *Phys. Stat. Sol. (b)* **140**, K107 (1987); K. P. Ghatak and M. Mondal, *J. Mag. Mag. Mat.* **74**, 203 (1988); K. P. Ghatak and M. Mondal, *Phys. Stat. Sol. (b)* **139**, 195 (1987); K. P. Ghatak and M. Mondal, *Phys. Stat. Sol. (b)* **148**, 645 (1988); K. P. Ghatak, B. Mitra and A. Ghoshal, *Phys. Stat. Sol. (b)* **154**, K121 (1989); K. P. Ghatak and S. N. Biswas, *J. Low Temp. Phys.* **78**, 219 (1990); K. P. Ghatak and M. Mondal, *Phys. Stat. Sol. (b)* **160**, 673 (1990); K. P. Ghatak and B. Mitra, *Phys. Letts. A* **156**, 233 (1991); K. P. Ghatak, A. Ghoshal and B. Mitra, *Nouvo Cimento D* **13D**, 867 (1991); K. P. Ghatak and M. Mondal, *Phys. Stat. Sol. (b)* **148**, 645 (1989); K. P. Ghatak and B. Mitra, *Int. J. Elect.* **70**, 345 (1991); K. P. Ghatak and S. N. Biswas, *J. Appl. Phys.* **70**, 299 (1991); K. P. Ghatak and A. Ghoshal, *Phys. Stat. Sol. (b)* **170**, K27 (1992); K. P. Ghatak, *Nuovo Cimento D* **13D**, 1321 (1992); K. P. Ghatak and B. Mitra, *Int. J. Elect.* **72**, 541 (1992); K. P. Ghatak and S. N. Biswas, *Nonlin. Opt.* **4**, 347 (1993); K. P. Ghatak and M. Mondal, *Phys. Stat. Sol. (b)* **175**, 113 (1993); K. P. Ghatak and S. N. Biswas, *Nonlin. Opt.* **4**, 39 (1993); K. P. Ghatak and B. Mitra, *Nuovo Cimento* **15D**, 97 (1993); K. P. Ghatak and S. N. Biswas, *Nanostruct. Mater.* **2**, 91 (1993); K. P. Ghatak and M. Mondal, *Phys. Stat. Sol. (b)* **185**, K5 (1994); K. P. Ghatak, B. Goswami, M. Mitra and B. Nag, *Nonlinear Optics* **16**, 9 (1996); K. P. Ghatak, M. Mitra, B. Goswami and B. Nag, *Nonlinear Optics* **16**, 167 (1996); K. P. Ghatak and B. Nag, *Nanostruct. Mater.* **10**, 923 (1998).

3. D. Roy Choudhury, A. K. Choudhury, K. P. Ghatak and A. N. Chakravarti, *Phys. Stat. Sol. (b)* **98**, K141 (1980); A. N. Chakravarti, K. P. Ghatak, A. Dhar and S. Ghosh, *Phys. Stat. Sol. (b)* **105**, K55 (1981); A. N. Chakravarti, A. K. Choudhury and K. P. Ghatak, *Phys. Stat. Sol. (a)* **63**, K97 (1981); A. N. Chakravarti, A. K. Choudhury, K. P. Ghatak, S. Ghosh and A. Dhar, *Appl. Phys.* **25**, 105 (1981); A. N. Chakravarti, K. P. Ghatak, G. B. Rao and K. K. Ghosh, *Phys. Stat. Sol. (b)* **112**, 75 (1982); A. N. Chakravarti, K. P. Ghatak, K. K. Ghosh and H. M. Mukherjee, *Phys. Stat. Sol. (b)* **116**, 17 (1983); M. Mondal and K. P. Ghatak, *Phys. Stat. Sol. (b)* **133**, K143 (1984); M. Mondal and K. P. Ghatak, *Phys. Stat. Sol. (b)* **126**, K47 (1984); M. Mondal and K. P. Ghatak, *Phys. Stat. Sol. (b)* **126**, K41 (1984); M. Mondal and K. P. Ghatak, *Phys. Stat. Sol. (b)* **129**, K745 (1985); M. Mondal and K. P. Ghatak, *Phys. Scr.* **31**, 615 (1985); M. Mondal and K. P. Ghatak, *Phys. Stat. Sol. (b)* **135**, 239 (1986); M. Mondal and K. P. Ghatak, *Phys. Stat. Sol. (b)* **93**, 377 (1986); M. Mondal and K. P. Ghatak, *Phys. Stat. Sol. (b)* **135**, K21 (1986); M. Mondal, S. Bhattacharyya and K. P. Ghatak, *Appl. Phys. A*

42A, 331 (1987); S. N. Biswas, N. Chattopadhyay and K. P. Ghatak, *Phys. Stat. Sol. (b)* **141**, K47 (1987); B. Mitra and K. P. Ghatak, *Phys. Stat. Sol. (b)* **149**, K117 (1988); B. Mitra, A. Ghoshal and K. P. Ghatak, *Phys. Stat. Sol. (b)* **150**, K67 (1988); M. Mondal and K. P. Ghatak, *Phys. Stat. Sol. (b)* **147**, K179 (1988); M. Mondal and K. P. Ghatak, *Phys. Stat. Sol. (b)* **146**, K97 (1988); B. Mitra, A. Ghoshal and K. P. Ghatak, *Phys. Stat. Sol. (b)* **153**, K209 (1989); B. Mitra and K. P. Ghatak, *Phys. Letts.* **142A**, 401 (1989); B. Mitra, A. Ghoshal and K. P. Ghatak, *Phys. Stat. Sol. (b)* **154**, K147 (1989); B. Mitra and K. P. Ghatak, *Sol. State Elect.* **32**, 515 (1989); B. Mitra, A. Ghoshal and K. P. Ghatak, *Phys. Stat. Sol. (b)* **155**, K23 (1989); B. Mitra and K. P. Ghatak, *Phys. Letts.* **135A**, 397 (1989); B. Mitra and K. P. Ghatak, *Phys. Letts. A* **146A**, 357 (1990); B. Mitra and K. P. Ghatak, *Phys. Stat. Sol. (b)* **164**, K13 (1991); S. N. Biswas and K. P. Ghatak, *Internat. Jour. of Elect.* **70**, 125 (1991).

4. P. R. Wallace, *Phys. Stat. Sol. (b)*, **92**, 49 (1979).
5. B. R. Nag, *Electron Transport in Compound Semiconductors*, Springer Series in Solid-State Sciences, Vol. 11 (Springer-Verlag, Germany, 1980).
6. K. P. Ghatak, S. Bhattacharya and D. De, *Einstein Relation in Compound Semiconductors and Their Nanostructures*, Springer Series in Materials Science, Vol. 116 (Springer-Verlag, Germany, 2009).
7. C. C. Wu and C. J. Lin, *J. Low Temp. Phys.* **57**, 469 (1984); M. H. Chen, C. C. Wu, C. J. Lin, *J. Low Temp. Phys.* **55**, 127 (1984).
8. E. Bangert and P. Kastner, *Phys. Stat. Sol (b)* **61**, 503 (1974).
9. G. M. T. Foley and P. N. Langenberg, *Phys. Rev. B* **15B**, 4850 (1977).
10. M. Singh, P. R. Wallace, S. D. Jog and E. Arushanov, *J. Phys. Chem. Solids* **45**, 409 (1984).
11. Y. Yamada, *Phys. Soc. Japan* **35**, 1600 (1973), **37**, 606 (1974).
12. L. A. Vassilev, *Phys. Stat. Sol. (b)* **121**, 203 (1984).
13. R. W. Cunningham, *Phys. Rev.* **167**, 761 (1968).
14. A. I. Yekimov, A. A. Onushchenko, A. G. Plyukhin, and Al, L. Efros, *J. Expt. Theor. Phys.* **88**, 1490 (1985).
15. B. J. Roman and A. W. Ewald, *Phys. Rev. B* **5**, 3914 (1972).
16. G. E. Stillman, C. M. Wolfe and J. C. Dinmockin, in Semiconductors and Semimetals, R. K. Willardson and A. C. Beer (eds.), **12**, 169 (1977).
17. E. D. Palik, G. S. Picus, S. Teither and R. F. Wallis, *Phys. Rev.* **122**, 475 (1961).

Chapter 13

The QC in Heavily Doped Quantum Well Transistors (HDQWT) Under Cross-Fields Configuration

Knowledge is proud, he knows too much, but the wise is humble, he knows no more.

13.1 Introduction

The electronic properties of bulk specimens of compound semiconductors have extensively been studied in the literature [1–14]. Incidentally, it appears from the literature that the QC in HDQWTs under cross-fields configurations has yet to be investigated. In this Chapter, in Section 13.2.1, of theoretical background, the QC in HDQWTs of tetragonal and nonlinear optical materials in the presence of crossed electric and quantizing magnetic fields has been investigated by formulating the electron dispersion relation. Section 13.2.2 reflects the study of the QC in HDQWTs of III–V, ternary and quaternary compounds as a special case of Section 13.2.1. Section 13.2.3 contains the study of the QC in HDQWTs of II–VI semiconductors in the present case. In Section 13.2.4, the QC under cross-field configuration in HDQWTs of IV–VI semiconductors has been investigated. In Section 13.2.5, the QC in HDQWTs of stressed Kane type semiconductors has been investigated. Section 13.3 contains summary and conclusion. Section 13.4 contains 11 open research problems.

13.2 Theoretical Background

13.2.1 *The QC in HDQWT of tetragonal and nonlinear optical materials in the presence of cross-fields configuration*

The DR of the conduction electrons in HD QWTs of tetragonal and nonlinear optical material in the presence of cross-fields configuration can be written from (10.4) as

$$T_{22}(E,\eta_g) = \left[\left(\left(n+\frac{1}{2}\right)\hbar\omega_{01}\right) + \left(\frac{[\hbar]^2}{2a(E,\eta_g)}\right)\left(\frac{\pi n_z}{d_z}\right)^2 - \left(\frac{E_0\hbar k_y \rho(E,\eta_g)}{B}\right) - \left(\frac{M_\perp \rho^2(E,\eta_g)E_0^2}{2B^2}\right)\right] \quad (13.1)$$

The use of (13.1) leads to the expression of EEM along y direction as

$$m_y^*(e_{fA1},\eta_g,n,E_0,n_z) = \text{Real Part of}\,(B/E_0)^2 T_{49}(e_{fA1},\eta_g,n_z) \times [T_{49}(e_{fA1},\eta_g,n_z)]' \quad (13.2)$$

where e_{fA1} is the Fermi energy in this case and

$$T_{49}(e_{fA1},\eta_g,n_z) = \left[[T_{22}(e_{fA1},\eta_g)] - \left[\left(\left(n+\frac{1}{2}\right)\hbar\omega_{01}\right) + \left(\frac{[\hbar]^2}{2a(e_{fA1},\eta_g)}\right)\left(\frac{\pi n_z}{d_z}\right)^2 - \left(\frac{M_\perp \rho^2(e_{fA1},\eta_g)E_0^2}{2B^2}\right)\right]\right] \rho(e_{fA1},\eta_g)^{-1}$$

The sub-band energy $E_{13,1}$ in this case assumes the from

$$T_{22}(E_{13,1},\eta_g) = \left[\left(\left(n+\frac{1}{2}\right)\hbar\omega_{01}\right) + \left(\frac{[\hbar]^2}{2a(E_{13,1},\eta_g)}\right)\left(\frac{\pi n_z}{d_z}\right)^2 - \left(\frac{M_\perp \rho^2(E_{13,1},\eta_g)E_0^2}{2B^2}\right)\right] \quad (13.3a)$$

The surface electron statistics can be written as

$$n_s = \text{Real part of}\,\bar{n}_s \quad (13.3b)$$

where

$$\bar{n}_s = \left(\frac{g_v eB}{\pi\hbar}\right)\sum_{n=0}^{n_{\max}}\sum_{n_z=1}^{n_{z\max}} F_{-1}(\eta_{13,2,1}), \tag{13.3c}$$

in which $\eta_{13,2,1} = (e_{fA1} - E_{13.1})(k_B T)^{-1}$.

Using (13.3b) and (11.7b) we can study the QC in this case.

13.2.2 *The QC in HDQWT of III–V, ternary and quaternary materials under cross-fields configuration*

(a) Three-band model of Kane

Under the conditions $\delta = 0, \Delta_{||} = \Delta_{\perp} = \Delta$ and $m_{||}^* = m_{\perp}^* = m_c$ (13.1) assumes the form

$$T_{33}(E,\eta_g) = \left(n+\frac{1}{2}\right)\hbar\omega_{01} + \frac{\left[\hbar\frac{\pi n_z}{d_z}\right]^2}{2m_c} - \frac{E_0}{B}\hbar k_y\{T_{33}(E,\eta_g)\}' - \frac{m_c E_0^2[\{T_{33}(E,\eta_g)\}']^2}{2B^2} \tag{13.3d}$$

The use of (13.3d) leads to the expression of EEM along y direction as

$$m_y^*(e_{fA2},\eta_g,n,E_0,n_z) = \text{Real Part of}\,(B/E_0)^2 T_{50}(e_{fA2},\eta_g,n_z) \times[T_{50}(e_{fA2},\eta_g,n_z)]' \tag{13.4a}$$

where e_{fA2} is the Fermi energy in this case and

$$T_{50}(e_{fA2},\eta_g,n_z) = \left[T_{30}(e_{fA2},\eta_g) - \left(n+\frac{1}{2}\right)\hbar\omega_0 - \frac{\left[\hbar\frac{\pi n_z}{d_z}\right]^2}{2m_c} + \frac{m_\perp E_0^2[\{T_{33}(e_{fA2},\eta_g)\}']^2}{2B^2}\right][\{T_{33}(e_{fA2},\eta_g)\}']^{-1}$$

The sub-band energy $E_{13,2}$ in this case assumes the from

$$T_{33}(E_{13,2},\eta_g) = \left(n+\frac{1}{2}\right)\hbar\omega_0 + \frac{\left[\hbar\frac{\pi n_z}{d_z}\right]^2}{2m_c} - \frac{m_c E_0^2[\{T_{33}(E_{13,2},\eta_g)\}']^2}{2B^2} \tag{13.4b}$$

The surface electron statistics can be written as

$$n_s = \text{Real part of}\, \bar{n}_s \tag{13.4c}$$

where

$$\bar{n}_s = \left(\frac{g_v eB}{\pi\hbar}\right) \sum_{n=0}^{n_{\max}} \sum_{n_z=1}^{n_{z\max}} F_{-1}(\eta_{13,2,2}) \tag{13.4d}$$

in which $\eta_{13,2,2} = (e_{fA2} - E_{13.2})(k_B T)^{-1}$.

Using (13.4c) and (11.7b) we can study the QC in this case.

(b) Two-band model of Kane

Under the condition $\Delta \gg E_g$, (13.3d) assumes the form

$$\gamma_2(E, \eta_g) = \left(n + \frac{1}{2}\right)\hbar\omega_0 - \frac{E_0}{B}\hbar k_y \gamma_2'(E, \eta_g) - \frac{m_c E_0^2}{2B^2}(\gamma_2'(E, \eta_g))^2 + \frac{\left[\hbar\frac{n_z \pi}{d_z}\right]^2}{2m_c} \tag{13.5}$$

The use of (13.5) leads to the expression of EEM along y direction as

$$m_y^*(e_{fA3}, \eta_g, n, E_0, n_z) = (B/E_0)^2 T_{51}(e_{fA3}, \eta_g, n_z)[T_{51}(e_{fA3}, \eta_g, n_z)]' \tag{13.6a}$$

where e_{fA3} is the Fermi energy in this case and

$$T_{51}(e_{fA3}, \eta_g, n_z) = \left[\gamma_2(e_{fA3}, \eta_g) - \left(n + \frac{1}{2}\right)\hbar\omega_0 - \frac{\left[\hbar\frac{\pi n_z}{d_z}\right]^2}{2m_c} + \frac{m_c E_0^2[\{\gamma_2(e_{fA3}, \eta_g)\}']^2}{2B^2}\right][\{\gamma_2(e_{fA3}, \eta_g)\}']^{-1}$$

The sub-band energy $E_{13,3}$ in this case assumes the from

$$\gamma_2(E_{13,3}, \eta_g) = \left(n + \frac{1}{2}\right)\hbar\omega_0 - \frac{m_c E_0^2}{2B^2}(\gamma_2'(E_{13,3}, \eta_g))^2 + \frac{\left[\hbar\frac{n_z \pi}{d_z}\right]^2}{2m_c} \tag{13.6b}$$

The surface electron statistics can be written as

$$\bar{n}_s = \left(\frac{g_v eB}{\pi\hbar}\right) \sum_{n=0}^{n_{\max}} \sum_{n_z=1}^{n_{z\max}} F_{-1}(\eta_{13,2,3}) \qquad (13.6c)$$
$$\eta_{13,2,3} = (e_{fA3} - E_{13.3})(k_B T)^{-1}$$

Using (13.6c) and (11.7b) we can study the QC in this case.

(c) Parabolic energy bands

The DR for this model under the condition $\alpha \to 0$ can be written as

$$\gamma_3(E, \eta_g) = \left(n + \frac{1}{2}\right)\hbar\omega_0 - \frac{E_0}{B}\hbar k_y \gamma_3'(E, \eta_g) - \frac{m_c E_0^2}{2B^2}(\gamma_3'(E, \eta_g))^2 + \frac{\left[\hbar\frac{n_z\pi}{d_z}\right]^2}{2m_c} \qquad (13.7)$$

The use of (13.7) leads to the expression of EEM along y direction as

$$m_y^*(e_{fA4}, \eta_g, n, E_0, n_z) = (B/E_0)^2 T_{52}(e_{fA4}, \eta_g, n_z)[T_{52}(e_{fA4}, \eta_g, n_z)]' \qquad (13.8a)$$

where e_{fA4} is the Fermi energy in this case and

$$T_{52}(e_{fA4}, \eta_g, n_z) = \left[\gamma_2(e_{fA4}, \eta_g) - \left(n + \frac{1}{2}\right)\hbar\omega_0 - \frac{\left[\hbar\frac{\pi n_z}{d_z}\right]^2}{2m_c} + \frac{m_c E_0^2[\{\gamma_3(e_{fA4}, \eta_g)\}']^2}{2B^2}\right][\{\gamma_3(e_{fA4}, \eta_g)\}']^{-1}$$

The sub-band energy $E_{13,4}$ in this case assumes the fro

$$\gamma_3(E_{13,4}, \eta_g) = \left(n + \frac{1}{2}\right)\hbar\omega_0 - \frac{m_c E_0^2}{2B^2}(\gamma_3'(E_{13,4}, \eta_g))^2 + \frac{\left[\hbar\frac{n_z\pi}{d_z}\right]^2}{2m_c} \qquad (13.8b)$$

The surface electron statistics can be written as

$$\bar{n}_s = \left(\frac{g_v eB}{\pi\hbar}\right) \sum_{n=0}^{n_{\max}} \sum_{n_z=1}^{n_{z\max}} F_{-1}(\eta_{13,2,4}) \qquad (13.8c)$$
$$\eta_{13,2,4} = (e_{fA4} - E_{13,4})(k_B T)^{-1}$$

Using (13.8c) and (11.7b) we can study the QC in this case.

13.2.3 The QC in HDQWT of II–VI materials under cross-fields configuration

The DR of the conduction electrons in HD QWTs of II–VI materials in the presence of cross-fields configuration can be written as

$$\gamma_3(E,\eta_g) = \beta_1(n,E_0) - \frac{E_0}{B}\hbar k_y \gamma_3'(E,\eta_g) - \frac{m_{||}^* E_0^2}{2B^2}(\gamma_3'(E,\eta_g))^2 + \frac{\left[\hbar \frac{n_z \pi}{d_z}\right]^2}{2m_{||}} \quad (13.9)$$

The use of (13.9) leads to the expression of EEM along y direction as

$$m_y^*(e_{fA4},\eta_g,n,E_0,n_z) = (B/E_0)^2 T_{53}(e_{fA5},\eta_g,n_z)[T_{53}(e_{fA5},\eta_g,n_z)]' \quad (13.10a)$$

where e_{fA5} is the Fermi energy in this case and

$$T_{53}(e_{fA5},\eta_g,n_z) = \left[\gamma_3(e_{fA5},\eta_g) - \beta_1(n,E_0) - \frac{\left[\hbar \frac{\pi n_z}{d_z}\right]^2}{2m_{||}^*} + \frac{m_{||}^* E_0^2[\{\gamma_3(e_{fA5},\eta_g)\}']^2}{2B^2}\right][\{\gamma_3(e_{fA5},\eta_g)\}']^{-1}$$

The sub-band energy $E_{13,5}$ in this case assumes the from

$$\gamma_3(E_{13,5},\eta_g) = \beta_1(n,E_0) - \frac{m_{||}^* E_0^2}{2B^2}(\gamma_3'(E_{13,5},\eta_g))^2 + \frac{\left[\hbar \frac{n_z \pi}{d_z}\right]^2}{2m_{||}^*} \quad (13.10b)$$

The surface electron statistics can be written as

$$\bar{n}_s = \left(\frac{g_v e B}{2\pi\hbar}\right) \sum_{n=0}^{n_{\max}} \sum_{n_z=1}^{n_{z\max}} F_{-1}(\eta_{13,2,5})$$

$$\eta_{13,2,5} = (e_{fA5} - E_{13,5})(k_B T)^{-1} \quad (13.10c)$$

Using (13.10c) and (11.7b) we can study the QC in this case.

13.2.4 *The QC in HDQWT of IV–VI materials under cross-fields configuration*

The DR of the conduction electrons in HD QWTs of IV–VI materials in the presence of cross-fields configuration can be written using (10.59a) as

$$g^*(E,\eta_g) = \left(n+\frac{1}{2}\right)\hbar\overline{\omega_{i1}}(E,\eta_g) + \frac{\hbar^2(\frac{n_z\pi}{d_z})^2}{2M_3^*(E,\eta_g)} - \frac{E_0}{B}\rho_1^*(E,\eta_g)\hbar k_y - \frac{E_0^2}{2B^2}[\rho_1^*(E,\eta_g)]^2 M_1^*(E,\eta_g) \quad (13.11a)$$

The use of (13.11a) leads to the expression of EEM along y direction as

$$m_y^*(e_{fA6},\eta_g,n,E_0,n_z) = \text{Real Part of}\,(B/E_0)^2 T_{54}(e_{fA6},\eta_g,n_z) \times [T_{54}(e_{fA6},\eta_g,n_z)]' \quad (13.11b)$$

where e_{fA6} is the Fermi energy in this case and

$$T_{54}(e_{fA6},\eta_g,n_z) = \left[g^*(e_{fA6},\eta_g) - \left(n+\frac{1}{2}\right)\hbar\overline{\omega_{i1}}(e_{fA6},\eta_g) - \frac{\hbar^2\left(\frac{n_z\pi}{d_z}\right)^2}{2M_3^*(e_{fA6},\eta_g)} + \frac{E_0^2}{2B^2}[\rho_1^*(e_{fA6},\eta_g)]^2 M_1^*(e_{fA6},\eta_g)\right] \times [\rho_1^*(e_{fA6},\eta_g)]^{-1} \quad (13.12a)$$

The sub-band energy $E_{13,6}$ in this case assumes the form

$$g^*(E_{13,6},\eta_g) - \left(n+\frac{1}{2}\right)\hbar\overline{\omega_{i1}}(E_{13,6},\eta_g) + \frac{\hbar^2\left(\frac{n_z\pi}{d_z}\right)^2}{2M_3^*(E_{13,6},\eta_g)} - \frac{E_0^2}{2B^2}[\rho_1^*(E_{13,6},\eta_g)]^2 M_1^*(E_{13,6},\eta_g) \quad (13.12b)$$

The surface electron statistics can be written as

$$n_s = \text{Real part of}\,\bar{n}_s \quad (13.12c)$$

where

$$\bar{n}_s = \left(\frac{g_v eB}{\pi\hbar}\right)\sum_{n=0}^{n_{\max}}\sum_{n_z=1}^{n_{z\max}} F_{-1}(\eta_{13,2,6}), \tag{13.12d}$$

in which $\eta_{13,2,6} = (e_{fA6} - E_{13.6})(k_B T)^{-1}$.

Using (13.12d) and (11.7b) we can study the QC in this case.

13.2.5 *The QC in HDQWT of stressed Kane type materials under cross-fields configuration*

Using (10.72), the DR in this case assumes the form

$$G^*(E,\eta_g) = \left(n+\frac{1}{2}\right)\hbar\overline{\omega_i}(E,\eta_g) + \frac{\hbar^2\left(\frac{n_z\pi}{d_z}\right)^2}{2m_3^*(E,\eta_g)} - \frac{E_0}{B}\rho^*(E,\eta_g)$$
$$\times\left[\frac{m_1^*(E,\eta_g)}{m_2^*(E,\eta_g)}\right]^{\frac{1}{2}}\hbar k_y - \frac{E_0^2}{2B^2}[\rho^*(E,\eta_g)]^2 m_1^*(E,\eta_g) \tag{13.13}$$

The use of (13.13) leads to the expression of EEM along y direction as

$$m_y^*(e_{fA7},\eta_g,n,E_0,n_z) = \text{Real Part of}\,(B/E_0)^2 T_{55}(e_{fA7},\eta_g,n_z)$$
$$\times[T_{55}(e_{fA7},\eta_g,n_z)]' \tag{13.14a}$$

where e_{fA7} is the Fermi energy in this case,

$$T_{55}(e_{fA7},\eta_g,n_z) = \left[G^*(e_{fA7},\eta_g) - \left(n+\frac{1}{2}\right)\hbar\overline{\omega_i}(e_{fA7},\eta_g)\right.$$
$$\left. - \frac{\hbar^2\left(\frac{n_z\pi}{d_z}\right)^2}{2m_3^*(e_{fA7},\eta_g)} + \frac{E_0^2}{2B^2}[\rho^*(E,\eta_g)]^2 m_1^*(E,\eta_g)\right]$$
$$\times[m_4^*(e_{fA7},\eta_g)]^{-1} \quad \text{and}$$

$$m_4^*(e_{fA7},\eta_g) = \rho^*(e_{fA7},\eta_g)\left[\frac{m_1^*(e_{fA7},\eta_g)}{m_2^*(e_{fA7},\eta_g)}\right]^{1/2}$$

The sub-band energy $E_{13,7}$ in this case assumes the from

$$G^*(E_{13,7},\eta_g) = \left(n+\frac{1}{2}\right)\hbar\overline{\omega_i}(E_{13,7},\eta_g) + \frac{\hbar^2\left(\frac{n_z\pi}{d_z}\right)^2}{2m_3^*(E_{13,7},\eta_g)} - \frac{E_0^2}{2B^2}[\rho^*(E_{13,7},\eta_g)]^2 m_1^*(E_{13,7},\eta_g) \quad (13.14b)$$

The surface electron statistics can be written as

$$n_s = \text{Real part of}\,\bar{n}_s \quad (13.14c)$$

where,

$$\bar{n}_s = \left(\frac{g_v eB}{\pi\hbar}\right)\sum_{n=0}^{n_{\max}}\sum_{n_z=1}^{n_{z\max}} F_{-1}(\eta_{13,2,7}), \quad (13.14d)$$

in which $\eta_{13,2,7} = (e_{fA7} - E_{13.7})(k_BT)^{-1}$.

Using (13.14d) and (11.7b) we can study the QC in this case.

13.3 Summary and Conclusion

From the theoretical background, we note that the expression for surface electron concentration of 2D HDQWTs operated under cross-fields configurations of nonlinear optical, III–V (three- and two-band models of Kane together with parabolic energy bands), II–VI, IV–VI and stressed Kane type materials are given by (13.3b), (13.4c), (13.6c), (13.8c), (13.10c), (13.12c) and (13.14c), respectively. By combining each of them with the (11.7b) we can study the QC in each case. Besides, the EEM for the aforementioned cases is given by (13.2), (13.4a), (13.6a), (13.8a), (13.10a), (13.11b) and (13.14a), respectively. In addition to Fermi energy, the EEMs' are functions of quantum numbers, electric field, magnetic field and scattering potential, respectively. We have not given the graphs of the QCs for various 2D HDQWTs made of many materials as discussed here. We have not given any graph of QC since we feel that our readers will perform the intricate numerical computation to obtain the oscillatory plots for better assessment and joy of understanding.

13.4 Open Research Problems

(R.13.1) Investigate the QCs in the presence of cross-fields for all the HD QWTs'of the materials whose bulk EMs are given in problems in

R.1.1 of Chapter 1. Draw the corresponding graphs and explain the results.

(R.13.2) Investigate the QCs in the presence of cross-fields in HD QWTs' of nonlinear optical semiconductors by including broadening and the electron spin. Study all the special cases for HD QWTs' of III–V, ternary and quaternary materials in this context. Draw the corresponding graphs and explain the results.

(R.13.3) Investigate the QCs for HD QWTs' of IV–VI, II–VI and stressed Kane type compounds in the presence of cross-fields by including broadening and electron spin. Draw the corresponding graphs and explain the results.

(R.13.4) Investigate the QCs in the presence of cross-fields in HD QWTs'of nonlinear optical semiconductors by including broadening and the electron spin. Study all the special cases for HD QWTs'of III–V, ternary and quaternary materials in this context. Draw the corresponding graphs and explain the results.

(R.13.5) Investigate the QCs for HD QWTs of IV–VI, II–VI and stressed Kane type compounds in the presence of cross-fields by including broadening and electron spin. Draw the corresponding graphs and explain the results.

(R.13.6) Investigate the QCs for all the HD QWTs of materials whose bulk DRs are stated in R.1.1 of Chapter 1 in the presence of an arbitrarily oriented quantizing magnetic field and crossed electric field by including broadening and electron spin under the condition of heavy doping. Draw the corresponding graphs and explain the results.

(R.13.7) Investigate the QC in HD QWTs of nonlinear optical, III–V, II–VI, IV–VI and stressed Kane type semiconductors in the presence of non-uniform strain. Draw the corresponding graphs and explain the results.

(R.13.8) Investigate the QCs in the presence of an arbitrarily oriented non-quantizing magnetic field for QWTs of HD nonlinear optical semiconductors by including the electron spin. Study all the special cases for III–V, ternary and quaternary materials in this context. Draw the corresponding graphs and explain the results.

(R.13.9) Investigate the QCs in HD QWTs' of IV–VI, II–VI and stressed Kane type compounds in the presence of an arbitrarily oriented non-quantizing magnetic field by including the electron spin. Draw the corresponding graphs and explain the results.

(R.13.10) Investigate the QCs for HDQWTs' of all the materials as stated in R.1.1 of Chapter 1 in the presence of an arbitrarily oriented magnetic field by including electron spin and broadening. Draw the corresponding graphs and explain the results.

(R.13.11) Investigate all the problems of this section by removing all the mathematical approximations and establishing the respective appropriate uniqueness conditions. Draw the corresponding graphs and explain the results.

References

1. W. Zawadzki and B. Lax, *Phys. Rev. Lett.* **16**, 1001 (1966).
2. M. J. Harrison, *Phys. Rev. A* **29**, 2272 (1984); J. Zak and W. Zawadzki, *Phys. Rev.* **145**, 536 (1966).
3. W. Zawadzki, Q. H. Vrehen and B. Lax, *Phys. Rev.* **148**, 849 (1966); Q. H. Vrehen, W. Zawadzki and M. Reine, *Phys. Rev.* **158**, 702 (1967); M. H. Weiler, W. Zawadzki and B. Lax, *Phys. Rev.* **163**, 733 (1967).
4. W. Zawadzki and J. Kowalski, *Phys. Rev. Lett.* **27**, 1713 (1971); C. Chu, M. S. Chu and T. Ohkawa, *Phys. Rev. Lett.* **41**, 653 (1978); P. Hu and C. S. Ting, *Phys. Rev. B* **36**, 9671 (1987).
5. E. I. Butikov, A. S. Kondratev and A. E. Kuchma, *Sov. Phys. Sol. State* **13**, 2594 (1972).
6. K. P. Ghatak, J. P. Banerjee, B. Goswami and B. Nag, *Non. Opt. Quant. Opt.*, **16**, 241 (1996); M. Mondal and K. P. Ghatak, *Phys. Stat. Sol. (b)* **133**, K67 (1986).
7. M. Mondal, N. Chattopadhyay and K. P. Ghatak, *J. Low Temp. Phys.* **66**, 131 (1987); K. P. Ghatak and M. Mondal, *Zeitschrift fur Physik B* **69**, 471 (1988).
8. M. Mondal and K. P. Ghatak, *Phys. Lett. A* **131A**, 529 (1988); M. Mondal and K. P. Ghatak, *Phys. Stat. Sol. (b) Germany* **147**, K179 (1988); B. Mitra and K. P. Ghatak, *Phys. Lett.* **137A**, 413 (1989).
9. B. Mitra, A. Ghoshal and K. P. Ghatak, *Phys. Stat. Soli. (b)* **154**, K147 (1989).
10. B. Mitra and K. P. Ghatak, *Phys. Stat. Sol. (b)*, **164**, K13 (1991); K. P. Ghatak and B. Mitra, *Int. J. Electron.* **70**, 345 (1991); S. M. Adhikari, D. De, J. K. Baruah, S. Chowdhury and K. P. Ghatak, *Adv. Sci. Focus* **1**, 57 (2013); S. Pahari, S. Bhattacharya, D. De, S. M. Adhikari, A. Niyogi, A. Dey, N. Paitya, S. C. Saha, K. P. Ghatak and P. K. Bose, *Physica B: Condensed Matter* **405**, 4064 (2010); S. Bhattaacharya, S. Choudhary, S. Ghoshal, S. K. Bishwas, D. De and K. P. Ghatak, *J. Comp. Theo. Nanosci.* **3**, 423 (2006); M. Mondal and K. P. Ghatak, *Ann. der Physik.* **46**, 502 (1989); K. P. Ghatak, B. De *MRS Proc.* **242**, 377 (1992); K. P. Ghatak and B. Mitra, *Int. J. Electron. Theo. Exp.* **70**, 343 (1991); B. Mitra and K. P. Ghatak, *Phys. Lett. A* **141**, 81 (1989); B. Mitra and K. P. Ghatak, *Phys. Lett. A* **137**, 413 (1989);

M. Mondal, N. Chattopadhyay and K. P. Ghatak, *J. Low Temp. Phys.* **73**, 321 (1988); K. P. Ghatak, N. Chattopadhyay and S. N. Biswas, *Int. Soc. Opt. Photon. Proc. Soc. Photo-Optical Instru. Engg., USA*, 203 (1987).

11. K. P. Ghatak, M. Mitra, B. Goswami and B. Nag, *Nonlin. Opt* **16**, 167 (1996); K. P. Ghatak, D. K. Basu, and B. Nag, *J. Phys. Chem. Sol.* **58**, 133 (1997).
12. S. Biswas, N. Chattopadhyay and K. P. Ghatak, *Int. Soc. Opt. Photon. Proc. Soc. Photo-Optical Instru. Engg. USA*, **836**, 175 (1987); K. P. Ghatak, M. Mondal and S. Bhattacharyya, *SPIE*, **1284**, 113 (1990); K. P. Ghatak, *SPIE*, **1280**, Photonic materials and optical bistability, **53** (1990).
13. K. P. Ghatak and S. N. Biswas, *SPIE*, Growth and Characterization of Materials for Infrared Detectors and *Nonlin. Opt. Switches*, **1484**, 149 (1991).
14. K. P. Ghatak, SPIE, *Fiber Opt. Laser Sens. IX*, **1584**, 435 (1992).

Chapter 14

The QC in Heavily Doped NWFETs of Important Materials

To be the best, I must extract the best from myself.

14.1 Introduction

The fabrication methods and the physics of NWs have extensively been investigated in the literature [1–27]. Incidentally, the QC in HDNWFETs of different important materials has yet to be studied. In this Chapter, in Sections 14.2.1–14.2.12, we have investigated the QCs' in HDNWFETs of tetragonal and nonlinear optical, III–V, ternary, quaternary, II–VI, IV–VI, stressed Kane type, Te, GaP, $PtSb_2$, Bi_2Te_3, Ge, GaAs and II–V materials, respectively. Section 14.3 contains the summary and conclusion pertaining to this chapter. Section 14.4 presents 17 open research problems.

14.2 Theoretical Background

14.2.1 *The QC in HD NWFETs of tetragonal and nonlinear optical materials*

The DR of the 1D electrons in this case can be written following (8.25) as

$$\frac{\hbar^2(n_z\pi/d_z)^2}{2m_{\parallel}^*T_{21}(E,\eta_g)}+\frac{\hbar^2(n_y\pi/d_y)^2}{2m_{\perp}^*T_{22}(E,\eta_g)}+\frac{\hbar^2k_x^2}{2m_{\perp}^*T_{22}(E,\eta_g)}=1 \tag{14.1}$$

where $n_z(=1,2,3,\ldots)$, d_z are the size quantum number and the nanothickness along the z-direction, respectively, $n_y(=1,2,3,\ldots)$ and d_y are the size quantum number and the nanothickness along the y-direction, respectively.

The 1D DOS function per sub-band is given by

$$N_{1D}(E) = \frac{2g_v}{\pi}\frac{\partial k_x}{\partial E} \tag{14.2}$$

Thus, by using (14.1) and (14.2) the total DOS function in this case can be written as

$$N_{1DHD\Gamma}(E,\eta_g) = \frac{2g_v}{\pi}\sum_{n_y=1}^{n_{y\max}}\sum_{n_z=1}^{n_{z\max}} T_1'(E,n_y,n_z,\eta_g)H(E-E_{1HDNW}') \tag{14.3}$$

where E_{1HDNW}' is the complex subband energy which can be expressed in this case as

$$\frac{\hbar^2(n_z\pi/d_z)^2}{2m_\parallel^* T_{21}(E_{1HDNW}',\eta_g)} + \frac{\hbar^2(n_y\pi/d_y)^2}{2m_\perp^* T_{22}(E_{1HDNW}',\eta_g)} = 1 \tag{14.4}$$

$$T_{1HDNW}(E,n_y,n_z,\eta_g) \equiv \left[\left[1-\left[\frac{\hbar^2(n_z\pi/d_z)^2}{2m_\parallel^* T_{21}(E,\eta_g)} + \frac{\hbar^2(n_y\pi/d_y)^2}{2m_\perp^* T_{22}(E,\eta_g)}\right]\right] \times \frac{2m_\perp^* T_{21}(E,\eta_g)}{\hbar^2}\right]^{1/2}$$

and

$$T_{1HDNW}'(E,n_y,n_z,\eta_g) = \frac{\partial}{\partial E}[T_{1HDNW}(E,n_y,n_z,\eta_g)]$$

The EEM in this case in given by

$$m^*(E_{F1HDNW},n_y,n_z,\eta_g) = \frac{\hbar^2}{2}\left[\text{Real part of } \frac{\partial}{\partial(E_{F1HDNW})}[T_{1HDNW}(E,n_y,n_z,\eta_g)]^2\right] \tag{14.5a}$$

where E_{F1HDNW} is the Fermi energy in this case.

Thus, we observe that the EEM is the function of size quantum numbers in both the directions and the Fermi energy due to the combined influence of the crystal filed splitting constant and the anisotropic spin-orbit splitting constants respectively. Besides it is a function of η_g due to which the EEM exists in the band gap, which is otherwise impossible.

The 1D electron statistics in HD NWFETs of tetragonal and nonlinear optical materials under the condition of extreme carrier degeneracy can be written as

$$n_{01D} = \text{Real part of } \bar{n}_{01D} \tag{14.5b}$$

where

$$\bar{n}_{01D} = \frac{2g_v}{\pi}\left[\sum_{n_y=1}^{n_{y\max}}\sum_{n_z=1}^{n_{z\max}}[T_{1HDNW}(E_{F1HDNW}, n_y, n_z, \eta_g)]\right] \tag{14.5c}$$

The QC in this case can be expressed as

$$C_g = e\frac{\partial n_{01D}}{\partial V_g} \tag{14.5d}$$

Thus, using (14.5b) and (14.5d) we can study the QC in this case.

14.2.2 *The QC in HD NWFETs of III–V, ternary and quaternary materials*

(i) Three-band model of Kane

The DR of the 1D electrons in this case can be written following (8.39) as

$$\frac{\hbar^2(n_z\pi/d_z)^2}{2m_c} + \frac{\hbar^2(n_y\pi/d_y)^2}{2m_c} + \frac{\hbar^2k_x^2}{2m_c} = T_{31}(E,\eta_g) + iT_{32}(E,\eta_g) \tag{14.6}$$

The DOS function in this case assume the form

$$N_{1DHD\Gamma}(E,\eta_g) = \frac{2g_v}{\pi}\sum_{n_y=1}^{n_{y\max}}\sum_{n_z=1}^{n_{z\max}} T_2'(E, n_y, n_z, \eta_g)H(E - E_{2HDNW}') \tag{14.7}$$

where

$$T_2(E, n_y, n_z, \eta_g) \equiv \left[\left[T_{31}(E,\eta_g) + iT_{32}(E,\eta_g) - \left[\frac{\hbar^2(n_z\pi/d_z)^2}{2m_c} + \frac{\hbar^2(n_y\pi/d_y)^2}{2m_c}\right]\right]\frac{2m_c}{\hbar^2}\right]^{1/2}$$

and

$$T_2'(E, n_y, n_z, \eta_g) = \frac{\partial}{\partial E}[T_2(E, n_y, n_z, \eta_g)]$$

In (14.7), E_{2HDNW}' is the sub-band energy in this case which can be expressed as

$$\frac{\hbar^2(n_z\pi/d_z)^2}{2m_c} + \frac{\hbar^2(n_y\pi/d_y)^2}{2m_c} = T_{31}(E_{2HDNW}', \eta_g) + iT_{32}(E_{2HDNW}', \eta_g) \tag{14.8}$$

The EEM in this case is given by

$$m^*(E_{F1HDNW}, \eta_g) = m_c[T'_{31}(E_{F1HDNW}, \eta_g)] \tag{14.9}$$

The electron concentration per unit length in HD NWFETs of III–V, ternary and quaternary materials under the condition of extreme carrier degeneracy can be written as

$$n_{01D} = \text{ Real part of } \bar{n}_{01D} \tag{14.10}$$

where

$$\bar{n}_{01D} = \frac{2g_v}{\pi}\left[\sum_{\bar{n}_y=1}^{n_{y\max}} \sum_{\bar{n}_z=1}^{n_{z\max}} [T_{2HDNW}(E_{F1HDNW}, n_y, n_z, \eta_g)]\right] \tag{14.11}$$

Thus, using (14.10) and (14.5d) we can study the QC in this case.

(ii) Two-band model of Kane

The DR under heavy doping of the 1D electrons in this case can be written from (8.50) as

$$\frac{\hbar^2(n_z\pi/d_z)^2}{2m_c} + \frac{\hbar^2(n_y\pi/d_y)^2}{2m_c} + \frac{\hbar^2 k_x^2}{2m_c} = \gamma_2(E, \eta_g) \tag{14.12}$$

The DOS function in this case assume the form

$$N_{1DHD\Gamma}(E, \eta_g) = \frac{2g_v}{\pi} \sum_{n_y=1}^{n_{y\max}} \sum_{n_z=1}^{n_{z\max}} T'_3(E, n_y, n_z, \eta_g) H(E - E'_{3HDNW}) \tag{14.13}$$

where

$$T_3(E, n_y, n_z, \eta_g) = \left[\left[\gamma_2(E, \eta_g) - \left[\frac{\hbar^2(n_z\pi/d_z)^2}{2m_c} + \frac{\hbar^2(n_y\pi/d_y)^2}{2m_c}\right]\right] \frac{2m_c}{\hbar^2}\right]^{1/2}$$

and

$$T'_3(E, n_y, n_z, \eta_g) = \frac{\partial}{\partial E}[T_3(E, n_y, n_z, \eta_g)]$$

In (14.13), E'_{3HDNW} is the sub-band energy in this case which can be expressed as

$$\frac{\hbar^2(n_z\pi/d_z)^2}{2m_c} + \frac{\hbar^2(n_y\pi/d_y)^2}{2m_c} = \gamma_2(E'_{3HDNW}, \eta_g) \tag{14.14}$$

The EEM in this case is given by

$$m^*(E_{F1HDNW}, \eta_g) = m_c[\gamma'_2(E_{F1HDNW}, \eta_g)] \tag{14.15}$$

The electron concentration per unit length in HD NWFETs of III–V, ternary and quaternary materials under the condition of extreme carrier degeneracy can be written as

$$n_{01D} = \frac{2g_v}{\pi} \sum_{n_y=1}^{n_{y\max}} \sum_{n_z=1}^{n_{z\max}} [T_3(E_{F1HDNW}, n_y, n_z, \eta_g)] \tag{14.16}$$

Thus, using (14.16) and (14.5d) we can study the QC in this case.

(iii) Parabolic energy bands

The DR under heavy doping of the 1D electrons in this case can be written from (8.60) as

$$\frac{\hbar^2 (n_z\pi/d_z)^2}{2m_c} + \frac{\hbar^2 (n_y\pi/d_y)^2}{2m_c} + \frac{\hbar^2 k_x^2}{2m_c} = \gamma_3(E, \eta_g) \tag{14.17}$$

The DOS function in this case assume the form

$$N_{1DHD\Gamma}(E, \eta_g) = \frac{2g_v}{\pi} \sum_{n_y=1}^{n_{y\max}} \sum_{n_z=1}^{n_{z\max}} T_4'(E, n_y, n_z, \eta_g) H(E - E'_{4HDNW}) \tag{14.18}$$

where

$$T_4(E, n_y, n_z, \eta_g) = \left[\left[\gamma_3(E, \eta_g) - \left[\frac{\hbar^2 (n_z\pi/d_z)^2}{2m_c} + \frac{\hbar^2 (n_y\pi/d_y)^2}{2m_c}\right]\right] \frac{2m_c}{\hbar^2}\right]^{1/2}$$

and

$$T_4'(E, n_y, n_z, \eta_g) = \frac{\partial}{\partial E}[T_4(E, n_y, n_z, \eta_g)]$$

In (14.18), E'_{4HDNW} is the sub-band energy in this case which can be expressed as

$$\frac{\hbar^2 (n_z\pi/d_z)^2}{2m_c} + \frac{\hbar^2 (n_y\pi/d_y)^2}{2m_c} = \gamma_3(E'_{4HDNW}, \eta_g) \tag{14.19}$$

The EEM in this case is given by

$$m^*(E_{F1HDNW}, \eta_g) = m_c[\gamma_3'(E_{F1HDNW}, \eta_g)] \tag{14.20}$$

The electron concentration per unit length in HD NWFETs of III–V, ternary and quaternary materials can be written as

$$n_{01D} = \frac{2g_v}{\pi} \sum_{n_y=1}^{n_{y\max}} \sum_{n_z=1}^{n_{z\max}} [T_4(E_{F1HDNW}, n_y, n_z, \eta_g)] \tag{14.21}$$

Thus, using (14.20) and (14.5d) we can study the QC in this case.

(iv) The Model of Stillman *et al.*

The DR of the 1D electrons under heavy doping in this case can be written from (11.23) as

$$\frac{\hbar^2(n_z\pi/d_z)^2}{2m_c}+\frac{\hbar^2(n_y\pi/d_y)^2}{2m_c}+\frac{\hbar^2k_x^2}{2m_c}=\theta_4(E,\eta_g) \tag{14.22}$$

where $\theta_4(E,\eta_g)=I_{12}(E,\eta_g)$.

The sub-band energy (E'_{9HDNW}) in this case can be expressed as

$$\frac{\hbar^2(n_z\pi/d_z)^2}{2m_c}+\frac{\hbar^2(n_y\pi/d_y)^2}{2m_c}=\theta_4(E'_{9HDNW},\eta_g) \tag{14.23}$$

The EEM in this case is given by

$$m^*(E_{F1HDNW},\eta_g)=m_c[\theta'_4(E_{F1HDNW},\eta_g)]. \tag{14.24}$$

where $\theta'_4(E_{F1HDNW},\eta_g)=\frac{\partial}{\partial E}[\theta_4(E_{F1HDNW},\eta_g)]$.

The 1D electron statistics in HD NWFETs of III–V, ternary and quaternary materials under the condition of extreme carrier degeneracy can be written as

$$n_{01D}=\left(\frac{2g_v}{\pi}\right)\sum_{n_y=1}^{n_{y\max}}\sum_{n_z=1}^{n_{z\max}}[T_{1HDNW}(E_{F1HDNW},n_y,n_z,\eta_g)] \tag{14.25}$$

where

$$T_{1HDNW}(E_{F1HDNW},n_y,n_z,\eta_g)$$
$$=\left[\left[\theta_4(E_{F1HDNW},\eta_g)-\frac{\hbar^2(n_z\pi/d_z)^2}{2m_c}-\frac{\hbar^2(n_y\pi/d_y)^2}{2m_c}\right]\frac{2m_c}{\hbar^2}\right]^{1/2}$$

Thus, using (14.25) and (14.5d) we can study the QC in this case.

(v) The Model of Palik *et al.*

The DR of the 1D electrons under heavy doping in this case can be written from (11.32) as

$$\frac{\hbar^2(n_z\pi/d_z)^2}{2m_c}+\frac{\hbar^2(n_y\pi/d_y)^2}{2m_c}+\frac{\hbar^2k_x^2}{2m_c}=\theta_5(E,\eta_g) \tag{14.26}$$

where $\theta_5(E,\eta_g)=I_{13}(E,\eta_g)$.

The sub-band energy (E'_{6HDNW}) in this case can be expressed as

$$\frac{\hbar^2(n_z\pi/d_z)^2}{2m_c}+\frac{\hbar^2(n_y\pi/d_y)^2}{2m_c}=\theta_5(E'_{6HDNW},\eta_g) \tag{14.27}$$

The EM in this case is given by

$$m^*(E_{F1HDNW},\eta_g)=m_c[\theta_5'(E_{F1HDNW},\eta_g)] \tag{14.28}$$

where

$$\theta_5'(E_{F1HDNW},\eta_g)=\frac{\partial}{\partial E}[\theta_5(E_{F1HDNW},\eta_g)]$$

The carrier statistics in HD NWFETs of III–V, ternary and quaternary materials under the condition of extreme carrier degeneracy can be written as

$$n_{01D}=\left(\frac{2g_v}{\pi}\right)\sum_{n_y=1}^{n_{y\max}}\sum_{n_z=1}^{n_{z\max}}[T_{6HDNW}(E_{F1HDNW},n_y,n_z,\eta_g)] \tag{14.29}$$

where

$$T_{6HDNW}(E_{F1HDNW},n_y,n_z,\eta_g)$$
$$=\left[\left[\theta_5(E_{F1HDNW},\eta_g)-\frac{\hbar^2(n_z\pi/d_z)^2}{2m_c}-\frac{\hbar^2(n_y\pi/d_y)^2}{2m_c}\right]\frac{2m_c}{\hbar^2}\right]^{1/2}$$

Thus, using (14.29) and (14.5d) we can study the QC in this case.

14.2.3 *The QC in HD NWFETs of II–VI materials*

The 1D electron dispersion law in HD NWFETs of II–VI materials can be written following (8.70) as

$$\gamma_3(E,\eta_g)=a_0'\left[\left(\frac{n_x\pi}{d_x}\right)^2+\left(\frac{n_y\pi}{d_y}\right)^2\right]\pm\bar{\lambda}_0\left[\left(\frac{n_x\pi}{d_x}\right)^2+\left(\frac{n_y\pi}{d_y}\right)^2\right]^{1/2}+\frac{\hbar^2k_z^2}{2m_\parallel^*} \tag{14.30}$$

The DOS function in this case assume the form

$$N_{1DHD\Gamma}(E,\eta_g)=\frac{g_v}{\pi}\sum_{n_x=1}^{n_{x\max}}\sum_{n_y=1}^{n_{y\max}}T_7'(E,n_x,n_y,\eta_g)H(E-E'_{13HDNW}) \tag{14.31}$$

where

$$T_7(E,n_x,n_y,\eta_g)\equiv\left[[\gamma_3(E,\eta_g)-[G_{3,\pm}(n_x,n_y)]]\frac{2m_\parallel^*}{\hbar^2}\right]^{1/2}$$

and

$$G_{3,\pm}(n_x,n_y)=a_0'\left[\left[\left(\frac{n_x\pi}{d_x}\right)^2+\left(\frac{n_y\pi}{d_y}\right)^2\right]\pm\bar{\lambda}_0\left[\left(\frac{n_x\pi}{d_x}\right)^2+\left(\frac{n_y\pi}{d_y}\right)^2\right]^{1/2}\right]$$

The sub-band energy (E'_{13HDNW}) in this case can be expressed as

$$\gamma_3(E'_{13HDNW},\eta_g)$$
$$=a_0'\left[\left(\frac{n_x\pi}{d_x}\right)^2+\left(\frac{n_y\pi}{d_y}\right)^2\right]\pm\bar{\lambda}_0\left[\left(\frac{n_x\pi}{d_x}\right)^2+\left(\frac{n_y\pi}{d_y}\right)^2\right]^{1/2} \quad (14.32)$$

The EEM in this case is given by

$$m^*(E_{F1HDNW},\eta_g)=m_{\|}^*\gamma_3'(E_{F1HDNW},\eta_g) \quad (14.33)$$

The carrier statistics in HD NWFETs of II–VI materials under the condition of extreme carrier degeneracy can be written as

$$n_{01D}=\frac{g_v}{\pi}\sum_{n_x=1}^{n_{x\max}}\sum_{n_y=1}^{n_{y\max}}[T_7(E_{F1HDNW},n_x,n_y,\eta_g)] \quad (14.34)$$

Thus, using (14.34) and (14.5d) we can study the QC in this case.

14.2.4 *The QC in HD NWFETs of IV–VI materials*

(i) Dimmock model

The 1D electron dispersion law in HD NWFETs of IV–VI materials can be expressed following (3.56) as

$$\gamma_2(E,\eta_g)+\alpha\gamma_3(E,\eta_g)\left(\frac{\hbar^2}{2x_4}\left(\frac{n_x\pi}{d_x}\right)^2+\frac{\hbar^2}{2x_5}\left(\frac{n_y\pi}{d_y}\right)^2\right)$$
$$+\alpha\gamma_3(E,\eta_g)\frac{\hbar^2}{2x_6}k_z^2-(1+\alpha\gamma_3(E,\eta_g))\left(\frac{\hbar^2}{2x_1}\left(\frac{n_x\pi}{d_x}\right)^2+\frac{\hbar^2}{2x_2}\left(\frac{n_y\pi}{d_y}\right)^2\right)$$
$$-\alpha\left(\frac{\hbar^2}{2\bar{x}_1}\left(\frac{n_x\pi}{d_x}\right)^2+\frac{\hbar^2}{2x_2}\left(\frac{n_y\pi}{d_y}\right)^2\right)-\alpha\left(\left(\frac{\hbar^2}{2x_1}\left(\frac{n_x\pi}{d_x}\right)^2\right.\right.$$

$$+\frac{\hbar^2}{2x_2}\left(\frac{n_y\pi}{d_y}\right)^2\Bigg)\frac{\hbar^2}{2\bar{x}_6}k_z^2-(1+\alpha\gamma_3(E,\eta_g))\frac{\hbar^2}{2x_3}k_z^2$$

$$-\alpha\frac{\hbar^2}{2x_3}k_z^2\left(\frac{\hbar^2}{2x_4}\left(\frac{n_x\pi}{d_x}\right)^2+\frac{\hbar^2}{2x_5}\left(\frac{n_y\pi}{d_y}\right)^2\right)-\alpha\frac{\hbar^4k_z^4}{4x_3x_6}=\frac{\hbar^2}{2m_1}\left(\frac{n_x\pi}{d_x}\right)^2$$

$$+\frac{\hbar^2}{2m_2}\left(\frac{n_y\pi}{d_y}\right)^2+\frac{\hbar^2}{2m_3}k_z^2 \tag{14.35}$$

(14.35) can be written as

$$k_z=T_{36}(E,\eta_g,n_x,n_y) \tag{14.36}$$

where

$$T_{36}(E,\eta_g,n_x,n_y)=[(2C_{22})^{-1}[-B_{HD}(E,\eta_g,n_x,n_y)$$
$$+\sqrt{B_{HD}^2(E,\eta_g,n_x,n_y)+4C_{22}A_{HD}(E,\eta_g,n_x,n_y)}\,]]^{1/2}$$

$$C_{22}=\left(\alpha\frac{\hbar^4}{4x_3x_6}\right),\quad B_{HD}(E,\eta_g,n_x,n_y)=\Bigg[\alpha\left(\frac{\hbar^2}{2x_1}\left(\frac{n_x\pi}{d_x}\right)^2\right.$$

$$\left.+\frac{\hbar^2}{2x_2}\left(\frac{n_y\pi}{d_y}\right)^2\right)\frac{\hbar^2}{2x_6}+(1+\alpha\gamma_3(E,\eta_g))\frac{\hbar^2}{2x_3}-\alpha\gamma_3(E,\eta_g)\frac{\hbar^2}{2x_6}$$

$$+\frac{\hbar^2}{2m_3}+\alpha\frac{\hbar^2}{2x_3}\left(\frac{\hbar^2}{2x_4}\left(\frac{n_x\pi}{d_x}\right)^2+\frac{\hbar^2}{2x_5}\left(\frac{n_y\pi}{d_y}\right)^2\right)\Bigg]$$

and

$$A_{HD}(E,\eta_g,n_x,n_y)$$
$$=\Bigg[-\left[\frac{\hbar^2}{2x_1}\left(\frac{n_x\pi}{d_x}\right)^2+\frac{\hbar^2}{2x_2}\left(\frac{n_y\pi}{d_y}\right)^2\right]\gamma_2(E,\eta_g)$$
$$+\alpha\gamma_3(E,\eta_g)\left(\frac{\hbar^2}{2x_4}\left(\frac{n_x\pi}{d_x}\right)^2+\frac{\hbar^2}{2x_5}\left(\frac{n_y\pi}{d_y}\right)^2\right)$$
$$-\alpha\left(\frac{\hbar^2}{2x_1}\left(\frac{n_x\pi}{d_x}\right)^2+\frac{\hbar^2}{2x_2}\left(\frac{n_y\pi}{d_y}\right)^2\right)$$

$$\times\left(\frac{\hbar^2}{2x_4}\left(\frac{n_x\pi}{d_x}\right)^2+\frac{\hbar^2}{2x_5}\left(\frac{n_y\pi}{d_y}\right)^2\right)$$
$$-(1+\alpha\gamma_3(E,\eta_g))\left(\frac{\hbar^2}{2x_1}\left(\frac{n_x\pi}{d_x}\right)^2+\frac{\hbar^2}{2x_2}\left(\frac{n_y\pi}{d_y}\right)^2\right)\Bigg]$$

The DOS function in this case assume the form

$$N_{1DHD\Gamma}(E)=\frac{2g_v}{\pi}\sum_{n_x=1}^{n_{x\max}}\sum_{n_y=1}^{n_{y\max}}T'_{36}(E,n_x,n_y,\eta_g)H(E-E'_{14HDNW}) \tag{14.37}$$

In (14.37), E'_{14HDNW} the sub-band energy in this case which can be expressed as

$$0=T_{36}(E'_{14HDNW},\eta_g,n_x,n_y) \tag{14.38}$$

The EEM in this case is given by

$$m^*(E_{F1HDNW},\eta_g,n_x,n_y)=\frac{\hbar^2}{2}\frac{\partial}{\partial E}[T^2_{36}(E_{F1HDNW},\eta_g,n_x,n_y)] \tag{14.39}$$

The 1D electron statistics in HD NWFETs of IV–VI materials under the condition of extreme carrier degeneracy can be written as

$$n_{01D}=\frac{2g_v}{\pi}\sum_{n_x=1}^{n_{x\max}}\sum_{n_y=1}^{n_{y\max}}[T_{36HDNW}(E_{F1HDNW},n_x,n_y,\eta_g)] \tag{14.40}$$

Thus, using (14.40) and (14.5d) we can study the QC in this case.

(ii) Bangert and Kastner model

The 1D DR in HD NWFETs of IV–VI materials in accordance with the present model can be written from (8.83) as

$$F_{1HD}(E,\eta_g)\left[\left(\frac{n_x\pi}{d_x}\right)^2+\left(\frac{n_y\pi}{d_y}\right)^2\right]+F_{2HD}(E,\eta_g)k_z^2=1 \tag{14.41}$$

The (14.41) can be written as

$$k_z=T_{60}(E,\eta_g,n_x,n_y) \tag{14.42}$$

where

$$T_{60}(E,\eta_g,n_x,n_y)$$
$$=\left[\left[1-F_{1HD}(E,\eta_g)\left[\left(\frac{n_x\pi}{d_x}\right)^2+\left(\frac{n_y\pi}{d_y}\right)^2\right]\right][F_{2HD}(E,\eta_g)]^{-1}\right]^{1/2}$$

The DOS function in this case assume the form

$$N_{1DHD\Gamma}(E) = \frac{2g_v}{\pi} \sum_{n_x=1}^{n_{x\max}} \sum_{n_y=1}^{n_{y\max}} T'_{40}(E, n_x, n_y, \eta_g) H(E - E'_{15HDNW}) \tag{14.43}$$

The sub-band energy (E'_{15HDNW}) can be expressed as

$$\text{Real part of } T_{60}(E'_{15HDNW}, \eta_g, n_x, n_y) = 0 \tag{14.44}$$

The EEM in this case is given by

$$m^*(E_{F1HDNW}, \eta_g, n_x, n_y) = \frac{\hbar^2}{2} \text{Real part of } \left[\frac{\partial}{\partial \bar{E}}[T_{60}^2(E_{F1HDNW}, \eta_g, n_x, n_y)]\right] \tag{14.45}$$

The 1D electron statistics in HD NWFETs of IV–VI materials under the condition of extreme carrier degeneracy can be written as

$$n_{01D} = \text{Real part of } \bar{n}_{01D} \tag{14.46}$$

where

$$\bar{n}_{01D} = \left(\frac{2g_v}{\pi}\right) \sum_{n_x=1}^{n_{x\max}} \sum_{n_y=1}^{n_{y\max}} [T_{40}(E_{F1HDNW}, n_x, n_y, \eta_g)] \tag{14.47}$$

Thus, using (14.46) and (14.5d) we can study the QC in this case.

14.2.5 *The QC in HD NWFETs of stressed Kane type materials*

The 1D DR in this case can be written can be written from (8.106) as

$$P_{11}(E, \eta_g)\left(\frac{\pi n_x}{d_x}\right)^2 + Q_{11}(E, \eta_g)\left(\frac{\pi n_y}{d_y}\right)^2 + S_{11}(E, \eta_g) k_z^2 = 1 \tag{14.48}$$

The (14.48) can be written as

$$k_z = T_{70}(E, \eta_g, n_x, n_y) \tag{14.49}$$

where

$$T_{70}(E, \eta_g, n_x, n_y) = \left[\left[1 - P_{11}(E, \eta_g)\left(\frac{\pi n_x}{d_x}\right)^2 + Q_{11}(E, \eta_g)\left(\frac{\pi n_y}{d_y}\right)^2\right] [S_{11}(E, \eta_g)]^{-1}\right]^{1/2}$$

The DOS function in this case assume the form

$$N_{1DHD\Gamma}(E) = \frac{2g_v}{\pi} \sum_{n_x=1}^{n_{x\max}} \sum_{n_y=1}^{n_{y\max}} T'_{70}(E, n_x, n_y, \eta_g) H(E - E'_{30HDNW}) \tag{14.50}$$

The sub-band energy (E'_{30HDNW}) in this case can be expressed as

$$0 = T_{70}(E'_{30HDNW}, \eta_g, n_x, n_y) \tag{14.51}$$

The EEM in this case is given by

$$m^*(E_{F1HDNW}, \eta_g, n_x, n_y) = \frac{\hbar^2}{2} \frac{\partial}{\partial E} [T^2_{70}(E_{F1HDNW}, \eta_g, n_x, n_y)] \tag{14.52}$$

The 1D electron statistics in HD NWFETs of stressed Kane type materials under the condition of extreme carrier degeneracy can be written as

$$n_{01D} = \frac{2g_v}{\pi} \sum_{n_x=1}^{n_{x\max}} \sum_{n_y=1}^{n_{y\max}} [T_{70}(E_{F1HDNW}, n_x, n_y, \eta_g)] \tag{14.53}$$

Thus, using (14.53) and (14.5d) we can study the QC in this case.

14.2.6 *The QC in HD NWFETs of Te*

The 1D DR in HD NWFETs of Te may be written in this case following (11.72) as

$$k_x = t_{72}(E, n_y, n_z, \eta_g) \tag{14.54}$$

where

$$t_{72}(E, n_y, n_z, \eta_g) = \left[-\left(\frac{n_y \pi}{d_y}\right)^2 + \psi_{5HD}(E, \eta_g) - \psi_6 \left(\frac{\pi n_z}{d_z}\right)^2 \pm \psi_7 \left[\psi^2_{8HD}(E, \eta_g) - \left(\frac{\pi n_z}{d_z}\right)^2 \right]^{1/2} \right]^{1/2}$$

The DOS function in this case assume the form

$$N_{1DHD\Gamma}(E) = \frac{g_v}{2\pi} \sum_{n_y=1}^{n_{x\max}} \sum_{n_z=1}^{n_{y\max}} t'_{72}(E, n_y, n_z, \eta_g) H(E - E'_{31HDNW}) \tag{14.55}$$

The sub-band energy (E'_{31HDNW}) in this case can be expressed as

$$0 = t_{72}(E'_{31HDNW}, \eta_g, n_y, n_z) \tag{14.56}$$

The EEM in this case is given by

$$m^*(E_{F1HDNW}, \eta_g, n_y, n_z) = \frac{\hbar^2}{2} \frac{\partial}{\partial E} [t^2_{72}(E_{F1HDNW}, \eta_g, n_y, n_z)] \tag{14.57}$$

The 1D electron statistics in HD NWFETs of Te under the condition of extreme carrier degeneracy can be written as

$$n_{01D} = \frac{g_v}{2\pi} \sum_{n_y=1}^{n_{x\max}} \sum_{n_z=1}^{n_{y\max}} [t_{72}(E_{F1HDNW}, n_y, n_z, \eta_g)] \tag{14.58}$$

Thus, using (14.58) and (14.5d) we can study the QC in this case.

14.2.7 *The QC in HD NWFETs of GaP*

The 1D DR in HD NWFETs of GaP may be written in this case following (11.77) as

$$k_x = u_{70}(E, n_y, n_z, \eta_g) \tag{14.59}$$

where

$$u_{70}(E, n_y, n_z, \eta_g) = \left[-\left(\frac{n_y \pi}{d_y}\right)^2 + t_{11}\gamma_3(E, \eta_g) + t_{21} - t_{31}\left(\frac{n_z \pi}{d_z}\right)^2 - t_{41}\left[\left(\frac{n_z \pi}{d_z}\right)^2 + t_5^2(E, \eta_g)\right]^{1/2}\right]^{1/2}$$

The DOS function in this case assume the form

$$N_{1DHD\Gamma}(E) = \frac{2g_v}{\pi} \sum_{n_y=1}^{n_{x\max}} \sum_{n_z=1}^{n_{y\max}} u'_{70}(E, n_y, n_z, \eta_g) H(E - E'_{32HDNW}) \tag{14.60}$$

The sub-band energy (E'_{32HDNW}) in this case can be expressed as

$$0 = u_{70}(E'_{32HDNW}, \eta_g, n_y, n_z) \tag{14.61}$$

The EEM in this case is given by

$$m^*(E_{F1HDNW}, \eta_g, n_y, n_z) = \frac{\hbar^2}{2} \frac{\partial}{\partial E} [u_{70}^2(E_{F1HDNW}, \eta_g, n_y, n_z)] \tag{14.62}$$

The 1D electron statistics in HD NWFETs of GaP under the condition of extreme carrier degeneracy can be written as

$$n_{01D} = \frac{2g_v}{\pi} \sum_{n_y=1}^{n_{y\max}} \sum_{n_z=1}^{n_{z\max}} [u_{70}(E_{F1HDNW}, \eta_g, n_y, n_z)] \tag{14.63}$$

Thus, using (14.63) and (14.5d) we can study the QC in this case.

14.2.8 The QC in HD NWFETs of $PtSb_2$

The 1D DR in HD NWFETs of $PtSb_2$ may be written in this case following (11.89) as

$$k_x = V_{70}(E, n_y, n_z, \eta_g) \tag{14.64}$$

where

$$V_{70}(E, n_y, n_z, \eta_g) = \left[-\left(\frac{n_y \pi}{d_y}\right)^2 + A_{60}(E, \eta_g, n_z)\right]^{1/2}$$

The DOS function in this case assume the form

$$N_{1DHD\Gamma}(E, \eta_g) = \frac{2g_v}{\pi} \sum_{n_y=1}^{n_{y\max}} \sum_{n_z=1}^{n_{z\max}} V'_{70}(E, n_y, n_z, \eta_g) H(E - E'_{34HDNW}) \tag{14.65}$$

The sub-band energy (E'_{34HDNW}) in this case can be expressed as

$$0 = V_{70}(E'_{34HDNW}, \eta_g, n_y, n_z) \tag{14.66}$$

The EEM in this case is given by

$$m^*(E_{F1HDNW}, n_y, n_z, \eta_g) = \frac{\hbar^2}{2} \frac{\partial}{\partial E}[V_{70}^2(E_{F1HDNW}, n_y, n_z, \eta_g)] \tag{14.67}$$

The carrier statistics in HD NWFETs of $PtSb_2$ under the condition of extreme carrier degeneracy can be written as

$$n_{01D} = \frac{2g_v}{\pi} \sum_{n_y=1}^{n_{y\max}} \sum_{n_z=1}^{n_{z\max}} [V_{70}(E_{F1HDNW}, n_y, n_z, \eta_g)] \tag{14.68}$$

Thus, using (14.68) and (14.5d) we can study the QC in this case.

14.2.9 The QC in HD NWFETs of bismuth telluride

The DR in HD NWFETs of bismuth telluride in this case can be written from (11.92) as

$$k_x = J_{70}(E, n_y, n_z, \eta_g) \tag{14.69}$$

where

$$J_{70}(E, n_y, n_z, \eta_g)$$
$$= \left[\left[\gamma_2(E, \eta_g) - I_2\left(\frac{n_y \pi}{d_y}\right)^2 - I_3\left(\frac{n_z \pi}{d_z}\right)^2 - 2I_4 \frac{n_y n_z \pi^2}{d_y d_z}\right](I_1)^{-1}\right]^{1/2}$$

The DOS function in this case assume the form

$$N_{1DHD\Gamma}(E,\eta_g) = \frac{2g_v}{\pi}\sum_{n_y=1}^{n_{y\max}}\sum_{n_z=1}^{n_{z\max}} J'_{70}(E,n_y,n_z,\eta_g)H(E-E'_{50HDNW}) \tag{14.70}$$

The sub-band energy (E'_{50HDNW}) in this case can be expressed as

$$0 = J_{70}(E'_{50HDNW},\eta_g,n_y,n_z) \tag{14.71}$$

The EEM in this case is given by

$$m^*(E_{F1HDNW},\eta_g,n_y,n_z) = \frac{\hbar^2}{2}\frac{\partial}{\partial E}[J^2_{70}(E_{F1HDNW},\eta_g,n_y,n_z)] \tag{14.72}$$

The carrier statistics in HD NWFETs of bismuth telluride under the condition of extreme carrier degeneracy which can, in turn, be written as

$$n_{01D} = \frac{2g_v}{\pi}\sum_{n_y=1}^{n_{y\max}}\sum_{n_z=1}^{n_{z\max}}[J_{70}(E_{F1HDNW},n_y,n_z,\eta_g)] \tag{14.73}$$

Thus, using (14.73) and (14.5d) we can study the QC in this case.

14.2.10 *The QC in HD NWFETs of germanium*

(a) Model of Cardona *et al.*

The DR in accordance with this model in the present case can be written from (11.96) as

$$k_x = L_{70}(E,n_y,n_z,\eta_g) \tag{14.74}$$

where

$$L_{70}(E,n_y,n_z,\eta_g) = \left[\left[\gamma_2(E,\eta_g) + \alpha\left[\frac{\hbar^2}{2m^*_{\|}}\left(\frac{n_z\pi}{d_z}\right)^2\right]^2 - (1+2\alpha\gamma_3(E,\eta_g))\frac{\hbar^2}{2m^*_{\|}}\left(\frac{n_z\pi}{d_z}\right)^2\right]\left(\frac{2m^*_{\|}}{\hbar^2}\right)\right]^{1/2}$$

The DOS function in this case assume the form

$$N_{1DHD\Gamma}(E,\eta_g) = \frac{2g_v}{\pi}\sum_{n_y=1}^{n_{y\max}}\sum_{n_z=1}^{n_{z\max}} L'_{70}(E,n_y,n_z,\eta_g)H(E-E'_{52HDNW}) \tag{14.75}$$

The sub-band energy (E'_{52HDNW}) in this case can be expressed as

$$0 = L_{70}(E'_{52HDNW}, \eta_g, n_y, n_z) \tag{14.76}$$

The EM in this case is given by

$$m^*(E_{F1HDNW}, \eta_g, n_y, n_z) = \frac{\hbar^2}{2}\frac{\partial}{\partial E}[L_{70}^2(E_{F1HDNW}, \eta_g, n_y, n_z)] \tag{14.77}$$

The carrier statistics in HD NWFETs of Ge under the condition of extreme carrier degeneracy can be written as

$$n_{01D} = \frac{2g_v}{\pi}\sum_{n_y=1}^{n_{y\max}}\sum_{n_z=1}^{n_{z\max}}[L_{70HDNW}(E_{F1HDNW}, n_y, n_z, \eta_g)] \tag{14.78}$$

Thus, using (14.78) and (14.5d) we can study the QC in this case.

(b) Model of Wang *et al.*

The DR in accordance with this model in the present case can be written from (11.99) as

$$k_x = \beta_{70}(E, n_y, n_z, \eta_g) \tag{14.79}$$

where

$$\beta_{70}(E, n_y, n_z, \eta_g) = \left[-\left(\frac{n_y\pi}{d_y}\right)^2 + \frac{2m_\perp^*}{\hbar^2}\left[\bar{\alpha}_8 - \bar{\alpha}_9\left(\frac{\pi n_z}{d_z}\right)^2 - \bar{\alpha}_{10}\left[\left(\frac{\pi n_z}{d_z}\right)^4 + \bar{\alpha}_{11}\left(\frac{\pi n_z}{d_z}\right)^2 + \bar{\alpha}_{12}(E, \eta_g)\right]^{1/2}\right]\right]^{1/2}$$

The DOS function in this case assume the form

$$N_{1DHD\Gamma}(E, \eta_g) = \frac{2g_v}{\pi}\sum_{n_y=1}^{n_{y\max}}\sum_{n_z=1}^{n_{z\max}}\beta'_{70}(E, n_y, n_z, \eta_g)H(E - E'_{54HDNW}) \tag{14.80}$$

The sub-band energy (E'_{54HDNW}) in this case can be expressed as

$$0 = \beta_{70}(E'_{54HDNW}, \eta_g, n_y, n_z) \tag{14.81}$$

The EEM in this case is given by

$$m^*(E_{F1HDNW}, \eta_g, n_y, n_z) = \frac{\hbar^2}{2}\frac{\partial}{\partial E}[\beta_{70}^2(E_{F1HDNW}, \eta_g, n_y, n_z)] \tag{14.82}$$

The carrier statistics in HD NWFETs of Ge under the condition of extreme carrier degeneracy can be written as

$$n_{01D} = \left(\frac{2g_v}{\pi}\right) \sum_{n_y=1}^{n_{y\max}} \sum_{n_z=1}^{n_{z\max}} [\beta_{70HDNW}(E_{F1HDNW}, n_y, n_z, \eta_g)] \tag{14.83}$$

Thus, using (14.83) and (14.5d) we can study the QC in this case.

14.2.11 *The QC in HD NWFETs of gallium antimonide*

The DR of the 1D electrons in this case can be written following (11.105) as

$$\frac{\hbar^2(n_z\pi/d_z)^2}{2m_c} + \frac{\hbar^2(n_y\pi/d_y)^2}{2m_c} + \frac{\hbar^2 k_x^2}{2m_c} = I_{36}(E, \eta_g) \tag{14.84}$$

The DOS function in this case assume the form

$$N_{1DHD\Gamma}(E, \eta_g) = \frac{2g_v}{\pi} \sum_{n_y=1}^{n_{y\max}} \sum_{n_z=1}^{n_{z\max}} N'_{100}(E, n_y, n_z, \eta_g) H(E - E'_{100HDNW}) \tag{14.85}$$

where

$$N_{100}(E, n_y, n_z, \eta_g) = \left(\frac{\sqrt{2m_c}}{\hbar}\right) \left[- \left\{ \frac{\hbar^2(n_z\pi/d_z)^2}{2m_c} + \frac{\hbar^2(n_y\pi/d_y)^2}{2m_c} \right\} + I_{36}(E, \eta_g) \right]^{1/2}$$

The sub-band energy ($E'_{100HDNW}$) in this case can be expressed as

$$\frac{\hbar^2(n_z\pi/d_z)^2}{2m_c} + \frac{\hbar^2(n_y\pi/d_y)^2}{2m_c} = I_{36}(E'_{100HDNW}, \eta_g) \tag{14.86}$$

The EEM in this case is given by

$$m^*(E_{F1HDNW}, \eta_g) = m_c [I'_{36}(E_{F1HDNW}, \eta_g)] \tag{14.87}$$

The 1D electron statistics in HD NWFETs of GaSb under the condition of extreme carrier degeneracy can be written as

$$n_{01D} = \frac{2g_v}{\pi} \sum_{n_y=1}^{n_{y\max}} \sum_{n_z=1}^{n_{z\max}} [N_{100}(E_{F1HDNW}, n_y, n_z, \eta_g)] \tag{14.88}$$

Thus, using (14.88) and (14.5d) we can study the QC in this case.

14.2.12 *The QC in HD NWFETs of II–V materials*

In the absence of band-tailing, the 1D hole energy spectrum in this case assumes the form [28]

$$E = A_{10}\left(\frac{n_x\pi}{d_x}\right)^2 + A_{11}\left(\frac{n_y\pi}{d_y}\right)^2 + A_{12}k_z^2 + A_{13}\left(\frac{n_x\pi}{d_x}\right)$$
$$\pm\left[\left(A_{14}\left(\frac{n_x\pi}{d_x}\right)^2 + A_{15}\left(\frac{n_y\pi}{d_y}\right)^2 + A_{16}k_z^2 + A_{17}\left(\frac{n_x\pi}{d_x}\right)\right)^2\right.$$
$$\left. + A_{18}\left(\frac{n_y\pi}{d_y}\right)^2 + A_{19}^2\right]^{1/2} \quad (14.89)$$

where from A_{10} to A_{19} are energy band cons tan ts.

The DR of the 1D holes in in HD NWFETs of II–V compounds can be expressed as

$$\gamma_3(E,\eta_g) = A_{10}\left(\frac{n_x\pi}{d_x}\right)^2 + A_{11}\left(\frac{n_y\pi}{d_y}\right)^2 + A_{12}k_z^2 + A_{13}\left(\frac{n_x\pi}{d_x}\right)$$
$$\pm\left[\left(A_{14}\left(\frac{n_x\pi}{d_x}\right)^2 + A_{15}\left(\frac{n_y\pi}{d_y}\right)^2 + A_{16}k_z^2 + A_{17}\left(\frac{n_x\pi}{\bar{d}_x}\right)^2\right)^2\right.$$
$$\left. + A_{18}\left(\frac{n_y\pi}{\bar{d}_y}\right)^2 + A_{19}^2\right]^{1/2} \quad (14.90)$$

where the numerical values of the energy band constants are given in Table 14.1.

Equation (14.90) can be expressed as

$$k_z = \Delta_{27}(E,\eta_g,n_x,n_y) \quad (14.91)$$

where

$$\Delta_{27}(E,\eta_g,n_x,n_y) = [(A_{12}^2 - A_{16}^2)^{-1}[\Delta_{21}(E,\eta_g,n_x,n_y)A_{12}$$
$$+ A_{16}\Delta_{22}(n_x,n_y)] + [[\Delta_{21}(E,\eta_g,n_x,n_y)A_{12}$$
$$+ A_{16}\Delta_{22}(n_x,n_y)]^2 - (A_{12}^2 - A_{16}^2)\Delta_{25}(n_x,n_y,E,\eta_g)]^{1/2}$$

$$\Delta_{21}(E,\eta_g,n_x,n_y) = \left[\gamma_3(E,\eta_g) - A_{10}\left(\frac{n_x\pi}{d_x}\right)^2 - A_{11}\left(\frac{n_y\pi}{d_y}\right)^2 - A_{13}\left(\frac{n_x\pi}{d_x}\right)\right],$$

$$\Delta_{22}(n_x,n_y) = \left[A_{14}\left(\frac{n_x\pi}{d_x}\right)^2 + A_{15}\left(\frac{n_y\pi}{d_y}\right)^2 + A_{17}\left(\frac{n_x\pi}{d_x}\right)\right],$$

$$\Delta_{25}(n_x,n_y,E,\eta_g) = [\Delta_{21}^2(E,\eta_g,n_x,n_y) - \Delta_{24}(n_x,n_y)],$$

$$\Delta_{24}(n_x,n_y) = [\Delta_{22}^2(E,\eta_g,n_x,n_y) + \Delta_{23}(n_y)] \quad \text{and}$$

$$\Delta_{23}(n_y) = \left[A_{18}\left(\frac{n_y\pi}{d_y}\right)^2 + A_{19}^2\right]$$

The DOS function in this case can be written as

$$N_{1DHD\Gamma}(E,\eta_g) = \frac{g_v}{\pi}\sum_{n_x=1}^{n_{x\max}}\sum_{n_y=1}^{n_{y\max}}\Delta_{27}'(E,n_x,n_y,\eta_g)H(E - E_{200HDNW}) \tag{14.92}$$

In (14.92), $E_{200HDNW}$ is the sub-band energy and in this case can be expressed as

$$0 = \Delta_{27}(E_{200HDNW},n_x,n_y,\eta_g) \tag{14.93}$$

The EEM in this case is given by

$$m^*(E_{F1HDNW},\eta_g,n_x,n_y) = \frac{\hbar^2}{2}\frac{\partial}{\partial E}[\Delta_{27}^2(E_{F1HDNW},\eta_g,n_x,n_y)] \tag{14.94}$$

The carrier statistics in HD NWFETs of II–V materials under the condition of extreme carrier degeneracy in this case can be written as

$$n_{1D} = \left(\frac{g_v}{\pi}\right)\sum_{n_x=1}^{n_{x\max}}\sum_{n_y=1}^{n_{y\max}}[\Delta_{27}(E_{F1HDNW},n_x,n_y,\eta_g)] \tag{14.95}$$

Thus, using (14.95) and (14.5d) we can study the QC in this case.

14.3 Summary and Conclusions

In Section 14.2.1, we have formulated the electron statistics in HDNWFETs of tetragonal and nonlinear optical materials as given by (14.5b) and by combining it with (14.5d) we can study the QC in this case. In Section 14.2.2, we have formulated the electron statistics in HDNWFETs of

III–V, ternary and quaternary materials in accordance with the three-band model of Kane, two-band model of Kane, isotropic parabolic energy bands together with the models of Stillman *et al.* and Palik *et al.* as given by (14.10), (14.16), (14.21), (14.25) and (14.29) and by combining them with (14.5d) we can study the QC for each model in this case. In Section 14.2.3, we have formulated the electron statistics in HDNWFETs of II–VI materials as given by (14.34) and by combining it with (14.5d) we can study the QC in this case.

In Section 14.2.4, we have formulated the electron statistics in HDNWFETs of IV–VI materials in accordance with the models of together with the models of Dimmock and Bangert and Kastner as given by (14.40), and (14.46) and by combining them with (14.5d) we can study the QC for each model in this case. In Section 14.2.5, we have formulated the electron statistics in HDNWFET of stressed Kane type as given by (14.53) and by combining it with (14.5d) we can study the QC in this case. In Section 14.2.6, we have formulated the electron statistics in HDNWFETs of Te as given by (14.58) and by combining it with (14.5d) we can study the QC in this case. In Section 14.2.7, we have formulated the electron statistics in HDNWFETs of GaP as given by (14.63) and by combining it with (14.5d) we can study the QC in this case. In Section 14.2.8, we have formulated the electron statistics in HDNWFETs of $PtSb_2$ as given by (14.68) and by combining it with (14.5d) we can study the QC in this case. In Section 14.2.9, we have formulated the electron statistics in HDNWFETs of Bi_2Te_3 as given by (14.73) and by combining it with (14.5d) we can study the QC in this case.

In Section 14.2.10, we have formulated the electron statistics in HDNWFETs of Ge in accordance with the models of together with the models of Cardona *et al.* and Wang and Ressler as given by (14.78) and (14.83) and by combining them with (14.5d) we can study the QC for each model in this case. In Section 14.2.11, we have formulated the electron statistics in HDNWFETs of GaSb as given by (14.88) and by combining it with (14.5d) we can study the QC in this case. In Section 14.2.11, we have formulated the electron statistics in HDNWFETs of II–V materials as given by (14.95) and by combining it with (14.5d) we can study the QC in this case.

We have not given the graphs of the QCs for 2D HD NWFETs made of many materials as discussed here since we feel that our readers will perform the intricate numerical computation to obtain the oscillatory plots for better assessment and joy of understanding. It is the band structure that changes in a fundamental way and consequently all the physical properties of all the electronic materials changes radically leading to new physical concepts.

14.4 Open Research Problems

(R14.1) Investigate the QC in HDNWFETs of the materials in the presence of Gaussian, exponential, Kane, Halperian, Lax and Bonch-Burevich types of band tails for all systems whose unperturbed carrier energy spectra are defined in R1.1. Draw the corresponding graphs and explain the results.

(R14.2) Investigate the QC in HDNWFETs of the negative refractive index, organic, magnetic and other advanced optical materials under strain. Draw the corresponding graphs and explain the results.

(R14.3) Investigate QC in HDNWFETs of the negative refractive index, organic, magnetic and other advanced optical materials in the presence of an arbitrarily oriented alternating electric field. Draw the corresponding graphs and explain the results.

(R14.4) Investigate the QC in HDNWFETs of the materials whose unperturbed carrier energy spectra are defined in R1.1. Draw the corresponding graphs and explain the results.

(R14.5) Investigate the QC in HDNWFETs of all the materials of this chapter in the presence of finite potential wells. Draw the corresponding graphs and explain the results.

(R14.6) Investigate the QC in HDNWFETs of all the materials of this chapter in the presence of parabolic potential wells. Draw the corresponding graphs and explain the results.

(R14.7) Investigate the QC in HDNWFETs of all the materials of negative refractive index and other advanced optical materials in the presence of an arbitrarily oriented alternating electric field and non-uniform light waves and in the presence of strain. Draw the corresponding graphs and explain the results.

(R14.8) Investigate the QC in HDNWFETs of all the materials of the advanced optical materials in the presence of an arbitrarily oriented alternating electric field in the presence of strain. Draw the corresponding graphs and explain the results.

(R14.9) Investigate the QC in HDNWFETs for all the problems of (R1.1) in the presence of arbitrarily oriented magnetic field. Draw the corresponding graphs and explain the results.

(R14.10) Investigate the QC in HDNWFETs for all the problems of (R1.1) in the presence of alternating electric field. Draw the corresponding graphs and explain the results.

(R14.11) Investigate the QC in HDNWFETs for all the problems of (R1.1) in the presence of alternating magnetic field. Draw the corresponding graphs and explain the results.

(R14.12) Investigate the QC in HDNWFETs for all the problems of (R1.1) in the presence of crossed electric field and quantizing magnetic fields. Draw the corresponding graphs and explain the results.

(R14.13) Investigate the QC in HDNWFETs for all the problems of (R1.1) in the presence of crossed alternating electric field and alternating quantizing magnetic fields. Draw the corresponding graphs and explain the results.

(R14.14) Investigate the QC in HDNWFETs of the negative refractive index, organic and magnetic materials. Draw the corresponding graphs and explain the results.

(R14.15) Investigate the QC in HDNWFETs of the magnetic materials in the presence of alternating time-dependent magnetic field. Draw the corresponding graphs and explain the results.

(R14.16) Investigate all the appropriate problems of this chapter by including the many body, image force, broadening and hot carrier effects, respectively. Draw the corresponding graphs and explain the results.

(R14.17) Investigate all the appropriate problems of this chapter by removing all the mathematical approximations and establishing the respective appropriate uniqueness conditions. Draw the corresponding graphs and explain the results.

References

1. P. Harrison, *Quantum Wells, Wires and Dots*, John Wiley and Sons, Ltd, (2002); B. K. Ridley, *Electrons and Phonons in Semiconductors Multilayers* (Cambridge University Press, Cambridge, 1997); G. Bastard, *Wave Mechanics Applied to Semiconductor Heterostructures*, Halsted; Les Ulis, *Les Editions de Physique*, New York (1988); V. V. Martin, A. A. Kochelap and M. A. Stroscio, *Quantum Heterostructures* (Cambridge University Press, Cambridge 1999).
2. C. S. Lent and D. J. Kirkner, *J. Appl. Phys.* **67**, 6353 (1990); F. Sols, M. Macucci, U. Ravaioli and K. Hess, *Appl. Phys. Lett.* **54**, 350 (1980).

3. C. S. Kim, A. M. Satanin, Y. S. Joe and R. M. Cosby, *Phys. Rev. B* **60**, 10962 (1999).
4. S. Midgley and J. B. Wang, *Phys. Rev. B* **64**, 153304 (2001).
5. T. Sugaya, J. P. Bird, M. Ogura, Y. Sugiyama, D. K. Ferry and K. Y. Jang, *Appl. Phys. Lett.* **80**, 434 (2002).
6. B. Kane, G. Facer, A. Dzurak, N. Lumpkin, R. Clark, L. PfeiKer and K. West, *Appl. Phys. Lett.* **72**, 3506 (1998).
7. C. Dekker, *Phys. Today,* **52**, 22 (1999).
8. A. Yacoby, H. L. Stormer, N. S. Wingreen, L. N. Pfeiffer, K. W. Baldwin and K. W. West, *Phys. Rev. Lett.* **77**, 4612 (1996).
9. Y. Hayamizu, M. Yoshita, S. Watanabe, H. A. L. PfeiKer and K. West, *Appl. Phys. Lett.* **81**, 4937 (2002).
10. S. Frank, P. Poncharal, Z. L. Wang and W. A. Heer, *Science* **280**, 1744 (1998).
11. I. Kamiya, I. I. Tanaka, K. Tanaka, F. Yamada, Y. Shinozuka and H. Sakaki, *Physica E* **13**, 131 (2002).
12. A. K. Geim, P. C. Main, N. LaScala, L. Eaves, T. J. Foster, P. H. Beton, J. W. Sakai, F. W. Sheard, M. Henini, G. Hill *et. al.*, *Phys. Rev. Lett.* **72**, 2061 (1994).
13. A. S. Melinkov and V. M. Vinokur, *Nature* **415**, 60(2002).
14. K. Schwab, E. A. henriksen, J. M. Worlock and M. L. Roukes, *Nature* **404**, 974 (2000).
15. L. Kouwenhoven, *Nature* **403**, 374 (2000).
16. S. Komiyama, O. Astafiev, V. Antonov and H. Hirai, *Nature* **403**, 405 (2000).
17. E. Paspalakis, Z. Kis, E. Voutsinas and A. F. Terziz, *Phys. Rev. B* **69**, 155316 (2004).
18. J. H. Jefferson, M. Fearn, D. L. J. Tipton and T. P. Spiller, *Phys. Rev. A* **66**, 042328 (2002).
19. J. Appenzeller, C. Schroer, T. Schapers, A. Hart, A. Froster, B. Lengeler and H. Luth, *Phys. Rev. B* **53**, 9959 (1996).
20. J. Appenzeller and C. Schroer, *J. Appl. Phys.* **87**, 31659 (2002).
21. P. Debray, O. E. Raichev, M. Rahman, R. Akis and W. C. Mitchel, *Appl. Phys. Lett.* **74**, 768 (1999).
22. P. M. Solomon, *Proc. IEEE,* **70**, 489 (1982); T. E. Schlesinger and T. Kuech, *Appl. Phys. Lett.* **49**, 519 (1986).
23. D. Kasemet, C. S. Hong, N. B. Patel and P. D. Dapkus, *Appl. Phys. Lett.* **41**, 912 (1982); K. Woodbridge, P. Blood, E. D. Pletcher and P. J. Hulyer, *Appl. Phys. Lett.* **45**, 16 (1984); S. Tarucha and H. O. Okamoto, *Appl. Phys. Lett.* **45**, 16 (1984); H. Heiblum, D. C. Thomas, C. M. Knoedler, and M. I Nathan, *Appl. Phys. Lett.* **47**, 1105 (1985).
24. O. Aina, M. Mattingly, F. Y. Juan and P. K. Bhattacharyya, *Appl. Phys. Lett.* **50**, 43 (1987).
25. I. Suemune and L. A. Coldren, *IEEE, J. Quant. Electron.* **24**, 1178 (1988).
26. D. A. B. Miller, D. S. Chemla, T. C. Damen, J. H. Wood, A. C. Burrus, A. C. Gossard and W. Weigmann, IEEE, *J. Quant. Electron.* **21**, 1462 (1985).

27. J. S. Blakemore, *Semiconductor Statistics* (Dover, New York, 1987); K. P. Ghatak, S. Bhattacharya, S. K. Biswas, A. Deyand and A. K. Dasgupta, *Phys. Scr.*, **75**, 820 (2007).
28. K. P. Ghatak and M. Mitra, *Elastic Constants in Heavily Doped Low Dimensional Materials* (World Scientific, 2021).

Chapter 15

The QC in Quantized Transistors in the Presence of Strong Light Waves

Bleesed are those who can give without remembering but take without forgetting.

15.1 Introduction

With the advent of nanophotonics, there has been a considerable interest in studying the optical processes in semiconductors and their nanostructures [1]. It appears from the literature, that the investigations have been carried out on the assumption that the carrier energy spectra are invariant quantities in the presence of intense light waves, *which is not fundamentally true*. The physical properties of semiconductors in the presence of light waves which change the basic dispersion relation have been relatively less investigated in the literature [2–5]. In this chapter, we shall study the QC in MOSFETs, QWFETs, NWFETs, CNTFETs and their heavily doped counter parts under different physical conditions made of III–V, ternary and quaternary materials on the basis of newly formulated electron dispersion law in the presence of strong light waves. In Section 15.2.1, of the theoretical background 15.2, we have formulated the DR of the conduction electrons of III–V, ternary and quaternary materials in the presence of strong light waves whose unperturbed electron energy spectrum is described by the three- and two-band models of Kane together with parabolic energy bands.

In Section 15.2.2, we have studied the QC of the said compounds in the presence of strong external photo-excitation. In Section 15.2.3, we have studied the opto QC in the said materials under magnetic quantization. In Section 15.2.4, we have studied the opto QC in the presence of crossed electric and quantizing magnetic fields. In Section 15.2.5, we have studied the opto QC in QWFETs in Kane type materials. In Section 15.2.6, we have studied the opto QC in QWFETs in Kane type materials in the presence of a quantizing magnetic field. In Section 15.2.7, we have studied the opto QC in QWFETs in Kane type materials in the presence of crossed electric and quantizing magnetic fields. In Section 15.2.8, we have studied the opto QC in NWFETs. In Section 15.2.9, we have studied the opto QC in CNTFETs. In Section 15.2.10, we have formulated the dispersion relations in HD III–V and optoelectronic materials in the presence of strong light waves. We have studied the opto QC in HD MOSFETs, HD QWFETs and HD CNTFETs of all the cases as have been studied in Sections 15.2.2 up to 15.2.9 chronologically in Sections 15.2.11–15.2.18, respectively. Few transport-related electronic quantities have also been discussed in this context. Section 15.3 contains the result and discussions pertinent to this chapter. Section 15.4 presents five open research problems.

15.2 Theoretical Background

15.2.1 *The DRs in III–V and opto-electronic materials in the presence of strong light waves*

The Hamiltonian ($\hat{H}$) of an electron in the presence of light wave characterized by the vector potential $\vec{A}$ can be written following [3] as

$$\hat{H} = [|(\hat{p} + |e|\vec{A})|^2/2m] + V(\bar{r}) \tag{15.1}$$

in which, $\hat{p}$ is the momentum operator, $V(\bar{r})$ is the crystal potential and m is the free electron mass (15.1) can be expressed as

$$\hat{H} = \hat{H}_0' + \hat{H}' \tag{15.2}$$

where

$$\hat{H}_0' = \frac{\hat{p}^2}{2m} + V(\bar{r})$$

and

$$\hat{H}' = \frac{|e|}{2m}\bar{A}.\hat{p} \tag{15.3}$$

The perturbed Hamiltonian $\hat{H}'$ can be written as

$$\hat{H}' = \left(\frac{-i\hbar|e|}{2m}\right)(\bar{A}.\nabla) \tag{15.4}$$

where $i = \sqrt{-1}$ and $\hat{p} = -i\hbar\nabla$.

The vector potential $((\vec{A}))$ of the monochromatic light of plane wave can be expressed as

$$\vec{A} = A_0 \vec{\varepsilon}_s \cos(\vec{s}_0 \cdot \vec{r} - \omega t) \tag{15.5}$$

where A_0 is the amplitude of the light wave, $\vec{\varepsilon}_s$ is the polarization vector, $\vec{s}_0$ is the momentum vector of the incident photon, $\vec{r}$ is the position vector, ω is the angular frequency of light wave and t is the time scale. The matrix element of $\hat{H}'_{nl}$ between initial state, $\psi_1(\vec{q}, \vec{r})$ and final state $\psi_n(\vec{k}, \vec{r})$ in different bands can be written as

$$\hat{H}'_{nl} = \frac{|e|}{2m} \left\langle n\vec{k} | \vec{A} \bullet \hat{p} | l\vec{q} \right\rangle \tag{15.6}$$

Using (15.4) and (15.5), we can re-write (15.6) as

$$\hat{H}'_{nl} = \left(\frac{-i\hbar|e|A_0}{4m} \right) \vec{\varepsilon}_s \bullet \left[\left\{ \left\langle n\vec{k} | e^{(i\vec{s}_0 \cdot \vec{r})} \nabla | l\vec{q} \right\rangle e^{-i\omega t} \right\} \right.$$
$$\left. + \left\{ \left\langle n\vec{k} | e^{(i\vec{s}_0 \cdot \vec{r})} \nabla | l\vec{q} \right\rangle e^{i\omega t} \right\} \right] \tag{15.7}$$

The first matrix element of (15.7) can be written as

$$\left\langle n\vec{k} | e^{(i\vec{s}_0 \cdot \vec{r})} \nabla | l\vec{q} \right\rangle = \int e^{\left(i[\vec{q} + \vec{s}_0 - \vec{k}] \cdot \vec{r}\right)} i\vec{q} u_n^*(\vec{k} \cdot \vec{r}) u_l(\vec{q}, \vec{r}) d^3 r$$
$$+ \int e^{\left(i[\vec{q} + \vec{s}_0 - \vec{k}] \cdot \vec{r}\right)} u_n^*(\vec{k}, \vec{r}) \nabla u_l(\vec{q}, \vec{r}) d^3 r \tag{15.8}$$

The functions $u_n^* u_l$ and $u_n^* \nabla u_l$ are periodic. The integral over all space can be separated into a sum over unit cells times an integral over a single unit cell. It is assumed that the wave length of the electromagnetic wave is sufficiently large so that if $\vec{k}$ and $\vec{q}$ are within the Brillouin zone, $(\vec{q} + \vec{s}_0 - \vec{k})$ is not a reciprocal lattice vector.

Therefore, we can write (15.8) as

$$\left\langle n\vec{k} \left| e^{(i\vec{s}_0 - \vec{r})} \nabla \right| l\vec{q} \right\rangle = \left[\frac{(2\pi)^3}{\Omega} \right] \left\{ i\vec{q} \delta(\vec{q} + \vec{s}_0 - \vec{k}) \delta_{nl} \right.$$
$$\left. + \delta(\vec{q} + \vec{s}_0 - \vec{k}) u_n^* \int_{cell} (\vec{k}, \vec{r}) \nabla u_1(\vec{q}, \vec{r}) d^3 r \right\}$$
$$= \left[\frac{(2\pi)^3}{\Omega} \right] \left\{ \delta(\vec{q} + \vec{s}_0 - \vec{k}) \int_{cell} u_n^*(\vec{k}, \vec{r}) \nabla u_1(\vec{q}, \vec{r}) d^3 r \right\} \tag{15.9}$$

where Ω is the volume of the unit cell and $\int u_n^*(\vec{k}, \vec{r}) u_l(\vec{q}, \vec{r}) d^3 r = \delta(\vec{q} - \vec{k}) \delta_{nl} = 0$, since $n \neq l$.

The delta function expresses the conservation of wave vector in the absorption of light wave and $\vec{s}_0$ is small compared to the dimension of a typical Brillouin zone and we set $\vec{q}=\vec{k}$.

From (15.8) and (15.9), we can write,

$$\hat{H}'_{nl} = \frac{|e|A_0}{2m}\vec{\varepsilon}_s \bullet \hat{p}_{nl}(\vec{k})\delta(\vec{q}-\vec{k})\cos(\omega t) \tag{15.10}$$

where $\hat{p}_{nl}(\vec{k}) = -i\hbar\int u_n^*\nabla u_1 d^3t = \int u_n^*(\vec{k},\vec{r})\hat{p}u_l(\vec{k},\vec{r})d^3r$

Therefore, we can write

$$\hat{H}'_{nl} = \frac{|e|A_0}{2m}\vec{\varepsilon}_s \bullet \hat{p}_{nl}(\vec{k}) \tag{15.11}$$

where $\vec{\varepsilon}=\vec{\varepsilon}_s\cos\omega t$.

When a photon interacts with a semiconductor, the carriers (i.e., electrons) are generated in the bands which are followed by the inter-band transitions. For example, when the carriers are generated in the valence band, the carriers then make an inter-band transition to the conduction band. The transition of the electrons within the same band, i.e., $\hat{H}'_{nn} = \langle n\vec{k}|\hat{H}'|n\vec{k}\rangle$ is neglected. Because, in such a case, i.e., when the carriers are generated within the same bands by photons, are lost by recombination within the aforementioned band resulting in zero carriers.

Therefore,

$$\left\langle n\vec{k}|\hat{H}'|n\vec{k}\right\rangle = 0 \tag{15.12}$$

With $n = c$ stands for conduction band and $l = v$ stands for valance band, the energy equation for the conduction electron can approximately be written as

$$I_{11}(E) = \left(\frac{\hbar^2k^2}{2m_c}\right) + \frac{\left(\frac{|e|A_0}{2m}\right)^2\left\langle|\vec{\varepsilon}\bullet\hat{p}_{cv}(\vec{k})|^2\right\rangle_{av}}{E_c(\vec{k})-E_v(\vec{k})} \tag{15.13}$$

where $I_{11}(E) \equiv E(aE+1)(bE+1)/(cE+1), a \equiv 1/E_{g0}, E_{g0}$ is the un-perturbed band-gap, $b \equiv 1/(E_{g0}+\Delta)$, $c \equiv 1/(E_{g0}+2\Delta/3)$, and $\langle|\vec{\varepsilon}\bullet\hat{p}_{cv}(\vec{k})|^2\rangle_{av}$ represents the average of the square of the optical matrix element (OME).

For the three-band model of Kane, we can write,

$$\xi_{1k} = E_c(\vec{k}) - E_v(\vec{k}) = (E_{g0}^2 + E_{g0}\hbar^2k^2/m_r)^{1/2} \tag{15.14}$$

where m_r is the reduced mass and is given by $m_r^{-1} = (m_c)^{-1} + m_v^{-1}$, and m_v is the effective mass of the heavy hole at the top of the valance band in the absence of any field.

The doubly degenerate wave functions $u_1(\vec{k},\vec{r})$ and $u_2(\vec{k},\vec{r})$ can be expressed as [5]

$$u_1(\vec{k},\vec{r}) = a_{k+}[(is)\downarrow'] + b_{k+}\left[\frac{X' - iY'}{\sqrt{2}}\uparrow'\right] + c_{k+}[Z'\downarrow'] \tag{15.15}$$

and

$$u_2(\vec{k},\vec{r}) = a_{k-}[(is)\uparrow'] - b_{k-}\left[\frac{X' + iY'}{\sqrt{2}}\downarrow'\right] + c_{k-}[Z'\uparrow'] \tag{15.16}$$

s is the s-type atomic orbital in both unprimed and primed coordinates, $\downarrow'$ indicates the spin down function in the primed coordinates,

$$a_{k\pm} \equiv \beta[E_{g0} - (\gamma_{0k\pm})^2(E_{g0} - \delta')]^{1/2}(E_{g0} + \delta')^{-1/2}$$

$$\beta \equiv [(6(E_{g0} + 2\Delta/3)(E_{g0} + \Delta))/\chi]^{1/2}, \quad \chi \equiv (6E_{g0}^2 + E_{g0}\Delta + 4\Delta^2),$$

$$\gamma_{0k\pm} \equiv \left[\frac{(\xi_{1k} \mp E_{g0})}{2(\xi_{1k} + \delta')}\right]^{1/2}$$

$$\xi_{1k} = E_c(\vec{k}) - E_v(\vec{k}) = E_{g0}\left[1 + 2\left(1 + \frac{m_c}{m_v}\right)\frac{I_{11}(E)}{E_{g0}}\right]^{1/2},$$

$$\delta' \equiv (E_{g0}^2\Delta)(\chi)^{-1},$$

X', Y', and Z' are the p-type atomic orbitals in the primed coordinates, $\uparrow'$ indicates the spin-up function in the primed coordinates,

$$b_{k\pm} \equiv \rho\gamma_{0k\pm}, \quad \rho \equiv (4\Delta^2/3\chi)^{1/2}, \quad c_{k\pm} \equiv t\gamma_{0k\pm} \quad \text{and}$$

$$t \equiv [6(E_{g0} + 2\Delta/3)^2/\chi]^{1/2}$$

We can, therefore, write the expression for the optical matrix element (OME) as

$$OME = \hat{p}_{cv}(\vec{k}) = \langle u_1(\vec{k},\vec{r})|\hat{p}|u_2(\vec{k},r')\rangle \tag{15.17}$$

Since the photon vector has no interaction in the same band for the study of inter-band optical transition, we can therefore write

$$\langle S|\hat{p}|S\rangle = \langle X|\hat{p}|X\rangle = \langle Y|\hat{p}|Y\rangle = \langle Z|\hat{p}|Z\rangle = 0$$

and

$$\langle X|\hat{p}|Y\rangle = \langle Y|\hat{p}|Z\rangle = \langle Z|\hat{p}|X\rangle = 0$$

There are finite interactions between the conduction band (CB) and the valance band (VB) and we can obtain

$$\langle S|\hat{p}|X\rangle = \hat{i} \bullet \hat{p} = \hat{i} \bullet \hat{p}_x$$
$$\langle S|\hat{p}|Y\rangle = \hat{j} \bullet \hat{p} = \hat{j} \bullet \hat{p}_x$$
$$\langle S|\hat{p}|Z\rangle = \hat{k} \bullet \hat{p} = \hat{k} \bullet \hat{p}_x$$

where $\hat{i}, \hat{j}$ and $\hat{k}$ are the unit vectors along x, y and z axes, respectively.

It is well known that

$$\begin{bmatrix} \uparrow' \\ \downarrow' \end{bmatrix} = \begin{bmatrix} e^{-i\phi/2}\cos(\theta/2) & e^{i\phi/2}\sin(\theta/2) \\ -e^{-i\phi/2}\sin(\theta/2) & e^{i\phi/2}\cos(\theta/2) \end{bmatrix} \begin{bmatrix} \uparrow \\ \downarrow \end{bmatrix}$$

and

$$\begin{bmatrix} X' \\ Y' \\ Z' \end{bmatrix} = \begin{bmatrix} \cos\theta\cos\phi & \cos\theta\sin\phi & -\sin\theta \\ -\sin\phi & \cos\phi & 0 \\ \sin\theta\cos\phi & \sin\theta\sin\phi & \cos\theta \end{bmatrix} \begin{bmatrix} X \\ Y \\ Z \end{bmatrix}$$

Besides, the spin vector can be written as

$$\vec{S} = \frac{\hbar}{2}\vec{\sigma}$$

where

$$\sigma_x = \begin{bmatrix} 0 & 1 \\ 1 & 0 \end{bmatrix}, \quad \sigma_y = \begin{bmatrix} 0 & -i \\ i & 0 \end{bmatrix} \text{ and } \sigma_z = \begin{bmatrix} 1 & 0 \\ 0 & -1 \end{bmatrix}$$

From above, we can write

$$\begin{aligned}\hat{P}_{CV}(\vec{k}) &= \left\langle u_1(\vec{k},\vec{r})|\hat{p}|u_2(\vec{k},r')\right\rangle \\ &= \left\langle \left\{ a_{k_+}[(iS)\downarrow'] + b_{k_+}\left[\left(\frac{X'-iY'}{\sqrt{2}}\right)\uparrow'\right] + c_{k_+}[Z'\downarrow']\right\} |\hat{P}| \right. \\ &\quad \left. \times \left\{ a_{k_-}(iS)\uparrow' - b_{k_-}\left[\left(\frac{X'+iY'}{\sqrt{2}}\right)\downarrow' + c_{k_-}[Z'\uparrow']\right]\right\}\right\rangle\end{aligned}$$

Using the above relations, we get

$$\begin{aligned}\hat{p}_{CV}(\vec{k}) &= \left\langle u_1(\vec{k},\vec{r})|\hat{P}|u_2(\vec{k},r')\right\rangle \\ &= \frac{b_{k_+}a_{k_-}}{\sqrt{2}}\left\{\left\langle (X'-iY')|\hat{P}|iS\right\rangle \langle\uparrow'|\uparrow'\rangle\right\} \\ &\quad + c_{k_+}a_{k_-}\left\{\left\langle Z'|\hat{P}|iS\right\rangle \langle\downarrow'|\uparrow'\rangle\right\}\end{aligned}$$

$$-\frac{a_{k_+}b_{k_-}}{\sqrt{2}}\left\{\left\langle iS|\hat{P}|(X'+iY')\right\rangle \langle\downarrow'|\downarrow'\rangle\right\}$$

$$+a_{k_+}c_{k_-}\left\{\left\langle iS|\hat{P}|Z'\right\rangle \langle\downarrow'|\uparrow'\rangle\right\} \tag{15.18}$$

We can also write

$$\begin{aligned}\left\langle (X'-iY')|\hat{P}|iS\right\rangle &= \left\langle (X')|\hat{P}|iS\right\rangle - \left\langle (iY')|\hat{P}|iS\right\rangle \\ &= i\int u^*_{X+}\hat{P}S - \int -iu^*_{Y+}\hat{P}iu_x \\ &= i\left\langle X'|\hat{P}|iS\right\rangle - \left\langle Y'|\hat{P}|iS\right\rangle\end{aligned}$$

From the above relations, for X', Y' and Z', we get

$$|X'\rangle = \cos\theta\cos\phi|X\rangle + \cos\theta\sin\phi|Y\rangle - \sin\theta|Z\rangle$$

Thus

$$\langle X'|\hat{P}|S\rangle = \cos\theta\cos\phi\langle X'|\hat{P}|S\rangle + \cos\theta\sin\phi\langle Y|\hat{P}|S\rangle - \sin\theta\langle Z|\hat{P}|S\rangle = \hat{P}\hat{r}_1$$

where $\hat{r}_1 = \hat{\imath}\cos\theta\cos\phi + \hat{\jmath}\cos\theta\sin\phi - \hat{k}\sin\theta$

$$|Y'\rangle = -\sin\phi|X\rangle + \cos\phi|Y\rangle + 0|Z\rangle$$

Thus

$$\langle Y'|\hat{P}|S\rangle = -\sin\phi\langle X|\hat{P}|S\rangle + \cos\phi\langle Y|\hat{P}|S\rangle + 0\langle Z|\hat{P}|S\rangle = \hat{P}\hat{r}_2$$

where

$$\hat{r}_2 = -\hat{\imath}\sin\phi + \hat{\jmath}\cos\phi$$

so that

$$\langle (X'-iY')|\hat{P}|S\rangle - \hat{P}(i\hat{r}_1 - \hat{r}_2)$$

Thus,

$$\frac{a_{k_-}b_{k_+}}{\sqrt{2}}\left\langle (X'-iY')|\hat{P}|S\right\rangle \langle\uparrow'|\uparrow'\rangle = \frac{a_{k_-}b_{k_+}}{\sqrt{2}}(i\hat{r}_1 - \hat{r}_2)\langle\uparrow'|\uparrow'\rangle \tag{15.19}$$

Now since,

$$\left\langle iS|\hat{P}|(X'+iY')\right\rangle = i\left\langle S|\hat{P}|X'\right\rangle - \left\langle S|\hat{P}|Y'\right\rangle = \hat{P}(i\hat{r}_1 - \hat{r}_2)$$

We can write,

$$-\left[\frac{a_{k_+}b_{k_-}}{\sqrt{2}}\left\{\left\langle iS|\hat{P}|(X'+iY')\right\rangle\langle\downarrow'|\downarrow'\rangle\right\}\right]=\left[\frac{a_{k_+}b_{k_-}}{\sqrt{2}}\hat{P}(i\hat{r}_1-\hat{r}_2)\langle\downarrow'|\downarrow'\rangle\right] \tag{15.20}$$

Similarly, we get

$$|Z'\rangle=\sin\theta\cos\phi|X\rangle+\sin\theta\sin\phi|Y\rangle+\cos\theta|Z\rangle$$

So that,

$$\left\langle Z'|\hat{P}|S\right\rangle=i\left\langle Z'|\hat{P}|S\right\rangle-i\hat{P}\{\sin\theta\cos\phi\hat{i}+\sin\theta\sin\phi\hat{j}+\cos\theta\hat{k}\}=i\hat{P}\hat{r}_3$$

where $\hat{r}_3=\hat{\imath}\sin\theta\cos\phi+\hat{\jmath}\sin\theta\sin\phi+\hat{k}\cos\theta$.

Thus,

$$c_{k_+}a_{k_-}\left\langle Z'|\hat{P}|S\right\rangle\langle\downarrow'|\uparrow'\rangle=c_{k_+}a_{k_-}i\hat{P}\hat{r}_3\langle\downarrow'|\uparrow'\rangle \tag{15.21}$$

similarly, we can write,

$$c_{k_+}a_{k_-}\left\langle iS|\hat{P}|Z'\right\rangle\langle\downarrow'|\uparrow'\rangle=c_{k_-}a_{k_+}i\hat{P}\hat{r}_3\langle\downarrow'|\uparrow'\rangle \tag{15.22}$$

Therefore, we obtain

$$\begin{aligned}&\frac{a_{k_-}b_{k_+}}{\sqrt{2}}\left\{\left\langle(X'+iY')|\hat{P}|S\right\rangle\langle\uparrow'|\uparrow'\rangle\right\}\\&\quad-\frac{a_{k_-}b_{k_+}}{\sqrt{2}}\left\{\left\langle iS|\hat{P}|(X'+iY')\right\rangle\langle\downarrow'|\downarrow'\rangle\right\}\\&=\frac{\hat{P}}{\sqrt{2}}(-a_{k_+}b_{k_-}\langle\downarrow'|\downarrow'\rangle+a_{k_-}b_{k_+}\langle\uparrow'|\uparrow'\rangle)(i\hat{r}_2-\hat{r}_2)\end{aligned} \tag{15.23}$$

Also, we can write,

$$\begin{aligned}&c_{k_+}a_{k_-}\left\langle Z'|\hat{P}|iS\right\rangle\langle\downarrow'|\uparrow'\rangle+c_{k_-}a_{k_+}\left\langle iS|\hat{P}|Z'\right\rangle\langle\downarrow'|\uparrow'\rangle\\&=i\hat{P}(c_{k_+}a_{k_-}+c_{k_-}a_{k_+})\hat{r}_3\langle\downarrow'|\downarrow'\rangle\end{aligned} \tag{15.24}$$

Combining (15.23) and (15.24), we find

$$\begin{aligned}\hat{p}_{CV}(\vec{k})=&\frac{\hat{p}}{\sqrt{2}}(i\hat{r}_1-\hat{r}_2)\left\{(b_{k_+}a_{k_-})\langle\uparrow'|\uparrow'\rangle-(b_{k_-}a_{k_+})\langle\downarrow'|\downarrow'\rangle\right\}\\&+i\hat{P}\hat{r}_3(c_{k_+}a_{k_-}-c_{k_-}c_{k_+})\langle\downarrow'|\uparrow'\rangle\end{aligned} \tag{15.25}$$

From the above relations, we obtain,

$$\left.\begin{aligned}\uparrow'&=e^{-i\phi/2}\cos(\theta/2)\uparrow+e^{i\phi/2}\sin(\theta/2)\downarrow\\\downarrow'&=e^{-i\phi/2}\sin(\theta/2)\uparrow+e^{i\phi/2}\cos(\theta/2)\downarrow\end{aligned}\right\} \tag{15.26}$$

Therefore,

$$\begin{aligned}\langle\downarrow'|\uparrow'\rangle_x &= -\sin(\theta/2)\cos(\theta/2)\,\langle\uparrow|\uparrow\rangle_x + e^{-i\theta}\cos^2(\theta/2)\,\langle\downarrow|\uparrow\rangle_x \\ &\quad - e^{i\phi}\sin^2(\theta/2)\,\langle\uparrow|\downarrow\rangle_x + \sin(\theta/2)\cos(\theta/2)\,\langle\downarrow|\downarrow\rangle_x \end{aligned} \tag{15.27}$$

But we know from above that

$$\langle\uparrow|\uparrow\rangle_x = 0, \langle\downarrow|\uparrow\rangle = \frac{1}{2}, \quad \langle\downarrow|\uparrow\rangle_x = \frac{1}{2} \text{ and } \langle\downarrow|\downarrow\rangle_x = 0$$

Thus, we get

$$\begin{aligned}\langle\downarrow'|\uparrow'\rangle_x &= \frac{1}{2}[e^{-i\phi}\cos^2(\theta/2) - e^{i\phi}\sin^2(\theta/2)] \\ &= \frac{1}{2}[(\cos\phi - i\sin\phi)\cos^2(\theta/2) - (\cos\phi + i\sin\phi)\sin^2(\theta/2)] \\ &= \frac{1}{2}[\cos\phi\cos\theta - i\sin\phi] \end{aligned} \tag{15.28}$$

Similarly, we obtain

$$\langle\downarrow'|\uparrow\rangle_y = \frac{1}{2}\,[i\cos\phi + \sin\phi\cos\theta] \quad \text{and}$$

$$\langle\downarrow'|\uparrow\rangle_z = \frac{1}{2}\,[-\sin\theta]\,\langle\downarrow'|\uparrow'\rangle = \hat{\imath}\,\langle\downarrow'|\uparrow'\rangle_x + \hat{\jmath}\,\langle\downarrow'|\uparrow'\rangle_y + \hat{k}\,\langle\downarrow'|\uparrow'\rangle$$

Therefore,

$$\begin{aligned}&\frac{1}{2}\left\{(\cos\theta\cos\phi - i\sin\phi)\hat{\imath} + (i\cos\phi + \sin\phi\cos\theta)\hat{\jmath} - \sin\theta\hat{k}\right\} \\ &= \frac{1}{2}\left[\left\{(\cos\theta\cos\phi)\hat{\imath} + (\sin\phi\cos\theta)\hat{\jmath} - \sin\theta\hat{k}\right\} + i\left\{-\hat{\imath}\sin\phi + \hat{\jmath}\cos\phi\right\}\right] \\ &= \frac{1}{2}[\hat{r}_1 + i\hat{r}_2] = -\frac{1}{2}i\,[i\hat{r}_1 - \hat{r}_2]\end{aligned}$$

Similarly, we can write

$$\langle\uparrow'|\uparrow'\rangle = \frac{1}{2}\left[\hat{\imath}\sin\theta\cos\phi + \hat{\jmath}\sin\theta\sin\phi + \hat{k}\cos\theta\right] = \frac{1}{2}\hat{r}_3 \quad \text{and}$$

$$\langle\downarrow'|\downarrow'\rangle = -\frac{1}{2}\hat{r}_3$$

Using the above results, we can write

$$\begin{aligned}\hat{p}_{CV}(\vec{k}) &= \frac{\hat{p}}{\sqrt{2}}(i\hat{r}_2 - \hat{r}_2)\left\{(a_{k_-}b_{k_+})\langle\uparrow'|\uparrow'\rangle - (b_{k_-}a_{k_+})\langle\downarrow'|\downarrow'\rangle\right\} \\ &\quad + i\hat{P}\hat{r}_3\left\{(c_{k_+}a_{k_-} - c_{k_-}a_{k_+})\langle\downarrow'|\uparrow'\rangle\right\}\end{aligned}$$

$$= \frac{\hat{P}}{2}\hat{r}_3(i\hat{r}_1 - \hat{r}_2)\left\{\left(\frac{a_{k_-}b_{k_+}}{\sqrt{2}} + \frac{b_{k_-}a_{k_+}}{\sqrt{2}}\right)\right\}$$

$$+ \frac{\hat{P}}{2}\hat{r}_3\,(i\hat{r}_2 - \hat{r}_2)\,\{(c_{k_+}a_{k_-} + c_{k_-}c_{k_+})\}$$

Thus,

$$\hat{p}_{CV}(\vec{k}) = \frac{\hat{p}}{2}\hat{r}_3(i\hat{r}_2 - \hat{r}_2)\left\{a_{k_+}\left(\frac{b_{k_-}}{\sqrt{2}} + c_{k_-}\right) + a_{k_-}\left(\frac{b_{k_+}}{\sqrt{2}} + c_{k_+}\right)\right\} \quad (15.29)$$

we can write that,

$$|\hat{r}_1| = |\hat{r}_2| = |\hat{r}_2| = 1, \quad \text{also,}$$

$$\hat{P}\hat{r}_3 = \hat{P}_x \sin\theta\cos\phi\hat{i} + \hat{P}_y \sin\theta\sin\phi\hat{j} + \hat{P}_z\cos\theta\hat{k}$$

where $\hat{P} = \langle S|P|X\rangle = \langle S|P|Y\rangle = \langle S|P|Z\rangle$,

$$\langle S|P'|X\rangle = \int u_C^*\,(0,\vec{r})\,\hat{P}u_{VX}\,(0,\vec{r})\,d^3r = \hat{P}_{CVX}(0) \quad \text{and}$$

$$\langle S|\hat{P}|Z\rangle = \hat{P}_{CVZ}(0)$$

Thus, $\hat{P} = \hat{P}_{CVX}(0) = \hat{P}_{CVY}(0) = \hat{P}_{CVZ}(0) = \hat{P}_{CV}(0)$
where $\hat{P}_{CV}(0) = \int u_C^*\,(0,\vec{r})\hat{P}_{u_V}(0,\vec{r})d^3r \equiv \hat{P}$.

For a plane polarized light wave, we have the polarization vector $\vec{\varepsilon}_s = \hat{k}$, when the light wave vector is traveling along the z-axis. Therefore, for a plane polarized light-wave, we have considered $\vec{\varepsilon}_s = \hat{k}$.

Then, from (15.29) we get

$$(\vec{\varepsilon} \bullet \hat{p}_{CV}(\vec{k})) = \vec{k} \bullet \frac{\hat{P}}{2}\hat{r}_3(i\hat{r}_1 - \hat{r}_2)[A(\vec{k}) + B(\vec{k})]\cos\omega t \quad (15.30)$$

and

$$\left.\begin{aligned} A(\vec{k}) &= a_{k_-}\left(\frac{b_{k_+}}{\sqrt{2}} + c_{k_+}\right) \\ B(\vec{k}) &= a_{k_+}\left(\frac{b_{k_-}}{\sqrt{2}} + c_{k_-}\right) \end{aligned}\right\} \quad (15.31)$$

Thus,

$$\left|\vec{\varepsilon} \bullet \hat{p}_{CV}(\vec{k})\right|^2 = \left|\vec{k} \bullet \frac{\hat{P}}{2}\hat{r}_3\right|^2 |i\hat{r}_1 - \hat{r}_2|^2[A(\vec{k}) + B(\vec{k})]^2\cos\omega t$$

$$= \frac{1}{4}\left|\hat{P}_z\cos\theta\right|^2[A(\vec{k}) + B(\vec{k})]^2\cos^2\omega t \quad (15.32)$$

So, the average value of $|\vec{\varepsilon} \bullet \hat{p}_{CV}(\vec{k})|^2$ for a plane polarized light-wave is given by

$$\left\langle \left|\vec{\varepsilon} \bullet \hat{p}_{CV}\left(\vec{k}\right)\right|^2 \right\rangle_{av}$$

$$= \frac{2}{4}\left|\hat{P}_z\right|^2 [A(\vec{k}) + B(\vec{k})]^2 \left(\int_0^{2\pi} d\phi \int_0^{\pi} \cos^2\theta \sin\theta d\theta\right)\left(\frac{1}{2}\right)$$

$$= \frac{2\pi}{3}\left|\hat{P}_z\right|^2 [A(\vec{k}) + B(\vec{k})]^2 \tag{15.33}$$

where

$$\left|\hat{P}_z\right|^2 = \left(\frac{1}{2}\right)\left|\vec{k} \bullet \hat{p}_{cv}(0)\right|^2$$

and

$$\left|\vec{k} \bullet \hat{p}_{cv}(0)\right|^2 = \frac{m^2}{4m_r}\frac{E_{g_0}(E_{g_0} + \Delta)}{(E_{g_0} + \frac{2}{3}\Delta)} \tag{15.34}$$

We shall express $A(\vec{k})$ and $B(\vec{k})$ in terms of constants of the energy spectra in the following way:

Substituting $a_{k\pm}, b_{k\pm}, c_{k\pm}$ and $\gamma_{0k\pm}$ in $A(\vec{k})$ and $B(\vec{k})$ in (1.31) we get

$$A(\vec{k}) = \beta\left(t + \frac{\rho}{\sqrt{2}}\right)\left\{\left(\frac{E_{g_0}}{E_{g_0} + \delta'}\right)\gamma_{0k_+}^2 - \gamma_{0k_+}^2\gamma_{0k_-}^2\left(\frac{E_{g_0} - \delta'}{E_{g_0} + \delta'}\right)\right\}^{1/2} \tag{15.35}$$

$$B(\vec{k}) = \beta\left(t + \frac{\rho}{\sqrt{2}}\right)\left\{\left(\frac{E_{g_0}}{E_{g_0} + \delta'}\right)\gamma_{0k_-}^2 - \gamma_{0k_+}^2\gamma_{0k_-}^2\left(\frac{E_{g_0} - \delta'}{E_{g_0} + \delta'}\right)\right\}^{1/2} \tag{15.36}$$

in which

$$\gamma_{0k_+}^2 = \frac{\xi_{1k} - E_{g_0}}{2(\xi_{1k} + \delta')} \equiv \frac{1}{2}\left[1 - \left(\frac{E_{g_0} + \delta'}{\xi_{1k} + \delta'}\right)\right] \quad \text{and}$$

$$\gamma_{0k_-}^2 = \frac{\xi_{1k} + E_{g_0}}{2(\xi_{1k} + \delta')} \equiv \frac{1}{2}\left[1 - \left(\frac{E_{g_0} - \delta'}{\xi_{1k} + \delta'}\right)\right]$$

Substituting $x \equiv \xi_{1k} + \delta'$ in $\gamma_{0k\pm}^2$, we can write,

$$A(\vec{k}) = \beta\left(t + \frac{\rho}{\sqrt{2}}\right)\left\{\left(\frac{E_{g_0}}{E_{g_0} + \delta'}\right)\frac{1}{2}\left(1 - \frac{E_{g_0} + \delta'}{x}\right)\right.$$

$$\left. - \frac{1}{4}\left(\frac{E_{g_0} - \delta'}{E_{g_0} + \delta'}\right)\left(1 - \frac{E_{g_0} + \delta'}{x}\right)\left(1 - \frac{E_{g_0} - \delta'}{x}\right)\right\}^{1/2}$$

Thus

$$A(\vec{k}) = \frac{\beta}{2}\left(t + \frac{\rho}{\sqrt{2}}\right)\left\{1 - \frac{2a_0}{x} + \frac{a_1}{x^2}\right\}^{1/2}$$

where

$$a_0 \equiv (E_{g_0}^2 + \delta'^2)(E_{g_0} + \delta')^{-1} \quad \text{and} \quad a_1 \equiv (E_{g_0} - \delta')^2.$$

After tedious algebra, one can show that

$$A(\vec{k}) = \frac{\beta}{2}\left(t + \frac{\rho}{\sqrt{2}}\right)(E_{g_0} - \delta')\left[\frac{1}{\xi_{1k} + \delta'} - \frac{1}{E_{g_0} + \delta'}\right]^{1/2} \times \left[\frac{1}{\xi_{1k} + \delta'}\frac{(E_{g_0} + \delta')}{(E_{g_0} - \delta')^2}\right]^{1/2} \tag{15.37}$$

Similarly, from (15.36), we can write,

$$B(\vec{k}) = \beta\left(t + \frac{\rho}{\sqrt{2}}\right)\left\{\left(\frac{E_{g_0}}{E_{g_0} + \delta'}\right)\frac{1}{2}\left(1 + \frac{E_{g_0} - \delta'}{x}\right) - \frac{1}{4}\left(\frac{E_{g_0} - \delta'}{E_{g_0} + \delta'}\right)\left(1 - \frac{E_{g_0} + \delta'}{x}\right)\left(1 + \frac{E_{g_0} - \delta'}{x}\right)\right\}^{1/2}$$

So that, finally we get,

$$B(\vec{k}) = \frac{\beta}{2}\left(t + \frac{\rho}{\sqrt{2}}\right)\left(1 + \frac{E_{g_0} - \delta'}{\xi_{1k} + \delta'}\right) \tag{15.38}$$

Using (15.33), (15.34), (15.37) and (15.38), we can write

$$\left(\frac{|e|A_0}{2m}\right)^2 \frac{\left\langle\left|\vec{\varepsilon} \bullet \hat{p}_{CV}\left(\vec{k}\right)\right|^2\right\rangle_{av}}{E_c(\vec{k}) - E_v(\vec{k})} = \left(\frac{|e|A_0}{2m}\right)^2 \frac{2\pi}{3}|\vec{k} \bullet \hat{p}_{CV}(0)|^2 \frac{\beta^2}{4}\left(t + \frac{\rho}{\sqrt{2}}\right)^2 \frac{1}{\xi_{1k}}\left\{\left(1 + \frac{E_{g_0} - \delta'}{\xi_{1k} + \delta'}\right) + (E_{g_0} - \delta')\left[\frac{1}{\xi_{1k} + \delta'} - \frac{1}{E_{g_0} + \delta'}\right]^{1/2} \times \left[\frac{1}{\xi_{1k} + \delta'} - \frac{E_{g_0} + \delta'}{(E_{g_0} - \delta')^2}\right]^{1/2}\right\}^2 \tag{15.39}$$

Following Nag [5], it can be shown that

$$A_0^2 = \frac{I\lambda^2}{2\pi^2 c^3 \sqrt{\varepsilon_{sc}\varepsilon_0}} \tag{15.40}$$

where I is the light intensity of wavelength λ, ε_0 is the permittivity of free space and c is the velocity of light. Thus, the simplified electron energy spectrum in III–V, ternary and quaternary materials in the presence of light waves can approximately be written as

$$\frac{\hbar^2 k^2}{2m_c} = \beta_0(E, \lambda) \tag{15.41}$$

where

$$\beta_0(E,\lambda) \equiv [I_{11}(E) - \theta_0(E,\lambda)],$$

$$\theta_0(E,\lambda) \equiv \frac{|e|^2}{96 m_r \pi c^3} \frac{I\lambda^2}{\sqrt{\varepsilon_{sc}\varepsilon_0}} \frac{E_{g_0}(E_{g_0}+\Delta)}{(E_{g_0}+\frac{2}{3}\Delta)} \frac{\beta^2}{4}\left(t+\frac{\rho}{\sqrt{2}}\right)^2 \frac{1}{\phi_0(E)}$$
$$\times \left\{\left(1+\frac{E_{g_0}-\delta'}{\phi_0(E)+\delta'}\right) + (E_{g_0}-\delta')\left[\frac{1}{\phi_0(E)+\delta'} - \frac{1}{E_{g_0}+\delta'}\right]^{1/2}\right.$$
$$\left. \times \left[\frac{1}{\phi_0(E)+\delta'} - \frac{E_{g_0}+\delta'}{(E_{g_0}-\delta')^2}\right]^{1/2}\right\}^2$$

and

$$\phi_0(E) \equiv E_{g_0}\left(1+2\left(1+\frac{m_c}{m_v}\right)\frac{I_{11}(E)}{E_{g_0}}\right)^{1/2}$$

Thus, under the limiting condition $\vec{k} \to 0$, from (1.41), we observe that $E \neq 0$ and is positive. Therefore, in the presence of external light waves, the energy of the electron does not tend to zero when $\vec{k} \to 0$, where as for the un-perturbed three-band model of Kane, $I_{11}(E) = [\hbar^2 k^2/(2m_c)]$ in which $E \to 0$ for $\vec{k} \to 0$. As the conduction band is taken as the reference level of energy, therefore the lowest positive value of E for $k' \to 0$ provides the increased band gap (ΔE_g) of the semiconductor due to photon excitation. The values of the increased band gap can be obtained by computer iteration processes for various values of I and λ, respectively.

Special Cases:

For the two-band model of Kane, we have $\Delta \to 0$. Under this condition,

$$I_{11}(E) \to E(1+aE) = \frac{\hbar^2 k^2}{2m_c}.$$

Since, $\beta \to 1, t \to 1, \rho \to 0, \delta' \to 0$ for $\Delta \to 0$ from (15.41), we can write the energy spectrum of III–V, ternary and quaternary materials in the presence of external photo-excitation whose unperturbed conduction electrons obey the two-band model of Kane as

$$\frac{\hbar^2 k^2}{2m_c} = \omega_0(E, \lambda) \tag{15.42}$$

where $\omega_0(E, \lambda) \equiv E(1 + aE) - B_0(E, \lambda)$,

$$B_0(E, \lambda) \equiv \frac{|e|^2}{384\pi c^3 m_r} \frac{I\lambda^2 E_{g_0}}{\sqrt{\varepsilon_{sc}\varepsilon_0}} \frac{1}{\phi_1(E)} \times \left\{\left(1 + \frac{E_{g_0}}{\phi_0(E)}\right) + E_{g_0}\left[\frac{1}{\phi_1(E)} - \frac{1}{E_{g_0}}\right]\right\}^2,$$

$$\phi_1(E) \equiv E_{g_0}\left\{1 + \frac{2m_c}{m_r} aE(1 + aE)\right\}^{1/2}$$

For relatively wide band gap materials, one can write, $a \to 0, b \to 0, c \to 0$ and $I_{11}(E) \to E$.

Thus, from (15.42), we get,

$$\frac{\hbar^2 k^2}{2m_c} = \rho_0(E, \lambda) \tag{15.43}$$

where

$$\rho_0(E, \lambda) \equiv E - \frac{|e|^2 I\lambda^2}{96\pi C^3 m_r \sqrt{\varepsilon_{sc}\varepsilon_0}} \left[1 + \left(\frac{2m_c}{m_r}\right) aE\right]^{-3/2}$$

15.2.2 *The QC in 2D MOSFETs in the presence of strong light waves*

(a) Perturbed three-band model of Kane

In the presence of a surface electric field F_s along z-direction and perpendicular to the surface, the (15.41) assumes the form

$$\frac{\hbar^2 k^2}{2m_c} = \beta_0(E - |e|F_s z, \lambda) \tag{15.44}$$

where, for this chapter, E represents the electron energy as measured from the edge of the conduction band at the surface in the vertically upward direction.

Using (15.44) and (1.7), the 2D DR in 2D MOSFETs of III–V, ternary and quaternary semiconductors (whose unperturbed electrons obey the three-band model of Kane) operated under inversion mode in the presence of strong light waves under the condition of weak electric field limit, assumes the form

$$\beta_0(E,\lambda) = \frac{\hbar^2 k_s^2}{2m_c} + S_i[\frac{\hbar|e|F_s[\beta_0(E,\lambda)]'}{\sqrt{2m_c}}]^{2/3} \tag{15.45}$$

The EEM can be expressed as

$$m^*(E_{Fw15,1},i) = \hbar^2 k_s \left.\frac{\partial k_s}{\partial E}\right|_{E=E_{Fw15,1}} = m_c\,[P_{3L}(E_{Fw15,1},i)]| \tag{15.46}$$

where $E_{Fw15,1}$ is the Fermi energy under the weak electric field limit as measured from the edge of the conduction band at the surface in the vertically upward direction in the presence of light waves and is equal to

$$E_{Fw15,1} = eV_g - \frac{e^2 n_{2Dw} d_{ox}}{\varepsilon_{ox}} - E_{fb}$$

and

$$P_{3L}(E,i) \equiv \left\{[\beta_0(E,i)]' - \left\{\frac{2}{3}S_i\left[\frac{\hbar|e|F_s}{\sqrt{2m_c}}\right]^{2/3}\{[\beta_0(E,i)]'\}^{-1/3}[\beta_0(E,i)]''\right\}\right\}$$

Thus, one can observe that the EEM is a function of the sub-band index, surface electric field, the Fermi energy and the other spectrum constants due to the combined influence of E_g, Δ and λ.

The sub-band energy ($E_{n_{iw2L}}$) in this case can be obtained from (15.45) as

$$\beta_0(E_{n_{iw2L}},\lambda) = S_i\left[\frac{\hbar|e|F_s[\beta_0(E_{n_{iw2L}},\lambda)]'}{\sqrt{2m_c}}\right]^{2/3} \tag{15.47}$$

The 2D total density-of-states function in weak electric field limit can be expressed as

$$N_{2D_i}(E) = \frac{m_c g_v}{\pi\hbar^2}\sum_{i=0}^{i_{\max}}[P_{3L}(E,i)H(E-E_{n_{iw2L}})] \tag{15.48}$$

Using (15.48) and the occupation probability, the n_{2Dw} in the present case can be written as

$$n_{2Dw} = \frac{g_v m_c}{\pi\hbar^2}\sum_{i=0}^{i_{\max}}[P_{4wL}(E_{Fw15,1},i)] \tag{15.49}$$

where

$$P_{4wL}(E_{Fw15,1}, i) \equiv \left\{ \beta_0(E_{Fw15,1}, \lambda) - S_i \left[\frac{\hbar e F_s [\beta_0(E_{Fw15,1}, \lambda)]'}{\sqrt{2m_c}} \right]^{2/3} \right\}$$

Using (15.44) and (1.7), the 2D electron DR in 2D MOSFETs of III–V, ternary and quaternary semiconductors operated under inversion mode in the presence of light waves under the condition of strong electric field limit, assumes the form

$$\left[\beta_0(E, \lambda) - \left\{ \frac{|e| F_s \hbar}{\sqrt{2m_c}} \left(\frac{2\sqrt{2}(S_i)^{3/2}}{3} \right) \sqrt{[\beta_0(E, \lambda)]''} \right\} \right] = \frac{\hbar^2 k_s^2}{2m_c} \quad (15.50)$$

The EEM can be expressed as

$$m^*(E_{Fs15,1}, i) = \hbar^2 k_s \left. \frac{\partial k_s}{\partial E} \right|_{E=E_{Fs15,1}} = m_c \left[P_{5L}(E, i) \right] |_{E=E_{Fs15,1}} \quad (15.51)$$

where $E_{Fs15,1}$ is the Fermi energy under the strong electric field limit as measured from the edge of the conduction band at the surface in the presence of light waves, $E_{Fs15,1} = eV_g - \frac{e^2 n_{2Ds} d_{ox}}{\varepsilon_{ox}} - E_{fb}$ and

$$P_{5L}(E, i) \equiv \left\{ \{\beta_0(E, i)\}' - \left\{ \frac{|e| \, F_s \hbar}{\sqrt{2m_c}} \left(\frac{\sqrt{2}(S_i)^{3/2}}{3} \right) \right. \right.$$
$$\left. \left. \times \left([\beta_0(E, i)]'' \right)^{-1/2} [\beta_0(E, i)]''' \right\} \right\}.$$

Thus, one can observe that the EEM is a function of the sub-band index, surface electric field, the Fermi energy and the other spectrum constants due to the combined influence of E_g, Δ and λ.

The sub-band energy ($E_{n_{is2L}}$) in this case can be obtained from (15.50) as

$$\beta_0(E_{n_{is2L}}, \lambda) - \left\{ \frac{|e| F_s \hbar}{\sqrt{2m_c}} \left(\frac{2\sqrt{2}(S_i)^{3/2}}{3} \right) \sqrt{[\beta_0(E_{n_{is2L}}, \lambda)]''} \right\} = 0 \quad (15.52)$$

The total 2D DOS function under the strong electric field limit can be expressed as

$$N_{2D_i}(E) = \frac{m_c g_v}{\pi \hbar^2} \sum_{i=0}^{i_{\max}} [P_{5L}(E, i) H(E - E_{n_{is2L}})] \quad (15.53)$$

Using (15.53) and the Fermi–Dirac occupation probability factor, the n_{2Ds} in the present case under the conditions of strong electric field and extreme carrier degeneracy can be written as

$$n_{2Ds} = \frac{g_v m^*}{\pi\hbar^2} \sum_{i=0}^{i_{\max}} [P_{6sL}(E_{Fs15,1}, i)] \tag{15.54}$$

where

$$P_{6sL}(E_{Fs15,1}, i) \equiv \left\{\beta_0(E_{Fs15,1}, i) - \left[\frac{2\sqrt{2}}{3}(S_i)^{3/2}\frac{\hbar|e|F_s[\beta_0(E_{Fs15,1}, i)]''}{\sqrt{2m_c}}\right]\right\}.$$

Using (15.49), (15.54) and (1.4) in accordance with the perturbed three-band model of Kane, the QC per unit area for low (C_{3g} low) and high (C_{3g} high) electric field limits can respectively, be expressed as

$$\begin{aligned}C_{3g_{low}} = e^2 &\left[\sum_{i=0}^{i_{\max}} \left[\beta_0'(E_{Fw15,1}, \lambda) - \frac{2}{3}S_i\left(\frac{\hbar e F_s}{\sqrt{2m_c}}\right)^{\frac{2}{3}}\right.\right.\\
&\times \left.\left.\left[\beta_0'(E_{Fw15,1}, \lambda)^{-\frac{1}{3}}\right] \cdot \beta_0''(E_{Fw15,1}, \lambda)\right]\right]\\
&\times \left[\frac{\pi\hbar^2}{m_c} + \sum_{i=0}^{i_{\max}}\left[\left[\beta_0'(E_{Fw15,1}, \lambda)\frac{e^2 d_{ox}}{\varepsilon_{ox}}\right.\right.\right.\\
&+ \frac{2}{3}\frac{S_i}{N_s}\left(\frac{\hbar e^2 N_s \beta_0'(E_{Fw15,1}, \lambda)}{\varepsilon_{sc}\sqrt{2m_c}}\right)^{\frac{2}{3}}\\
&- \left.\left.\left.\left(\frac{e^2 d_{ox}}{\varepsilon_{ox}}\right)\left(\frac{2S_i\beta_0''(E_{Fw15,1}, \lambda)}{3\beta_0'(E_{Fw15,1}, \lambda)}\right)\left(\frac{\hbar e F_s \beta_0''(E_{Fw15,1}, \lambda)}{\sqrt{2m_c}}\right)^{\frac{2}{3}}\right]\right]\right]^{-1}\end{aligned} \tag{15.55}$$

and

$$\begin{aligned}C_{3g_{high}} = e^2 &\left[\sum_{i=0}^{i_{\max}}\left[\left(\beta_0'(E_{Fs15,1}, \lambda)\frac{e^2 d_{ox}}{\varepsilon_{ox}}\right.\right.\right.\\
&+ \left.\frac{2}{3}\frac{S_i}{N_s}\left(\frac{\hbar e^2 N_s \beta_0'(E_{Fs15,1}, \lambda)}{\varepsilon_{sc}\sqrt{2m_c}}\right)^{\frac{2}{3}}, \lambda\right)\\
&- \left.\left.\frac{N_s \hbar e^2 S_i^{\frac{3}{2}} \beta_0'''(E_{Fs15,1}, \lambda)}{3\varepsilon_{sc}\sqrt{m_c}\beta_0''(E_{Fs15,1}, \lambda)}\right]\right]\\
&\times \left[\frac{\pi\hbar^2}{m_c} + \sum_{i=0}^{i_{\max}}\left[\beta_0'(E_{Fs15,1}, \lambda)\frac{e^2 d_{ox}}{E_{ox}}\right.\right.\end{aligned}$$

$$+\frac{2}{3}\frac{e^2\hbar S_i^{\frac{3}{2}}}{\varepsilon_{sc}\sqrt{m_c}}(\beta_0''(E_{Fs15,1},\lambda))^{\frac{1}{2}}$$

$$-\frac{N_s\hbar e^2 S_i^{\frac{3}{2}}}{3\varepsilon_{sc}\sqrt{m_c}}\cdot\frac{e^2 d_{ox}}{\varepsilon_{ox}}\left(\frac{\beta_0'''(E_{Fs15,1},\lambda)}{\sqrt{\beta_0''(E_{Fs15,1},\lambda)}}\right)\Bigg]\Bigg]^{-1} \quad (15.56)$$

where

$$\beta_0(E,\lambda) = [I_{11}(E) - \theta(E,\lambda)],$$

$$I_{11}(E) = \left[\frac{E(E+E_{g0})(E+E_{g0}+\Delta)(E_g+\frac{2}{3}\Delta)}{E_{g0}(E_{g0}+\Delta)(E+E_{g0}+\frac{2}{3}\Delta)}\right]$$

$$\theta_0(E) = J_0\left[\frac{H^2(E,\lambda)}{\varphi_0(E)}\right],$$

$$J_0 = \left[\left(\frac{I_0(e\lambda)^2}{96m_r\pi c^3}\right)(\varepsilon_0\varepsilon_{sc})^{-\frac{1}{2}}\frac{E_{g0}(E_{g0}+\Delta)}{(E_{g0}+\frac{2}{3}\Delta)}\left(\frac{\beta^2}{4}\right)\times\left(\bar{t}+\frac{P}{\sqrt{2}}\right)^2\right],$$

$$\beta = \left[\frac{6(E_{g0}+\frac{2}{3}\Delta)(E_{g0}+\Delta)}{\chi}\right]^{\frac{1}{2}}, \quad \chi = [6E_{g0}^2 + 9\Delta E_{g0} + 4\Delta^2],$$

$$\bar{t} = \left[\frac{6\left(E_{g0}+\frac{2\Delta}{3}\right)^2}{\chi}\right], \quad P = [(4\Delta^2)/3\chi],$$

$$H(E,\lambda) = \left[\left(1+\frac{E_{g0}-\delta'}{\varphi_0(E)+\delta'}\right) + (E_{g0}-\delta')g_1(E)g_2(E)\right],$$

$$\delta' = \left(\frac{\Delta E_{g0}^2}{\chi}\right), \quad \varphi_0(E) = E_{g0}\left[1+2\left(1+\frac{m_c}{m_v}\right)\frac{I(E)}{E_{g0}}\right]^{\frac{1}{2}}$$

$$g_1^2(E) = \left[\frac{1}{(\varphi_0(E)+\delta')} - \frac{1}{(E_{g0}+\delta')}\right],$$

$$g_2^2(E) = \left[\frac{1}{\varphi_0(E)+\delta'} - \frac{(E_{g0}+\delta')}{(E_{g0}-\delta')^2}\right],$$

$$\beta_0'(E,\lambda) = \beta'(E,\lambda) = I_{11}'(E) - \theta_0'(E,\lambda),$$

$$I_{11}'(E) = I_{11}(E)\left[\frac{1}{E} + \frac{1}{E+E_{g0}} + \frac{1}{E+E_{g0}+\Delta} + \frac{1}{E+E_{g0}+\frac{2}{3}\Delta}\right]$$

$$\theta_0'(E,\lambda) = [\varphi_0(E)]^{-1}[-\theta_0(E,\lambda)\,\varphi_0'(E) + 2J_0 H(E,\lambda)H'(E,\lambda)],$$

$$\varphi_0'(E) = [\varphi_0(E)]^{-1}[E_{g0}(1+\frac{m_{sc}}{m_2})I_{11}'(E)],$$

$$H'(E,\lambda) = \left[-\frac{\varphi_0'(E)[E_{g0}-\delta']}{[E_{g0}+\delta']^2} + (E_{g0}-\delta')[g_1(E)g_2'(E)+g_2(E)g_1'(E)]\right]$$

$$g_1'(E) = -[2g_1(E)]^{-1}[\varphi_0'(E)] \times [\varphi_0(E)+\delta']^{-1},$$

$$g_2'(E) = -[2g_2(E)]^{-1}[\varphi_0'(E)] \times [\varphi_0(E)+\delta']^{-1},$$

$$\beta_0''(E,\lambda) = [I_{11}''(E) - \theta_0''(E,\lambda)]$$

$$I_{11}''(E) = \left[\frac{[I_{11}'(E)]^2}{I_{11}(E)} - I_{11}(E)\right.$$
$$\left.\times\left[\frac{1}{E^2} + \frac{1}{(E+E_{g0})^2} + \frac{1}{(E+E_{g0}+\Delta)^2} - \frac{1}{(E+E_{g0}+\frac{2}{3}\Delta)^2}\right]\right]$$

$$\theta_0''(E,\lambda) = [\varphi_0(E)]^{-1}\left[-\theta_0'(E,\lambda)\varphi_0'(E) - \theta_0(E,\lambda)\varphi_0''(E)\right.$$
$$+\frac{\theta_0(E,\lambda)[\varphi_0'(E)]^2}{\theta_0(E)} + 2J_0[H'(E,\lambda)]^2$$
$$\left.+2J_0H(E,\lambda)H''(E,\lambda) - \frac{2J_0H(E,\lambda)H'(E,\lambda)\varphi_0'(E)}{\theta_0(E)}\right]$$

$$\varphi_0''(E) = [\varphi_0(E)]^{-1}\left[E_{g0}\left(1+\frac{m_{sc}}{m_v}\right)I''(E) - [\varphi_0'(E)]^2\right],$$

$$H''(E,\lambda) = \left[-\frac{\varphi_0''(E)(E_{g0}-\delta')}{[\varphi_0(E)+\delta']^2} + \frac{2(\varphi_0'(E))^2(E_{g0}-\delta')}{[\varphi_0(E)+\delta']^3}\right.$$
$$\left.+[E_{g0}-\delta'] \times [g_1(E)g_2''(E) + g_2(E)g_1''(E) + 2g_1'(E)g_2'(E)]\right]$$

$$g_1''(E) = [2g_1(E)]^{-1}\left[-\frac{\varphi_0''(E)}{[\varphi_0(E)+\delta']} + \frac{[\varphi_0'(E)]^2}{[\varphi_0(E)+\delta']^2}\right],$$

$$g_2''(E) = [2g_2(E)]^{-1}\left[-\frac{\varphi_0''(E)}{[\varphi_0(E)+\delta']} + \frac{[\varphi_0'(E)]^2}{[\varphi_0(E)+\delta']^2}\right]$$

$$\beta_0'''(E,\lambda) = [I_{11}'''(E) - \theta_0'''(E,\lambda)],$$

$$I_{11}'''(E) = \left[\frac{2I_{11}'(E)I_{11}''(E)}{I_{11}(E)} - \frac{[I_{11}(E)]^3}{I_{11}^2(E)}\right.$$

$$\left.\times \left[\frac{1}{E^2} + \frac{1}{(E+E_{g0})^2} + \frac{1}{(E+E_{g0}+\Delta)^2} - \frac{1}{(E+E_{g0}+\frac{2}{3}\Delta)^2}\right]\right]$$

$$+ 2I(E)\left[\frac{1}{E^3} + \frac{1}{(E+E_{g0})^3}\right.$$

$$\left. + \frac{1}{(E+E_{g0}+\Delta)^3} - \frac{1}{(E+E_{g0}+\frac{2}{3}\Delta)^3}\right],$$

$$\theta_0'''(E,\lambda) = \left[-\frac{\theta_0''(E,\lambda)\varphi_0'(E)}{\varphi_0(E)} - \frac{2\theta_0'(E)\varphi_0''(E)}{\varphi_0(E)} + \frac{2\theta_0'(E)[\varphi_0'(E)]^2}{\varphi_0^2(E)}\right.$$

$$- \frac{\theta_0(E)\varphi_0'''(E)}{\varphi_0(E)} + \frac{3\theta_0(E)\varphi_0'(E)\varphi_0''(E)}{\varphi_0^2(E)} - \frac{2\theta_0(E)[\varphi_0'(E)]^3}{[\varphi_0(E)]^{33}}$$

$$+ \frac{6J_0H'(E,\lambda)H''(E,\lambda)}{\varphi_0(E)} - \frac{4J_0\varphi_0'(E)[H'(E,\lambda)]^2}{\varphi_0^2(E)}$$

$$+ \frac{2J_0H(E,\lambda)H''(E,\lambda)}{\varphi_0(E)} - \frac{4J_0H(E,\lambda)\varphi_0'(E)H''(E,\lambda)}{\varphi_0^2(E)}$$

$$+ \frac{4J_0H(E,\lambda)H'(E,\lambda)[\varphi_0'(E)][\varphi_0'(E)]]^2}{\varphi_0^3(E)}$$

$$\left. - \frac{2J_0H(E,\lambda)H'(E,\lambda)\varphi_0''(E)}{\varphi_0^2(E)}\right],$$

$$\varphi_0'''(E) = \left[-\frac{2\varphi_0'(E)\varphi_0''(E)}{\varphi_0(E)} + \frac{[\varphi_0'(E)]^3}{\varphi_0^2(E)}\right.$$

$$\left. + E_g\left(1+\frac{m_c}{m_v}\right)\left\{\frac{I_{11}'''(E)}{\varphi_0(E)} - \frac{I_{11}''(E)\varphi_0'(E)}{\varphi_0^2(E)}\right\}\right]$$

$$H'''(E,\lambda) = \left[-\frac{\varphi_0'''(E)(E_{g0}-\delta')}{[\varphi_0(E)+\delta']^2} + \frac{6\varphi_0'(E)\varphi_0''(E)(E_{g0}-\delta')}{[\varphi_0(E)+\delta']^3}\right.$$

$$- \frac{6[\varphi_0'(E)]^3(E_{g0}-\delta')}{[\varphi_0(E)+\delta']^4} + (E_{g0}-\delta')$$

$$\times [3g_1''(E)g_1''(E) + 3g_1'(E)g''(E)$$

$$\left. + g_1(E)g_2'''(E) + g_2(E)g_1'''(E)]\right]$$

$$g_1'''(E) = [2g_1(E)]^{-1}[g_3(E) - 6g_1'(E)g_1''(E)],$$

$$g_3(E) = \left[-\frac{\varphi_0'''(E)}{[[\varphi_0(E)+\delta']]} + \frac{3\varphi_0'(E)\varphi_0''(E)}{[\varphi_0(E)+\delta']^2} - \frac{2[\varphi_0'(E)]^3}{[\varphi_0(E)+\delta']^3}\right]$$

$$g_2'''(E) = [2g_1(E)]^{-1}\left[g_3(E) - 6g_2'(E)g_2''(E)\right]$$

(b) Perturbed two-band model of Kane

Again for this section, the (15.42) assumes the form

$$\frac{\hbar^2 k^2}{2m_c} = \omega_0(E - |e|F_s z, \lambda) \tag{15.57}$$

Using (15.57) and (1.7), the 2D DR in 2D MOSFETs of III–V, ternary and quaternary semiconductors (whose unperturbed electrons obey the two-band model of Kane) operated under inversion mode in the presence of strong light waves under the condition of weak electric field limit, assumes the form

$$\omega_0(E,\lambda) = \frac{\hbar^2 k_s^2}{2m_c} + S_i\left[\frac{\hbar|e|F_s[\omega_0(E,\lambda)]'}{\sqrt{2m_c}}\right]^{2/3} \tag{15.58}$$

The EEM can be expressed as

$$m^*(E_{Fw15,2}, i) = m_c[P_{3L2}(E_{Fw15,2}, i)]\Big| \tag{15.59}$$

where $E_{Fw15,2}$ is the Fermi energy under the weak electric field limit as measured from the edge of the conduction band at the surface in the vertically upward direction in the presence of light waves in this case and

$$P_{3L2}(E,i) \equiv \left\{[\omega_0(E,\lambda)]' - \left\{\frac{2}{3}S_i\left[\frac{\hbar|e|F_s}{\sqrt{2m_c}}\right]^{2/3}\{[\omega_0(E,\lambda)]'\}^{-1/3}[\omega_0(E,\lambda)]''\right\}\right\}.$$

Thus, one can observe that the EEM is a function of the sub-band index, surface electric field, the Fermi energy and the other spectrum constants due to the combined influence of E_g and λ.

The sub-band energy $(E_{n_{iw2L2}})$ in this case can be obtained from (15.58) as

$$\omega_0(E_{n_{iw2L2}},\lambda) = S_i\left[\frac{\hbar\,|e|\,F_s[\omega_0(E_{n_{iw2L2}},\lambda)]'}{\sqrt{2m_c}}\right]^{2/3} \tag{15.60a}$$

The 2D total density-of-states function in weak electric field limit can be expressed as

$$N_{2D_i}(E) = \frac{m_c g_v}{\pi\hbar^2}\sum_{i=0}^{i_{\max}}[P_{3L2}(E,i)H(E-E_{n_{iw2L2}})] \tag{15.60b}$$

Using (15.60b) and the occupation probability, the n_{2Dw} in the present case can be written as

$$n_{2Dw} = \frac{g_v m_c}{\pi\hbar^2}\sum_{i=0}^{i_{\max}}[P_{4wL2}(E_{Fw15,2},i)] \tag{15.61}$$

where

$$P_{4wL2}(E_{Fw15,2},i) \equiv \left\{\omega_0(E_{Fw15,2},\lambda) - S_i\left[\frac{\hbar e F_s[\omega_0(E_{Fw15,2},\lambda)]'}{\sqrt{2m_c}}\right]^{2/3}\right\}$$

Using the appropriate equations, the 2D DR in 2D MOSFETs of III–V, ternary and quaternary semiconductors (whose unperturbed electrons obey the two-band model of Kane) operated under inversion mode in the presence of strong light waves under the condition of strong electric field limit, assumes the form

$$\left[\omega_0(E,\lambda) - \left\{\frac{|e|F_s\hbar}{\sqrt{2m_c}}\left(\frac{2\sqrt{2}(S_i)^{3/2}}{3}\right)\sqrt{[\omega_0(E,\lambda)]''}\right\}\right] = \frac{\hbar^2 k_s^2}{2m_c} \tag{15.62}$$

The EEM can be expressed as

$$m^*(E_{Fs15,2},i) = m_c[P_{5L2}(E,i)]\big|_{E=E_{Fs15,2}} \tag{15.63}$$

where $E_{Fs15,2}$ is the Fermi energy under the strong electric field limit as measured from the edge of the conduction band at the surface in the presence of light waves in the present case and

$$P_{5L2}(E,i) \equiv \left\{\{\omega_0(E,\lambda)\}' - \left\{\frac{|e|F_s\hbar}{\sqrt{2m_c}}\left(\frac{\sqrt{2}(S_i)^{3/2}}{3}\right)([\omega_0(E,\lambda)]'')^{-1/2}[\omega_0(E,\lambda)]'''\right\}\right\}.$$

Thus, one can observe that the EEM is a function of the sub-band index, surface electric field, the Fermi energy and the other spectrum constants due to the combined influence of E_g and λ.

The sub-band energy ($E_{n_{is2L2}}$) in this case can be obtained from (15.62) as

$$\omega_0(E_{n_{is2L2}}, \lambda) - \left\{ \frac{|e|F_s\hbar}{\sqrt{2m_c}} \left(\frac{2\sqrt{2}(S_i)^{3/2}}{3} \right) \sqrt{[\omega_0(E_{n_{is2L2}}, \lambda)]''} \right\} = 0 \quad (15.64)$$

The total 2D density-of-states function under the strong electric field limit can be expressed as

$$N_{2D_i}(E) = \frac{m_c g_v}{\pi\hbar^2} \sum_{i=0}^{i_{\max}} [P_{5L2}(E, i) H(E - E_{n_{is2L2}})] \quad (15.65)$$

Using (15.65) and the Fermi–Dirac occupation probability factor, the n_{2Ds} in the present case under the strong electric field can be written as

$$n_{2Ds} = \frac{g_v m_c}{\pi\hbar^2} \sum_{i=0}^{i_{\max}} [P_{6sL2}(E_{Fs15,2}, i)] \quad (15.66)$$

where

$$P_{6sL2}(E_{Fs15,2}, i) \equiv \left\{ \omega_0(E_{Fs15,2}, \lambda) - \left[\frac{2\sqrt{2}}{3} (S_i)^{3/2} \frac{\hbar|e|F_s[\omega_0(E_{Fs15,2}, \lambda)]''}{\sqrt{2m_c}} \right] \right\}.$$

In this case, the forms of the (15.65) and (15.66) remain as it is where

$$\beta_0(E, \lambda), \ \beta_0'(E, \lambda), \ \beta_0''(E, \lambda) \quad \text{and} \quad \beta_0'''(E, \lambda)$$

are replaced by $\omega_0(E, \lambda)$, $\omega_0'(E, \lambda)$, $\omega_0''(E, \lambda)$ and $\omega_0'''(E, \lambda)$, respectively where, $\omega_0(E, \lambda) = [E(1 + \alpha E) - B_0(E, \lambda)]$, $\alpha = \frac{1}{E_{g0}}$, $B_0(E, \lambda) = \left[\frac{J_1}{\varphi_1^3(E)}\right]$,

$$J_1 = \left[\frac{4e^2 I_0 \lambda^2 E_{g0}^3}{384\pi c^3 m_r (\varepsilon_0 \varepsilon_{sc})^{1/2}} \right],$$

$$\varphi_1(E) = E_{g0} \left[1 + \frac{2m_c}{m_r} \frac{E(1 + \alpha E)}{E_{g0}} \right]^{1/2},$$

$$\omega_0'(E, \lambda) = [1 + 2\alpha E - B_0'(E, \lambda)],$$

$$B_0'(E,\lambda) = \left[-\frac{3J_1\varphi_1'(E)}{\varphi_1^4(E)}\right], \quad \varphi_1'(E) = [\varphi_1(E)]^{-1}\left[\frac{m_c}{m_r}\right][1+2\alpha E],$$
$$\omega_0''(E,\lambda) = [2\alpha - B_0''(E,\lambda)],$$
$$B_0''(E,\lambda) = [\varphi_1(E)]^{-5}[-3J_1\varphi_1(E)\varphi_1''(E) + 12J_1[\varphi_1'(E)]^2],$$
$$\varphi_1''(E) = \left[-\frac{[\varphi_1'(E)]^2}{\varphi_1(E)} + \frac{2\alpha m_c}{m_r}[\varphi_1(E)]^{-1}\right],$$
$$\omega_0'''(E_F,\lambda) = [-B_0'''(E,\lambda)],$$
$$B_0'''(E,\lambda) = [\varphi_1^5(E)]^{-1} \times \left[-3J_1\varphi_1(E)\varphi_1'''(E) + 36J_1\varphi_1'(E)\varphi_1''(E) - \frac{60J_1[\varphi_1'(E)]^3}{\varphi_1(E)}\right]$$
$$\varphi_1'''(E) = \left[-\frac{2\varphi_1'(E)\varphi_1''(E)}{\varphi_1(E)} + \frac{[\varphi_1'(E)]^3}{\varphi_1^2(E)} - \frac{2\alpha m_c\varphi_1'(E)}{m_r[\varphi_1(E)]^2}\right]$$

(c) Perturbed parabolic band model

Besides for this section, the (15.43) assumes the form

$$\frac{\hbar^2k^2}{2m_c} = \rho_0(E - |e|F_s z, \lambda) \tag{15.67}$$

Using (15.67) and (1.7), the 2D DR in 2D MOSFETs of III–V, ternary and quaternary semiconductors (whose unperturbed electrons obey the parabolic energy bands) operated under inversion mode in the presence of strong light waves under the condition of weak electric field limit, assumes the form

$$\rho_0(E,\lambda) = \frac{\hbar^2k_s^2}{2m_c} + S_i\left[\frac{\hbar|e|F_s[\rho_0(E,\lambda)]'}{\sqrt{2m^*}}\right]^{2/3} \tag{15.68}$$

The EEM can be expressed as

$$m^*(E_{Fw15,1}, i) = m_c[P_{3L1}(E_{Fw15,1}, i)] \tag{15.69}$$

$E_{Fw15,1}$ is the Fermi energy under the weak electric field limit as measured from the edge of the conduction band at the surface in the vertically upward direction in the presence of light waves in this case and

$$P_{3L1}(E,i) \equiv \left\{[\rho_0(E,i)]' - \left\{\frac{2}{3}S_i\left[\frac{\hbar|e|F_s}{\sqrt{2m_c}}\right]^{2/3}\{[\rho_0(E,i)]'\}^{-1/3}[\rho_0(E,i)]''\right\}\right\}.$$

Thus, one can observe that the EEM is a function of the sub-band index, surface electric field, the Fermi energy and the other spectrum constants due to the combined influence of E_g and λ.

The sub-band energy ($E_{n_{iw2L1}}$) in this case can be obtained from (15.69) as

$$\rho_0(E_{n_{iw2L1}}, \lambda) = S_i \left[\frac{\hbar|e|F_s[\rho_0(E_{n_{iw2L1}}, \lambda)]'}{\sqrt{2m_c}} \right]^{2/3} \tag{15.70a}$$

The 2D total density-of-states function in weak electric field limit can be expressed as

$$N_{2D_i}(E) = \frac{m_c g_v}{\pi\hbar^2} \sum_{i=0}^{i_{\max}} [P_{3L1}(E, i) H(E - E_{n_{iw2L1}})] \tag{15.70b}$$

Using (15.70b) and the occupation probability, the n_{2Dw} in the present case can be written as

$$n_{2Dw} = \frac{g_v m_c}{\pi\hbar^2} \sum_{i=0}^{i_{\max}} [P_{4wL1}(E_{Fw15,1}, i)] \tag{15.71}$$

where

$$P_{4wL1}(E_{Fw15,1}, i) \equiv \left\{ \rho_0(E_{Fw15,1}, \lambda) - S_i \left[\frac{\hbar e F_s[\rho_0(E_{Fw15,1}, \lambda)]'}{\sqrt{2m_c}} \right]^{2/3} \right\}$$

Using the appropriate equations, the 2D DR in 2D MOSFETs of III–V, ternary and quaternary semiconductors (whose unperturbed electrons obey the two-band model of Kane) operated under inversion mode in the presence of strong light waves under the condition of strong electric field limit, assumes the form

$$\left[\rho_0(E, \lambda) - \left\{ \frac{|e|F_s\hbar}{\sqrt{2m_c}} \left(\frac{2\sqrt{2}(S_i)^{3/2}}{3} \right) \sqrt{[\rho_0(E, \lambda)]''} \right\} \right] = \frac{\hbar^2 k_s^2}{2m_c} \tag{15.72}$$

The EEM can be expressed as

$$m^*(E_{Fs15,2}, i) = m_c[P_{5L2}(E, i)] \Big|_{E=E_{Fs15,2}} \tag{15.73}$$

where $E_{Fs15,2}$ is the Fermi energy under the strong electric field limit as measured from the edge of the conduction band at the surface in the presence of light waves in the present case and

$$P_{5L2}(E,i) \equiv \left\{ \{\rho_0(E,i)\}' - \left\{ \frac{|e|F_s\hbar}{\sqrt{2m_c}} \left(\frac{\sqrt{2}(S_i)^{3/2}}{3} \right) ([\rho_0(E,i)]'')^{-1/2} [\rho_0(E,i)]''' \right\} \right\}.$$

Thus, one can observe that the EEM is a function of the sub-band index, surface electric field, the Fermi energy and the other spectrum constants due to the combined influence of E_g and λ.

The sub-band energy ($E_{n_{is2L2}}$) in this case can be obtained from (15.72) as

$$\rho_0(E_{n_{is2L2}},\lambda) - \left\{ \frac{|e|F_s\hbar}{\sqrt{2m_c}} \left(\frac{2\sqrt{2}(S_i)^{3/2}}{3} \right) \sqrt{[\rho_0(E_{n_{is2L2}},\lambda)]''} \right\} = 0 \quad (15.74)$$

The total 2D density-of-states function under the strong electric field limit can be expressed as

$$N_{2D_i}(E) = \frac{m_c g_v}{\pi\hbar^2} \sum_{i=0}^{i_{\max}} [P_{5L2}(E,i)H(E - E_{n_{is2L2}})] \quad (15.75)$$

Using (15.75) and the Fermi–Dirac occupation probability factor, the n_{2Ds} in the present case under the strong electric field can be written as

$$n_{2Ds} = \frac{g_v m_c}{\pi\hbar^2} \sum_{i=0}^{i_{\max}} [P_{6sL2}(E_{Fs15,2},i)] \quad (15.76)$$

where

$$P_{6sL2}(E_{Fs15,2},i) \equiv \left\{ \rho_0(E_{Fs15,2},\lambda) - \left[\frac{2\sqrt{2}}{3}(S_i)^{3/2} \frac{\hbar|e|F_s[\rho_0(E_{Fs15,2},\lambda)]''}{\sqrt{2m_c}} \right] \right\}$$

In this case the forms of the (15.77) and (15.78) remain as it is where

$$\beta_0(E,\lambda),\ \beta_0'(E,\lambda),\ \beta_0''(E,\lambda) \text{ and } \beta_0'''(E,\lambda)$$

are replaced by $\rho_0(E,\lambda)$, $\rho_0'(E,\lambda)$, $\rho_0''(E,\lambda)$ and $\rho_0'''(E,\lambda)$, respectively, where

$$\rho_0(E,\lambda) = [E - J_2[1+J_3E]^{-3/2}], \quad J_2 = \left[\frac{I_0e^2\lambda^2}{96\pi c^3 m_r\sqrt{\varepsilon_0\varepsilon sc}}\right],$$

$$J_3 = \left[\frac{2m_c}{m_rE_{g0}}\right]$$

$$\rho_0'(E,\lambda) = \left[1+\frac{3}{2}J_2J_3(1+J_3E)^{-5/2}\right],$$

$$|\rho_0''(E,\lambda)| = \left[\frac{15}{4}J_2J_3^2(1+J_3E)^{-\frac{7}{2}}\right],$$

$$\rho_0'''(E_F,\lambda) = \left[\frac{105}{8}J_2J_3^3(1+J_3E)^{-9/2}\right]$$

15.2.3 *The QC in 2D MOSFETs in the presence of strong light waves and quantizing magnetic field*

The generalized expressions of the surface electron concentration and the magneto gate capacitance for both low and high electric field limits for perturbed three, two and parabolic energy bands can, respectively, be expressed as

$$N_s = \left(\frac{2eBg_v}{h}\right)\sum_{n=0}^{n_{\max}}\sum_{i=0}^{i_{\max}} F_{-1}(\eta_{j,15,w,s}) \tag{15.77}$$

and

$$C_{gB} = \left[e^2\sum_{n=0}^{n_{\max}}\sum_{i=0}^{i_{\max}} F_{-2}(\eta_{j,15,w,s})\right] \times\left[\frac{hk_BT}{2cBg_v}+\frac{e^2d_{0x}}{c_{0x}}\sum_{n=0}^{n_{\max}}\sum_{i=0}^{i_{\max}}F_{-2}(\eta_{j,15,w,s})+\sum_{n=0}^{n_{\max}}\sum_{i=0}^{i_{\max}}t_iF_{-2}(\eta_{j,15,w,s})\right]^{-1} \tag{15.78}$$

where

$$\eta_{j,15,w,s} = \left[\frac{E_{FB,w,s}-E_j}{k_BT}\right], \quad E_{FB,w,s} = \left[eV_g - \frac{d_{0x}e^2n_{2D,w,s}}{\varepsilon_{0x}} - E_{fBB}\right],$$

E_{fBB} is the Fermi energy in bulk materials in the present case under magnetic quantization, $J = 1$ to 6, E_1 and E_2 are the totally quantized energy for perturbed three-band model of Kane for both low and high electric field limits and are given by

$$\left(n+\frac{1}{2}\right)\hbar\omega_0 + S_i\left[\frac{\hbar eF_s\beta_0'(E_1,\lambda)}{\sqrt{2m_c}}\right]^{\frac{2}{3}} = \beta_0(E_1,\lambda) \tag{15.79}$$

and

$$\left(n+\frac{1}{2}\right)\hbar\omega_0 + \frac{2\hbar}{3}\left(\frac{eF_s}{\sqrt{m_c}}\right)S_i^{3/2}[\beta_0''(E_2,\lambda)]^{1/2} = \beta_0(E_2,\lambda) \tag{15.80}$$

where

$$\omega_0 = \frac{eB}{m_c}$$

Besides for low and high electric field limits the corresponding t_1 and t_2 can, respectively, be written as

$$t_1 = \frac{\left[\left(\frac{2S_i}{3}(n_{2Dw})^{-1/3}\right)\left[\hbar e^2\frac{\beta_0'(E_1,\lambda)}{\varepsilon_{sc}\sqrt{2m_c}}\right]^{2/3}\right]}{\left[\beta_0'(E_1,\lambda) - \frac{2}{3}S_i\left(\frac{\hbar eF_s}{\sqrt{2m_c}}\right)^{2/3}[\beta_0'(E_1,\lambda)]^{-1/3}\beta_0''(E_1,\lambda)\right]} \tag{15.81}$$

and

$$t_2 = \frac{\left[\hbar\left(\frac{2}{3}\frac{e^2}{\varepsilon_{sc}}\right)\left(\frac{S_i^{3/2}}{\sqrt{m_c}}\right)[\beta_0''(E_2,\lambda)]^{1/2}\right]}{\left[\beta_0'(E_2,\lambda) - \frac{\hbar S_i^{3/2}}{3}\left(\frac{eF_s}{\sqrt{m_c}}\right)[\beta_0''(E_2,\lambda)]^{-1/2}\cdot\beta_0'''(E_2,\lambda)\right]} \tag{15.82}$$

Besides E_3 and E_4 are the totally quantized energy for perturbed two-band model of Kane for both low and high electric field limits are given by

$$\left(n+\frac{1}{2}\right)\hbar\omega_0 + S_i\left[\frac{\hbar eF_s\omega_0'(E_3,\lambda)}{\sqrt{2m_c}}\right]^{2/3} = \omega_0(E_3,\lambda) \tag{15.83}$$

and

$$\left(n+\frac{1}{2}\right)\hbar\omega_0 + \frac{2\hbar}{3}\left(\frac{eF_s}{\sqrt{m_c}}\right)S_i^{3/2}[\omega_0''(E_4,\lambda)]^{1/2} = \omega_0(E_4,\lambda) \tag{15.84}$$

The corresponding t_3 and t_4 can, respectively be expressed as

$$t_3 = \frac{\left[\left(\frac{2S_i}{3}(n_{2Dw})^{-1/3}\right)\left[\hbar e^2\frac{\omega_0'(E_3,\lambda)}{\varepsilon_{sc}\sqrt{2m_c}}\right]^{2/3}\right]}{\left[\omega_0'(E_3,\lambda) - \frac{2}{3}S_i\left(\frac{\hbar eF_s}{\sqrt{2m_c}}\right)^{2/3}[\omega_0'(E_3,\lambda)]^{-1/3}\omega_0''(E_3,\lambda)\right]} \tag{15.85}$$

and

$$t_4 = \frac{\left[\hbar\left(\frac{2}{3}\frac{e^2}{\varepsilon_{sc}}\right)\left(\frac{S_i^{3/2}}{\sqrt{m_c}}\right)[\omega_0''(E_4,\lambda)]^{1/2}\right]}{\left[\omega_0'(E_4,\lambda) - \frac{\hbar S_i^{3/2}}{3}\left(\frac{eF_s}{\sqrt{m_c}}\right)[\omega_0''(E_4,\lambda)]^{-1/2}\cdot\omega_0'''(E_4,\lambda)\right]} \tag{15.86}$$

Additionally, the E_5 and E_6 are the totally quantized energy for the perturbed parabolic energy band model and are given by

$$\left(n+\frac{1}{2}\right)\hbar\omega_0 + S_i\left[\frac{\hbar eF_s\rho_0'(E_5,\lambda)}{\sqrt{2m_c}}\right]^{2/3} = \rho_0(E_5,\lambda) \tag{15.87}$$

and

$$\left(n+\frac{1}{2}\right)\hbar\omega_0 + \frac{2\hbar}{3}\left(\frac{eF_s}{\sqrt{m_c}}\right)S_i^{3/2}[\rho_0''(E_6,\lambda)]^{3/2} = \rho_0(E_6,\lambda) \tag{15.88}$$

where the corresponding t_5 and t_6 are given by

$$t_5 = \frac{\left[\left(\frac{2S_i}{3}(n_{2Dw})^{-1/3}\right)\left[\hbar e^2\frac{\rho_0'(E_5,\lambda)}{\varepsilon_{sc}\sqrt{2m_c}}\right]^{2/3}\right]}{\left[\rho_0'(E_5,\lambda) - \frac{2}{3}S_i\left(\frac{\hbar eF_s}{\sqrt{2m_c}}\right)^{2/3}[\rho_0'(E_5,\lambda)]^{-1/3}\rho_0''(E_5,\lambda)\right]} \tag{15.89}$$

and

$$t_6 = \frac{\left[\hbar\left(\frac{2}{3}\frac{e^2}{\varepsilon_{sc}}\right)\left(\frac{s_i^{3/2}}{\sqrt{m_c}}\right)[\rho_0''(E_6,\lambda)]^{1/2}\right]}{\left[\rho_0'(E_6,\lambda) - \frac{\hbar S_i^{3/2}}{3}\left(\frac{eF_s}{\sqrt{m_c}}\right)[\rho_0''(E_6,\lambda)]^{-1/2}\cdot\rho_0'''(E_6,\lambda)\right]} \tag{15.90}$$

15.2.4 *The QC in 2D MOSFETs in the presence of strong light waves and the crossed fields configuration*

(a) Perturbed three-band model of Kane

Using the method as given in Chapter 6, the DR in this case can be written as

$$\beta_0(E,\lambda) = \left(n+\frac{1}{2}\right)\hbar\omega_0 + \frac{\hbar^2[k_z(E)]^2}{2m_c} - \frac{E_0}{B}\hbar k_y\{\beta_0'(E,\lambda)\} - \frac{m_cE_0^2\{\beta_0'(E,\lambda)\}^2}{2B^2} \tag{15.91}$$

In the presence of a surface electric field F_s along z-axis, from (15.91) we get

$$\frac{\hbar^2 k^2}{2m_c} = \alpha_{15,1}(E, k_y) - eF_s z\beta_{15,1}(E, k_y) \tag{15.92}$$

where

$$\alpha_{15,1}(E, k_y) = \left[\beta_0(E, \lambda) - \left(n + \frac{1}{2}\right)\hbar\omega_0 + \frac{E_0}{B}\hbar k_y \beta_0'(E, \lambda) - \frac{m_c E_0^2[\beta_0'(E, \lambda)^2]}{2B^2}\right]$$

and

$$\beta_{15,1}(E, k_y) = \left[\beta_0'(E, \lambda) + \frac{E_0}{B}\hbar k_y \beta_0''(E, \lambda) + \frac{m_c E_0^2}{B^2}\beta_0'(E, \lambda)\beta_0''(E, \lambda)\right]$$

The DR in this case is given by

$$\alpha_{15,1}(E, k_y) = \left(\frac{\hbar e F_s}{\sqrt{2m_c}}\right)^{2/3} S_i[\beta_{15,1}(E, k_y)]^{2/3} \tag{15.93}$$

The electron concentration can be expressed as

$$N_s = \frac{g_s g_v eB}{2\pi\hbar}\sum_{n=0}^{n_{\max}}\sum_{i=0}^{i_{\max}} F_{-1}(\eta_{15,CF1}) \tag{15.94}$$

where $\eta_{15,CF1} = [E_{FCF} - E_{15,1}(0)][k_B T]^{-1}$, E_{FCF} is the Fermi energy in this case and $E_{15,1}(0)$ is the positive root of the equation

$$\alpha_{15,1}(E_{15,1}(0), 0) = \left(\frac{\hbar e F_s}{\sqrt{2m_c}}\right)^{2/3} S_i[\beta_{15,1}(E_{15,1}(0), 0)]^{2/3} \tag{15.95}$$

Thus, using the appropriate equations and (1.4) we can study the QC in this case.

(b) Perturbed two-band model of Kane

Using the method as given in Chapter 6, the DR in this case can be written as

$$\omega_0(E, \lambda) = \left(n + \frac{1}{2}\right)\hbar\omega_0 + \frac{\hbar^2[k_z(E)]^2}{2m_c} - \frac{E_0}{B}\hbar k_y\{\omega_0'(E, \lambda)\} - \frac{m_c E_0^2\{\omega_0'(E, \lambda)\}^2}{2B^2} \tag{15.96}$$

In the presence of an a surface electric field F_s along z-axis, from (15.96) we get

$$\frac{\hbar^2 k^2}{2m_c} = \alpha_{15,2}(E, k_y) - eF_s z \omega_{15,2}(E, k_y) \tag{15.97}$$

where

$$\alpha_{15,2}(E, k_y) = \left[\omega_0(E, \lambda) - \left(n + \frac{1}{2}\right)\hbar\omega_0 + \frac{E_0}{B}\hbar k_y \omega_0'(E, \lambda) - \frac{m_c E_0^2[\omega_0'(E, \lambda)]^2}{2B^2}\right]$$

and

$$\beta_{15,2}(E, k_y) = \left[\omega_0'(E, \lambda) + \frac{E_0}{B}\hbar k_y \omega_0''(E, \lambda) + \frac{m_c E_0^2}{B^2}\omega_0'(E, \lambda)\omega_0''(E, \lambda)\right]$$

The DR in this case is given by

$$\alpha_{15,2}(E, k_y) = \left(\frac{\hbar e F_s}{\sqrt{2m_c}}\right)^{2/3} S_i[\beta_{15,2}(E, k_y)]^{2/3} \tag{15.98}$$

The electron concentration can be expressed as

$$N_s = \frac{g_s g_v eB}{2\pi\hbar} \sum_{n=0}^{n_{\max}} \sum_{i=0}^{i_{\max}} F_{-1}(\eta_{15,CF2}) \tag{15.99}$$

where $\eta_{15,CF2} = [E_{FCF} - E_{15,2}(0)][k_B T]^{-1}$, E_{FCF} is the Fermi energy in this case and $E_{15,2}(0)$ is the positive root of the equation

$$\alpha_{15,2}(E_{15,2}(0), 0) = \left(\frac{\hbar e F_s}{\sqrt{2m_c}}\right)^{2/3} S_i[\beta_{15,2}(E_{15,2}(0), 0)]^{2/3} \tag{15.100}$$

Thus, using the appropriate equations and (1.4) we can study the QC in this case.

(c) Perturbed parabolic band model

Using the method as given in Chapter 6, the DR in this case can be written as

$$\rho_0(E, \lambda) = \left(n + \frac{1}{2}\right)\hbar\omega_0 + \frac{\hbar^2[k_z(E)]^2}{2m_c} - \frac{E_0}{B}\hbar k_y\{\rho_0'(E, \lambda)\} - \frac{m_c E_0^2\{\rho_0'(E, \lambda)\}^2}{2B^2} \tag{15.101}$$

In the presence of a surface electric field F_s along z-axis, from (15.101) we get

$$\frac{\hbar^2 k^2}{2m_c} = \alpha_{15,3}(E, k_y) - eF_s z\omega_{15,3}(E, k_y) \tag{15.102}$$

where

$$\alpha_{15,3}(E, k_y) = \left[\rho_0(E, \lambda) - \left(n + \frac{1}{2}\right)\hbar\omega_0 + \frac{E_0}{B}\hbar k_y \rho_0'(E, \lambda) - \frac{m_c E_0^2 [\rho_0'(E, \lambda)]^2}{2B^2}\right]$$

and

$$\beta_{15,3}(E, k_y) = \left[\rho_0'(E, \lambda) + \frac{E_0}{B}\hbar k_y \rho_0''(E, \lambda) + \frac{m_c E_0^2}{B^2}\rho_0'(E, \lambda)\rho_0''(E, \lambda)\right]$$

The DR in this case is given by

$$\alpha_{15,3}(E, k_y) = \left(\frac{\hbar e F_s}{\sqrt{2m_c}}\right)^{2/3} S_i[\beta_{15,3}(E, k_y)]^{2/3} \tag{15.103}$$

The electron concentration can be expressed as

$$N_s = \frac{g_s g_v eB}{2\pi\hbar} \sum_{n=0}^{n_{\max}} \sum_{i=0}^{i_{\max}} F_{-1}(\eta_{15,CF3}) \tag{15.104}$$

where $\eta_{15,CF3} = [E_{FCF} - E_{15,3}(0)][k_B T]^{-1}$, E_{FCF} is the Fermi energy in this case and $E_{15,3}(0)$ is the positive root of the equation

$$\alpha_{15,3}(E_{15,3}(0), 0) = \left(\frac{\hbar e F_s}{\sqrt{2m_c}}\right)^{2/3} S_i[\beta_{15,3}(E_{15,3}(0), 0)]^{2/3} \tag{15.105}$$

Thus, using the appropriate equations and (1.4) we can study the QC in this case.

15.2.5 *The QC in QWFETs in the presence of light waves*

The 2D DR in QWFETs of III–V, ternary and quaternary materials, whose unperturbed band structure is defined by the three- and two-band models of Kane together with parabolic energy bands can, respectively, be written using (15.41), (15.42) and (15.43) as

$$\frac{\hbar^2 k_s^2}{2m_c} + \frac{\hbar^2}{2m_c}\left(\frac{n_z \pi}{d_z}\right)^2 = \beta_0(E, \lambda) \tag{15.106}$$

$$\frac{\hbar^2 k_s^2}{2m_c} + \frac{\hbar^2}{2m_c}\left(\frac{n_z \pi}{d_z}\right)^2 = \omega_0(E, \lambda) \tag{15.107}$$

$$\frac{\hbar^2 k_s^2}{2m_c} + \frac{\hbar^2}{2m_c}\left(\frac{n_z \pi}{d_z}\right)^2 = \rho_0(E, \lambda) \tag{15.108}$$

The surface electron concentration for the above cases can, respectively, be written under the condition of extreme carrier degeneracy as

$$N_S = \frac{m_c}{\pi \hbar^2} \sum_{n_z=1}^{n_{z\max}} \left[\beta_0(E_{F15,5}, \lambda) - \frac{\hbar^2}{2m_c}\left(\frac{n_z \pi}{d_z}\right)^2\right] \tag{15.109}$$

$$N_S = \frac{m_c}{\pi \hbar^2} \sum_{n_z=1}^{n_{z\max}} \left[\omega_0(E_{F15,5}, \lambda) - \frac{\hbar^2}{2m_c}\left(\frac{n_z \pi}{d_z}\right)^2\right] \tag{15.110}$$

$$N_S = \frac{m_c}{\pi \hbar^2} \sum_{n_z=1}^{n_{z\max}} \left[\rho_0(E_{F15,5}, \lambda) - \frac{\hbar^2}{2m_c}\left(\frac{n_z \pi}{d_z}\right)^2\right] \tag{15.111}$$

where

$$E_{F15,5} = eV_g - \frac{e^2 N_S d_{0x}}{\varepsilon_{0x}} - E_{fB}$$

Using (15.109), (15.110), (15.111) and (1.4) the QC can, respectively, be written as

$$C_g = \left[e^2 \sum_{n_z=1}^{n_{z\max}} \beta_0'(E_{F15,5}, \lambda)\right]\left[\frac{\pi \hbar^2}{m_c} + \frac{e^2 d_{0x}}{\varepsilon_{0x}} \sum_{n_z=1}^{n_{z\max}} \beta_0'(E_{F15,5}, \lambda)\right]^{-1} \tag{15.112}$$

$$C_g = \left[e^2 \sum_{n_z=1}^{n_{z\max}} \omega_0'(E_{F15,5}, \lambda)\right]\left[\frac{\pi \hbar^2}{m_c} + \frac{e^2 d_{0x}}{\varepsilon_{0x}} \sum_{n_z=1}^{n_{z\max}} \omega_0'(E_{F15,5}, \lambda)\right]^{-1} \tag{15.113}$$

$$C_g = \left[e^2 \sum_{n_z=1}^{n_{z\max}} \rho_0'(E_{F15,5}, \lambda)\right]\left[\frac{\pi \hbar^2}{m_c} + \frac{e^2 d_{0x}}{\varepsilon_{0x}} \sum_{n_z=1}^{n_{z\max}} \rho_0'(E_{F15,5}, \lambda)\right]^{-1} \tag{15.114}$$

The EEM in this case can, respectively, be written as

$$m^*(E_{F15,5},\lambda) = m_c\beta_0'(E_{F15,5},\lambda) \tag{15.115}$$

$$m^*(E_{F15,5},\lambda) = m_c\omega_0'(E_{F15,5},\lambda) \tag{15.116}$$

$$m^*(E_{F15,5},\lambda) = m_c\rho_0'(E_{F15,5},\lambda) \tag{15.117}$$

The ER in this case can, respectively, be written as

$$\frac{D}{\mu} = \frac{1}{e}\left[\sum_{n_z=1}^{n_{z\max}}\left[\beta_0(E_{F15,5},\lambda) - \frac{\hbar^2}{2m_c}\left(\frac{n_z\pi}{d_z}\right)^2\right]\right]\left[\sum_{n_z=1}^{n_{z\max}}[\beta_0'(E_{F15,5},\lambda)]\right]^{-1} \tag{15.118}$$

$$\frac{D}{\mu} = \frac{1}{e}\left[\sum_{n_z=1}^{n_{z\max}}\left[\omega_0(E_{F15,5},\lambda) - \frac{\hbar^2}{2m_c}\left(\frac{n_z\pi}{d_z}\right)^2\right]\right]\left[\sum_{n_z=1}^{n_{z\max}}[\omega_0'(E_{F15,5},\lambda)]\right]^{-1} \tag{15.119}$$

$$\frac{D}{\mu} = \frac{1}{e}\left[\sum_{n_z=1}^{n_{z\max}}\left[\rho_0(E_{F15,5},\lambda) - \frac{\hbar^2}{2m_c}\left(\frac{n_z\pi}{d_z}\right)^2\right]\right]\left[\sum_{n_z=1}^{n_{z\max}}[\rho_0'(E_{F15,5},\lambda)]\right]^{-1} \tag{15.120}$$

15.2.6 *The QC in QWFETs in the presence of light waves and quantizing magnetic field*

The E_7, E_8 and E_9 are the totally quantized energies for the perturbed three, two and parabolic band models in this case can be written using (15.106), (15.107) and (15.108) as

$$\left(n+\frac{1}{2}\right)\hbar\omega_0 + \frac{\hbar^2}{2m_c}\left(\frac{n_z\pi}{d_z}\right)^2 = \beta_0(E_7,\lambda) \tag{15.121}$$

$$\left(n+\frac{1}{2}\right)\hbar\omega_0 + \frac{\hbar^2}{2m_c}\left(\frac{n_z\pi}{d_z}\right)^2 = \omega_0(E_8,\lambda) \tag{15.122}$$

and

$$\left(n+\frac{1}{2}\right)\hbar\omega_0 + \frac{\hbar^2}{2m_c}\left(\frac{n_z\pi}{d_z}\right)^2 = \rho_0(E_9,\lambda) \tag{15.123}$$

The generalized expressions of the surface electron concentration and the QC can, respectively, be written as

$$N_s = \frac{2eBg_v}{h}\sum_{n_z=1}^{n_{z\max}}\sum_{n=0}^{n_{\max}} F_{-1}(\zeta_u) \tag{15.124}$$

and

$$C_{gB} = \left[e^2 \sum_{n=0}^{n_{\max}} \sum_{n_z=1}^{n_{z\max}} F_{-2}(\zeta_u)\right] \left[\frac{hK_BT}{2eBg_v} + \left(\frac{e^2 d_{0x}}{\varepsilon_{0x}}\right) \sum_{n=0}^{n_{\max}} \sum_{n_z=1}^{n_{z\max}} F_{-2}(\zeta_u)\right]^{-1} \tag{15.125}$$

where

$$\zeta_u = \frac{\bar{E}_{FB} - E_u}{K_BT}, \quad \bar{E}_{FB} = \left[eV_g - \frac{e^2 d_{0x}}{\varepsilon_{0x}} - E_{JBB}\right], \quad u = 7, 8 \text{ and } 9$$

15.2.7 *The QC in QWFETs in the presence of light waves and the crossed fields configuration*

(a) Perturbed three-band model of Kane

From (15.91), the DR in this case can be written as

$$\beta_0(E, \lambda) = \left(n + \frac{1}{2}\right) \hbar\omega_0 + \frac{\hbar^2 (n_z\pi/d_z)^2}{2m_c} - \frac{E_0}{B} \hbar k_y \{\beta_0'(E, \lambda)\} - \frac{m_c E_0^2 \{\beta_0'(E, \lambda)\}^2}{2B^2} \tag{15.126}$$

The sub-band energy ($E_{15,71}$) can be written from (15.126) as

$$\beta_0(E_{15,71}, \lambda) = \left(n + \frac{1}{2}\right) \hbar\omega_0 + \frac{\hbar^2 (n_z\pi/d_z)^2}{2m_c} - \frac{m_c E_0^2 \{\beta_0'(E_{15,71}, \lambda)\}^2}{2B^2} \tag{15.127}$$

The surface electron concentration can be written as

$$N_s = \frac{g_s g_v eB}{2\pi\hbar} \sum_{n=0}^{n_{\max}} \sum_{n_z=1}^{n_{z\max}} F_{-1}(\eta_{15,71}) \tag{15.128a}$$

where $\eta_{15,71} = (E_{F15,71} - E_{15,71})(k_BT)^{-1}$ and $E_{F15,71}$ is the Fermi energy in this case.

The ER can be written as

$$\frac{D}{\mu} = \frac{k_BT}{e} \left[\sum_{n=0}^{n_{\max}} \sum_{n_z=1}^{n_{z\max}} F_{-1}(\eta_{15,71})\right] \left[\sum_{n=0}^{n_{\max}} \sum_{n_z=1}^{n_{z\max}} F_{-2}(\eta_{15,71})\right]^{-1} \tag{15.128b}$$

Using the appropriate equations and (1.4) we can study the QC in this case.

(b) Perturbed two-band model of Kane

From (15.96), the DR in this case can be written as

$$\omega_0(E,\lambda) = \left(n+\frac{1}{2}\right)\hbar\omega_0 + \frac{\hbar^2(n_z\pi/d_z)^2}{2m_c} - \frac{E_0}{B}\hbar k_y\{\omega_0'(E,\lambda)\} - \frac{m_c E_0^2\{\omega_0'(E,\lambda)\}^2}{2B^2} \tag{15.129}$$

The sub-band energy ($E_{15,72}$) can be written from (15.129) as

$$\omega_0(E_{15,72},\lambda) = \left(n+\frac{1}{2}\right)\hbar\omega_0 + \frac{\hbar^2(n_z\pi/d_z)^2}{2m_c} - \frac{m_c E_0^2\{\omega_0'(E_{15,72},\lambda)\}^2}{2B^2} \tag{15.130}$$

The surface electron concentration can be written as

$$N_s = \frac{g_s g_v eB}{2\pi\hbar}\sum_{n=0}^{n_{\max}}\sum_{n_z=1}^{n_{z\max}} F_{-1}(\eta_{15,72}) \tag{15.131}$$

where $\eta_{15,72} = (E_{F15,72} - E_{15,72})(k_B T)^{-1}$ and $E_{F15,72}$ is the Fermi energy in this case.

The ER can be written as

$$\frac{D}{\mu} = \frac{k_B T}{e}\left[\sum_{n=0}^{n_{\max}}\sum_{n_z=1}^{n_{z\max}} F_{-1}(\eta_{15,72})\right]\left[\sum_{n=0}^{n_{\max}}\sum_{n_z=1}^{n_{z\max}} F_{-2}(\eta_{15,72})\right]^{-1} \tag{15.132}$$

Using the appropriate equations and (1.4) we can study the QC in this case.

(c) Perturbed parabolic band model

From (15.101), the DR in this case can be written as

$$\rho_0(E,\lambda) = \left(n+\frac{1}{2}\right)\hbar\omega_0 + \frac{\hbar^2(n_z\pi/d_z)^2}{2m_c} - \frac{E_0}{B}\hbar k_y\{\rho_0'(E,\lambda)\} - \frac{m_c E_0^2\{\rho_0'(E,\lambda)\}^2}{2B^2} \tag{15.133}$$

The sub-band energy ($E_{15,73}$) can be written from (15.133) as

$$\rho_0(E_{15,73},\lambda) = \left(n+\frac{1}{2}\right)\hbar\omega_0 + \frac{\hbar^2(n_z\pi/d_z)^2}{2m_c} - \frac{m_c E_0^2\{\rho_0'(E_{15,73},\lambda)\}^2}{2B^2} \tag{15.134}$$

The surface electron concentration can be written as

$$N_s = \frac{g_s g_v eB}{2\pi\hbar}\sum_{n=0}^{n_{\max}}\sum_{n_z=1}^{n_{z\max}} F_{-1}(\eta_{15,73}) \tag{15.135}$$

where $\eta_{15,73} = (E_{F15,73} - E_{15,73})(k_B T)^{-1}$ and $E_{F15,73}$ is the Fermi energy in this case.

The ER can be written as

$$\frac{D}{\mu} = \frac{k_B T}{e}\left[\sum_{n=0}^{n_{\max}}\sum_{n_z=1}^{n_{z\max}} F_{-1}(\eta_{15,73})\right]\left[\sum_{n=0}^{n_{\max}}\sum_{n_z=1}^{n_{z\max}} F_{-2}(\eta_{15,73})\right]^{-1} \quad (15.136)$$

Using the appropriate equations and (1.4) we can study the QC in this case.

15.2.8 *The QC in NWFETs in the presence of strong light waves*

(a) Perturbed three-band model of Kane

The 1D DR in NWFETs of III–V, ternary and quaternary materials, whose unperturbed band structure is defined by the three-band model of Kane, in the presence of strong light waves can be expressed from Equation (15.41) as

$$\frac{\hbar^2 k_x^2}{2m_c} = \beta_0(E,\lambda) - \left[\frac{\hbar^2}{2m_c}\left(\frac{n_y\pi}{d_y}\right)^2 + \frac{\hbar^2}{2m_c}\left(\frac{n_z\pi}{d_z}\right)^2\right] \quad (15.137)$$

where $n_y(=1,2,3,\ldots)$ and $n_z(=1,2,3,\ldots)$ are the size quantum numbers along y and z directions while d_y and d_z are the nanothickness along the y and z directions, respectively.

The sub-band energies $(E_{n_{l10}})$ can be expressed as

$$\beta_0(E_{n_{l10}},\lambda) = \left[\frac{\hbar^2}{2m_c}(n_y\pi/d_y)^2 + \frac{\hbar^2}{2m_c}(n_z\pi/d_z)^2\right] \quad (15.138)$$

The EMM in this case can be written as

$$m^*(E_{F1DL}, n_y, n_z, \lambda) = m_c\{\beta_0(E_{F1DL},\lambda)\}' \quad (15.139)$$

where E_{F1DL} is the Fermi energy in the present case as measured from the edge of the conduction band in the vertically upward direction in the absence of any quantization.

The 1D density-of-states function $(N_{1D}(E,\lambda))$ is given by

$$N_{1D}(E,\lambda) = \left(\frac{g_v\sqrt{2m_c}}{\pi\hbar}\right)\sum_{n_y=1}^{n_{y\max}}\sum_{n_z=1}^{n_{z\max}}\{\beta_0(E,\lambda)\}' \times [\beta_0(E,\lambda) - \phi(n_y,n_z)]^{-1/2} H(E - E_{n_{l10}}) \quad (15.140)$$

where

$$\phi(n_y, n_z) = \frac{\hbar^2\pi^2}{2m_c}\left[\left(\frac{n_y}{d_y}\right)^2 + \left(\frac{n_z}{d_z}\right)^2\right]$$

Combining Equation (15.140), with the Fermi–Dirac occupation probability factor, the 1D electron concentration (n_{1D}) can thus be written under the condition of extreme carrier degeneracy as

$$n_{1D} = \frac{2g_v}{\pi}\frac{\sqrt{2m_c}}{\hbar}\sum_{n_y=1}^{n_{y\max}}\sum_{n_z=1}^{n_{z\max}}[M_{21}(n_y, n_z, E_{F1DL}, \lambda)] \tag{15.141}$$

where $M_{21}(n_y, n_z, E_{F1DL}, \lambda) \equiv [\beta_0(E_{F1DL}, \lambda) - \phi(n_y, n_z)]^{1/2}$.

The ER can be written as

$$\frac{D}{\mu} = \frac{1}{e}\frac{\partial n_{1D}}{\partial E_{F1DL}} \tag{15.142}$$

The use of (15.141) and (15.142) leads to the expression of the 1D-ER in this case as

$$\frac{D}{\mu} = \frac{1}{|e|}\frac{\sum_{n_y=1}^{n_{y\max}}\sum_{n_z=1}^{n_{z\max}}[M_{21}(n_y, n_z, E_{F1DL}, \lambda)]}{\sum_{n_y=1}^{n_{y\max}}\sum_{n_z=1}^{n_{z\max}}[\{M_{21}(n_y, n_z, E_{F1DL}, \lambda)\}']} \tag{15.143}$$

Using (15.143) and (1.4) we can study the QC in this case.

(b) Perturbed two-band model of Kane

Using Equation (15.42), the expressions for the 1D DR, the sub-band energies, the EMM, the density-of-states function and the electron concentration for NWFETs of III–V, ternary and quaternary materials, whose unperturbed band structure is defined by the two-band model of Kane, can, respectively, be written in the presence of strong external photo-excitation as

$$\frac{\hbar^2 k_x^2}{2m_c} = \{\omega_0(E, \lambda) - \phi(n_y, n_z)\} \tag{15.144}$$

$$\omega_0(E_{n_{l11}}, \lambda) = \phi(n_y, n_z) \tag{15.145}$$

$$m^*(E_{F1DL}, n_y, n_z, \lambda) = m_c\{\omega_0(E_{F1DL}, \lambda)\}' \tag{15.146}$$

$$N_{1D}(E, \lambda) = \left(\frac{g_v\sqrt{2m_c}}{\pi\hbar}\right)\sum_{n_y=1}^{n_{y\max}}\sum_{n_z=1}^{n_{z\max}}\{\omega_0(E, \lambda)\}' \times [\omega_0(E, \lambda) - \phi(n_y, n_z)]^{-1/2} H(E - E_{n_{l11}}) \tag{15.147}$$

$$n_{1D} = \frac{2g_v}{\pi}\frac{\sqrt{2m_c}}{\hbar}\sum_{n_y=1}^{n_{y\max}}\sum_{n_z=1}^{n_{z\max}}[M_{22}(n_y, n_z, E_{F1DL}, \lambda)] \tag{15.148}$$

where $M_{22}(n_y, n_z, E_{F1DL}, \lambda) \equiv [\omega_0(E_{F1DL}, \lambda) - \phi(n_y, n_z)]^{1/2}$.

The use of Equations (15.148) and (15.142) leads to the expression of the ER in this case as

$$\frac{D}{\mu} = \frac{1}{|e|}\frac{\sum_{n_y=1}^{n_{y\max}}\sum_{n_z=1}^{n_{z\max}}[M_{22}(n_y, n_z, E_{F1DL}, \lambda)]}{\sum_{n_y=1}^{n_{y\max}}\sum_{n_z=1}^{n_{z\max}}[\{M_{22}(n_y, n_z, E_{F1DL}, \lambda)\}']} \tag{15.149}$$

Using (15.140) and (1.4) we can study the QC in this case.

(c) Perturbed parabolic band model

Using Equation (15.43), the expressions for the 1D DR, the sub-band energies, the EMM, the density of states function and the electron concentration for NWFETs of III–V, ternary and quaternary materials, whose unperturbed band structure is defined by the parabolic energy bands, can respectively be written in the presence of strong external photo-excitation as

$$\frac{\hbar^2 k_x^2}{2m_c} = [\rho_0(E, \lambda) - \phi(n_y, n_z)] \tag{15.150}$$

$$\rho_0(E_{n_{l11}}, \lambda) = \phi(n_y, n_z) \tag{15.151}$$

$$m^*(E_{F1DL}, n_y, n_z, \lambda) = m_c\{\rho_0(E_{F1DL}, \lambda)\}' \tag{15.152}$$

$$N_{1D}(E, \lambda) = \left(\frac{g_v\sqrt{2m_c}}{\pi\hbar}\right)\sum_{n_y=1}^{n_{y\max}}\sum_{n_z=1}^{n_{z\max}}\{\rho_0(E, \lambda)\}' \times [\rho_0(E, \lambda) - \phi(n_y, n_z)]^{-1/2}H(E - E_{n_{l11}}) \tag{15.153}$$

$$n_{1D} = \frac{2g_v\sqrt{2m_c}}{\pi\hbar}\sum_{n_y=1}^{n_{y\max}}\sum_{n_z=1}^{n_{z\max}}[M_{23}(n_y, n_z, E_{F1DL}, \lambda)] \tag{15.154}$$

where $M_{23}(n_y, n_z, E_{F1DL}, \lambda) \equiv [\rho_0(E_{F1DL}, \lambda) - \phi(n_y, n_z)]^{1/2}$.

The use of Equations (15.154) and (15.142) leads to the expression of the ER in this case as

$$\frac{D}{\mu} = \frac{1}{|e|}\frac{\sum_{n_y=1}^{n_{y\max}}\sum_{n_z=1}^{n_{z\max}}[M_{23}(n_y, n_z, E_{F1DL}, \lambda)]}{\sum_{n_y=1}^{n_{y\max}}\sum_{n_z=1}^{n_{z\max}}[\{M_{23}(n_y, n_z, E_{F1DL}, \lambda)\}']} \tag{15.155}$$

Using (15.155) and (1.4) we can study the QC in this case.

15.2.9 *The QC in CNTFETs in the presence of light waves*

In the presence of light waves the DRs arm chair and zigzag CNTs can, respectively, be written using (7.170) and (7.171) as

$$\rho_0(E,\lambda) = \pm t_c\sqrt{1+4\cos\frac{m\pi}{n}\cos\frac{k_y a_c}{2}+4\cos^2\frac{k_y a_c}{2}},$$
$$-\frac{\pi}{a_c} < k_y < \frac{\pi}{a_c} \text{ and } m(=1,2,\ldots,2n) \tag{15.156}$$

$$\rho_0(E,\lambda) = \pm t_c\sqrt{1+4\cos\frac{\sqrt{3}k_x a_c}{n}\cos\frac{m\pi}{n}+4\cos^2\frac{m\pi}{n}},$$
$$-\frac{\pi}{\sqrt{3}a_c} < k_x < \frac{\pi}{\sqrt{3}a_c} \text{ and } m(=1,2,\ldots,2n) \tag{15.157}$$

where $\rho_0(E,\lambda)$ is defined in (15.43).

The above equations can be expressed as

$$k_y = \frac{2}{a_{cc}\sqrt{3}}A_1(E,E_i,\lambda) \tag{15.158}$$

and

$$k_y = \frac{2}{3a_{cc}}A_2(E,E_i,\lambda) \tag{15.159}$$

where

$$A_1(E,E_i,\lambda) = \cos^{-1}[f_1(E,E_i,\lambda)],$$

$$f_1(E,E_i,\lambda) = \left[+\frac{1}{2}\left[\frac{-\rho_0^2(E_i,\lambda)}{t^2}+5\right]\right.$$
$$\left.-\frac{1}{2}\left[\left(\frac{\rho_0^2(E_i,\lambda)}{t^2}-5\right)^2-\left(1-\frac{\rho_0^2(E_i,\lambda)}{t^2}\right)\right]^{1/2}\right],$$

$$E_i = \frac{|3i-n_1+n_2|}{2}|t_c|\frac{d}{r},$$

d and r are the system constants of the nanotube, i, n_1, n_2, are integers

$$A_2(E,E_i,\lambda) = \cos^{-1}[f_2(E,E_i,\lambda)],$$

$$f_2(E,E_i,\lambda) = \left[\bar{\phi}_0(E_i,\lambda)+\frac{1}{4\bar{\phi}_0(E_i,\lambda)}-\frac{\rho_0^2(E,\lambda)}{4t_c^2\bar{\phi}_0(E_i,\lambda)}\right]$$

and

$$\bar{\phi}_0(E_i,\lambda) = \left[\frac{1}{2}+\frac{1}{2t_c}\rho_0(E_i,\lambda)\right]$$

At low temperatures, where that quantum effects become prominent the electron concentration per unit length in arm chair and zig-zag carbon nanotubes in the presence of light waves can, respectively, be written as

$$n_{1D} = \frac{8}{a_{cc}\pi\sqrt{3}} \sum_{i=0}^{i_{\max}} [A_1(E_{F15,9}, E_i, \lambda)] \tag{15.160}$$

$$n_{1D} = \frac{8}{3a_{cc}\pi} \sum_{i=0}^{i_{\max}} [A_2(E_{F15,9}, E_i, \lambda)] \tag{15.161}$$

where $E_{F15,9}$ is the Fermi energy in this case.

The QC in arm chair and zig-zag carbon nanotubes in the presence of light waves can, respectively, be expressed as

$$C_g = e^2 \left[\sum_{i=1}^{i_{\max}} f_3(E_f, E_i, \lambda)\right] \left[\frac{a_{cc}\pi\sqrt{3}}{4} + \frac{e^2}{\varepsilon_{OX}} \sum_{i=1}^{i_{\max}} f_3(E_f, E_i, \lambda)\right]^{-1} \tag{15.162}$$

$$C_g = e^2 \left[\sum_{i=1}^{i_{\max}} f_4(E_f, E_i, \lambda)\right] \left[\frac{3a_{cc}\pi}{4} + \frac{e^2}{\varepsilon_{OX}} \sum_{i=1}^{i_{\max}} f_4(E_f, E_i, \lambda)\right]^{-1} \tag{15.163}$$

where

$$f_3(E_f, E_i, \lambda) = \Bigg[[1 - f_1^2(E_f, E_i, \lambda)]^{-1/2} \times [\rho_0(E_f, \lambda)\rho_0'(E_f, \lambda) \times \left[\left(\frac{\rho_0^2(E_f, \lambda)}{t^2} - 5\right)^2 - \left(1 - \frac{\rho_0^2(E_f, \lambda)}{t^2}\right)\right]^{-1/2}\Bigg]$$

$$f_4(E_f, E_i, \lambda) = \frac{\rho_0(E_f, \lambda)\rho_0'(E_f, \lambda)}{4t^2\bar{\phi}(E_i, \lambda)[1 - f_2^2(E_f, E_i, \lambda)]^{1/2}}$$

15.2.10 *The dispersion relations in HD III–V and optoelectronic materials in the presence of strong light waves*

The (15.41), (15.42) and (15.43) can approximately be written as

$$\frac{\hbar^2 k^2}{2m_c} = U_\lambda I_{11}(E) - P_\lambda \tag{15.164}$$

$$\frac{\hbar^2 k^2}{2m_c} = t_{1\lambda}E + t_{2\lambda}E^2 - \delta_\lambda \tag{15.165}$$

and

$$\frac{\hbar^2 k^2}{2m_c} = t_{1\lambda}E - \delta_\lambda \tag{15.166}$$

where

$$U_\lambda = (1+\theta_\lambda), \quad \theta_\lambda = \frac{C_0}{A}\left(t_\lambda + \frac{BJ_\lambda}{A}\right),$$

$$C_0 = \left[\frac{|e|^2}{96\pi c^3 m_r}\frac{I\lambda^2 E_{g_0}(E_{g_0}+\Delta)}{\sqrt{\varepsilon_{sc}\varepsilon_0}(E_{g_0}+\frac{2}{3}\Delta)}\frac{\beta^2}{4}(1+\frac{\rho}{\sqrt{2}})^2\right],$$

$$A = E_{g_0}, \quad B = \left[1+\frac{m^*}{m_V}\right], \quad G_\lambda = \left[\frac{2B}{(A+\delta')^3} - \frac{BC_\lambda}{(A+\delta')}\right],$$

$$C_\lambda = [(E_{g_0}+\delta')^{-1} + (E_{g_0}+\delta')(E_{g_0}-\delta')^{-2}](A+\delta')^{-1},$$

$$P_\lambda = \frac{C_0}{A}J_\lambda, \quad J_\lambda = (D_\lambda + 2(E_{g_0}-\delta')\sqrt{f_\lambda}),$$

$$D_\lambda = \left(1+\frac{2(E_{g_0}-\delta')}{(A+\delta')}\right), \quad f_\lambda = \left[\frac{1}{(A+\delta')^2} + \frac{1}{(E_{g_0}-\delta')^2} - C_\lambda\right],$$

$$t_{1\lambda} = \left(1+\frac{3m_c}{m_r}\alpha\delta_\lambda\right), \quad \alpha = \frac{1}{E_{g_0}},$$

$$\delta_\lambda = \frac{|e|^2 I\lambda^2}{96 m_r \pi c^3\sqrt{\varepsilon_{sc}\varepsilon_0}} \quad \text{and} \quad t_{2\lambda} = \alpha t_{1\lambda}$$

Under the condition of heavy doping, the DR in III–V and optoelectronic materials in the presence of strong light can be written by combining (15.164), (15.165), (15.166) and (8.2) as

$$\frac{\hbar^2 k^2}{2m_c} = T_1(E,\eta_g,\lambda) \tag{15.167}$$

$$\frac{\hbar^2 k^2}{2m_c} = T_2(E,\eta_g,\lambda) \tag{15.168}$$

and

$$\frac{\hbar^2 k^2}{2m_c} = T_3(E,\eta_g,\lambda) \tag{15.169}$$

where

$$T_1(E,\eta_g,\lambda) = [U_\lambda[T_{31}(E,\eta_g) + iT_{32}(E,\eta_g)] - P_\lambda]$$

$$T_{31}(E,\eta_g) \equiv \left(\frac{2}{1+Erf(E/\eta_g)}\right)$$

$$\times\left[\frac{\alpha b}{c}\theta_0(E,\eta_g)+\left[\frac{\alpha c+bc-\alpha b}{c^2}\right]\gamma_0(E,\eta_g)\right.$$
$$+\frac{1}{c}\left(1-\frac{\alpha}{c}\right)\left(1-\frac{b}{c}\right)\frac{1}{2}\left[1+Erf\left(\frac{E}{\eta_g}\right)\right]$$
$$-\frac{1}{c}\left(1-\frac{\alpha}{c}\right)\left(1-\frac{b}{c}\right)\frac{2}{c\eta_g\sqrt{\pi}}\exp(-u_2^2)$$
$$\left.\times\left[\sum_{p=1}^{\infty}\frac{\exp(-p^2/4)}{p}sinh(pu_2)\right]\right],$$

$$b\equiv\left(\frac{1}{E_g+\Delta}\right),\quad c\equiv\left(\frac{1}{E_g+\frac{2}{3}\Delta}\right),\quad u_2\equiv\frac{1+cE}{c\eta_g}$$

$$T_{32}(E,\eta_g)\equiv\left(\frac{2}{1+Erf(E/\eta_g)}\right)\frac{1}{c}\left(1-\frac{\alpha}{c}\right)\left(1-\frac{b}{c}\right)\frac{\sqrt{\pi}}{c\eta_g}\exp(-u_2^2)$$

$$\theta_0(E,\eta_g)=\frac{\eta_g E}{2\sqrt{\pi}}\exp\left(\frac{-E^2}{\eta_g^2}\right)+\frac{1}{4}(\eta_g^2+2E^2)\left[1+Erf\left(\frac{E}{\eta_g}\right)\right]$$

$$\gamma_0(E,\eta_g)=\eta_g\exp(-E^2/\eta_g^2)(2\sqrt{\pi})^{-1}+\frac{E}{2}(1+Erf(E/\eta_g))$$

$$T_2(E,\eta_g,\lambda)=[U_{1\lambda}\gamma_3(E,\eta_g)+(t_{2\lambda})2\theta_0(E,\eta_g)[1+Erf(E/\eta_g)]^{-1}-\delta_\lambda]$$

$$\gamma_3(E,\eta_g)\equiv\left[\frac{2}{(1+Erf(E/\eta_g))}\right]\gamma_0(E,\eta_g)\text{ and}$$

$$T_3(E,\eta_g,\lambda)=[t_{1\lambda}\gamma_3(E,\eta_g)-\delta_\lambda]$$

15.2.11 *The QC in HD MOSFETs in the presence of strong light waves under accumulation mode of operation*

(a) Perturbed three-band model of Kane

In the presence of a surface electric field F_s along z-direction and perpendicular to the surface, the (15.157) assumes the form

$$T_1(E-|e|F_s z,\eta_g,\lambda)=\frac{\hbar^2k^2}{2m_c}\tag{15.170}$$

where, for this section, E represents the electron energy as measured from the edge of the conduction band at the surface in the vertically upward direction.

Using (15.170) and (1.7), under the weak electric field limit, the 2D DR in HD MOSFETs of HD III–V, ternary and quaternary materials operated under accumulation mode, whose unperturbed band structure is defined by the three-band model of Kane, in the presence of light waves can be expressed as

$$T_1(E,\eta_g,\lambda) = \frac{\hbar^2 k_s^2}{2m_c} + S_i\left[\frac{\hbar|e|F_s[T_1(E,\eta_g,\lambda)]'}{\sqrt{2m_c}}\right]^{2/3} \tag{15.171}$$

Equation (15.171) represents the DR of the 2D electrons in HD MOSFETs III–V, ternary and quaternary materials operated with accumulation mode under the weak electric field limit in the presence of light waves whose bulk electrons obey the HD three-band model of Kane. Since the DR in accordance with the HD three-band model of Kane is complex in nature, the (15.171) will also be complex. Both complexities occur due to the presence of poles in the finite complex plane of the dispersion relation of the materials in the presence of band tails.

The EEM can be expressed as

$$m_L^*(E'_{fL},i,\eta_g) = m_c \text{ Real part of } [P_{3HDL}(E'_{fL},i,\eta_g,\lambda)]' \tag{15.172}$$

where

$$P_{3HDL}(E'_{fL},i,\eta_g,\lambda) = \left[T_1(E'_{fL},\eta_g,\lambda) - S_i\left[\frac{\hbar|e|F_2[T_1(E'_{fL},\eta_g,\lambda)]'}{\sqrt{2m_c}}\right]^{2/3}\right],$$

the Fermi energy in this case is E'_{fL} which can be expressed as

$$E'_{fL} = eV_g - \frac{e^2 n_s d_{ex}}{\varepsilon_{ox}},$$

n_s is the total surface electron concentration and E_{FB} should be, under the condition extreme degeneracy, determined from the equation

$$n_B = \frac{4\pi}{3}\frac{2g_v}{(2\pi)^3}\left(\frac{2m_c}{\hbar^2}\right)^{3/2} \text{ Real Part of } [T_1(E_{FB},\eta_g,\lambda)]^{3/2} \tag{15.173}$$

and n_B is the bulk electron concentration.

Thus, one can observe that the EEM is a function of light intensity, scattering potential, the sub-band index, surface electric field, the Fermi energy and the other spectrum constants due to the combined influence of E_g and Δ.

The sub-band energy E'_{i1L} is given by

$$0 = \text{ Real part of } \left[T_1(E'_{i1L}, \eta_g, \lambda) - S_i \left[\frac{\hbar|e|F_s[T_1(E'_{i1L}, \eta_g, \lambda)]'}{\sqrt{2m_c}}\right]^{2/3}\right] \tag{15.174}$$

The DOS function can be written as

$$N_{2DiL}(E) = \frac{m_c g_v}{\pi\hbar^2} \sum_{i=0}^{i_{\max}} [P_{3HDL}(E, i, \eta_g, \lambda)]' H(E - E'_{i1L}) \tag{15.175}$$

Thus, the DOS function is complex in nature.

The surface concentration can be written as

$$\bar{n}_{S1} = g_v \text{ Real part of } \sum_{i=0}^{i_{\max}} \left[\frac{m_c}{\pi\hbar^2} P_{3HDL}(E'_{fL}, i, \eta_g, \lambda)\right] \tag{15.176}$$

The total surface electron concentration is given by

$$n_S = \text{Real part of } \bar{n}_S \tag{15.177}$$

where

$$\bar{n}_S = g_v \sum_{i=0}^{i_{\max}} \left[\left[\frac{m_c}{\pi\hbar^2} P_{3HDL}(E'_{fL}, i, \eta_g, \lambda)\right] + \frac{1}{3\pi^2}\left(\frac{2m_c}{\hbar^2}\right)^{3/2} t_i [T_1(E_{FB}, \eta_g, \lambda)]^{3/2}\right] \tag{15.178}$$

in which

$$t_i = \frac{E_{Li1\max}}{eF_s(1 + i_{\max})},$$

$E_{Li1\max}$ is the root of the Real part of the equation

$$0 = \text{Real part of } \left[T_1(E_{Li_{\max}}, \eta_g, \lambda) - S_{i_{\max}} \left[\frac{\hbar|e|F_s[T_1(E_{Li_{\max}}, \eta_g, \lambda)]'}{\sqrt{2m_c}}\right]^{2/3}\right] \tag{15.179}$$

Thus, combining the appropriate equations with (1.4) we can study the QC in this case.

(b) Perturbed two-band model of Kane

In the presence of a surface electric field F_s along z-direction and perpendicular to the surface, the (15.42) assumes the form

$$T_2(E - |e|F_s z, \eta_g, \lambda) = \frac{\hbar^2 k^2}{2m_c} \tag{15.180}$$

Using (15.180) and (1.7), under the weak electric field limit, the 2D DR in HD MOSFETs of HD III–V, ternary and quaternary materials operated under accumulation mode, whose unperturbed band structure is defined by the two-band model of Kane, in the presence of light waves can be expressed as

$$T_2(E,\eta_g,\lambda) = \frac{\hbar^2 k_s^2}{2m_c} + S_i\left[\frac{\hbar|e|F_s[T_2(E,\eta_g,\lambda)]'}{\sqrt{2m_c}}\right]^{2/3} \tag{15.181}$$

Equation (15.181) represents the DR of the 2D electrons in HD MOSFETs III–V, ternary and quaternary materials operated with accumulation mode under the weak electric field limit in the presence of light waves whose bulk electrons obey the HD two-band model of Kane. Since the DR in accordance with the HD two-band model of Kane is real in nature, the (15.181) will also be real.

The EEM can be expressed as

$$m_L^*(E'_{fL2},i,\eta_g) = m_c\,[P_{3HDL2}(E'_{fL2},i,\eta_g,\lambda)]' \tag{15.182}$$

where

$$P_{3HDL2}(E'_{fL2},i,\eta_g,\lambda) = \left[T_2(E'_{fL2},\eta_g,\lambda) - S_i\left[\frac{\hbar|e|F_2[T_2(E'_{fL2},\eta_g,\lambda)]'}{\sqrt{2m_c}}\right]^{2/3}\right],$$

the Fermi energy in this case is E'_{fL2} which can be expressed as

$$E'_{fL2} = eV_g - \frac{e^2 n_s d_{ex}}{\varepsilon_{ox}},$$

and E_{FB2} should be, under the condition of extreme degeneracy, determined from the equation as follows:

$$n_B = \frac{4\pi}{3}\frac{2g_v}{(2\pi)^3}\left(\frac{2m_c}{\hbar^2}\right)^{3/2}[T_2(E_{FB2},\eta_g,\lambda)]^{3/2} \tag{15.183}$$

The sub-band energy E'_{i1L2} is given by

$$0 = \left[T_2(E'_{i1L2},\eta_g,\lambda) - S_i\left[\frac{\hbar|e|F_s[T_2(E'_{i1L2},\eta_g,\lambda)]'}{\sqrt{2m_c}}\right]^{2/3}\right] \tag{15.184}$$

The DOS function can be written as

$$N_{2DiL}(E) = \frac{m_c g_v}{\pi\hbar^2}\sum_{i=0}^{i_{\max}}[P_{3HDL2}(E,i,\eta_g,\lambda)]'H(E-E'_{i1L2}) \tag{15.185}$$

The surface concentration can be written as

$$\bar{n}_{S2} = g_v \sum_{i=0}^{i_{\max}} \left[\frac{m_c}{\pi\hbar^2} P_{3HDL2}(E'_{fL2}, i, \eta_g, \lambda) \right] \tag{15.186}$$

The total surface electron concentration is given by

$$n_S = g_v \sum_{i=0}^{i_{\max}} \left[\left[\frac{m_c}{\pi\hbar^2} P_{3HDL2}(E'_{fL2}, i, \eta_g, \lambda) \right] + \frac{1}{3\pi^2} \left(\frac{2m_c}{\hbar^2} \right)^{3/2} t_i [T_2(E_{FB2}, \eta_g, \lambda)]^{3/2} \right] \tag{15.187}$$

where

$$t_i = \frac{E_{Li2\max}}{eF_s(1 + i_{\max})}$$

$E_{Li2\max}$ is the root of the Real part of the equation

$$0 = \left[T_2(E_{Li2_{\max}}, \eta_g, \lambda) - S_{i_{\max}} \left[\frac{\hbar|e|F_s[T_2(E_{Li2_{\max}}, \eta_g, \lambda)]'}{\sqrt{2m_c}} \right]^{2/3} \right] \tag{15.188}$$

Thus, combining the appropriate equations with (1.4) we can study the QC in this case.

(c) Perturbed parabolic band model

In the presence of a surface electric field F_s along z-direction and perpendicular to the surface, the (15.43) assumes the form

$$T_3(E - |e|F_s z, \eta_g, \lambda) = \frac{\hbar^2 k^2}{2m_c} \tag{15.189}$$

Using (15.189) and (1.7), under the weak electric field limit, the 2D DR in HD MOSFETs of HD III–V, ternary and quaternary materials operated under accumulation mode, whose unperturbed band structure is defined by the parabolic band model, in the presence of light waves can be expressed as

$$T_3(E, \eta_g, \lambda) = \frac{\hbar^2 k_s^2}{2m_c} + S_i \left[\frac{\hbar|e|F_s[T_3(E, \eta_g, \lambda)]'}{\sqrt{2m_c}} \right]^{2/3} \tag{15.190}$$

Equation (15.190) represents the DR of the 2D electrons in HD MOSFETs III–V, ternary and quaternary materials operated with accumulation mode under the weak electric field limit in the presence of light waves whose bulk electrons obey the HD parabolic band model. Since the DR in accordance with the HD parabolic band model is real in nature, the (15.190) will also be real.

The EEM can be expressed as

$$m_L^*(E'_{fL23}, i, \eta_g) = m_c\ [P_{3HDL23}(E'_{fL23}, i, \eta_g, \lambda)]' \tag{15.191}$$

where

$$P_{3HDL23}(E'_{fL23}, i, \eta_g, \lambda) = \left[T_3(E'_{fL23}, \eta_g, \lambda) - S_i\left[\frac{\hbar|e|F_2[T_3(E'_{fL23}, \eta_g, \lambda)]'}{\sqrt{2m_c}}\right]^{2/3}\right],$$

the Fermi energy in this case is E'_{fL23} which can be expressed as

$$E'_{fL23} = eV_g - \frac{e^2 n_s d_{ex}}{\varepsilon_{ox}},$$

and E_{FB3} should be, under the condition extreme degeneracy, determined from the equation

$$n_B = \frac{4\pi}{3}\frac{2g_v}{(2\pi)^3}\left(\frac{2m_c}{\hbar^2}\right)^{3/2}[T_3(E_{FB3}, \eta_g, \lambda)]^{3/2} \tag{15.192}$$

The sub-band energy E'_{i1L23} is given by

$$0 = \left[T_3(E'_{i1L3}, \eta_g, \lambda) - S_i\left[\frac{\hbar|e|F_s[T_3(E'_{i1L3}, \eta_g, \lambda)]'}{\sqrt{2m_c}}\right]^{2/3}\right] \tag{15.193}$$

The DOS function can be written as

$$N_{2DiL}(E) = \frac{m_c g_v}{\pi\hbar^2}\sum_{i=0}^{i_{\max}}[P_{3HDL3}(E, i, \eta_g, \lambda)]' H(E - E'_{i1L3}) \tag{15.194}$$

The surface concentration can be written as

$$\bar{n}_{S3} = g_v\sum_{i=0}^{i_{\max}}\left[\frac{m_c}{\pi\hbar^2}P_{3HDL3}(E'_{fL3}, i, \eta_g, \lambda)\right] \tag{15.195}$$

The total surface electron concentration is given by

$$n_S = g_v\sum_{i=0}^{i_{\max}}\left[\left[\frac{m_c}{\pi\hbar^2}P_{3HDL3}(E'_{fL3}, i, \eta_g, \lambda)\right] + \frac{1}{3\pi^2}\left(\frac{2m_c}{\hbar^2}\right)^{3/2} t_i[T_3(E_{FB3}, \eta_g, \lambda)]^{3/2}\right] \tag{15.196}$$

where

$$t_i = \frac{E_{Li3\max}}{eF_s(1+i_{\max})},$$

$E_{Li3\max}$ is the root of the Real part of the equation

$$0 = \left[T_3(E_{Li3_{\max}}, \eta_g, \lambda) - S_{i_{\max}}\left[\frac{\hbar|e|F_s[T_3(E_{Li3_{\max}}, \eta_g, \lambda)]'}{\sqrt{2m_c}}\right]^{2/3}\right] \qquad (15.197)$$

Thus, combining the appropriate equations with (1.4) we can study the QC in this case.

15.2.12 *The QC in HD magneto MOSFETs in the presence of strong light waves under accumulation mode of operation*

(i) Perturbed three-band model of Kane

Using (15.41), the magneto-dispersion law, in the absence of spin, of III–V, ternary and quaternary materials, in the presence of photo-excitation, whose unperturbed conduction electrons obey the three-band model of Kane, is given by

$$T_1(E, \eta_g, \lambda) = \left(n + \frac{1}{2}\right)\hbar\omega_0 + \frac{\hbar^2 k_z^2}{2m_c} \qquad (15.198)$$

Using (15.198), the DOS function in the present case can be expressed as

$$D_g(E, \eta_g, \lambda) = \frac{g_v|e|\sqrt{2m_c}}{2\pi^2\hbar^2}\sum_{n=0}^{n_{\max}}\left[\left[\{T_1(E, \eta_g, \lambda)\}'\left\{T_1(E, \eta_g, \lambda) - \left(n + \frac{1}{2}\right)\hbar\omega_0\right\}^{-1/2}\right]H(E - E_{l1})\right] \qquad (15.199)$$

where E_{l1} is the Landau sub-band energies in this case and is given as

$$T_1(E_{l1}, \eta_g, \lambda) = \left(n + \frac{1}{2}\right)\hbar\omega_0 \qquad (15.200)$$

The EEM in this case assumes the form

$$m^*(E_{FHDLB}, \eta_g, \lambda) = m_c \text{ Real part of } \{T_1(E_{FHDLB}, \eta_g, \lambda)\}' \qquad (15.201)$$

where E_{FHDLB} is the Fermi energy under quantizing magnetic field in the presence of light waves as measured from the edge of the conduction band in the vertically upward direction in the absence of any quantization.

The electron concentration is given by

$$n_0 = \text{Real part of } \bar{n}_0 \tag{15.202}$$

where,

$$\bar{n}_0 = \frac{g_v|e|\sqrt{2m_c}}{\pi^2\hbar^2}\sum_{n=0}^{n_{\max}}\left[\left\{T_1(E_{FHDLB},\eta_g,\lambda) - \left(n+\frac{1}{2}\right)\hbar\omega_0\right\}^{1/2}\right]$$

The 2D DR in HD magneto MOSFETs of III–V, ternary and quaternary materials operated under accumulation mode, whose unperturbed band structure is defined by the three-band model of Kane, in the presence of light waves under magnetic quantization, can be expressed as

$$T_1(E_{17,30},\eta_g,\lambda) = \left(n+\frac{1}{2}\right)\hbar\omega_0 + S_i\left[\frac{\hbar|e|F_s[T_1(E_{17,30},\eta_g,\lambda)]'}{\sqrt{2m_c}}\right]^{\frac{2}{3}} \tag{15.203}$$

where $E_{17,30}$ is the totally quantized energy in this case (15.203) represents the magneto DR of the 2D electrons of HD magneto MOSFETs of III–V, ternary and quaternary materials operated under accumulation mode under the weak electric field limit in the presence of light waves whose bulk electrons obey the HD three-band model of Kane. Since the DR in accordance with the HD three-band model of Kane is complex in nature, the (15.203) will also be complex in the energy plane. The total energy is quantized since the wave vector space is totally quantized.

The DOS function is given by

$$N_{2DALB}(E) = \frac{g_v eB}{\pi\hbar}\sum_{n=0}^{n_{\max}}\sum_{i=0}^{i_{\max}}\delta'(E - E_{17,30}) \tag{15.204}$$

The 2D surface electron concentration can be written as

$$n_0 = \text{Real part of } \bar{n}_0 \tag{15.205}$$

where

$$\bar{n}_0 = \frac{g_v eB}{\pi\hbar}\sum_{n=0}^{n_{\max}}\sum_{i=0}^{i_{\max}}F_{-1}(\eta_{17,30}) \tag{15.206}$$

in which

$$\eta_{17,30} = \frac{E_{F17,30} - E_{17,30}}{k_B T}$$

and $E_{F17,30}$ is the Fermi energy in this case and can be written as

$$E_{F17,30} \equiv eV_g - \frac{e^2 N_s}{C_{OX}}$$

The total surface concentration in HD magneto MOSFET in the presence of strong light waves under accumulation mode of operation in this case and can be written as

$$N_s = \text{Real part of } g_v \sum_{n=0}^{n_{\max}} \sum_{i=0}^{i_{\max}} \left[\frac{eB}{\pi\hbar}[F_{-1}(\eta_{17,30})] + \left(\frac{|e|\sqrt{2m_c}}{\pi^2\hbar^2} t_{i1BHD} \right) \times \left[\left\{ T_1(E_{FHDLB}, \eta_g, \lambda) - \left(n+\frac{1}{2}\right)\hbar\omega_0 \right\} \right]^{1/2} \right]$$

where

$$t_{i1BHD} = \frac{E_{15,1\max HD}}{eF_s(1+i_{\max})},$$

$E_{15,1\max HD}$ is the maximum value of the real part of the equation

$$T_1(E_{15,1\max HD}, \eta_g, \lambda) = \left(n+\frac{1}{2}\right)\hbar\omega_0 + S_{i_{\max}} \left[\frac{\hbar|e|F_s[T_1(E_{15,1\max HD}, \eta_g, \lambda)]'}{\sqrt{2m_c}} \right]^{2/3} \tag{15.207}$$

Thus, combining the appropriate equations with (1.4) we can study the QC in this case.

(ii) Perturbed two-band model of Kane

Using (15.42), the magneto-dispersion law, in the absence of spin, for the carriers of III–V, ternary and quaternary materials, in the presence of photo-excitation, whose unperturbed conduction electrons obey the two-band model of Kane, is given by

$$T_2(E, \eta_g, \lambda) = \left(n+\frac{1}{2}\right)\hbar\omega_0 + \frac{\hbar^2 k_z^2}{2m_c} \tag{15.208}$$

Using (15.208), the DOS function in the present case can be expressed as

$$D_g(E, \eta_g, \lambda) = \frac{g_v|e|\sqrt{2m_c}}{2\pi^2\hbar^2} \sum_{n=0}^{n_{\max}} \left[\left[\{T_2(E, \eta_g, \lambda)\}' \left\{ T_2(E, \eta_g, \lambda) - \left(n+\frac{1}{2}\right)\hbar\omega_0 \right\}^{-1/2} \right] H(E - E_{l1}) \right] \tag{15.209}$$

where E_{l2} is the Landau sub-band energies in this case and is given as

$$T_2(E_{l2}, \eta_g, \lambda) = \left(n + \frac{1}{2}\right) \hbar\omega_0 \tag{15.210}$$

The EEM in this case assumes the form

$$m^*(E_{FHDLB}, \eta_g, \lambda) = m_c\{T_2(E_{FHDLB}, \eta_g, \lambda)\}' \tag{15.211}$$

where E_{FHDLB} is the Fermi energy under quantizing magnetic field in the presence of strong light waves as measured from the edge of the conduction band in the vertically upward direction in the absence of any quantization.

The electron concentration is given by

$$n_0 = \bar{n}_0 \tag{15.212}$$

where,

$$\bar{n}_0 = \frac{g_v|e|\sqrt{2m_c}}{\pi^2\hbar^2} \sum_{n=0}^{n_{\max}} \left[\left\{T_2(E_{FHDLB}, \eta_g, \lambda) - \left(n + \frac{1}{2}\right) \hbar\omega_0\right\}^{1/2}\right]$$

The 2D DR in HD magneto MOSFETs of III–V, ternary and quaternary materials operated under accumulation mode, whose unperturbed band structure is defined by the two-band model of Kane, in the presence of light waves under magnetic quantization, can be expressed as

$$T_2(E_{17,32}, \eta_g, \lambda) = \left(n + \frac{1}{2}\right) \hbar\omega_0 + S_i \left[\frac{\hbar|e|F_s[T_2(E_{17,32}, \eta_g, \lambda)]'}{\sqrt{2m_c}}\right]^{\frac{2}{3}} \tag{15.213}$$

where $E_{17,31}$ is the totally quantized energy in this case (15.205) represents the magneto DR of the 2D electrons of HD magneto MOSFETs of III–V, ternary and quaternary materials operated under accumulation mode under the weak electric field limit in the presence of light waves whose bulk electrons obey the HD two-band model of Kane.

The DOS function is given by

$$N_{2DALB}(E) = \frac{g_v eB}{\pi\hbar} \sum_{n=0}^{n_{\max}} \sum_{i=0}^{i_{\max}} \delta'(E - E_{17,32}) \tag{15.214}$$

The 2D surface electron concentration can be written as

$$n_0 = \bar{n}_0 \tag{15.215}$$

where

$$\bar{n}_0 = \frac{g_v eB}{\pi\hbar} \sum_{n=0}^{n_{\max}} \sum_{i=0}^{i_{\max}} F_{-1}(\eta_{17,32}) \tag{15.216}$$

in which

$$\eta_{17,32} = \frac{E_{F17,32} - E_{17,32}}{k_B T}$$

and $E_{\mathrm{F}17,32}$ is the Fermi energy in this case and can be written as

$$E_{F17,31} \equiv eV_g - \frac{e^2 N_s}{C_{OX}}$$

The total surface concentration in HD magneto MOSFET of in the presence of strong light waves under accumulation mode of operation in this case and can be written as

$$N_s = g_v \sum_{n=0}^{n_{\max}} \sum_{i=0}^{i_{\max}} \left[\frac{eB}{\pi\hbar}[F_{-1}(\eta_{17,32})] + \left(\frac{|e|\sqrt{2m_c}}{\pi^2\hbar^2} t_{i2BHD}\right) \right.$$

$$\left. \times \left[\left\{T_2(E_{FHDLB}, \eta_g, \lambda) - \left(n+\frac{1}{2}\right)\hbar\omega_0\right\}\right]^{1/2} \right] \quad (15.217)$$

where

$$t_{i2BHD} = \frac{E_{15,2\max HD}}{eF_s(1+i_{\max})},$$

$E_{15,2\max HD}$ is the maximum value of the real part of the equation

$$T_2(E_{15,2\max HD}, \eta_g, \lambda)$$

$$= \left(n+\frac{1}{2}\right)\hbar\omega_0 + S_{i_{\max}} \left[\frac{\hbar|e|F_s[T_2(E_{15,2\max HD}, \eta_g, \lambda)]'}{\sqrt{2m_c}}\right]^{\frac{2}{3}} \quad (15.218)$$

Thus, combining the appropriate equations with (1.4) we can study the QC in this case.

(iii) Perturbed parabolic band model

Using (15.43), the magneto-dispersion law, in the absence of spin, for the carriers of III–V, ternary and quaternary materials, in the presence of photo-excitation, whose unperturbed conduction electrons obey the three-band model of Kane, is given by

$$T_3(E, \eta_g, \lambda) = \left(n+\frac{1}{2}\right)\hbar\omega_0 + \frac{\hbar^2 k_z^2}{2m_c} \quad (15.219)$$

Using (15.219), the DOS function in the present case can be expressed as

$$D_g(E, \eta_g, \lambda) = \frac{g_v|e|\sqrt{2m_c}}{2\pi^2\hbar^2} \sum_{n=0}^{n_{\max}} \left[\left[\{T_3(E, \eta_g, \lambda)\}' \left\{T_3(E, \eta_g, \lambda) \right.\right.\right.$$

$$\left.\left.\left. - \left(n+\frac{1}{2}\right)\hbar\omega_0\right\}^{-1/2}\right] H(E - E_{l3})\right] \quad (15.220)$$

where E_{l3} is the Landau sub-band energies in this case and is given as

$$T_3(E_{l3}, \eta_g, \lambda) = \left(n + \frac{1}{2}\right) \hbar\omega_0 \tag{15.221}$$

The EEM in this case assumes the form

$$m^*(E_{FHDLB}, \eta_g, \lambda) = m_c\{T_3(E_{FHDLB}, \eta_g, \lambda)\}' \tag{15.222}$$

where E_{FHDLB} is the Fermi energy under quantizing magnetic field in the presence of light waves as measured from the edge of the conduction band in the vertically upward direction in the absence of any quantization.

The electron concentration is given by

$$n_0 = \bar{n}_0 \tag{15.223}$$

where,

$$\bar{n}_0 = \frac{g_v|e|\sqrt{2m_c}}{\pi^2\hbar^2} \sum_{n=0}^{n_{\max}} \left[\left\{T_3(E_{FHDLB}, \eta_g, \lambda) - \left(n + \frac{1}{2}\right) \hbar\omega_0\right\}^{1/2}\right]$$

The 2D DR in HD magneto MOSFETs of III–V, ternary and quaternary materials operated under accumulation mode, whose unperturbed band structure is defined by the parabolic energy bands, in the presence of light waves under magnetic quantization, can be expressed as

$$T_3(E_{17,34}, \eta_g, \lambda) = \left(n + \frac{1}{2}\right) \hbar\omega_0 + S_i \left[\frac{\hbar|e|F_s[T_3(E_{17,34}, \eta_g, \lambda)]'}{\sqrt{2m_c}}\right]^{\frac{2}{3}} \tag{15.224}$$

where $E_{17,31}$ is the totally quantized energy in this case (15.205) represents the magneto DR of the 2D electrons of HD magneto MOSFETs of III–V, ternary and quaternary materials operated under accumulation mode under the weak electric field limit in the presence of light waves whose bulk electrons obey the HD parabolic model.

The DOS function is given by

$$N_{2DALB}(E) = \frac{g_v eB}{\pi\hbar} \sum_{n=0}^{n_{\max}} \sum_{i=0}^{i_{\max}} \delta'(E - E_{17,34}) \tag{15.225}$$

The 2D surface electron concentration can be written as

$$n_0 = \bar{n}_0 \tag{15.226}$$

where

$$\bar{n}_0 = \frac{g_v eB}{\pi\hbar} \sum_{n=0}^{n_{\max}} \sum_{i=0}^{i_{\max}} F_{-1}(\eta_{17,34}) \tag{15.227}$$

in which

$$\eta_{17,34} = \frac{E_{F17,34} - E_{17,34}}{k_B T}$$

and $E_{F17,34}$ is the Fermi energy in this case and can be written as

$$E_{F17,34} \equiv eV_g - \frac{e^2 N_s}{C_{OX}}$$

The total surface concentration in HD magneto MOSFET of in the presence of strong light waves under accumulation mode of operation in this case and can be written as

$$N_s = g_v \sum_{n=0}^{n_{\max}} \sum_{i=0}^{i_{\max}} \left[\frac{eB}{\pi\hbar} [F_{-1}(\eta_{17,34})] + \left(\frac{|e|\sqrt{2m_c}}{\pi^2 \hbar^2} t_{i3BHD} \right) \right.$$
$$\left. \times \left[\left\{ T_3(E_{FHDLB}, \eta_g, \lambda) - \left(n + \frac{1}{2}\right) \hbar\omega_0 \right\} \right]^{1/2} \right] \quad (15.228)$$

where

$$t_{i3BHD} = \frac{E_{15,3\max HD}}{eF_s(1 + i_{\max})},$$

$E_{15,3\max HD}$ is the maximum value of the real part of the equation

$$T_3(E_{15,3\max HD}, \eta_g, \lambda)$$
$$= \left(n + \frac{1}{2}\right) \hbar\omega_0 + S_{i_{\max}} \left[\frac{\hbar|e|F_s[T_2(E_{15,3\max HD}, \eta_g, \lambda)]'}{\sqrt{2m_c}} \right]^{\frac{2}{3}} \quad (15.229)$$

Thus, combining the appropriate equations with (1.4) we can study the QC in this case.

15.2.13 *The QC in 2D HD MOSFETs in the presence of strong light waves and the crossed fields configurations under accumulation mode of operation*

(i) Perturbed three-band model of Kane

The electron dispersion law in the present case is given by

$$T_1(E, \eta_g, \lambda) = \left(n + \frac{1}{2}\right) \hbar\omega_0 + \frac{[\hbar k_z(E)]^2}{2m_c} - \frac{E_0}{B} \hbar k_y \{T_1(E, \eta_g, \lambda)\}'$$
$$- \left\{ \frac{m_c E_0^2 [\{T_1(E, \eta_g, \lambda)\}']^2}{2B^2} \right\} \quad (15.230)$$

The use of (15.230) leads to the expressions of the EEMs' along z and y directions as

$$m_z^*(E_{F_{BLHDC}}, n, E_0, B, \lambda)$$
$$= \text{Real part of } m_c \left[\{T_1(E_{F_{BLHDC}}, \eta_g, \lambda)\}'' + \frac{m_c E_0^2 \{T_1(E_{F_{BLHDC}}, \eta_g, \lambda)\}' \{T_1(E_{F_{BLHDC}}, \eta_g, \lambda)\}''}{B^2} \right] \quad (15.231)$$

$$m_y^*(E_{F_{BLHDC}}, n, E_0, B, \lambda)$$
$$= \text{Real part of } \left(\frac{B}{E_0}\right)^2 [\{T_1(E_{F_{BLHDC}}, \eta_g, \lambda)\}']^{-1} \left[T_1(E_{F_{BLHDC}}, \eta_g, \lambda) - \left(n + \frac{1}{2}\right)\hbar\omega_0 + \frac{m_c E_0^2 \{T_1(E_{F_{BLHDC}}, \eta_g, \lambda)\}'}{2B^2} \right]$$
$$\times \left[\frac{\{T_1(E_{F_{BLHDC}}, \eta_g, \lambda)\}''}{[\{T_1(E_{F_{BLHDC}}, \eta_g, \lambda)\}']^2} \left[T_1(E_{F_{BLHDC}}, \eta_g, \lambda) - \left(n + \frac{1}{2}\right)\hbar\omega_0 + \frac{m_c E_0^2 [\{T_1(E_{F_{BLHDC}}, \eta_g, \lambda)\}']^2}{2B^2} \right] + 1 + \frac{m_c E_0^2 \{T_1(E_{F_{BLHDC}}, \eta_g, \lambda)\}''}{B^2} \right] \quad (15.232)$$

where $E_{F_{BLHDC}}$ is the Fermi energy in this case.

The Landau energy $(E_{n_{l41}})$ can be written as

$$T_1(E_{n_{l41}}, \eta_g, \lambda) = \left(n + \frac{1}{2}\right)\hbar\omega_0 - \left\{ \frac{m_c E_0^2 [\{T_1(E_{n_{l41}}, \eta_g, \lambda)\}']^2}{2B^2} \right\} \quad (15.233)$$

The electron concentration in this case can be expressed as

$$n_0 = \text{Real part of } \bar{n}_0 \quad (15.234)$$

where

$$n_0 = \frac{2g_v B\sqrt{2m_c}}{3L_x \pi^2 \hbar^2 E_0} \sum_{n=0}^{n_{\max}} [M_{161}(E_{F_{BLHDC}}, n, E_0, B, \lambda) + N_{161}(E_{F_{BLHDC}}, n, E_0, B, \lambda)] \quad (15.235)$$

in which

$$M_{161}(n, E_{F_{BL}}, \lambda) = \left[\left[T_1(E_{F_{BLHDC}}, n, E_0, B, \lambda) - \left(n + \frac{1}{2}\right)\hbar\omega_0 - \frac{m_c E_0^2}{2B^2} \right] \times [\{T_1(E_{F_{BLHDC}}, n, E_0, B, \lambda)\}']^2 \right.$$

$$+|e|E_0L_x[\{T_1(E_{F_{BLHDC}},n,E_0,B,\lambda)\}']^{3/2}$$
$$-\left[T_1(E_{F_{BLHDC}},n,E_0,B,\lambda)-\left(n+\frac{1}{2}\right)\hbar\omega_0\right.$$
$$\left.\left.-\frac{m_cE_0^2}{2B^2}[\{T_1(E_{F_{BLHDC}},n,E_0,B,\lambda)\}']^2\right]^{3/2}\right]$$
$$\times\frac{1}{[\{T_1(E_{F_{BLHDC}},n,E_0,B,\lambda)\}']} \tag{15.236}$$

and

$$N_{161}(n,E_{F_{BL}},\lambda)=\sum_{r=1}^{s}[L(r)M_{161}(n,E_{F_{BL}},\lambda)]. \tag{15.237}$$

In the presence of a surface electric field F_s along z-direction from (15.220) we can write that

$$k_z=\frac{\sqrt{2m_c}}{\hbar}[\alpha_{15,13,1}-eF_sz\beta_{15,13,1}]^{1/2} \tag{15.238}$$

where

$$\alpha_{15,13,1}=\left[T_1(E,\eta_g,\lambda)-\left(n+\frac{1}{2}\right)\hbar\omega_0+\frac{E_0}{B}\hbar k_y\{T_1(E,\eta_g,\lambda)\}'\right.$$
$$\left.+\left\{\frac{m_cE_0^2[\{T_1(E,\eta_g,\lambda)\}']^2}{2B^2}\right\}\right]$$

and

$$\beta_{15,13,1}=\left[T_1'(E,\eta_g,\lambda)+\frac{E_0}{B}\hbar k_y\{T_1(E,\eta_g,\lambda)\}'\right.$$
$$\left.+\left\{\frac{m_cE_0^2[\{T_1(E,\eta_g,\lambda)\}'][\{T_1(E,\eta_g,\lambda)\}'']}{B^2}\right\}\right]$$

Using (15.238), the 2D DR in this case assumes the form

$$\left[\left[T_1(E,\eta_g,\lambda)-\left(n+\frac{1}{2}\right)\hbar\omega_0\right.\right.$$
$$\left.+\frac{E_0}{B}\hbar k_y\{T_1(E,\eta_g,\lambda)\}'+\left\{\frac{m_cE_0^2[\{T_1(E,\eta_g,\lambda)\}']^2}{2B^2}\right\}\right]$$
$$=S_i\left(\frac{e^2\hbar}{\varepsilon_{sc}\sqrt{2m_c}}\right)^{2/3}N_s^{2/3}\times\left[T_1'(E,\eta_g,\lambda)+\frac{E_0}{B}\hbar k_y\{T_1(E,\eta_g,\lambda)\}'\right.$$
$$\left.\left.+\left\{\frac{m_cE_0^2[\{T_1(E,\eta_g,\lambda)\}'][\{T_1(E,\eta_g,\lambda)\}'']}{B^2}\right\}\right]^{2/3}\right] \tag{15.239}$$

The sub-band energy in this case ($E_{15,13,1s}$) can be written as

$$\left[\left[T_1(E_{15,13,1s},\eta_g,\lambda)-\left(n+\frac{1}{2}\right)\hbar\omega_0+\left\{\frac{m_c E_0^2[\{T_1(E_{15,13,1s},\eta_g,\lambda)\}']^2}{2B^2}\right\}\right]\right.$$

$$= S_i\left(\frac{e^2\hbar}{\varepsilon_{sc}\sqrt{2m_c}}\right)^{2/3} N_s^{2/3}$$

$$\times\left[T_1'(E_{15,13,1s},\eta_g,\lambda)\right.$$

$$\left.+\left\{\frac{m_c E_0^2[\{T_1(E_{15,13,1s},\eta_g,\lambda)\}'][\{T_1(E_{15,13,1s},\eta_g,\lambda)\}'']}{B^2}\right\}\right]^{2/3}\Bigg] \tag{15.240}$$

The surface electron concentration can be written as

$$\bar{N}_s = g_v \text{ Real part of } \sum_{i=0}^{i_{\max}}\sum_{n=0}^{n_{\max}}\frac{g_s eB}{2\pi\hbar}F_{-1}(\eta_{15,13,1})] \tag{15.241}$$

where

$$\eta_{15,13,1}=(k_BT)^{-1}(E_{F15,13,1}-E_{15,13,1s}) \quad \text{and} \quad E_{F15,13,1}=eV_g-\frac{e^2N_s}{C_{OX}}$$

The total surface concentration per unit area in this case can be written as

$$N_s = g_v \text{ Real part of } \sum_{i=0}^{i_{\max}}\sum_{n=0}^{n_{\max}}\left[\frac{2B\sqrt{2m_c}}{3L_x\pi^2\bar{\hbar}^2E_0}\bar{t}_{15,13,1\max}\right.$$

$$\times[M_{161}(E_{F_{BLHDC}},n,E_0,B,\lambda)+N_{161}(E_{F_{BLHDC}},n,E_0,B,\lambda)]$$

$$\left.+\frac{g_s eB}{2\pi\hbar}F_{-1}(\eta_{15,13,1})\right] \tag{15.242a}$$

where

$$\bar{t}_{15,13,1\max}=\left[\frac{\varepsilon_{sc}E_{15,13,1s\max}}{N_s e^2(1+i_{\max})}\right]$$

and $E_{15,13,1s\max}$ is the maximum value of the positive real part of the equation

$$\left[\left[T_1(E_{15,13,1s\max},\eta_g,\lambda)\right.\right.$$

$$\left.-\left(n+\frac{1}{2}\right)\hbar\omega_0+\left\{\frac{m_c E_0^2[\{T_1(E_{15,13,1s\max},\eta_g,\lambda)\}']^2}{2B^2}\right\}\right]$$

$$= S_i \left(\frac{e^2\hbar}{\varepsilon_{sc}\sqrt{2m_c}}\right)^{2/3} N_s^{2/3}$$
$$\times \left[T_1'(E_{15,13,1s\max}, \eta_g, \lambda)\right.$$
$$\left. + \left\{\frac{m_c E_0^2[\{T_1(E_{15,13,1s\max}, \eta_g, \lambda)\}'][\{T_1(E_{15,13,1s\max}, \eta_g, \lambda)\}'']}{B^2}\right\}\right]^{2/3}$$

Using (15.242) and (1.4) we can study the QC in this case.

(ii) Perturbed two-band model of Kane

The electron dispersion law in the present case is given by

$$T_2(E, \eta_g, \lambda) = \left(n + \frac{1}{2}\right)\hbar\omega_0 + \frac{[\hbar k_z(E)]^2}{2m_c} - \frac{E_0}{B}\hbar k_y\{T_2(E, \eta_g, \lambda)\}' - \left\{\frac{m_c E_0^2[\{T_2(E, \eta_g, \lambda)\}']^2}{2B^2}\right\} \quad (15.242b)$$

The use of (15.242) leads to the expressions of the EEMs' along z and y directions as

$$m_z^*(E_{F_{BLHDC}}, n, E_0, B, \lambda) = m_c\left[\{T_2(E_{F_{BLHDC}}, \eta_g, \lambda)\}'' + \frac{m_c E_0^2\{T_2(E_{F_{BLHDC}}, \eta_g, \lambda)\}'\{T_2(E_{F_{BLHDC}}, \eta_g, \lambda)\}''}{B^2}\right] \quad (15.243)$$

$$m_y^*(E_{F_{BLHDC}}, n, E_0, B, \lambda)$$
$$= \left(\frac{B}{E_0}\right)^2 [\{T_2(E_{F_{BLHDC}}, \eta_g, \lambda)\}']^{-1}\left[T_2(E_{F_{BLHDC}}, \eta_g, \lambda) - \left(n + \frac{1}{2}\right)\hbar\omega_0\right.$$
$$\left. + \frac{m_c E_0^2\{T_2(E_{F_{BLHDC}}, \eta_g, \lambda)\}'}{2B^2}\right]\left[\frac{\{T_2(E_{F_{BLHDC}}, \eta_g, \lambda)\}''}{[\{T_2(E_{F_{BLHDC}}, \eta_g, \lambda)\}']^2}\right.$$
$$\times \left[T_2(E_{F_{BLHDC}}, \eta_g, \lambda) - \left(n + \frac{1}{2}\right)\hbar\omega_0 + \frac{m_c E_0^2[\{T_2(E_{F_{BLHDC}}, \eta_g, \lambda)\}']^2}{2B^2}\right]$$
$$\left. + 1 + \frac{m_c E_0^2\{T_2(E_{F_{BLHDC}}, \eta_g, \lambda)\}''}{B^2}\right] \quad (15.244)$$

where $E_{F_{BLHDC}}$ is the Fermi energy in this case.

The Landau energy ($E_{n_{l41}}$) can be written as

$$T_2(E_{n_{l42}}, \eta_g, \lambda) = \left(n + \frac{1}{2}\right)\hbar\omega_0 - \left\{\frac{m_c E_0^2[\{T_2(E_{n_{l42}}, \eta_g, \lambda)\}']^2}{2B^2}\right\} \tag{15.245}$$

The electron concentration in this case can be expressed as

$$n_0 = \bar{n}_0 \tag{15.246}$$

where

$$n_0 = \frac{2g_v B\sqrt{2m_c}}{3L_x\pi^2\bar{h}^2 E_0}\sum_{n=0}^{n_{\max}}[M_{162}(E_{F_{BLHDC}}, n, E_0, B, \lambda) + N_{162}(E_{F_{BLHDC}}, n, E_0, B, \lambda)] \tag{15.247}$$

in which

$$M_{162}(n, E_{F_{BL}}, \lambda) = \Bigg[\Bigg[T_2(E_{F_{BLHDC}}, n, E_0, B, \lambda) - \left(n + \frac{1}{2}\right)\hbar\omega_0 - \frac{m_c E_0^2}{2B^2}\Bigg] \times [\{T_2(E_{F_{BLHDC}}, n, E_0, B, \lambda)\}']^2 + |e|E_0 L_x[\{T_2(E_{F_{BLHDC}}, n, E_0, B, \lambda)\}']^{3/2} - [T_2(E_{F_{BLHDC}}, n, E_0, B, \lambda) - \left(n + \frac{1}{2}\right)\hbar\omega_0 - \frac{m_c E_0^2}{2B^2}[\{T_2(E_{F_{BLHDC}}, n, E_0, B, \lambda)\}']^2\Bigg]^{3/2}\Bigg] \times \frac{1}{[\{T_2(E_{F_{BLHDC}}, n, E_0, B, \lambda)\}']} \tag{15.248}$$

and

$$N_{162}(n, E_{F_{BL}}, \lambda) = \sum_{r=1}^{s}[L(r)M_{162}(n, E_{F_{BL}}, \lambda)]. \tag{15.249}$$

In the presence of a surface electric field Fs along z direction from (15.242) we can write that

$$k_z = \frac{\sqrt{2m_c}}{\hbar}[\alpha_{15,13,2} - eF_s z\beta_{15,13,2}]^{1/2} \tag{15.250}$$

where

$$\alpha_{15,13,2} = \left[T_2(E, \eta_g, \lambda) - \left(n + \frac{1}{2}\right)\hbar\omega_0 + \frac{E_0}{B}\hbar k_y\{T_2(E, \eta_g, \lambda)\}' + \left\{\frac{m_c E_0^2[\{T_2(E, \eta_g, \lambda)\}']^2}{2B^2}\right\}\right]$$

and

$$\beta_{15,13,2} = \left[T_2'(E,\eta_g,\lambda) + \frac{E_0}{B}\hbar k_y\{T_2(E,\eta_g,\lambda)\}' + \left\{\frac{m_c E_0^2[\{T_2(E,\eta_g,\lambda)\}'][\{T_2(E,\eta_g,\lambda)\}'']}{B^2}\right\}\right]$$

Using (15.250), the 2D DR in this case assumes the form

$$\left[\left[T_2(E,\eta_g,\lambda) - \left(n+\frac{1}{2}\right)\hbar\omega_0 + \frac{E_0}{B}\hbar k_y\{T_2(E,\eta_g,\lambda)\}' + \left\{\frac{m_c E_0^2[\{T_2(E,\eta_g,\lambda)\}']^2}{2B^2}\right\}\right] = S_i\left(\frac{e^2\hbar}{\varepsilon_{sc}\sqrt{2m_c}}\right)^{2/3} N_s^{2/3} \times \left[T_2'(E,\eta_g,\lambda) + \frac{E_0}{B}\hbar k_y\{T_2(E,\eta_g,\lambda)\}' + \left\{\frac{m_c E_0^2[\{T_2(E,\eta_g,\lambda)\}'][\{T_2(E,\eta_g,\lambda)\}'']}{B^2}\right\}\right]^{2/3}\right] \quad (15.251)$$

The sub-band energy in this case ($E_{15,13,1s}$) can be written as

$$\left[\left[T_2(E_{15,13,2s},\eta_g,\lambda) - \left(n+\frac{1}{2}\right)\hbar\omega_0 + \left\{\frac{m_c E_0^2[\{T_2(E_{15,13,2s},\eta_g,\lambda)\}']^2}{2B^2}\right\}\right] = S_i\left(\frac{e^2\hbar}{\varepsilon_{sc}\sqrt{2m_c}}\right)^{2/3} N_s^{2/3} \times [T_2'(E_{15,13,2s},\eta_g,\lambda) + \left\{\frac{m_c E_0^2[\{T_2(E_{15,13,2s},\eta_g,\lambda)\}'][\{T_2(E_{15,13,2s},\eta_g,\lambda)\}'']}{B^2}\right\}\right]^{2/3}\right] \quad (15.252)$$

The surface electron concentration can be written as

$$\bar{N}_s = g_v \sum_{i=0}^{i_{\max}} \sum_{n=0}^{n_{\max}} \left[\frac{g_s eB}{2\pi\hbar} F_{-1}(\eta_{15,13,2})\right] \quad (15.253)$$

where

$$\eta_{15,13,2} = (k_B T)^{-1}(E_{F15,13,2} - E_{15,13,2s}) \quad \text{and}$$

$$E_{F15,13,2} = eV_g - \frac{e^2 N_s}{C_{OX}}$$

The total surface concentration per unit area in this case can be written as

$$N_s = g_v \sum_{i=0}^{i_{\max}} \sum_{n=0}^{n_{\max}} \left[\frac{2B\sqrt{2m_c}}{3L_x\pi^2\bar{\hbar}^2E_0} \bar{t}_{15,13,2\max}[M_{162}(E_{F_{BLHDC}}, n, E_0, B, \lambda) + N_{162}(E_{F_{BLHDC}}, n, E_0, B, \lambda)] + \frac{g_s eB}{2\pi\hbar} F_{-1}(\eta_{15,13,2}) \right] \quad (15.254)$$

where

$$\bar{t}_{15,13,2\max} = \left[\frac{\varepsilon_{sc} E_{15,13,2s\max}}{N_s e^2(1+i_{\max})} \right]$$

and $E_{15,13,2s\max}$ is the maximum value of the positive real part of the equation

$$\left[\left[T_2(E_{15,13,2s\max}, \eta_g, \lambda) - \left(n+\frac{1}{2}\right)\hbar\omega_0 + \left\{ \frac{m_c E_0^2[\{T_2(E_{15,13,2s\max}, \eta_g, \lambda)\}']^2}{2B^2} \right\}\right]\right.$$
$$= S_i \left(\frac{e^2\hbar}{\varepsilon_{sc}\sqrt{2m_c}} \right)^{2/3} N_s^{2/3} \times \left[T_2'(E_{15,13,2s\max}, \eta_g, \lambda) + \left\{ \frac{m_c E_0^2[\{T_2(E_{15,13,2s\max}, \eta_g, \lambda)\}'][\{T_2(E_{15,13,2s\max}, \eta_g, \lambda)\}'']}{B^2} \right\} \right]^{2/3}$$

Using (15.254) and (1.4) wecan study the QC in this case.

(iii) Perturbed parabolic band model

The electron dispersion law in the present case is given by

$$T_3(E, \eta_g, \lambda) = \left(n+\frac{1}{2}\right)\hbar\omega_0 + \frac{[\hbar k_z(E)]^2}{2m_c} - \frac{E_0}{B}\hbar k_y\{T_3(E, \eta_g, \lambda)\}' - \left\{ \frac{m_c E_0^2[\{T_3(E, \eta_g, \lambda)\}']^2}{2B^2} \right\} \quad (15.255)$$

The use of (15.255) leads to the expressions of the EEMs' along z and y directions as

$$m_z^*(E_{F_{BLHDC}}, n, E_0, B, \lambda) = m_c \left[\{T_2(E_{F_{BLHDC}}, \eta_g, \lambda)\}'' + \frac{m_c E_0^2\{T_2(E_{F_{BLHDC}}, \eta_g, \lambda)\}'\{T_2(E_{F_{BLHDC}}, \eta_g, \lambda)\}''}{B^2} \right] \quad (15.256)$$

$$m_y^*(E_{F_{BLHDC}}, n, E_0, B, \lambda)$$
$$= \left(\frac{B}{E_0}\right)^2 [\{T_2(E_{F_{BLHDC}}, \eta_g, \lambda)\}']^{-1}$$
$$\times \left[T_2(E_{F_{BLHDC}}, \eta_g, \lambda) - \left(n + \frac{1}{2}\right)\hbar\omega_0 + \frac{m_c E_0^2 \{T_2(E_{F_{BLHDC}}, \eta_g, \lambda)\}'}{2B^2}\right]$$
$$\times \left[\frac{\{T_2(E_{F_{BLHDC}}, \eta_g, \lambda)\}''}{[\{T_2(E_{F_{BLHDC}}, \eta_g, \lambda)\}']^2}\right.$$
$$\left[T_2(E_{F_{BLHDC}}, \eta_g, \lambda) - \left(n + \frac{1}{2}\right)\hbar\omega_0 + \frac{m_c E_0^2 [\{T_2(E_{F_{BLHDC}}, \eta_g, \lambda)\}']^2}{2B^2}\right]$$
$$\left. + 1 + \frac{m_c E_0^2 \{T_2(E_{F_{BLHDC}}, \eta_g, \lambda)\}''}{B^2}\right] \tag{15.257}$$

where $E_{F_{BLHDC}}$ is the Fermi energy in this case.

The Landau energy ($E_{n_{l43}}$) can be written as

$$T_3(E_{n_{l43}}, \eta_g, \lambda) = \left(n + \frac{1}{2}\right)\hbar\omega_0 - \left\{\frac{m_c E_0^2 [\{T_3(E_{n_{l43}}, \eta_g, \lambda)\}']^2}{2B^2}\right\} \tag{15.258}$$

The electron concentration in this case can be expressed as

$$n_0 = \bar{n}_0 \tag{15.259}$$

where

$$n_0 = \frac{2g_v B\sqrt{2m_c}}{3L_x \pi^2 \hbar^2 E_0} \sum_{n=0}^{n_{\max}} [M_{163}(E_{F_{BLHDC}}, n, E_0, B, \lambda)$$
$$+ N_{163}(E_{F_{BLHDC}}, n, E_0, B, \lambda)] \tag{15.260}$$

in which

$$M_{163}(n, E_{F_{BL}}, \lambda) = \left[\left[T_3(E_{F_{BLHDC}}, n, E_0, B, \lambda) - \left(n + \frac{1}{2}\right)\hbar\omega_0 - \frac{m_c E_0^2}{2B^2}\right]\right.$$
$$[\{T_3(E_{F_{BLHDC}}, n, E_0, B, \lambda)\}']^2 + |e|E_0 L_x [\{T_3(E_{F_{BLHDC}}, n, E_0, B, \lambda)\}']^{3/2}$$
$$- \left[T_3(E_{F_{BLHDC}}, n, E_0, B, \lambda) - \left(n + \frac{1}{2}\right)\hbar\omega_0\right.$$
$$\left.\left. - \frac{m_c E_0^2}{2B^2} [\{T_3(E_{F_{BLHDC}}, n, E_0, B, \lambda)\}']^2\right]^{3/2}\right]$$
$$\frac{1}{[\{T_3(E_{F_{BLHDC}}, n, E_0, B, \lambda)\}']} \tag{15.261}$$

and

$$N_{163}(n, E_{F_{BL}}, \lambda) = \sum_{r=1}^{s} [L(r) M_{163}(n, E_{F_{BL}}, \lambda)]. \quad (15.262)$$

In the presence of a surface electric field F_s along z-direction from (15.255) we can write that

$$k_z = \frac{\sqrt{2m_c}}{\hbar}[\alpha_{15,13,3} - eF_s z \beta_{15,13,3}]^{1/2} \quad (15.263)$$

where

$$\alpha_{15,13,3} = \left[T_3(E, \eta_g, \lambda) - \left(n + \frac{1}{2}\right)\hbar\omega_0 + \frac{E_0}{B}\hbar k_y \{T_3(E, \eta_g, \lambda)\}' + \left\{\frac{m_c E_0^2 [\{T_3(E, \eta_g, \lambda)\}']^2}{2B^2}\right\}\right]$$

and

$$\beta_{15,13,3} = \left[T_3'(E, \eta_g, \lambda) + \frac{E_0}{B}\hbar k_y \{T_3(E, \eta_g, \lambda)\}' + \left\{\frac{m_c E_0^2 [\{T_3(E, \eta_g, \lambda)\}'][\{T_3(E, \eta_g, \lambda)\}'']}{B^2}\right\}\right]$$

Using (15.263), the 2D DR in this case assumes the form

$$\left[\left[T_3(E, \eta_g, \lambda) - \left(n + \frac{1}{2}\right)\hbar\omega_0 + \frac{E_0}{B}\hbar k_y \{T_3(E, \eta_g, \lambda)\}' + \left\{\frac{m_c E_0^2 [\{T_3(E, \eta_g, \lambda)\}']^2}{2B^2}\right\}\right] = S_i \left(\frac{e^2 \hbar}{\varepsilon_{sc}\sqrt{2m_c}}\right)^{2/3} N_s^{2/3} \times \left[T_3'(E, \eta_g, \lambda) + \frac{E_0}{B}\hbar k_y \{T_3(E, \eta_g, \lambda)\}' + \left\{\frac{m_c E_0^2 [\{T_3(E, \eta_g, \lambda)\}'][\{T_3(E, \eta_g, \lambda)\}'']}{B^2}\right\}\right]^{2/3}\right] \quad (15.264)$$

The sub-band energy in this case ($E_{15,13,1s}$) can be written as

$$\left[\left[T_3(E_{15,13,3s}, \eta_g, \lambda) - \left(n + \frac{1}{2}\right)\hbar\omega_0 + \left\{\frac{m_c E_0^2 [\{T_3(E_{15,13,3s}, \eta_g, \lambda)\}']^2}{2B^2}\right\}\right] = S_i \left(\frac{e^2 \hbar}{\varepsilon_{sc}\sqrt{2m_c}}\right)^{2/3} N_s^{2/3}\right.$$

$$\times\ [T_3'(E_{15,13,3s},\eta_g,\lambda) + \left\{\frac{m_c E_0^2[\{T_3(E_{15,13,3s},\eta_g,\lambda)\}'][\{T_3(E_{15,13,3s},\eta_g,\lambda)\}'']}{B^2}\right\}\Big]^{2/3}\Big] \tag{15.265}$$

The surface electron concentration can be written as

$$\bar{N}_s = g_v \sum_{i=0}^{i_{\max}} \sum_{n=0}^{n_{\max}} \frac{g_s eB}{2\pi\hbar} F_{-1}(\eta_{15,13,3})] \tag{15.266}$$

where $\eta_{15,13,3} = (k_B T)^{-1}(E_{F15,13,3} - E_{15,13,3s})$ and

$$E_{F15,13,3} = eV_g - \frac{e^2 N_s}{C_{OX}}$$

The total surface concentration per unit area in this case can be written as

$$N_s = g_v \sum_{i=0}^{i_{\max}} \sum_{n=0}^{n_{\max}} \left[\frac{2B\sqrt{2m_c}}{3L_x\pi^2\hbar^2 E_0} \bar{t}_{15,13,3\max}[M_{163}(E_{F_{BLHDC}}, n, E_0, B, \lambda) + N_{163}(E_{F_{BLHDC}}, n, E_0, B, \lambda)] + \frac{g_s eB}{2\pi\hbar} F_{-1}(\eta_{15,13,3})\right] \tag{15.267}$$

where

$$\bar{t}_{15,13,3\max} = \left[\frac{\varepsilon_{sc} E_{15,13,3s\max}}{N_s e^2(1 + i_{\max})}\right]$$

and $E_{15,13,3s\max}$ is the maximum value of the positive real part of the equation

$$\Big[\Big[T_3(E_{15,13,3s\max},\eta_g,\lambda) - \left(n + \frac{1}{2}\right)\hbar\omega_0 + \left\{\frac{m_c E_0^2[\{T_3(E_{15,13,3s\max},\eta_g,\lambda)\}']^2}{2B^2}\right\}\Big] = S_i\left(\frac{e^2\hbar}{\varepsilon_{sc}\sqrt{2m_c}}\right)^{2/3} N_s^{2/3}$$
$$\times\ [T_3'(E_{15,13,3s\max},\eta_g,\lambda) + \left\{\frac{m_c E_0^2[\{T_3(E_{15,13,3s\max},\eta_g,\lambda)\}'][\{T_3(E_{15,13,3s\max},\eta_g,\lambda)\}'']}{B^2}\right\}\Big]^{2/3}\Big]$$

Using (15.257) and (1.4) we can study the QC in this case.

15.2.14 *The QC in HDQWFETs in the presence of light waves*

(i) Perturbed three-band model of Kane

The 2D DR in HDQWFETs of III–V, ternary and quaternary materials, whose unperturbed band structure is defined by the three-band model of Kane, in the presence of light waves, can be expressed as

$$\frac{\hbar^2 k_s^2}{2m_c} + \frac{\hbar^2}{2m_c}\left(\frac{n_z\pi}{d_z}\right)^2 = T_1(E, \eta_g, \lambda) \tag{15.268}$$

The sub-band energies ($E_{n_{l7HD}}$) can be written as

$$T_1(E_{n_{l7HD}}, \eta_g, \lambda) = \frac{\hbar^2}{2m_c}(n_z\pi/d_z)^2 \tag{15.269}$$

The expression of the EEM in this case is given by

$$m^*(E_{F2DLHD}, n_z, \lambda) = m_c \text{ Real part of } \{T_1(E_{F2DLHD}, \eta_g, \lambda)\}' \tag{15.270}$$

where E_{F2DLHD} is the Fermi energy in the present case as measured from the edge of the conduction band in the vertically upward direction in the absence of any quantization.

The DOS function can be written as

$$N(E, \eta_g, \lambda) = \left(\frac{m_c g_v}{\pi\hbar^2}\right)\sum_{n_z=1}^{n_{z\max}}[T_1(E, \eta_g, \lambda)]' H(E - E_{n_{l7HD}}) \tag{15.271}$$

The electron concentration can be written as

$$n_0 = \text{Real part of } \bar{n}_0 \tag{15.272}$$

where

$$n_0 = \left(\frac{m_c g_v}{\pi\hbar^2}\right)\sum_{n_z=1}^{n_{z\max}}\left[\left[T_1(E_{F2DLHD}, n_z, \lambda) - \frac{\hbar^2}{2m_c}\left(\frac{\pi n_z}{d_z}\right)^2\right] + \sum_{r=1}^{s} L(r)[T_1(E_{F2DLHD}, n_z, \lambda)]\right] \tag{15.273}$$

Using (15.263) and (1.4) we can study the QC in this case.

(ii) Perturbed two-band model of Kane

The 2D DR in HD*QWFETs of* III–V, ternary and quaternary materials, whose unperturbed band structure is defined by the two-band model of

Kane, in the presence of light waves, can be expressed as

$$\frac{\hbar^2 k_s^2}{2m_c} + \frac{\hbar^2}{2m_c}\left(\frac{n_z \pi}{d_z}\right)^2 = T_2(E, \eta_g, \lambda) \qquad (15.274)$$

The sub-band energies ($E_{n_{l7HD}}$) an be written as

$$T_2(E_{n_{l7HD}}, \eta_g, \lambda) = \frac{\hbar^2}{2m_c}(n_z \pi / d_z)^2 \qquad (15.275)$$

The expression of the EM in this case is given by

$$m^*(E_{F2DLHD}, n_z, \lambda) = m_c\{T_2(E_{F2DLHD}, \eta_g, \lambda)\}' \qquad (15.276)$$

where E_{F2DLHD} is the Fermi energy in the present case as measured from the edge of the conduction band in the vertically upward direction in the absence of any quantization.

The DOS function can be written as

$$N(E, \eta_g, \lambda) = \left(\frac{m_c g_v}{\pi \hbar^2}\right) \sum_{n_z=1}^{n_{z\max}} [T_2(E, \eta_g, \lambda)]' H(E - E_{n_{l7HD}}) \qquad (15.277)$$

The electron concentration can be written as

$$n_0 = \left(\frac{m_c g_v}{\pi \hbar^2}\right) \sum_{n_z=1}^{n_{z\max}} \left[\left[T_2(E_{F2DLHD}, n_z, \lambda) - \frac{\hbar^2}{2m_c}\left(\frac{\pi n_z}{d_z}\right)^2\right] + \sum_{r=1}^{s} L(r)[T_2(E_{F2DLHD}, n_z, \lambda)]\right] \qquad (15.278)$$

Using (15.278) and (1.4) we can study the QC in this case.

(iii) Perturbed parabolic band model

The 2D DR in HDQWFETs of III–V, ternary and quaternary materials, whose unperturbed band structure is defined by the parabolic energy bands in the presence of light waves can be expressed as

$$\frac{\hbar^2 k_s^2}{2m_c} + \frac{\hbar^2}{2m_c}\left(\frac{n_z \pi}{d_z}\right)^2 = T_3(E, \eta_g, \lambda) \qquad (15.279)$$

The sub-band energies ($E_{n_{l7HD1}}$) can be written as

$$T_3(E_{n_{l7HD1}}, \eta_g, \lambda) = \frac{\hbar^2}{2m_c}(n_z \pi / d_z)^2 \qquad (15.280)$$

The expression of the EM in this case is given by

$$m^*(E_{F2DLHD}, n_z, \lambda) = m_c \{T_3(E_{F2DLHD}, \eta_g, \lambda)\}' \tag{15.281}$$

The DOS function can be written as

$$N(E, \eta_g, \lambda) = \left(\frac{m_c g_v}{\pi\hbar^2}\right) \sum_{n_z=1}^{n_{z\max}} [T_3(E, \eta_g, \lambda)]' H(E - E_{n_{l7HD1}}) \tag{15.282}$$

The electron concentration can be written as

$$n_0 = \left(\frac{m_c g_v}{\pi\hbar^2}\right) \sum_{n_z=1}^{n_{z\max}} \left[\left[T_3(E_{F2DLHD}, n_z, \lambda) - \frac{\hbar^2}{2m_c}\left(\frac{\pi n_z}{d_z}\right)^2\right] + \sum_{r=1}^{s} L(r)[T_3(E_{F2DLHD}, n_z, \lambda)]\right] \tag{15.283}$$

Using (15.283) and (1.4) we can study the QC in this case.

15.2.15 *The QC in HDQWFETs in the presence of light waves and quantizing magnetic field*

(i) Perturbed three-band model of Kane

The magneto DR in HDQWFETs of III–V, ternary and quaternary materials, whose unperturbed band structure is defined by the three-band model of Kane, in the presence of light waves can be expressed as

$$\left(n + \frac{1}{2}\right)\hbar\omega_0 + \frac{\hbar^2}{2m_c}\left(\frac{n_z\pi}{d_z}\right)^2 = T_1(E_{17,8}, \eta_g, \lambda) \tag{15.284}$$

where $E_{17,8}$ is the totally quantized energy in this case.

The DOS function is given by

$$N_{0DT}(E) = \frac{eBg_v}{\pi\hbar} \sum_{n_z=1}^{n_{z\max}} \sum_{n=0}^{n_{\max}} \delta'(E - E_{17,8}) \tag{15.285}$$

The electron concentration can be written as

$$n_0 = \text{Real part of } \bar{n}_0 \tag{15.286}$$

$$n_0 = \frac{eBg_v}{\pi\hbar} \sum_{n=0}^{n_{\max}} \sum_{n_z=1}^{n_{z\max}} F_{-1}(\eta_{128}) \tag{15.287}$$

where

$$\eta_{128} = \frac{E_{F128} - E_{17.8}}{k_B T}$$

and E_{F128} is the Fermi energy in this case.

Using (15.286) and (1.4) we can study the QC in this case.

(ii) Perturbed two-band model of Kane

The magneto DR *in* HD QWFETs of HD III–V, ternary and quaternary materials, whose unperturbed band structure is defined by the two-band model of Kane, in the presence of light waves can be expressed as

$$\left(n + \frac{1}{2}\right)\hbar\omega_0 + \frac{\hbar^2}{2m_c}\left(\frac{n_z\pi}{d_z}\right)^2 = T_2(E_{17,9}, \eta_g, \lambda) \tag{15.288}$$

where $E_{17,9}$ is the totally quantized energy in this case.

The DOS function is given by

$$N_{0DT}(E) = \frac{eBg_v}{\pi\hbar} \sum_{n_z=1}^{n_{z\max}} \sum_{n=0}^{n_{\max}} \delta'(E - E_{17,9}) \tag{15.289}$$

The electron concentration can be written as

$$n_0 = \frac{eBg_v}{\pi\hbar} \sum_{n=0}^{n_{\max}} \sum_{n_z=1}^{n_{z\max}} F_{-1}(\eta_{129}) \tag{15.290}$$

where

$$\eta_{129} = \frac{E_{F129} - E_{17.9}}{k_B T}$$

and E_{F129} is the Fermi energy in this case.

Using (15.290) and (1.4) we can study the QC in this case.

(iii) Perturbed parabolic band model

The magneto DR in HDQWFETs of III–V, ternary and quaternary materials, whose unperturbed band structure is defined by the parabolic energy bands in the presence of light waves, can be expressed as

$$\left(n + \frac{1}{2}\right)\omega_0 + \frac{\hbar^2}{2m_c}\left(\frac{n_z\pi}{d_z}\right)^2 = T_3(E_{17,10}, \eta_g, \lambda) \tag{15.291}$$

where $E_{17,10}$ is the totally quantized energy in this case.

The DOS function is given by

$$N_{0DT}(E) = \frac{eBg_v}{\pi\hbar} \sum_{n_z=1}^{n_{z\max}} \sum_{n=0}^{n_{\max}} \delta'(E - E_{17,10}) \tag{15.292}$$

The electron concentration can be written as

$$n_0 = \frac{eBg_v}{\pi\hbar} \sum_{n=0}^{n_{\max}} \sum_{n_z=1}^{n_{z\max}} F_{-1}(\eta_{1210}) \tag{15.293}$$

where

$$\eta_{1210} = \frac{E_{F1210} - E_{17.10}}{k_B T}$$

and E_{F1210} is the Fermi energy in this case.

Using (15.293) and (1.4) we can study the QC in this case.

15.2.16 *The QC in HDQWFETs in the presence of strong light waves and the crossed fields configurations*

(i) Perturbed three-band model of Kane

The DR in HDQWFETs of III–V, ternary and quaternary materials, whose unperturbed band structure is defined by the three-band model of Kane, in the presence of light waves and under crossed fields configuration can be expressed as

$$T_1(E_{15,16}, \eta_g, \lambda) = \left(n + \frac{1}{2}\right)\hbar\omega_0 + \frac{[\hbar(n_z\pi/d_z)]^2}{2m_c} - \frac{E_0}{B}\hbar k_y \{T_1(E_{15,16}, \eta_g, \lambda)\}' - \left\{\frac{m_c E_0^2[\{T_1(E_{15,16}, \eta_g, \lambda)\}']^2}{2B^2}\right\} \tag{15.294}$$

where $E_{15,16}$ is the totally quantized energy in this case.

The DOS function can be expressed as

$$N_{DOS} = \frac{g_s g_v eB}{2\pi\hbar} \sum_{n=0}^{n_{\max}} \sum_{n_z=1}^{n_{z\max}} \delta'(E - E_{15,16}) \tag{15.295}$$

The surface electron concentration can be written as

$$N_s = \frac{g_s g_v eB}{2\pi\hbar} \text{Real part of} \sum_{n=0}^{n_{\max}} \sum_{n_z=1}^{n_{z\max}} F_{-1}(\eta_{15,16}) \tag{15.296}$$

where $\eta_{15,16} = (k_B T)^{-1}[E_{F15,16} - E_{15,16}]$ and $E_{F15,16}$ is the Fermi energy in this case.

Using (15.296) and (1.4) we can study the QC in this case.

(ii) Perturbed two-band model of Kane

The DR in HDQWFETs of III–V, ternary and quaternary materials, whose unperturbed band structure is defined by the three-band model of Kane, in the presence of light waves and under crossed fields configuration can be expressed as

$$T_2(E_{15,17}, \eta_g, \lambda) = \left(n + \frac{1}{2}\right)\hbar\omega_0 + \frac{[\hbar(n_z\pi/d_z)]^2}{2m_c} - \frac{E_0}{B}\hbar k_y\{T_2(E_{15,17}, \eta_g, \lambda)\}' - \left\{\frac{m_c E_0^2[\{T_2(E_{15,17}, \eta_g, \lambda)\}']^2}{2B^2}\right\} \tag{15.297}$$

where $E_{15,17}$ is the totally quantized energy in this case.

The DOS function can be expressed as

$$N_{DOS} = \frac{g_s g_v eB}{2\pi\hbar}\sum_{n=0}^{n_{\max}}\sum_{n_z=1}^{n_{z\max}}\delta'(E - E_{15,17}) \tag{15.298}$$

The surface electron concentration can be written as

$$N_s = \frac{g_s g_v eB}{2\pi\hbar}\sum_{n=0}^{n_{\max}}\sum_{n_z=1}^{n_{z\max}}F_{-1}(\eta_{15,17}) \tag{15.299}$$

where $\eta_{15,17} = (k_B T)^{-1}[E_{F15,17} - E_{15,17}]$ and $E_{F15,17}$ is the Fermi energy in this case.

Using (15.299) and (1.4) we can study the QC in this case.

(iii) Perturbed parabolic band model

The DR in HDQWFETs of III–V, ternary and quaternary materials, whose unperturbed band structure is defined by the three-band model of Kane, in the presence of light waves and under crossed fields configuration can be expressed as

$$T_3(E_{15,18}, \eta_g, \lambda) = \left(n + \frac{1}{2}\right)\hbar\omega_0 + \frac{[\hbar(n_z\pi/d_z)]^2}{2m_c} - \frac{E_0}{B}\hbar k_y\{T_3(E_{15,18}, \eta_g, \lambda)\}' - \left\{\frac{m_c E_0^2[\{T_3(E_{15,18}, \eta_g, \lambda)\}']^2}{2B^2}\right\} \tag{15.300}$$

where $E_{15,17}$ is the totally quantized energy in this case.

The DOS function can be expressed as

$$N_{DOS} = \frac{g_s g_v eB}{2\pi\hbar} \sum_{n=0}^{n_{\max}} \sum_{n_z=1}^{n_{z\max}} \delta'(E - E_{15,18}) \tag{15.301}$$

The surface electron concentration can be written as

$$N_s = \frac{g_s g_v eB}{2\pi\hbar} \sum_{n=0}^{n_{\max}} \sum_{n_z=1}^{n_{z\max}} F_{-1}(\eta_{15,18}) \tag{15.302}$$

where $\eta_{15,18} = (k_B T)^{-1}[E_{F15,18} - E_{15,18}]$ and $E_{F15,18}$ is the Fermi energy in this case.

Using (15.302) and (1.4) we can study the QC in this case.

15.2.17 *The QC in HDNWFETs in the presence of strong light waves*

(a) Perturbed three-band model of Kane

The 1D DR in HDNWFETs of III–V, ternary and quaternary materials, whose unperturbed band structure is defined by the three-band model of Kane in the absence of any field, in the presence of light waves can be expressed as

$$\frac{\hbar^2 (n_z \pi/d_z)^2}{2m_c} + \frac{\hbar^2 (n_y \pi/d_y)^2}{2m_c} + \frac{\hbar^2 k_x^2}{2m_c} = T_1(E, \eta_g, \lambda) \tag{15.303}$$

The DOS function in this case assumes the form

$$N_{1DHD\Gamma L}(E, \eta_g) = \frac{2g_v}{\pi} \sum_{n_y=1}^{n_{y\max}} \sum_{n_z=1}^{n_{z\max}} T'_{3L1}(E, n_y, n_z, \eta_g, \lambda) \times H(E - E'_{3HDNWL1}) \tag{15.304}$$

where

$$T_{3L1}(E, n_y, n_z, \eta_g, \lambda) = \left[\left[T_1(E, \eta_g, \lambda) - \left[\frac{\hbar^2 (n_z \pi/d_z)^2}{2m_c} + \frac{\hbar^2 (n_y \pi/d_y)^2}{2m_c}\right]\right] \frac{2m_c}{\hbar^2}\right]^{\frac{1}{2}}$$

In (1.418), $E'_{3HDNWL1}$ is the sub-band energy in this case which can be expressed as

$$\frac{\hbar^2 (n_z \pi/d_z)^2}{2m_c} + \frac{\hbar^2 (n_y \pi/d_y)^2}{2m_c} + \frac{\hbar^2 k_x^2}{2m_c} = T_1(E, \eta_g, \lambda) \tag{15.305}$$

The EEM in this case is given by

$$m^*(E_{F1HDNWL1}, \eta_g, \lambda) = m_c \text{ Real part of } [T_1(E_{F1HDNWL1}, \eta_g, \lambda)] \tag{15.306}$$

The electron statistics per unit length can be written as

$$n_0 = \text{Real Part of } \bar{n}_0 \tag{15.307}$$

where

$$\bar{n}_0 = \frac{2g_v}{\pi} \sum_{n_y=1}^{n_{y\max}} \sum_{n_z=1}^{n_{z\max}} \left[T_{3L1}(E_{F1HDNWL1}, n_y, n_z, \eta_g, \lambda) + \sum_{r=1}^{r=s} L(r)[T_{3L1}(E_{F1HDNWL1}, n_y, n_z, \eta_g, \lambda)] \right]$$

Using (15.307) and (1.4) we can study the QC in this case.

(b) Perturbed two-band model of Kane

The 1D DR in NWs HDNWFETs of III–V, ternary and quaternary materials, whose unperturbed band structure is defined by the two-band model of Kane in the absence of any field, in the presence of light waves can be expressed as

$$\frac{\hbar^2 (n_z\pi/d_z)^2}{2m_c} + \frac{\hbar^2 (n_y\pi/d_y)^2}{2m_c} + \frac{\hbar^2 k_x^2}{2m_c} = T_2(E, \eta_g, \lambda) \tag{15.308}$$

The DOS function in this case assumes the form

$$N_{1DIIDFL}(E, \eta_y) - \frac{2g_v}{\pi} \sum_{n_y=1}^{n_{y\max}} \sum_{n_z=1}^{n_{z\max}} T'_{3L2}(E, n_y, n_z, \eta_g, \lambda) \times H(E - E'_{3HDNWL2}) \tag{15.309}$$

where

$$T_{3L2}(E, n_y, n_z, \eta_g, \lambda) = \left[\left[T_2(E, \eta_g, \lambda) - \left[\frac{\hbar^2 (n_z\pi/d_z)^2}{2m_c} + \frac{\hbar^2 (n_y\pi/d_y)^2}{2m_c} \right] \right] \frac{2m_c}{\hbar^2} \right]^{\frac{1}{2}}$$

The $E'_{3HDNWL2}$ is the sub-band energy in this case which can be expressed as

$$\frac{\hbar^2 (n_z\pi/d_z)^2}{2m_c} + \frac{\hbar^2 (n_y\pi/d_y)^2}{2m_c} = T_2(E'_{3HDNWL2,}\eta_g \lambda) \tag{15.310}$$

The EEM in this case is given by

$$m^*(E_{F1HDNWL2}, \eta_g, \lambda) = m_c[T_2'(E_{F1HDNWL2}, \eta_g, \lambda)] \quad (15.311)$$

The electron statistics per unit length can be written as

$$n_0 = \frac{2g_v}{\pi} \sum_{n_y=1}^{n_{y\max}} \sum_{n_z=1}^{n_{z\max}} \left[T_{3L2}(E_{F1HDNWL2}, n_y, n_z, \eta_g, \lambda) + \sum_{r=1}^{r=s} L(r)[T_{3L2}(E_{F1HDNWL2}, n_y, n_z, \eta_g, \lambda)] \right] \quad (15.312)$$

Using (15.312) and (1.4) we can study the QC in this case.

(c) Perturbed parabolic band model

The 1D DR in NWs HDNWFETs of III–V, ternary and quaternary materials, whose unperturbed band structure is defined by the parabolic energy bands in the absence of any field, in the presence of light waves can be expressed as

$$\frac{\hbar^2(n_z\pi/d_z)^2}{2m_c} + \frac{\hbar^2(n_y\pi/d_y)^2}{2m_c} + \frac{\hbar^2 k_x^2}{2m_c} = T_3(E, \eta_g, \lambda) \quad (15.313)$$

The DOS function in this case assumes the form

$$N_{1DHD\Gamma L}(E, \eta_g) = \frac{2g_v}{\pi} \sum_{n_y=1}^{n_{y\max}} \sum_{n_z=1}^{n_{z\max}} T_{3L3}'(E, n_y, n_z, \eta_g, \lambda) \times H(E - E_{3HDNWL3}') \quad (15.314)$$

where

$$T_{3L3}(E, n_y, n_z, \eta_g, \lambda) = \left[\left[T_3(E, \eta_g, \lambda) - \left[\frac{\hbar^2(n_z\pi/d_z)^2}{2m_c} + \frac{\hbar^2(n_y\pi/d_y)^2}{2m_c} \right] \right] \frac{2m_c}{\hbar^2} \right]^{\frac{1}{2}}$$

In (15.305), $E_{3HDNWL3}'$ is the sub-band energy in this case which can be expressed as

$$\frac{\hbar^2(n_z\pi/d_z)^2}{2m_c} + \frac{\hbar^2(n_y\pi/d_y)^2}{2m_c} = T_3(E_{3HDNWL3,}'\eta_g\lambda) \quad (15.315)$$

The EEM in this case is given by

$$m^*(E_{F1HDNWL3}, \eta_g, \lambda) = m_c[T_3'(E_{F1HDNWL3}, \eta_g, \lambda)] \quad (15.316)$$

The electron statistics per unit length can be written as

$$n_0 = \frac{2g_v}{\pi} \sum_{n_y=1}^{n_{y\max}} \sum_{n_z=1}^{n_{z\max}} \left[T_{3L3}(E_{F1HDNWL3}, n_y, n_z, \eta_g, \lambda) \right.$$
$$\left. + \sum_{r=1}^{r=s} L(r) \left[T_{3L3}(E_{F1HDNWL3}, n_y, n_z, \eta_g, \lambda) \right] \right] \quad (15.317)$$

Using (15.308) and (1.4) we can study the QC in this case.

15.2.18 *The QC in HDCNTFETs in the presence of light waves*

The exponential-type distribution $F(V)$ of the impurity potential is given by [10]

$$F(V) = \theta_0 \exp(-V/\theta_0) \quad (15.318)$$

where θ_0 is the band talling parameter of exponential type band tails.

Under heavy doping in the presence of exponential-type band tails, the use of (15.309) leads to the dispersion relations of arm chair and zigzag CNTs as

$$\gamma_0(E, \theta_0) = \left[\pm t_c 1 + 4\cos\left(\frac{\pi\nu}{n}\right) \cos\left(\frac{k_y a_{cc}\sqrt{3}}{2}\right) + 4\cos^2\left(\frac{k_y a_{cc}\sqrt{3}}{2}\right) \right]^{1/2} \quad (15.319)$$

and

$$\gamma_0(E, \theta_0) = \left[\pm t_c 1 + 4\cos\left(\frac{3k_y a_{cc}}{2}\right) \cos\left(\frac{\pi\nu}{n}\right) + 4\cos^2\left(\frac{\pi\nu}{n}\right) \right]^{1/2} \quad (15.320)$$

where

$$\gamma_0(E, \theta_0) = \left[E + \frac{\left[\exp\left(-\frac{E}{\theta_0}\right) - \theta_0\right]}{\left[1 - \exp\left(-\frac{E}{\theta_0}\right)\right]} \right]$$

The DRs of heavily doped arm-chair and zig-zag CNTs can, respectively, be expressed as

$$k_y = \left(\frac{2}{a_{cc}\sqrt{3}}\right) (A_7(E, E_i, \theta_0)) \quad (15.321)$$

$$k_y = \left(\frac{2}{3a_{cc}}\right) (A_8(E, E_i, \theta_0)) \quad (15.322)$$

where

$$A_7(E,E_i,\theta_0)=\cos^{-1}[f_9(E,E_i,\theta_0)],$$

$$f_9(E,E_i,\theta_0)=\left[+\frac{1}{2}\left[\frac{-\gamma_0^2(E_i,\theta_0)}{t^2}+5\right]-\frac{1}{2}\left[\left(\frac{\gamma_0^2(E_i,\theta_0)}{t^2}-5\right)^2-\left(1-\frac{\gamma_0^2(E_i,\theta_0)}{t^2}\right)\right]^{1/2}\right],$$

$$A_8(E,E_i,\theta_0)=\cos^{-1}[f_{10}(E,E_i,\theta_0)],$$

$$f_{10}(E,E_i,\theta_0)=\left[\bar{\phi}_2(E_i,\theta_0)+\frac{1}{4\bar{\phi}_2(E_i,\theta_0)}-\frac{\gamma_0^2(E_i,\theta_0)}{4t_c^2\bar{\phi}_2(E_i,\theta_0)}\right],$$

$$\bar{\phi}_2(E_i,\theta_0)=\left[\frac{1}{2}+\frac{1}{2t_c}\gamma_0(E_i,\theta_0)\right]$$

The electron concentration per unit length for both types of CNTs under the condition of extreme carrier degeneracy can, in turn, be expressed as

$$n_{1D}=\frac{8}{\pi a_{cc}\sqrt{3}}\sum_{i=1}^{i_{\max}}A_7(E,E_i,\theta_0)\tag{15.323}$$

$$n_{1D}=\frac{8}{\pi a_{cc}\sqrt{3}}\sum_{i=1}^{i_{\max}}A_8(E,E_i,\theta_0)\tag{15.324}$$

The gate capacitance per unit length for heavily doped arm-chair and zig-zag CNTs can, respectively, be expressed as

$$C_g=\frac{e^2\sum_{i=1}^{i_{\max}}f_{11}(E_f,E_i,\theta_0)}{\left[\frac{a_{cc}\pi\sqrt{3}}{4}+\left(\frac{e^2}{\varepsilon_{0x}}\right)\sum_{i=1}^{i_{\max}}f_{11}(E_f,E_i,\theta_0)\right]}\tag{15.325}$$

and

$$C_g=\frac{e^2\sum_{i=1}^{i_{\max}}f_{12}(E_f,E_i,\theta_0)}{\left[\frac{3\pi a_{cc}}{4}+\left(\frac{e^2}{\varepsilon_{0x}}\right)\sum_{i=1}^{i_{\max}}f_{12}(E_f,E_i,\theta_0)\right]}\tag{15.326}$$

where

$$f_{11}(E_f,E_i,\theta_0)=[[1-f_9^2(E_f,E_i,\theta_0)]^{-1/2}][\gamma_0(E_f,\theta_0)\gamma_0'(E_f,\theta_0)]$$
$$\cdot\left[\left(\frac{\gamma_0^2(E_i,\theta_0)}{t_2}-5\right)^2-\left(1-\frac{\gamma_0^2(E_i,\theta_0)}{t^2}\right)\right]^{-1/2},$$

$$\gamma_0'(E_f,\theta_0) = \Bigg[1 + [1-\exp(-E_f/\theta_0)]^{-2} \times \Bigg[(1-\exp(-E_f/\theta_0))$$
$$\cdot\left(\exp(-E_f/\theta_0) - \frac{E_f}{\theta}\exp(-E_f/\theta_0) - \exp(-E_f/\theta_0)\right)$$
$$- [\{\exp(-E_f/\theta_0)(E_f+\theta) - \theta\}(\theta)^{-1}\cdot(\exp(-E_f/\theta_0))]\Bigg]\Bigg]$$

$$f_{12}(E_f,E_i,\theta_0) = \left[\left[\frac{\gamma_0(E_f/\theta_0).\gamma_0'(E_f/\theta_0)}{4t^2\bar{\varphi}_2(E_i,\theta_0)}\right] \times [1 - f_{10}^2(E_f,E_i,\theta_0)]^{-1/2}\right]$$

15.3 Result and Discussions

Using the appropriate equations together with the parameters as given in Chapter 18 we have plotted the normalized QC as a function of surface electron concentration in QWFETs of InSb, InAs, $Hg_{1-x}Cd_xTe$, $Ga_{1-x}Al_xAs_yP_{1-y}$ lattice matched to InP as shown in Figs. 15.1–15.4 for (a) perturbed three-band model of Kane, (b) perturbed two-band model of Kane and (c) perturbed parabolic energy bands, respectively, in the presence of strong light waves. It appears that QC increases with increasing

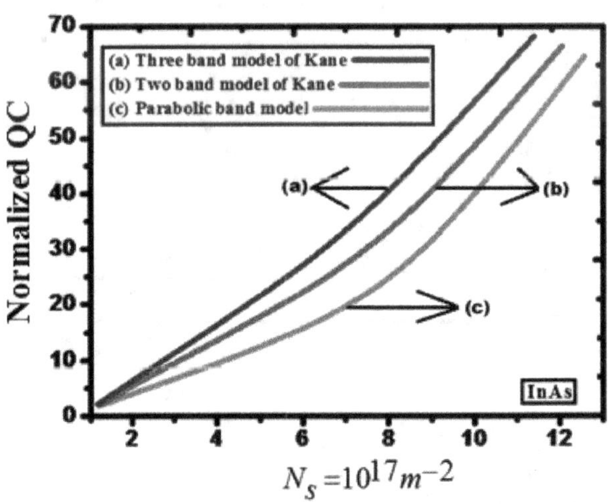

Fig. 15.1. Plot of the normalized QC versus surface electron concentration per unit area in the presence of light waves in QWFETs of InAs in accordance with (a) perturbed three-band model of Kane, (b) perturbed two-band model of Kane and (c) perturbed parabolic energy bands, respectively.

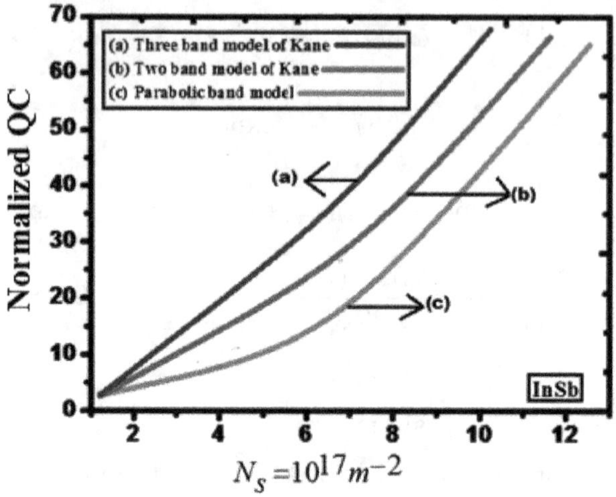

Fig. 15.2. Plot of the normalized QC versus surface electron concentration per unit area in the presence of light waves in QWFETs of InSb in accordance with (a) perturbed three-band model of Kane, (b) perturbed two-band model of Kane and (c) perturbed parabolic energy bands, respectively.

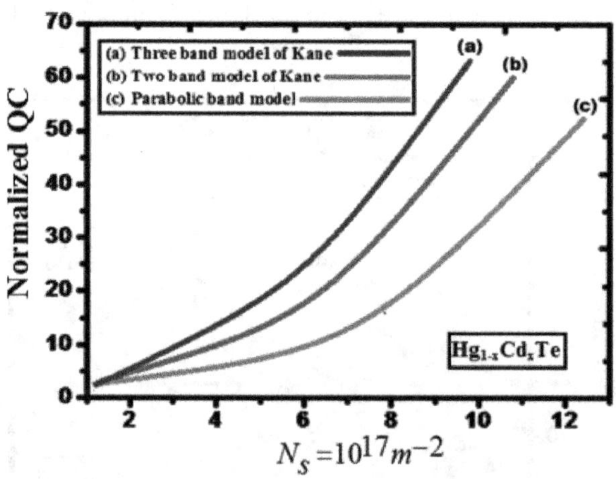

Fig. 15.3. Plot of the normalized QC versus surface electron concentration per unit area in the presence of light waves in QWFETs of $Hg_{1-x}Cd_xTe$ in accordance with (a) perturbed three-band model of Kane, (b) perturbed two-band model of Kane and (c) perturbed parabolic energy bands, respectively.

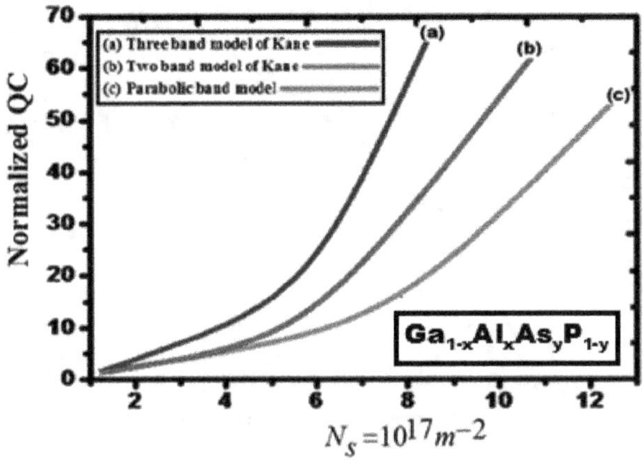

Fig. 15.4. Plot of the normalized QC versus surface electron concentration per unit area in the presence of light waves in QWFETs of $Ga_{1-x}Al_xAs_yP_{1-y}$ lattice matched to InP in accordance with (a) perturbed three-band model of Kane, (b) perturbed two-band model of Kane and (c) perturbed parabolic energy bands, respectively.

N_s for all the models for QWFETs of the said materials. The numerical values of QC are the greatest for perturbed three-band model of Kane and least for perturbed parabolic energy bands for all the cases.

We have plotted the normalized QC as a function of film thickness in QWFETs of InSb, InAs, $Hg_{1-x}Cd_xTe$, $Ga_{1-x}Al_xAs_yP_{1-y}$ lattice matched to InP as shown in Figs. 15.5–15.8 for (a) perturbed three-band model of Kane, (b) perturbed two-band model of Kane and (c) perturbed parabolic energy bands, respectively, in the presence of strong light waves. It appears that QC increases with increasing d_z in an oscillatory way for all the models for QWFETs of the said materials. The numerical values of oscillatory QC is the greatest for perturbed three-band model of Kane and least for perturbed parabolic energy bands for all the cases. We have plotted the normalized QC as a function of normalized intensity in QWFETs of InSb, InAs, $Hg_{1-x}Cd_xTe$, $Ga_{1-x}Al_xAs_yP_{1-y}$ lattice matched to InP as shown in Figs. 15.9–15.12 for (a) perturbed three-band model of Kane, (b) perturbed two-band model of Kane and (c) perturbed parabolic energy bands, respectively, in the presence of strong light waves. It appears that QC decreases with increasing intensity in an oscillatory way for all the models for QWFETs of the said materials. The numerical values of oscillatory QC are the greatest for perturbed three-band model of Kane and least for perturbed parabolic energy bands for all the cases. We have plotted

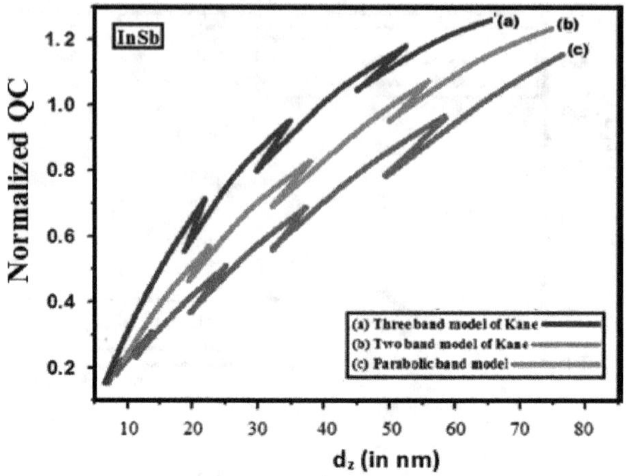

Fig. 15.5. Plot of the normalized QC per unit area versus film thickness (d_z) in the presence of light waves in QWFETs of InSb in accordance with (a) perturbed three-band model of Kane, (b) perturbed two-band model of Kane and (c) perturbed parabolic energy bands, respectively.

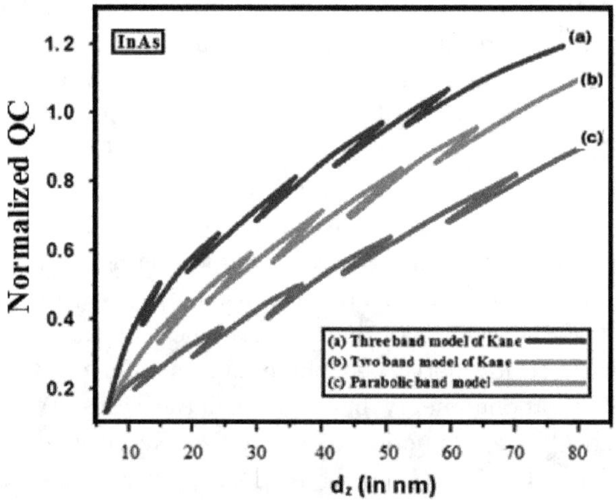

Fig. 15.6. Plot of the normalized QC per unit area versus film thickness (d_z) in the presence of light waves in QWFETs of InAs in accordance with (a) perturbed three-band model of Kane, (b) perturbed two-band model of Kane and (c) perturbed parabolic energy bands, respectively.

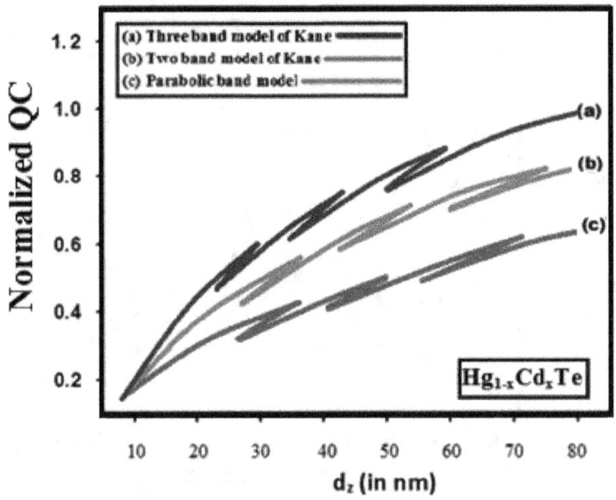

Fig. 15.7. Plot of the normalized QC per unit area versus film thickness (d_z) in the presence of light waves in QWFETs of $\text{Hg}_{1-x}\text{Cd}_x\text{Te}$ in accordance with (a) perturbed three-band model of Kane, (b) perturbed two-band model of Kane and (c) perturbed parabolic energy bands, respectively.

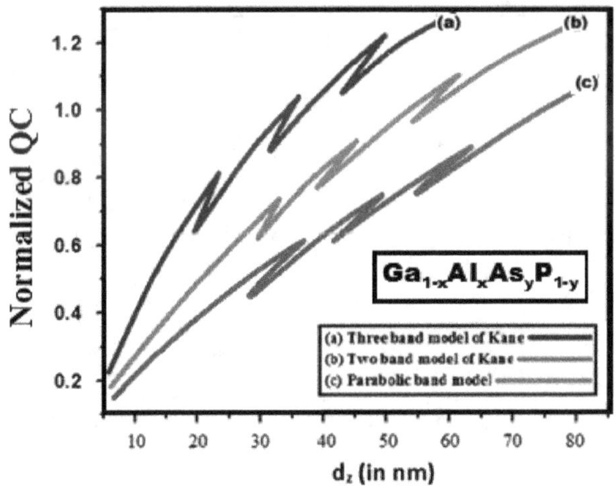

Fig. 15.8. Plot of the normalized QC per unit area versus film thickness (d_z) in the presence of light waves in QWFETs of $\text{Ga}_{1-x}\text{Al}_x\text{As}_y\text{P}_{1-y}$ lattice matched to InP in accordance with (a) perturbed three-band model of Kane, (b) perturbed two-band model of Kane and (c) perturbed parabolic energy bands, respectively.

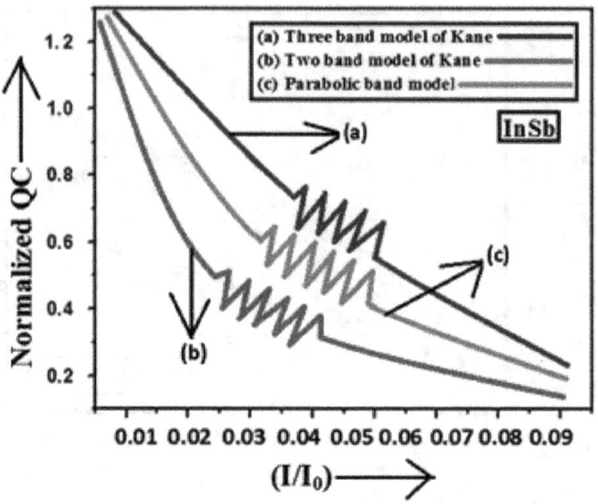

Fig. 15.9. Plot of the normalized QC per unit area versus normalized light intensity in QWFETs of InSb in accordance with (a) perturbed three-band model of Kane, (b) perturbed two-band model of Kane and (c) perturbed parabolic energy bands, respectively.

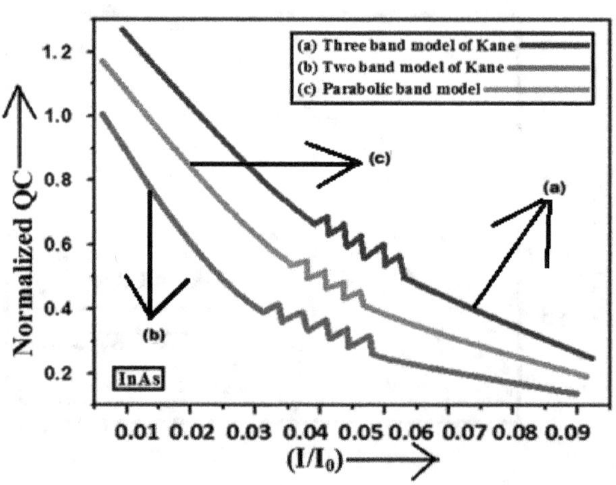

Fig. 15.10. Plot of the normalized QC per unit area versus normalized light intensity in QWFETs of InAs in accordance with (a) perturbed three-band model of Kane, (b) perturbed two-band model of Kane and (c) perturbed parabolic energy bands, respectively.

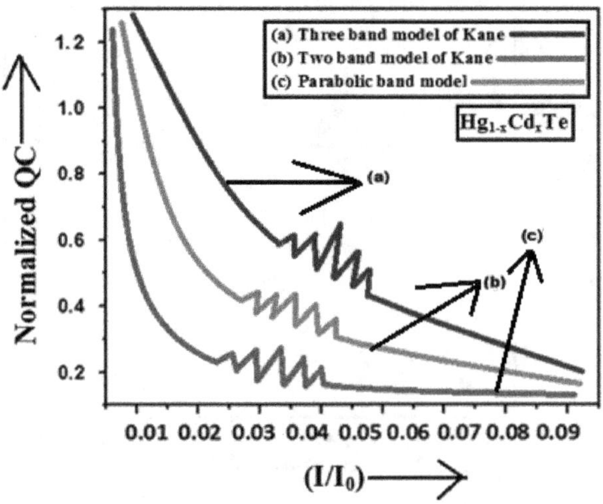

Fig. 15.11. Plot of the normalized QC per unit area versus normalized light intensity in QWFETs of $Hg_{1-x}Cd_xTe$ in accordance with (a) perturbed three-band model of Kane, (b) perturbed two-band model of Kane and (c) perturbed parabolic energy bands, respectively.

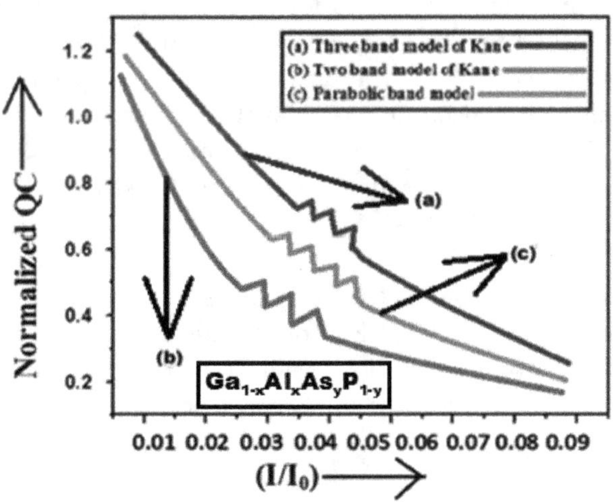

Fig. 15.12. Plot of the normalized QC per unit area versus normalized light intensity in QWFETs of $Ga_{1-x}Al_xAs_yP_{1-y}$ lattice matched to InP in accordance with (a) perturbed three-band model of Kane, (b) perturbed two-band model of Kane and (c) perturbed parabolic energy bands.

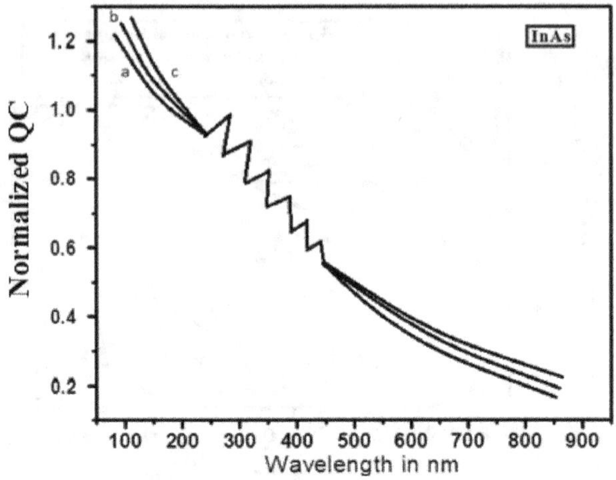

Fig. 15.13. Plot of the normalized QC per unit area versus wavelength (λ) in QWFETs of InAs in accordance with (a) perturbed three-band model of Kane, (b) perturbed two-band model of Kane and (c) perturbed parabolic energy bands.

the normalized QC as a function of wave length in QWFETs of InSb and InAs as shown in Figs. 15.13–15.14 for (a) perturbed three-band model of Kane, (b) perturbed two-band model of Kane and (c) perturbed parabolic energy bands, respectively, in the presence of strong light waves. We have also plotted the normalized QC as a function of wavelength in QWFETs of $Hg_{1-x}Cd_xTe$, $Ga_{1-x}Al_xAs_yP_{1-y}$ lattice matched to InP as shown in Figs. 15.15 and 15.16 in accordance with perturbed three- and two-band models of Kane for QWFETs of $Hg_{1-x}Cd_xTe$ and perturbed three-band model of Kane for QWFETs of $Ga_{1-x}Al_xAs_yP_{1-y}$ lattice matched to InP, respectively. It appears that QC decreases with increasing λ in an oscillatory way for all the models for QWFETs of the said materials.

Using the appropriate equations we have plotted the normalized ER as a function of the film thickness in the presence of strong light waves for QWFETs of InAs, InSb, $Hg_{1-x}Cd_xTe$ and $In_{1-x}Ga_xAs_yP_{1-y}$, as shown in Figs. 15.17–15.20 in which the curves (a), (b) and (c) represents the respective ER for the perturbed three- and the two models of Kane and the parabolic energy bands. It has been observed that the ER in this case also decreases with the increase in film thickness in a step wise manner. Figures 15.21–15.24 exhibit the plot of ER as a function of electron

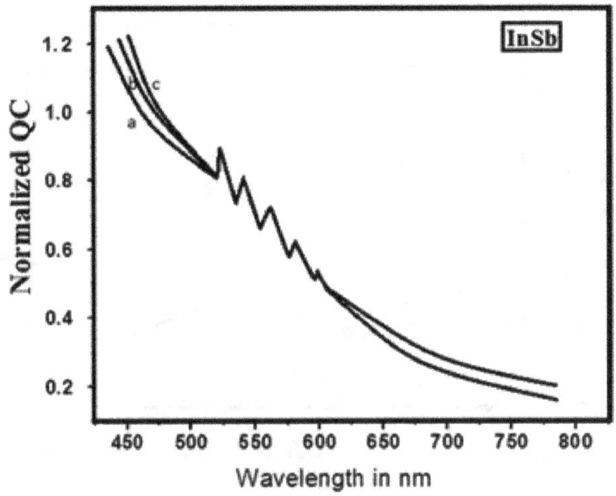

Fig. 15.14. Plot of the normalized QC per unit area versus wavelength (λ) in QWFETs of InSb in accordance with (a) perturbed three-band model of Kane, (b) perturbed two-band model of Kane and (c) perturbed parabolic energy bands.

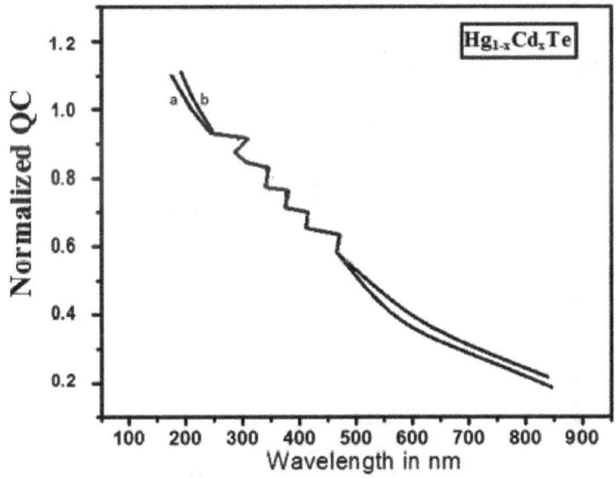

Fig. 15.15. Plot of the normalized QC per unit area versus wavelength (λ) in QWFETs of $Hg_{1-x}Cd_xTe$ in accordance with (a) perturbed three-band model of Kane and (b) perturbed two-band model of Kane.

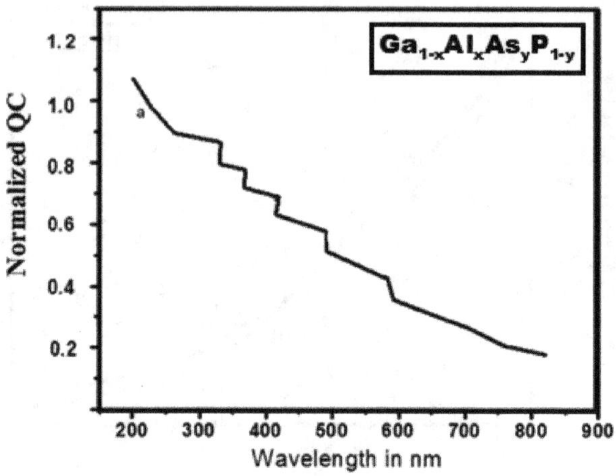

Fig. 15.16. Plot of the normalized QC per unit area versus wavelength (λ) in QWFETs of $Ga_{1-x}Al_xAs_yP_{1-y}$ lattice matched to InP in accordance with perturbed three-band model of Kane.

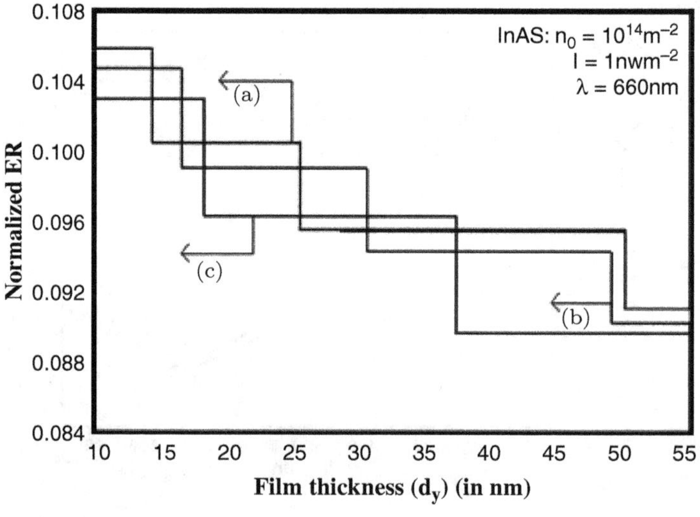

Fig. 15.17. Plot of the normalized 2D ER as a function of film thickness for QWFETs of InAs in the presence of light waves in which the curves (a), (b) and (c) represent the perturbed three- and two-band models of Kane and that of the parabolic energy bands, respectively.

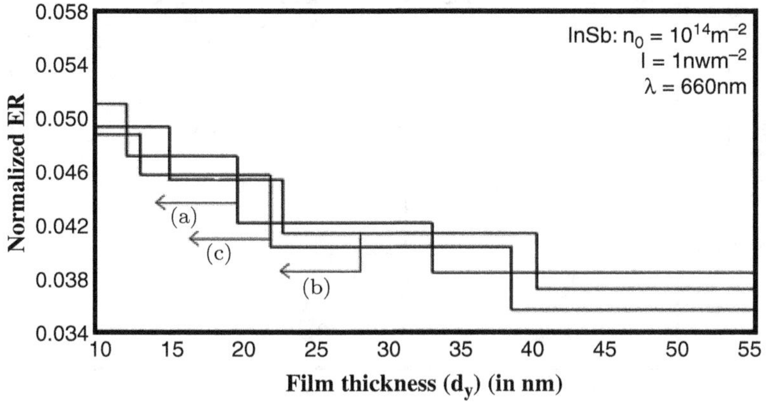

Fig. 15.18. Plot of the normalized 2D ER as a function of film thickness for QWFETs of InSb in the presence of light waves in which the curves (a), (b) and (c) represent the perturbed three- and two-band models of Kane and that of the parabolic energy bands, respectively.

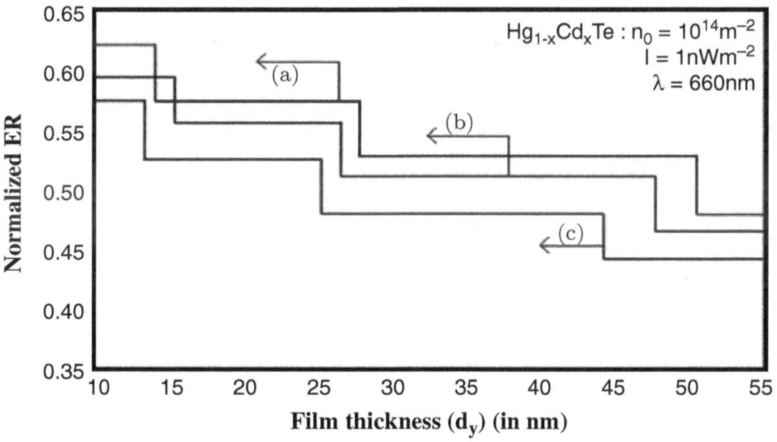

Fig. 15.19. Plot of the normalized 2D ER as a function of film thickness for QWFETs of $Hg_{1-x}Cd_xTe$ in the presence of light waves in which the curves (a), (b) and (c) represent the perturbed three- and two-band models of Kane and that of the parabolic energy bands, respectively.

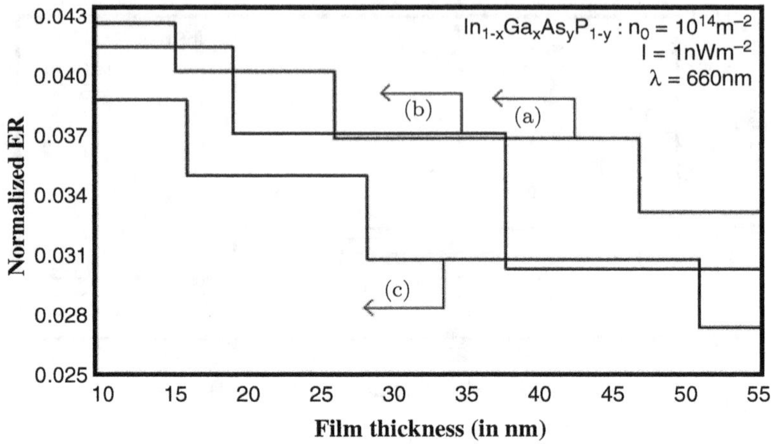

Fig. 15.20. Plot of the normalized 2D ER as a function of film thickness for QWFETs of $In_{1-x}Ga_xAs_yP_{1-y}$ lattice matched to InP in the presence of light waves in which the curves (a), (b) and (c) represent the perturbed three- and two-band models of Kane and that of the parabolic energy bands, respectively.

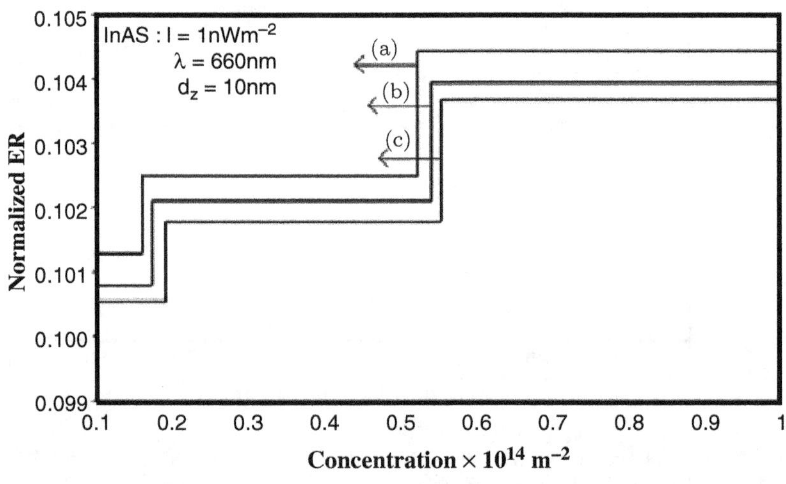

Fig. 15.21. Plot of the normalized 2D ER as a function of surface electron concentration for QWFETs of InAs in the presence of light waves in which the curves (a), (b) and (c) represent the perturbed three- and two-band models of Kane and that of the parabolic energy bands, respectively.

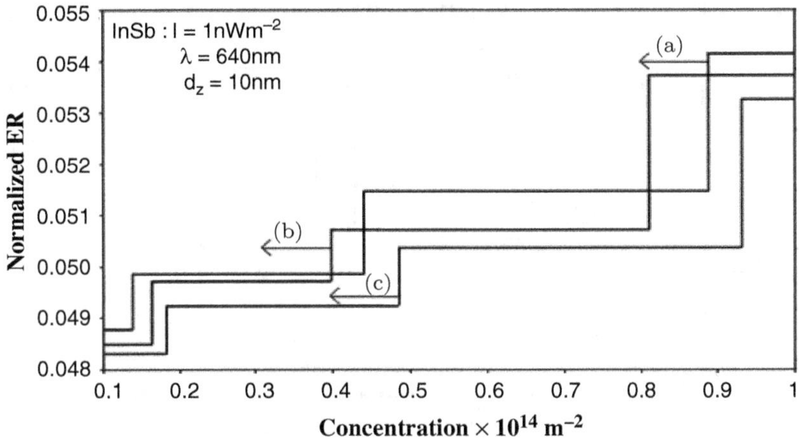

Fig. 15.22. Plot of the normalized 2D ER as a function of surface electron concentration for QWFETs of InSb in the presence of light waves in which the curves (a), (b) and (c) represent the perturbed three- and two-band models of Kane and that of the parabolic energy bands, respectively.

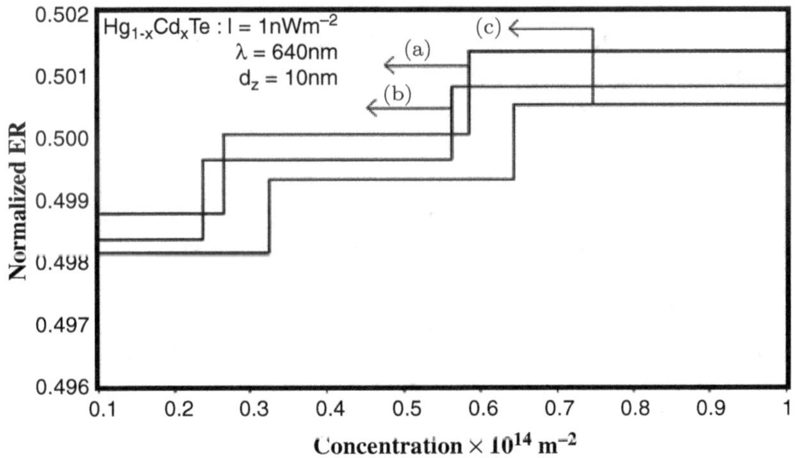

Fig. 15.23. Plot of the normalized 2D ER as a function of surface electron concentration for QWFETs of $Hg_{1-x}Cd_xTe$ in the presence of light waves in which the curves (a), (b) and (c) represent the perturbed three- and two-band models of Kane and that of the parabolic energy bands, respectively.

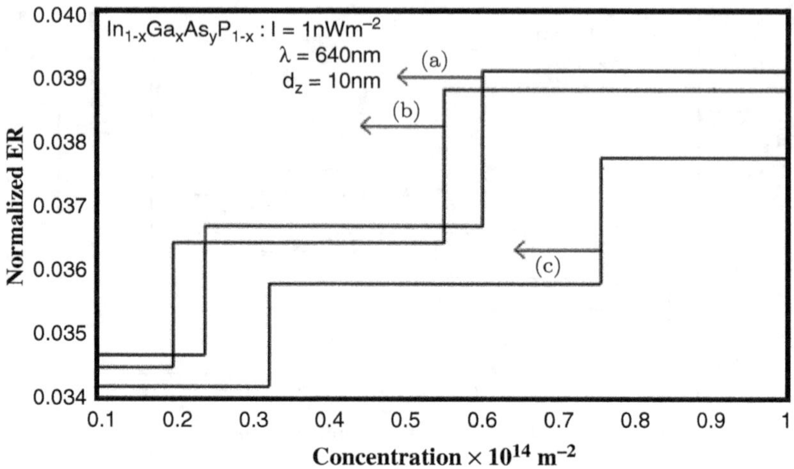

Fig. 15.24. Plot of the normalized 2D ER as a function of surface electron concentration for QWFETs of $In_{1-x}Ga_xAs_yP_{1-y}$ lattice matched to InP in the presence of light waves in which the curves (a), (b) and (c) represent the perturbed three- and two-band models of Kane and that of the parabolic energy bands, respectively.

concentration per unit area under strong light waves for QWFETs of InAs, InSb, $Hg_{1-x}Cd_xTe$ and $In_{1-x}Ga_xAs_yP_{1-y}$. It appears from the said figures that the ER increases with increasing surface electron statistics in a step-wise manner.

Figures 15.25–15.28 exhibit the plot of ER as a function of light intensity for QWFETs of InAs, InSb, $Hg_{1-x}Cd_xTe$ and $In_{1-x}Ga_xAs_yP_{1-y}$. It appears from the said figures that the ER decreases with increasing light intensity in a step-wise manner. Figures 15.29–15.32 exhibit the variation of the ER as functions of wavelength.

The ER in this case decreases as the wavelength increases in a step-like manner. Figures 15.33 and 15.34 show that the ER decreases with increasing alloy composition for the QWFETs of ternary and quaternary materials. Using the numerical values of the energy band constants from Chapter 18, we have plotted the normalized EEM as functions of film thickness, surface electron concentration, intensity and wavelength at $T = 4.2\,\mathrm{K}$ by taking QWFETs of ternary materials which are used for the purpose of numerical computations in accordance with the perturbed three- and two-band models of Kane together with perturbed parabolic energy bands, as shown in Figs. 15.35–15.38, respectively.

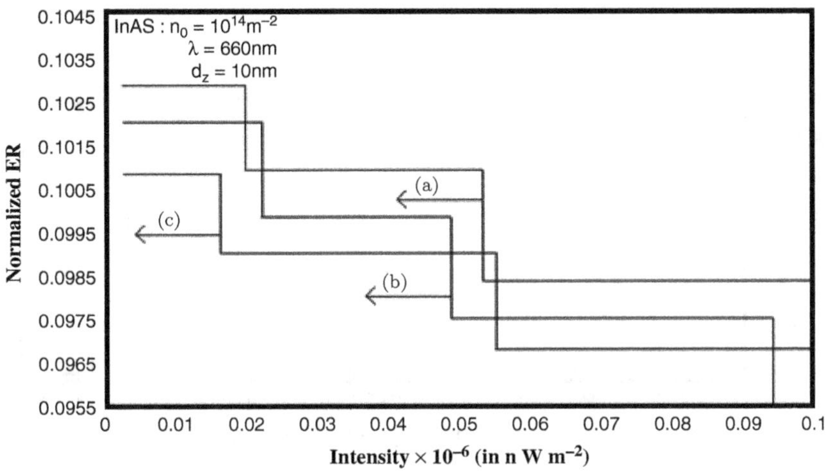

Fig. 15.25. Plot of the normalized 2D ER as a function of light intensity for QWFETs of InAs in which the curves (a), (b) and (c) represent the perturbed three- and two-band models of Kane and that of the parabolic energy bands, respectively.

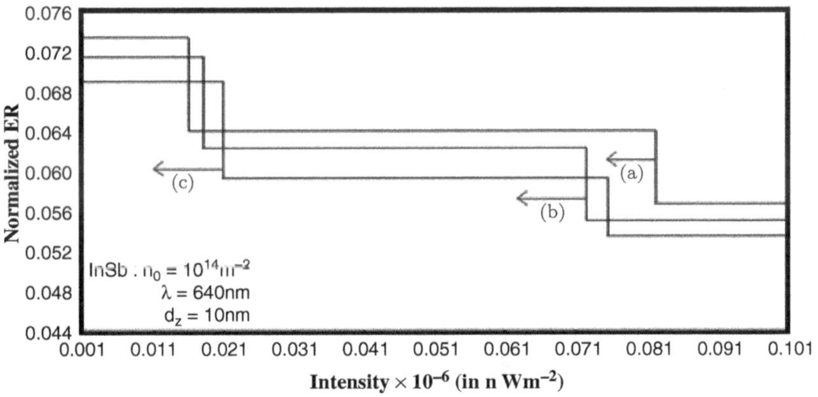

Fig. 15.26. Plot of the normalized 2D ER as a function of light intensity for QWFETs of InSb in which the curves (a), (b) and (c) represent the perturbed three- and two-band models of Kane and that of the parabolic energy bands, respectively.

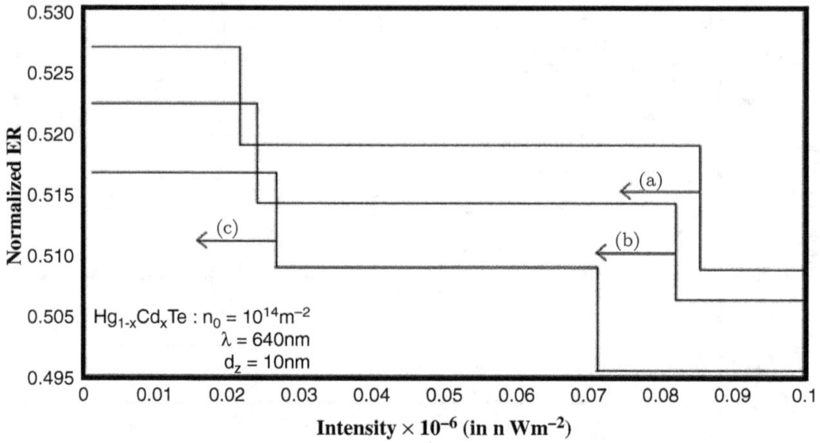

Fig. 15.27. Plot of the normalized 2D ER as a function of light intensity for QWFETs of $Hg_{1-x}Cd_xTe$ in which the curves (a), (b) and (c) represent the perturbed three- and two-band models of Kane and that of the parabolic energy bands, respectively.

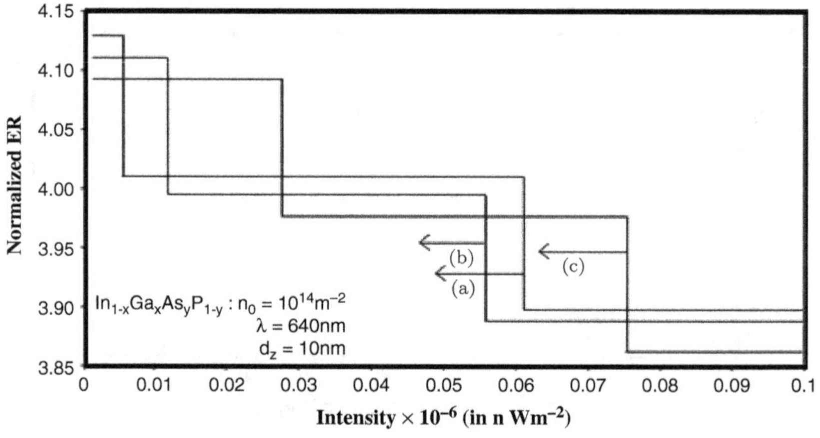

Fig. 15.28. Plot of the normalized 2D ER as a function of light intensity for QWFETs of $In_{1-x}Ga_xAs_yP_{1-y}$ lattice matched to InP in which the curves (a), (b) and (c) represent the perturbed three- and two-band models of Kane and that of the parabolic energy bands, respectively.

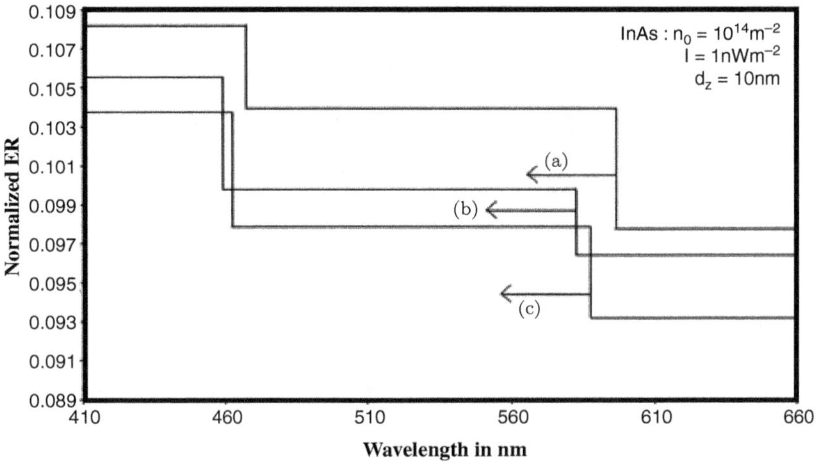

Fig. 15.29. Plot of the normalized 2D ER as a function of wave length (λ) for QWFETs of InAs in which the curves (a), (b) and (c) represent the perturbed three- and two-band models of Kane and that of the parabolic energy bands, respectively.

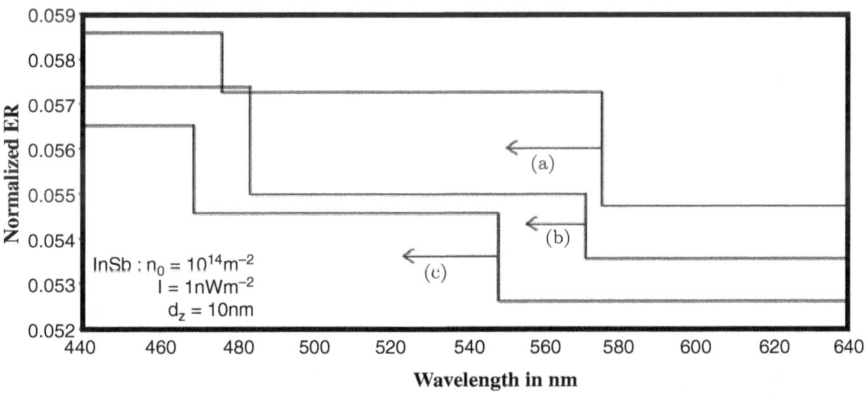

Fig. 15.30. Plot of the normalized 2D ER as a function of wave length (λ) for QWFETs of InSb in which the curves (a), (b) and (c) represent the perturbed three- and two-band models of Kane and that of the parabolic energy bands, respectively.

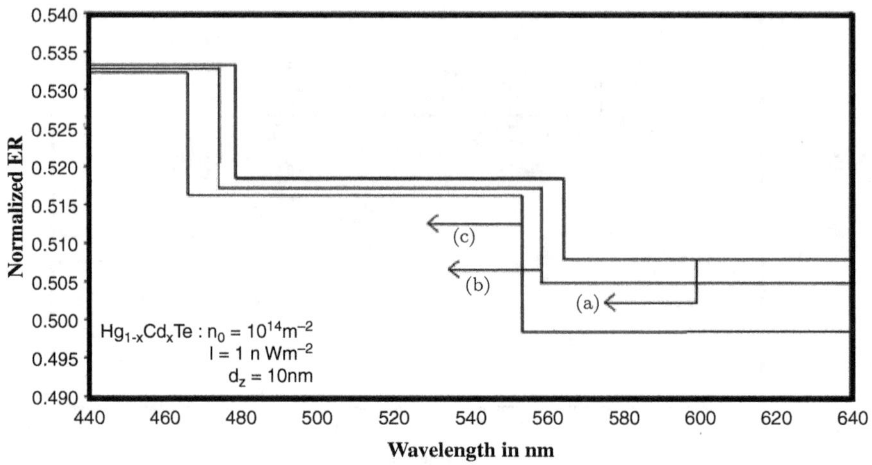

Fig. 15.31. Plot of the normalized 2D ER as a function of wave length (λ) for QWFETs of $Hg_{1-x}Cd_xTe$ in which the curves (a), (b) and (c) represent the perturbed three- and two-band models of Kane and that of the parabolic energy bands, respectively.

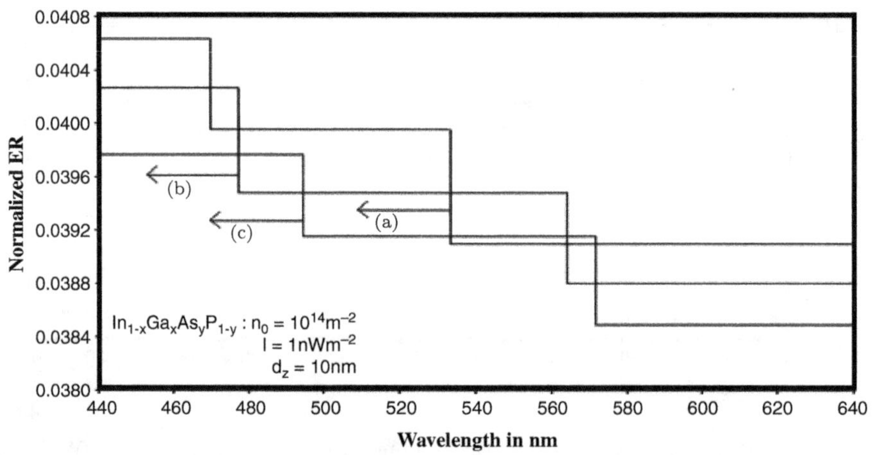

Fig. 15.32. Plot of the normalized 2D ER as a function of wave length (λ) for QWFETs of $In_{1-x}Ga_xAs_yP_{1-y}$ lattice matched to InP in which the curves (a), (b) and (c) represent the perturbed three- and two-band models of Kane and that of the parabolic energy bands, respectively.

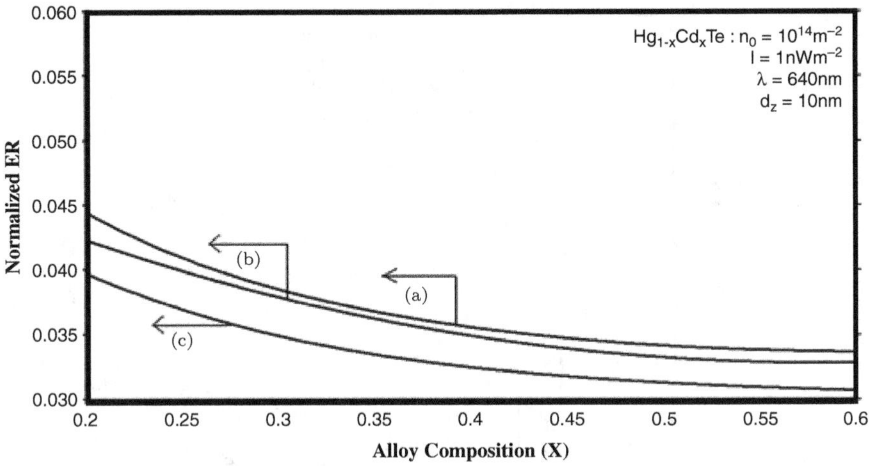

Fig. 15.33. Plot of the normalized 2D ER as a function of Alloy composition (x) for QWFETs of $Hg_{1-x}Cd_xTe$ in the presence of light waves in which the curves (a), (b) and (c) represent the perturbed three- and two-band models of Kane and that of the parabolic energy bands, respectively.

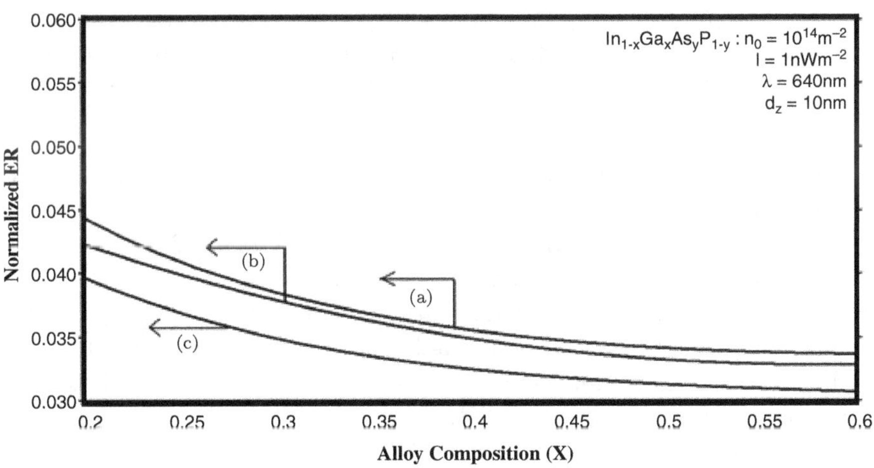

Fig. 15.34. Plot of the normalized 2D ER as a function of Alloy composition (x) for QWFETs of $In_{1-x}Ga_xAs_yP_{1-y}$ lattice matched to InP in the presence of light waves in which the curves (a), (b) and (c) represent the perturbed three- and two-band models of Kane and that of the parabolic energy bands, respectively.

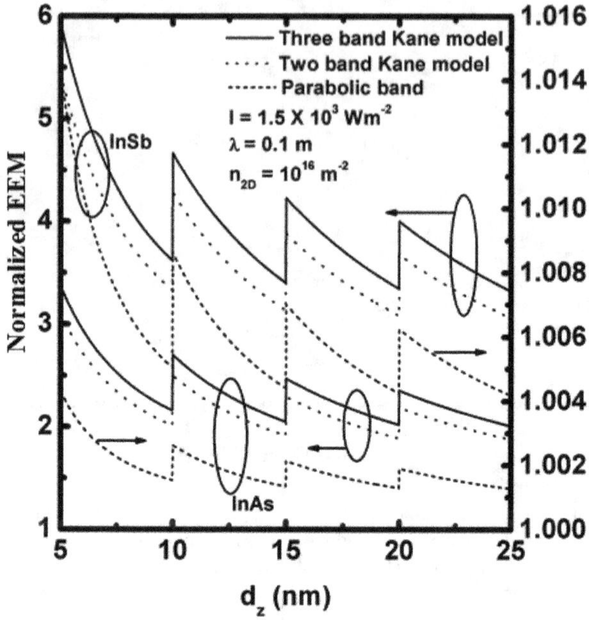

Fig. 15.35. Plot of the normalized EEM as a function of film thickness for QWFETs of InSb and InAs in the presence of light waves in accordance with the perturbed three-the two-band models of Kane and the parabolic energy band model, respectively.

Using the appropriate equations we have plotted the normalized EEM as a function of different variables in the presence of strong light waves for QWFETs of InAs, InSb, $Hg_{1-x}Cd_xTe$ and $In_{1-x}Ga_xAs_yP_{1-y}$, as shown in Figs. 15.35–15.38 in accordance with various purterbed band models, respectively. The influence of carrier confinement in 2D under the presence of an external photo-excitation on the behavior of normalized EEM can be understood from Figs. 15.35 and 15.36. The effect of quantum confinement is immediately apparent from all the curves of Fig. 15.35, since, the 2D normalized EEM depends strongly on the nanothickness, which is in direct contrast with the corresponding bulk specimens which is also the *direct signature of quantum confinement*. It appears from the said figures that the normalized EEM in this case decreases with the increasing film thickness in a step-like manner as considered here although the numerical values vary widely and are determined by the constants of the energy.

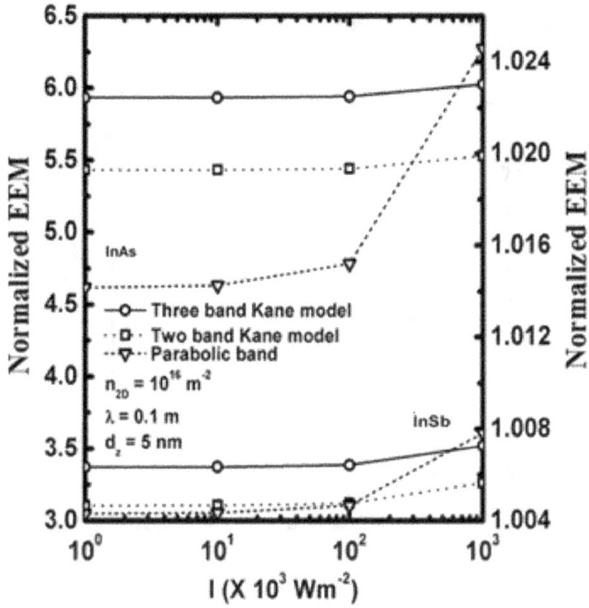

Fig. 15.36. Plot of the normalized EEM as a function of surface electron concentration for QWFETs of InSb and InAs in the presence of light waves in accordance with the perturbed three- the two-band models of Kane and the parabolic energy band model, respectively, in the presence of external photo-excitation.

The oscillatory dependence is due to the crossing over of the Fermi level by the size quantized levels. For each coincidence of a size quantized level with the Fermi level, there would be a discontinuity in the density-of-states function resulting in a peak of oscillations. With large values of film thickness, the height of the steps decreases and the normalized EEM decreases with increasing film thickness in non-oscillatory manner and exhibits monotonic decreasing dependence. The height of step size and the rate of decrement are totally dependent on the band structure. The influence of energy band non-parabolicity is immediately apparent by comparing the curves of the said figures. The energy band non-parabolicity and the spin-orbit splitting constant significantly enhances the numerical values of the normalized EEM both the cases of the materials. The numerical values of the normalized EEM in accordance with the three-band model of Kane are different as compared with the corresponding two-band

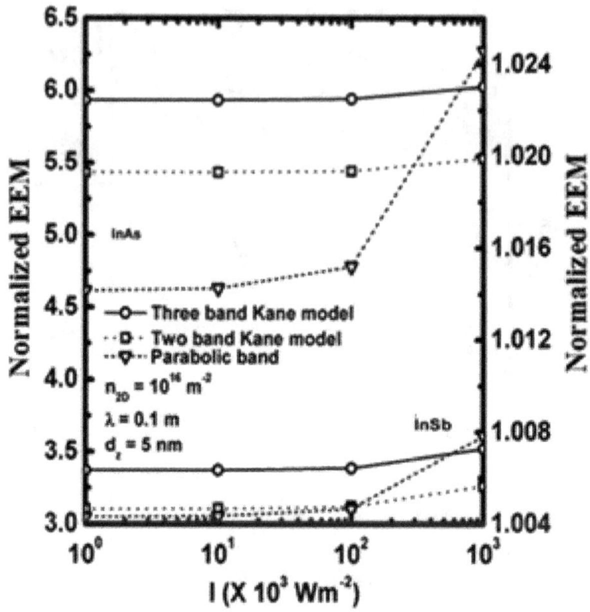

Fig. 15.37. Plot of the normalized EEM as a function of light intensity for QWFETs of InSb and InAs in the presence of light waves in accordance with the perturbed three-the two-band models of Kane and the parabolic energy band model, respectively.

model, which reflects that fact that the presence of the spin-orbit splitting constant changes the magnitude of the normalized EEM. It may be noted that the presence of the band non-parabolicity in accordance with the two-band model of Kane further changes the peaks of the oscillatory normalized EEM for all cases of quantum confinements.

In Fig. 15.36, we have plotted the normalized EEM as a function of surface electron concentration per unit area for all cases of Fig. 15.35. It appears that the normalized EEM increases with increasing carrier degeneracy and also reflects the signature of the 1D confinement through the nonlinear dependence with the 2D electron statistics. Since most of the electrons at low temperatures occupy the lowest sub-band level, we have plotted the normalized EEM by considering the lowest sub-band energy in Figs. 15.36 and 15.37. If more sub-bands were considered, the oscillatory dependence will be less and less prominent with increasing carrier concentration and ultimately, for bulk specimens of the same material, the normalized EEM will be found to increase continuously with

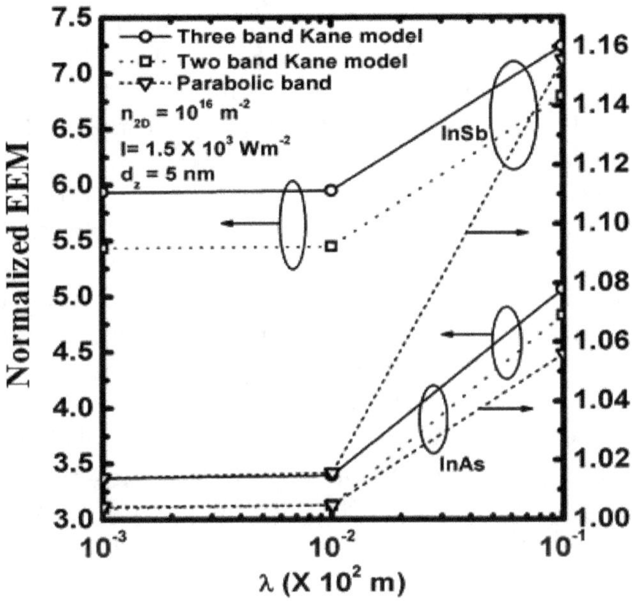

Fig. 15.38. Plot of the normalized EEM as a function of wavelength for QWFETs of InSb and InAs in the presence of light waves in accordance with the perturbed three-the two-band models of Kane and the parabolic energy band model, respectively.

increasing electron concentration in a non-oscillatory manner. The effects of the light intensity and wavelength on the normalized EEM have been exhibited in Figs. 15.37 and 15.38, respectively in the regime of very low temperatures. The normalized EEM increases with both the variables with different slopes. It appears that the normalized EEM in the case of parabolic energy band varies extremely slowly both with the light intensity and wavelength, although the sharp and significant variations are exhibited for both three- and the two-band energy models. It may be noted that the basic aim of this chapter is not solely to demonstrate the influence of quantum confinement in the presense of light waves on the QC and other band structure-dependent electronic quantities but also to formulate the appropriate carrier statistics in the most generatized form, since the transport and other phenomena in modern nanostructured devices having different band structures and the derivation of the expressions of many important carrier properties are based on the temperature-dependent carrier statistics in such quantum devices in the presence of strong light waves.

15.4 Open Research Problems

(R15.1) Investigate the QC for the QWFETs made of optoelectronic materials in the presence of arbitrarily oriented photo-excitation for all the cases of this chapter. Draw the corresponding graphs and explain the results.

(R15.2) Investigate the QC for the NWFETs made of magnetic and advanced materials in the presence of arbitrarily oriented photo-excitation. Draw the corresponding graphs and explain the results.

(R15.3) Investigate the QC for NWFETs of the optoelectronic materials in the presence of arbitrarily oriented photo-excitation. Draw the corresponding graphs and explain the results.

(R15.4) Investigate the QC for NWFETs of the optoelectronic materials in the presence of arbitrarily oriented alternating photo-excitation. Draw the corresponding graphs and explain the results.

(R15.5) Investigate the QC for CNTFETs in the presence of arbitrary photo-excitation. Draw the corresponding graphs and explain the results.

References

1. P. K. Basu, *Theory of Optical Processes in Materials, Bulk and Microstructures* (Oxford University Press, Oxford, 1997).
2. K. P. Ghatak, S. Bhattacharya, S. Bhowmik, R. Benedictus and S. Chowdhury, *J. Appl. Phys.* **103**, 094314 (2008); K. P. Ghatak and S. Bhattacharya, *J. Appl. Phys.* **102**, 073704 (2007); K. P. Ghatak, S. Bhattacharya, S. K. Biswas, A. De and A. K. Dasgupta, *Phys. Scr.* **75**, 820 (2007); P. K. Bose, N. Paitya, S. Bhattacharya, D. De, S. Saha, K. M. Chatterjee, S. Pahari and K. P. Ghatak, *Quant. Matt.* **1**, 89 (2012); K. P. Ghatak, S. Bhattacharya, A. Mondal, S. Debbarma, P. Ghorai and A. Bhattacharjee, *Quant. Matt.* **2**, 25 (2013); S. Bhattacharya, D. De, S. Ghosh, J. P. Bannerje, M. Mitra, B. Nag, S. Saha, S. K. Bishwas and M. Paul, *J. Comp. Theo. Nanosci.* **7**, 1066 (2010).
3. K. P. Ghatak, S. Bhattacharya, S. Pahari, S. N. Mitra and P. K. Bose, *J. Phys. Chem. Solids* **70**, 122 (2009), S. Bhattacharya, D. De, R. Sarkar, S. Pahari, A. De, A. K. Dasgupta, S. N. Biswas and K. P. Ghatak, *J. Comp. Theo. Nanosci.* **5**, 1345 (2008); S. Mukherjee, D. De, D. Mukherjee, S. Bhattacharya, A. Sinha, and K. P. Ghatak *Physica B* **393**, 347 (2007).
4. K. Seeger, *Semiconductor Physics* 7th Edition. (Springer-Verlag, 2006).
5. B. R. Nag, *Physics of Quantum Well Devices* (Kluwer Academic Publishers, The Netherlands, 2000).

Chapter 16

The QC in Quantized Transistors in the Presence of Strong Electric Field

16.1 Introduction

With the advent of modern nano devices, there has been considerable interest in studying the electric-field-induced processes in semiconductors having different band structures. *It appears from the literature that the studies have been made on the assumption that the carrier dispersion laws are invariant quantities in the presence of intense electric field, which is not fundamentally true.* In this chapter, we shall study the quantum capacitance (QC) in quantum confined optoelectronic materials under strong electric fields. In Section 16.2.1, on the theoretical background, we formulate the dispersion relation (DR) in III-V and optoelectronic compounds in the presence of intense electric field. In Section 16.2.2, we study the QC in quantum wire field-effect transistors (QWFETs) in the presence of electric fields. In Section 16.2.3, we investigate the QC in carbon nanotube field-effect transistors (CNTFETs) under electric fields, and in Section 16.2.4, we study the QC in nanowire field-effect transistors (NWFETs) in the present case. Section 16.3 presents summary and conclusion pertinent to this chapter. The last Section 16.4, contains few open research problems.

16.2 Theoretical Background

16.2.1 *The dispersion relations in III-Vand optoelectronic materials in the presence of strong electric fields*

The expression of the inter-band transition matrix element (X_{12}) in this case can be written as

$$X_{12} = i \int u_{\bar{k}1}^{*}(\bar{r}) \cdot \frac{\partial}{\partial k_x} u_{\bar{k}2}(\bar{r}) d^3 r \tag{16.1a}$$

where $u_{\bar{k}1}(\bar{r}) \equiv u_1(\bar{k}, \bar{r})$ and $u_{\bar{k}2}(\bar{r}) \equiv u_2(\bar{k}, \bar{r})$ in which $u_1(\bar{k}, \bar{r})$ and $u_2(\bar{k}, \bar{r})$ are given by (15.15) and (15.16). In the presence of an external electric field, F_s along x axis the X_{12} has finite interaction from one band to the same band, e.g.,

$$
\begin{aligned}
&áS \mid S\tilde{n} = áX \mid X\tilde{n} = áY \mid Y\tilde{n} = áZ \mid Z\tilde{n} = 1,\\
&áX \mid Y\tilde{n} = áY \mid Z\tilde{n} = áZ \mid X\tilde{n} = 0,\\
&áS \mid X\tilde{n} = áX \mid S\tilde{n} = 0, \quad áS \mid Y\tilde{n} = áY \mid S\tilde{n} = 0 \quad \text{and}\\
&áS \mid Z\tilde{n} = áZ \mid S\tilde{n} = 0.
\end{aligned}
$$

Using the appropriate equations, we can write

$$
\begin{aligned}
X_{12} &= i\int d^3r \left\{ a_{k_+}[(iS)\downarrow'] + b_{k_+}\left[\left(\frac{X' - iY'}{\sqrt{2}}\right)\uparrow'\right] + c_{k_+}[Z'\downarrow'] \right\}^* \frac{\partial}{\partial k_x}\\
&\quad \times \left\{ a_{k_-}[(iS)\uparrow'] - b_{k_-}\left[\left(\frac{X' + iY'}{\sqrt{2}}\downarrow'\right)\right] + c_{k_-}[Z'\uparrow'] \right\}\\
&= i\int d^3r \left[\left\{\left(a_{k_+}\frac{\partial}{\partial k_x}a_{k_-}\right)\cdot[(iS)\downarrow']^*\cdot([(iS)\uparrow'])\right.\right.\\
&\quad + \left(\frac{b_{k_+}}{\sqrt{2}}\frac{\partial}{\partial k_x}a_{k_-}\right)[(X' - iY')\uparrow']^*\cdot[(iS)\uparrow']\\
&\quad \left.\left. + \left(c_{k_+}\frac{\partial}{\partial k_x}a_{k_-}\right)[(Z'\downarrow')]^*\cdot[(iS)\uparrow']\right\}\right]\\
&\quad + i\int d^3r \left[\left\{\left(-\frac{a_{k_+}}{\sqrt{2}}\frac{\partial}{\partial k_x}b_{k_-}\right)\cdot[(iS)\downarrow']^*\cdot[(X' + iY')\downarrow']\right.\right.\\
&\quad - \left(\frac{b_{k_+}\dfrac{\partial}{\partial k_x}b_{k_-}}{2}\right)[(X' - iY')\uparrow']^*\cdot[(X' + iY')\downarrow']\\
&\quad \left.\left. - \left(\frac{c_{k_+}\frac{\partial}{\partial k_x}b_{k_-}}{\sqrt{2}}\right)[(Z'\downarrow')]^*\cdot[(X' + iY')\downarrow']\right\}\right]\\
&= i\int d^3r \left[\left\{\left(a_{k_+}\frac{\partial}{\partial k_x}c_{k_-}\right)\cdot([(iS)\downarrow']^*\cdot[Z'\uparrow'])\right.\right.\\
&\quad + \left(\frac{b_{k_+}}{\sqrt{2}}\frac{\partial}{\partial k_x}c_{k_-}\right)[(X' - iY')\uparrow']^*\cdot[Z'\uparrow']\\
&\quad \left.\left. + \left(c_{k_+}\frac{\partial}{\partial k_x}c_{k_-}\right)[(Z'\downarrow')]^*\cdot[Z'\uparrow']\right\}\right]
\end{aligned}
\tag{16.2}
$$

Therefore,

$$\begin{aligned}
\frac{X_{12}}{i} = & \left[\left\{\left[\left(a_{k_+}\frac{\partial}{\partial k_x}a_{k_-}\right)\cdot\langle iS|iS\rangle\langle\downarrow'\mid\uparrow'\rangle\right]\right.\right.\\
& + \left[\left(\frac{b_{k_+}}{\sqrt{2}}\frac{\partial}{\partial k_x}a_{k_-}\right)\langle(X'-iY')|iS\rangle\langle\uparrow'\mid\downarrow'\rangle\right]\\
& \left.\left. + \left[\left(c_{k_+}\frac{\partial}{\partial k_x}a_{k_-}\right)\langle Z'|iS\rangle\langle\downarrow'\mid\uparrow'\rangle\right]\right\}\right]\\
& - \left[\left\{\left[\left(\frac{a_{k_+}}{\sqrt{2}}\frac{\partial}{\partial k_x}b_{k_-}\right)\cdot\langle iS|(X'+iY')\rangle\langle\downarrow'\mid\downarrow'\rangle\right]\right.\right.\\
& + \left[\left(\frac{b_{k_+}}{2}\frac{\partial}{\partial k_x}b_{k_+}\right)\langle(X'-iY')|(X'+iY')\rangle\langle\uparrow'\mid\downarrow'\rangle\right]\\
& \left.\left. + \left[\left(\frac{c_k}{\sqrt{2}}\frac{\partial}{\partial k_x}b_{k_-}\right)\langle Z'|(X'+iY')\rangle\langle\downarrow'\mid\downarrow'\rangle\right]\right\}\right]\\
& + \left\{\left[\left(a_k\frac{\partial}{\partial k_x}c_{k_-}\right)\cdot\langle iS|Z'\rangle\langle\downarrow'\mid\uparrow'\rangle\right]\right.\\
& + \left[\left(\frac{b_{k_+}}{\sqrt{2}}\frac{\partial}{\partial k_x}c_{k_-}\right)\cdot\langle(X'-iY')|Z'\rangle\langle\uparrow'\mid\uparrow'\rangle\right]\\
& \left. + \left[\left(c_k\frac{\partial}{\partial k_x}c_{k_-}\right)\cdot\langle Z'|Z'\rangle\langle\downarrow'\mid\uparrow'\rangle\right]\right\}
\end{aligned} \tag{16.3}$$

Therefore we can write,

$$X_{12} = i\left\{-\left(a_{k_+}\frac{\partial}{\partial k_x}a_{k_-}\right)\langle\downarrow'\mid\uparrow'\rangle + \left(c_{k_+}\frac{\partial}{\partial k_x}c_{k_-}\right)\langle\downarrow'\mid\uparrow'\rangle\right\} \tag{16.4}$$

We can prove that

$$\langle\downarrow'\mid\uparrow'\rangle = \frac{1}{2}(\hat{r}_1 + i\hat{r}_2) \tag{16.5}$$

Therefore using (16.4) and (16.5), we get

$$\begin{aligned}
X_{12} &= i\left\{-\left(a_{k+}\frac{\partial}{\partial k_x}a_{k-}\right) + \left(c_{k+}\frac{\partial}{\partial k_x}c_{k-}\right)\right\}\langle\downarrow'\mid\uparrow'\rangle\\
&= -i\left\{\left(a_{k_+}\frac{\partial}{\partial k_x}a_{k-}\right) - \left(c_{k+}\frac{\partial}{\partial k_x}c_{k-}\right)\right\}\cdot\frac{1}{2}(\hat{r}_1 + i\hat{r}_2)\\
&= \frac{-iA(k)}{2}(\hat{r}_1 + i\hat{r}_2)
\end{aligned} \tag{16.6}$$

where

$$A(k) = \left(a_{k+}\frac{\partial}{\partial k_x}a_{k-}\right) - \left(c_{k+}\frac{\partial}{dk_x}c_{k-}\right) \tag{16.7}$$

From (16.6), we find

$$|X_{12}|^2 = \frac{1}{4}A^2(k)(1+1) \quad [\text{since, } |\hat{r}_1| = |\hat{r}_2| = 1] \tag{16.8}$$

Considering spin-up and spin-down, we have to multiply by 2

$$|X_{12}|^2 = 2 \times \frac{1}{2}A^2(k) = A^2(k) \tag{16.9}$$

We can evaluate X_{11} and X_{22}. in the following way:

$$\begin{aligned}
X_{11} &= i\int u^*_{\bar{k}1}(\bar{r}) \cdot \frac{\partial}{\partial k_x}u_{\bar{k}1}(\bar{r}) \cdot d^3r \\
&= i\int d^3r \left\{\left(a_{k_+}\frac{\partial}{\partial k_x}a_{k_+}\right) + \left(b_{k_+}\frac{\partial}{\partial k_x}b_{k_+}\right) + \left(c_{k_+}\frac{\partial}{\partial k_x}c_{k_+}\right)\right\} \\
&= \frac{1}{2}i\int d^3r \left\{\frac{\partial}{\partial k_x}(a_k^2 + b_k^2 + c_k^2)\right\} = \frac{1}{2}i\int d^3r\left\{\frac{\partial}{\partial k_x}(1)\right\} = 0, \\
&\quad \text{since } a_{k+}^2 + b_{k+}^2 + c_{k+}^2 = 1.
\end{aligned}$$

Therefore, $X_{11} = 0$, and similarly we can prove $X_{22} = 0$. Thus, we conclude that intraband momentum matrix element due to external electric field (X_{CC}) is zero. From the expression of $a_{k\pm}$, we can write

$$a_{k_+}^2 = r_0^2\left[\frac{E_g - \gamma_{k_+}^2(E_g - \delta')}{E_g + \delta'}\right]^2 \quad \text{and} \quad a_{k-}^2 = r_0^2\left[\frac{E_g - \gamma_{k-}^2(E_g - \delta')}{E_g + \delta'}\right]^2$$

Therefore,

$$2a_{k-}\frac{\partial}{\partial k_x}a_{k-} = r_0^2\left[-\left(\frac{E_g - \delta'}{E_g + \delta'}\right)\right]\frac{\partial \gamma_{k-}^2}{\partial k_x} \quad \text{and}$$

$$\frac{\partial a_{k-}}{\partial k_x} = -\frac{r_0^2}{2}\left(\frac{E_g - \delta'}{E_g + \delta'}\right)\frac{1}{a_{k-}}\frac{\partial \gamma_{k-}^2}{\partial k_x}$$

Combining we can write

$$a_{k+}\frac{\partial a_{k-}}{\partial k_x} = -\frac{r_0^2}{2}\left(\frac{E_g - \delta'}{E_g + \delta'}\right)\frac{a_{k+}}{a_{k-}}\frac{\partial \gamma_{k-}^2}{\partial k_x}$$

Similarly, $c_{k_+} = t\gamma_{k_+}$ and $c_{k_-} = t\gamma_{k_-}$.

Therefore,

$$c_{k+}\frac{\partial}{\partial k_x}c_{k-} = \frac{t^2}{2}\frac{c_{k+}}{c_{k-}}\frac{\partial \gamma_{k-}^2}{\partial k_x}$$

$$A(k) = \left\{-\frac{r_0^2}{2}\left[\frac{E_g-\delta'}{E_g+\delta'}\right]\frac{a_{k+}}{a_{k-}} - \frac{t^2}{2}\frac{c_{k+}}{c_{k-}}\right\}\frac{\partial \gamma_{k-}^2}{\partial k_x}$$

Now,

$$\left(\frac{a_{k+}}{a_{k-}}\right)^2 = \frac{E_g-\gamma_{k+}^2(E_g-\delta')}{E_g-\gamma_{k-}^2(E_g-\delta')} = \frac{E_g-\frac{\eta-E_g}{2(\eta+\delta')}(E_g-\delta')}{E_g-\frac{\eta+E_g}{2(\eta+\delta')}(E_g-\delta')}$$

$$= \frac{2E_g-(\eta+\delta')-(\eta-E_g)(E_g-\delta')}{2E_g(\eta+\delta')-(\eta+E_g)(E_g-\delta')}$$

$$= \frac{\eta(E_g+\delta')+E_g(E_g+\delta')}{\eta(E_g+\delta')-E_g(E_g-3\delta')}$$

Therefore,

$$\left(\frac{a_{k+}}{a_{k-}}\right)^2 = \frac{\eta+E_g}{\eta-E_g\left(\frac{E_g-3\delta'}{E_g+\delta'}\right)} = \frac{\eta+E_g}{\eta-E_g'}$$

where

$$E_g' = \frac{E_g(E_g-3\delta')}{E_g+\delta'}$$

Thus,

$$\frac{a_{k+}}{a_{k-}} = \sqrt{\frac{\eta+E_g}{\eta-E_g'}}.$$

Similarly,

$$\frac{c_{k+}}{c_{k-}} = \frac{\gamma_{k+}}{\gamma_{k-}} = \sqrt{\frac{\eta-E_g}{\eta+E_g'}}$$

and thus,

$$A(k) = -\left\{P\left(\frac{\eta+E_g}{\eta-E_g'}\right)^{1/2} + Q\left(\frac{\eta-E_g}{\eta+E_g}\right)^{1/2}\right\}\frac{\partial \gamma_{k-}^2}{\partial k_x} \tag{16.10}$$

where

$$P = \frac{r_0^2}{2}\left(\frac{E_g-\delta'}{E_g+\delta'}\right) \quad \text{and} \quad Q = t^2/2.$$

Now,

$$\gamma_{k-}^2 = \frac{\eta + E_g}{2(\eta+\delta')} \quad \text{so that} \quad \frac{\partial \gamma_{k-}^2}{\partial k_x} = \frac{1}{2}\left[\frac{\partial\eta/\partial k_x}{(\eta+\delta')} - \frac{\eta+E_g}{(\eta+\delta')^2}\frac{\partial\eta}{\partial k_x}\right]$$

Thus,

$$\frac{\partial \gamma_{k-}^2}{\partial k_x} = \frac{1}{2}\left[\frac{\eta+\delta'-\eta-E_g}{(\eta+\delta')}\right]\frac{\partial\eta}{\partial k_x} = -\frac{1}{2}\frac{(E_g-\delta')}{(\eta+\delta')^2}\cdot\frac{\partial\eta}{\partial k_x} \tag{16.11}$$

From (16.10) and (16.11), we get

$$A(k) = \frac{1}{2}\frac{(E_g-\delta')}{(\eta+\delta')^2}\cdot\frac{\partial\eta}{\partial k_u}\cdot\left\{P\left(\frac{\eta+E_g}{\eta-E_g'}\right)^{1/2} + Q\left(\frac{\eta-E_g}{\eta+E_g}\right)^{1/2}\right\} \tag{16.12}$$

This implies

$$\frac{\partial\eta}{\partial k_x} = \frac{E_g\hbar^2}{m_r}\cdot\frac{k_x}{\eta} \tag{16.13}$$

From (16.12) and (16.13), one can write

$$A(k) = \frac{E_g\hbar^2}{2m_r}\cdot\frac{k_u}{\eta}\cdot\frac{(E_g-\delta')}{(\eta+\delta')^2}\cdot\left\{P\left(\frac{\eta+E_g}{\eta-E_g'}\right)^{1/2} + Q\left(\frac{\eta-E_g}{\eta+E_g}\right)^{1/2}\right\} \tag{16.14}$$

Thus,

$$|A(k)|^2 = \frac{E_g^2(E_g-\delta')\hbar^2}{4m_r}\frac{\hbar^2k_x^2}{m_r}\frac{1}{\eta^2}\frac{1}{(\eta+\delta')^4}\left\{P\left(\frac{\eta+E_g}{\eta-E_g'}\right)^{1/2} + Q\left(\frac{\eta-E_g}{\eta+E_g}\right)^{1/2}\right\}^2 \tag{16.15}$$

and

$$|X_{12}|^2 = |A('k)|^2 \tag{16.16}$$

From (16.15) and (16.16), we can write the square of the magnitude of the inter-band transition matrix element due to external electric field ($|X_{CV}|^2$).

It is well known that the energy Eigen value, $E_n^{(2)}(\bar{k})$, in the presence of a perturbed Hamiltonian, H', is given by [1]

$$E_n^{(2)}(\bar{k}) = E_n(\bar{k}) + \langle n\bar{k}|H'|n\bar{k}\rangle + \{|\langle n\bar{k}|H'|n\bar{k}\rangle|^2[E_n(\bar{k}) - E_m(\bar{k})]\} \tag{16.17}$$

where

$$H\psi_n(\bar{k},\bar{r}) = E\psi(\bar{k},\bar{r}) \tag{16.18}$$

$$H = H_0 + H' \tag{16.19}$$

in which H is the total Hamiltonian, $\psi(\bar{k},\bar{r})$ is the wave function, where $u_n(\bar{k},\bar{r})$ is the periodic function of it, H_0 is the unperturbed Hamiltonian, n is the band index, and $E_n(\bar{k})$ is the energy of an electron in the periodic lattice.

For an external electric field (F_S) applied along the x-axis, the perturbed Hamiltonian (H') can be written as

$$H' = -F \cdot x \tag{16.20}$$

where

$F(= eF_s)$.

Therefore, we get

$$E_n^{(2)}(\bar{k}) = E_n(\bar{k}) - F\langle n\bar{k}|H'|n\bar{k}\rangle + F^2\{|\langle n\bar{k}|H'|n\bar{k}\rangle|^2/[E_n(\bar{k}) - E_m(\bar{k})]\} \tag{16.21}$$

In (2.22), the second and the third terms are due to the perturbation factor.

For

$$X_{nm}(\bar{k}) = \langle n\bar{k}|x|m\bar{k}\rangle \tag{16.22}$$

we find

$$X_{nm}(\bar{k}) = i\int u_n^*(\bar{k},\bar{r})(\partial/\partial u)[u_m(\bar{k},\bar{r})]d^3r \tag{16.23}$$

where k_x is the x component of the $\bar{k}$ and the integration in (16.23) extends over the unit cell. From (16.21), (16.22), and (16.23), with the n corresponding to the conduction band (C) and m corresponding to the valance band (V), we get

$$E_c^{(2)}(\bar{k}) = E_c(\bar{k}) - FX_{cc} + \{F^2|X_{CV}|^2/E_C(\bar{k}) - E_V(\bar{k})]\} \tag{16.24}$$

Thus, combining the appropriate equations, the DR of the conduction electrons in the presence of electric field along x-axis can be written as

$$\begin{aligned} I_{11}(E) &= \frac{\hbar^2k_x^2}{2m_c} + \frac{\hbar^2k_y^2}{2m_c} + \frac{\hbar^2k_z^2}{2m_c} + \frac{F^2|X_{12}|^2}{\eta} = \left[\frac{\hbar^2k_x^2}{2m_c} + \frac{\hbar^2k_y^2}{2m_c} + \frac{\hbar^2k_z^2}{2m_c}\right. \\ &\quad + \left\{\frac{\hbar^2k_x^2}{2m_c}\cdot\frac{2m_c}{m_r}\cdot\frac{F^2\hbar^2E_g^2(E_g-\delta')^2}{4m_r}\frac{1}{\eta^3}\frac{1}{(\eta+\delta')^4}\right. \\ &\quad \times\left.\left.\left[P\left(\frac{\eta+E_g}{\eta-E_g'}\right)^{1/2} + Q\left(\frac{\eta-E_g}{\eta+E_g}\right)^{1/2}\right]^2\right\}\right] \end{aligned} \tag{16.25}$$

When $F \to 0$, we have from (2.26),

$$k^2 \to \frac{2m_c}{\hbar^2} I_{11}(E) \quad \text{and} \quad \eta_1^2 = \left[E_g^2 + E_g \frac{2m_c}{m_r} I_{11}(E)\right].$$

Using the method of successive approximation, one can write

$$1 = \frac{\hbar^2 k_x^2}{2m_c I_{11}(E)} + \frac{\hbar^2 k_y^2}{2m_c I_{11}(E)} + \frac{\hbar^2 k_z^2}{2m_c I_{11}(E)} + \frac{\hbar^2 k_x^2}{2m_c I_{11}(E)} \cdot \Phi(E, F) \tag{16.26}$$

where

$$\Phi(E, F) = \frac{2m_c}{m_r} \frac{F^2 \hbar^2 E_g^2 (E_g - \delta')^2}{4m_r} \frac{1}{\eta_1^3} \frac{1}{(\eta_1 + \delta')^4} \times \left[P\left(\frac{\eta_1 + E_g}{\eta_1 - E_g'}\right)^{1/2} + Q\left(\frac{\eta_1 - E_g}{\eta_1 + E_g}\right)^{1/2}\right]^2$$

Therefore, the $E - k$ DR in the presence of an external electric field for III-V, ternary and quaternary materials whose unperturbed energy band structures are defined by the three-band model of Kane can be expressed as

$$\frac{k_x^2}{\frac{2m_c}{\hbar^2}\left[\frac{I_{11}(E)}{1+\Phi(E,F)}\right]} + \frac{k_y^2}{\frac{2m_c}{\hbar^2} I_{11}(E)} + \frac{k_z^2}{\frac{2m_c}{\hbar^2} I_{11}(E)} = 1 \tag{16.27}$$

In (16.27), the coefficients of k_x, k_y, and k_z are not same and, for this reason, this basic equation is "anisotropic" in nature together with the fact that the anisotropic dispersion relation is the ellipsoid of revolution in the k-space.

From (16.27), the expressions of the effective electron masses along x, y, and z directions can, respectively, be written as

$$m_x^*(E, F) = \hbar^2 k_x \left.\frac{\partial k_x}{\partial E}\right|_{k_y=0, k_z=0} = m_c[1 + \Phi(E, F)]^{-2}[[1 + \Phi(E, F)] I_{11}'(E) - I_{11}(E)(\Phi'(E, F))] \tag{16.28}$$

$$m_y^*(E, F) = \hbar^2 k_y \left.\frac{\partial k_y}{\partial E}\right|_{k_x=0, k_z=0} = m_0 I_{11}'(E) \tag{16.29}$$

$$m_z^*(E, F) = \hbar^2 k_z \left.\frac{\partial k_z}{\partial E}\right|_{k_x=0, k_y=0} = m_0 I_{11}'(E) \tag{16.30}$$

where

$$I'_{11}(E) = \frac{\partial}{\partial E}(I_{11}(E)) \quad \text{and} \quad \Phi'(E, F) = \frac{\partial}{\partial E}[\Phi(E, F)].$$

It may be noted from (16.28) that the effective mass along x-direction is a function of both electron energy and electric field, respectively, whereas from (16.29) and (16.30) we can infer that the expressions of the effective masses along y and z directions are same and they depend on the electron energy only. Thus, in the presence of an electric field, the mass anisotropy for Kane type semiconductors depends both on electron energy and electric field, respectively.

The use of the usual approximation [2]

$$k_x^2 \approx \frac{1}{3}k^2 \tag{16.31}$$

in (16.27) leads to the simplified expression of the electron energy spectrum in the present case as

$$I_{11}(E) = \frac{\hbar^2 k^2}{2m_c} + \frac{F^2\hbar^2 E_g^2(E_g - \delta')^2}{12m_r} \frac{2m_c}{m_r} I_{11}(E) \frac{1}{\eta_1} \frac{1}{(\eta_1 + \delta')^4} \times \left\{ P\left(\frac{\eta_1 + E_g}{\eta_1 - E'_g}\right)^{1/2} + Q\left(\frac{\eta_1 - E_g}{\eta_1 + E_g}\right)^{1/2} \right\}^2 \tag{16.32}$$

Equation (16.32) can approximately be written as

$$\frac{\hbar^2 k^2}{2m_c} = \beta_{11}(E, F) \tag{16.33}$$

where

$$\beta_{11}(E, F) = \left[I_{11}(E) - \left[\frac{F^2\hbar^2 E_g^2(E_g - \delta')^2}{12m_r} \frac{2m_c}{m_r} I_{11}(E) \frac{1}{\eta_1} \frac{1}{(\eta_1 + \delta')^4} \times \left\{ P\left(\frac{\eta_1 + E_g}{\eta_1 - E'_g}\right)^{1/2} + Q\left(\frac{\eta_1 - E_g}{\eta_1 + E_g}\right)^{1/2} \right\}^2 \right] \right]$$

(I) Under the condition $\Delta \to 0$, Eq. (16.32) assumes the form

$$\frac{k_x^2}{\frac{2m_c}{\hbar^2}\left[\frac{\gamma_0(E)}{1+\Phi_1(E,F)}\right]} + \frac{k_y^2}{\frac{2m_c}{\hbar^2}\gamma_0(E)} + \frac{k_z^2}{\frac{2m_c}{\hbar^2}\gamma_0(E)} = 1 \tag{16.34}$$

where

$$\gamma_0(E) = E(1 + \alpha E)$$

and

$$\Phi_1(E,F) = \frac{\hbar^2 F^2}{4m_r E_g^2 \gamma_0(E)}\left[1+\frac{2m_c}{m_r}\frac{\gamma_0(E)}{E_g}\right]^{-5/2} \tag{16.35}$$

Equation (16.34) represents the electron energy spectrum of III-V, ternary and quaternary materials in the presence of an external electric field whose unperturbed band structures are defined by the two-band model of Kane.

From (16.34) along with the substitution $\Delta \to 0$, we get

$$\frac{\hbar^2 k^2}{2m_c} = \gamma_{11}(E,F) \tag{16.36}$$

where

$$\gamma_{11}(E,F) = \gamma_0(E)\left[1-\frac{1}{3}\Phi_1(E,F)\right]$$

Equation (16.36) represents the approximate $\boldsymbol{E}-\boldsymbol{k}$ DR of III-V, ternary and quaternary compounds in the presence of an external electric field whose unperturbed band structures are defined by the two-band model of Kane.

16.2.2 *The QC in QWFETs in the presence of electric field*

The DR of the conduction electrons in this case can be written using the perturbed three-band Kane model and the two-band Kane model as

$$\frac{\hbar^2 k_s^2}{2m_c} = \beta_{11}(E,F) - \frac{\hbar^2}{2m_c}\left(\frac{n_z \pi}{d_z}\right)^2 \tag{16.37}$$

$$\frac{\hbar^2 k^2}{2m_c} = \gamma_{11}(E,F) - \frac{\hbar^2}{2m_c}\left(\frac{n_z \pi}{d_z}\right)^2 \tag{16.38}$$

The electron concentration per unit area under the condition of extreme carrier degeneracy can, respectively, be written as

$$N_S = \frac{m_c}{\pi\hbar^2}\sum_{n_z=1}^{n_{z\max}}\left[\beta_{11}(E_{F16,1},F) - \frac{\hbar^2}{2m_c}\left(\frac{n_z \pi}{d_z}\right)^2\right] \tag{16.39}$$

$$N_S = \frac{m_c}{\pi\hbar^2}\sum_{n_z=1}^{n_{z\max}}\left[\gamma_{11}(E_{F16,1},F) - \frac{\hbar^2}{2m_c}\left(\frac{n_z \pi}{d_z}\right)^2\right] \tag{16.40}$$

Using (16.39), (16.40), and (1.4), the QC can be written as

$$C_g = \left[e^2 \sum_{n_z=1}^{n_{z\max}} \beta'_{11}(E_{F16,1},F)\right]\left[+\frac{e^2 d_{0x}}{\varepsilon_{0x}}\sum_{n_z=1}^{n_{z\max}} \beta'_{11}(E_{F16,1},F)\right]^{-1} \tag{16.41}$$

and

$$C_g = \left[e^2 \sum_{n_z=1}^{n_{z\max}} \gamma'_{11}(E_{F16,1}, F)\right] \left[\frac{\pi\hbar^2}{m_c} + \frac{e^2 d_{0x}}{\varepsilon_{0x}} \sum_{n_z=1}^{n_{z\max}} \gamma'_{11}(E_{F16,1}, F)\right]^{-1} \tag{16.42}$$

16.2.3 The QC in CNTFETs in the presence of electric field

The dispersion relations in arm-chair and zigzag nanotubes can be written following (7.170) and (7.171) as

$$\delta_{11}(E,F) = \pm t_c\sqrt{1 + 4\cos\frac{m\pi}{n}\cos\frac{k_y a_c}{2} + 4\cos^2\frac{k_y a_c}{2}},$$
$$-\frac{\pi}{a_c} < k_y < \frac{\pi}{a_c} \quad \text{and} \quad m(=1,2,\ldots,2n) \tag{16.43}$$

$$\delta_{11}(E,F) = \pm t_c\sqrt{1 + 4\cos\frac{\sqrt{3}k_x a_c}{n}\cos\frac{m\pi}{n} + 4\cos^2\frac{m\pi}{n}},$$
$$-\frac{\pi}{\sqrt{3}a_c} < k_x < \frac{\pi}{\sqrt{3}a_c} \quad \text{and} \quad m(=1,2,\ldots,2n) \tag{16.44}$$

where

$$\delta_{11}(E,F) = \left[E - \bar{C}_2(F)\left[E + \frac{m_r E_g}{2m_c}\right]^{-\frac{5}{2}}\right]$$

and

$$\bar{C}_2(F) = \left[\frac{(\hbar eF)^2 m_r^{3/2}(E_g)^{1/2}}{12(2m_c)^{5/2}}\right]$$

(16.43) and (16.44) can be written as

$$k_y = \left(\frac{2}{a_{cc}\sqrt{3}}\right)\cos^{-1}[f_5(E, E_i, F)] \tag{16.45}$$

and

$$k_y = \left(\frac{2}{3a_{cc}}\right)\cos^{-1}[f_6(E, E_i, F)] \tag{16.46}$$

where

$$f_5(E,E_i,F) = \left[\frac{1}{2}\left[\frac{-\delta_{11}^2(E_i,F)}{t_c^2}+5\right] - \frac{1}{2}\left[\left(\frac{\delta_{11}^2(E_i,F)}{t_c^2}-5\right)^2 - \left(1-\frac{\delta_{11}^2(E_i,F)}{t_c^2}\right)\right]^{\frac{1}{2}}\right]$$

$$f_6(E,E_i,F) = \left[\bar{\varphi}_1(E_i,F) + \frac{1}{4\bar{\varphi}_1(E_i,F)} - \frac{\delta_{11}^2(E_i,F)}{4t_c^2\bar{\varphi}_1(E_i,F)}\right],$$

$$\bar{\varphi}_1(E_i,F) = \left[\frac{1}{2}+\frac{1}{2t_c}\delta_{11}(E_i,F)\right].$$

The electron concentration per unit length for arm-chair and Zig-Zag CNT FET in the presence of an electric field F along x direction can, respectively be expressed under the condition of extreme carrier degeneracy as

$$n_{1D} = \left(\frac{8}{\pi a_{cc}\sqrt{3}}\right)\sum_{i=1}^{i_{\max}} \cos^{-1}[f_5(E_{F16,3},E_i,F)] \tag{16.47}$$

and

$$n_{1D} = \left(\frac{8}{3\pi a_{cc}}\right)\sum_{i=1}^{i_{\max}} \cos^{-1}[f_6(E_{F16,3},E_i,F)] \tag{16.48}$$

where $E_{F16,3}$ is the Fermi energy in this case.

The QC per unit length for arm-chair and Zig-Zag CNT FETs can, respectively, be expressed as

$$C_g = \frac{e^2\sum_{i=1}^{i_{\max}} f_7(E_{F16,3},E_i,F)}{\left[\frac{a_{cc}\pi\sqrt{3}}{4} + \frac{e^2}{\varepsilon_{0x}}\sum_{i=1}^{i_{\max}} f_7(E_{F16,3},E_i,F)\right]} \tag{16.49}$$

and

$$C_g = \frac{e^2\sum_{i=1}^{i_{\max}} f_8(E_{F16,3},E_i,F)}{\left[\frac{3\pi a_{cc}}{4} + \frac{e^2}{\varepsilon_{0x}}\sum_{i=1}^{i_{\max}} f_8(E_{F16,3},E_i,F)\right]} \tag{16.50}$$

where

$$f_7(E_f,E_i,F_x) = [[1-f_5^2(E_{F16,3},E_i,F)]^{-1/2}][\delta_{11}(E_{F16,3},F)\delta_{11}'(E_{F16,3},F)] \cdot\left[\left(\frac{\delta_{11}^2(E_{F16,3},F)}{t_c^2}-5\right)^2 - \left(1-\frac{\delta_{11}^2(E_{F16,3},F)}{t_c^2}\right)\right]^{-1/2}$$

F_x is the electric field along x direction,

$$\delta_{11}(E, F_x) = \left[E - \bar{C}_2(F_x)\left[E + \frac{m_r E_g}{2m_c}\right]^{-5/2}\right],$$

$$\delta_{11}(E, F_x) = \left[E_f - \bar{C}_2\left[E_f + \frac{m_r E_g}{2m_c}\right]^{-5/2}\right]$$

$$f_8(E_f, F_i, F) = \frac{\delta_{11}(E_{F16,3}, F)\delta'_{11}(E_{F16,3}, F)}{[4t^2\bar{\varphi}_1(E_i, F)][1 - f_6^2(E_f, E_i, F)]^{1/2}},$$

$$\delta'_{11}(E, F) = \left[1 + \frac{5}{2}\bar{C}_2(F)\left[E + \frac{m_r E_g}{2m_c}\right]^{-\frac{7}{2}}\right]$$

16.2.4 *The QC in NWFETs in the presence of strong electric field*

The one-dimensional (1D) electron energy spectrum in NWFETs of III-V, ternary and quaternary materials, whose unperturbed band structure is defined by the three-band model of Kane, in the presence of electric field can be expressed from equation (16.33) as

$$\frac{\hbar^2 k_x^2}{2m_c} = \beta_{11}(E, F) - \left[\frac{\hbar^2}{2m_c}\left(\frac{n_y\pi}{d_y}\right)^2 + \frac{\hbar^2}{2m_c}\left(\frac{n_z\pi}{d_z}\right)^2\right] \tag{16.51}$$

where $n_y(= 1, 2, 3, \ldots)$ and $n_z(= 1, 2, 3, \ldots)$ are the size quantum numbers along y and z directions while d_y and d_z are the nanothickness along the y and z directions, respectively.

The sub-band energies ($E_{16,1}$) can be expressed as

$$\beta_{11}(E_{16,1}, F) = \left[\frac{\hbar^2}{2m_c}(n_y\pi/d_y)^2 + \frac{\hbar^2}{2m_c}(n_z\pi/d_z)^2\right] \tag{16.52}$$

The EEM in this case can be written as

$$m^*(E_{F16,4}, n_y, n_z, F) = m_c\{\beta_{11}(E_{16,4}, F)\}' \tag{16.53}$$

where $E_{16,4}$ is the Fermi energy in the present case as measured from the edge of the conduction band in the vertically upward direction in the absence of any quantization.

The 1D density-of-states function ($N_{1D}(E,F)$) is given by

$$N_{1D}(E,F) = \left(\frac{g_v\sqrt{2m_c}}{\pi\hbar}\right)\sum_{n_y=1}^{n_{y\max}}\sum_{n_z=1}^{n_{z\max}}\{\beta_{11}(E,F)\}' \times [\beta_{11}(E,F) - \phi(n_y,n_z)]^{-1/2}H(E - E_{16,4}) \quad (16.54)$$

where

$$\phi(n_y,n_z) = \frac{\hbar^2\pi^2}{2m_c}\left[\left(\frac{n_y}{d_y}\right)^2 + \left(\frac{n_z}{d_z}\right)^2\right]$$

Combining (16.54), with the Fermi Dirac occupation probability factor, the 1D electron concentrations (n_{1D}) under the condition of extreme carrier degeneracy can thus be written as

$$n_{1D} = \frac{2g_v}{\pi}\frac{\sqrt{2m_c}}{\hbar}\sum_{n_y=1}^{n_{y\max}}\sum_{n_z=1}^{n_{z\max}}[M_{216}(n_y,n_z,E_{16,4},F)] \quad (16.55)$$

where

$$M_{216}(n_y,n_z,E_{F16,4},F) \equiv [\beta_{11}(E_{F16,4},F) - \phi(n_y,n_z)]^{1/2}$$

The Einstein Relation can be written as

$$\frac{D}{\mu} = \frac{1}{e}\frac{\partial n_{1D}}{\partial E_{F16,4}} \quad (16.56)$$

The use of (16.55) and (16.56) leads to the expression of the 1D-ER in this case as

$$\frac{D}{\mu} = \frac{1}{|e|}\frac{\sum_{n_y=1}^{n_{y\max}}\sum_{n_z=1}^{n_{z\max}}[M_{216}(n_y,n_z,E_{F16,4},F)]}{\sum_{n_y=1}^{n_{y\max}}\sum_{n_z=1}^{n_{z\max}}[\{M_{216}(n_y,n_z,E_{F16,4},F)\}']} \quad (16.57)$$

Using (16.55) and (1.4), we can study the QC in this case.

(a) Perturbed two-band model of Kane

Using (16.36), the expressions for the 1D dispersion relation, the sub-band energies, the EMM, the density-of-states function, and the electron concentration for NWFETs of III-V, ternary and quaternary materials, whose unperturbed band structure is defined by the two-band model of

Kane, can, respectively, be written in the presence of photo-excitation as

$$\frac{\hbar^2 k_x^2}{2m_c} = \{\gamma_{11}(E,F) - \phi(n_y,n_z)\} \tag{16.58}$$

$$\gamma_{11}(E_{16,5},F) = \phi(n_y,n_z) \tag{16.59}$$

$$m^*(E_{F16,5},n_y,n_z,F) = m_c\{\gamma_{11}(E_{F16,5},F)\}' \tag{16.60}$$

$$N_{1D}(E,\lambda) = \left(\frac{g_v\sqrt{2m_c}}{\pi\hbar}\right)\sum_{n_y=1}^{n_{y\max}}\sum_{n_z=1}^{n_{z\max}}\{\gamma_{11}(E,F)\}' \times[\gamma_{11}(E,F) - \phi(n_y,n_z)]^{-1/2}H(E-E_{16,5}) \tag{16.61}$$

$$n_{1D} = \frac{2g_v}{\pi}\frac{\sqrt{2m_c}}{\hbar}\sum_{n_y=1}^{n_{y\max}}\sum_{n_z=1}^{n_{z\max}}[M_{226}(n_y,n_z,E_{F16,5},F)] \tag{16.62}$$

where

$$M_{226}(n_y,n_z,E_{F16,5},F) \equiv [\gamma_{11}(E_{F16,5},F) - \phi(n_y,n_z)]^{1/2}$$

The use of (16.56) and (16.61) leads to the expression of the ER in this case as

$$\frac{D}{\mu} = \frac{1}{|e|}\frac{\sum_{n_y=1}^{n_{y\max}}\sum_{n_z=1}^{n_{z\max}}[M_{226}(n_y,n_z,E_{F16,5},F)]}{\sum_{n_y=1}^{n_{y\max}}\sum\limits_{n_z=1}^{n_{z\max}}[\{M_{226}(n_y,n_z,E_{F16,5},F)\}']} \tag{16.63}$$

Using (16.62) and (1.4), we can study the QC in this case.

16.3 Summary and Conclusions

In Section 16.2.1, we have derived the DRs in III-V and optoelectronic materials in the presence of strong electric fields as given by (16.33) and (16.36), respectively, whose energy band spectra in the absence of any field are described by three- and two-band models of Kane, respectively. In Section 16.2.2, we have formulated the surface electron concentration in QWFETs under strong electric fields as given by (16.39) and (16.40) in accordance with both types of band models, and by combining each of them with (1.4), we can study QC for the respective models. The QC per unit length for arm-chair and Zig-Zag CNT FETs under strong electric field can, respectively, be expressed as (16.49) and (16.50) and these are derived in Section 16.2.3. In Section 16.2.4, we have formulated the expression for surface electron concentration in NWFETs under strong electric fields as given by (16.55), and by combining the same with (1.4), we can study the QC in this case.

We have not given the graphs of the QCs for any QWFETs, CNTFETs, and NWFETs under strong electric field configurations as discussed here since we feel that our readers will perform the intricate numerical computation to obtain the oscillatory plots for better assessment and joy of understanding. Incidentally, one can easily infer how little is presented and how much more there is yet to be investigated in the research field of QCs in this case for different FETs having different band structures in general, which is the signature of the co-existence of new physics and advanced mathematics. Finally, we again realize that it is the band structure which changes in a fundamental way and consequently all the physical properties of all the electronic materials change radically leading to new physical concepts.

16.4 Open Research Problems

(R.16.1) Investigate the QCs in the presence of intense external non-uniform electric field for all the FETs whose respective dispersion relations of the carriers are given in this chapter. Draw the corresponding graphs and explain the results.

(R.16.2) Investigate the QCs for the heavily doped FETs in the presences of Gaussian, exponential, Kane, Halperian, Lax, and Bonch-Burevich types of band tails as discussed in this chapter in the presence of external oscillatory and non-uniform electric fields. Draw the corresponding graphs and explain the results.

(R.16.3) Investigate the QCs in the presence of external oscillatory electric field for all types of FETs as discussed here in the presence of an arbitrarily oriented alternating electric field. Draw the corresponding graphs and explain the results.

(R.16.4) Investigate all the appropriate problems of this chapter for a Dirac electron. Draw the corresponding graphs and explain the results.

(R.16.5) Investigate all the appropriate problems of this chapter by including the many-body, broadening, and hot-carrier effects, respectively. Draw the corresponding graphs and explain the results.

(R.16.6) Investigate all the appropriate problems of this chapter by removing all the mathematical approximations and establishing the respective appropriate uniqueness conditions. Draw the corresponding graphs and explain the results.

References

1. L. Schiff, *Quantum Mechanics* (McGraw-Hill, USA, 1968).
2. B. R. Nag, *Electron Transport in Compound Semiconductors* (Springer, Heidelberg, 1980).
3. H. Sasaki, *Phys. Rev. B* **30**, 7016 (1984).
4. H. X. Jiang, J. Y. Lin, *J. Appl. Phys.* **61**, 624 (1987).

Chapter 17

Conclusion and Scope for Future Research

When my inner landscape is full of new beautiful thoughts, everything I do brings real pleasure

This monograph deals with the quantum capacitance (QC) in quantized transistors (QTs) and few electron statistics (ES) dependent electronic properties in various types of QTs made of different materials under different physical conditions. We have studied the dispersion relation, the effective electron mass, the sub-band energy, the density-of-states function, Einstein relation, Einstein's photoemission, the elastic constants and the field emission for various QTs. The QTs considered are metal oxide semiconductor field effect transistors (MOSFETs), quantum well field effect transistors (QWFETs), nano wire field effect transistors (NWFETs), made of nonlinear optical, III–V, II–VI, IV–VI, stressed Kane type, Bi_2Te_3, Ge, GaSb, GaAs, GaP, InSb, $PtSb_2$ and II–V materials, respectively *We have also considered the Heavily Doped MOSFETs, QWFETs and NWFETs for investigating the respective QCs. We have studied the influences of strong magnetic quantization and cross-fields configuration on the QC of the said QTs having various band structures.*

The external photo excitation, quantization and strong electric field alter profoundly the basic band structures, which, in turn, generate pinpointed knowledge regarding different electronic properties in different types of QTs made of the said materials and we have also considered these in studying the QC in QTs in this context. *We have studied the $I_D - V_D$ equations for ballistic NWFETs, HDNWFETs, MOSFETs, QWFETs, HDQWFWTs, Magneto QWFETs, Magneto HDQWFETs and magneto MOSFETs, respectively.* It may be noted that the in-depth experimental investigations covering the whole spectrum of QTs in general, are extremely important to uncover the underlying physics and the related mathematics in this particular pinpointed topic. Our investigations are based on the simplified $k.p$ formalism of solid-state science without incorporating the advanced field theoretic techniques. In spite of such constraints, the role of band structure, which generates, in turn, new concepts is truly amazing and discussed throughout the text.

We further present few challenging open research problems in this pin pointed topic of research of quantized structures.

(R17.1) Investigate the dispersion relation, the sub-band energy, the EEM, the density-of-states function, the ES, the ER, the TPSM, the ΔC_{44}, the ΔC_{456}, the different scattering mechanisms and the corresponding mobilities in the presence of a quantizing magnetic field for different QTs under exponential, Kane, Halperin, Lax and Bonch-Bruevich band tails [1] for all the problems of this monograph whose unperturbed carrier energy spectra are defined in Chapter 1 by including spin and broadening effects. Draw the corresponding graphs and explain the results.

(R17.2) Investigate the influence of defect traps and surface states separately on the dispersion relation, the sub-band energy, the EEM, the density-of-states function, the ES, the ER, the TPSM, the ΔC_{44}, the ΔC_{456}, the different scattering mechanisms and the corresponding mobilities of the QTs for all the appropriate problems of all the chapters after proper modifications. Draw the corresponding graphs and explain the results.

(R17.3) Investigate the dispersion relation, the sub-band energy, the EEM, the density-of-states function, the ES, the ER, the TPSM, the ΔC_{44}, the ΔC_{456}, the different scattering mechanisms and the corresponding mobilities of the QTs under the condition of non-equilibrium of the carrier states for all the appropriate problems of this monograph. Draw the corresponding graphs and explain the results.

(R17.4) Investigate the dispersion relation, the sub-band energy, the EEM, the density-of-states function, the ES, the ER, the TPSM, the ΔC_{44}, the ΔC_{456}, the different scattering mechanisms and the corresponding mobilities for all the appropriate problems of this monograph for the corresponding HD QTs. Draw the corresponding graphs and explain the results.

(R17.5) Investigate the dispersion relation, the sub-band energy, the EEM, the density-of-states function, the ES, the ER, the TPSM, the ΔC_{44}, the ΔC_{456}, the different scattering mechanisms and the corresponding mobilities for all the appropriate problems of this monograph for all types of QTs under arbitrarily oriented electric field in the presence of strain. Draw the corresponding graphs and explain the results.

(R17.6) Investigate the dispersion relation, the sub-band energy, the EEM, the density-of-states function, the ES, the ER, the TPSM, the ΔC_{44}, the ΔC_{456}, the different scattering mechanisms and the corresponding mobilities for all the appropriate problems of this monograph for all types of HD QTs in the presence of hot electron effects. Draw the corresponding graphs and explain the results.

(R17.7) Investigate the dispersion relation, the sub-band energy, the EEM, the density-of-states function, the ES, the ER, the TPSM, the ΔC_{44}, the ΔC_{456}, the different scattering mechanisms and the corresponding mobilities for all the appropriate problems of this monograph for all types of QTs for nonlinear charge transport. Draw the corresponding graphs and explain the results.

(R17.8) Investigate the dispersion relation, the sub-band energy, the EEM, the density-of-states function, the ES, the ER, the TPSM, the ΔC_{44}, the ΔC_{456}, the different scattering mechanisms and the corresponding mobilities for all the appropriate problems of this monograph for all types of HD QTs in the presence of strain in an arbitrary direction. Draw the corresponding graphs and explain the results.

(R17.9) Investigate all the appropriate problems of this monograph for strongly correlated electronic HD QTs in the presence of strain. Draw the corresponding graphs and explain the results.

(R17.10) Investigate all the appropriate problems of this chapter in the presence of arbitrarily oriented photon field and strain. Draw the corresponding graphs and explain the results.

(R17.11) Investigate the influence of the localization of carriers on the dispersion relation, the sub-band energy, the EEM, the density-of-states function, the ES, the ER, the TPSM, the ΔC_{44}, the ΔC_{456}, the different scattering mechanisms and the corresponding mobilities in HD QTs for all the appropriate problems of this monograph in the presence of crossed fields. Draw the corresponding graphs and explain the results.

(R17.12) Investigate the dispersion relation, the sub-band energy, the EEM, the density-of-states function, the ES, the ER, the TPSM, the ΔC_{44}, the ΔC_{456}, the different scattering mechanisms and the corresponding mobilities for QTs made of amorphous materials under different appropriate physical conditions as discussed in this monograph in the presence of strain and crossed fields. Draw the corresponding graphs and explain the results.

(R17.13) Investigate the dispersion relation, the sub-band energy, the EEM, the density-of-states function, the ES, the ER, the TPSM, the ΔC_{44}, the ΔC_{456}, the different scattering mechanisms and the corresponding mobilities for all types of QTs under different appropriate physical conditions as discussed in this monograph in the presence of strain and crossed fields. Draw the corresponding graphs and explain the results.

(R17.14) Investigate all the appropriate problems of this monograph in the presence of crossed electric and alternating quantizing magnetic fields. Draw the corresponding graphs and explain the results.

(R17.15) Investigate all the appropriate problems of this monograph in the presence of crossed alternating electric and quantizing magnetic fields. Draw the corresponding graphs and explain the results.

(R17.16) Investigate all the appropriate problems of this monograph in the presence of crossed alternating nonuniform electric and alternating quantizing magnetic fields. Draw the corresponding graphs and explain the results.

(R17.17) Investigate all the appropriate problems of this monograph in the presence of alternating crossed electric and alternating quantizing magnetic fields. Draw the corresponding graphs and explain the results.

(R17.18) Investigate all the appropriate problems of this monograph in the presence of arbitrarily oriented pulsed electric and quantizing magnetic fields. Draw the corresponding graphs and explain the results.

(R17.19) Investigate all the appropriate problems of this monograph in the presence of arbitrarily oriented alternating electric and quantizing magnetic fields. Draw the corresponding graphs and explain the results.

(R17.20) Investigate all the appropriate problems of this monograph in the presence of crossed in homogeneous electric and alternating quantizing magnetic fields. Draw the corresponding graphs and explain the results.

(R17.21) Investigate all the appropriate problems of this monograph in the presence of arbitrarily oriented electric and alternating quantizing magnetic fields under strain. Draw the corresponding graphs and explain the results.

(R17.22) Investigate all the appropriate problems of this monograph in the presence of arbitrarily oriented electric and alternating quantizing magnetic fields under strong arbitrarily oriented light waves. Draw the corresponding graphs and explain the results.

(R17.23) Investigate the dispersion relation, the sub-band energy, the EEM, the density-of-states function, the ES, the ER, the TPSM, the ΔC_{44}, the ΔC_{456}, the different scattering mechanisms and the corresponding mobilities for all types of QTs in this monograph in the presence of many-body effects, strain and arbitrarily oriented alternating light waves respectively. Draw the corresponding graphs and explain the results.

(R17.24) Investigate all the problems of this monograph by removing all the physical and mathematical approximations and establishing the respective appropriate uniqueness conditions. Draw the corresponding graphs and explain the results.

The formulation of the QCs and the transport properties for all types of QTs made of different materials considering the influence of all the bands created due to all types of quantizations after removing all the assumptions and establishing the respective appropriate uniqueness conditions are, in general, an extremely difficult problem. The **circa 200** open research problems have been presented in this monograph and we hope that the readers will not only solve them but also will generate new concepts, both theoretical and experimental. Incidentally, we can easily infer how little is presented and how much more is yet to be investigated in this exciting topic which is the signature of the coexistence of new physics, advanced mathematics combined with the inner fire for performing creative research in this context from the young scientists since like Kikoin [5] we firmly believe that *"A young scientist is no good if his teacher learns nothing new from him and gives his teacher nothing to be proud of"*. In the meantime, our research interest has been shifted and we are leaving this particular beautiful topic with the hope that (R17.24) alone is sufficient to draw the attention of researchers the from diverse fields and our readers are surely in tune with the fact that *"Exposition, criticism, appreciation is the work for second-rate minds"* [6].

References

1. B. R. Nag, *Electron Transport in Compound Semiconductors*, Springer Series in Solid State Sciences, Vol. 11 (Springer-Verlag, Germany, 1980).
2. R. W. Cunningham, *Phys. Rev.* **167**, 761 (1968).
3. A. I. Yekimov, A. A. Onushchenko, A. G. Plyukhin and Al L. Efros, *J. Expt. Theor. Phys.* **88**, 1490 (1985).
4. B. J. Roman and A. W. Ewald, *Phys. Rev. B* **5**, 3914 (1972).
5. I. K. Kikoin, *Science for Everyone: Encounters with Physicists and Physics* (Mir Publishers, Russia, 1989), p. 154.
6. G. H. Hardy, *A Mathematician's Apology* (Cambridge University Press, 1990), p. 61.

Chapter 18

The Numerical Values of the Energy Band Constants of Few Materials

Most important to recognize is that we have learned enough to understand how little we know and to see dimly how much more there is yet to study in Quantum Capacitance in Quantized Transistors.

	Materials	Numerical values of the energy band constants
1	The conduction electrons of n-cadmium germanium arsenide can be described by three types of band models	1. The values of the energy band constants in accordance with the generalized electron dispersion relation of nonlinear optical materials (as given by (2.1)) are as follows: $E_{g_0} = 0.57eV$, $\Delta_{\|} = 0.30eV$, $\Delta_{\perp} = 0.36eV$, $m^*_{\|} = 0.034m_0$, $m^*_{\perp} = 0.039m_0$, $T = 4K, \delta = -0.21eV$, $g_v = 1[1, 2]$, $\varepsilon_{sc} = 18.4\varepsilon_0$ [3] (ε_{sc} and ε_0 are the permittivity of the semiconductor material and free space respectively) and $W = 4eV$ [4]. 2. In accordance with the three-band model of Kane (as given by (2.19)), the spectrum constants are given by $\Delta = (\Delta_{\|} + \Delta_{\perp})/2 = 0.33eV$, $E_{g_0} = 0.57eV$, $m^* = (m^*_{\|} + m^*_{\perp})/2 = 0.0365m_0$ and $\delta = 0eV$. 3. In accordance with two-band model of Kane, $E_{g_0} = 0.57eV$ and $m^* = 0.0365m_0$.
2	n-Indium arsenide	The values $E_{g_0} = 0.36eV$, $\Delta = 0.43eV$, $m^* = 0.026m_0$, $g_v = 1$, $\varepsilon_{sc} = 12.25\varepsilon_0$ [5] and $W = 5.06eV$ [6] are valid for three-band model of Kane as given by (2.19).

(*Continued*)

(*Continued*)

	Materials	Numerical values of the energy band constants
3	n-Gallium arsenide	The values $E_{g_0} = 1.55eV$, $\Delta = 0.35eV$, $m^* = 0.07m_0$, $g_v = 1$, $\varepsilon_{sc} = 12.9\varepsilon_0$ [5] and $W = 4.07eV$ [7] are valid for three-band model of Kane as given by (2.19). The values $\alpha_{13} = -1.97 \times 10^{-37}\, eVm^4$ and $a_{15} = -2.3 \times 10^{-34} eVm^4$ [8] are valid for (3.25). The values $\alpha_{11} = -2132 \times 10^{-40} eVm^4$, $\alpha_{12} = 9030 \times 10^{-50} eVm^5$, $\beta_{11} = -2493 \times 10^{-40} eVm^4$, $\beta_{12} = 12594 \times 10^{-50} eVm^5$, $\gamma_{11} = 30 \times 10^{-30} eVm^3$, $\gamma_{12} = -154 \times 10^{-42} eVm^4$ [9] are valid for (3.30).
4	n-Gallium aluminium arsenide	$E_{g_0} = (1.424 + 1.266x + 0.26x^2)eV$, $\Delta = (0.34 - 0.5x)eV$, $m^* = [0.066 + 0.088x]m_0$, $g_v = 1$, $\varepsilon_{sc} = [13.18 - 3.12x]\varepsilon_0$ [10] and $W = (3.64 - 0.14x)eV$ [11].
5	n-Mercury cadmium telluride	$E_{g_0} = (-0.302 + 1.93x + 5.35 \times 10^{-4}(1-2x)T - 0.810x^2 + 0.832x^3)eV$, $\Delta = (0.63 + 0.24x - 0.27x^2)eV$, $m^* = 0.1m_0 E_{g_0}(eV)^{-1}$, $g_v = 1$, $\varepsilon_{sc} = [20.262 - 14.812x + 5.22795x^2]\varepsilon_0$ [12] and $W = (4.23 - 0.813(E_{g_0} - 0.083))eV$ [13].
6	n-Indium gallium arsenide phosphide lattice matched to Indium phosphide	$E_{g_0} = (1.337 - 0.73y + 0.13y^2)eV$, $\Delta = (0.114 + 0.26y - 0.22y^2)eV$, $m^* = (0.08 - 0.039y)m_0$, $y = (0.1896 - 0.4052x)/(0.1896 - 0.0123x)$, $g_v = 1$, $\varepsilon_{sc} = [10.65 + 0.1320y]\varepsilon_0$ and $W(x,y) = [5.06(1-x)y + 4.38(1-x)(1-y) + 3.64xy + 3.75\{x(1-y)\}]eV$ [14].
7	n-Indium antimonide	$E_{g_0} = 0.2352eV$, $\Delta = 0.81eV$, $m^* = 0.01359m_0$, $g_v = 1$, $\varepsilon_{sc} = 15.56\varepsilon_0$ [5] and $W = 4.72eV$ [6].
8	n-Gallium antimonide	The values of $E_{g_0} = 0.81eV$, $\Delta = 0.80eV$, $P = 9.48 \times 10^{-10} eVm$, $\bar{\varsigma}_0 = -2.1$, $\bar{v}_0 = -1.49$, $\bar{\omega}_0 = 0.42$, $g_v = 1$ [15] and $\varepsilon_{sc} = 15.85\varepsilon_0$ [15, 16] are valid for the model of Seiler *et al.* [15] as given by (3.113). The values $\overline{E_1} = 1.024eV$, $\overline{E_2} = 0eV$, $\overline{E_3} = -1.132eV$, $\overline{E_4} = 0.05eV$, $\overline{E_5} = 1.107eV$, $\overline{E_6} = -0.113eV$ and $\overline{E_7} = -0.0072eV$ [16] are valid for the model of Zhang [16] as given by (3.128).
9	n-Cadmium sulphide	$m^*_{\parallel} = 0.7m_0$, $m^*_{\perp} = 1.5m_0$, $\bar{\lambda}_0 = 1.4 \times 10^{-8} eVm$ (value changed), $g_v = 1$ [5], $\varepsilon_{sc} = 15.5\varepsilon_0$ [17] and $W = 4.5eV$ [6]
10	n-Lead telluride	The values $m_t^- = 0.070m_0$, $m_l^- = 0.54m_0$, $m_t^+ = 0.010m_0$, $m_l^+ = 1.4m_0$, $P_{\|} = 141meVnm$, $P_{\perp} = 486meVnm$, $E_{g_0} = 190meV$, $g_v = 4$ [5], $\varepsilon_{sc} = 33\varepsilon_0$ [5, 18] and $W = 4.6eV$ [19] are valid for the Dimmock model [20] as given by (3.165).

(*Continued*)

	Materials	Numerical values of the energy band constants
		The values $(\overline{R})^2 = 2.3 \times 10^{-19}\,(eVm)^2$, $E_{g_0} = 0.16eV$, $(\overline{s})^2 = 4.6\,(\overline{R})^2$, $\Delta_c' = 3.07eV$, $(\overline{Q})^2 = 1.3\,(\overline{R})^2$, $\Delta_c'' = 3.28eV$, $(\overline{A})^2 = 0.83 \times 10^{-19}\,(eVm)^2$ [21] and $W = 4.21eV$ [6] are valid for the model of Bangert and Kastner [21] as given by (3.172).
		The values $m_{tv} = 0.0965m_0$, $m_{lv} = 1.33m_0$, $m_{tc} = 0.088m_0$, $m_{lc} = 0.83m_0$ [19] are valid for the model of Foley *et al.* [19] as given by (3.179). The values $m_1 = 0.0239m_0$, $m_2 = 0.024m_0$, $m_2' = 0.31m_0$, $m_3 = 0.24m_0$ [22] are valid for the Cohen model [23] as given by (2.144).
11	Stressed n-indium antimonide	The values $m^* = 0.048m_o$, $E_{g_0} = 0.081eV$, $B_2 = 9 \times 10^{-10}eVm$, $C_1^c = 3eV$, $C_2^c = 2eV$, $a_0 = -10eV$, $b_0 = -1.7eV$, $d = -4.4eV$, $S_{xx} = 0.6 \times 10^{-3}(kbar)^{-1}$, $S_{yy} = 0.42 \times 10^{-3}(kbar)^{-1}$, $S_{zz} = 0.39 \times 10^{-3}(kbar)^{-1}$, $S_{xy} = 0.5 \times 10^{-3}(kbar)^{-1}$, $\varepsilon_{xx} = \sigma S_{xx}$, $\varepsilon_{yy} = \sigma S_{yy}$, $\varepsilon_{zz} = \sigma S_{zz}$, $\varepsilon_{xy} = \sigma S_{xy}$, σ is the stress in kilobar, $g_v = 1$[24] are valid for the model of Seiler *et al.* [24] as given by (2.102).
12	Bismuth	$E_{g_0} = 0.0153eV$, $m_1 = 0.00194m_0$, $m_2 = 0.313m_0$, $m_3 = 0.00246m_0$, $m_2' = 0.36m_0$, $g_v = 3$, $g_s = 2$ [25], $M_2 = 0.128m_0$, $M_2' = 0.80m_0$ [26] and $W = 4.34eV$.
13	Mercury telluride	$m_v^* = 0.028m_0$, $g_v = 1$, $\varepsilon_\infty = 15.2\varepsilon_0$ [27] and $W = 5.5eV$ [28].
14	Platinum antimonide	For valence bands, along $\langle\ 111\ \rangle$ direction, $\lambda_1 = 0.33eV$, $l_1 = 1.09eV$, $\nu_1 = 0.17eV$, $\overline{n} = 0.22eV$, $\overline{\overline{a}} = 0.643\,nm$, $I_0 = 0.30(eV)^2$, $\delta_0' = 0.33eV$, $g_v = 8$ [29], $\varepsilon_{sc} = 30\varepsilon_0$ [30] and $\phi \approx 3.0eV$ [31].
15	n-Gallium phosphide	$m_{\|}^* = 0.92m_0$, $m_\perp^* = 0.25m_0$, $k_0 = 1.7 \times 10^{19}m^{-1}$, $\|V_G\| = 0.21eV$, $g_v = 6$, $g_s = 2$ [32] and $W = 3.75eV$ [6].
16	Germanium	$E_{g_0} = 0.785eV$, $m_{\|}^* = 1.57m_0$, $m_\perp^* = 0.0807m_0$ [7] and $W = 4.14eV$ [6]

(*Continued*)

(*Continued*)

	Materials	Numerical values of the energy band constants
17	Tellurium	The values $A_6 = 6.7 \times 10^{-16} meVm^2$, $A_7 = 4.2 \times 10^{-16} meVm^2$, $A_8 = (6 \times 10^{-8} meVm)^2$ and $A_9 = (3.6 \times 10^{-8} meVm)^2$ [33] are valid for the model of Bouat *et al.* [33] as given by (3.76). The values $t_1 = 0.06315eV$, $t_2 = -10.0\hbar^2/2m_0$, $t_3 = -5.55\hbar^2/2m_0$, $t_4 = 0.3 \times 10^{-36} eVm^4$, $t_5 = 0.3 \times 10^{-36} eVm^4$, $t_6 = -5.55\hbar^2/2m_0$, $t_7 = 6.18 \times 10^{-20}(eVm)^2$ [34] and $W = 1.9708eV$ [35] are valid for the model of Ortenberg and Button [34] as given by (3.82).
18	Graphite	The values $\Delta_1 = -0.0002eV$, $\gamma_1 = 0.392eV$, $\gamma_5 = 0.194eV$, $\bar{c} = 0.674nm$, $\gamma_2 = -0.019eV$, $\bar{a} = 0.246nm$, $\gamma_0 = 3eV$, $\gamma_4 = 0.193eV$ [36] and $W = 4.6eV$ [37] are valid for the model of Brandt *et al.* [36] as given by (3.88).
19	Lead germanium telluride	The values $\overline{E_{g_0}} = 0.21eV$, $g_v = 4$ [38] and $\phi \approx 6eV$ [39] are valid for the model of Vassilev [38] as given by (3.107a).
20	Cadmium antimonide	The values $A_{10} = -4.65 \times 10^{-19} eVm^2$, $A_{11} = -2.035 \times 10^{-19} eVm^2$, $A_{12} = -5.12 \times 10^{-19} eVm^2$, $A_{13} = -0.25 \times 10^{-10} eVm$, $A_{14} = 1.42 \times 10^{-19} eVm^2$, $A_{15} = 0.405 \times 10^{-19} eVm^2$, $A_{16} = -4.07 \times 10^{-19} eVm^2$, $A_{17} = 3.22 \times 10^{-10} eVm$, $A_{18} = 1.69 \times 10^{-20}(eVm)^2$, $A_{19} = 0.070eV$ [40] and $\phi \approx 2eV$ [41] are valid for the model of Yamada [40] as given by (3.186).
21	Cadmium diphosphide	The values $\beta_1 = 8.6 \times 10^{-21} eVm^2$, $\beta_2 = 1.8 \times 10^{-21}(eVm)^2$, $\beta_4 = 0.0825eV$, $\beta_5 = -1.9 \times 10^{-19} eVm^2$ [42] and $\phi \approx 5eV$ [43] are valid for the model of Chuiko [42] and are given by (3.193).
22	Zinc diphosphide	The values $\beta_1 = 8.7 \times 10^{-21} eVm^2$, $\beta_2 = 1.9 \times 10^{-21}(eVm)^2$, $\beta_4 = 0.0875eV$, $\beta_5 = -1.9 \times 10^{-19} eVm^2$ [42] and $W \approx 3.9eV$ [43] are valid for the model of Chuiko [42] and are given by (3.193).
23	Bismuth telluride	The values $E_{g_0} = 0.145eV$, $\alpha_{11} = 3.25$, $\alpha_{22} = 4.81$, $\alpha_{33} = 9.02$, $\alpha_{23} = 4.15$, $g_s = 2$, $g_v = 6$ [44] and $\varphi = 5.3eV$ [45] are valid for the model of Stordeur *et al.* [44] using (3.200).
24	Carbon nanotube	The values $a_c = 0.144nm$ [46], $t_c = 2.7eV$ [47], $r_0 = 0.7nm$ [48] and $W = 3.2eV$ [49] are valid for graphene band structure realization of carbon nanotube [48] and are given by (2.167) and (2.168).

(*Continued*)

	Materials	Numerical values of the energy band constants
25	Antimony	The values $\bar{\alpha}_{11} = 16.7$, $\bar{\alpha}_{22} = 5.98$, $\bar{\alpha}_{33} = 11.61$, $\bar{\alpha}_{23} = 7.54$ [50] and $W = 4.63eV$ [6] are valid for the model of Ketterson [50] and are given by (3.207), (3.208) and (3.209), respectively.
26	Zinc selenide	$m_2^* = 0.16m_0$, $\Delta_2 = 0.42eV$, $E_{g0_2} = 2.82eV$ [7] and $W = 3.2eV$ [51].
27	Lead selenide	$m_t^- = 0.23m_0$, $m_l^- = 0.32m_0$, $m_t^+ = 0.115m_0$, $m_l^+ = 0.303m_0$, $P_{\|\|} \approx 138meVnm$, $P_\perp = 471meVnm$, $E_{g0} = 0.28eV$ [52], $\varepsilon_{sc} = 21.0\varepsilon_0$ [32] and $W = 4.2eV$ [53].

References

1. J. L. Shay and J. W. Wernik, *Ternary Chalcoprite Semiconductors: Growth, Electronic Properties and Applications* (Pergamon Press, London, 1975).
2. E. A. Arushanov, A. A. Kaynzev, A. N. Natepov and S. I. Radautsan, *Sov. Phys. Semicond.* **15**, 828 (1981).
3. K. S. Hong, R. F. Speyer and R. A. Condrate, *J. Phys. Chem. Solids* **51**, 969 (1990).
4. M. Mondal, S. Banik and K. P. Ghatak, *J. Low. Temp. Phys.* **74**, 423 (1989).
5. B. R. Nag, *Electron Transport in Compound Semiconductor* (Springer-Verlag, Germany, 1980); M. Krieehbaum, P. Kocevar, H. Pascher and G. Bauer, *IEEE QE* **24** 1727 (1988); J. J. Hopfield, *J. Appl. Phys.* **32**, 2277 (1961).
6. S. Adachi, *Properties of Group-IV, III–V and II–VI Semiconductors* (John Wiley and Sons, 2005).
7. O. Madelung, *Semiconductors: Data Handbook*, 3rd ed. (Springer-Verlag, Germany, 2003).
8. D. J. Newson, A. Kurobe, *Semicond. Sci. Technol.* **3**, 786 (1988).
9. U. Rossler, *Solid State Commun.* **49**, 943 (1984).
10. S. Adachi, *J. Appl. Phys.* **58**, R1 (1985).
11. S. Adachi, *GaAs and Related Materials: Bulk Semiconductors and Superlattice Properties*, (World Scientific, USA, 1994).
12. G. L. Hansen, J. L. Schmit and T. N. Casselman, *J. Appl. Phys.* **63**, 7079 (1982)
13. J. Wenus, J. Rutkowski, and A. Rogalski, *IEEE Trans. Elect. Dev.* **48**, 1326 (2001)
14. S. Adachi, *J. Appl. Phys.* **53**, 8775 (1982).
15. D. G. Seiler, W. M. Beeker and L. M. Roth, *Phys. Rev.* **1**, 764 (1970).
16. I. V. Skryabinskii, Yu. I. Ukhanov, *Sov. Phys. Solid State*, **14**, 2838 (1973); H. I. Zhang, *Phys. Rev. B*, **1**, 3450 (1970); P. C. Mathur and S. Jain, *Phys. Rev.* **19**, 1359 (1979).

17. S. Tiwari and S. Tiwari, *Cryst. Res. Technol.* **41**, 78 (2006).
18. J. R. Lowney and S. D. Senturia, *J. Appl. Phys.* **47**, 1771 (1976).
19. W. E. Spicer and G. J. Lapeyre, *Phys. Rev.* **139**, A565(1965); G. M. T. Foley and P. N. Langenberg, *Phys. Rev. B* **15**, 4850 (1977).
20. J. O. Dimmock, in *The Physics of Semimetals and Narrowgap Semiconductors* by D. L. Carter and R. T. Bates (eds.) (Pergamon Press, Oxford, 1971).
21. E. Bangert and P. Kastner, *Phys. Stat. Sol. (b)* **61**, 503 (1974).
22. D. R. Lovett, in *Semimetals and Narrow Band Gap Semiconductors* (Pion Limited, London, 1977).
23. M. H. Cohen, *Phys. Rev.* **121**, 387 (1961).
24. D. G. Seiler, B. D. Bajaj and A. E. Stephens, *Phys. Rev. B* **16**, 2822 (1977); A. V. Germaneko and G. M. Minkov, *Phys. Stat. Sol.* (b) **184**, 9 (1994); G. L. Bir and G. E. Pikus, *Symmetry and Strain-Induced Effects in Semiconductors* (Nauka, Russia, 1972). (in Russian); M. Mondal and K. P. Ghatak, *Phys. Stat. Sol. (b)* **135**, K21 (1986).
25. C. C. Wu and C. J. Lin, *J. Low. Temp. Phys.* **57**, 469 (1984).
26. S. Takaoka, H. Kawamura, K. Murasa, S. Takano, *Phys. Rev. B* **13**, 1428 (1976).
27. V. I. Ivanov-Omskii, A. Sh. Mekhtisev, S. A. Rustambekova and E. N. Ukraintsev, *Phys. Stat. Sol. (b)* **119**, 159 (1983).
28. H. Kim, K. Cho, H. Song, B. Min, J. Lee, G. Kim, S. Kim, S. H. Kim and T. Noh, *Appl. Phys. Lett.* **83**, 4619 (2003).
29. P. R. Emtage, *Phys. Rev. A* **246**, 138 (1965).
30. R. A. Reynolds, M. J. Brau, R. A. Chapman, *J. Phys. Chem. Solids* **29**, 755 (1968).
31. J. O'Shaughnessy and C. Smith, *Solid State Commun.* **8**, 481 (1970).
32. G. J. Rees, Phys in of Compounds, in Proceedings of the 13th International Nature Conference F. G. Fumi (ed.) (North Holland Company, 1976), p. 1166.
33. J. Bouat and J. C. Thuillier, *Surface Sci.* **73**, 528 (1978).
34. M. V. Ortenberg and K. J. Button, *Phys. Rev. B* **16**, 2618 (1977).
35. G. Haeffler, A. E. Klinkmüller, J. Rangell, U. Berzinsh and D. Hanstorp, *Z. Phys. D* **38**, 211 (1996).
36. N. B. Brandt, V. N. Davydov, V. A. Kulbachinskii and O. M. Nikitina, *Sov. Phys. Sol. Stat.* **29**, 1014 (1987).
37. L. M. Viculis, J. J. Mack, O. M. Mayer, H. T. Hahn and R. B. Kaner, *J. Mater. Chem.* **15**, 974 (2005).
38. L. A. Vassilev, Phys. Stat. Sol. (b), **121**, 203 (1984); S. Takaoka and K. Murase, *Phys. Rev. B, Phys. Rev. B* **20**, 2823 (1979).
39. D. L. Partin, *Superlatt. Microstruct.*, **1**, 131 (1985).
40. Y. Yamada, J. Phys. Japan, **35**, 1600 (1973); M. Singh, P. R. Wallace, S. D. Jog and E. Arushanov, *J. Phys. Chem. Solids* **45**, 409 (1984).
41. W. J. Turner, A. S. Fischler, and W. E. Reese, *Phys. Rev.* **121**, 759 (1961).
42. G. P. Chuiko, Sov. *Phys. Semi.* **19**(12), 1381 (1985).
43. W. E. Swank, P. G.Le Comber, *Phys. Rev.* **153**, 844 (1967).

44. M. Stordeur and W. Kuhnberger, *Phys. Stat. Sol. (b)*, **69**, 377 (1975); D. R. Lovett, *Semimetals and Narrow-Bandgap Semiconductors*, (Pion Limited, 1977), **185**; H. Köhler, *Phys. Stat. Sol. (b)* **74**, 591 (1976).
45. D. Haneman, *J. Phys. Chem. Solids*, **11**, 205 (1959).
46. N. Wei, G. Wu and J. Dong, *Phys. Lett. A* **325**, 403 (2004).
47. S. Reich, J. Maultzsch, C. Thomsen and P. Ordejón, *Phys. Rev. B* **66**, 035412 (2006).
48. M. S. Lundstrom, J. Guo, *Nanoscale Transistors: Device Physics, Modeling and Simulation*, (Springer, USA, 2006); J. W. Mintmire and C. T. White, *Phys. Rev. Lett.* **81**, 2506 (1998).
49. F Buonocore, F Trani, D Ninno, A Di Matteo, G Cantele and G Iadonisi, *Nanotechnology* **19**, 025711 (2008).
50. J. B. Ketterson, *Phys. Rev.* **129**, 18 (1963).
51. R. C. Vilão, J. M. Gil, A. Weidinger, H. V. Alberto, J. Piroto Duarte, N. A. de Campos, R. L. Lichti, K. H. Chow, S. P. Cottrell and S. F. J. Cox, *Phys. Rev. B* **77**, 235212 (2008).
52. I. Kang and F. W. Wise, *Phys. Rev. B, J. Opt. Soc. Am. B* **14**, 1632 (1997).
53. D. Cui, J. Xu, S.-Y. Xu, G. Paradee, B. A. Lewis and M. D. Gerhold, *IEEE. Trans. Elect. Dev.* **5**, 362 (2006).

Chapter 19

$I_D - V_D$ Equation for Ballistic NWFETs

Well-done is better than well-said

19.1 Introduction

The I_D–V_D equation for ballistic NWFETs has relatively been less investigated in the literature. In this appendix, we shall formulate the $I_D - V_D$ equations for ballistic NWFETs made of tetragonal and nonlinear optical, III–V ternary and quaternary, II–VI, IV–VI, stressed Kane type, Bi_2Te_3, Ge, GaSb, GaP, $PtSb_2$ and II–V materials from Section 19.2.1 up to 19.2.11, respectively. Section 19.3 contains summary and conclusion and Section 19.4 presents one multidimensional open research problem.

19.2 Theoretical Background

19.2.1 *The I_D–V_D equation for ballistic NWFETs of tetragonal and nonlinear optical materials*

The I_D is given by

$$I_D = I_+ - I_- \tag{19.1a}$$

where

$$I_+ = eg_\nu \sum_{n_x=1}^{n_{x\max}} \sum_{n_y=1}^{n_{y\max}} \int_{E_{11}}^{\infty} \left[\frac{1}{2}\frac{2}{\pi}\frac{\partial k_z}{\partial E}\frac{1}{\hbar}\frac{\partial E}{\partial k_z}\right] f(E)dE, \tag{19.1b}$$

E_{11} is the sub-band energy in the case which is defined in (7.2) and $f(E)$ is the Fermi Dirac occupation probablity factor.

From (19.1b) we can write that

$$I_+ = \frac{ek_BTg_v}{\pi\hbar}\sum_{n_x=1}^{n_{x\max}}\sum_{n_y=1}^{n_{y\max}} F_0(\eta_{19,1}) \tag{19.2}$$

where $\eta_{19,1} = (k_BT)^{-1}[(E_{F19,1} - E_{11})]$ and $E_{F19,1}$ is the Fermi energy in this case.

From (19.2) we can write that

$$I_- = \frac{ek_BTg_v}{\pi\hbar}\sum_{n_x=1}^{n_{x\max}}\sum_{n_y=1}^{n_{y\max}} F_0(\eta_{19,1} - U_D) \tag{19.3}$$

where $U_D = (eV_D/k_BT)$.
Using (19.1a), (19.2) and (19.3) we get

$$I_D = \frac{ek_BTg_v}{\pi\hbar}\sum_{n_x=1}^{n_{x\max}}\sum_{n_y=1}^{n_{y\max}} [F_0(\eta_{19,1}) - F_0(\eta_{19,1} - U_D)] \tag{19.4}$$

It appears than that the computation of I_D needs an expression of electron concentration per unit length which can, in turn, be expressed as

$$n_{1D} = \frac{g_v}{\pi}\sum_{n_x=1}^{n_{x\max}}\sum_{n_y=1}^{n_{y\max}}\left[\left[\sqrt{A_{11}(E_{F19,1}, n_x, n_y)} + \sqrt{A_{11}((E_{F19,1} - eV_D), n_x, n_y)}\right]\right.$$
$$\left. + \sum_{r=1}^{s} L(r)\left[\left[\sqrt{A_{11}(E_{F19,1}, n_x, n_y)} + \sqrt{A_{11}((E_{F19,1} - eV_D), n_x, n_y)}\right]\right]\right] \tag{19.5}$$

where $A_{11}(E, n_x, n_y)$ has been defined in (7.1) of Chapter 7 and $L(r) = \left[2(k_BT)^{2r}(1 - 2^{1-2r})\xi(2r)\frac{\partial^{2r}}{\partial E_F^{2r}}\right]$.

19.2.2 *The $I_D - V_D$ equation for ballistic NWFETs of III–V, ternary and quaternary materials*

(i) Three-band model of Kane

From (19.2), in this case we can write that

$$I_+ = \frac{ek_BTg_v}{\pi\hbar}\sum_{n_x=1}^{n_{x\max}}\sum_{n_y=1}^{n_{y\max}} F_0(\eta_{19,2}) \tag{19.6}$$

where $\eta_{19,2} = (k_BT)^{-1}[(E_{F19,2} - E_{12})]$, $E_{F19,2}$ is the Fermi energy in this case and E_{12} is the sub-band energy which is defined in (7.17) of Chapter 7.

From (19.6) we can write that

$$I_- = \frac{ek_BTg_v}{\pi\hbar}\sum_{n_x=1}^{n_{x\max}}\sum_{n_y=1}^{n_{y\max}} F_0(\eta_{19,2} - U_D) \tag{19.7}$$

Using (19.6), (19.7) and (19.1a) we get

$$I_D = \frac{ek_BTg_v}{\pi\hbar}\sum_{n_x=1}^{n_{x\max}}\sum_{n_y=1}^{n_{y\max}} [F_0(\eta_{19,2}) - F_0(\eta_{19,2} - U_D)] \tag{19.8}$$

The n_{1D} can be expressed as

$$n_{1D} = \frac{g_v}{\pi}\sum_{n_x=1}^{n_{x\max}}\sum_{n_y=1}^{n_{y\max}}\Bigg[\Big[\sqrt{A_{12}(E_{F19,2}, n_x, n_y)} + \sqrt{A_{12}((E_{F19,2} - eV_D), n_x, n_y)}\Big] + \sum_{r=1}^{s} L(r)\Big[\Big[\sqrt{A_{12}(E_{F19,2}, n_x, n_y)} + \sqrt{A_{12}((E_{F19,2} - eV_D), n_x, n_y)}\Big]\Big]\Bigg] \tag{19.9}$$

where $A_{12}(E, n_x, n_y)$ has been defined in (7.16a) of Chapter 7.

(ii) Two-band model of Kane

From (19.2), in this case we can write that

$$I_+ = \frac{ek_BTg_v}{\pi\hbar}\sum_{n_x=1}^{n_{x\max}}\sum_{n_y=1}^{n_{y\max}} F_0(\eta_{19,3}) \tag{19.10}$$

where $\eta_{19,3} = (k_BT)^{-1}[(E_{F19,3} - E_{13})]$, $E_{F19,3}$ is the Fermi energy in this case and E_{13} is the sub-band energy which is defined in (7.26) of Chapter 7.

From (19.10) we can write that

$$I_- = \frac{ek_BT}{\pi\hbar}\sum_{n_x=1}^{n_{x\max}}\sum_{n_y=1}^{n_{y\max}} F_0(\eta_{19,3} - U_D) \tag{19.11}$$

Using (19.10), (19.11) and (19.1a) we get

$$I_D = \frac{ek_BTg_v}{\pi\hbar}\sum_{n_x=1}^{n_{x\max}}\sum_{n_y=1}^{n_{y\max}} [F_0(\eta_{19,3}) - F_0(\eta_{19,3} - U_D)] \tag{19.12}$$

The n_{1D} can be expressed as

$$n_{1D} = \frac{g_v}{\pi} \sum_{n_x=1}^{n_{x\max}} \sum_{n_y=1}^{n_{y\max}} \left[\left[\sqrt{A_{13}(E_{F19,3}, n_x, n_y)} + \sqrt{A_{13}((E_{F19,3} - eV_D), n_x, n_y)} \right] \right.$$
$$\left. + \sum_{r=1}^{s} L(r) \left[\left[\sqrt{A_{13}(E_{F19,3}, n_x, n_y)} + \sqrt{A_{13}((E_{F19,3} - eV_D), n_x, n_y)} \right] \right] \right] \tag{19.13}$$

where $A_{13}(E, n_x, n_y)$ has been defined in (7.25) of chapter 7

(iii) Parabolic band model

From (19.2), in this case we can write that

$$I_+ = \frac{ek_B T g_v}{\pi\hbar} \sum_{n_x=1}^{n_{x\max}} \sum_{n_y=1}^{n_{y\max}} F_0(\bar{\eta}_{19,3}) \tag{19.14}$$

where $\bar{\eta}_{19,3} = (k_B T)^{-1}[(\bar{E}_{F19,3} - \bar{E}_{13})]$, $\bar{E}_{F19,3}$ is the Fermi energy in this case and $\bar{E}_{13}$ is the sub-band energy which is defined in (7.35) of Chapter 7.

From (19.14) we can write that

$$I_- = \frac{ek_B T g_v}{\pi\hbar} \sum_{n_x=1}^{n_{x\max}} \sum_{n_y=1}^{n_{y\max}} F_0(\bar{\eta}_{19,3} - U_D) \tag{19.15}$$

Using (19.14), (19.15) and (19.1a) we get

$$I_D = \frac{ek_B T g_v}{\pi\hbar} \sum_{n_x=1}^{n_{x\max}} \sum_{n_y=1}^{n_{y\max}} [F_0(\bar{\eta}_{19,3}) - F_0(\bar{\eta}_{19,3} - U_D)] \tag{19.16}$$

The n_{1D} can be expressed as

$$n_{1D} = \frac{g_v}{\pi} \sum_{n_x=1}^{n_{x\max}} \sum_{n_y=1}^{n_{y\max}} \left[\left[\sqrt{\bar{A}_{13}(\bar{E}_{F19,3}, n_x, n_y)} + \sqrt{\bar{A}_{13}((\bar{E}_{F19,3} - eV_D), n_x, n_y)} \right] \right.$$
$$\left. + \sum_{r=1}^{s} L(r) \left[\left[\sqrt{\bar{A}_{13}(\bar{E}_{F19,3}, n_x, n_y)} + \sqrt{\bar{A}_{13}((\bar{E}_{F19,3} - eV_D), n_x, n_y)} \right] \right] \right] \tag{19.17}$$

where $\bar{A}_{13}(E, n_x, n_y)$ has been defined in (7.34) of Chapter 7.

(iv) The model of Stillman *et al.*

From (19.2), in this case we can write that

$$I_+ = \frac{ek_BTg_v}{\pi\hbar}\sum_{n_x=1}^{n_{x\max}}\sum_{n_y=1}^{n_{y\max}} F_0(\eta_{19,4}) \tag{19.18}$$

where $\eta_{19,4} = (k_BT)^{-1}[(E_{F19,4} - E_{14})]$, $E_{F19,4}$ is the Fermi energy in this case and E_{14} is the sub-band energy which is defined in (7.44) of Chapter 7.

From (19.18) we can write that

$$I_- = \frac{ek_BTg_v}{\pi\hbar}\sum_{n_x=1}^{n_{x\max}}\sum_{n_y=1}^{n_{y\max}} F_0(\eta_{19,4} - U_D) \tag{19.19}$$

Using (19.18), (19.19) and (19.1a), we get

$$I_D = \frac{ek_BTg_v}{\pi\hbar}\sum_{n_x=1}^{n_{x\max}}\sum_{n_y=1}^{n_{y\max}} [F_0(\eta_{19,4}) - F_0(\eta_{19,4} - U_D)] \tag{19.20}$$

The n_{1D} can be expressed as

$$n_{1D} = \frac{g_v}{\pi}\sum_{n_x=1}^{n_{x\max}}\sum_{n_y=1}^{n_{y\max}}\left[\left[\sqrt{A_{14}(E_{F19,4}, n_x, n_y)} + \sqrt{A_{14}((E_{F19,4} - eV_D), n_x, n_y)}\right]\right.$$
$$\left. + \sum_{r=1}^{s} L(r)\left[\left[\sqrt{A_{14}(E_{F19,4}, n_x, n_y)} + \sqrt{A_{14}((E_{F19,4} - eV_D), n_x, n_y)}\right]\right]\right] \tag{19.21}$$

where,

$$A_{14}(E_{F19,4}, n_x, n_y) = \left[\frac{2m_c}{\hbar^2}\gamma_5(E_{F19,4}) - \theta_1(n_x, n_y)\right] \tag{19.22}$$

$$\theta_1(n_x, n_y) = \left(\frac{n_x\pi}{d_x}\right)^2 + \left(\frac{n_y\pi}{d_y}\right)^2 \tag{19.23}$$

$$L(\gamma) = 2(k_BT)^{2r}(1 - 2^{1-2r})\xi(2r)\left(\frac{\partial^{2r}}{\partial E_{F19,4}^{2r}}\right) \tag{19.24}$$

(v) The model of Newson and Kourbe *et al.*

From (19.2), in this case, we can write that

$$I_+ = \frac{ek_BTg_v}{\pi\hbar}\sum_{n_x=1}^{n_{x\max}}\sum_{n_y=1}^{n_{y\max}} F_0(\eta_{19,5}) \tag{19.25}$$

where $\eta_{19,5} = (k_B T)^{-1}[(E_{F19,5} - E_{16})]$, $E_{F19,5}$ is the Fermi energy in this case and E_{16} is the sub-band energy which is defined in (7.53) of Chapter 7.

From (19.25) we can write that

$$I_- = \frac{ek_B T g_v}{\pi\hbar} \sum_{n_x=1}^{n_{x\max}} \sum_{n_y=1}^{n_{y\max}} F_0(\eta_{19,5} - U_D) \tag{19.26}$$

Using (19.25), (19.26) and (19.1a), we get

$$I_D = \frac{ek_B T g_v}{\pi\hbar} \sum_{n_x=1}^{n_{x\max}} \sum_{n_y=1}^{n_{y\max}} [F_0(\eta_{19,5}) - F_0(\eta_{19,5} - U_D)] \tag{19.27}$$

The n_{1D} can be expressed as

$$\begin{aligned} n_{1D} = \frac{g_v}{\pi} \sum_{n_x=1}^{n_{x\max}} \sum_{n_y=1}^{n_{y\max}} \Bigg[& \left[\sqrt{A_{15}(E_{F19,5}, n_x, n_y)} + \sqrt{A_{15}((E_{F19,5} - eV_D), n_x, n_y)}\right] \\ & + \sum_{r=1}^{s} L(r) \left[\left[\sqrt{A_{15}(E_{F19,5}, n_x, n_y)} + \sqrt{A_{15}((E_{F19,5} - eV_D), n_x, n_y)}\right]\right]\Bigg] \end{aligned} \tag{19.28}$$

where $A_{15}(E, n_x, n_y)$ has been defined in (7.52) of Chapter 7.

(vi) The model of Palik *et al.*

From (19.2), in this case we can write that

$$I_+ = \frac{ek_B T g_v}{\pi\hbar} \sum_{n_x=1}^{n_{x\max}} \sum_{n_y=1}^{n_{y\max}} F_0(\bar{\eta}_{19,5}) \tag{19.29}$$

where $\bar{\eta}_{19,5} = (k_B T)^{-1}[(\bar{E}_{F19,5} - \bar{E}_{16})]$, $\bar{E}_{F19,5}$ is the Fermi energy in this case and $\bar{E}_{16}$ is the sub-band energy which is defined in (7.61) of Chapter 7.

From (19.29) we can write that

$$I_- = \frac{ek_B T g_v}{\pi\hbar} \sum_{n_x=1}^{n_{x\max}} \sum_{n_y=1}^{n_{y\max}} F_0(\bar{\eta}_{19,5} - U_D) \tag{19.30}$$

Using (19.29), (19.30) and (19.1a) we get

$$I_D = \frac{ek_B T g_v}{\pi\hbar} \sum_{n_x=1}^{n_{x\max}} \sum_{n_y=1}^{n_{y\max}} [F_0(\bar{\eta}_{19,5}) - F_0(\bar{\eta}_{19,5} - U_D)] \tag{19.31}$$

The n_{1D} can be expressed as

$$n_{1D} = \frac{g_v}{\pi} \sum_{n_x=1}^{n_{x\max}} \sum_{n_y=1}^{n_{y\max}} \left[\left[\sqrt{\bar{A}_{15}(\bar{E}_{F19,5}, n_x, n_y)} + \sqrt{\bar{A}_{15}((\bar{E}_{F19,5} - eV_D), n_x, n_y)} \right] \right.$$

$$\left. + \sum_{r=1}^{s} L(r) \left[\left[\sqrt{\bar{A}_{15}(\bar{E}_{F19,5}, n_x, n_y)} + \sqrt{\bar{A}_{15}((\bar{E}_{F19,5} - eV_D), n_x, n_y)} \right] \right] \right] \tag{19.32}$$

where $\bar{A}_{15}(E, n_x, n_y)$ has been defined in (7.60) of Chapter 7.

19.2.3 *The $I_D - V_D$ equation for ballistic NWFETs of II–VI materials*

From (19.2), in this case we can write that

$$I_+ = \frac{ek_BTg_v}{2\pi\hbar} \sum_{n_x=1}^{n_{x\max}} \sum_{n_y=1}^{n_{y\max}} F_0(\eta_{19,6}) \tag{19.33}$$

where $\eta_{19,6} = (k_BT)^{-1}[(E_{F19,6} - \bar{E}_{161})]$, $E_{F19,6}$ is the Fermi energy in this case and $\bar{E}_{161}$ is the sub-band energy which is defined in (7.75) of Chapter 7.

From (19.33) we can write that

$$I_- = \frac{ek_BTg_v}{2\pi\hbar} \sum_{n_x=1}^{n_{x\max}} \sum_{n_y=1}^{n_{y\max}} F_0(\eta_{19,6} - U_D) \tag{19.34}$$

Using (19.33), (19.34) and (19.1a) we get

$$I_D = \frac{ek_BTg_v}{2\pi\hbar} \sum_{n_x=1}^{n_{x\max}} \sum_{n_y=1}^{n_{y\max}} [F_0(\eta_{19,6}) - F_0(\eta_{19,6} - U_D)] \tag{19.35}$$

The n_{1D} can be expressed as

$$n_{1D} = \frac{g_v}{2\pi\sqrt{B_0}} \sum_{n_x=1}^{n_{x\max}} \sum_{n_y=1}^{n_{y\max}} \left[[t_7(E_{F19,6}, n_x, n_y) + t_7((E_{F19,6} - eV_D), n_x, n_y)] \right.$$

$$\left. + \sum_{r=1}^{s} L(r) \left[[t_7(E_{F19,6}, n_x, n_y) + t_7((E_{F19,6} - eV_D), n_x, n_y)] \right] \right] \tag{19.36}$$

where $t_7(E, n_x, n_y)$ has been defined in (7.70) of Chapter 7.

19.2.4 The $I_D - V_D$ equation for ballistic NWFETs of IV–VI materials

From (19.2), in this case we can write that

$$I_+ = \frac{ek_BTg_v}{\pi\hbar}\sum_{n_x=1}^{n_{x\max}}\sum_{n_y=1}^{n_{y\max}} F_0(\eta_{19,7}) \tag{19.37}$$

where $\eta_{19,7} = (k_BT)^{-1}[(E_{F19,7} - \bar{E}_{22})]$, $E_{F19,7}$ is the Fermi energy in this case and $\bar{E}_{22}$ is the sub-band energy which is defined in (7.78) of Chapter 7.

From (19.37) we can write that

$$I_- = \frac{ek_BTg_v}{\pi\hbar}\sum_{n_x=1}^{n_{x\max}}\sum_{n_y=1}^{n_{y\max}} F_0(\eta_{19,7} - U_D) \tag{19.38}$$

Using (19.37), (19.38) and (19.1a) we get

$$I_D = \frac{ek_BTg_v}{\pi\hbar}\sum_{n_x=1}^{n_{x\max}}\sum_{n_y=1}^{n_{y\max}} [F_0(\eta_{19,7}) - F_0(\eta_{19,7} - U_D)] \tag{19.39}$$

The n_{1D} can be expressed as

$$n_{1D} = \frac{g_v}{\pi}\sum_{n_x=1}^{n_{x\max}}\sum_{n_y=1}^{n_{y\max}}\Bigg[\left[\sqrt{A_{23}(E_{F19,7}, n_x, n_y)} + \sqrt{A_{23}((E_{F19,7} - eV_D), n_x, n_y)}\right] + \sum_{r=1}^{s} L(r)\left[\left[\sqrt{A_{23}(E_{F19,7}, n_x, n_y)} + \sqrt{A_{23}((E_{F19,7} - eV_D), n_x, n_y)}\right]\right]\Bigg] \tag{19.40}$$

where $A_{23}(E, n_x, n_y)$ has been defined in (7.76) of Chapter 7.

19.2.5 The $I_D - V_D$ equation for ballistic NWFETs of stressed Kane type materials

From (19.2), in this case we can write that

$$I_+ = \frac{ek_BTg_v}{\pi\hbar}\sum_{n_x=1}^{n_{x\max}}\sum_{n_y=1}^{n_{y\max}} F_0(\eta_{19,8}) \tag{19.41}$$

where $\eta_{19,8} = (k_BT)^{-1}[(E_{F19,8} - \bar{E}_{23})]$, $E_{F19,8}$ is the Fermi energy in this case and $\bar{E}_{23}$ is the sub-band energy which is defined in (7.85) of Chapter 7.

From (19.41) we can write that

$$I_- = \frac{ek_BTg_v}{\pi\hbar}\sum_{n_x=1}^{n_{x\max}}\sum_{n_y=1}^{n_{y\max}} F_0(\eta_{19,8} - U_D) \tag{19.42}$$

Using (19.41), (19.42) and (19.1a) we get

$$I_D = \frac{ek_BTg_v}{\pi\hbar}\sum_{n_x=1}^{n_{x\max}}\sum_{n_y=1}^{n_{y\max}} [F_0(\eta_{19,8}) - F_0(\eta_{19,8} - U_D)] \tag{19.43}$$

The n_{1D} can be expressed as

$$n_{1D} = \frac{g_v}{\pi}\sum_{n_x=1}^{n_{x\max}}\sum_{n_y=1}^{n_{y\max}}\left[\left[\sqrt{A_{24}(E_{F19,8}, n_x, n_y)} + \sqrt{A_{24}((E_{F19,8} - eV_D), n_x, n_y)}\right]\right.$$
$$\left.+\sum_{r=1}^{s} L(r)\left[\left[\sqrt{A_{24}(E_{F19,8}, n_x, n_y)} + \sqrt{A_{24}((E_{F19,8} - eV_D), n_x, n_y)}\right]\right]\right] \tag{19.44}$$

where $A_{24}(E, n_x, n_y)$ has been defined in (7.84) of Chapter 7.

19.2.6 *The $I_D - V_D$ equation for ballistic NWFETs of bismuth telluride*

From (19.2), in this case we can write that

$$I_+ = \frac{ek_BTg_v}{\pi\hbar}\sum_{n_y=1}^{n_{y\max}}\sum_{n_z=1}^{n_{z\max}} F_0(\eta_{19,9}) \tag{19.45}$$

where $\eta_{19,9} = (k_BT)^{-1}[(E_{F19,9} - E_{30})]$, $E_{F19,9}$ is the Fermi energy in this case and E_{30} is the sub-band energy which is defined in (7.93) of Chapter 7.

From (19.45) we can write that

$$I_- = \frac{ek_BTg_v}{\pi\hbar}\sum_{n_y=1}^{n_{y\max}}\sum_{n_z=1}^{n_{z\max}} F_0(\eta_{19,9} - U_D) \tag{19.46}$$

Using (19.45), (19.46) and (19.1a), we get

$$I_D = \frac{ek_BTg_v}{\pi\hbar}\sum_{n_y=1}^{n_{y\max}}\sum_{n_z=1}^{n_{z\max}} [F_0(\eta_{19,9}) - F_0(\eta_{19,9} - U_D)] \tag{19.47}$$

The n_{1D} can be expressed as

$$n_{1D} = \frac{g_v}{\pi} \sum_{n_y=1}^{n_{y\max}} \sum_{n_z=1}^{n_{z\max}} \left[\left[\sqrt{A_{28}(E_{F19,9}, n_y, n_z)} + \sqrt{A_{28}((E_{F19,9} - eV_D), n_y, n_z)} \right] + \sum_{r=1}^{s} L(r) \left[\left[\sqrt{A_{28}(E_{F19,9}, n_y, n_z)} + \sqrt{A_{28}((E_{F19,9} - eV_D), n_y, n_z)} \right] \right] \right] \tag{19.48}$$

where $A_{28}(E, n_x, n_y)$ has been defined in (7.92) of Chapter 7.

19.2.7 The $I_D - V_D$ equation for ballistic NWFETs of germanium

(i) Model of Cardona *et al.*

From (19.2), in this case we can write that

$$I_+ = \frac{ek_BTg_v}{\pi\hbar} \sum_{n_x=1}^{n_{x\max}} \sum_{n_z=1}^{n_{z\max}} F_0(\eta_{19,10}) \tag{19.49}$$

where $\eta_{19,10} = (k_BT)^{-1}[(E_{F19,10} - E_{31})]$, $E_{F19,10}$ is the Fermi energy in this case and E_{31} is the sub-band energy which is defined in (7.101) of Chapter 7.

From (19.45) we can write that

$$I_- = \frac{ek_BTg_v}{\pi\hbar} \sum_{n_x=1}^{n_{x\max}} \sum_{n_z=1}^{n_{z\max}} F_0(\eta_{19,10} - U_D) \tag{19.50}$$

Using (19.49), (19.50) and (19.1a) we get

$$I_D = \frac{ek_BTg_v}{\pi\hbar} \sum_{n_x=1}^{n_{x\max}} \sum_{n_z=1}^{n_{z\max}} [F_0(\eta_{19,10}) - F_0(\eta_{19,10} - U_D)] \tag{19.51}$$

The n_{1D} can be expressed as

$$n_{1D} = \frac{g_v}{\pi} \sum_{n_x=1}^{n_{x\max}} \sum_{n_z=1}^{n_{z\max}} \left[\left[\sqrt{A_{29}(E_{F19,10}, n_x, n_z)} + \sqrt{A_{29}((E_{F19,10} - eV_D), n_x, n_z)} \right] + \sum_{r=1}^{s} L(r) \left[\left[\sqrt{A_{29}(E_{F19,10}, n_x, n_z)} + \sqrt{A_{29}((E_{F19,10} - eV_D), n_x, n_z)} \right] \right] \right] \tag{19.52}$$

where $A_{29}(E, n_x, n_y)$ has been defined in (7.100) of Chapter 7.

(ii) Model of Wang and Ressler

From (19.2), in this case we can write that

$$I_+ = \frac{ek_BTg_v}{\pi\hbar}\sum_{n_x=1}^{n_{x\max}}\sum_{n_z=1}^{n_{z\max}} F_0(\eta_{19,11}) \tag{19.53}$$

where $\eta_{19,11} = (k_BT)^{-1}[(E_{F19,11} - E_{32})]$, $E_{F19,11}$ is the Fermi energy in this case and E_{32} is the sub-band energy which is defined in (7.109) of Chapter 7.

From (19.53) we can write that

$$I_- = \frac{ek_BTg_v}{\pi\hbar}\sum_{n_x=1}^{n_{x\max}}\sum_{n_z=1}^{n_{z\max}} F_0(\eta_{19,11} - U_D) \tag{19.54}$$

Using (19.53), (19.54) and (19.1a) we get

$$I_D = \frac{ek_BTg_v}{\pi\hbar}\sum_{n_x=1}^{n_{x\max}}\sum_{n_z=1}^{n_{z\max}} [F_0(\eta_{19,11}) - F_0(\eta_{19,11} - U_D)] \tag{19.55}$$

The n_{1D} can be expressed as

$$n_{1D} = \frac{g_v}{\pi}\sum_{n_x=1}^{n_{x\max}}\sum_{n_z=1}^{n_{z\max}}\left[\left[\sqrt{A_{30}(E_{F19,11}, n_x, n_z)} + \sqrt{A_{30}((E_{F19,11} - eV_D), n_x, n_z)}\right]\right.$$
$$\left. + \sum_{r=1}^{s} L(r)\left[\left[\sqrt{A_{30}(E_{F19,11}, n_x, n_z)} + \sqrt{A_{30}((E_{F19,11} - eV_D), n_x, n_z)}\right]\right]\right] \tag{19.56}$$

where $A_{30}(E, n_x, n_y)$ has been defined in (7.108) of Chapter 7.

19.2.8 *The $I_D - V_D$ equation for ballistic NWFETs of gallium antimonide*

(i) Model of Sailer *et al.*

From (19.2), in this case we can write that

$$I_+ = \frac{ek_BTg_v}{\pi\hbar}\sum_{n_x=1}^{n_{x\max}}\sum_{n_y=1}^{n_{y\max}} F_0(\eta_{19,12}) \tag{19.57}$$

where $\eta_{19,12} = (k_BT)^{-1}[(E_{F19,12} - \bar{E}_{32})]$, $E_{F19,12}$ is the Fermi energy in this case and $\bar{E}_{32}$ is the sub-band energy which is defined in (7.123) of Chapter 7.

From (19.57) we can write that

$$I_- = \frac{ek_B T g_v}{\pi\hbar} \sum_{n_x=1}^{n_{x\max}} \sum_{n_y=1}^{n_{y\max}} F_0(\eta_{19,12} - U_D) \tag{19.58}$$

Using (19.57), (19.58) and (19.1a) we get

$$I_D = \frac{ek_B T g_v}{\pi\hbar} \sum_{n_x=1}^{n_{x\max}} \sum_{n_z=1}^{n_{z\max}} [F_0(\eta_{19,12}) - F_0(\eta_{19,12} - U_D)] \tag{19.59}$$

The n_{1D} can be expressed as

$$n_{1D} = \frac{g_v}{\pi} \sum_{n_x=1}^{n_{x\max}} \sum_{n_y=1}^{n_{y\max}} \left[\left[\sqrt{\bar{\gamma}_6(E_{F19,12}, n_x, n_y)} + \sqrt{\bar{\gamma}_6((E_{F19,12} - eV_D), n_x, n_y)} \right] \right.$$
$$\left. + \sum_{r=1}^{s} L(r) \left[\left[\sqrt{\bar{\gamma}_6(E_{F19,12}, n_x, n_y)} + \sqrt{\bar{\gamma}_6((E_{F19,12} - eV_D), n_x, n_y)} \right] \right] \right] \tag{19.60}$$

where $\bar{\gamma}_6(E, n_x, n_y)$ has been defined in (7.116) of Chapter 7.

(ii) Model of Mathur and Jain

From (19.2), in this case we can write that

$$I_+ = \frac{ek_B T g_v}{\pi\hbar} \sum_{n_x=1}^{n_{x\max}} \sum_{n_y=1}^{n_{y\max}} F_0(\eta_{19,13}) \tag{19.61}$$

where $\eta_{19,13} = (k_B T)^{-1}[(E_{F19,13} - \bar{E}_{321})]$, $E_{F19,13}$ is the Fermi energy in this case and $\bar{E}_{321}$ is the sub-band energy which is defined in (7.131) of Chapter 7.

From (19.61) we can write that

$$I_- = \frac{ek_B T g_v}{\pi\hbar} \sum_{n_x=1}^{n_{x\max}} \sum_{n_y=1}^{n_{y\max}} F_0(\eta_{19,13} - U_D) \tag{19.62}$$

Using (19.61), (19.62) and (19.1a) we get

$$I_D = \frac{ek_B T g_v}{\pi\hbar} \sum_{n_x=1}^{n_{x\max}} \sum_{n_z=1}^{n_{z\max}} [F_0(\eta_{19,13}) - F_0(\eta_{19,13} - U_D)] \tag{19.63}$$

The n_{1D} can be expressed as

$$n_{1D} = \frac{g_v}{\pi} \sum_{n_x=1}^{n_{x\max}} \sum_{n_y=1}^{n_{y\max}} \left[\left[\sqrt{\bar{\gamma}_7(E_{F19,13}, n_x, n_y)} + \sqrt{\bar{\gamma}_7((E_{F19,13} - eV_D), n_x, n_y)} \right] \right.$$
$$\left. + \sum_{r=1}^{s} L(r) \left[\left[\sqrt{\bar{\gamma}_7(E_{F19,13}, n_x, n_y)} + \sqrt{\bar{\gamma}_7((E_{F19,13} - eV_D), n_x, n_y)} \right] \right] \right] \tag{19.64}$$

where $\bar{\gamma}_7(E, n_x, n_y)$ has been defined in (7.125) of Chapter 7.

(iii) Model of Zhang

From (19.2), in this case we can write that

$$I_+ = \frac{ek_BTg_v}{\pi\hbar} \sum_{n_x=1}^{n_{x\max}} \sum_{n_y=1}^{n_{y\max}} F_0(\eta_{19,14}) \tag{19.65}$$

where $\eta_{19,14} = (k_BT)^{-1}[(E_{F19,14} - \bar{E}_{322})]$, $E_{F19,14}$ is the Fermi energy in this case and $\bar{E}_{322}$ is the sub-band energy which is defined in (7.139) of Chapter 7.

From (19.65) we can write that

$$I_- = \frac{ek_BTg_v}{\pi\hbar} \sum_{n_x=1}^{n_{x\max}} \sum_{n_y=1}^{n_{y\max}} F_0(\eta_{19,14} - U_D) \tag{19.66}$$

Using (19.65), (19.66) and (19.1a) we get

$$I_D = \frac{ek_BTg_v}{\pi\hbar} \sum_{n_x=1}^{n_{x\max}} \sum_{n_y=1}^{n_{y\max}} [F_0(\eta_{19,14}) - F_0(\eta_{19,14} - U_D)] \tag{19.67}$$

The n_{1D} can be expressed as

$$n_{1D} = \frac{g_v}{\pi} \sum_{n_x=1}^{n_{x\max}} \sum_{n_y=1}^{n_{y\max}} \left[\left[\sqrt{\bar{\gamma}_8(E_{F19,14}, n_x, n_y)} + \sqrt{\bar{\gamma}_8((E_{F19,14} - eV_D), n_x, n_y)} \right] \right.$$
$$\left. + \sum_{r=1}^{s} L(r) \left[\left[\sqrt{\bar{\gamma}_8(E_{F19,14}, n_x, n_y)} + \sqrt{\bar{\gamma}_7((E_{F19,14} - eV_D), n_x, n_y)} \right] \right] \right] \tag{19.68}$$

where $\bar{\gamma}_8(E, n_x, n_y)$ has been defined in (7.133) of Chapter 7.

19.2.9 The $I_D - V_D$ equation for ballistic NWFETs of GaP

From (19.2), in this case we can write that

$$I_+ = \frac{ek_BTg_v}{\pi\hbar}\sum_{n_x=1}^{n_{x\max}}\sum_{n_y=1}^{n_{y\max}} F_0(\eta_{19,15}) \tag{19.69}$$

where $\eta_{19,15} = (k_BT)^{-1}[(E_{F19,15} - \bar{E}_{27})]$, $E_{F19,15}$ is the Fermi energy in this case and $\bar{E}_{27}$ is the sub-band energy which is defined in (7.146) of Chapter 7.

From (19.65) we can write that

$$I_- = \frac{ek_BTg_v}{\pi\hbar}\sum_{n_x=1}^{n_{x\max}}\sum_{n_y=1}^{n_{y\max}} F_0(\eta_{19,15} - U_D) \tag{19.70}$$

Using (19.65), (19.66) and (19.1a) we get

$$I_D = \frac{ek_BTg_v}{\pi\hbar}\sum_{n_x=1}^{n_{x\max}}\sum_{n_y=1}^{n_{y\max}} [F_0(\eta_{19,15}) - F_0(\eta_{19,15} - U_D)] \tag{19.71}$$

The n_{1D} can be expressed as

$$n_{1D} = \frac{g_v}{\pi}\sum_{n_x=1}^{n_{x\max}}\sum_{n_y=1}^{n_{y\max}} \left[\left[\sqrt{A_{26}(E_{F19,15}, n_x, n_y)} + \sqrt{A_{26}((E_{F19,15} - eV_D), n_x, n_y)}\right]\right.$$
$$\left. + \sum_{r=1}^{s} L(r)\left[\left[\sqrt{A_{26}(E_{F19,15}, n_x, n_y)} + \sqrt{A_{26}((E_{F19,15} - eV_D), n_x, n_y)}\right]\right]\right] \tag{19.72}$$

where $A_{26}(E, n_x, n_y)$ has been defined in (7.145) of Chapter 7.

19.2.10 The $I_D - V_D$ equation for ballistic NWFETs of $PtSb_2$

From (19.2), in this case we can write that

$$I_+ = \frac{ek_BTg_v}{\pi\hbar}\sum_{n_y=1}^{n_{y\max}}\sum_{n_z=1}^{n_{z\max}} F_0(\eta_{19,16}) \tag{19.73}$$

where $\eta_{19,16} = (k_BT)^{-1}[(E_{F19,16} - \bar{E}_{28})]$, $E_{F19,16}$ is the Fermi energy in this case and $\bar{E}_{28}$ is the sub-band energy which is defined in (7.160) of Chapter 7.

From (19.65) we can write that

$$I_- = \frac{ek_BTg_v}{\pi\hbar}\sum_{n_y=1}^{n_{y\max}}\sum_{n_z=1}^{n_{z\max}} F_0(\eta_{19,16} - U_D) \tag{19.74}$$

Using (19.73), (19.74) and (19.1a) we get

$$I_D = \frac{ek_BTg_v}{\pi\hbar}\sum_{n_y=1}^{n_{y\max}}\sum_{n_z=1}^{n_{z\max}} [F_0(\eta_{19,16}) - F_0(\eta_{19,16} - U_D)] \tag{19.75}$$

The n_{1D} can be expressed as

$$n_{1D} = \frac{g_v}{\pi}\sum_{n_y=1}^{n_{y\max}}\sum_{n_z=1}^{n_{z\max}}\Bigg[\left[\sqrt{A_{27}(E_{F19,16}, n_y, n_z)} + \sqrt{A_{27}((E_{F19,16} - eV_D), n_y, n_z)}\right]$$
$$+ \sum_{r=1}^{s} L(r)\left[\left[\sqrt{A_{27}(E_{F19,16}, n_y, n_z)} + \sqrt{A_{27}((E_{F19,16} - eV_D), n_y, n_z)}\right]\right]\Bigg] \tag{19.76}$$

where $A_{27}(E, n_y, n_z)$ has been defined in (7.154) of Chapter 7.

19.2.11 *The $I_D - V_D$ equation for ballistic NWFETs of II–V materials*

From (19.2), in this case we can write that

$$I_+ = \frac{ek_BTg_v}{2\pi\hbar}\sum_{n_x=1}^{n_{x\max}}\sum_{n_y=1}^{n_{y\max}} F_0(\eta_{19,17,\pm}) \tag{19.77}$$

where $\eta_{19,17,\pm} = (k_BT)^{-1}[(E_{F19,17} - E_{34\pm})]$, $E_{F19,17}$ is the Fermi energy in this case and $E_{34\pm}$ is the sub-band energy which is defined in (7.163) of Chapter 7.

From (19.77) we can write that

$$I_- = \frac{ek_BTg_v}{2\pi\hbar}\sum_{n_x=1}^{n_{x\max}}\sum_{n_y=1}^{n_{y\max}} F_0(\eta_{19,17\pm} - U_D) \tag{19.78}$$

Using (19.77), (19.78) and (19.1a) we get

$$I_D = \frac{ek_BTg_v}{2\pi\hbar}\sum_{n_x=1}^{n_{x\max}}\sum_{n_y=1}^{n_{y\max}} [F_0(\eta_{19,17\pm}) - F_0(\eta_{19,17\pm} - U_D)] \tag{19.79}$$

The n_{1D} can be expressed as

$$n_{1D} = \frac{g_v}{2\pi} \sum_{n_x=1}^{n_{x\max}} \sum_{n_y=1}^{n_{y\max}} \left[\left[\sqrt{A_{32\pm}(E_{F19,17\pm}, n_x, n_y)} \right.\right.$$

$$\left. + \sqrt{A_{32\pm}(E_{F19,17\pm} - eV_D), n_x, n_y)} \right]$$

$$+ \sum_{r=1}^{s} L(r) \left[\left[\sqrt{A_{32\pm}(E_{F19,17\pm}, n_x, n_y)} \right.\right.$$

$$\left.\left.\left. + \sqrt{A_{32\pm}(E_{F19,17\pm} - eV_D), n_x, n_y)} \right] \right] \right] \tag{19.80}$$

where $A_{32\pm}(E, n_x, n_y)$ has been defined in (7.162) of Chapter 7.

19.3 Summary and Conclusion

In this chapter, in Section 19.2.1, we have formulated $I_D - V_D$ for ballistic NWFETs made of tetragonal and nonlinear optical materials as given by (19.4) for the computation of which we need the expression of the 1D electron statistics which is given in (19.5). The $I_D - V_D$ equations for NWFETs made of III–V, ternary and quaternary materials has been investigated in Section 19.2.2 in accordance with the three-band model of Kane, two-band model of Kane, Parabolic energy bands, model of Stillman *et al.*, the model of Newson and Kurobe and the model of Palik *et al.* are given by (19.8), (19.12), (19.16), (19.20), (19.27) and (19.31) and the corresponding electron statistics are expressed through (19.9), (19.13), (19.17), (19.21), (19.28) and (19.32) respectively. In Section 19.2.3, we have studied the $I_D - V_D$ equation for NWFETs of II–VI materials which is given by (19.35) and the corresponding electron statistics has been represented in (19.36). In Section 19.2.4, we have studied the $I_D - V_D$ equation in NWFETs of IV–VI materials which is given by (19.39) and the corresponding electron statistics has been represented in (19.40). In Section 19.2.5, we have investigated the $I_D - V_D$ equation in NWFETs of stressed Kane type materials which is given by (19.43) and the corresponding electron statistics has been represented in (19.44). In Section 19.2.6, we have studied the $I_D - V_D$ equation in NWFETs of Bismuth Telluride materials which is given by (19.47) and the corresponding electron statistics has been represented in (19.48). In Section 19.2.7, we have studied the $I_D - V_D$ equation in NWFETs of Germanium in accordance with the models of

Cardona *et al.* and Wang *et al.* which are given by (19.51) and (19.55) and the corresponding electron statistics have been represented in (19.52) and (19.56) respectively. In Section 19.2.8, we have studied the $I_D - V_D$ equation in NWFETs of gallium antimonide in accordance with the models of Sailer *et al.*, Mathur *et al.* and Zhang which are given by (19.59), (19.63) and (19.67) and the corresponding electron statistics have been represented in (19.60), (19.64) and (19.68) respectively. In Section 19.2.9, we have investigated the $I_D - V_D$ equation in NWFETs of GaP which is given by (19.71) and the corresponding electron statistics has been represented in (19.72). In Section 19.2.10, we have investigated the $I_D - V_D$ equation in NWFETs of $PtSb_2$ which is given by (19.75) and the corresponding electron statistics has been represented in (19.76). In Section 19.2.11, we have investigated the $I_D - V_D$ equation in NWFETs of class II–V materials which is given by (19.79) and the corresponding electron statistics has been represented in (19.80).

We have not given the graphs of the $I_D - V_D$ equations for ballistic NWFETs made of many materials as discussed here since we feel that our readers will perform the intricate numerical computation to obtain the oscillatory plots for better assessment and joy of understanding. It is the band structure which changes in a fundamental way and consequently all the physical properties of all the electronic materials changes radically leading to new physical concepts.

19.4 Open Research Problem

(R19.1) Derive $I_D - V_D$ equations for ballistic NWFETs made of all the materials whose DRs are given in (R1.1). Draw the corresponding graphs and explain the results.

Chapter 20

The $I_D - V_D$ Equation for Ballistic HDNWFETs

The greatest wealth in life is time

20.1 Introduction

It appears from the literature that the $I_D - V_D$ equations for ballistic heavily doped HDNWFETs have yet to be investigated. In this appendix, we shall formulate the $I_D - V_D$ equations for ballistic HD NWFETs made of HD tetragonal and nonlinear optical, III–V ternary and quaternary, II–VI, IV–VI, stressed Kane type, Te, GaP, $PtSb_2$, Bi_2Te_3, Ge, GaSb, and II–V materials from Section 20.2.1 up to 20.2.12, respectively. Section 20.3 contains summary and conclusion and Section 20.4 presents one multidimensional open research problem.

20.2 Theoretical Background

20.2.1 *The $I_D - V_D$ equation for ballistic HD NWFETs of tetragonal and nonlinear optical materials*

The I_D is given by

$$I_D = I_+ - I_- \tag{20.1}$$

where

$$I_+ = eg_\nu \text{ Real part of } \sum_{n_y=1}^{n_{y\max}} \sum_{n_z=1}^{n_{z\max}} \int_{E'_{1HDNW}}^{\infty} \left[\left[\frac{1}{2}\frac{2}{\pi}\frac{\partial k_z}{\partial E}\frac{1}{\hbar}\frac{\partial E}{\partial k_z}\right] f(E)dE\right], \tag{20.2}$$

and E'_{1HDNW} is the sub-band energy in the case which is defined in (14.4) of Chapter 14.

From (20.2) we can write that

$$I_+ = \frac{ek_BTg_\nu}{\pi\hbar} \text{ Real part of } \sum_{n_y=1}^{n_{y\max}} \sum_{n_z=1}^{n_{z\max}} F_0(\eta_{20,1}) \tag{20.3}$$

where $\eta_{20,1} = (k_BT)^{-1}[(E_{F20,1} - E'_{1HDNW})]$ and $E_{F20,1}$ is the Fermi energy in this case.

From (20.3) we can write that

$$I_- = \frac{ek_BTg_\nu}{\pi\hbar} \text{ Real Part of } \sum_{n_y=1}^{n_{y\max}} \sum_{n_z=1}^{n_{z\max}} F_0(\eta_{20,1} - U_D) \tag{20.4}$$

where $U_D = (eV_D/k_BT)$.

Using (20.1), (20.3) and (20.4) we get

$$I_D = \frac{ek_BTg_\nu}{\pi\hbar} \text{ Real part of } \sum_{n_y=1}^{n_{y\max}} \sum_{n_z=1}^{n_{z\max}} [F_0(\eta_{20,1}) - F_0(\eta_{20,1} - U_D)] \tag{20.5}$$

It appears than that the computation of I_D needs an expression of electron concentration per unit length which can, in turn, be expressed as

$$n_{1D} = \frac{g_\nu}{\pi} \text{ Real part of} \sum_{n_y=1}^{n_{y\max}} \sum_{n_z=1}^{n_{z\max}} \Bigg[[T_1(E_{F20,1}, n_y, n_z, \eta_g) + T_1((E_{F20,1} - eV_D), n_y, n_z, \eta_g)] + \sum_{r=1}^{s} L(r) \, [[T_1(E_{F20,1}, n_y, n_z, \eta_g) + T_1((E_{F20,1} - eV_D), n_y, n_z, \eta_g)]] \Bigg] \tag{20.6}$$

where $T_1(E, n_x, n_y, \eta_g)$ has been defined in (14.4) of Chapter 14.

20.2.2 *The $I_D - V_D$ equation for ballistic HDNWFETs of III–V, ternary and quaternary materials*

(i) Three-band-model of Kane

From (20.2) in this case we can write

$$I_+ = \frac{ek_BTg_\nu}{\pi\hbar} \text{ Real part of } \sum_{n_y=1}^{n_{y\max}} \sum_{n_z=1}^{n_{z\max}} F_0(\eta_{20,2}) \tag{20.7}$$

where $\eta_{20,2} = (k_B T)^{-1}[(E_{F20,2} - E'_{2HDNW})]$, $E_{F20,2}$ is the Fermi energy in this case and E'_{2HDNW} is the sub-band energy in the case which is defined in (14.8) of Chapter 14.

From (20.7) we can write that

$$I_- = \frac{ek_B T g_\nu}{\pi\hbar} \text{ Real Part of } \sum_{n_y=1}^{n_{y\max}} \sum_{n_z=1}^{n_{z\max}} F_0(\eta_{20,2} - U_D) \tag{20.8}$$

Using (20.7), (20.8) and (20.1) we get

$$I_D = \frac{ek_B T g_\nu}{\pi\hbar} \text{ Real part of } \sum_{n_y=1}^{n_{y\max}} \sum_{n_z=1}^{n_{z\max}} [F_0(\eta_{20,2}) - F_0(\eta_{20,2} - U_D)] \tag{20.9}$$

The n_{1D} in this case can be expressed as

$$n_{1D} = \frac{g_\nu}{\pi} \text{ Real part of } \sum_{n_y=1}^{n_{y\max}} \sum_{n_z=1}^{n_{z\max}} \Bigg[[T_2(E_{F20,2}, n_y, n_z, \eta_g) + T_2((E_{F20,2} - eV_D), n_y, n_z, \eta_g)] + \sum_{r=1}^{s} L(r) \, [[T_2(E_{F20,2}, n_y, n_z, \eta_g) + T_2((E_{F20,2} - eV_D), n_y, n_z, \eta_g)]] \Bigg] \tag{20.10}$$

where $T_2(E, n_x, n_y, \eta_g)$ has been defined in (14.7) of Chapter 14.

(ii) Two-band model of Kane

From (20.2) in this case we can write

$$I_+ = \frac{ek_B T g_\nu}{\pi\hbar} \sum_{n_y=1}^{n_{y\max}} \sum_{n_z=1}^{n_{z\max}} F_0(\eta_{20,3}) \tag{20.11}$$

where $\eta_{20,3} = (k_B T)^{-1}[(E_{F20,3} - E'_{3HDNW})]$, $E_{F20,3}$ is the Fermi energy in this case and E'_{3HDNW} is the sub-band energy in the case which is defined in (14.14) of Chapter 14.

From (20.11) we can write that

$$I_- = \frac{ek_BTg_\nu}{\pi\hbar}\sum_{n_y=1}^{n_{y\max}}\sum_{n_z=1}^{n_{z\max}} F_0(\eta_{20,3} - U_D) \tag{20.12}$$

Using (20.11), (20.12) and (20.1) we get

$$I_D = \frac{ek_BTg_\nu}{\pi\hbar}\sum_{n_y=1}^{n_{y\max}}\sum_{n_z=1}^{n_{z\max}} [F_0(\eta_{20,3}) - F_0(\eta_{20,3} - U_D)] \tag{20.13}$$

The n_{1D} in this case can be expressed as

$$n_{1D} = \frac{g_\nu}{\pi}\sum_{n_y=1}^{n_{y\max}}\sum_{n_z=1}^{n_{z\max}}\Bigg[[T_3(E_{F20,3}, n_y, n_z, \eta_g) + T_3((E_{F20,3} - eV_D), n_y, n_z, \eta_g)] + \sum_{r=1}^{s} L(r)[[T_3(E_{F20,3}, n_y, n_z, \eta_g) + T_3((E_{F20,3} - eV_D), n_y, n_z, \eta_g)]]\Bigg] \tag{20.14}$$

where $T_3(E, n_y, n_z, \eta_g)$ has been defined in (14.13) of Chapter 14.

(iii) Parabolic band model

From (20.2) in this case we can write

$$I_+ = \frac{ek_BTg_\nu}{\pi\hbar}\sum_{n_y=1}^{n_{y\max}}\sum_{n_z=1}^{n_{z\max}} F_0(\eta_{20,4}) \tag{20.15}$$

where $\eta_{20,4} = (k_BT)^{-1}[(E_{F20,4} - E'_{4HDNW})]$, $E_{F20,4}$ is the Fermi energy in this case and E'_{4HDNW} is the sub-band energy in the case which is defined in (14.19) of Chapter 14.

From (20.15) we can write that

$$I_- = \frac{ek_BTg_\nu}{\pi\hbar}\sum_{n_y=1}^{n_{y\max}}\sum_{n_z=1}^{n_{z\max}} F_0(\eta_{20,4} - U_D) \tag{20.16}$$

Using (20.15), (20.16) and (20.1) we get

$$I_D = \frac{ek_BTg_\nu}{\pi\hbar}\sum_{n_y=1}^{n_{y\max}}\sum_{n_z=1}^{n_{z\max}} [F_0(\eta_{20,4}) - F_0(\eta_{20,4} - U_D)] \tag{20.17}$$

The n_{1D} in this case can be expressed as

$$n_{1D} = \frac{g_\nu}{\pi} \sum_{n_y=1}^{n_{y\max}} \sum_{n_z=1}^{n_{z\max}} \Bigg[[T_4(E_{F20,4}, n_y, n_z, \eta_g) + T_4((E_{F20,4} - eV_D), n_y, n_z, \eta_g)] + \sum_{r=1}^{s} L(r)[[T_4(E_{F20,4}, n_y, n_z, \eta_g) + T_4((E_{F20,4} - eV_D), n_y, n_z, \eta_g)]] \Bigg] \tag{20.18}$$

where $T_4(E, n_y, n_z, \eta_g)$ has been defined in (14.18) of Chapter 14.

(iv) Stillman *et al.*

From (20.2) in this case we can write

$$I_+ = \frac{ek_BTg_\nu}{\pi\hbar} \sum_{n_y=1}^{n_{y\max}} \sum_{n_z=1}^{n_{z\max}} F_0(\eta_{20,5}) \tag{20.19}$$

where $\eta_{20,5} = (k_BT)^{-1}[(E_{F20,5} - E'_{9HDNW})]$, $E_{F20,5}$ is the Fermi energy in this case and E'_{9HDNW} is the sub-band energy in the case which is defined in (14.23) of Chapter 14.

From (20.19) we can write that

$$I_- = \frac{ek_BTg_\nu}{\pi\hbar} \sum_{n_y=1}^{n_{y\max}} \sum_{n_z=1}^{n_{z\max}} F_0(\eta_{20,5} - U_D) \tag{20.20}$$

Using (20.19), (20.20) and (20.1) we get

$$I_D = \frac{ek_BTg_\nu}{\pi\hbar} \sum_{n_y=1}^{n_{y\max}} \sum_{n_z=1}^{n_{z\max}} [F_0(\eta_{20,5}) - F_0(\eta_{20,5} - U_D)] \tag{20.21}$$

The n_{1D} in this case can be expressed as

$$n_{1D} = \frac{g_\nu}{\pi} \sum_{n_y=1}^{n_{y\max}} \sum_{n_z=1}^{n_{z\max}} \Bigg[[T_{11}(E_{F20,5}, n_y, n_z, \eta_g) + T_{11}((E_{F20,5} - eV_D), n_y, n_z, \eta_g) \Bigg] + \sum_{r=1}^{s} L(r)[[T_{11}(E_{F20,5}, n_y, n_z, \eta_g) + T_{11}((E_{F20,5} - eV_D), n_y, n_z, \eta_g)]] \Bigg] \tag{20.22}$$

where $T_{11}(E, n_y, n_z, \eta_g)$ has been defined in (14.25) of Chapter 14.

(v) Palik *et al.*

From (20.2) in this case we can write

$$I_+ = \frac{ek_BTg_\nu}{\pi\hbar}\sum_{n_y=1}^{n_{y\max}}\sum_{n_z=1}^{n_{z\max}}F_0(\eta_{20,6}) \tag{20.23}$$

where $\eta_{20,6} = (k_BT)^{-1}[(E_{F20,6} - E'_{6HDNW})]$, $E_{F20,6}$ is the Fermi energy in this case and E'_{6HDNW} is the sub-band energy in the case which is defined in (14.27) of Chapter 14.

From (20.23) we can write that

$$I_- = \frac{ek_BTg_\nu}{\pi\hbar}\sum_{n_y=1}^{n_{y\max}}\sum_{n_z=1}^{n_{z\max}}F_0(\eta_{20,6} - U_D) \tag{20.24}$$

Using (20.23), (20.24) and (20.1) we get

$$I_D = \frac{ek_BTg_\nu}{\pi\hbar}\sum_{n_y=1}^{n_{y\max}}\sum_{n_z=1}^{n_{z\max}}[F_0(\eta_{20,6}) - F_0(\eta_{20,6} - U_D)] \tag{20.25}$$

The n_{1D} in this case can be expressed as

$$n_{1D} = \frac{g_\nu}{\pi}\sum_{n_y=1}^{n_{y\max}}\sum_{n_z=1}^{n_{z\max}}\Bigg[[T_6(E_{F20,6}, n_y, n_z, \eta_g) + T_6((E_{F20,6} - eV_D), n_y, n_z, \eta_g)] + \sum_{r=1}^{s}L(r)[[T_6(E_{F20,6}, n_y, n_z, \eta_g) + T_6((E_{F20,6} - eV_D), n_y, n_z, \eta_g)]]\Bigg] \tag{20.26}$$

where $T_6(E, n_y, n_z, \eta_g)$ has been defined in (14.29) of Chapter 14.

20.2.3 The $I_D - V_D$ equation for ballistic HDNWFETs of II–VI materials

From (20.2) in this case we can write

$$I_+ = \frac{ek_BTg_\nu}{2\pi\hbar}\sum_{n_x=1}^{n_{x\max}}\sum_{n_y=1}^{n_{y\max}}F_0(\eta_{20,7}) \tag{20.27}$$

where $\eta_{20,7} = (k_BT)^{-1}[(E_{F20,7} - E'_{13HDNW})]$, $E_{F20,7}$ is the Fermi energy in this case and E'_{13HDNW} is the sub-band energy in the case which is defined in (14.32) of Chapter 14. From (20.27) we can write that

$$I_- = \frac{ek_BTg_\nu}{2\pi\hbar}\sum_{n_x=1}^{n_{x\max}}\sum_{n_y=1}^{n_{y\max}} F_0(\eta_{20,7} - U_D) \tag{20.28}$$

Using (20.27), (20.28) and (20.1) we get

$$I_D = \frac{ek_BTg_\nu}{2\pi\hbar}\sum_{n_x=1}^{n_{x\max}}\sum_{n_y=1}^{n_{y\max}} [F_0(\eta_{20,7}) - F_0(\eta_{20,7} - U_D)] \tag{20.29}$$

The n_{1D} in this case can be expressed as

$$n_{1D} = \frac{g_\nu}{2\pi}\sum_{n_x=1}^{n_{x\max}}\sum_{n_y=1}^{n_{y\max}}\Bigg[[T_7(E_{F20,7}, n_x, n_y, \eta_g) + T_7((E_{F20,7} - eV_D), n_x, n_y, \eta_g)] + \sum_{r=1}^{s} L(r)[[T_7(E_{F20,7}, n_x, n_y, \eta_g) + T_7((E_{F20,7} - eV_D), n_x, n_y, \eta_g)]]\Bigg] \tag{20.30}$$

where $T_7(E, n_x, n_y, \eta_g)$ has been defined in (14.31) of Chapter 14.

20.2.4 *The $I_D - V_D$ equation for ballistic HDNWFETs of IV–VI materials*

(i) Dimmock model

From (20.2) in this case we can write

$$I_+ = \frac{ek_BTg_\nu}{\pi\hbar}\sum_{n_x=1}^{n_{x\max}}\sum_{n_y=1}^{n_{y\max}} F_0(\eta_{20,8}) \tag{20.31}$$

where $\eta_{20,8} = (k_BT)^{-1}[(E_{F20,8} - E'_{14HDNW})]$, $E_{F20,8}$ is the Fermi energy in this case and E'_{14HDNW} is the sub-band energy in the case which is defined in (14.38) of Chapter 14.

From (20.31) we can write that

$$I_- = \frac{ek_BTg_\nu}{\pi\hbar}\sum_{n_x=1}^{n_{x\max}}\sum_{n_y=1}^{n_{y\max}} F_0(\eta_{20,8} - U_D) \tag{20.32}$$

Using (20.31), (20.32) and (20.1) we get

$$I_D = \frac{ek_BTg_\nu}{\pi\hbar}\sum_{n_x=1}^{n_{x\max}}\sum_{n_y=1}^{n_{y\max}} [F_0(\eta_{20,8}) - F_0(\eta_{20,8} - U_D)] \tag{20.33}$$

The n_{1D} in this case can be expressed as

$$n_{1D} = \frac{g_\nu}{\pi} \sum_{n_x=1}^{n_{x\max}} \sum_{n_y=1}^{n_{y\max}} \left[[T_{36}(E_{F20,8}, n_x, n_y, \eta_g) + T_{36}((E_{F20,8} - eV_D), n_x, n_y, \eta_g)] + \sum_{r=1}^{s} L(r)[[T_{36}(E_{F20,8}, n_x, n_y, \eta_g) + T_{36}((E_{F20,8} - eV_D), n_x, n_y, \eta_g)]] \right] \tag{20.34}$$

where $T_{36}(E, n_x, n_y, \eta_g)$ has been defined in (14.36) of Chapter 14.

(ii) Bangert and Kastner model

From (20.2) in this case we can write

$$I_+ = \frac{ek_BTg_\nu}{\pi\hbar} \text{ Real part of } \sum_{n_x=1}^{n_{x\max}} \sum_{n_y=1}^{n_{y\max}} F_0(\eta_{20,9}) \tag{20.35}$$

where $\eta_{20,9} = (k_BT)^{-1}[(E_{F20,9} - E'_{15HDNW})]$, $E_{F20,9}$ is the Fermi energy in this case and E'_{15HDNW} is the sub-band energy in the case which is defined in (14.44) of Chapter 14.

From (20.35) we can write that

$$I_- = \frac{ek_BTg_\nu}{\pi\hbar} \text{ Real part of } \sum_{n_x=1}^{n_{x\max}} \sum_{n_y=1}^{n_{y\max}} F_0(\eta_{20,9} - U_D) \tag{20.36}$$

Using (20.35), (20.36) and (20.1) we get

$$I_D = \frac{ek_BTg_\nu}{\pi\hbar} \text{ Real part of } \sum_{n_x=1}^{n_{x\max}} \sum_{n_y=1}^{n_{y\max}} [F_0(\eta_{20,9}) - F_0(\eta_{20,9} - U_D)] \tag{20.37}$$

The n_{1D} in this case can be expressed as

$$n_{1D} = \frac{g_\nu}{\pi} \text{ Real part of } \sum_{n_x=1}^{n_{x\max}} \sum_{n_y=1}^{n_{y\max}} \left[[T_{60}(E_{F20,9}, n_x, n_y, \eta_g) + T_{60}((E_{F20,9} - eV_D), n_x, n_y, \eta_g)] + \sum_{r=1}^{s} L(r)[[T_{60}(E_{F20,9}, n_x, n_y, \eta_g) + T_{60}((E_{F20,9} - eV_D), n_x, n_y, \eta_g)]] \right] \tag{20.38}$$

where $T_{60}(E, n_x, n_y, \eta_g)$ has been defined in (14.42) of Chapter 14.

20.2.5 The $I_D - V_D$ equation for ballistic HDNWFETs of stressed Kane type materials

From (20.2) in this case we can write

$$I_+ = \frac{ek_BTg_\nu}{\pi\hbar} \sum_{n_x=1}^{n_{x\max}} \sum_{n_y=1}^{n_{y\max}} F_0(\eta_{20,10}) \tag{20.39}$$

where $\eta_{20,10} = (k_BT)^{-1}[(E_{F20,10} - E'_{30HDNW})]$, $E_{F20,10}$ is the Fermi energy in this case and E'_{30HDNW} is the sub-band energy in the case which is defined in (14.51) of Chapter 14.

From (20.39) we can write that

$$I_- = \frac{ek_BTg_\nu}{\pi\hbar} \sum_{n_x=1}^{n_{x\max}} \sum_{n_y=1}^{n_{y\max}} F_0(\eta_{20,10} - U_D) \tag{20.40}$$

Using (20.39), (20.40) and (20.1) we get

$$I_D = \frac{ek_BTg_\nu}{\pi\hbar} \sum_{n_x=1}^{n_{x\max}} \sum_{n_y=1}^{n_{y\max}} [F_0(\eta_{20,10}) - F_0(\eta_{20,10} - U_D)] \tag{20.41}$$

The n_{1D} in this case can be expressed as

$$n_{1D} = \frac{g_\nu}{\pi} \sum_{n_x=1}^{n_{x\max}} \sum_{n_y=1}^{n_{y\max}} \Bigg[[T_{70}(E_{F20,10}, n_x, n_y, \eta_g) + T_{70}((E_{F20,10} - eV_D), n_x, n_y, \eta_g)] + \sum_{r=1}^{s} L(r)[[T_{70}(E_{F20,10}, n_x, n_y, \eta_g) + T_{70}((E_{F20,10} - eV_D), n_x, n_y, \eta_g)]] \Bigg] \tag{20.42}$$

where $T_{70}(E, n_x, n_y, \eta_g)$ has been defined in (14.49) of Chapter 14.

20.2.6 The $I_D - V_D$ equation for ballistic HDNWFETs of Te

From (20.2) in this case we can write

$$I_+ = \frac{ek_BTg_\nu}{2\pi\hbar} \sum_{n_y=1}^{n_{x\max}} \sum_{n_z=1}^{n_{y\max}} F_0(\eta_{20,11}) \tag{20.43}$$

where $\eta_{20,11} = (k_BT)^{-1}[(E_{F20,11} - E'_{31HDNW})]$, $E_{F20,11}$ is the Fermi energy in this case and E'_{31HDNW} is the sub-band energy in the case which is defined in (14.55) of Chapter 14.

From (20.43) we can write that

$$I_- = \frac{ek_BTg_\nu}{2\pi\hbar}\sum_{n_y=1}^{n_{y\max}}\sum_{n_z=1}^{n_{z\max}} F_0(\eta_{20,11} - U_D) \tag{20.44}$$

Using (20.43), (20.44) and (20.1) we get

$$I_D = \frac{ek_BTg_\nu}{2\pi\hbar}\sum_{n_y=1}^{n_{y\max}}\sum_{n_z=1}^{n_{z\max}} [F_0(\eta_{20,11}) - F_0(\eta_{20,11} - U_D)] \tag{20.45}$$

The n_{1D} in this case can be expressed as

$$n_{1D} = \frac{g_\nu}{2\pi}\sum_{n_y=1}^{n_{y\max}}\sum_{n_z=1}^{n_{z\max}}\Bigg[[t_{72}(E_{F20,11}, n_y, n_z, \eta_g) + t_{72}((E_{F20,11} - eV_D), n_y, n_z, \eta_g)] + \sum_{r=1}^{s} L(r)[[t_{72}(E_{F20,11}, n_y, n_z, \eta_g) + t_{72}((E_{F20,11} - eV_D), n_y, n_z, \eta_g)]]\Bigg] \tag{20.46}$$

where $t_{72}(E, n_y, n_z, \eta_g)$ has been defined in (14.54) of Chapter 14.

20.2.7 The $I_D - V_D$ equation for ballistic HDNWFETs of GaP

From (20.2) in this case we can write

$$I_+ = \frac{ek_BTg_\nu}{\pi\hbar}\sum_{n_y=1}^{n_{y\max}}\sum_{n_z=1}^{n_{z\max}} F_0(\eta_{20,12}) \tag{20.47}$$

where $\eta_{20,12} = (k_BT)^{-1}[(E_{F20,12} - E'_{32HDNW})]$, $E_{F20,12}$ is the Fermi energy in this case and E'_{32HDNW} is the sub-band energy in the case which is defined in (14.61) of Chapter 14.

From (20.47) we can write that

$$I_- = \frac{ek_BTg_\nu}{\pi\hbar}\sum_{n_y=1}^{n_{y\max}}\sum_{n_z=1}^{n_{z\max}} F_0(\eta_{20,12} - U_D) \tag{20.48}$$

Using (20.47), (20.48) and (20.1) we get

$$I_D = \frac{ek_BTg_\nu}{\pi\hbar}\sum_{n_y=1}^{n_{y\max}}\sum_{n_z=1}^{n_{z\max}} [F_0(\eta_{20,12}) - F_0(\eta_{20,12} - U_D)] \tag{20.49}$$

The n_{1D} in this case can be expressed as

$$n_{1D} = \frac{g_\nu}{\pi} \sum_{n_y=1}^{n_{y\max}} \sum_{n_z=1}^{n_{z\max}} \Bigg[[u_{70}(E_{F20,12}, n_y, n_z, \eta_g) + u_{70}((E_{F20,12} - eV_D), n_y, n_z, \eta_g)] + \sum_{r=1}^{s} L(r)[[u_{70}(E_{F20,12}, n_y, n_z, \eta_g) + u_{70}((E_{F20,12} - eV_D), n_y, n_z, \eta_g)]] \Bigg] \tag{20.50}$$

where $u_{70}(E, n_y, n_z, \eta_g)$ has been defined in (14.59) of Chapter 14.

20.2.8 The $I_D - V_D$ equation for ballistic HDNWFETs of $PtSb_2$

From (20.2) in this case we can write

$$I_+ = \frac{ek_BTg_\nu}{\pi\hbar} \sum_{n_y=1}^{n_{y\max}} \sum_{n_z=1}^{n_{z\max}} F_0(\eta_{20,13}) \tag{20.51}$$

where $\eta_{20,13} = (k_BT)^{-1}[(E_{F20,13} - E'_{34HDNW})]$, $E_{F20,13}$ is the Fermi energy in this case and E'_{34HDNW} is the sub-band energy in the case which is defined in (14.66) of Chapter 14.

From (20.51) we can write that

$$I_- = \frac{ek_BTg_\nu}{\pi\hbar} \sum_{n_y=1}^{n_{y\max}} \sum_{n_z=1}^{n_{z\max}} F_0(\eta_{20,13} - U_D) \tag{20.52}$$

Using (20.51), (20.52) and (20.1) we get

$$I_D = \frac{ek_BTg_\nu}{\pi\hbar} \sum_{n_y=1}^{n_{y\max}} \sum_{n_z=1}^{n_{z\max}} [F_0(\eta_{20,13}) - F_0(\eta_{20,13} - U_D)] \tag{20.53}$$

The n_{1D} in this case can be expressed as

$$n_{1D} = \frac{g_\nu}{\pi} \sum_{n_y=1}^{n_{y\max}} \sum_{n_z=1}^{n_{z\max}} \Bigg[[V_{70}(E_{F20,13}, n_y, n_z, \eta_g) + V_{70}((E_{F20,13} - eV_D), n_y, n_z, \eta_g)] + \sum_{r=1}^{s} L(r)[[V_{70}(E_{F20,13}, n_y, n_z, \eta_g) + V_{70}((E_{F20,13} - eV_D), n_y, n_z, \eta_g)]] \Bigg] \tag{20.54}$$

where $V_{70}(E, n_y, n_z, \eta_g)$ has been defined in (14.64) of Chapter 14.

20.2.9 The $I_D - V_D$ equation for ballistic HDNWFETs of bismuth telluride

From (20.2) in this case we can write

$$I_+ = \frac{ek_BTg_\nu}{\pi\hbar}\sum_{n_y=1}^{n_{y\max}}\sum_{n_z=1}^{n_{z\max}} F_0(\eta_{20,14}) \tag{20.55}$$

where $\eta_{20,14} = (k_BT)^{-1}[(E_{F20,14} - E'_{50HDNW})]$, $E_{F20,14}$ is the Fermi energy in this case and E'_{50HDNW} is the sub-band energy in the case which is defined in (14.71) of Chapter 14.

From (20.55) we can write that

$$I_- = \frac{ek_BTg_\nu}{\pi\hbar}\sum_{n_y=1}^{n_{y\max}}\sum_{n_z=1}^{n_{z\max}} F_0(\eta_{20,14} - U_D) \tag{20.56}$$

Using (20.55), (20.56) and (20.1) we get

$$I_D = \frac{ek_BTg_\nu}{\pi\hbar}\sum_{n_y=1}^{n_{y\max}}\sum_{n_z=1}^{n_{z\max}} [F_0(\eta_{20,14}) - F_0(\eta_{20,14} - U_D)] \tag{20.57}$$

The n_{1D} in this case can be expressed as

$$n_{1D} = \frac{g_\nu}{\pi}\sum_{n_y=1}^{n_{y\max}}\sum_{n_z=1}^{n_{z\max}}\Bigg[[J_{70}(E_{F20,14}, n_y, n_z, \eta_g) + J_{70}((E_{F20,14} - eV_D), n_y, n_z, \eta_g)] + \sum_{r=1}^{s} L(r)[[J_{70}(E_{F20,14}, n_y, n_z, \eta_g) + J_{70}((E_{F20,14} - eV_D), n_y, n_z, \eta_g)]]\Bigg] \tag{20.58}$$

where $J_{70}(E, n_y, n_z, \eta_g)$ has been defined in (14.69) of Chapter 14.

20.2.10 The $I_D - V_D$ equation for ballistic HDNWFETs of germanium

(i) Model of Cardona *et al.*

From (20.2) in this case we can write

$$I_+ = \frac{ek_BTg_\nu}{\pi\hbar}\sum_{n_y=1}^{n_{y\max}}\sum_{n_z=1}^{n_{z\max}} F_0(\eta_{20,15}) \tag{20.59}$$

where $\eta_{20,15} = (k_BT)^{-1}[(E_{F20,15} - E'_{52HDNW})]$, $E_{F20,15}$ is the Fermi energy in this case and E'_{52HDNW} is the sub-band energy in the case which is defined in (14.76) of Chapter 14.

From (20.59) we can write that

$$I_- = \frac{ek_BTg_\nu}{\pi\hbar} \sum_{n_y=1}^{n_{y\max}} \sum_{n_z=1}^{n_{z\max}} F_0(\eta_{20,15} - U_D) \tag{20.60}$$

Using (20.59), (20.60) and (20.1) we get

$$I_D = \frac{ek_BTg_\nu}{\pi\hbar} \sum_{n_y=1}^{n_{y\max}} \sum_{n_z=1}^{n_{z\max}} [F_0(\eta_{20,15}) - F_0(\eta_{20,15} - U_D)] \tag{20.61}$$

The n_{1D} in this case can be expressed as

$$\begin{aligned} n_{1D} = \frac{g_\nu}{\pi} \sum_{n_y=1}^{n_{y\max}} \sum_{n_z=1}^{n_{z\max}} \Bigg[& [L_{70}(E_{F20,15}, n_y, n_z, \eta_g) \\ & + L_{70}((E_{F20,15} - eV_D), n_y, n_z, \eta_g)] \\ & + \sum_{r=1}^{s} L(r)[[L_{70}(E_{F20,15}, n_y, n_z, \eta_g) \\ & + L_{70}((E_{F20,15} - eV_D), n_y, n_z, \eta_g)]] \Bigg] \end{aligned} \tag{20.62}$$

where $L_{70}(E, n_y, n_z, \eta_g)$ has been defined in (14.74) of Chapter 14.

(ii) Model of Wang *et al.*

From (20.2) in this case we can write

$$I_+ = \frac{ek_BTg_\nu}{\pi\hbar} \sum_{n_y=1}^{n_{y\max}} \sum_{n_z=1}^{n_{z\max}} F_0(\eta_{20,16}) \tag{20.63}$$

where $\eta_{20,16} = (k_BT)^{-1}[(E_{F20,16} - E'_{54HDNW})]$, $E_{F20,16}$ is the Fermi energy in this case and E'_{54HDNW} is the sub-band energy in the case which is defined in (14.81) of Chapter 14.

From (20.63) we can write that

$$I_- = \frac{ek_BTg_\nu}{\pi\hbar} \sum_{n_y=1}^{n_{y\max}} \sum_{n_z=1}^{n_{z\max}} F_0(\eta_{20,16} - U_D) \tag{20.64}$$

Using (20.63), (20.64) and (20.1) we get

$$I_D = \frac{ek_BTg_\nu}{\pi\hbar} \sum_{n_y=1}^{n_{y\max}} \sum_{n_z=1}^{n_{z\max}} [F_0(\eta_{20,16}) - F_0(\eta_{20,16} - U_D)] \tag{20.65}$$

The n_{1D} in this case can be expressed as

$$n_{1D} = \frac{g_\nu}{\pi} \sum_{n_y=1}^{n_{y\max}} \sum_{n_z=1}^{n_{z\max}} \Bigg[[\beta_{70}(E_{F20,16}, n_y, n_z, \eta_g) + \beta_{70}((E_{F20,16} - eV_D), n_y, n_z, \eta_g)] + \sum_{r=1}^{s} L(r)[[\beta_{70}(E_{F20,16}, n_y, n_z, \eta_g) + \beta_{70}((E_{F20,16} - eV_D), n_y, n_z, \eta_g)]] \Bigg] \tag{20.66}$$

where $\beta_{70}(E, n_y, n_z, \eta_g)$ has been defined in (14.79) of Chapter 14.

20.2.11 The $I_D - V_D$ equation for ballistic HDNWFETs of gallium antimonide

From (20.2) in this case we can write

$$I_+ = \frac{ek_BTg_\nu}{\pi\hbar} \sum_{n_y=1}^{n_{y\max}} \sum_{n_z=1}^{n_{z\max}} F_0(\eta_{20,17}) \tag{20.67}$$

where $\eta_{20,17} = (k_BT)^{-1}[(E_{F20,17} - E'_{100HDNW})]$, $E_{F20,17}$ is the Fermi energy in this case and $E'_{100HDNW}$ is the sub-band energy in the case which is defined in (14.86) of Chapter 14.

From (20.67) we can write that

$$I_- = \frac{ek_BTg_\nu}{\pi\hbar} \sum_{n_y=1}^{n_{y\max}} \sum_{n_z=1}^{n_{z\max}} F_0(\eta_{20,17} - U_D) \tag{20.68}$$

Using (20.67), (20.68) and (20.1) we get

$$I_D = \frac{ek_BTg_\nu}{\pi\hbar} \sum_{n_y=1}^{n_{y\max}} \sum_{n_z=1}^{n_{z\max}} [F_0(\eta_{20,17}) - F_0(\eta_{20,17} - U_D)] \tag{20.69}$$

The n_{1D} in this case can be expressed as

$$n_{1D} = \frac{g_\nu}{\pi} \sum_{n_y=1}^{n_{y\max}} \sum_{n_z=1}^{n_{z\max}} \Bigg[[N_{100}(E_{F20,17}, n_y, n_z, \eta_g) + N_{100}((E_{F20,17} - eV_D), n_y, n_z, \eta_g)] + \sum_{r=1}^{s} L(r)[[N_{100}(E_{F20,17}, n_y, n_z, \eta_g) + N_{100}((E_{F20,17} - eV_D), n_y, n_z, \eta_g)]] \Bigg] \quad (20.70)$$

where $N_{100}(E, n_y, n_z, \eta_g)$ has been defined in (14.85) of Chapter 14.

20.2.12 *The $I_D - V_D$ equation for ballistic HDNWFETs of II–V materials*

From (20.2) in this case we can write

$$I_+ = \frac{ek_BTg_\nu}{2\pi\hbar} \sum_{n_x=1}^{n_{x\max}} \sum_{n_y=1}^{n_{y\max}} F_0(\eta_{20,18}) \quad (20.71)$$

where $\eta_{20,18} = (k_BT)^{-1}[(E_{F20,18} - E_{200_{HDNW}})]$, $E_{F20,17}$ is the Fermi energy in this case and $E_{200_{HDNW}}$ is the sub-band energy in the case which is defined in (14.94) of Chapter 14.

From (20.71) we can write that

$$I_- = \frac{ek_BTg_\nu}{2\pi\hbar} \sum_{n_x=1}^{n_{x\max}} \sum_{n_y=1}^{n_{y\max}} F_0(\eta_{20,18} - U_D) \quad (20.72)$$

Using (20.71), (20.72) and (20.1) we get

$$I_D = \frac{ek_BTg_\nu}{2\pi\hbar} \sum_{n_x=1}^{n_{x\max}} \sum_{n_y=1}^{n_{y\max}} [F_0(\eta_{20,18}) - F_0(\eta_{20,18} - U_D)] \quad (20.73)$$

The n_{1D} in this case can be expressed as

$$n_{1D} = \frac{g_\nu}{2\pi} \sum_{n_y=1}^{n_{y\max}} \sum_{n_z=1}^{n_{z\max}} \Bigg[[\Delta_{27}(E_{F20,18}, n_x, n_y, \eta_g) + \Delta_{27}((E_{F20,18} - eV_D), n_x, n_y, \eta_g)]$$

$$+\sum_{r=1}^{s} L(r)[[\Delta_{27}(E_{F20,18}, n_x, n_y, \eta_g) + \Delta_{27}((E_{F20,18} - eV_D), n_x, n_y, \eta_g)]]\Bigg] \tag{20.74}$$

where $\Delta_{27}(E, n_x, n_y, \eta_g)$ has been defined in (14.92) of Chapter 14.

20.3 Summary and Conclusion

In this chapter, in Section 20.2.1, we have formulated $I_D - V_D$ for ballistic HDNWFETs made of nonlinear optical materials as given by (20.5) for the computation of which we need the expression of the 1D electron statistics which is given in (20.6). The $I_D - V_D$ equations for HDNWFETs made of III–V, ternary and quaternary materials has been investigated in Section 20.2.2 in accordance with the three-band model of Kane, two-band model of Kane, Parabolic energy bands, model of Stillman *et al.* and the model of Palik *et al.* are given by (20.9), (20.13), (20.17), (20.21) and (20.25) and the corresponding electron statistics are expressed through (20.10), (20.14), (20.18), (20.22) and (20.26), respectively. In Section 20.2.3, we have studied the $I_D - V_D$ equation for HDNWFETs of II–VI materials which is given by (20.29) and the corresponding electron statistics has been represented in (20.30). In Section 20.2.4, we have studied the $I_D - V_D$ equation in HDNWFETs of IV–VI materials in accordance with the models of Dimmock and Bangert and Kastner which are given by (20.33) and (20.37) and the corresponding electron statistics have been represented in (20.34) and (20.38), respectively. In Section 20.2.5, we have investigated the $I_D - V_D$ equation in HDNWFETs of stressed Kane type materials which is given by (20.41) and the corresponding electron statistics has been represented in (20.42). In Section 20.2.6, we have studied the $I_D - V_D$ equation in HDNWFETs of Te which is given by (20.45) and the corresponding electron statistics has been represented in (20.46). In Section 20.2.7, we have studied the $I_D - V_D$ equation in HDNWFETs of GaP which is given by (20.49) and the corresponding electron statistics has been represented in (20.50). In Section 20.2.8, we have investigated the $I_D - V_D$ equation in HDNWFETs of $PtSb_2$ which is given by (20.53) and the corresponding electron statistics has been represented in (20.54). In Section 20.2.9, we have studied the $I_D - V_D$ equation in HDNWFETs of bismuth telluride materials which is given by (20.57) and the corresponding electron statistics has been represented in (20.58). In Section 20.2.10, we have studied the $I_D - V_D$ equation in HDNWFETs of Germanium in accordance with the models of Cardona *et al.* and Wang *et al.* which are given by (20.61) and (20.65) and the corresponding electron statistics have been represented in (20.62) and (20.66), respectively. In Section 20.2.11, we have studied the $I_D - V_D$ equation in HDNWFETs of gallium antimonide which is given by (20.69) and the corresponding electron statistics has been represented in (20.70). In Section 20.2.12, we have investigated the $I_D - V_D$ equation in HDNWFETs of class II–V materials which is given by (20.73) and the corresponding electron statistics has been represented in (20.74).

We have not given the graphs of the $I_D - V_D$ equations for ballistic HDNWFETs made of many materials as discussed here since we feel that our readers will perform the intricate numerical computation to obtain the oscillatory plots for better assessment and joy of understanding. It is the band structure that changes in a fundamental way and consequently all the physical properties of all the electronic materials changes radically leading to new physical concepts.

20.4 Open Research Problem

(R20.1) Derive $I_D - V_D$ equations for ballistic HDNWFETs made of all the materials whose DRs are given in (R1.1). Draw the corresponding graphs and explain the results.

Chapter 21

$I_D - V_D$ Equation for Ballistic MOSFETs

Time is the ultimate judge. If you give time to time, then time can tell you whether you are right or wrong.

21.1 Introduction

This is completely a new chapter where from Sections 21.2.1 up to 21.2.11 we study the $I_D - V_D$ equation in MOSFETs of tetragonal and nonlinear optical, III–V, ternary and quaternary, II–VI, IV–VI, stressed Kane type, Bi_2Te_3, Ge, GaAs, GaSb, and InSb materials, respectively. In Section 21.3, we present summary and conclusion and last Section 21.4 we offer one multidimensional open research problem.

21.2 Theoretical Background

21.2.1 *The $I_D - V_D$ equation for Ballistic MOSFETs of tetragonal and nonlinear optical materials*

The current density can be expressed as

$$J_D = W[J_+(E_F) - J_-(E_F - eV_D)] \tag{21.1}$$

where W is the width,

$$J_+(E_F) = \frac{eg_v}{\pi} \sum_{i=0}^{i_{\max}} \int_{E_i}^{\infty} \frac{V_x N_{2D_i}(E)dE}{[1 + \exp(\frac{E-E_F}{k_B T})]} \tag{21.2}$$

and

$$J_-(E_F - eV_D) = \frac{eg_v}{\pi} \sum_{i=0}^{i_{\max}} \int_{E_i}^{\infty} \frac{V_x N_{2D_i}(E)dE}{[1 + \exp(\frac{E-E_F+eV_D}{k_B T})]} \tag{21.3}$$

where E_i is the sub-band energy of the i-th sub-band, $N_{2D_i}(E)$ is the 2D density-of-states function per sub-band and V_x is the velocity along z-direction.

Therefore, it appears that the determination of J_D, requires an equation for E_F which can be determine form

$$n_s^+(E_F) + n_s^-(E_F - eV_D) = \frac{C_{ins}}{e}(V_T - V_{GS}) \tag{21.4}$$

where for a given device C_{ins}, V_T and V_{GS} are known quantities.

From (21.4) we can write that

$$\frac{g_v}{(2\pi)^2}\sum_{i=0}^{i_{\max}}\left[[(A(E_F) + A(E_F - eV_D)] + \sum_{r=1}^{s} L(r)[(A(E_F) + A(E_F - eV_D)]\right] = \frac{C_{ins}}{e}(V_T - V_{GS}) \tag{21.5}$$

(i) Weak electric field limit

From (1.6) we can write

$$k_x = g_{21,1}(E) \tag{21.6}$$

where

$$g_{21,1}(E) = \left[\frac{\gamma(E)}{f_1(E)}\right]^{1/2}$$

From (21.6) we can write

$$V_x = \frac{1}{\hbar}\frac{\partial E}{\partial k_x} = [\hbar g'_{21,1}(E)]^{-1} \tag{21.7}$$

Therefore, using (21.1), (21.2), (21.3) and (1.12) we get

$$(J_D/W) = \frac{eg_v}{2\pi^2\hbar}\sum_{i=0}^{i_{\max}}\left[\int_{E_{niw1}}^{\infty}\left[\frac{G_7(E,i)}{g'_{21,1}(E)}\right]\left[\frac{1}{1+\exp(\frac{E-E_{F21,1}}{k_BT})} - \frac{1}{1+\exp\left(\frac{E-E_{F21,1}+eV_D}{k_BT}\right)}\right]dE\right] \tag{21.8}$$

where $E_{F21,1}$ is the Fermi energy in the present case.

The $E_{F21,1}$ should be determined from the following equation

$$\frac{g_v}{4\pi}\sum_{i=0}^{i_{\max}}\left[\left[\frac{\psi_1(E_{F21,1})-Q_7(E_{F21,1},i)}{P_7(E_{F21,1},i)} + \frac{\psi_1(E_{F21,1}-eV_D)-Q_7(E_{F21,1}-eV_D,i)}{P_7(E_{F21,1}-eV_D,i)}\right] + \sum_{r=1}^{s}L(r)\left[\frac{\psi_1(E_{F21,1})-Q_7(E_{F21,1},i)}{P_7(E_{F21,1},i)} + \frac{\psi_1(E_{F21,1}-eV_D)-Q_7(E_{F21,1}-eV_D,i)}{P_7(E_{F21,1}-eV_D,i)}\right]\right] = \frac{C_{ins}}{e}(V_T-V_{GS}) \tag{21.9}$$

(ii) Strong electric field limit

Therefore, using (21.1), (21.2), (21.3) and (1.56) we get

$$(J_D/W) = \frac{eg_v}{2\pi^2\hbar}\sum_{i=0}^{i_{\max}}\left[\int_{E_{niw1}}^{\infty}\left[\frac{P_2(E,i)}{g'_{21,1}(E)}\right] \times\left[\frac{1}{1+\exp\left(\frac{E-E_{F21,5}}{k_BT}\right)} - \frac{1}{1+\exp\left(\frac{E-E_{F21,5}+eV_D}{k_BT}\right)}\right]dE\right] \tag{21.10}$$

where $E_{F21,5}$ is the Fermi energy in the present case.

The $E_{F21,5}$ should be determined from the following equation:

$$\frac{g_v}{4\pi}\sum_{i=0}^{i_{\max}}\left[[P_2(E_{F21,5},i)+P_2((E_{F21,5}-eV_D),i)] + \sum_{r=1}^{s}L(r)[P_2(E_{F21,5},i)+P_2((E_{F21,5}-eV_D),i)]\right] = \frac{C_{ins}}{e}(V_T-V_{GS}) \tag{21.11}$$

21.2.2 The $I_D - V_D$ equation for Ballistic MOSFETs of III–V, ternary and quaternary materials

Special Cases:

(i) Three-band model of Kane

(a) Weak electric field limit

Using (1.23) we can write

$$V_x = \left[\sqrt{\frac{2}{m_c}}\right][I'_{11}(E)]^{-1}[\sqrt{I_{11}(E)}] \tag{21.12}$$

Therefore, using (21.1), (21.2), (21.3) and (1.26) we get

$$(J_D/W) = \frac{eg_v}{\pi^2\hbar^2}\sqrt{2m_c}\sum_{i=0}^{i_{\max}}\left[\int_{E_{niw2}}^{\infty}(P_3(E,i))\left[\frac{\sqrt{I_{11}(E)}}{I'_{11}(E)}\right]\right.$$
$$\left.\times\left[\frac{1}{1+\exp\left(\frac{E-E_{F21,2}}{k_BT}\right)}-\frac{1}{1+\exp\left(\frac{E-E_{F21,2}+eV_D}{k_BT}\right)}\right]dE\right] \tag{21.13}$$

where $E_{F21,2}$ is the Fermi energy in the present case and can be determined from the equation as follows:

$$\frac{g_v m_c}{2\pi\hbar^2}\sum_{i=0}^{i_{\max}}\left[[P_{4w}(E_{F21,2},i)+P_{4w}((E_{F21,2}-eV_D),i)]\right.$$
$$\left.+\sum_{r=1}^{s}L(r)[P_{4w}(E_{F21,2},i)+P_{4w}((E_{F21,2}-eV_D),i)]\right]$$
$$=\frac{C_{ins}}{e}(V_T-V_{GS}) \tag{21.14}$$

(b) Two-band model of Kane

Using (1.33a) we can write

$$V_x = \left[\sqrt{\frac{2}{m_c}E(1+\alpha E)}\right][1+2\alpha E]^{-1} \tag{21.15}$$

Therefore, using (21.1), (21.2), (21.3) and (1.36) we get

$$(J_D/W) = \frac{eg_v}{\pi^2\hbar^2}\sqrt{2m_c}\sum_{i=0}^{i_{\max}}\left[\int_{E_{niw3}}^{\infty}\left[\left[1+2\alpha E-\frac{4\alpha}{3}S_i\left[\frac{\hbar\,|e|\,F_s}{\sqrt{2m_c}}\right]^{2/3}\right.\right.\right.$$

$$\left.\times(1+2\alpha E)^{-1/3}\right]\left[\frac{\sqrt{E(1+\alpha E)}}{(1+2\alpha E)}\right]$$

$$\left.\times\left[\frac{1}{1+\exp\left(\frac{E-E_{F21,3}}{k_BT}\right)}-\frac{1}{1+\exp\left(\frac{E-E_{F21,3}+eV_D}{k_BT}\right)}\right]dE\right] \tag{21.16}$$

where $E_{F21,3}$ is the Fermi energy in the present case and can be determined from the equation as follows:

$$\frac{g_v m_c}{2\pi\hbar^2}\sum_{i=0}^{i_{\max}}\left[[P_{5w}(E_{F21,3},i)+P_{5w}((E_{F21,3}-eV_D),i)]\right.$$

$$\left.+\sum_{r=1}^{s}L(r)[P_{5w}(E_{F21,3},i)+P_{5w}((E_{F21,3}-eV_D),i)]\right]$$

$$=\frac{C_{ins}}{e}(V_T-V_{GS}) \tag{21.17}$$

(c) Parabolic band model

In this case we can write we can write

$$V_x = \left[\sqrt{\frac{2}{m_c}E}\right] \tag{21.18}$$

Therefore, using (21.1), (21.2), (21.3) and (1.48) we get

$$(J_D/W) = \frac{eg_v}{\pi^2\hbar^2}\sqrt{2m_c}\sum_{i=0}^{i_{\max}}\left[\int_{E_{niw4}}^{\infty}[\sqrt{E}]\right.$$

$$\left.\times\left[\frac{1}{1+\exp\left(\frac{E-E_{F21,4}}{k_BT}\right)}-\frac{1}{1+\exp\left(\frac{E-E_{F21,4}+eV_D}{k_BT}\right)}\right]dE\right] \tag{21.19}$$

where $E_{F21,4}$ is the Fermi energy in the present case and can be determined from the equation as follows:

$$
\begin{aligned}
n_s = \frac{m_c g_v}{2\pi\hbar^2}\sum_{i=0}^{i_{\max}}\Bigg[\Bigg[E_{F21,4} - S_i\left[\frac{\hbar\,|e|\,F_s}{\sqrt{2m_c}}\right]^{2/3}\Bigg]
&+ \Bigg[E_{F21,4} - eV_D - S_i\left[\frac{\hbar\,|e|\,F_s}{\sqrt{2m_c}}\right]^{2/3}\Bigg]\\
&+ \sum_{r=1}^{s} L(r)\Bigg[\Bigg[E_{F21,4} - S_i\left[\frac{\hbar\,|e|\,F_s}{\sqrt{2m_c}}\right]^{2/3}\Bigg]\\
&+ \Bigg[E_{F21,4} - eV_D - S_i\left[\frac{\hbar\,|e|\,F_s}{\sqrt{2m_c}}\right]^{2/3}\Bigg]\\
= \frac{C_{ins}}{e}(V_T - V_{GS})
\end{aligned}
\tag{21.20}
$$

(ii) Strong electric field limit

(a) Three-band model of Kane

Therefore, using (21.1), (21.2), (21.3) and (1.65) we get

$$
\begin{aligned}
(J_D/W) = \frac{eg_v}{\pi^2\hbar^2}\sqrt{2m_c}\sum_{i=0}^{i_{\max}}\Bigg[\int_{E_{nis2}}^{\infty}\left(P_5(E,i)\left[\frac{\sqrt{I_{11}(E)}}{I'_{11}(E)}\right]\right)\\
\times\left[\frac{1}{1+\exp\left(\frac{E-E_{F21,6}}{k_BT}\right)} - \frac{1}{1+\exp\left(\frac{E-E_{F21,6}+eV_D}{k_BT}\right)}\right] dE\Bigg]
\end{aligned}
\tag{21.21}
$$

where $E_{F21,6}$ is the Fermi energy in the present case and can be determined from the equation as follows:

$$
\begin{aligned}
\frac{g_v m_c}{2\pi\hbar^2}\Bigg[\sum_{i=0}^{i_{\max}}[P_{6s}(E_{F21,6},i) + P_{6s}((E_{F21,6} - eV_D),i)]\\
+ \sum_{r=1}^{s} L(r)[P_{6s}(E_{F21,6},i) + P_{6s}((E_{F21,6} - eV_D),i)]\Bigg]\\
= \frac{C_{ins}}{e}(V_T - V_{GS})
\end{aligned}
\tag{21.22}
$$

(b) Two-band model of Kane

Therefore, using (21.1), (21.2), (21.3) and (1.75) we get

$$(J_D/W) = \frac{eg_v}{\pi^2\hbar^2}\sqrt{2m_c}\sum_{i=0}^{i_{\max}}\left[\int_{E_{n_{is5}}}^{\infty}[\sqrt{E(1+\alpha E)}]\right.$$

$$\left.\times\left[\frac{1}{1+\exp\left(\frac{E-E_{F21,7}}{k_BT}\right)} - \frac{1}{1+\exp\left(\frac{E-E_{F21,7}+eV_D}{k_BT}\right)}\right]dE\right] \tag{21.23a}$$

where $E_{F21,7}$ is the Fermi energy in the present case and can be determined from the equation as follows:

$$\frac{g_v m_c}{2\pi\hbar^2}\sum_{i=0}^{i_{\max}}\left[[\bar{P}_5(E_{F21,7},i) + \bar{P}_5((E_{F21,7} - eV_D),i)]\right.$$

$$\left. + \sum_{r=1}^{s} L(r)[\bar{P}_5(E_{F21,7},i) + \bar{P}_5(E_{F21,7} - eV_D),i)]\right]$$

$$= \frac{C_{ins}}{e}(V_T - V_{GS}) \tag{21.23b}$$

where

$$\bar{P}_5(E_{F21,7},i) = \left[E_{F21,7}(1+\alpha E_{F21,7}) - \frac{2\sqrt{2}}{3}(S_i)^{3/2}\left[\frac{\hbar\,|e|\,F_s}{\sqrt{2m_cE_g}}\right]\right]$$

21.2.3 *The $I_D - V_D$ equation for Ballistic MOSFETs of II–VI materials*

From (2.1) we can write

$$V_x = \left[\sqrt{C_0^2 + \frac{2E\hbar^2}{m_\perp^*}}\right](\hbar)^{-1} \tag{21.24}$$

Using (21.1), (21.2), (21.3), (21.24), (2.4a) and (2.4b) we can write

$$\frac{J_D}{W} = \frac{2g_v m_\perp^* e}{\pi^2\hbar^3}\sum_{i=0}^{i_{\max}}\int_{E_{i1}}^{\infty}\left[\left[C_0^2 + \frac{2E\hbar^2}{m_\perp^*}\right]^{1/2}\left[\frac{1}{1+\exp\left(\frac{E-E_{F21,8}}{k_BT}\right)}\right.\right.$$

$$\left.\left. - \frac{1}{1+\exp\left(\frac{E-E_{F21,8}+eV_D}{k_BT}\right)}\right]\right]dE \tag{21.25}$$

where $E_{F21,8}$ is the Fermi energy in the present case and can be determined from the equation as follows:

$$\frac{C_{ins}}{e}(V_T - V_{GS}) = \frac{1}{2}N_{c2D}\sum_{i=0}^{i_{\max}}[F_0(\bar{\eta}_{i21}) + F_0(\theta_{21,1})] \tag{21.26}$$

where

$$\theta_{21,1} = \bar{\eta}_{i21} - \frac{eV_D}{k_BT}, \quad \bar{\eta}_{i21} = \frac{E_{F21,8} - E_{i1}}{k_BT}$$

21.2.4 *The $I_D - V_D$ equation for Ballistic MOSFETs of IV–VI materials*

(i) Weak electric field limit

From (2.7) we can write

$$V_x = \sqrt{\frac{2}{m_1}}\left[\frac{\sqrt{E(1+\alpha E)}}{1+2\alpha E}\right] \tag{21.27}$$

Using (21.1), (21.2), (21.3), (21.27), and (2.13) we can write

$$\frac{J_D}{W} = \left(\frac{g_v e}{2\pi^3}\right)\sqrt{\frac{2}{m_1}}\sum_{i=0}^{i_{\max}}\int_{E_{n_{iw7}}}^{\infty}\left[[R_{71}(E,i)]\left[\frac{\sqrt{E(1+\alpha E)}}{(1+2\alpha E)}\right]\right.$$
$$\left.\times\left[\frac{1}{1+\exp\left(\frac{E-E_{F21,9}}{k_BT}\right)} - \frac{1}{1+\exp\left(\frac{E-E_{F21,9}+eV_D}{k_BT}\right)}\right]\right]dE \tag{21.28}$$

where $E_{F21,9}$ is the Fermi energy in this case and can be expressed through the following equation:

$$\frac{C_{ins}}{e}(V_T - V_{GS}) = \frac{g_v}{3\pi^2}\left(\frac{r_{71}}{p_{71}}\right)^{1/2}\left[\sum_{i=0}^{i_{\max}}[P_{7w}(E_{F21,9},i)\right.$$
$$+ P_{7w}((E_{F21,9} - eV_D),i)] + \sum_{r=1}^{s}L(r)[P_{7w}(E_{F21,9},i)$$
$$\left. + P_{7w}((E_{F21,9} - eV_D),i)]\right] \tag{21.29}$$

(ii) Strong electric field limit

Using (21.1), (21.2), (21.3) and (2.18e) we get

$$\frac{J_D}{W} = \left(\frac{g_v e}{2\pi^3}\right)\sqrt{\frac{2}{m_1}}\sum_{i=0}^{i_{\max}}\int_{\bar{E}_{n_{is8}}}^{\infty}\left[R_{72}(E,i)\left[\frac{\sqrt{E(1+\alpha E)}}{(1+2\alpha E)}\right]\right.$$
$$\left.\times\left[\frac{1}{1+\exp\left(\frac{E-E_{F21,10}}{k_BT}\right)} - \frac{1}{1+\exp\left(\frac{E-E_{F21,10}+eV_D}{k_BT}\right)}\right]\right]dE \tag{21.30}$$

where $E_{F21,10}$ is the Fermi energy in this case and can be expressed through the following equation:

$$\frac{C_{ins}}{e}(V_T - V_{GS})$$
$$= \frac{g_v}{3\pi^2}\left(\frac{r_{72}}{p_{72}}\right)^{1/2}\left\{\sum_{i=0}^{i_{\max}}[P_{8s}(E_{F21,10},i) + P_{8s}((E_{F21,10} - eV_D),i)]\right\}$$
$$+\sum_{r=1}^{s}L(r)[P_{8s}(E_{F21,10},i) + P_{8s}((E_{F21,10} - eV_D),i)]] \tag{21.31}$$

21.2.5 *The $I_D - V_D$ equation for Ballistic MOSFETs of stressed Kane type materials*

(i) Weak electric field limit

From (2.21) we get

$$V_x = [\hbar[\bar{a}_0(E)]']^{-1} \tag{21.32}$$

Using (21.1), (21.2), (21.3), (21.32) and (2.27) we get

$$\frac{J_D}{W} = \frac{g_v e}{2\pi^2\hbar}\sum_{i=0}^{i_{\max}}\int_{E_{n_{iw8}}}^{\infty}\left[\frac{L_{47}(E,i)}{[\bar{a}_0(E)]'}\right]$$
$$\times\left[\left[\frac{1}{1+\exp\left(\frac{E-E_{F21,11}}{k_BT}\right)} - \frac{1}{1+\exp\left(\frac{E-E_{F21,11}+eV_D}{k_BT}\right)}\right]\right]dE \tag{21.33}$$

where $E_{F21,11}$ is the Fermi energy in this case and can be expressed through the following equation:

$$\frac{C_{ins}}{e}(V_T - V_{GS}) = \frac{g_v}{4\pi}\sum_{i=0}^{i_{\max}}[P_{8w}(E_{F21,11}, i) + P_{8w}((E_{F21,11} - eV_D), i)] + \sum_{r=1}^{s} L(r)[P_{8w}(E_{F21,11}, i) + P_{8w}((E_{F21,11} - eV_D), i)]] \tag{21.34}$$

(ii) Strong electric field limit

Using (21.1), (21.2), (21.3), (21.32) and (2.35) we get

$$\frac{J_D}{W} = \frac{g_v e}{2\pi^2\hbar}\sum_{i=0}^{i_{\max}}\int_{E_{n_{is8}}}^{\infty}\left[\frac{T_{147}(E,i)}{[\bar{a}_0(E)]'}\right]\left[\left[\frac{1}{1+\exp\left(\frac{E-E_{F21,12}}{k_B T}\right)} - \frac{1}{1+\exp\left(\frac{E-E_{F21,12}+eV_D}{k_B T}\right)}\right]\right] dE \tag{21.35}$$

where $E_{F21,12}$ is the Fermi energy in this case and can be expressed through the equation:

$$\frac{C_{ins}}{e}(V_T - V_{GS}) = \frac{g_v}{4\pi}\left[\sum_{i=0}^{i_{\max}}[P_{9s}(E_{F21,12}, i) + P_{9s}((E_{F21,12} - eV_D), i)] + \sum_{r=1}^{s} L(r)[P_{9s}(E_{F21,12}, i) + P_{9s}((E_{F21,12} - eV_D), i)]\right] \tag{21.36}$$

21.2.6 *The $I_D - V_D$ equation for Ballistic MOSFETs of bismuth telluride*

(i) Weak electric field limit

From (2.39) we can write

$$V_x = \left[\frac{2\sqrt{I_1 E(1+\alpha E)}}{\hbar(1+2\alpha E)}\right] \tag{21.37}$$

Using (21.1), (21.2), (21.3) and (2.41) we can write

$$\frac{J_D}{W} = \frac{g_v e}{\pi^2 \hbar}(\sqrt{I_1})(I_2 I_3 - I_4^2)^{-1/2} \sum_{i=0}^{i_{\max}} \int_{E_{46}}^{\infty} \left[T'_{46}(E,i)\left(\frac{\sqrt{E(1+\alpha E)}}{(1+2\alpha E)}\right)\right]$$
$$\times \left[\left[\frac{1}{1+\exp\left(\frac{E-E_{F21,13}}{k_B T}\right)} - \frac{1}{1+\exp\left(\frac{E-E_{F21,13}+eV_D}{k_B T}\right)}\right]\right] dE \tag{21.38}$$

where $E_{F21,13}$ is the Fermi energy in this case and can be expressed through the following equation:

$$\frac{C_{ins}}{e}(V_T - V_{GS}) = \frac{g_v}{4\pi}(I_2 I_3 - I_4^2)^{-1/2} \sum_{i=0}^{i_{\max}} [[[E_{F21,13}(1+\alpha E_{F21,13})$$
$$- S_i(eF_S(1+2\alpha E_{F21,13}))^{2/3}(I_1)^{1/3}]$$
$$+ [(E_{F21,13} - eV_D)(1+\alpha(E_{F21,13} - eV_D))$$
$$- S_i(eF_S(1+2\alpha(E_{F21,13} - eV_D))^{2/3}(I_1)^{1/3}]]$$

$$\sum_{r=1}^{s} L(r)[[E_{F21,13}(1+\alpha E_{F21,13}) - S_i(eF_S(1+2\alpha E_{F21,13}))^{2/3}(I_1)^{1/3}]$$
$$+ [(E_{F21,13} - eV_D)(1+\alpha(E_{F21,13} - eV_D))$$
$$- S_i(eF_S(1+2\alpha(E_{F21,13} - eV_D))^{2/3}(I_1)^{1/3}]]] \tag{21.39}$$

(ii) Strong electric field limit

Using (21.1), (21.2), (21.3), (21.37) and (2.44) we can write

$$\frac{J_D}{W} = \frac{g_v e}{\pi^2 \hbar}(\sqrt{I_1})(I_2 I_3 - I_4^2)^{-1/2} \sum_{i=0}^{i_{\max}} \int_{E_{47}}^{\infty} \left[T'_{47}(E,i)\left(\frac{\sqrt{E(1+\alpha E)}}{(1+2\alpha E)}\right)\right]$$
$$\times \left[\left[\frac{1}{1+\exp\left(\frac{E-E_{F21,14}}{k_B T}\right)} - \frac{1}{1+\exp\left(\frac{E-E_{F21,14}+eV_D}{k_B T}\right)}\right]\right] dE \tag{21.40}$$

where $E_{F21,14}$ is the Fermi energy in this case and can be expressed through the following equation:

$$\frac{C_{ins}}{e}(V_T - V_{GS}) = \frac{g_v}{4\pi}(I_2 I_3 - I_4^2)^{-1/2} \sum_{i=0}^{i_{\max}} \left[\left[\left[E_{F21,14}(1+\alpha E_{F21,14}) - \frac{4}{3}S_i^{3/2} eF_s(I_1\alpha)^{1/2}\right] + \left[(E_{F21,14} - eV_D) \times (1+\alpha(E_{F21,14} - eV_D)) - \frac{4}{3}S_i^{3/2} eF_s(I_1\alpha)^{1/2}\right]\right]\right.$$
$$\sum_{r=1}^{s} L(r)\left[\left[E_{F21,14}(1+\alpha E_{F21,14}) - \frac{4}{3}S_i^{3/2} eF_s(I_1\alpha)^{1/2}\right] + \left[(E_{F21,14} - eV_D)(1+\alpha(E_{F21,14} - eV_D)) - \frac{4}{3}S_i^{3/2} eF_s(I_1\alpha)^{1/2}\right]\right]\Bigg] \tag{21.41}$$

21.2.7 The $I_D - V_D$ equation for ballistic MOSFETs of germanium

From (2.47) we can write

$$V_x = \left[\sqrt{\frac{2}{m_\perp^*}} \frac{\sqrt{E(1+\alpha E)}}{(1+2\alpha E)}\right] \tag{21.42}$$

Using (21.1), (21.2), (21.3), (21.42) and (2.51) we can write

$$\frac{J_D}{W} = \frac{e g_v}{(2\pi)^2} \cdot \frac{4\sqrt{m_1 m_2}}{\hbar^2} \sqrt{\frac{2}{m_\perp^*}} \sum_{i=0}^{i_{\max}} \int_{E_{i2}}^{\infty} \left(\frac{\sqrt{E(1+\alpha E)}}{(1+2\alpha E)}\right)[1+2\alpha E - 2\alpha E_{i2}] \times \left[\left[\frac{1}{1+\exp\left(\frac{E-E_{F21,15}}{k_B T}\right)} - \frac{1}{1+\exp\left(\frac{E-E_{F21,15}+eV_D}{k_B T}\right)}\right]\right] dE \tag{21.43}$$

where $E_{F21,15}$ is the Fermi energy in this case and can be expressed through the following equation:

$$\frac{C_{ins}}{e}(V_T - V_{GS}) = \frac{g_v k_B T}{2\pi} \cdot \frac{\sqrt{m_1 m_2}}{\hbar^2} \sum_{i=0}^{i_{\max}} [\tau_i F_0(\bar{\eta}_{4i21}) + 2\alpha k_B T F_1(\bar{\eta}_{4i21})] + [\tau_i F_0(\bar{\eta}_{4i2}) + 2\alpha k_B T F_1(\bar{\eta}_{4i2})] \tag{21.44}$$

where $\bar{\eta}_{4i21} = (E_{F21,15} - E_{i2})(k_B)^{-1}$ and $\bar{\eta}_{4i2} = ((E_{F21,15} - eV_D) - E_{i2}))(k_BT)^{-1}$

21.2.8 *The $I_D - V_D$ equation for ballistic MOSFETs of gallium arsenide*

(a) Rossler model

(i) Weak electric field limit

From (2.57) we can write

$$V_x = \left[\frac{2\sqrt{\gamma_{4,\pm}(E)}}{\hbar\gamma'_{4,\pm}(E)}\right] \tag{21.45}$$

Using (21.1), (21.2), (21.3), (21.45) and (2.57) we can write

$$\frac{J_D}{W} = \frac{2g_ve}{4\pi^2\hbar}\sum_{i=0}^{i_{\max}}\int_{\bar{E}_{31,\pm}}^{\infty}[T'_{31,\pm}(E,i)]\left[\frac{\sqrt{\gamma_{4,\pm}(E)}}{\gamma'_{4,\pm}(E)}\right] \times\left[\left[\frac{1}{1+\exp\left(\frac{E-E_{F21,16}}{k_BT}\right)} - \frac{1}{1+\exp\left(\frac{E-E_{F21,16}+eV_D}{k_BT}\right)}\right]\right]dE \tag{21.46}$$

where $E_{F21,16}$ is the Fermi energy in this case and can be expressed through the following equation:

$$\frac{C_{ins}}{e}(V_T - V_{GS}) = \frac{g_v}{8\pi}\sum_{i=0}^{i_{\max}}\left[[T_{31,\pm}(E_{F21,16},i)] + [T_{31,\pm}((E_{F21,16} - eV_D),i))] + \sum_{r=1}^{s}L(r)[T_{31,\pm}(E_{F21,16},i)] + [T_{31,\pm}((E_{F21,16} - eV_D),i))]\right] \tag{21.47}$$

(ii) Strong electric field limit

Using (21.1), (21.2), (21.3), (21.45) and (2.62) we can write

$$\frac{J_D}{W} = \frac{2g_ve}{4\pi^2\hbar}\sum_{i=0}^{i_{\max}}\int_{\bar{E}_{32,\pm}}^{\infty}[T_{32,\pm}(E,i)]'\left[\frac{\sqrt{\gamma'_{4,\pm}(E)}}{\gamma'_{4,\pm}(E)}\right]\left[\left[\frac{1}{1+\exp\left(\frac{E-E_{F21,17}}{k_BT}\right)} - \frac{1}{1+\exp\left(\frac{E-E_{F21,17}+eV_D}{k_BT}\right)}\right]\right]dE \tag{21.48}$$

where $E_{F21,17}$ is the Fermi energy in this case and can be expressed through the following equation:

$$\frac{C_{ins}}{e}(V_T - V_{GS}) = \frac{g_v}{8\pi}\sum_{i=0}^{i_{\max}}\left[[T_{32,\pm}(E_{F21,17},i)] + [T_{32,\pm}((E_{F21,17} - eV_D),i))] + \sum_{r=1}^{s} L(r)[T_{32,\pm}(E_{F21,17},i)] + [T_{32,\pm}((E_{F21,17} - eV_D),i))]\right] \quad (21.49)$$

(b) Stillman *et al.*

(i) Weak electric field limit

From (2.65) we can write

$$V_x = \left[\frac{2\sqrt{\gamma_5(E)}}{\hbar\gamma_5'(E)}\right] \quad (21.50)$$

Using (21.1), (21.2), (21.3), (21.50) and (2.67) we can write

$$\frac{J_D}{W} = \frac{g_v e}{\pi^2\hbar}\sum_{i=0}^{i_{\max}}\int_{\bar{E}_{33}}^{\infty}[(T_{33}(E,i)]'\left[\frac{\sqrt{\gamma_5(E)}}{\gamma_5'(E)}\right]\left[\left[\frac{1}{1+\exp\left(\frac{E-E_{F21,18}}{k_BT}\right)} - \frac{1}{1+\exp\left(\frac{E-E_{F21,18}+eV_D}{k_BT}\right)}\right]\right]dE \quad (21.51)$$

where $E_{F21,18}$ is the Fermi energy in this case and can be expressed through the following equation:

$$\frac{C_{ins}}{e}(V_T - V_{GS}) = \frac{g_v}{4\pi}\sum_{i=0}^{i_{\max}}\left[[T_{33}(E_{F21,18},i)] + [T_{33}((E_{F21,18} - eV_D),i))] + \sum_{r=1}^{s} L(r)[T_{33}(E_{F21,18},i)] + [T_{33}((E_{F21,18} - eV_D),i))]\right] \quad (21.52a)$$

(ii) Strong electric field limit

Using (21.1), (21.2), (21.3), (21.50) and (2.70) we can write

$$\frac{J_D}{W} = \frac{g_v e}{\pi^2 \hbar} \sum_{i=0}^{i_{\max}} \int_{\bar{E}_{34}}^{\infty} [(T_{34}(E,i)]' \left[\frac{\sqrt{\gamma_5(E)}}{\gamma_5'(E)}\right] \left[\left[\frac{1}{1+\exp\left(\frac{E-E_{F21,19}}{k_B T}\right)} - \frac{1}{1+\exp\left(\frac{E-E_{F21,19}+eV_D}{k_B T}\right)}\right]\right] dE \qquad (21.52b)$$

where $E_{F21,17}$ is the Fermi energy in this case and can be expressed through the following equation:

$$\frac{C_{ins}}{e}(V_T - V_{GS}) = \frac{g_v}{4\pi}\left[\sum_{i=0}^{i_{\max}} [T_{34}(E_{F21,19}, i)] + [T_{34}((E_{F21,19} - eV_D), i))] + \sum_{r=1}^{s} L(r)[T_{34}(E_{F21,19}, i)] + [T_{34}((E_{F21,19} - eV_D), i))]\right] \qquad (21.53)$$

21.2.9 *The $I_D - V_D$ equation for ballistic MOSFETs of gallium antimonide*

(a) Seiler model

(i) Weak electric field limit

From (2.65) we can write

$$V_x = \left[\frac{2\sqrt{\gamma_6(E)}}{\hbar \gamma_6'(E)}\right] \qquad (21.54)$$

Using (21.1), (21.2), (21.3), (21.54) and (2.79) we can write

$$\frac{J_D}{W} = \frac{g_v e}{2\pi^2 \hbar} \sum_{i=0}^{i_{\max}} \int_{\bar{E}_{35}}^{\infty} [T_{35}(E,i)]' \left[\frac{\sqrt{\gamma_6(E)}}{\gamma_6'(E)}\right] \left[\left[\frac{1}{1+\exp\left(\frac{E-E_{F21,20}}{k_B T}\right)} - \frac{1}{1+\exp\left(\frac{E-E_{F21,20}+eV_D}{k_B T}\right)}\right]\right] dE \qquad (21.55)$$

where $E_{F21,20}$ is the Fermi energy in this case and can be expressed through the following equation:

$$\frac{C_{ins}}{e}(V_T - V_{GS}) = \frac{g_v}{8\pi}\sum_{i=0}^{i_{\max}}\left[[T_{35}(E_{F21,20},i)] + [T_{35}((E_{F21,20} - eV_D),i))] + \sum_{r=1}^{s} L(r)[T_{35}(E_{F21,20},i)] + [T_{35}((E_{F21,20} - eV_D),i))]\right]\right] \quad (21.56)$$

(ii) Strong electric field limit

Using (21.1), (21.2), (21.3), (21.54) and (2.82) we can write

$$\frac{J_D}{W} = \frac{g_v e}{\pi^2\hbar}\sum_{i=0}^{i_{\max}}\int_{\bar{E}_{36}}^{\infty}[T_{36}(E,i)]'\left[\frac{\sqrt{\gamma_6(E)}}{\gamma_6'(E)}\right]\left[\left[\frac{1}{1+\exp\left(\frac{E-E_{F21,21}}{k_B T}\right)} - \frac{1}{1+\exp\left(\frac{E-E_{F21,21}+eV_D}{k_B T}\right)}\right]\right] dE \quad (21.57)$$

where $E_{F21,21}$ is the Fermi energy in this case and can be expressed through the following equation:

$$\frac{C_{ins}}{e}(V_T - V_{GS}) = \frac{g_v}{8\pi}\left[\sum_{i=0}^{i_{\max}}[T_{36}(E_{F21,21},i)] + [T_{36}((E_{F21,21} - eV_D),i))] + \sum_{r=1}^{s} L(r)[T_{36}(E_{F21,21},i)] + [T_{36}((E_{F21,21} - eV_D),i))]\right] \quad (21.58)$$

(b) Model of Mathur and Jain *et al.*

(i) Weak electric field limit

From (2.86) we can write

$$V_x = \left[\frac{2\sqrt{\gamma_7(E)}}{\hbar\gamma_7'(E)}\right] \quad (21.59)$$

Using (21.1), (21.2), (21.3), (21.59) and (2.88) we can write

$$\frac{J_D}{W} = \frac{g_v e}{\pi^2 \hbar} \sum_{i=0}^{i_{\max}} \int_{\bar{E}_{37}}^{\infty} [T_{37}(E,i)]' \left[\frac{\sqrt{\gamma_7(E)}}{\gamma_7'(E)}\right] \left[\left[\frac{1}{1+\exp\left(\frac{E-E_{F21,22}}{k_B T}\right)} - \frac{1}{1+\exp\left(\frac{E-E_{F21,22}+eV_D}{k_B T}\right)}\right]\right] dE \tag{21.60}$$

where $E_{F21,22}$ is the Fermi energy in this case and can be expressed through the following equation:

$$\frac{C_{ins}}{e}(V_T - V_{GS}) = \frac{g_v}{4\pi}\left[\sum_{i=0}^{i_{\max}} [T_{37}(E_{F21,22}, i)] + [T_{37}((E_{F21,22} - eV_D), i))] + \sum_{r=1}^{s} L(r)[T_{37}(E_{F21,22}, i)] + [T_{37}((E_{F21,22} - eV_D), i))]\right] \tag{21.61}$$

(ii) Strong electric field limit

Using (21.1), (21.2), (21.3), (21.54) and (2.70) we can write

$$\frac{J_D}{W} = \frac{g_v e}{\pi^2 \hbar} \sum_{i=0}^{i_{\max}} \int_{\bar{E}_{38}}^{\infty} [T_{38}(E,i)]' \left[\frac{\sqrt{\gamma_7(E)}}{\gamma_7'(E)}\right] \left[\left[\frac{1}{1+\exp\left(\frac{E-E_{F21,23}}{k_B T}\right)} - \frac{1}{1+\exp\left(\frac{E-E_{F21,23}+eV_D}{k_B T}\right)}\right]\right] dE \tag{21.62}$$

where $E_{F21,23}$ is the Fermi energy in this case and can be expressed through the following equation:

$$\frac{C_{ins}}{e}(V_T - V_{GS}) = \frac{g_v}{4\pi}\left[\sum_{i=0}^{i_{\max}} [T_{38}(E_{F21,23}, i)] + [T_{38}((E_{F21,23} - eV_D), i))] + \sum_{r=1}^{s} L(r)[T_{38}(E_{F21,23}, i)] + [T_{38}((E_{F21,23} - eV_D), i))]\right] \tag{21.63}$$

(c) Model of Zhang

(i) Weak electric field limit

From (2.95) we can write

$$V_x = \left[\frac{2\sqrt{\gamma_8(E)}}{\hbar\gamma_8'(E)}\right] \tag{21.64}$$

Using (21.1), (21.2), (21.3), (21.64) and (2.97) we can write

$$\frac{J_D}{W} = \frac{g_v e}{\pi^2\hbar}\sum_{i=0}^{i_{\max}}\int_{\bar{E}_{39}}^{\infty}[T_{39}(E,i)]'\left[\frac{\sqrt{\gamma_8(E)}}{\gamma_8'(E)}\right]\left[\left[\frac{1}{1+\exp\left(\frac{E-E_{F21,24}}{k_BT}\right)}\right.\right.$$
$$\left.\left.-\frac{1}{1+\exp\left(\frac{E-E_{F21,24}+eV_D}{k_BT}\right)}\right]\right]dE \tag{21.65}$$

where $E_{F21,24}$ is the Fermi energy in this case and can be expressed through the following equation:

$$\frac{C_{ins}}{e}(V_T - V_{GS})$$
$$= \frac{g_v}{4\pi}\left[\sum_{i=0}^{i_{\max}}[T_{39}(E_{F21,24},i)] + [T_{39}((E_{F21,24}-eV_D),i)]\right.$$
$$\left.+\sum_{r=1}^{s}L(r)[T_{39}(E_{F21,24},i)] + [T_{39}((E_{F21,24}-eV_D),i))]\right] \tag{21.66}$$

(ii) Strong electric field limit

Using (21.1), (21.2), (21.3), (21.54) and (2.100) we can write

$$\frac{J_D}{W} = \frac{g_v e}{\pi^2\hbar}\sum_{i=0}^{i_{\max}}\int_{\bar{E}_{40}}^{\infty}[T_{40}(E,i)]'\left[\frac{\sqrt{\gamma_8(E)}}{\gamma_8'(E)}\right]$$
$$\times\left[\left[\frac{1}{1+\exp\left(\frac{E-E_{F21,25}}{k_BT}\right)} - \frac{1}{1+\exp\left(\frac{E-E_{F21,25}+eV_D}{k_BT}\right)}\right]\right]dE \tag{21.67}$$

where $E_{F21,25}$ is the Fermi energy in this case and can be expressed through the following equation:

$$\frac{C_{ins}}{e}(V_T - V_{GS}) = \frac{g_v}{4\pi}\left[\sum_{i=0}^{i_{\max}}[T_{40}(E_{F21,25},i)] + [T_{40}((E_{F21,25} - eV_D),i)] + \sum_{r=1}^{s} L(r)[T_{40}(E_{F21,25},i)] + [T_{40}((E_{F21,25} - eV_D),i))]\right] \tag{21.68}$$

21.2.10 *The $I_D - V_D$ equation for ballistic MOSFETs of indium antimonide*

(a) Palik *et al.*

(i) Weak electric field limit

From (2.104) we can write

$$V_x = \left[\frac{2\sqrt{\gamma_9(E)}}{\hbar\gamma_9'(E)}\right] \tag{21.69}$$

Using (21.1), (21.2), (21.3), (21.69) and (2.106) we can write

$$\frac{J_D}{W} = \frac{g_v e}{\pi^2\hbar}\sum_{i=0}^{i_{\max}}\int_{E_{401}}^{\infty}[T_{401}(E,i)]'\left[\frac{\sqrt{\gamma_9(E)}}{\gamma_9'(E)}\right] \times \left[\left[\frac{1}{1+\exp\left(\frac{E-E_{F21,26}}{k_BT}\right)} - \frac{1}{1+\exp\left(\frac{E-E_{F21,26}+eV_D}{k_BT}\right)}\right]\right] dE \tag{21.70}$$

where $E_{F21,26}$ is the Fermi energy in this case and can be expressed through the following equation:

$$\frac{C_{ins}}{e}(V_T - V_{GS}) = \frac{g_v}{4\pi}\left[\sum_{i=0}^{i_{\max}}[T_{401}(E_{F21,26},i)] + [T_{401}((E_{F21,26} - eV_D),i))] + \sum_{r=1}^{s} L(r)[T_{401}(E_{F21,26},i)] + [T_{401}((E_{F21,26} - eV_D),i))]\right] \tag{21.71}$$

(ii) Strong electric field limit

Using (21.1), (21.2), (21.3), (21.69) and (2.110) we can write

$$\frac{J_D}{W} = \frac{g_v e}{\pi^2 \hbar} \sum_{i=0}^{i_{\max}} \int_{E_{410}}^{\infty} [T_{410}(E,i)]' \left[\frac{\sqrt{\gamma_9(E)}}{\gamma_9'(E)}\right] \times \left[\left[\frac{1}{1+\exp\left(\frac{E-E_{F21,27}}{k_B T}\right)} - \frac{1}{1+\exp\left(\frac{E-E_{F21,27}+eV_D}{k_B T}\right)}\right]\right] dE \tag{21.72}$$

where $E_{F21,27}$ is the Fermi energy in this case and can be expressed through the following equation:

$$\frac{C_{ins}}{e}(V_T - V_{GS}) = \frac{g_v}{4\pi}\left[\sum_{i=0}^{i_{\max}} [T_{410}(E_{F21,27},i)] + [T_{410}((E_{F21,27}-eV_D),i)] + \sum_{r=1}^{s} L(r)[T_{410}(E_{F21,27},i)] + [T_{410}((E_{F21,27}-eV_D),i)]\right] \tag{21.73}$$

(b) Model of Johnson *et al.*

(i) Weak electric field limit

From (2.114) we can write

$$V_x = \left[\frac{2\sqrt{\gamma_{10}(E)}}{\hbar \gamma_{10}'(E)}\right] \tag{21.74}$$

Using (21.1), (21.2), (21.3), (21.74) and (2.116) we can write

$$\frac{J_D}{W} = \frac{g_v e}{\pi^2 \hbar} \sum_{i=0}^{i_{\max}} \int_{E_{41}}^{\infty} [T_{41}(E,i)]' \left[\frac{\sqrt{\gamma_{10}(E)}}{\gamma_{10}'(E)}\right] \times \left[\left[\frac{1}{1+\exp\left(\frac{E-E_{F21,28}}{k_B T}\right)} - \frac{1}{1+\exp\left(\frac{E-E_{F21,28}+eV_D}{k_B T}\right)}\right]\right] dE \tag{21.75}$$

where $E_{F21,28}$ is the Fermi energy in this case and can be expressed through the following equation:

$$\frac{C_{ins}}{e}(V_T - V_{GS})$$
$$= \frac{g_v}{4\pi}\left[\sum_{i=0}^{i_{\max}}[T_{41}(E_{F21,28}, i)] + [T_{41}((E_{F21,28} - eV_D), i)]\right.$$
$$\left. + \sum_{r=1}^{s} L(r)[T_{41}(E_{F21,28}, i)] + [T_{41}((E_{F21,28} - eV_D), i)]\right] \quad (21.76)$$

(ii) Strong electric field limit

Using (21.1), (21.2), (21.3), (21.74) and (2.120) we can write

$$\frac{J_D}{W} = \frac{g_v e}{\pi^2 \hbar}\sum_{i=0}^{i_{\max}}\int_{E_{42}}^{\infty}[T_{42}(E,i)]'\left[\frac{\sqrt{\gamma_{10}(E)}}{\gamma_{10}'(E)}\right]\left[\left[\frac{1}{1+\exp\left(\frac{E-E_{F21,29}}{k_B T}\right)}\right.\right.$$
$$\left.\left. - \frac{1}{1+\exp\left(\frac{E-E_{F21,29}+eV_D}{k_B T}\right)}\right]\right] dE \quad (21.77)$$

where $E_{F21,29}$ is the Fermi energy in this case and can be expressed through the following equation:

$$\frac{C_{ins}}{e}(V_T - V_{GS})$$
$$= \frac{g_v}{4\pi}\left[\sum_{i=0}^{i_{\max}}[T_{42}(E_{F21,29}, i)] + [T_{42}((E_{F21,29} - eV_D), i)]\right.$$
$$\left. + \sum_{r=1}^{s} L(r)[T_{42}(E_{F21,29}, i)] + [T_{42}((E_{F21,29} - eV_D), i)]\right] \quad (21.78)$$

(iii) Model of Agafonov *et al.*

(a) Weak electric field limit

From (2.124) we can write

$$V_x = \left[\frac{2\sqrt{\gamma_{11}(E)}}{\hbar\gamma_{11}'(E)}\right] \quad (21.79)$$

Using (21.1), (21.2), (21.3), (21.79) and (2.126) we can write

$$\frac{J_D}{W} = \frac{g_v e}{\pi^2 \hbar} \sum_{i=0}^{i_{\max}} \int_{E_{43}}^{\infty} [T_{43}(E,i)]' \left[\frac{\sqrt{\gamma_{11}(E)}}{\gamma'_{11}(E)}\right] \left[\left[\frac{1}{1+\exp\left(\frac{E-E_{F21,30}}{k_B T}\right)} - \frac{1}{1+\exp\left(\frac{E-E_{F21,30}+eV_D}{k_B T}\right)}\right]\right] dE \tag{21.80}$$

where $E_{F21,30}$ is the Fermi energy in this case and can be expressed through the following equation:

$$\frac{C_{ins}}{e}(V_T - V_{GS}) = \frac{g_v}{4\pi}\left[\sum_{i=0}^{i_{\max}} [T_{43}(E_{F21,30},i)] + [T_{43}((E_{F21,30}-eV_D),i)] + \sum_{r=1}^{s} L(r)[T_{43}(E_{F21,30},i)] + [T_{43}((E_{F21,30}-eV_D),i)]\right] \tag{21.81}$$

(b) Strong electric field limit

Using (21.1), (21.2), (21.3), (21.79) and (2.130) we can write

$$\frac{J_D}{W} = \frac{g_v e}{\pi^2 \hbar} \sum_{i=0}^{i_{\max}} \int_{E_{45}}^{\infty} [T_{45}(E,i)]' \left[\frac{\sqrt{\gamma_{11}(E)}}{\gamma'_{11}(E)}\right] \left[\left[\frac{1}{1+\exp\left(\frac{E-E_{F21,31}}{k_B T}\right)} - \frac{1}{1+\exp\left(\frac{E-E_{F21,31}+eV_D}{k_B T}\right)}\right]\right] dE \tag{21.82}$$

where $E_{F21,31}$ is the Fermi energy in this case and can be expressed through the following equation:

$$\frac{C_{ins}}{e}(V_T - V_{GS}) = \frac{g_v}{4\pi}\left[\sum_{i=0}^{i_{\max}} [T_{45}(E_{F21,31},i)] + [T_{45}((E_{F21,31}-eV_D),i)] + \sum_{r=1}^{s} L(r)[T_{45}(E_{F21,31},i)] + [T_{45}((E_{F21,31}-eV_D),i)]\right] \tag{21.83}$$

21.3 Summary and Conclusions

In this chapter, in Section 21.2.1, we have formulated $I_D - V_D$ for ballistic MOSFETs made of nonlinear optical materials for both and strong electric field limits as given by (21.8) and (21.10), respectively and for the computation of which we need the corresponding expression of the 2D electron statistics for both the limits which are given in (21.9) and (21.11), respectively. The $I_D - V_D$ equations for MOSFETs made of III–V, ternary and quaternary materials has been investigated in Section 21.2.2 under weak electric field limit in accordance with the three-band model of Kane, two-band model of Kane together with Parabolic energy bands, are given by (21.13), (21.16) and (21.19) and the corresponding 2D electron statistics are expressed through (21.14), (21.17) and (21.20), respectively. Besides the $I_D - V_D$ equations in this case for strong electric field limit in accordance with the three-band model of Kane and two-band model of Kane are given by (21.21) and (21.23a) and the corresponding 2D electron statistics are expressed through (21.22) and (21.23b), respectively. In Section 21.2.3, we have studied the $I_D - V_D$ equation for ballistic MOSFETs of II–VI materials which is given by (21.25) and the corresponding 2D electron statistics has been represented in (21.26). In Section 21.2.4, we have studied the $I_D - V_D$ equation in MOSFETs of IV–VI materials for both weak and strong electric field limits which are given by (21.28) and (21.30), respectively and the corresponding 2D electron statistics have been represented in (21.29) and (21.31), respectively. In Section 21.2.5, we have investigated the $I_D - V_D$ equation in ballistic MOSFETs of stressed Kane type materials for both weak and strong electric field limits which are given by (21.33) and (21.35), respectively and the corresponding 2D electron statistics have been represented in (21.34) and (21.36), respectively. In Section 21.2.6, we have studied the $I_D - V_D$ equation in ballistic MOSFETs of bismuth telluride for both weak and strong electric field limits which are given by (21.38) and (21.40), respectively and the corresponding 2D electron statistics have been represented in (21.39) and (21.41), respectively. In Section 21.2.7, we have studied the $I_D - V_D$ equation in ballistic MOSFETs of germanium which is given by (21.43) and the corresponding 2D electron statistics has been represented in (21.44). In Section 21.2.8, we have investigated the $I_D - V_D$ equation in MOSFETs of gallium arsenide in accordance with the models of Rossler and Stillman *et al.* for weak electric field limits which are given by (21.46) and (21.51) and the corresponding 2D electron statistics has been represented in (21.47) and (21.52a), respectively. In the same section, we have also investigated the $I_D - V_D$ equation in this case in accordance with the said models for strong electric field limits which are given by (21.48) and (21.52b) and the corresponding 2D electron statistics has been represented in (21.49) and (21.53), respectively. In Section 21.2.9, we have studied the $I_D - V_D$ equation in ballistic MOSFETs of gallium antimonide in accordance with the models of Seiler, Mathur *et al.* and Zhang. for weak electric field limits which are given by (21.55), (21.60) and (21.65) and the corresponding 2D electron statistics has been represented in (21.56), (21.61) and (21.66), respectively. In the same section, we have also investigated the $I_D - V_D$ equation in this case in accordance with the said models for strong electric field limits which are given by (21.57), (21.62) and (21.67) and the corresponding 2D electron statistics has been represented in (21.58), (21.63) and (21.68), respectively. In Section 21.2.10, we have studied the $I_D - V_D$ equation in ballistic MOSFETs of indium antimonide in accordance with the models of Palik, Johnson *et al.* and Agafonov *et al.* for weak electric field limits which are given by (21.70), (21.75) and (21.80) and the corresponding 2D electron statistics has been represented in (21.71), (21.76) and (21.81), respectively. In the same section, we have also investigated the $I_D - V_D$ equation in this case in accordance with the said models for strong electric field limits which are given by (21.72), (21.77) and (21.82) and the corresponding 2D electron statistics has been represented in (21.73), (21.78) and (21.83), respectively.

We have not given the graphs of the $I_D - V_D$ equations for ballistic MOSFETs made of many materials as discussed here since we feel that our readers will perform the intricate numerical computation to obtain the oscillatory plots for better assessment and joy of understanding. It is the band structure that changes in a fundamental way and consequently all the physical properties of all the electronic materials changes radically leading to new physical concepts.

21.4 Open Research Problem

(R21.1) Derive $I_D - V_D$ equations for ballistic MOSFETs made of all the materials whose DRs are given in (R1.1). Draw the corresponding graphs and explain the results.

Chapter 22

$I_D - V_D$ Equation for Ballistic QWFETs

The young thinks about the utility of a thing, but the old knows the futility of utility.

22.1 Introduction

This is completely a new chapter where from Sections 22.2.1 up to 22.2.13 we study the $I_D - V_D$ equation in ballistic QWFETs of tetragonal and nonlinear optical, III–V ternary and quaternary, II–VI, IV–VI, stressed Kane type, Te, GaP, $PtSb_2$, Bi_2Te_3, Ge, GaAs, GaSb, and InSb materials, respectively. In Section 22.3, we present summary and conclusion and last Section 22.4, we offer one multidimensional open research problems.

22.2 Theoretical Background

22.2.1 *The $I_D - V_D$ equation for ballistic QWFETs of tetragonal and nonlinear optical materials*

Using (21.7), (21.1), (21.2), (21.3) and (3.5) we can write

$$(J_D/W) = \frac{eg_v}{2\pi^2\hbar}\sum_{n_z=1}^{n_{z\max}}\left[\int_{E_{n_{z1}}}^{\infty}\left[\frac{1}{g'_{21,1}(E)}\right]\left[\frac{1}{1+\exp\left(\frac{E-E_{F22,1}}{k_BT}\right)}\right.\right.$$

$$\left.-\frac{1}{1+\exp\left(\frac{E-E_{F22,1}+eV_D}{k_BT}\right)}\right][\psi_2(E)]^{-2}\left[\psi_2(E)\left\{\{\psi_1(E)\}'\right.\right.$$

$$- \{\psi_3(E)\}' \left(\frac{n_z\pi}{d_z}\right)^2 \Bigg\} - \left\{\psi_1(E) - \psi_3(E)\left(\frac{n_z\pi}{d_z}\right)^2\right\}$$

$$\times \{\psi_2(E)\}' \Bigg]\Bigg] dE \tag{22.1}$$

where $E_{F22,1}$ is the Fermi energy in the present case.

The $E_{F22,1}$ should be determined form the following equation:

$$\frac{g_v}{4\pi}\sum_{n_z=1}^{n_{z\max}}\Bigg[[T_{51}(E_{F22,1},n_z) + T_{51}((E_{F22,1} - eV_D), n_z)]$$

$$+ \sum_{r=1}^{s} L(r)[T_{51}(E_{F22,1},n_z) + T_{51}((E_{F22,1} - eV_D), n_z)]\Bigg]$$

$$= \frac{C_{ins}}{e}(V_T - V_{GS}) \tag{22.2}$$

22.2.2 *The $I_D - V_D$ equation for ballistic QWFETs of III–V, ternary and quaternary materials*

(a) Three-band model of Kane

Using (21.12), (21.1), (21.2), (21.3) and (3.10) we can write

$$(J_D/W) = \frac{eg_v}{\pi^2\hbar^2}\sqrt{2m_c}\sum_{n_z=1}^{n_{z\max}}\left[\int_{E_{n_{z2}}}^{\infty}\sqrt{I_{11}(E)}\left[\frac{1}{1+\exp\left(\frac{E-E_{F22,2}}{k_BT}\right)}\right.\right.$$

$$\left.\left. - \frac{1}{1+\exp\left(\frac{E-E_{F22,2}+eV_D}{k_BT}\right)}\right] dE\right] \tag{22.3}$$

where $E_{F22,2}$ is the Fermi energy in the present case and is given by

$$\frac{m_c g_v}{2\pi\hbar^2}\sum_{n_z=1}^{n_{z\max}}\Bigg[[T_{53}(E_{F22,2},n_z) + T_{53}((E_{F22,2} - eV_D), n_z)]$$

$$+ \sum_{r=1}^{s} L(r)[T_{53}(E_{F22,2},n_z) + T_{53}((E_{F22,2} - eV_D), n_z)]\Bigg]$$

$$= \frac{C_{ins}}{e}(V_T - V_{GS}) \tag{22.4}$$

(b) Two-band model of Kane

Using (21.12), (21.1), (21.2), (21.3) and (3.15) we can write

$$(J_D/W) = \frac{eg_v}{\pi^2\hbar^2}\sqrt{2m_c}\sum_{n_z=1}^{n_{z\max}}\left[\int_{E_{n_{z3}}}^{\infty}\left[\sqrt{E(1+\alpha E)}\right]\left[\frac{1}{1+\exp\left(\frac{E-E_{F22,3}}{k_BT}\right)} - \frac{1}{1+\exp\left(\frac{E-E_{F22,3}+eV_D}{k_BT}\right)}\right]dE\right] \quad (22.5)$$

where $E_{F22,3}$ is the Fermi energy in the present case and is given by

$$\frac{C_{ins}}{e}(V_T - V_{GS}) = \frac{m_c k_B T g_v}{2\pi\hbar^2}\sum_{n_z=1}^{n_{z\max}}[(1+2\alpha E_{n_{z3}})F_0(\eta_{n_{1,22}}) + 2\alpha k_B T F_1(\eta_{n_{1,22}}) + (1+2\alpha E_{n_{z3}})F_0(\bar{\eta}_{n_1}) + 2\alpha k_B T F_1(\bar{\eta}_{n_1})] \quad (22.6)$$

where

$$\bar{\eta}_{n_1} = [\eta_{n_{1,22}} - (eV_D/k_BT)] \quad \text{and} \quad \eta_{n_{1,22}} = \frac{E_{F22,3} - E_{n_{z3}}}{k_BT}$$

(c) Parabolic band model

Using (21.12), (21.1), (21.2), (21.3) and (3.18b) we can write

$$(J_D/W) = \frac{eg}{\pi^2\hbar^2}\sqrt{2m_c}\sum_{n_z=1}^{n_{z\max}}\left[\int_{E_{n_{zp}}}^{\infty}[\sqrt{E}]\left[\frac{1}{1+\exp\left(\frac{E-E_{F22,4}}{k_BT}\right)} - \frac{1}{1+\exp\left(\frac{E-E_{F22,4}+eV_D}{k_BT}\right)}\right]dE\right] \quad (22.7)$$

where $E_{F22,4}$ is the Fermi energy in the present case and is given by

$$\frac{C_{ins}}{e}(V_T - V_{GS}) = \frac{m_c k_B T g_v}{2\pi\hbar^2}\sum_{n_z=1}^{n_{z\max}}[F_0(\eta_{n_{2,22}}) + F_0(\bar{\eta}_{n_2})] \quad (22.8)$$

where

$$\bar{\eta}_{n_2} = [\eta_{n2,22} - (eV_D/k_BT)] \quad \text{and} \quad \eta_{n_{2,22}} = \frac{E_{F22,4} - E_{n_{zp}}}{k_BT}$$

22.2.3 The $I_D - V_D$ equation for ballistic QWFETs of II–VI materials

Using (21.24), (21.1), (21.2), (21.3) and (3.25) we can write

$$(J/W) = \frac{eg_v m_\perp^*}{2\pi^2\hbar^3}\left[\sum_{n_z=1}^{n_{z\max}} \int_{E_{n_{z5}}}^{\infty} \left[1 - \frac{(C_0/2\sqrt{a_0'})}{\sqrt{E+\delta_{51}(n_z)}}\right]\left[\sqrt{C_0^2 + \frac{2E\hbar^2}{m_\perp^*}}\right.\right.$$
$$\left.\left.\times\left[\frac{1}{1+\exp\left(\frac{E-E_{F22,5}}{k_BT}\right)} - \frac{1}{1+\exp\left(\frac{E-E_{F22,5}+eV_D}{k_BT}\right)}\right] dE\right]\right] \tag{22.9}$$

where $E_{F22,5}$ is the Fermi energy in the present case and is given by

$$\frac{C_{ins}}{e}(V_T - V_{GS}) = \frac{g_v m_\perp^* k_BT}{2\pi\hbar^2}\sum_{n_z=1}^{n_{z\max}}\left[F_0(\eta_{n_{z51}}) - \frac{C_0 f_s(E_{F22,5}, n_z)}{2\sqrt{a_0'}k_BT}\right]$$
$$+\left[F_0(\bar{\eta}_{n_{z51}}) - \frac{C_0 f_s(E_{F22,5} - eV_D), n_z)}{2\sqrt{a_0'}k_BT}\right] \tag{22.10}$$

where

$$\eta_{n_{z51}} \equiv (k_BT)^{-1}[E_{F22,5} - b_0'(n_z\pi/d_z)^2] \quad \text{and}$$
$$\bar{\eta}_{n_{z51}} = [\eta_{n_{z51}} - (eV_D/k_BT)]$$

$$f_s((E_{F22,5} - eV_D), n_z) \equiv \left[2\left(\sqrt{\bar{\eta}_{n_{z51}} + \delta_{52}(n_z)} - \sqrt{\delta_{52}(n_z)}\right)\right.$$
$$\left.+\sum_{r=1}^{s}\left[2(1-2^{1-2r})\zeta(2r)\frac{(-1)^{2r-1}(2r-1)!}{(\bar{\eta}_{n_{z51}} + \delta_{52}(n_z))^{2r}}\right]\right]$$

22.2.4 The $I_D - V_D$ equation for ballistic QWFETs of IV–VI materials

(a) McClure and Choi model

From (3.27) we get

$$V_x = \sqrt{\frac{2}{m_1}}\left[\frac{\sqrt{E(1+\alpha E)}}{1+2\alpha E}\right] \tag{22.11}$$

Using (22.11), (21.1), (21.2), (21.3) and (3.33) we can write

$$\frac{J_D}{W} = \left(\frac{2g_v e}{3\pi^3}\right)\sqrt{\frac{2}{m_1 p_1}}\sum_{n_z=1}^{n_{z\max}}\int_{E_{n_{z6}}}^{\infty}\left[R_3(E,n_z)\left[\frac{\sqrt{E(1+\alpha E)}}{(1+2\alpha E)}\right]\right.$$
$$\left.\times\left[\frac{1}{1+\exp\left(\frac{E-E_{F22,6}}{k_B T}\right)} - \frac{1}{1+\exp\left(\frac{E-E_{F22,6}+eV_D}{k_B T}\right)}\right]\right] dE \tag{22.12}$$

where $E_{F22,6}$ is the Fermi energy in this case and can be expressed through the following equation:

$$\frac{C_{ins}}{e}(V_T - V_{GS})$$
$$= \left(\frac{g_v}{3\pi^2\sqrt{p_1}}\right)\sum_{n_z=1}^{n_{z\max}}\left[[\theta_1(E_{F22,6},n_z) + \theta_1((E_{F22,6}-eV_D),n_z)]\right.$$
$$\left. + \sum_{r=1}^{s} L(r)[\theta_1(E_{F22,6},n_z) + \theta_1((E_{F22,6}-eV_D),n_z)]\right] \tag{22.13}$$

(b) Hybrid model

From (3.36) we get

$$V_x = \sqrt{\frac{2}{m_1}}\left[\frac{\sqrt{E(1+\alpha E)}}{1+2\alpha E}\right] \tag{22.14}$$

Using (22.14), (21.1), (21.2), (21.3) and (3.41) we can write

$$\frac{J_D}{W} = \frac{eg_v\sqrt{2m_3}}{\pi^2\hbar^2}\left[\sum_{n_y=1}^{n_{y\max}}\int_{E_{n_y}}^{\infty}[t_{29}(E,n_y)]'\left[\frac{\sqrt{E(1+\alpha E)}}{(1+2\alpha E)}\right]\right.$$
$$\left.\times\left[\frac{1}{1+\exp\left(\frac{E-E_{F22,7}}{k_BT}\right)}-\frac{1}{1+\exp\left(\frac{E-E_{F22,7}+eV_D}{k_BT}\right)}\right]\right]dE \quad (22.15)$$

where $E_{F22,7}$ is the Fermi energy in this case and can be expressed through the following equation:

$$\frac{C_{ins}}{e}(V_T - V_{GS})$$
$$= \frac{g_v\sqrt{m_1m_3}}{2\pi\hbar^2}\sum_{n_y=1}^{n_{y\max}}\left[[t_{29}(E_{F22,7},n_y)+t_{29}(E_{F22,7}-eV_D),n_y]\right.$$
$$\left.+\sum_{r=1}^{s}L(r)[t_{29}(E_{F22,7},n_y)+t_{29}(E_{F22,7}-eV_D),n_y]\right] \quad (22.16)$$

(c) Cohen model

Using (3.43), (21.1), (21.2), (21.3) and (3.46) we can write

$$\frac{J_D}{W} = \left(\frac{2g_ve}{3\pi^3}\right)\sqrt{\frac{2}{m_1p_1}}\left[\sum_{n_z=1}^{n_{z\max}}\int_{E_{nz7}}^{\infty}[R_4(E,n_z)]\left[\frac{\sqrt{E(1+\alpha E)}}{(1+2\alpha E)}\right]\right.$$
$$\left.\times\left[\frac{1}{1+\exp\left(\frac{E-E_{F22,8}}{k_BT}\right)}-\frac{1}{1+\exp\left(\frac{E-E_{F22,8}+eV_D}{k_BT}\right)}\right]dE\right]$$
$$(22.17)$$

where $E_{F22,8}$ is the Fermi energy in this case and can be expressed through the following equation:

$$\frac{C_{ins}}{e}(V_T - V_{GS})$$
$$= \left(\frac{g_v}{3\pi^2\sqrt{p_1}}\right)\sum_{n_z=1}^{n_{z\max}}\left[[\theta_3(E_{F22,8},n_z)+\theta_3((E_{F22,8}-eV_D),n_z)]\right.$$
$$\left.+\sum_{r=1}^{s}L(r)[\theta_3(E_{F22,8},n_z)+\theta_3((E_{F22,8}-eV_D),n_z)]\right] \quad (22.18)$$

(d) Lax model

Using (21.1), (21.2), (21.3) and (3.49) we can write

$$(J_D/W) = \frac{eg_v}{\pi^2\hbar^2}\sqrt{2m_2}\sum_{n_z=1}^{n_{z\max}}\left[\int_{E_{n_{z8}}}^{\infty}\left[\sqrt{E(1+\alpha E)}\right]\left[\frac{1}{1+\exp\left(\frac{E-E_{F22,9}}{k_BT}\right)} - \frac{1}{1+\exp\left(\frac{E-E_{F22,9}+eV_D}{k_BT}\right)}\right]dE\right] \quad (22.19)$$

where $E_{F22,9}$ is the Fermi energy in the present case and is given by

$$\frac{C_{ins}}{e}(V_T - V_{GS}) = \frac{\sqrt{m_1m_2}k_BTg_v}{2\pi\hbar^2}\sum_{n_z=1}^{n_{z\max}}[(1+2\alpha E_{n_{z8}})F_0(\eta_{y2,22}) + 2\alpha k_BTF_1(\eta_{y2,22}) + (1+2\alpha E_{n_{z8}})F_0(\bar{\eta}_{y2}) + 2\alpha k_BTF_1(\bar{\eta}_{y2})] \quad (22.20)$$

where

$$\bar{\eta}_{y2} = [\eta_{y2,22} - (eV_D/k_BT)] \quad \text{and} \quad \eta_{y2,22} = \frac{E_{F22,9} - E_{n_{z8}}}{k_BT}$$

(e) Dimmock model

From (3.56) we get

$$V_x = \frac{1}{\phi'_{227}(E)} \quad (22.21)$$

where

$$\phi'_{227}(E) = \left[\left[2\left[-\left[\frac{1}{m_t^*} + \frac{\alpha E}{m_t^+} - \frac{1+\alpha E}{m_t^-}\right] + \left(\frac{1}{m_t^*} + \frac{\alpha E}{m_t^+} - \frac{1+\alpha E}{m_t^-}\right)^2 - 4\frac{\alpha}{m_t^- m_t^+}E(1+\alpha E)\right]\right]\left[\frac{2\alpha}{m_t^- m_t^+}\right]^{-1}\right]^{1/2}$$

Using (22.21), (21.1), (21.2), (21.3) and (3.68) we can write

$$(J_D/W) = \frac{eg_v}{2\pi^2} \sum_{n_z=1}^{n_{z\max}} \left[\int_{E_{n_{z10}}}^{\infty} \left[\frac{\theta_5(E, n_z)}{\phi'_{227}(E)} \right] \left[\frac{1}{1 + \exp\left(\frac{E - E_{F22,10}}{k_B T}\right)} - \frac{1}{1 + \exp\left(\frac{E - E_{F22,10} + eV_D}{k_B T}\right)} \right] dE \right] \tag{22.22}$$

where $E_{F22,10}$ is the Fermi energy in the present case and is given by

$$\frac{C_{ins}}{e}(V_T - V_{GS}) = \frac{g_v}{4\pi} \sum_{n_z=1}^{n_{z\max}} [T_{59}(E_{F22,10}, n_z) + T_{59}((E_{F22,10} - eV_D), n_z)] + \sum_{r=1}^{s} L(r)[T_{59}(E_{F22,10}, n_z) + T_{59}((E_{F22,10} - eV_D), n_z)] \tag{22.23}$$

(f) Ellipsoidal parabolic model

Using (21.1), (21.2), (21.3) and (3.73) we can write

$$(J_D/W) = \frac{eg_v}{\pi^2 \hbar^2} \sqrt{2m_2} \sum_{n_z=1}^{n_{z\max}} \left[\int_{E_{n_{z9}}}^{\infty} [\sqrt{E}] \left[\frac{1}{1 + \exp\left(\frac{E - E_{F22,11}}{k_B T}\right)} - \frac{1}{1 + \exp\left(\frac{E - E_{F22,11} + eV_D}{k_B T}\right)} \right] dE \right] \tag{22.24}$$

where $E_{F22,11}$ is the Fermi energy in the present case and is given by

$$\frac{C_{ins}}{e}(V_T - V_{GS}) = \frac{\sqrt{m_1 m_2} k_B T g_v}{2\pi\hbar^2} \sum_{n_z=1}^{n_{z\max}} [F_0(\eta_{y3,22}) + F_0(\bar{\eta}_{y3})] \tag{22.25}$$

where $\bar{\eta}_{y3} = [\eta_{y3,22} - (eV_D/k_B T)]$ and $\eta_{y3,22} = \frac{E_{F22,11} - E_{n_{z9}}}{k_B T}$

22.2.5 The $I_D - V_D$ equation for ballistic QWFETs of stressed Kane type materials

Using (21.1), (21.2), (21.3), (21.32) and (3.79) we get

$$\frac{J_D}{W} = \frac{g_v e}{2\pi^2 \hbar} \sum_{n_z=1}^{n_{z\max}} \int_{E_{n_z 11}}^{\infty} \left[\frac{\theta_6(E,n_z)}{[\bar{a}_0(E)]'}\right] \left[\left[\frac{1}{1+\exp\left(\frac{E-E_{F22,12}}{k_B T}\right)} - \frac{1}{1+\exp\left(\frac{E-E_{F22,12}+eV_D}{k_B T}\right)}\right]\right] dE \tag{22.26}$$

where $E_{F22,12}$ is the Fermi energy in this case and can be expressed through the following equation:

$$\frac{C_{ins}}{e}(V_T - V_{GS}) = \frac{g_v}{4\pi}\left[\sum_{n_z=1}^{n_{z\max}} [T_{61}(E_{F22,12}, n_z) + T_{61}((E_{F22,12} - eV_D), n_z)] + \sum_{r=1}^{s} L(r)[T_{61}(E_{F22,12}, n_z) + T_{61}((E_{F22,12} - eV_D), n_z)]\right] \tag{22.27}$$

22.2.6 The $I_D - V_D$ equation for ballistic QWFETs of Tellurium

From (2.133) we can write

$$V_x = (\hbar)^{-1}\sqrt{\psi_4^2 + 4\psi_2 E} \tag{22.28}$$

Using (21.1), (21.2), (21.3), (22.28) and (3.84) we get

$$\frac{J_D}{W} - \frac{g_v e}{2\pi^2 \hbar} \sum_{n_z=1}^{n_{z\max}} \int_{E_{n_z 12}}^{\infty} \left[t'_{40}(E,n_z)\sqrt{\psi_4^2 + 4\psi_2 E}\right] \times \left[\left[\frac{1}{1+\exp\left(\frac{E-E_{F22,13}}{k_B T}\right)} - \frac{1}{1+\exp\left(\frac{E-E_{F22,13}+eV_D}{k_B T}\right)}\right]\right] dE \tag{22.29}$$

where $E_{F22,13}$ is the Fermi energy in this case and can be expressed through the following equation:

$$\frac{C_{ins}}{e}(V_T - V_{GS})$$
$$= \frac{g_v}{2\pi}\sum_{n_z=1}^{n_{z\max}}[t_{40}(E_{F22,13}, n_z) + t_{40}((E_{F22,13} - eV_D), n_z)]$$
$$\left. + \sum_{r=1}^{s} L(r)[t_{40}(E_{F22,13}, n_z) + t_{40}((E_{F22,13} - eV_D), n_z)]\right] \quad (22.30)$$

22.2.7 *The $I_D - V_D$ equation for ballistic QWFETs of gallium phosphide*

From (2.72) we can write

$$V_x = \frac{2}{\hbar}\left[\frac{\sqrt{\phi_{444}(E)}}{\phi'_{444}(E)}\right] \quad (22.31)$$

where

$$\phi_{444}(E) = (2a^2)^{-1}[c + 2a(E - |V_G|)$$
$$\pm\sqrt{[c + 2(E - |V_G|)a]^2 - 4a^2(E - |V_G|)^2 - |V_G|}]$$

Using (21.1), (21.2), (21.3), (22.31) and (3.91) we get

$$\frac{J_D}{W} = \frac{g_v e}{2\pi^2 a^2 \hbar}\sum_{n_z=1}^{n_{z\max}}\int_{E_{n_z13}}^{\infty}\left[t'_{42}(E, n_z)\left[\frac{\sqrt{\phi_{444}(E)}}{\phi'_{444}(E)}\right]\right]$$
$$\times\left[\left[\frac{1}{1 + \exp\left(\frac{E - E_{F22,14}}{k_B T}\right)} - \frac{1}{1 + \exp\left(\frac{E - E_{F22,14} + eV_D}{k_B T}\right)}\right]\right] dE$$
$$(22.32)$$

where $E_{F22,14}$ is the Fermi energy in this case and can be expressed through the following equation:

$$\frac{C_{ins}}{e}(V_T - V_{GS})$$
$$= \frac{g_v}{2\pi}\left[\sum_{n_z=1}^{n_{z\max}}[t_{42}(E_{F22,14}, n_z) + t_{42}((E_{F22,14} - eV_D), n_z)]\right.$$
$$\left. + \sum_{r=1}^{s} L(r)[t_{42}(E_{F22,14}, n_z) + t_{42}((E_{F22,14} - eV_D), n_z)]\right] \quad (22.33)$$

22.2.8 The $I_D - V_D$ equation for ballistic QWFETs of platinum antimonide

From (R1.11) we can write

$$V_x = [\hbar t'_{222}(E)]^{-1} \quad (22.34)$$

where

$$t_{222}(E) = [E[\theta_{22} + \theta_{221}] \pm [\sqrt{(\theta_{22} + \theta_{221})^2 E^2 + 4E^2(\theta_{222} - \theta_{22}\theta_{221})}]]^{1/2},$$

$$\theta_{22} = \left(\frac{\bar{a}}{2}\right)^2 [\bar{\lambda}_0 - \bar{l}], \theta_{221} = \bar{\delta}_0 - \left(\frac{\bar{a}}{2}\right)^2 [\bar{\upsilon} + \bar{n}] \quad \text{and} \quad \theta_{222} = I\left(\frac{\bar{a}}{2}\right)^2$$

Using (21.1), (21.2), (21.3), (22.34) and (3.99) we get

$$\frac{J_D}{W} = \frac{g_v e}{4\pi^2 A_9 \hbar} \sum_{n_z=1}^{n_{z\max}} \int_{E_{n_{z14}}}^{\infty} [t'_{44}(E, n_z)[t'_{222}(E)]^{-1}] \times \left[\left[\frac{1}{1 + \exp\left(\frac{E - E_{F22,15}}{k_B T}\right)} - \frac{1}{1 + \exp\left(\frac{E - E_{F22,15} + eV_D}{k_B T}\right)}\right]\right] dE \quad (22.35)$$

where $E_{F22,15}$ is the Fermi energy in this case and can be expressed through the following equation:

$$\frac{C_{ins}}{e}(V_T - V_{GS}) = \frac{g_v}{8\pi A_9} \sum_{n_z=1}^{n_{z\max}} [t_{44}(E_{F22,15}, n_z) + t_{44}((E_{F22,15} - eV_D), n_z)] + \sum_{r=1}^{s} L(r)[t_{44}(E_{F22,15}, n_z) + t_{44}((E_{F22,15} - eV_D), n_z)]\Bigg] \quad (22.36)$$

22.2.9 The $I_D - V_D$ equation for ballistic QWFETs of bismuth telluride

Using (21.1), (21.2), (21.3), (21.37) and (3.105) we can write

$$\frac{J_D}{W} = \frac{2g_v e m_0}{\pi^2 \hbar^3} \left(\sqrt{I_1}\right) (\alpha_{22}\alpha_{33} - 4\alpha_{23}^2)^{-1/2} \sum_{n_x=1}^{n_{x\max}} \int_{E_{n_{x15}}}^{\infty} \left[\sqrt{E(1+\alpha E)}\right] \times \left[\left[\frac{1}{1 + \exp\left(\frac{E - E_{F22,16}}{k_B T}\right)} - \frac{1}{1 + \exp\left(\frac{E - E_{F22,16} + eV_D}{k_B T}\right)}\right]\right] dE \quad (22.37)$$

where $E_{F22,16}$ is the Fermi energy in this case and can be expressed through the following equation:

$$\frac{C_{ins}}{e}(V_T - V_{GS}) = \frac{k_B T g_v}{2\pi\hbar^2}\left(\frac{m_0}{\sqrt{\alpha_{22}\alpha_{33} - 4\alpha_{23}^2}}\right)$$
$$\times \left[\sum_{n_x=1}^{n_{x\max}} [(1 + 2\alpha E_{n_{x15}})F_0(\eta_{n_{15,22}}) + 2\alpha k_B T F_1(\eta_{n_{15,22}})]\right]$$
$$+ [[(1 + 2\alpha E_{n_{x15}})F_0(\bar{\eta}_{n_{15}}) + 2\alpha k_B T F_1(\bar{\eta}_{n_{15}})]] \qquad (22.38)$$

where $\bar{\eta}_{n_{15}} = [\eta_{n_{15,22}} - (eV_D/k_B T)]$ and $\eta_{n_{15,22}} = \frac{E_{F22,16} - E_{n_{x15}}}{k_B T}$

22.2.10 *The $I_D - V_D$ equation for ballistic QWFETs of germanium*

(a) The model of Cardona *et al.*

Using (21.1), (21.2), (21.3), (21.42) and (3.112) we can write

$$\frac{J_D}{W} = \frac{4e}{\pi^2\hbar^2}\sqrt{m_1^* m_2^*}\sqrt{\frac{2}{m_\perp^*}}\sum_{n_z=1}^{n_{z\max}}\int_{E_{n_{z16}}}^{\infty}\left[1 + 2\alpha E - 2\alpha\left(\frac{\hbar^2}{2m_3^*}\left(\frac{\pi n_z}{d_z}\right)^2\right)\right]$$
$$\times \left[\frac{\sqrt{E(1+\alpha E)}}{(1+2\alpha E)}\right]\left[\left[\frac{1}{1+\exp\left(\frac{E - E_{F22,17}}{k_B T}\right)}\right.\right.$$
$$\left.\left. - \frac{1}{1+\exp\left(\frac{E - E_{F22,17} + eV_D}{k_B T}\right)}\right]\right] dE \qquad (22.39)$$

where $E_{F22,17}$ is the Fermi energy in this case and can be expressed through the following equation:

$$\frac{C_{ins}}{e}(V_T - V_{GS}) = \frac{2\sqrt{m_1^* m_2^*}k_B T}{\hbar^2}\sum_{n_z=1}^{n_{z\max}}[(A_1(n_z) + 2\alpha E_{n_{z16}})F_0(\eta_{n_{z16,22}})$$
$$+ 2\alpha k_B T F_1(\eta_{n_{z16,22}})] + [(A_1(n_z) + 2\alpha E_{n_{z16}})F_0(\bar{\eta}_{n_{z16}})$$
$$+ 2\alpha k_B T F_1(\bar{\eta}_{n_{z16}})] \qquad (22.40)$$

where in this case

$$\eta_{n_{z16,22}} = \frac{1}{k_B T}[E_{F22,17} - E_{n_{z16}}] \quad \text{and} \quad \bar{\eta}_{n_{z16}} = [\eta_{n_{z16,22}} - (eV_D/k_B T)]$$

(b) The model of Wang and Ressler *et al.*

From (2.53) we can write

$$V_x = [\omega'_{222}(E)]^{-1} \tag{22.41}$$

where

$$\omega_{222}(E) = \left[\left[\frac{1-\sqrt{1-4\bar{c}_1 E}}{2\bar{c}_1}\right] 2m_\perp^*\right]^{1/2}$$

Using (21.1), (21.2), (21.3), (22.41) and (3.120) we get

$$\frac{J_D}{W} = \frac{4e\sqrt{m_1^* m_2^*}}{\pi^2\hbar^2} \sum_{n_z=1}^{n_{z\max}} \int_{E_{n_z 17}}^{\infty} [I_1'(E, n_z)[[\omega'_{222}(E)]^{-1} \times \left[\left[\frac{1}{1+\exp\left(\frac{E-E_{F22,19}}{k_B T}\right)} - \frac{1}{1+\exp\left(\frac{E-E_{F22,19}+eV_D}{k_B T}\right)}\right]\right] dE \tag{22.42}$$

where $E_{F22,19}$ is the Fermi energy in this case and can be expressed through the following equation:

$$\frac{C_{ins}}{e}(V_T - V_{GS}) = \frac{2\sqrt{m_1^* m_2^*}}{\pi\hbar^2} \sum_{n_z=1}^{n_{z\max}} \left[[t_{46}(E_{F22,19}, n_z) + t_{46}((E_{F22,19} - eV_D), n_z)] + \sum_{r=1}^{s} L(r)[t_{46}(E_{F22,19}, n_z) + t_{46}((E_{F22,19} - eV_D), n_z)]\right] \tag{22.43}$$

22.2.11 *The $I_D - V_D$ equation for ballistic QWFETs of gallium arsenide*

(a) Rossler model

Using (21.1), (21.2), (21.3), (2.57) and (3.125) we get

$$\frac{J_D}{W} = \frac{2g_v e}{4\pi^2\hbar} \sum_{n_z=1}^{n_{z\max}} \int_{E_{n_z 18}}^{\infty} \left[\sqrt{\gamma_{4,\pm}(E)}\right] \left[\left[\frac{1}{1+\exp\left(\frac{E-E_{F22,20}}{k_B T}\right)} - \frac{1}{1+\exp\left(\frac{E-E_{F22,20}+eV_D}{k_B T}\right)}\right]\right] dE \tag{22.44}$$

where $E_{F22,20}$ is the Fermi energy in this case and can be expressed through the following equation:

$$\frac{C_{ins}}{e}(V_T - V_{GS}) = \frac{g_v}{8\pi}\sum_{n_z=1}^{n_{z\max}}\left[[T_{47,\pm}(E_{F22,20}, n_z)] + [T_{47,\pm}(E_{F22,20} - eV_D), n_z))] + \sum_{r=1}^{s} L(r)[T_{47,\pm}(E_{F22,20}, n_z)] + [T_{47,\pm}(E_{F22,20} - eV_D), n_z))]\right] \tag{22.45}$$

(b) Stillman *et al.*

Using (21.1), (21.2), (21.3), (21.50) and (3.130) we get

$$\frac{J_D}{W} = \frac{g_v e}{\pi^2\hbar}\sum_{n_z=1}^{n_{z\max}}\int_{E_{n_{z19}}}^{\infty}\left[\sqrt{\gamma_5(E)}\right]\left[\left[\frac{1}{1+\exp\left(\frac{E-E_{F22,21}}{k_BT}\right)} - \frac{1}{1+\exp\left(\frac{E-E_{F22,21}+eV_D}{k_BT}\right)}\right]\right]dE \tag{22.46}$$

where $E_{F22,21}$ is the Fermi energy in this case and can be expressed through the following equation:

$$\frac{C_{ins}}{e}(V_T - V_{GS}) = \frac{g_v}{4\pi}\sum_{n_z=1}^{n_{z\max}}\left[[T_{48}(E_{F22,21}, n_z)] + [T_{48}(E_{F22,21} - eV_D), n_z)] + \sum_{r=1}^{s} L(r)[T_{48}(E_{F22,21}, n_z)] + [T_{48}(E_{F22,21} - eV_D), n_z)]\right] \tag{22.47}$$

22.2.12 *The $I_D - V_D$ equation for ballistic QWFETs of gallium antimonide*

(a) Seiler *et al.*

Using (21.1), (21.2), (21.3), (21.54) and (3.135) we get

$$\frac{J_D}{W} = \frac{g_v e}{2\pi^2\hbar}\sum_{n_z=1}^{n_{z\max}}\int_{E_{n_{z20}}}^{\infty}\left[\sqrt{\gamma_6(E)}\right]\left[\left[\frac{1}{1+\exp\left(\frac{E-E_{F22,22}}{k_BT}\right)} - \frac{1}{1+\exp\left(\frac{E-E_{F22,22}+eV_D}{k_BT}\right)}\right]\right]dE \quad (22.48)$$

where $E_{F22,22}$ is the Fermi energy in this case and can be expressed through the following equation:

$$\frac{C_{ins}}{e}(V_T - V_{GS}) = \frac{g_v}{8\pi}\sum_{n_z=1}^{n_{z\max}}\left[[T_{49}(E_{F22,22}, n_z)] + [T_{49}(E_{F22,22} - eV_D), n_z)] + \sum_{r=1}^{s} L(r)[T_{49}(E_{F22,22}, n_z)] + [T_{49}(E_{F22,22} - eV_D), n_z)]\right] \quad (22.49)$$

(b) Mathur *et al.*

Using (21.1), (21.2), (21.3), (21.59) and (3.140) we get

$$\frac{J_D}{W} = \frac{g_v e}{\pi^2\hbar}\sum_{n_z=1}^{n_{z\max}}\int_{E_{n_{z21}}}^{\infty}\left[\sqrt{\gamma_7(E)}\right]\left[\left[\frac{1}{1+\exp\left(\frac{E-E_{F22,23}}{k_BT}\right)} - \frac{1}{1+\exp\left(\frac{E-E_{F22,23}+eV_D}{k_BT}\right)}\right]\right]dE \quad (22.50a)$$

where $E_{F22,23}$ is the Fermi energy in this case and can be expressed through the following equation:

$$\frac{C_{ins}}{e}(V_T - V_{GS}) = \frac{g_v}{4\pi}\sum_{n_z=1}^{n_{z\max}}\left[[T_{50}(E_{F22,23}, n_z)] + [T_{50}(E_{F22,23} - eV_D), n_z)] + \sum_{r=1}^{s} L(r)[T_{50}(E_{F22,23}, n_z)] + [T_{50}(E_{F22,23} - eV_D), n_z)]\right] \quad (22.50b)$$

(c) Zhang *et al.*

Using (21.1), (21.2), (21.3), (21.64) and (3.145) we get

$$\frac{J_D}{W} = \frac{g_v e}{\pi^2 \hbar} \sum_{n_z=1}^{n_{z\max}} \int_{\overline{E}_{n_{z22}}}^{\infty} \left[\sqrt{\gamma_8(E)}\right] \left[\left[\frac{1}{1+\exp\left(\frac{E-E_{F22,24}}{k_B T}\right)} - \frac{1}{1+\exp\left(\frac{E-E_{F22,24}+eV_D}{k_B T}\right)}\right]\right] dE \tag{22.51a}$$

where $E_{F22,24}$ is the Fermi energy in this case and can be expressed through the following equation:

$$\begin{aligned}\frac{C_{ins}}{e}(V_T - V_{GS}) &= \frac{g_v}{4\pi} \sum_{n_z=1}^{n_{z\max}} \Bigg[[T_{51}(E_{F22,24}, n_z)] + [T_{51}(E_{F22,24} - eV_D), n_z)] \\ &\quad + \sum_{r=1}^{s} L(r) [T_{51}(E_{F22,24}, n_z)] + [T_{51}(E_{F22,24} - eV_D), n_z)] \Bigg]\end{aligned} \tag{22.51b}$$

22.2.13 *The $I_D - V_D$ equation for ballistic QWFETs of indium antimonide*

(a) Model of Palik *et al.*

Using (21.1), (21.2), (21.3), (21.69) and (3.150) we get

$$\frac{J_D}{W} = \frac{g_v e}{\pi^2 \hbar} \sum_{n_z=1}^{n_{z\max}} \int_{E_{n_{z23}}}^{\infty} \left[\sqrt{\gamma_9(E)}\right] \left[\left[\frac{1}{1+\exp\left(\frac{E-E_{F22,25}}{k_B T}\right)} - \frac{1}{1+\exp\left(\frac{E-E_{F22,25}+eV_D}{k_B T}\right)}\right]\right] dE \tag{22.52}$$

where $E_{F22,25}$ is the Fermi energy in this case and can be expressed through the following equation:

$$\begin{aligned}\frac{C_{ins}}{e}(V_T - V_{GS}) \\ = \frac{g_v}{4\pi}\sum_{n_z=1}^{n_{z\max}}\Bigg[[T_{52}(E_{F22,25}, n_z)] + [T_{52}(E_{F22,25} - eV_D), n_z)] \\ + \sum_{r=1}^{s} L(r)[T_{52}(E_{F22,25}, n_z)] + [T_{52}(E_{F22,25} - eV_D), n_z)]\Bigg] \quad (22.53)\end{aligned}$$

(b) Model of Johnson *et al.*

Using (21.1), (21.2), (21.3), (21.74) and (3.155) we get

$$\begin{aligned}\frac{J_D}{W} = \frac{g_v e}{\pi^2\hbar}\sum_{n_z=1}^{n_{z\max}}\int_{E_{n_{z24}}}^{\infty}\left[\sqrt{\gamma_{10}(E)}\right]\left[\left[\frac{1}{1+\exp\left(\frac{E-E_{F22,26}}{k_BT}\right)}\right.\right. \\ \left.\left. - \frac{1}{1+\exp\left(\frac{E-E_{F22,26}+eV_D}{k_BT}\right)}\right]\right] dE \quad (22.54)\end{aligned}$$

where $E_{F22,26}$ is the Fermi energy in this case and can be expressed through the following equation:

$$\begin{aligned}\frac{C_{ins}}{e}(V_T - V_{GS}) \\ = \frac{g_v}{4\pi}\sum_{n_z=1}^{n_{z\max}}\Bigg[[T_{53}(E_{F22,26}, n_z)] + [T_{53}(E_{F22,26} - eV_D), n_z)] \\ + \sum_{r=1}^{s} L(r)[T_{53}(E_{F22,26}, n_z)] + [T_{53}(E_{F22,26} - eV_D), n_z)]\Bigg] \quad (22.55)\end{aligned}$$

(c) Model of Agafanov *et al.*

Using (21.1), (21.2), (21.3), (21.79) and (3.160) we get

$$\frac{J_D}{W} = \frac{g_v e}{\pi^2 \hbar} \sum_{n_z=1}^{n_{z\max}} \int_{E_{n_{z25}}}^{\infty} \left[\sqrt{\gamma_{11}(E)}\right] \left[\left[\frac{1}{1+\exp\left(\frac{E-E_{F22,27}}{k_B T}\right)} - \frac{1}{1+\exp\left(\frac{E-E_{F22,27}+eV_D}{k_B T}\right)}\right]\right] dE \quad (22.56)$$

where $E_{F22,27}$ is the Fermi energy in this case and can be expressed through the following equation:

$$\frac{C_{ins}}{e}(V_T - V_{GS}) = \frac{g_v}{4\pi} \sum_{n_z=1}^{n_{z\max}} \left[[T_{54}(E_{F22,27}, n_z)] + [T_{54}(E_{F22,27} - eV_D), n_z)] + \sum_{r=1}^{s} L(r)[T_{54}(E_{F22,27}, n_z)] + [T_{54}(E_{F22,27} - eV_D), n_z)] \right] \quad (22.57)$$

22.3 Summary and Conclusions

In this chapter, in Section 22.2.1, we have formulated $I_D - V_D$ for ballistic QWFETs made of nonlinear optical materials as given by (22.1) for the computation of which we need the expression of the 2D electron statistics which is given in (22.2). The $I_D - V_D$ equations for ballistic QWFETs made of III–V, ternary and quaternary materials has been investigated in Section 22.2.2 in accordance with the three-band model of Kane, two-band model of Kane and parabolic energy bands are given by (22.3), (22.5) and (22.7) and the corresponding 2D electron statistics are expressed through (22.4), (22.6) and (22.8), respectively. In Section 22.2.3, we have studied the $I_D - V_D$ equation for QWFETs of II–VI materials which is given by (22.9) and the corresponding 2D electron statistics has been represented in (22.10). In Section 22.2.4, we have studied the $I_D - V_D$ equation in QWFETs of IV–VI materials in accordance with the models of McClure and Choi, Hybrid, Cohen, Lax, Dimmock and ellipsoidal parabolic bands which are given by (22.12), (22.15), (22.17), (22.19), (22.22) and (22.24) and the corresponding 2D electron statistics have been represented in (22.13),

(22.16), (22.18), (22.20), (22.23) and (22.25), respectively. In Section 22.2.5, we have investigated the $I_D - V_D$ equation in QWFETs of stressed Kane type materials which is given by (22.26) and the corresponding 2D electron statistics has been represented in (22.27). In Section 22.2.6, we have studied the $I_D - V_D$ equation in QWFETs of Te which is given by (22.29) and the corresponding 2D electron statistics has been represented in (22.30). In Section 22.2.7, we have studied the $I_D - V_D$ equation in QWFETs of GaP which is given by (22.32) and the corresponding 2D electron statistics has been represented in (22.33). In Section 22.2.8, we have investigated the $I_D - V_D$ equation in QWFETs of $PtSb_2$ which is given by (22.35) and the corresponding 2D electron statistics has been represented in (22.36). In Section 22.2.9, we have studied the $I_D - V_D$ equation in QWFETs of bismuth telluride materials which is given by (22.37) and the corresponding 2D electron statistics has been represented in (22.38). In Section 22.2.10, we have studied the $I_D - V_D$ equation in QWFETs of germanium in accordance with the models of Cardona *et al.* and Wang *et al.* which are given by (22.39) and (22.42) and the corresponding 2D electron statistics have been represented in (22.40) and (22.43), respectively. In Section 22.2.11, we have studied the $I_D - V_D$ equation in QWFETs of gallium arsenide in accordance with the models of Rossler *et al.* and Stillman *et al.* which are given by (22.44) and (22.46) and the corresponding 2D electron statistics have been represented in (22.45) and (22.47), respectively. In Section 22.2.12, we have studied the $I_D - V_D$ equation in QWFETs of gallium antimonide in accordance with the models of Seiler *et al.*, Mathur *et al.* and Zhang *et al.* which are given by (22.48), (22.50a) and (22.51a) and the corresponding 2D electron statistics have been represented in (22.49), (22.50b) and (22.51b), respectively. In Section 22.2.13, we have studied the $I_D - V_D$ equation in QWFETs of indium antimonide in accordance with the models of Palik *et al.*, Johnson *et al.* and Agafonov *et al.* which are given by (22.52), (22.54) and (22.56) and the corresponding 2D electron statistics have been represented in (22.53), (22.55) and (22.57), respectively.

We have not given the graphs of the $I_D - V_D$ equations for ballistic QWFETs made of many materials as discussed here since we feel that our readers will perform the intricate numerical computation to obtain the oscillatory plots for better assessment and joy of understanding. It is the band structure which changes in a fundamental way and consequently all the physical properties of all the electronic materials changes radically leading to new physical concepts.

22.4 Open Research Problem

(R22.1) Derive $I_D - V_D$ equations for ballistic QWFETs made of all the materials whose DRs are given in (R1.1). Draw the corresponding graphs and explain the results.

Chapter 23

The $I_D - V_D$ Equation for Ballistic HDQWFETs

I am trying to accomplish something totally new without realizing that life is made up of little known things.

23.1 Introduction

This is completely a new chapter where from Sections 23.2.1 up to 23.2.11 we study the equation in ballistic HDQWFETs of tetragonal and nonlinear optical, III–V ternary and quaternary, II–VI, IV–VI, stressed Kane type, Te, GaP, $PtSb_2$, Bi_2Te_3, Ge and GaSb materials, respectively. In Section 23.3, we present summary and conclusion and last Section 23.4, we offer one multidimensional open research problems.

23.2 Theoretical Background

23.2.1 *The $I_D - V_D$ equation for ballistic HDQWFETs of tetragonal and nonlinear optical materials*

Using (8.25) we get

$$V_x = \sqrt{\frac{2}{m_\perp^*}} \frac{\sqrt{T_{22}(E,\eta_g)}}{T'_{22}(E,\eta_g)} \tag{23.1}$$

Using (21.1), (21.2), (21.3), (11.3) and (23.1) we get

$$\frac{J_D}{W} = \left[\frac{eg_v}{\pi^2\hbar^2}\sqrt{2m_\perp^*}\right] \text{ Real part of } \sum_{n_z=1}^{n_{z\max}} \int_{E_{n_zD1}}^{\infty} \left[\frac{\sqrt{T_{22}(E,\eta_g)}}{T'_{22}(E,\eta_g)} T'_{1D}(E,\eta_g,n_z)\right]$$
$$\times \left[\frac{1}{1+\exp\left(\frac{E-E_{F23,1}}{k_BT}\right)} - \frac{1}{1+\exp\left(\frac{E-E_{F23,1}+eV_D}{k_BT}\right)}\right] dE \tag{23.2}$$

where $E_{F23,1}$ is the Fermi energy in this case which can be determined form

$$\frac{C_{ins}}{e}(V_T - V_{GS})$$
$$= \frac{m_\perp^* g_v}{2\pi\hbar^2} \text{ Real part of } \sum_{n_z=1}^{n_{z\max}} \Bigg[[T_{1D}(E_{F23,1}, \eta_g, n_z)$$
$$+ T_{1D}((E_{F23,1} - eV_D), \eta_g, n_z)] + \sum_{r=1}^{s} L(r)[T_{1D}(E_{F23,1}, \eta_g, n_z)$$
$$+ T_{1D}((E_{F23,1} - eV_D), \eta_g, n_z)] \Bigg] \tag{23.3}$$

23.2.2 *The $I_D - V_D$ equation for ballistic HDQWFETs of III–V, ternary and quaternary materials*

(i) Three-band model of Kane

From (8.39) we get

$$V_x = \sqrt{\frac{2}{m_c}} \left[\frac{\sqrt{T_{31}(E, \eta_g) + iT_{32}(E, \eta_g)}}{T'_{31}(E, \eta_g) + iT'_{32}(E, \eta_g)} \right] \tag{23.4}$$

Using (21.1), (21.2), (21.3), (11.8) and (23.4) we get

$$\frac{J_D}{W} = \left[\frac{eg_v}{\pi^2\hbar^2} \sqrt{2m_c} \right] \text{ Real part of } \sum_{n_z=1}^{n_{z\max}} \int_{E_{n_z D5}}^{\infty}$$
$$\times \left[\left[\frac{\sqrt{T_{31}(E, \eta_g) + iT_{32}(E, \eta_g)}}{T'_{31}(E, \eta_g) + iT'_{32}(E, \eta_g)} \right] [T'_{5D}(E, \eta_g, n_z)] \right.$$
$$\left. \times \left[\frac{1}{1 + \exp\left(\frac{E - E_{F23,2}}{k_B T} \right)} - \frac{1}{1 + \exp\left(\frac{E - E_{F23,2} + eV_D}{k_B T} \right)} \right] dE \right] \tag{23.5}$$

where $E_{F23,2}$ is the Fermi energy in this case which can be determined form

$$\frac{C_{ins}}{e}(V_T - V_{GS})$$
$$= \frac{m_c g_v}{2\pi\hbar^2} \text{ Real part of } \sum_{n_z=1}^{n_{z\max}} \Bigg[[T_{5D}(E_{F23,2}, \eta_g, n_z) + T_{5D}((E_{F23,2}$$

$$- eV_D), \eta_g, n_z)] + \sum_{r=1}^{s} L(r)[T_{5D}(E_{F23,2}, \eta_g, n_z) + T_{5D}((E_{F23,2} - eV_D), \eta_g, n_z)]\Bigg] \quad (23.6)$$

(ii) Two-band model of Kane

From (8.50) we get

$$V_x = \sqrt{\frac{2}{m_c}} \left[\frac{\sqrt{\gamma_2(E, \eta_g)}}{\gamma_2'(E, \eta_g)} \right] \quad (23.7)$$

Using (21.1), (21.2), (21.3), (11.15) and (23.7) we get

$$\frac{J_D}{W} = \left[\frac{eg_v}{\pi^2 \hbar^2} \sqrt{2m_c} \right] \sum_{n_z=1}^{n_{z\max}} \int_{E_{n_z D7}}^{\infty} \left[\left[\frac{\sqrt{\gamma_2(E, \eta_g)}}{\gamma_2'(E, \eta_g)} \right] [T_{7D}'(E, \eta_g, n_z)] \times \left[\frac{1}{1 + \exp\left(\frac{E - E_{F23,3}}{k_B T} \right)} - \frac{1}{1 + \exp\left(\frac{E - E_{F23,3} + eV_D}{k_B T} \right)} \right] dE \right] \quad (23.8)$$

where $E_{F23,3}$ is the Fermi energy in this case which can be determined form

$$\frac{C_{ins}}{e}(V_T - V_{GS}) = \frac{m_c g_v}{2\pi\hbar^2} \sum_{n_z=1}^{n_{z\max}} \left[[T_{7D}(E_{F23,3}, \eta_g, n_z) + T_{7D}((E_{F23,3} - eV_D), \eta_g, n_z)] + \sum_{r-1}^{s} L(r)[T_{7D}(E_{F23,3}, \eta_g, n_z) + T_{7D}((E_{F23,3} - eV_D), \eta_g, n_z)] \right] \quad (23.9)$$

(iii) Parabolic band model of Kane

From (8.60) we get

$$V_x = \sqrt{\frac{2}{m_c}} \left[\frac{\sqrt{\gamma_3(E, \eta_g)}}{\gamma_3'(E, \eta_g)} \right] \quad (23.10)$$

Using (21.1), (21.2), (21.3), (11.20) and (23.10) we get

$$\frac{J_D}{W} = \left[\frac{eg_v}{\pi^2\hbar^2}\sqrt{2m_c}\right]\sum_{n_z=1}^{n_{z\max}}\int_{E_{n_zD9}}^{\infty}\left[\left[\frac{\sqrt{\gamma_3(E,\eta_g)}}{\gamma_3'(E,\eta_g)}\right][T_{9D}'(E,\eta_g,n_z)]\right.$$
$$\left.\times\left[\frac{1}{1+\exp\left(\frac{E-E_{F23,4}}{k_BT}\right)}-\frac{1}{1+\exp\left(\frac{E-E_{F23,4}+eV_D}{k_BT}\right)}\right]dE\right] \quad (23.11)$$

where $E_{F23,4}$ is the Fermi energy in this case which can be determined form

$$\frac{C_{ins}}{e}(V_T - V_{GS})$$
$$= \frac{m_c g_v}{2\pi\hbar^2}\sum_{n_z=1}^{n_{z\max}}\left[[T_{9D}(E_{F23,4},\eta_g,n_z)+T_{9D}((E_{F23,4}-eV_D),\eta_g,n_z)]\right.$$
$$\left.+\sum_{r=1}^{s}L(r)[T_{9D}(E_{F23,4},\eta_g,n_z)+T_{9D}((E_{F23,4}-eV_D),\eta_g,n_z)]\right]$$
$$(23.12)$$

(iv) Model of Stillman *et al.*

From (11.23) we get

$$V_x = \sqrt{\frac{2}{m_c}}\left[\frac{\sqrt{I_{12}(E,\eta_g)}}{I_{12}'(E,\eta_g)}\right] \quad (23.13)$$

Using (21.1), (21.2), (21.3), (11.25) and (23.13) we get

$$\frac{J_D}{W} = \left[\frac{eg_v}{\pi^2\hbar^2}\sqrt{2m_c}\right]\sum_{n_z=1}^{n_{z\max}}\int_{E_{n_zD11}}^{\infty}\left[\left[\frac{\sqrt{I_{12}(E,\eta_g)}}{I_{12}'(E,\eta_g)}\right][T_{11D}'(E,\eta_g,n_z)]\right.$$
$$\left.\times\left[\frac{1}{1+\exp(\frac{E-E_{F23,5}}{k_BT})}-\frac{1}{1+\exp\left(\frac{E-E_{F23,5}+eV_D}{k_BT}\right)}\right]dE\right] \quad (23.14)$$

where $E_{F23,5}$ is the Fermi energy in this case which can be determined form

$$\frac{C_{ins}}{e}(V_T - V_{GS}) = \frac{m_c g_v}{2\pi\hbar^2}\sum_{n_z=1}^{n_{z\max}}\left[[T_{11D}(E_{F23,5},\eta_g,n_z)+T_{11D}((E_{F23,5}-eV_D),\eta_g,n_z)] + \sum_{r=1}^{s} L(r)[T_{11D}(E_{F23,5},\eta_g,n_z)+T_{11D}((E_{F23,5}-eV_D),\eta_g,n_z)]\right] \quad (23.15)$$

(v) Model of Palik *et al.*

From (11.31) we get

$$V_x = \sqrt{\frac{2}{m_c}}\left[\frac{\sqrt{I_{13}(E,\eta_g)}}{I'_{13}(E,\eta_g)}\right] \quad (23.16)$$

Using (21.1), (21.2), (21.3), (11.33) and (23.16) we get

$$\frac{J_D}{W} = \left[\frac{eg_v}{\pi^2\hbar^2}\sqrt{2m_c}\right]\sum_{n_z=1}^{n_{z\max}}\int_{E_{n_zD13}}^{\infty}\left[\left[\frac{\sqrt{I_{13}(E,\eta_g)}}{I'_{13}(E,\eta_g)}\right][T'_{13D}(E,\eta_g,n_z)] \times\left[\frac{1}{1+\exp\left(\frac{E-E_{F23,6}}{k_BT}\right)} - \frac{1}{1+\exp\left(\frac{E-E_{F23,6}+eV_D}{k_BT}\right)}\right]dE\right] \quad (23.17)$$

where $E_{F23,6}$ is the Fermi energy in this case which can be determined form

$$\frac{C_{ins}}{e}(V_T - V_{GS}) = \frac{m_c g_v}{2\pi\hbar^2}\sum_{n_z=1}^{n_{z\max}}[[T_{13D}(E_{F23,6},\eta_g,n_z)+T_{13D}((E_{F23,6}-eV_D),\eta_g,n_z)] + \sum_{r=1}^{s} L(r)[T_{13D}(E_{F23,6},\eta_g,n_z)+T_{13D}((E_{F23,6}-eV_D),\eta_g,n_z)]\Bigg] \quad (23.18)$$

23.2.3 The $I_D - V_D$ equation for ballistic HDQWFETs of II–VI materials

From (8.70) we get

$$V_x = \frac{1}{\hbar}\left[\frac{\sqrt{(\bar{\lambda}_0)^2 + 4a_0'\gamma_3(E,\eta_g)}}{\gamma_3'(E,\eta_g)}\right] \tag{23.19}$$

Using (21.1), (21.2), (21.3), (11.38) and (23.19) we get

$$\frac{J_D}{W} = \left[\frac{eg_v}{2\pi^2\hbar^3}m_\perp^*\right]\sum_{n_z=1}^{n_{z\max}}\int_{E_{n_zD14}}^{\infty}\left[\left[\left[\frac{\sqrt{(\bar{\lambda}_0)^2 + 4a_0'\gamma_3(E,\eta_g)}}{\gamma_3'(E,\eta_g)}\right]\right.\right.$$
$$\times\left[1 \mp \frac{(\bar{\lambda}_0)}{\left[(\bar{\lambda}_0)^2 - 4a_0'b_0'\left(\frac{n_z\pi}{d_z}\right)^2 + 4a_0'\gamma_3(E_{F1HD},\eta_g)\right]^{1/2}}\right][\gamma_3'(E_{F23,7},\eta_g)]$$
$$\left.\left.\times\left[\frac{1}{1+\exp\left(\frac{E-E_{F23,7}}{k_BT}\right)} - \frac{1}{1+\exp\left(\frac{E-E_{F23,7}+eV_D}{k_BT}\right)}\right]dE\right]\right] \tag{23.20}$$

where $E_{F23,7}$ is the Fermi energy in this case which can be determined form

$$\frac{C_{ins}}{e}(V_T - V_{GS}) = \frac{m_\perp^* g_v}{2\pi\hbar^2}\sum_{n_z=1}^{n_{z\max}}[(\gamma_3(E_{F23,7},\eta_g) - E_{n_ZD_{14}} + (\bar{\lambda})^2 m_\perp^*\hbar^{-2})]$$
$$+ [(\gamma_3(E_{F23,7} - eV_D),\eta_g) - E_{n_ZD_{14}} + (\bar{\lambda})^2 m_\perp^*\hbar^{-2})]$$
$$+ \sum_{r=1}^{s} L(r)[(\gamma_3(E_{F23,7},\eta_g) - E_{n_ZD_{14}} + (\bar{\lambda})^2 m_\perp^*\hbar^{-2})]$$
$$+ [(\gamma_3(E_{F23,7} - eV_D),\eta_g) - E_{n_ZD_{14}} + (\bar{\lambda})^2 m_\perp^*\hbar^{-2})]] \tag{23.21}$$

23.2.4 The $I_D - V_D$ equation for ballistic HDQWFETs of IV–VI materials

From (8.85) we get

$$V_x = \frac{2}{\hbar}\left[\frac{[F_{1HD}(E,\eta_g)]^{3/2}}{F_{1HD}'(E,\eta_g)}\right] \tag{23.22}$$

Using (21.1), (21.2), (21.3), (11.62) and (23.22) we get

$$\frac{J_D}{W} = \left[\frac{eg_v}{\pi^2\hbar}\right] \text{ Real part of } \sum_{n_z=1}^{n_{z\max}} \int_{E_{n_z 71HD}}^{\infty} \times \left[\left[\left[\frac{[F_{1HD}(E,\eta_g)]^{3/2}}{F'_{1HD}(E,\eta_g)}\right] [F'_6(E,\eta_g,n_z)] \left[\frac{1}{1+\exp\left(\frac{E-E_{F23,8}}{k_BT}\right)} - \frac{1}{1+\exp\left(\frac{E-E_{F23,8}+eV_D}{k_BT}\right)}\right] dE\right]\right] \tag{23.23}$$

where $E_{F23,8}$ is the Fermi energy in this case which can be determined form

$$\frac{C_{ins}}{e}(V_T - V_{GS}) = \left[\frac{g_v}{4\pi}\right] \text{ Real part of } \sum_{n_z=1}^{n_{z\max}} [F_6(E_{F23,8},\eta_g,n_z)] + [F_6((E_{F23,8} - eV_D),\eta_g,n_z)] + \sum_{r=1}^{s} L(r)[F_6(E_{F23,8},\eta_g,n_z)] + [F_6((E_{F23,8} - eV_D),\eta_g,n_z)] \tag{23.24}$$

23.2.5 *The $I_D - V_D$ equation for ballistic HDQWFETs of stressed Kane type materials*

From (8.106) we get

$$V_x = \frac{2}{\hbar}\left[\frac{[P_{11}(E,\eta_g)]^{3/2}}{P'_{11}(E,\eta_g)}\right] \tag{23.25}$$

Using (21.1), (21.2), (21.3), (11.68) and (23.25) we get

$$\frac{J_D}{W} = \left[\frac{eg_v}{\pi^2\hbar}\right] \sum_{n_z=1}^{n_{z\max}} \int_{E_{n_z 8HD}}^{\infty} \left[\left[\left[\frac{[P_{11}(E,\eta_g)]^{3/2}}{P'_{11}(E,\eta_g)}\right] [A'_{56}(E,\eta_g,n_z)] \times \left[\frac{1}{1+\exp\left(\frac{E-E_{F23,9}}{k_BT}\right)} - \frac{1}{1+\exp\left(\frac{E-E_{F23,9}+eV_D}{k_BT}\right)}\right] dE\right]\right] \tag{23.26}$$

where $E_{F23,9}$ is the Fermi energy in this case which can be determined form

$$\frac{C_{ins}}{e}(V_T - V_{GS}) = \left[\frac{g_v}{4\pi}\right] \sum_{n_z=1}^{n_{z\max}} [T_{57HD}(E_{F23,9}, \eta_g, n_z)] + [T_{57HD}((E_{F23,9} - eV_D), \eta_g, n_z)] + \sum_{r=1}^{s} L(r)[T_{57HD}(E_{F23,9}, \eta_g, n_z)] + [T_{57HD}((E_{F23,9} - eV_D), \eta_g, n_z)] \tag{23.27}$$

23.2.6 *The $I_D - V_D$ equation for ballistic HDQWFETs of tellurium*

From (11.71) we get

$$V_x = \frac{1}{\hbar}\left[\frac{\sqrt{\psi_4^2 + 4\psi_2\gamma_3(E, \eta_g)}}{\gamma_3'(E, \eta_g)}\right] \tag{23.28}$$

Using (21.1), (21.2), (21.3), (11.74) and (23.28) we get

$$\frac{J_D}{W} = \left[\frac{eg_v}{\pi^2\hbar}\right] \sum_{n_z=1}^{n_{z\max}} \int_{E_{n_z 59HD}}^{\infty} \left[\left[\left[\frac{\sqrt{\psi_4^2 + 4\psi_2\gamma_3(E, \eta_g)}}{\gamma_3'(E, \eta_g)}\right] [\psi_{5HD}(E, \eta_g, n_z)]' \times \left[\frac{1}{1 + \exp\left(\frac{E - E_{F23,10}}{k_B T}\right)} - \frac{1}{1 + \exp\left(\frac{E - E_{F23,10} + eV_D}{k_B T}\right)}\right] dE\right]\right] \tag{23.29}$$

where $E_{F23,10}$ is the Fermi energy in this case which can be determined form

$$\frac{C_{ins}}{e}(V_T - V_{GS}) = \left[\frac{g_v}{2\pi}\right] \sum_{n_z=1}^{n_{z\max}} [t_{1HDTE}(E_{F23,10}, \eta_g, n_z)] + [t_{1HDTE}((E_{F23,10} - eV_D), \eta_g, n_z)] + \sum_{r=1}^{s} L(r)[t_{1HDTE}(E_{F23,10}, \eta_g, n_z)] + [t_{1HDTE}((E_{F23,10} - eV_D), \eta_g, n_z)] \tag{23.30}$$

23.2.7 The $I_D - V_D$ equation for ballistic HDQWFETs of GaP

From (11.77) we get

$$V_x = \frac{2}{\hbar}\left[\frac{\sqrt{\psi_{444HD}(E,\eta_g)}}{\psi'_{444HD}(E,\eta_g)}\right] \tag{23.31}$$

where

$$\psi_{444HD}(E,\eta_g) = (2a^2)^{-1}[c + 2a((\gamma_3(E,\eta_g) - |V_G|) \pm\sqrt{[c + 2a((\gamma_3(E,\eta_g) - |V_G|)a]^2 - 4a^2[((\gamma_3(E,\eta_g) - |V_G|)^2 - |V_G|]]},$$

Using (21.1), (21.2), (21.3), (11.80) and (23.31) we get

$$\frac{J_D}{W} = \left[\frac{eg_v}{\pi^2\hbar}\right]\sum_{n_z=1}^{n_{z\max}}\int_{E_{n_z8THD}}^{\infty}\left[\left[\left[\frac{\sqrt{\psi_{444HD}(E,\eta_g)}}{\psi'_{444HD}(E,\eta_g)}\right]\left[t_{11}\gamma'_3(E_{F23,11},\eta_g) - t_{41}t_5(E_{F23,11},\eta_g)t'_5(E_{F23,11},\eta_g)\times\left[\left(\frac{\pi n_z}{d_z}\right)^2 + t_5^2(E_{F23,11},\eta_g)\right]^{-1/2}\right]\right.\right.$$
$$\left.\left.\times\left[\frac{1}{1+\exp\left(\frac{E-E_{F23,11}}{k_BT}\right)} - \frac{1}{1+\exp\left(\frac{E-E_{F23,11}+eV_D}{k_BT}\right)}\right]dE\right]\right] \tag{23.32}$$

where $E_{F23,11}$ is the Fermi energy in this case which can be determined form

$$\frac{C_{ins}}{e}(V_T - V_{GS}) = \left[\frac{g_v}{2\pi}\right]\sum_{n_z=1}^{n_{z\max}}[t_{3HDGaP}(E_{F23,11},\eta_g,n_z)] + [t_{3HDGaP}((E_{F23,11} - eV_D),\eta_g,n_z)] + \sum_{r=1}^{s}L(r)[t_{3HDGaP}(E_{F23,11},\eta_g,n_z)] + [t_{3HDGaP}((E_{F23,11} - eV_D),\eta_g,n_z)] \tag{23.33}$$

23.2.8 The $I_D - V_D$ equation for ballistic HDQWFETs of $PtSb_2$

From (11.84) we get

$$V_x = \frac{2}{\hbar}\left[\frac{\sqrt{\beta_{23}(E,\eta_g)}}{\beta'_{23}(E,\eta_g)}\right] \tag{23.34}$$

where $\beta_{23}(E,\eta_g) = [2T_{11}]^{-1}[T_{21}(E,\eta_g) \pm \sqrt{T_{21}^2(E,\eta_g) + 4T_{11}T_{61}(E,\eta_g)}]$.

Using (21.1), (21.2), (21.3), (11.88b) and (23.34) we get

$$\frac{J_D}{W} = \left[\frac{eg_v}{\pi^2\hbar}\right]\sum_{n_z=1}^{n_{z\max}}\int_{E_{n_{z9THD}}}^{\infty}\left[\left[\left[\frac{\sqrt{\beta_{23}(E,\eta_g)}}{\beta'_{23}(E,\eta_g)}\right][A_{60}(E,\eta_g,n_z)]'\right.\right.$$
$$\left.\left.\times\left[\frac{1}{1+\exp\left(\frac{E-E_{F23,12}}{k_BT}\right)} - \frac{1}{1+\exp\left(\frac{E-E_{F23,12}+eV_D}{k_BT}\right)}\right]dE\right]\right] \tag{23.35}$$

where $E_{F23,12}$ is the Fermi energy in this case which can be determined form

$$\frac{C_{ins}}{e}(V_T - V_{GS}) = \left[\frac{g_v}{4\pi}\right]\sum_{n_z=1}^{n_{z\max}}[A_{60}(E_{F23,12},\eta_g,n_z)] + [A_{60}((E_{F23,12}$$
$$- eV_D),\eta_g,n_z)] + \sum_{r=1}^{s}L(r)[A_{60}(E_{F23,12},\eta_g,n_z)]$$
$$+ [A_{60}((E_{F23,12} - eV_D),\eta_g,n_z)] \tag{23.36}$$

23.2.9 The $I_D - V_D$ equation for ballistic HDQWFETs of bismuth telluride

From (11.92) we can write

$$V_x = \left[\frac{2\sqrt{I_1}}{\hbar}\right]\left[\frac{\sqrt{\gamma_2(E,\eta_g)}}{\gamma'_2(E,\eta_g)}\right] \tag{23.37}$$

Using (21.1), (21.2), (21.3), (11.94b) and (23.37) we get

$$\frac{J_D}{W} = \left[\frac{eg_v}{\pi^2\hbar}\right]\sqrt{I_1}\sum_{n_z=1}^{n_{z\max}}\int_{E_{n_z 10THD}}^{\infty}\left[\left[\left[\frac{\sqrt{\gamma_2(E,\eta_g)}}{\gamma_2'(E,\eta_g)}\right][R_{60}(E,\eta_g,n_z)]'\right.\right.$$

$$\left.\left.\times\left[\frac{1}{1+\exp\left(\frac{E-E_{F23,13}}{k_BT}\right)} - \frac{1}{1+\exp\left(\frac{E-E_{F23,13}+eV_D}{k_BT}\right)}\right]dE\right]\right] \tag{23.38}$$

where $E_{F23,13}$ is the Fermi energy in this case which can be determined form

$$\frac{C_{ins}}{e}(V_T - V_{GS}) = \left[\frac{g_v}{4\pi}\right]\sum_{n_z=1}^{n_{z\max}}[R_{60}(E_{F23,13},\eta_g,n_z)] + [R_{60}((E_{F23,13}$$

$$- eV_D),\eta_g,n_z)] + \sum_{r=1}^{s}L(r)[R_{60}(E_{F23,13},\eta_g,n_z)]$$

$$+ [R_{60}((E_{F23,13} - eV_D),\eta_g,n_z)] \tag{23.39}$$

23.2.10 *The $I_D - V_D$ equation for ballistic HDQWFETs of germanium*

From (2.47) we can write

$$V_x = \left[\sqrt{\frac{2}{m_\perp^*}}\frac{\sqrt{\gamma_2(E,\eta_g)}}{\gamma_2'(E,\eta_g)}\right] \tag{23.40}$$

Using (21.1), (21.2), (21.3), (11.96c) and (23.40) we get

$$\frac{J_D}{W} = \left[\frac{eg_v}{\pi^2\hbar^2}\right]\sqrt{2m_\perp^*}\sum_{n_z=1}^{n_{z\max}}\int_{E_{n_z 11THD}}^{\infty}\left[\left[\left[\frac{\sqrt{\gamma_2(E,\eta_g)}}{\gamma_2'(E,\eta_g)}\right][R_1(E,\eta_g,n_z)]'\right.\right.$$

$$\left.\left.\times\left[\frac{1}{1+\exp\left(\frac{E-E_{F23,14}}{k_BT}\right)} - \frac{1}{1+\exp\left(\frac{E-E_{F23,14}+eV_D}{k_BT}\right)}\right]dE\right]\right] \tag{23.41}$$

where $E_{F23,13}$ is the Fermi energy in this case which can be determined form

$$\frac{C_{ins}}{e}(V_T - V_{GS}) = \left[\frac{g_v}{4\pi}\right]\sum_{n_z=1}^{n_{z\max}}[R_1(E_{F23,14}, \eta_g, n_z)] + [R_1((E_{F23,14} - eV_D), \eta_g, n_z)] + \sum_{r=1}^{s} L(r)[R_1(E_{F23,14}, \eta_g, n_z)] + [R_1((E_{F23,14} - eV_D), \eta_g, n_z)] \quad (23.42)$$

23.2.11 The $I_D - V_D$ equation for ballistic HDQWFETs of gallium antimonide

From (11.105) we get

$$V_x = \sqrt{\frac{2}{m_c}}\left[\frac{\sqrt{I_{36}(E, \eta_g)}}{I'_{36}(E, \eta_g)}\right] \quad (23.43)$$

Using (21.1), (21.2), (21.3), (11.107) and (23.43) we get

$$\frac{J_D}{W} = \left[\frac{eg_v}{\pi^2\hbar^2}\sqrt{2m_c}\right]\sum_{n_z=1}^{n_{z\max}}\int_{E_{n_zD119}}^{\infty}\left[\left[\frac{\sqrt{I_{36}(E, \eta_g)}}{I'_{36}(E, \eta_g)}\right][T'_{119D}(E, \eta_g, n_z)] \times \left[\frac{1}{1+\exp\left(\frac{E-E_{F23,15}}{k_BT}\right)} - \frac{1}{1+\exp\left(\frac{E-E_{F23,15}+eV_D}{k_BT}\right)}\right] dE\right] \quad (23.44)$$

where $E_{F23,15}$ is the Fermi energy in this case which can be determined form

$$\frac{C_{ins}}{e}(V_T - V_{GS}) = \frac{m_c g_v}{2\pi\hbar^2}\sum_{n_z=1}^{n_{z\max}}[[T_{119D}(E_{F23,15}, \eta_g, n_z) + T_{119D}((E_{F23,15} - eV_D), \eta_g, n_z)] + \sum_{r=1}^{s} L(r)[T_{119D}(E_{F23,15}, \eta_g, n_z) + T_{119D}((E_{F23,15} - eV_D), \eta_g, n_z)]] \quad (23.45)$$

23.3 Summary and Conclusions

In this chapter, in Section 23.2.1, we have formulated $I_D - V_D$ for ballistic HDQWFETs made of nonlinear optical materials as given by (23.2) for the computation of which we need the expression of the 2D electron statistics which is given in (23.3). The $I_D - V_D$ equations for ballistic HDQWFETs made of III–V, ternary and quaternary materials has been investigated in Section 23.2.2 in accordance with the models of three-band model of Kane, two-band model of Kane, parabolic energy bands, Stillman *et al.* and Palik *et al.* are given by (23.5), (23.8), (23.11), (23.14) and (23.17) and the corresponding 2D electron statistics are expressed through (23.6), (23.9), (23.12), (23.15) and (23.18), respectively. In Section 23.2.3, we have studied the $I_D - V_D$ equation for HDQWFETs of II–VI materials which is given by (23.20) and the corresponding 2D electron statistics has been represented in (23.21). In Section 23.2.4, we have studied the $I_D - V_D$ equation in HDQWFETs of IV–VI materials which is given by (23.23) and the corresponding 2D electron statistics has been represented in (23.24). In Section 23.2.5, we have investigated the $I_D - V_D$ equation in HDQWFETs of stressed Kane type materials which is given by (23.26) and the corresponding 2D electron statistics has been represented in (23.27). In Section 23.2.6, we have studied the $I_D - V_D$ equation in HDQWFETs of Te which is given by (23.29) and the corresponding 2D electron statistics has been represented in (23.30). In Section 23.2.7, we have studied the $I_D - V_D$ equation in QWFETs of GaP which is given by (23.32) and the corresponding 2D electron statistics has been represented in (23.33). In Section 23.2.8, we have investigated the $I_D - V_D$ equation in HDQWFETs of $PtSb_2$ which is given by (23.35) and the corresponding 2D electron statistics has been represented in (23.36). In Section 23.2.9, we have studied the $I_D - V_D$ equation in HDQWFETs of bismuth telluride materials which is given by (23.38) and the corresponding 2D electron statistics has been represented in (23.39). In Section 23.2.10, we have studied the $I_D - V_D$ equation in HDQWFETs of germanium materials which is given by (23.41) and the corresponding 2D electron statistics has been represented in (23.42). In Section 23.2.11, we have studied the $I_D - V_D$ equation in HDQWFETs of gallium antimonide materials which is given by (23.44) and the corresponding 2D electron statistics has been represented in (23.45).

We have not given the graphs of the $I_D - V_D$ equations for ballistic HDQWFETs made of many materials as discussed here since we feel that our readers will perform the intricate numerical computation to obtain the oscillatory plots for better assessment and joy of understanding. It is the band structure that changes in a fundamental way and consequently all the physical properties of all the electronic materials changes radically leading to new physical concepts.

23.4 Open Research Problem

(R23.1) Derive equations for ballistic HDQWFETs made of all the materials whose DRs are given in (R1.1). Draw the corresponding graphs and explain the results.

Chapter 24

$I_D - V_D$ Equation for Ballistic MAGNETO QWFETs

To see the world in a grain of sand, and the heaven in a wild flower, hold the Infinity in the palm of your hands and the Eternity in an hour.

24.1 Introduction

This is completely a new chapter where from Sections 24.2.1 up to 24.2.11 we study the equation in ballistic Magneto QWFETs of tetragonal and nonlinear optical, III–V ternary and quaternary, II–VI, IV–VI, stressed Kane type, Bi_2Te_3, Ge, GaAs, GaSb, InSb, Te, $PtSb_2$ and GaP materials, respectively. In Section 24.3, we present summary and conclusion and last Section 24.4 we offer one multidimensional open research problems.

24.2 Theoretical Background

24.2.1 *The $I_D - V_D$ equation for ballistic magneto QWFETs of tetragonal and nonlinear optical materials*

Using (21.7), (21.1), (21.2), (21.3) and (4.101) we get

$$\frac{J_D}{W} - \frac{e^2 B g_v}{\pi^2 \hbar^2} \sum_{n=0}^{n_{\max}} \sum_{i=0}^{i_{\max}} [[g'_{21,1}(E_{77})]^{-1}[F_{-1}(\eta_{24,4,1}) \quad F_{-1}(\theta_{24,4,1})]] \tag{24.1}$$

where

$$\eta_{24,4,1} = (k_B T)^{-1}[E_{F24,1} - E_{77}], \quad \theta_{24,4,1} = \left[\eta_{24,4,1} - \frac{eV_D}{k_B T}\right]$$

and $E_{F24,1}$ is the Fermi energy in the present case which can be determined from

$$\frac{C_{ins}}{e}(V_T - V_{GS}) = \frac{g_v eB}{2\pi\hbar}\sum_{n=0}^{n_{\max}}\sum_{i=0}^{i_{\max}}[F_{-1}(\eta_{24,4,1}) + F_{-1}(\theta_{24,4,1})] \quad (24.2)$$

24.2.2 *The $I_D - V_D$ equation for ballistic magneto QWFETs of III–V, ternary and quaternary materials*

(a) Three-band model of Kane

Using (21.12), (21.1), (21.2), (21.3) and (4.104) we get

$$\frac{J_D}{W} = \frac{e^2 B g_v}{\pi^2\hbar}\sqrt{\frac{2}{m_c}}\sum_{n=0}^{n_{\max}}\sum_{i=0}^{i_{\max}}\left[\left[\sqrt{I_{11}(E_{78})}\right][I'_{11}(E_{78})]^{-1}\right.$$
$$\left.\times\,[F_{-1}(\eta_{24,4,3}) - F_{-1}(\theta_{24,4,3})]\right] \quad (24.3)$$

where

$$\eta_{24,4,3} = (k_B T)^{-1}[E_{F24,3} - E_{78}], \quad \theta_{24,4,3} = \left[\eta_{24,4,3} - \frac{eV_D}{k_B T}\right]$$

and $E_{F24,3}$ is the Fermi energy in the present case which can be determined from

$$\frac{C_{ins}}{e}(V_T - V_{GS}) = \frac{g_v eB}{2\pi\hbar}\sum_{n=0}^{n_{\max}}\sum_{i=0}^{i_{\max}}[F_{-1}(\eta_{24,4,3}) + F_{-1}(\theta_{24,4,3})] \quad (24.4)$$

(b) Two-band model of Kane

Using (21.15), (21.1), (21.2), (21.3) and (4.107) we get

$$\frac{J_D}{W} = \frac{e^2 B g_v}{\pi^2\hbar}\sqrt{\frac{2}{m_c}}\sum_{n=0}^{n_{\max}}\sum_{i=0}^{i_{\max}}\left[\left[\sqrt{E_{79}(1+\alpha E_{79})}\right]\right.$$
$$\left.\times\,[1+2\alpha E_{79}]^{-1}[F_{-1}(\eta_{24,4,4}) - F_{-1}(\theta_{24,4,4})]\right] \quad (24.5)$$

where

$$\eta_{24,4,4} = (k_B T)^{-1}[E_{F24,4} - E_{79}], \quad \theta_{24,4,4} = \left[\eta_{24,4,4} - \frac{eV_D}{k_B T}\right]$$

and $E_{F24,4}$ is the Fermi energy in the present case which can be determined from

$$\frac{C_{ins}}{e}(V_T - V_{GS}) = \frac{g_v eB}{2\pi\hbar} \sum_{n=0}^{n_{\max}} \sum_{i=0}^{i_{\max}} [F_{-1}(\eta_{24,4,4}) + F_{-1}(\theta_{24,4,4})] \tag{24.6}$$

(c) Parabolic band model

Using (21.18), (21.1), (21.2), (21.3) and (4.110) we get

$$\frac{J_D}{W} = \frac{e^2 B g_v}{\pi^2 \hbar} \sqrt{\frac{2}{m_c}} \sum_{n=0}^{n_{\max}} \sum_{i=0}^{i_{\max}} \left[\left[\sqrt{E_{80}} \right] [F_{-1}(\eta_{24,4,5}) - F_{-1}(\theta_{24,4,5})] \right] \tag{24.7}$$

where

$$\eta_{24,4,5} = (k_B T)^{-1}[E_{F24,5} - E_{80}], \quad \theta_{24,4,5} = \left[\eta_{24,4,5} - \frac{eV_D}{k_B T} \right]$$

and $E_{F24,5}$ is the Fermi energy in the present case which can be determined from

$$\frac{C_{ins}}{e}(V_T - V_{GS}) = \frac{g_v eB}{2\pi\hbar} \sum_{n=0}^{n_{\max}} \sum_{i=0}^{i_{\max}} [F_{-1}(\eta_{24,4,5}) + F_{-1}(\theta_{24,4,5})] \tag{24.8}$$

24.2.3 *The $I_D - V_D$ equation for ballistic magneto QWFETs of II–VI materials*

Using (21.24), (21.1), (21.2), (21.3) and (4.113) we get

$$\frac{J_D}{W} = \frac{e^2 B g_v}{\pi^2 \hbar^2} \sum_{n=0}^{n_{\max}} \sum_{i=0}^{i_{\max}} \left[\left[\sqrt{C_0^2 + \frac{2E_{81}\hbar^2}{m_\perp^*}} \right] [F_{-1}(\eta_{24,4,6}) - F_{-1}(\theta_{24,4,6})] \right] \tag{24.9}$$

where

$$\eta_{24,4,6} - (k_B T)^{-1}[E_{F24,6} - E_{81}], \quad \theta_{24,4,6} = \left[\eta_{24,4,6} - \frac{eV_D}{k_B T} \right]$$

and $E_{F24,6}$ is the Fermi energy in the present case which can be determined from

$$\frac{C_{ins}}{e}(V_T - V_{GS}) = \frac{g_v eB}{4\pi\hbar} \sum_{n=0}^{n_{\max}} \sum_{i=0}^{i_{\max}} [F_{-1}(\eta_{24,4,6}) + F_{-1}(\theta_{24,4,6})] \tag{24.10}$$

24.2.4 The $I_D - V_D$ equation for ballistic magneto QWFETs of IV–VI materials

(i) McClure and Choi model

Using (22.11), (21.1), (21.2), (21.3) and (4.116) we get

$$\frac{J_D}{W} = \frac{e^2 B g_v}{\pi^2 \hbar}\sqrt{\frac{2}{m_1}}\sum_{n=0}^{n_{\max}}\sum_{i=0}^{i_{\max}}\left[\left[\sqrt{E_{82}(1+\alpha E_{82})}\right]\right.$$
$$\left.\times\,[1+2\alpha E_{82}]^{-1}[F_{-1}(\eta_{24,4,7}) - F_{-1}(\theta_{24,4,7})]\right] \quad (24.11)$$

where

$$\eta_{24,4,7} = (k_B T)^{-1}[E_{F24,7} - E_{82}], \quad \theta_{24,4,7} = \left[\eta_{24,4,7} - \frac{eV_D}{k_B T}\right]$$

and $E_{F24,7}$ is the Fermi energy in the present case which can be determined from

$$\frac{C_{ins}}{e}(V_T - V_{GS}) = \frac{g_v e B}{2\pi\hbar}\sum_{n=0}^{n_{\max}}\sum_{i=0}^{i_{\max}}[F_{-1}(\eta_{24,4,7}) + F_{-1}(\theta_{24,4,7})] \quad (24.12)$$

(ii) Hybrid model

Using (22.14), (21.1), (21.2), (21.3) and (4.119) we get

$$\frac{J_D}{W} = \frac{e^2 B g_v}{\pi^2 \hbar}\sqrt{\frac{2}{m_1}}\sum_{n=0}^{n_{\max}}\sum_{i=0}^{i_{\max}}\left[\left[\sqrt{E_{83}(1+\alpha E_{83})}\right]\right.$$
$$\left.\times\,[1+2\alpha E_{83}]^{-1}[F_{-1}(\eta_{24,4,8}) - F_{-1}(\theta_{24,4,8})]\right] \quad (24.13a)$$

where

$$\eta_{24,4,8} = (k_B T)^{-1}[E_{F24,8} - E_{83}], \quad \theta_{24,4,8} = \left[\eta_{24,4,8} - \frac{eV_D}{k_B T}\right]$$

and $E_{F24,8}$ is the Fermi energy in the present case which can be determined from

$$\frac{C_{ins}}{e}(V_T - V_{GS}) = \frac{g_v e B}{2\pi\hbar}\sum_{n=0}^{n_{\max}}\sum_{i=0}^{i_{\max}}[F_{-1}(\eta_{24,4,8}) + F_{-1}(\theta_{24,4,8})] \quad (24.13b)$$

(iii) Cohen model

Using (3.43), (21.1), (21.2), (21.3) and (4.122) we get

$$\frac{J_D}{W} = \frac{e^2 B g_v}{\pi^2 \hbar}\sqrt{\frac{2}{m_1}}\sum_{n=0}^{n_{\max}}\sum_{i=0}^{i_{\max}}\left[\left[\sqrt{E_{84}(1+\alpha E_{84})}\right]\right.$$
$$\left.\times\,[1+2\alpha E_{84}]^{-1}[F_{-1}(\eta_{24,4,9}) - F_{-1}(\theta_{24,4,9})]\right] \quad (24.14a)$$

where

$$\eta_{24,4,9} = (k_B T)^{-1}[E_{F24,9} - E_{84}], \quad \theta_{24,4,9} = \left[\eta_{24,4,9} - \frac{eV_D}{k_B T}\right]$$

and $E_{F24,9}$ is the Fermi energy in the present case which can be determined from

$$\frac{C_{ins}}{e}(V_T - V_{GS}) = \frac{g_v e B}{2\pi\hbar}\sum_{n=0}^{n_{\max}}\sum_{i=0}^{i_{\max}}[F_{-1}(\eta_{24,4,9}) + F_{-1}(\theta_{24,4,9})] \quad (24.14b)$$

(iv) Lax model

Using (3.49), (21.1), (21.2), (21.3) and (4.125) we get

$$\frac{J_D}{W} = \frac{e^2 B g_v}{\pi^2 \hbar}\sqrt{\frac{2}{m_1}}\sum_{n=0}^{n_{\max}}\sum_{i=0}^{i_{\max}}\left[\left[\sqrt{E_{85}(1+\alpha E_{85})}\right]\right.$$
$$\left.\times\,[1+2\alpha E_{85}]^{-1}[F_{-1}(\eta_{24,4,10}) - F_{-1}(\theta_{24,4,10})]\right] \quad (24.15)$$

where

$$\eta_{24,4,10} = (k_B T)^{-1}[E_{F24,10} - E_{85}], \quad \theta_{24,4,10} = \left[\eta_{24,4,10} - \frac{eV_D}{k_D T}\right]$$

and $E_{F24,10}$ is the Fermi energy in the present case which can be determined from

$$\frac{C_{ins}}{e}(V_T - V_{GS}) = \frac{g_v e B}{2\pi\hbar}\sum_{n=0}^{n_{\max}}\sum_{i=0}^{i_{\max}}[F_{-1}(\eta_{24,4,10}) + F_{-1}(\theta_{24,4,10})] \quad (24.16)$$

(v) Dimmock model

Using (22.21), (21.1), (21.2), (21.3) and (4.128) we get

$$\frac{J_D}{W} = \frac{e^2 B g_v}{\pi^2 \hbar}\sum_{n=0}^{n_{\max}}\sum_{i=0}^{i_{\max}}[[\phi'_{227}(E_{86})]^{-1}[F_{-1}(\eta_{24,4,11}) - F_{-1}(\theta_{24,4,11})]] \quad (24.17)$$

where

$$\eta_{24,4,11} = (k_BT)^{-1}[E_{F24,11} - E_{86}], \quad \theta_{24,4,11} = \left[\eta_{24,4,11} - \frac{eV_D}{k_BT}\right]$$

and $E_{F24,11}$ is the Fermi energy in the present case which can be determined from

$$\frac{C_{ins}}{e}(V_T - V_{GS}) = \frac{g_v eB}{2\pi\hbar}\sum_{n=0}^{n_{\max}}\sum_{i=0}^{i_{\max}}[F_{-1}(\eta_{24,4,11}) + F_{-1}(\theta_{24,4,11})] \quad (24.18)$$

(vi) Ellipsoidal parabolic model

Using (21.1), (21.2), (21.3) and (4.131) we get

$$\frac{J_D}{W} = \frac{e^2Bg_v}{\pi^2\hbar}\sqrt{\frac{2}{m_1}}\sum_{n=0}^{n_{\max}}\sum_{i=0}^{i_{\max}}\left[\left[\sqrt{E_{87}}\right][F_{-1}(\eta_{24,4,12}) - F_{-1}(\theta_{24,4,12})]\right] \quad (24.19)$$

where

$$\eta_{24,4,12} = (k_BT)^{-1}[E_{F24,12} - E_{87}], \quad \theta_{24,4,12} = \left[\eta_{24,4,12} - \frac{eV_D}{k_BT}\right]$$

and $E_{F24,12}$ is the Fermi energy in the present case which can be determined from

$$\frac{C_{ins}}{e}(V_T - V_{GS}) = \frac{g_v eB}{2\pi\hbar}\sum_{n=0}^{n_{\max}}\sum_{i=0}^{i_{\max}}[F_{-1}(\eta_{24,4,12}) + F_{-1}(\theta_{24,4,12})] \quad (24.20)$$

24.2.5 *The $I_D - V_D$ equation for ballistic magneto QWFETs of stressed Kane type materials*

Using (21.32), (21.1), (21.2), (21.3) and (4.134) we get

$$\frac{J_D}{W} = \frac{e^2Bg_v}{\pi^2\hbar^2}\sum_{n=0}^{n_{\max}}\sum_{i=0}^{i_{\max}}[[[\bar{a}_0(E_{88})]']^{-1}[F_{-1}(\eta_{24,4,13}) - F_{-1}(\theta_{24,4,13})]] \quad (24.21)$$

where

$$\eta_{24,4,13} = (k_BT)^{-1}[E_{F24,13} - E_{88}], \quad \theta_{24,4,13} = \left[\eta_{24,4,13} - \frac{eV_D}{k_BT}\right]$$

and $E_{F24,13}$ is the Fermi energy in the present case which can be determined from

$$\frac{C_{ins}}{e}(V_T - V_{GS}) = \frac{g_v eB}{2\pi\hbar} \sum_{n=0}^{n_{\max}} \sum_{i=0}^{i_{\max}} [F_{-1}(\eta_{24,4,13}) + F_{-1}(\theta_{24,4,13})] \quad (24.22)$$

24.2.6 The $I_D - V_D$ equation for ballistic magneto QWFETs of bismuth telluride

Using (21.37), (21.1), (21.2), (21.3) and (4.137) we get

$$\frac{J_D}{W} = \frac{2e^2 B g_v}{\pi^2 \hbar^2} \sqrt{I_1} \sum_{n=0}^{n_{\max}} \sum_{i=0}^{i_{\max}} \Big[\Big[\sqrt{E_{89}(1+\alpha E_{89})} \Big] \times [1 + 2\alpha E_{89}]^{-1} [F_{-1}(\eta_{24,4,14}) - F_{-1}(\theta_{24,4,14})] \Big] \quad (24.23)$$

where

$$\eta_{24,4,14} = (k_B T)^{-1} [E_{F24,14} - E_{89}], \quad \theta_{24,4,14} = \left[\eta_{24,4,14} - \frac{eV_D}{k_B T} \right]$$

and $E_{F24,14}$ is the Fermi energy in the present case which can be determined from

$$\frac{C_{ins}}{e}(V_T - V_{GS}) = \frac{g_v eB}{2\pi\hbar} \sum_{n=0}^{n_{\max}} \sum_{i=0}^{i_{\max}} [F_{-1}(\eta_{24,4,14}) + F_{-1}(\theta_{24,4,14})] \quad (24.24)$$

24.2.7 The $I_D - V_D$ equation for ballistic magneto QWFETs of germanium

(i) Model of Cardona *et al.*

Using (21.42), (21.1), (21.2), (21.3) and (4.140) we get

$$\frac{J_D}{W} = \frac{e^2 B g_v}{\pi^2 \hbar} \sqrt{\frac{2}{m_\perp^*}} \sum_{n=0}^{n_{\max}} \sum_{i=0}^{i_{\max}} \Big[\Big[\sqrt{E_{90}(1+\alpha E_{90})} \Big] \times [1 + 2\alpha E_{90}]^{-1} [F_{-1}(\eta_{24,4,15}) - F_{-1}(\theta_{24,4,15})] \Big] \quad (24.25)$$

where

$$\eta_{24,4,15} = (k_B T)^{-1} [E_{F24,15} - E_{90}], \quad \theta_{24,4,15} = \left[\eta_{24,4,15} - \frac{eV_D}{k_B T} \right]$$

and $E_{F24,15}$ is the Fermi energy in the present case which can be determined from

$$\frac{C_{ins}}{e}(V_T - V_{GS}) = \frac{g_v eB}{2\pi\hbar} \sum_{n=0}^{n_{\max}} \sum_{i=0}^{i_{\max}} [F_{-1}(\eta_{24,4,15}) + F_{-1}(\theta_{24,4,15})] \quad (24.26)$$

(ii) Model of Wang *et al.*

Using (22.41), (21.1), (21.2), (21.3) and (4.142) we get

$$\frac{J_D}{W} = \frac{e^2 Bg_v}{\pi^2\hbar} \sum_{n=0}^{n_{\max}} \sum_{i=0}^{i_{\max}} [[[\omega'_{222}(E_{91})]^{-1}][F_{-1}(\eta_{24,4,16}) - F_{-1}(\theta_{24,4,16})]] \quad (24.27)$$

where

$$\eta_{24,4,16} = (k_B T)^{-1}[E_{F24,16} - E_{91}], \quad \theta_{24,4,16} = \left[\eta_{24,4,16} - \frac{eV_D}{k_B T}\right]$$

and $E_{F24,16}$ is the Fermi energy in the present case which can be determined from

$$\frac{C_{ins}}{e}(V_T - V_{GS}) = \frac{g_v eB}{2\pi\hbar} \sum_{n=0}^{n_{\max}} \sum_{i=0}^{i_{\max}} [F_{-1}(\eta_{24,4,16}) + F_{-1}(\theta_{24,4,16})] \quad (24.28)$$

24.2.8 *The $I_D - V_D$ equation for ballistic magneto QWFETs of gallium arsenide*

(i) Rossler model

Using (21.45), (21.1), (21.2), (21.3) and (4.146) we get

$$\frac{J_D}{W} = \frac{e^2 Bg_v}{\pi^2\hbar^2} \sum_{n=0}^{n_{\max}} \sum_{i=0}^{i_{\max}} \left[\left[\frac{\sqrt{\gamma_{4,\pm}(E_{92})}}{\gamma'_{4,\pm}(E_{92})}\right][F_{-1}(\eta_{24,4,17}) - F_{-1}(\theta_{24,4,17})]\right] \quad (24.29)$$

where

$$\eta_{24,4,17} = (k_B T)^{-1}[E_{F24,17} - E_{92}], \quad \theta_{24,4,17} = \left[\eta_{24,4,17} - \frac{eV_D}{k_B T}\right]$$

and $E_{F24,17}$ is the Fermi energy in the present case which can be determined from

$$\frac{C_{ins}}{e}(V_T - V_{GS}) = \frac{g_v eB}{4\pi\hbar} \sum_{n=0}^{n_{\max}} \sum_{i=0}^{i_{\max}} [F_{-1}(\eta_{24,4,17}) + F_{-1}(\theta_{24,4,17})] \quad (24.30)$$

(ii) Stillman model

Using (21.50), (21.1), (21.2), (21.3) and (4.149) we get

$$\frac{J_D}{W} = \frac{2e^2 B g_v}{\pi^2 \hbar^2} \sum_{n=0}^{n_{\max}} \sum_{i=0}^{i_{\max}} \left[\left[\frac{\sqrt{\gamma_5(E_{93})}}{\gamma_5'(E_{93})} \right] [F_{-1}(\eta_{24,4,18}) - F_{-1}(\theta_{24,4,18})] \right] \tag{24.31}$$

where

$$\eta_{24,4,18} = (k_B T)^{-1} [E_{F24,18} - E_{93}], \quad \theta_{24,4,18} = \left[\eta_{24,4,18} - \frac{eV_D}{k_B T} \right]$$

and $E_{F24,18}$ is the Fermi energy in the present case which can be determined from

$$\frac{C_{ins}}{e}(V_T - V_{GS}) = \frac{g_v e B}{2\pi\hbar} \sum_{n=0}^{n_{\max}} \sum_{i=0}^{i_{\max}} [F_{-1}(\eta_{24,4,18}) + F_{-1}(\theta_{24,4,18})] \tag{24.32}$$

24.2.9 *The $I_D - V_D$ equation for ballistic magneto QWFETs of gallium antimonide*

(a) Model of Sailer *et al.*

Using (21.54), (21.1), (21.2), (21.3) and (4.152) we get

$$\frac{J_D}{W} = \frac{e^2 B g_v}{\pi^2 \hbar^2} \sum_{n=0}^{n_{\max}} \sum_{i=0}^{i_{\max}} \left[\left[\frac{\sqrt{\gamma_6(E_{94})}}{\gamma_6'(E_{94})} \right] [F_{-1}(\eta_{24,4,19}) - F_{-1}(\theta_{24,4,19})] \right] \tag{24.33}$$

where

$$\eta_{24,4,19} = (k_B T)^{-1} [E_{F24,19} - E_{94}], \quad \theta_{24,4,19} = \left[\eta_{24,4,19} - \frac{eV_D}{k_B T} \right]$$

and $E_{F24,19}$ is the Fermi energy in the present case which can be determined from

$$\frac{C_{ins}}{e}(V_T - V_{GS}) = \frac{g_v e B}{4\pi\hbar} \sum_{n=0}^{n_{\max}} \sum_{i=0}^{i_{\max}} [F_{-1}(\eta_{24,4,19}) + F_{-1}(\theta_{24,4,19})] \tag{24.34}$$

(b) Model of Mathur and Jain

Using (21.59), (21.1), (21.2), (21.3) and (4.155) we get

$$\frac{J_D}{W} = \frac{2e^2 B g_v}{\pi^2 \hbar^2} \sum_{n=0}^{n_{\max}} \sum_{i=0}^{i_{\max}} \left[\left[\frac{\sqrt{\gamma_7(E_{95})}}{\gamma_7'(E_{95})} \right] [F_{-1}(\eta_{24,4,20}) - F_{-1}(\theta_{24,4,20})] \right] \tag{24.35}$$

where

$$\eta_{24,4,20} = (k_B T)^{-1}[E_{F24,20} - E_{95}], \quad \theta_{24,4,20} = \left[\eta_{24,4,20} - \frac{eV_D}{k_B T}\right]$$

and $E_{F24,20}$ is the Fermi energy in the present case which can be determined from

$$\frac{C_{ins}}{e}(V_T - V_{GS}) = \frac{g_v eB}{2\pi\hbar}\sum_{n=0}^{n_{\max}}\sum_{i=0}^{i_{\max}}[F_{-1}(\eta_{24,4,20}) + F_{-1}(\theta_{24,4,20})] \quad (24.36)$$

(c) Model of Zhang

Using (21.64), (21.1), (21.2), (21.3) and (4.158) we get

$$\frac{J_D}{W} = \frac{2e^2 B g_v}{\pi^2\hbar^2}\sum_{n=0}^{n_{\max}}\sum_{i=0}^{i_{\max}}\left[\left[\frac{\sqrt{\gamma_8(E_{96})}}{\gamma_8'(E_{96})}\right][F_{-1}(\eta_{24,4,21}) - F_{-1}(\theta_{24,4,21})]\right] \quad (24.37)$$

where

$$\eta_{24,4,21} = (k_B T)^{-1}[E_{F24,21} - E_{96}], \quad \theta_{24,4,21} = \left[\eta_{24,4,21} - \frac{eV_D}{k_B T}\right]$$

and $E_{F24,21}$ is the Fermi energy in the present case which can be determined from

$$\frac{C_{ins}}{e}(V_T - V_{GS}) = \frac{g_v eB}{2\pi\hbar}\sum_{n=0}^{n_{\max}}\sum_{i=0}^{i_{\max}}[F_{-1}(\eta_{24,4,21}) + F_{-1}(\theta_{24,4,21})] \quad (24.38)$$

24.2.10 *The $I_D - V_D$ equation for ballistic magneto QWFETs of indium antimonide*

(a) Model of Palik *et al.*

Using (21.69), (21.1), (21.2), (21.3) and (4.161) we get

$$\frac{J_D}{W} = \frac{2e^2 B g_v}{\pi^2\hbar^2}\sum_{n=0}^{n_{\max}}\sum_{i=0}^{i_{\max}}\left[\left[\frac{\sqrt{\gamma_9(E_{97})}}{\gamma_9'(E_{97})}\right][F_{-1}(\eta_{24,4,22}) - F_{-1}(\theta_{24,4,22})]\right] \quad (24.39)$$

where

$$\eta_{24,4,22} = (k_B T)^{-1}[E_{F24,22} - E_{97}], \quad \theta_{24,4,22} = \left[\eta_{24,4,21} - \frac{eV_D}{k_B T}\right]$$

and $E_{F24,22}$ is the Fermi energy in the present case which can be determined from

$$\frac{C_{ins}}{e}(V_T - V_{GS}) = \frac{g_v eB}{2\pi\hbar}\sum_{n=0}^{n_{\max}}\sum_{i=0}^{i_{\max}}[F_{-1}(\eta_{24,4,22}) + F_{-1}(\theta_{24,4,22})] \quad (24.40)$$

(b) Model of Johnson *et al.*

Using (21.74), (21.1), (21.2), (21.3) and (4.164) we get

$$\frac{J_D}{W} = \frac{2e^2 Bg_v}{\pi^2\hbar^2}\sum_{n=0}^{n_{\max}}\sum_{i=0}^{i_{\max}}\left[\left[\frac{\sqrt{\gamma_{10}(E_{98})}}{\gamma'_{10}(E_{98})}\right][F_{-1}(\eta_{24,4,23}) - F_{-1}(\theta_{24,4,23})]\right] \quad (24.41)$$

where

$$\eta_{24,4,23} = (k_B T)^{-1}[E_{F24,23} - E_{98}], \quad \theta_{24,4,23} = \left[\eta_{24,4,23} - \frac{eV_D}{k_B T}\right]$$

and $E_{F24,23}$ is the Fermi energy in the present case which can be determined from

$$\frac{C_{ins}}{e}(V_T - V_{GS}) = \frac{g_v eB}{2\pi\hbar}\sum_{n=0}^{n_{\max}}\sum_{i=0}^{i_{\max}}[F_{-1}(\eta_{24,4,23}) + F_{-1}(\theta_{24,4,23})] \quad (24.42)$$

(c) Model of Agafonov *et al.*

Using (21.79), (21.1), (21.2), (21.3) and (4.167) we get

$$\frac{J_D}{W} = \frac{2e^2 Bg_v}{\pi^2\hbar^2}\sum_{n=0}^{n_{\max}}\sum_{i=0}^{i_{\max}}\left[\left[\frac{\sqrt{\gamma_{11}(E_{99})}}{\gamma'_{11}(E_{99})}\right][F_{-1}(\eta_{24,4,24}) - F_{-1}(\theta_{24,4,24})]\right] \quad (24.43)$$

where

$$\eta_{24,4,24} = (k_B T)^{-1}[E_{F24,24} - E_{99}], \quad \theta_{24,4,24} = \left[\eta_{24,4,24} - \frac{eV_D}{k_B T}\right]$$

and $E_{F24,24}$ is the Fermi energy in the present case which can be determined from

$$\frac{C_{ins}}{e}(V_T - V_{GS}) = \frac{g_v eB}{2\pi\hbar}\sum_{n=0}^{n_{\max}}\sum_{i=0}^{i_{\max}}[F_{-1}(\eta_{24,4,24}) + F_{-1}(\theta_{24,4,24})] \quad (24.44)$$

24.2.11 The $I_D - V_D$ equation for ballistic magneto QWFETs of tellurium

Using (22.28), (21.1), (21.2), (21.3) and (4.170) we get

$$\frac{J_D}{W}=\frac{e^2Bg_v}{2\pi^2\hbar^2}\sum_{n=0}^{n_{\max}}\sum_{i=0}^{i_{\max}}\left[\left[\sqrt{\psi_4^2+4\psi_2E_{100}}\right][F_{-1}(\eta_{24,4,25})-F_{-1}(\theta_{24,4,25})]\right] \tag{24.45}$$

where

$$\eta_{24,4,25}=(k_BT)^{-1}[E_{F24,25}-E_{100}],\quad \theta_{24,4,25}=\left[\eta_{24,4,25}-\frac{eV_D}{k_BT}\right]$$

and $E_{F24,25}$ is the Fermi energy in the present case which can be determined from

$$\frac{C_{ins}}{e}(V_T-V_{GS})=\frac{g_veB}{4\pi\hbar}\sum_{n=0}^{n_{\max}}\sum_{i=0}^{i_{\max}}[F_{-1}(\eta_{24,4,25})+F_{-1}(\theta_{24,4,25})] \tag{24.46}$$

24.2.12 The $I_D - V_D$ equation for ballistic magneto QWFETs of $PtSb_2$

Using (22.34), (21.1), (21.2), (21.3) and (4.173) we get

$$\frac{J_D}{W}=\frac{e^2Bg_v}{2\pi^2\hbar^2}\sum_{n=0}^{n_{\max}}\sum_{i=0}^{i_{\max}}[[[t'_{222}(E_{101})]^{-1}][F_{-1}(\eta_{24,4,26})-F_{-1}(\theta_{24,4,26})]] \tag{24.47}$$

where

$$\eta_{24,4,26}=(k_BT)^{-1}[E_{F24,26}-E_{101}],\quad \theta_{24,4,26}=\left[\eta_{24,4,26}-\frac{eV_D}{k_BT}\right]$$

and $E_{F24,26}$ is the Fermi energy in the present case which can be determined from

$$\frac{C_{ins}}{e}(V_T-V_{GS})=\frac{g_veB}{4\pi\hbar}\sum_{n=0}^{n_{\max}}\sum_{i=0}^{i_{\max}}[F_{-1}(\eta_{24,4,26})+F_{-1}(\theta_{24,4,26})] \tag{24.48}$$

24.2.13 The $I_D - V_D$ equation for ballistic magneto QWFETs of GaP

Using (22.31), (21.1), (21.2), (21.3) and (4.176) we get

$$\frac{J_D}{W}=\frac{e^2Bg_v}{\pi^2\hbar^2}\sum_{n=0}^{n_{\max}}\sum_{i=0}^{i_{\max}}[[\sqrt{\frac{\psi_{444}(E_{102})}{\psi'_{444}(E_{102})}}][F_{-1}(\eta_{24,4,27})-F_{-1}(\theta_{24,4,27})]] \tag{24.49}$$

where

$$\psi_{444}(E_{102}) = \left[\frac{c + 2a(E - |V_G|) \pm \sqrt{[c + 2a(E - |V_G|)a]^2 - 4a^2[(E - |V_G|)^2 - |V_G|]}}{2a^2}\right],$$

$$\eta_{24,4,27} = (k_B T)^{-1}[E_{F24,27} - E_{102}], \quad \theta_{24,4,27} = \left[\eta_{24,4,27} - \frac{eV_D}{k_B T}\right]$$

and $E_{F24,27}$ is the Fermi energy in the present case which can be determined from

$$\frac{C_{ins}}{e}(V_T - V_{GS}) = \frac{g_v eB}{4\pi\hbar}\sum_{n=0}^{n_{\max}}\sum_{i=0}^{i_{\max}}[F_{-1}(\eta_{24,4,27}) + F_{-1}(\theta_{24,4,27})] \quad (24.50)$$

24.3 Summary and Conclusions

In this chapter, in Section 24.2.1, we have formulated $I_D - V_D$ for ballistic magneto QWFETs made of nonlinear optical materials as given by (24.1) for the computation of which we need the expression of the 2D electron statistics which is given in (24.2). The $I_D - V_D$ equations for ballistic magneto QWFETs made of III–V, ternary and quaternary materials has been investigated in Section 24.2.2 in accordance with the models of three-band model of Kane, two-band model of Kane and parabolic energy bands are given by (24.3), (24.5) and (24.7) and the corresponding 2D electron statistics are expressed through (24.4), (24.6) and (24.8), respectively. In Section 24.2.3, we have studied the $I_D - V_D$ equation for magneto QWFETs of II–VI materials which is given by (24.9) and the corresponding 2D electron statistics has been represented in (24.10). In Section 20.2.4, we have studied the $I_D - V_D$ equation in magneto QWFETs of IV–VI materials in accordance with the models of McClure-Choi, Hybrid, Cohen, Lax, Dimmock and ellipsoidal are given by (24.11), (24.13a), (24.14a), (24.15), (24.17) and (24.19) and the corresponding 2D electron statistics are expressed through (24.12), (24.13b), (24.14b), (24.16), (24.18) and (24.20), respectively. In Section 24.2.5, we have investigated the $I_D - V_D$ equation in magneto QWFETs of stressed Kane type materials which is given by (24.21) and the corresponding 2D electron statistics has been represented in (24.22). In Section 24.2.6, we have studied the $I_D - V_D$ equation in magneto QWFETs of bismuth telluride which is given by (24.23) and the corresponding 2D electron statistics has been represented in (24.24). In Section 23.2.7, we have studied the $I_D - V_D$ equation in QWFETs of germanium materials in accordance with the models of Cardona *et al.* and Wang *et al.* are given by (24.25) and (24.27) and the corresponding 2D electron statistics are expressed through (24.26) and

(24.28), respectively. In Section 23.2.8, we have investigated the $I_D - V_D$ equation in magneto QWFETs of gallium arsenide materials in accordance with the models of Rossler *et al.* and Stillman *et al.* are given by (24.29) and (24.31) and the corresponding 2D electron statistics are expressed through (24.30) and (24.32), respectively. In Section 24.2.9, we have studied the $I_D - V_D$ equation in magneto QWFETs of gallium antimonide materials in accordance with the models of Sailer *et al.*, Mathur *et al.* and Zhang *et al.* are given by (24.33), (24.35) and (24.37) and the corresponding 2D electron statistics are expressed through (24.34), (24.36) and (24.38), respectively. In Section 24.2.10, we have studied the $I_D - V_D$ equation in magneto QWFETs of indium antimonide materials in accordance with the models of Palik *et al.*, Johnson *et al.* and Agafonov *et al.* are given by (24.39), (24.41) and (24.43) and the corresponding 2D electron statistics expressed through (24.40), (24.42) and (24.44), respectively. In Section 24.2.11, we have studied the $I_D - V_D$ equation in magneto QWFETs of tellurium materials which is given by (24.45) and the corresponding 2D electron statistics has been represented in (24.46). In Section 24.2.12, we have studied the $I_D - V_D$ equation in magneto QWFETs of $PtSb_2$ materials which is given by (24.47) and the corresponding 2D electron statistics has been represented in (24.48). In Section 24.2.13, we have studied the $I_D - V_D$ equation in magneto QWFETs of GaP materials which is given by (24.49) and the corresponding 2D electron statistics has been represented in (24.50).

We have not given the graphs of the $I_D - V_D$ equations for ballistic magneto QWFETs made of many materials as discussed here since we feel that our readers will perform the intricate numerical computation to obtain the oscillatory plots for better assessment and joy of understanding. It is the band structure that changes in a fundamental way and consequently all the physical properties of all the electronic materials changes radically leading to new physical concepts.

24.4 Open Research Problem

(R24.1) Derive equations for ballistic Magneto QWFETs made of all the materials whose DRs are given in (R1.1). Draw the corresponding graphs and explain the results.

Chapter 25

$I_D - V_D$ Equation for Ballistic MAGNETO HDQWFETs

Begin at the beginning and go on till you come to an end remembering that the end is where we start from.

25.1 Introduction

This is completely a new chapter where from Sections 25.2.1 up to 25.2.11 we study the $I_D - V_D$ equation in ballistic magneto HDQWFETs of nonlinear optical, III–V, ternary and quaternary, II–VI, IV–VI, stressed Kane type, Te, GaP, $PtSb_2$, Bi_2Te_3, Ge and GaSb, respectively. In Section 25.3 we present summary and conclusion and in Section 25.4, we offer one multidimensional open research problem.

25.2 Theoretical Background

25.2.1 *The $I_D - V_D$ equation for ballistic magneto HDQWFETs of nonlinear optical materials*

Using (21.7), (21.1), (21.2), (21.3) and (12.5) we get

$$\frac{J_D}{W} = \frac{e^2 B g_v}{2\pi^2 \hbar}\sqrt{\frac{2}{m_\perp^*}} \text{ Real part of} \sum_{n=0}^{n_{\max}} \sum_{n_z=1}^{n_{z\max}} \left[\frac{\sqrt{T_{22}(e_{81}, \eta_g)}}{T'_{22}(e_{81}, \eta_g)}\right] \times [F_{-1}(\bar{\eta}_{25,1}) - F_{-1}(\bar{\theta}_{25,1})] \tag{25.1}$$

where

$$\bar{\eta}_{25,1} = (k_B T)^{-1}[E_{F25,1} - e_{81}], \quad \bar{\theta}_{25,1} = \left[\bar{\eta}_{25,1} - \frac{eV_D}{k_B T}\right]$$

and $E_{F25,1}$ is the Fermi energy in the present case which can be determined from

$$\frac{C_{ins}}{e}(V_T - V_{GS}) = \frac{g_v eB}{4\pi\hbar} \text{ Real part of} \sum_{n=0}^{n_{\max}} \sum_{i=0}^{i_{\max}} [F_{-1}(\bar{\eta}_{25,1}) + F_{-1}(\bar{\theta}_{25,1})] \tag{25.2}$$

25.2.2 *The $I_D - V_D$ equation for ballistic magneto HDQWFETs of III–V, ternary and quaternary materials*

(i) Three-band model of Kane

Using (23.4), (21.1), (21.2), (21.3) and (12.12b) we get

$$\frac{J_D}{W} = \frac{e^2 B g_v}{2\pi^2\hbar}\sqrt{\frac{2}{m_c}} \text{ Real part of} \sum_{n=0}^{n_{\max}} \sum_{i=0}^{i_{\max}} \left[\left[\frac{\sqrt{T_{31}(e_{82},\eta_g) + iT_{32}(e_{82},\eta_g)}}{T'_{31}(e_{82},\eta_g) + iT'_{32}(e_{82},\eta_g)}\right]\right.$$

$$\left.\times [F_{-1}(\bar{\eta}_{25,2}) - F_{-1}(\bar{\theta}_{25,2})]\right] \tag{25.3}$$

where

$$\bar{\eta}_{25,2} = (k_B T)^{-1}[E_{F25,2} - e_{82}], \quad \bar{\theta}_{25,2} = \left[\bar{\eta}_{25,2} - \frac{eV_D}{k_B T}\right]$$

and $E_{F25,2}$ is the Fermi energy in the present case which can be determined from

$$\frac{C_{ins}}{e}(V_T - V_{GS}) = \frac{g_v eB}{4\pi\hbar} \text{Real part of} \sum_{n=0}^{n_{\max}} \sum_{i=0}^{i_{\max}} [F_{-1}(\bar{\eta}_{25,2}) + F_{-1}(\bar{\theta}_{25,2})] \tag{25.4}$$

(ii) Two-band model of Kane

Using (21.15), (21.1), (21.2), (21.3) and (12.15) we get

$$\frac{J_D}{W} = \frac{e^2 B g_v}{2\pi^2\hbar}\sqrt{\frac{2}{m_c}} \sum_{n=0}^{n_{\max}} \sum_{i=0}^{i_{\max}} \left[\left[\frac{\sqrt{\gamma_2(e_{83},\eta_g)}}{\gamma'_2(e_{83},\eta_g)}\right] [F_{-1}(\bar{\eta}_{25,3}) - F_{-1}(\bar{\theta}_{25,3})]\right] \tag{25.5}$$

where

$$\bar{\eta}_{25,3} = (k_B T)^{-1}[E_{F25.3} - e_{83}], \;\; \bar{\theta}_{25,3} = \left[\bar{\eta}_{25,3} - \frac{eV_D}{k_B T}\right]$$

and $E_{F25,3}$ is the Fermi energy in the present case which can be determined from

$$\frac{C_{ins}}{e}(V_T - V_{GS}) = \frac{g_v e B}{4\pi\hbar}\sum_{n=0}^{n_{\max}}\sum_{i=0}^{i_{\max}}[F_{-1}(\bar{\eta}_{25,3}) + F_{-1}(\bar{\theta}_{25,3})] \tag{25.6}$$

(iii) Parabolic band model

Using (21.18), (21.1), (21.2), (21.3) and (12.18) we get

$$\frac{J_D}{W} = \frac{e^2 B g_v}{2\pi^2\hbar}\sqrt{\frac{2}{m_c}}\sum_{n=0}^{n_{\max}}\sum_{i=0}^{i_{\max}}\left[\left[\frac{\sqrt{\gamma_2(e_{84},\eta_g)}}{\gamma_2'(e_{84},\eta_g)}\right][F_{-1}(\bar{\eta}_{25,4}) - F_{-1}(\bar{\theta}_{25,4})]\right] \tag{25.7}$$

where

$$\bar{\eta}_{25,4} = (k_B T)^{-1}[E_{F25.4} - e_{84}], \;\; \bar{\theta}_{25,4} = \left[\bar{\eta}_{25,4} - \frac{eV_D}{k_B T}\right]$$

and $E_{F25,4}$ is the Fermi energy in the present case which can be determined from

$$\frac{C_{ins}}{e}(V_T - V_{GS}) = \frac{g_v e B}{4\pi\hbar}\sum_{n=0}^{n_{\max}}\sum_{i=0}^{i_{\max}}[F_{-1}(\bar{\eta}_{25,4}) + F_{-1}(\bar{\theta}_{25,4})] \tag{25.8}$$

(iv) Model of Stillman *et al.*

Using (23.13), (21.1), (21.2), (21.3) and (12.23) we get

$$\frac{J_D}{W} = \frac{e^2 B g_v}{\pi^2\hbar}\sqrt{\frac{2}{m_c}}\sum_{n=0}^{n_{\max}}\sum_{i=0}^{i_{\max}}\left[\left[\frac{\sqrt{I_{12}(e_{85},\eta_g)}}{I_{12}'(e_{85},\eta_g)}\right][F_{-1}(\bar{\eta}_{25,5}) - F_{-1}(\bar{\theta}_{25,5})]\right] \tag{25.9}$$

where

$$\bar{\eta}_{25,5} = (k_B T)^{-1}[E_{F25.5} - e_{85}], \;\; \bar{\theta}_{25,5} = \left[\bar{\eta}_{25,5} - \frac{eV_D}{k_B T}\right]$$

and $E_{F25,5}$ is the Fermi energy in the present case which can be determined from

$$\frac{C_{ins}}{e}(V_T - V_{GS}) = \frac{g_v e B}{4\pi\hbar}\sum_{n=0}^{n_{\max}}\sum_{i=0}^{i_{\max}}[F_{-1}(\bar{\eta}_{25,5}) + F_{-1}(\bar{\theta}_{25,5})] \tag{25.10}$$

(v) Model of Palik *et al.*

Using (23.16), (21.1), (21.2), (21.3) and (12.29) we get

$$\frac{J_D}{W} = \frac{e^2 B g_v}{\pi^2 \hbar} \sqrt{\frac{2}{m_c}} \sum_{n=0}^{n_{\max}} \sum_{i=0}^{i_{\max}} \left[\left[\frac{\sqrt{I_{13}(e_{86}, \eta_g)}}{I'_{13}(e_{86}, \eta_g)} \right] [F_{-1}(\bar{\eta}_{25,6}) - F_{-1}(\bar{\theta}_{25,6})] \right] \tag{25.11}$$

where

$$\bar{\eta}_{25,6} = (k_B T)^{-1} [E_{F25.6} - e_{86}], \;\; \bar{\theta}_{25,6} = \left[\bar{\eta}_{25,6} - \frac{eV_D}{k_B T} \right]$$

and $E_{F25,6}$ is the Fermi energy in the present case which can be determined from

$$\frac{C_{ins}}{e}(V_T - V_{GS}) = \frac{g_v e B}{4\pi\hbar} \sum_{n=0}^{n_{\max}} \sum_{i=0}^{i_{\max}} [F_{-1}(\bar{\eta}_{25,6}) + F_{-1}(\bar{\theta}_{25,6})] \tag{25.12}$$

25.2.3 *The equation for ballistic magneto HDQWFETs of II–VI materials*

Using (23.19), (21.1), (21.2), (21.3) and (12.32) we get

$$\frac{J_D}{W} = \frac{e^2 B g_v}{2\pi^2 \hbar^2} \sqrt{\frac{2}{m_c}} \sum_{n=0}^{n_{\max}} \sum_{i=0}^{i_{\max}} \left[\left[\frac{\sqrt{(\bar{\lambda}_0)^2 + 4 a'_0 \gamma_3(e_{87}, \eta_g)}}{\gamma'_3(e_{87}, \eta_g)} \right] \times [F_{-1}(\bar{\eta}_{25,7}) - F_{-1}(\bar{\theta}_{25,7})] \right] \tag{25.13}$$

where

$$\bar{\eta}_{25,7} = (k_B T)^{-1} [E_{F25.7} - e_{87}], \;\; \bar{\theta}_{25,7} = \left[\bar{\eta}_{25,7} - \frac{eV_D}{k_B T} \right]$$

and $E_{F25,7}$ is the Fermi energy in the present case which can be determined from

$$\frac{C_{ins}}{e}(V_T - V_{GS}) = \frac{g_v e B}{4\pi\hbar} \sum_{n=0}^{n_{\max}} \sum_{i=0}^{i_{\max}} [F_{-1}(\bar{\eta}_{25,7}) + F_{-1}(\bar{\theta}_{25,7})] \tag{25.14}$$

25.2.4 The $I_D - V_D$ equation for ballistic magneto HDQWFETs of IV–VI materials

Using (23.22), (21.1), (21.2), (21.3) and (12.52) we get

$$\frac{J_D}{W} = \frac{2e^2Bg_v}{\pi^2\hbar^2} \text{ Real part of} \sum_{n=0}^{n_{\max}} \sum_{i=0}^{i_{\max}} \left[\left[\frac{[F_{1HD}(e_{91}, \eta_g)]^{3/2}}{F'_{1HD}(e_{91}, \eta_g)} \right] \times [F_{-1}(\bar{\eta}_{25,8}) - F_{-1}(\bar{\theta}_{25,8})] \right] \quad (25.15)$$

where

$$\bar{\eta}_{25,8} = (k_BT)^{-1}[E_{F25.8} - e_{91}], \; \bar{\theta}_{25,8} = \left[\bar{\eta}_{25,8} - \frac{eV_D}{k_BT}\right]$$

and $E_{F25,8}$ is the Fermi energy in the present case which can be determined from

$$\frac{C_{ins}}{e}(V_T - V_{GS}) = \frac{g_veB}{2\pi\hbar} \text{ Real part of} \sum_{n=0}^{n_{\max}} \sum_{i=0}^{i_{\max}} [F_{-1}(\bar{\eta}_{25,8}) + F_{-1}(\bar{\theta}_{25,8})] \quad (25.16)$$

25.2.5 The $I_D - V_D$ equation for ballistic magneto HDQWFETs of stressed Kane type materials

Using (23.25), (21.1), (21.2), (21.3) and (12.62) we get

$$\frac{J_D}{W} - \frac{2e^2Bg_v}{\pi^2\hbar^2} \sum_{n=0}^{n_{\max}} \sum_{i=0}^{i_{\max}} \left[\left[\frac{[P_{11}(e_{93}, \eta_q)]^{3/2}}{P'_{11}(e_{93}, \eta_g)} \right] [F_{-1}(\bar{\eta}_{25,9}) - F_{-1}(\bar{\theta}_{25,9})] \right] \quad (25.17a)$$

where

$$\bar{\eta}_{25,9} - (k_BT)^{-1}[E_{F25.9} - e_{93}] \; \bar{\theta}_{25,9} = [\bar{\eta}_{25,9} - \frac{eV_D}{k_BT}]$$

and $E_{F25,9}$ is the Fermi energy in the present case which can be determined from

$$\frac{C_{ins}}{e}(V_T - V_{GS}) = \frac{g_veB}{2\pi\hbar} \sum_{n=0}^{n_{\max}} \sum_{i=0}^{i_{\max}} [F_{-1}(\bar{\eta}_{25,9}) + F_{-1}(\bar{\theta}_{25,9})] \quad (25.17b)$$

25.2.6 The $I_D - V_D$ equation for ballistic magneto HDQWFETs of tellurium

Using (23.28), (21.1), (21.2), (21.3) and (12.65) we get

$$\frac{J_D}{W} = \frac{e^2 B g_v}{2\pi^2\hbar^2} \sum_{n=0}^{n_{\max}} \sum_{i=0}^{i_{\max}} \left[\left[\frac{\sqrt{\psi_4^2 + 4\psi_2\gamma_3(\bar{e}_{94}, \eta_g)}}{\gamma_3'(\bar{e}_{94}, \eta_g)} \right] \times [F_{-1}(\bar{\eta}_{25,10}) - F_{-1}(\bar{\theta}_{25,10})] \right] \quad (25.18a)$$

where

$$\bar{\eta}_{25,10} = (k_B T)^{-1}[E_{F25.10} - \bar{e}_{94}], \ \bar{\theta}_{25,10} = \left[\bar{\eta}_{25,10} - \frac{eV_D}{k_B T}\right]$$

and $E_{F25,10}$ is the Fermi energy in the present case which can be determined from

$$\frac{C_{ins}}{e}(V_T - V_{GS}) = \frac{g_v e B}{4\pi\hbar} \sum_{n=0}^{n_{\max}} \sum_{i=0}^{i_{\max}} [F_{-1}(\bar{\eta}_{25,10}) + F_{-1}(\bar{\theta}_{25,10})] \quad (25.18b)$$

25.2.7 The $I_D - V_D$ equation for ballistic magneto HDQWFETs of GaP

Using (23.31), (21.1), (21.2), (21.3) and (12.69) we get

$$\frac{J_D}{W} = \frac{2e^2 B g_v}{\pi^2\hbar^2} \sum_{n=0}^{n_{\max}} \sum_{i=0}^{i_{\max}} \left[\left[\frac{\sqrt{\psi_{444HD}(\bar{e}_{941}, \eta_g)}}{\psi_{444HD}'(\bar{e}_{941}, \eta_g)} \right] \times [F_{-1}(\bar{\eta}_{25,11}) - F_{-1}(\bar{\theta}_{25,11})] \right] \quad (25.19)$$

where

$$\bar{\eta}_{25,11} = (k_B T)^{-1}[E_{F25.11} - \bar{e}_{941}], \ \bar{\theta}_{25,11} = \left[\bar{\eta}_{25,11} - \frac{eV_D}{k_B T}\right]$$

and $E_{F25,11}$ is the Fermi energy in the present case which can be determined from

$$\frac{C_{ins}}{e}(V_T - V_{GS}) = \frac{g_v e B}{4\pi\hbar} \sum_{n=0}^{n_{\max}} \sum_{i=0}^{i_{\max}} [F_{-1}(\bar{\eta}_{25,11}) + F_{-1}(\bar{\theta}_{25,11})] \quad (25.20)$$

25.2.8 *The $I_D - V_D$ equation for ballistic magneto HDQWFETs of $PtSb_2$*

Using (23.34), (21.1), (21.2), (21.3) and (12.73) we get

$$\frac{J_D}{W} = \frac{2e^2 B g_v}{\pi^2 \hbar^2} \sum_{n=0}^{n_{\max}} \sum_{i=0}^{i_{\max}} \left[\left[\frac{\sqrt{\beta_{23}(e_{95}, \eta_g)}}{\beta'_{23}(e_{95}, \eta_g)} \right] [F_{-1}(\bar{\eta}_{25,12}) - F_{-1}(\bar{\theta}_{25,12})] \right] \tag{25.21}$$

where

$$\bar{\eta}_{25,12} = (k_B T)^{-1}[E_{F25.12} - e_{95}], \ \ \bar{\theta}_{25,12} = \left[\bar{\eta}_{25,12} - \frac{eV_D}{k_B T}\right]$$

and $E_{F25,12}$ is the Fermi energy in the present case which can be determined from

$$\frac{C_{ins}}{e}(V_T - V_{GS}) = \frac{g_v e B}{2\pi\hbar} \sum_{n=0}^{n_{\max}} \sum_{i=0}^{i_{\max}} [F_{-1}(\bar{\eta}_{25,12}) + F_{-1}(\bar{\theta}_{25,12})] \tag{25.22}$$

25.2.9 *The $I_D - V_D$ equation for ballistic magneto HDQWFETs of bismuth telluride*

Using (23.37), (21.1), (21.2), (21.3) and (12.78) we get

$$\frac{J_D}{W} = \frac{2e^2 B g_v \sqrt{I_1}}{\pi^2 \hbar^2} \sum_{n=0}^{n_{\max}} \sum_{i=0}^{i_{\max}} \left[\left[\frac{\sqrt{\gamma_2(e_{96}, \eta_g)}}{\gamma'_2(e_{96}, \eta_g)} \right] [F_{-1}(\bar{\eta}_{25,13}) - F_{-1}(\bar{\theta}_{25,13})] \right] \tag{25.23a}$$

where

$$\bar{\eta}_{25,13} = (k_B T)^{-1}[E_{F25.13} - e_{96}] \tag{25.23b}$$

and

$$\bar{\theta}_{25,13} = \left[\bar{\eta}_{25,13} - \frac{eV_D}{k_B T}\right] \tag{25.24}$$

and $E_{F25,13}$ is the Fermi energy in the present case which can be determined from

$$\frac{C_{ins}}{e}(V_T - V_{GS}) = \frac{g_v e B}{2\pi\hbar} \sum_{n=0}^{n_{\max}} \sum_{i=0}^{i_{\max}} [F_{-1}(\bar{\eta}_{25,13}) + F_{-1}(\bar{\theta}_{25,13})] \tag{25.25}$$

25.2.10 *The $I_D - V_D$ equation for ballistic magneto HDQWFETs of germanium*

(i) Model of Cardona *et al.*

Using (23.40), (21.1), (21.2), (21.3) and (12.83) we get

$$\frac{J_D}{W} = \frac{e^2 B g_v}{\pi^2 \hbar} \sqrt{\frac{2}{m_\perp^*}} \sum_{n=0}^{n_{\max}} \sum_{i=0}^{i_{\max}} \left[\left[\frac{\sqrt{\gamma_2(e_{97}, \eta_g)}}{\gamma_2'(e_{97}, \eta_g)} \right] [F_{-1}(\bar{\eta}_{25,14}) - F_{-1}(\bar{\theta}_{25,14})] \right] \tag{25.26}$$

where

$$\bar{\eta}_{25,14} = (k_B T)^{-1} [E_{F25.14} - e_{97}] \ \bar{\theta}_{25,14} = \left[\bar{\eta}_{25,14} - \frac{eV_D}{k_B T} \right]$$

and $E_{F25,14}$ is the Fermi energy in the present case which can be determined from

$$\frac{C_{ins}}{e}(V_T - V_{GS}) = \frac{g_v e B}{2\pi\hbar} \sum_{n=0}^{n_{\max}} \sum_{i=0}^{i_{\max}} [F_{-1}(\bar{\eta}_{25,14}) + F_{-1}(\bar{\theta}_{25,14})] \tag{25.27}$$

(ii) Model of Wang and Ressler

From (11.99) we can write

$$V_x = \frac{\omega_{25}'(E, \eta_g)}{\sqrt{2m_\perp^*}} \tag{25.28}$$

where

$$\omega_{25}(E, \eta_g) = \left[\frac{1 - \sqrt{1 - 4\bar{\alpha}_4 \gamma_3(E, \eta_g)}}{2\alpha_4} \right]^{1/2}$$

Using (25.28), (21.1), (21.2), (21.3) and (12.88) we get

$$\frac{J_D}{W} = \frac{e^2 B g_v}{\pi^2 \hbar} \sqrt{\frac{1}{2m_\perp^*}} \sum_{n=0}^{n_{\max}} \sum_{i=0}^{i_{\max}} [[\omega_{25}'(e_{98}, \eta_g)][F_{-1}(\bar{\eta}_{25,15}) - F_{-1}(\bar{\theta}_{25,15})]] \tag{25.29}$$

where

$$\bar{\eta}_{25,15} = (k_B T)^{-1} [E_{F25.15} - e_{98}] \ \bar{\theta}_{25,15} = \left[\bar{\eta}_{25,15} - \frac{eV_D}{k_B T} \right]$$

and $E_{F25,15}$ is the Fermi energy in the present case which can be determined from

$$\frac{C_{ins}}{e}(V_T - V_{GS}) = \frac{g_v e B}{2\pi\hbar} \sum_{n=0}^{n_{\max}} \sum_{i=0}^{i_{\max}} [F_{-1}(\bar{\eta}_{25,15}) + F_{-1}(\bar{\theta}_{25,15})] \tag{25.30}$$

25.2.11 The $I_D - V_D$ equation for ballistic magneto HDQWFETs of gallium antimonide

Using (23.43), (21.1), (21.2), (21.3) and (12.93) we get

$$\frac{J_D}{W} = \frac{e^2 B g_v}{\pi^2 \hbar} \sqrt{\frac{2}{m_\perp^*}} \sum_{n=0}^{n_{\max}} \sum_{i=0}^{i_{\max}} \left[\left[\frac{\sqrt{I_{36}(e_{99}, \eta_g)}}{I'_{36}(e_{99}, \eta_g)} \right] \times [F_{-1}(\bar{\eta}_{25,15}) - F_{-1}(\bar{\theta}_{25,15})] \right] \tag{25.31}$$

where

$$\bar{\eta}_{25,15} = (k_B T)^{-1}[E_{F25.15} - e_{99}], \quad \bar{\theta}_{25,15} = \left[\bar{\eta}_{25,15} - \frac{eV_D}{k_B T} \right]$$

and $E_{F25,15}$ is the Fermi energy in the present case which can be determined from

$$\frac{C_{ins}}{e}(V_T - V_{GS}) = \frac{g_v eB}{2\pi\hbar} \sum_{n=0}^{n_{\max}} \sum_{i=0}^{i_{\max}} [F_{-1}(\bar{\eta}_{25,15}) + F_{-1}(\bar{\theta}_{25,15})] \tag{25.32}$$

25.3 Summary and conclusions

In this chapter, in Section 25.2.1, we have formulated $I_D - V_D$ for ballistic magneto HDQWFETs made of nonlinear optical materials as given by (25.1) for the computation of which we need the expression of the 2D electron statistics which is given in (25.2). The equations for $I_D - V_D$ for ballistic magneto HDQWFETs made of III–V, ternary and quaternary materials have been investigated in Section 25.2.2, in accordance with the models of three-band model of Kane, two-band model of Kane, parabolic energy bands, Stillman *et al.* and Palik *et al.* are given by (25.3), (25.5), (25.7), (25.9) and (25.11) and the corresponding 2D electron statistics are expressed through (25.4), (25.6), (25.8), (25.10) and (25.12), respectively. In Section 25.2.3, we have studied the $I_D - V_D$ equation for magneto HDQWFETs of II–VI materials which is given by (25.13) and the corresponding 2D electron statistics has been represented in (25.14). In Section 25.2.4, we have studied the $I_D - V_D$ equation in magneto HDQWFETs of IV–VI materials which is given by (25.15) and the corresponding 2D electron statistics has been represented in (25.16). In Section 25.2.5, we have investigated the $I_D - V_D$ equation in magneto HDQWFETs of stressed Kane type materials which is given by (25.17a) and the corresponding 2D electron statistics has been represented in (25.17b). In Section 25.2.6, we have studied the $I_D - V_D$ equation in magneto HDQWFETs of Te which is given by (25.18a) and the corresponding 2D electron statistics has been represented in (25.18b). In section 25.2.7, we have studied the $I_D - V_D$ equation in QWFETs of GaP which is given by (25.19) and the corresponding 2D electron statistics has been represented in (25.20). In Section 25.2.8, we have investigated the $I_D - V_D$ equation in magneto HDQWFETs of $PtSb_2$ which is given by (25.21) and the corresponding 2D electron statistics has been represented in (25.22). In Section 25.2.9, we have studied the $I_D - V_D$ equation in magneto HDQWFETs of bismuth telluride materials which is given by (25.23a) and the corresponding 2D electron statistics has been represented in (25.25). In Section 25.2.10, we have studied the $I_D - V_D$ equation in magneto HDQWFETs of germanium materials in accordance with the models of Cardona *et al.* and Wang *et al.* are given by (25.26) and (25.29) and the corresponding 2D electron statistics are expressed through (25.27) and (25.30), respectively. In Section 25.2.11, we have studied the $I_D - V_D$ equation in magneto HDQWFETs of gallium antimonide materials which is given by (25.31) and the corresponding 2D electron statistics has been represented in (25.32).

We have not given the graphs of the $I_D - V_D$ equations for ballistic magneto HDQWFETs made of many materials as discussed here since we feel that our readers will perform the intricate numerical computation to obtain the oscillatory plots for better assessment and joy of understanding. It is the band structure that changes in a fundamental way and consequently all the physical properties of all the electronic materials changes radically leading to new physical concepts.

25.4 Open Research problem

(R25.1) Derive equations for ballistic magneto HDQWFETs made of all the materials whose DRs are given in (R1.1). Draw the corresponding graphs and explain the results.

Chapter 26

The $I_D - V_D$ Equation in Magneto MOSFETs

Everything is predetermined in accordance with eastern philosophy. Answer to all the problems is the 360 degree mental attachment to the lotus feet of SADGURU.

26.1 Introduction

This is completely a new chapter where from Sections 26.2.1 up to 26.2.11, we study the $I_D - V_D$ equation in magneto MOSFETs of tetragonal and nonlinear optical, III–V, ternary and quaternary, II–VI, IV–VI, stressed Kane type, Bi_2Te_3, Ge, GaAs, GaSb, InSb materials and Te, respectively. In Section 26.3, we present summary and conclusion and last Section 26.4 we offer one multidimensional open research problem.

26.2 Theoretical Background

26.2.1 *The $I_D - V_D$ equation in magneto MOSFETs of tetragonal and nonlinear optical materials*

(a) Weak electric field limit

Using (21.7), (21.1), (21.2), (21.3) and (4.2) we get

$$\frac{J_D}{W} = \frac{e^2 B g_v}{\pi^2 \hbar^2} \sum_{n=0}^{n_{\max}} \sum_{i=0}^{i_{\max}} [[g'_{21,1}(E_{41})]^{-1}[F_{-1}(\bar{\eta}_{4,1}) - F_{-1}(\bar{\theta}_{4,1})]] \tag{26.1}$$

where

$$\bar{\eta}_{4,1} = (k_B T)^{-1}[E_{F26,1} - E_{41}] \tag{26.2}$$

$$\bar{\theta}_{4,1} = \left[\bar{\eta}_{4,1} - \frac{eV_D}{k_B T}\right]$$

and $E_{F26,1}$ is the Fermi energy in the present case which can be determined from

$$\frac{C_{ins}}{e}(V_T - V_{GS}) = \frac{g_v eB}{2\pi\hbar}\sum_{n=0}^{n_{\max}}\sum_{i=0}^{i_{\max}}[F_{-1}(\bar{\eta}_{4,1}) + F_{-1}(\bar{\theta}_{4,1})] \tag{26.3}$$

(b) Strong electric field limit

Using (21.7), (21.1), (21.2), (21.3) and (4.5) we get

$$\frac{J_D}{W} = \frac{e^2 B g_v}{\pi^2\hbar^2}\sum_{n=0}^{n_{\max}}\sum_{i=0}^{i_{\max}}[[g'_{21,1}(E_{42})]^{-1}[F_{-1}(\bar{\eta}_{4,2}) - F_{-1}(\bar{\theta}_{4,2})]] \tag{26.4}$$

where

$$\bar{\eta}_{4,2} = (k_B T)^{-1}[E_{F26,2} - E_{42}], \quad \bar{\theta}_{4,2} = \left[\bar{\eta}_{4,2} - \frac{eV_D}{k_B T}\right]$$

and $E_{F26,2}$ is the Fermi energy in the present case which can be determined from

$$\frac{C_{ins}}{e}(V_T - V_{GS}) = \frac{g_v eB}{2\pi\hbar}\sum_{n=0}^{n_{\max}}\sum_{i=0}^{i_{\max}}[F_{-1}(\bar{\eta}_{4,2}) + F_{-1}(\bar{\theta}_{4,2})] \tag{26.5a}$$

26.2.2 *The $I_D - V_D$ equation in magneto MOSFETs of III–V, ternary and quaternary materials*

(a) Three band model of Kane

(i) Weak electric field limit

Using (21.12), (21.1), (21.2), (21.3) and (4.8) we get

$$\frac{J_D}{W} = \frac{e^2 B g_v}{\pi^2\hbar^2}\sqrt{\frac{2}{m_c}}\sum_{n=0}^{n_{\max}}\sum_{i=0}^{i_{\max}}\left[\left[\sqrt{I_{11}(E_{43})}\right][I'_{11}(E_{43})]^{-1}\right] \times [F_{-1}(\bar{\eta}_{4,3}) - F_{-1}(\bar{\theta}_{4,3})]] \tag{26.5b}$$

where

$$\bar{\eta}_{4,3} = (k_B T)^{-1}[E_{F26,3} - E_{43}], \; \bar{\theta}_{4,3} = \left[\bar{\eta}_{4,3} - \frac{eV_D}{k_B T}\right]$$

and $E_{F26,3}$ is the Fermi energy in the present case which can be determined from

$$\frac{C_{ins}}{e}(V_T - V_{GS}) = \frac{g_v eB}{2\pi\hbar} \sum_{n=0}^{n_{\max}} \sum_{i=0}^{i_{\max}} [F_{-1}(\bar{\eta}_{4,3}) + F_{-1}(\bar{\theta}_{4,3})] \tag{26.6a}$$

(b) Two-band model of Kane

(i) Weak electric field limit

Using (21.15), (21.1), (21.2), (21.3) and (4.11) we get

$$\frac{J_D}{W} = \frac{e^2 B g_v}{\pi^2\hbar^2}\sqrt{\frac{2}{m_c}} \sum_{n=0}^{n_{\max}} \sum_{i=0}^{i_{\max}} \left[\left[\sqrt{E_{44}(1+\alpha E_{44})}\right][1+2\alpha E_{44}]^{-1}\right] \times [F_{-1}(\bar{\eta}_{4,4}) - F_{-1}(\bar{\theta}_{4,4})] \tag{26.6b}$$

where

$$\bar{\eta}_{4,4} = (k_B T)^{-1}[E_{F26,4} - E_{44}], \quad \bar{\theta}_{4,4} = \left[\bar{\eta}_{4,4} - \frac{eV_D}{k_B T}\right]$$

and $E_{F26,4}$ is the Fermi energy in the present case which can be determined from

$$\frac{C_{ins}}{e}(V_T - V_{GS}) = \frac{g_v eB}{2\pi\hbar} \sum_{n=0}^{n_{\max}} \sum_{i=0}^{i_{\max}} [F_{-1}(\bar{\eta}_{4,4}) + F_{-1}(\bar{\theta}_{4,4})] \tag{26.7}$$

(c) Parabolic band model

Using (21.18), (21.1), (21.2), (21.3) and (4.14) we get

$$\frac{J_D}{W} = \frac{e^2 B g_v}{\pi^2\hbar^2}\sqrt{\frac{2}{m_c}} \sum_{n=0}^{n_{\max}} \sum_{i=0}^{i_{\max}} \left[\sqrt{E_{45}}\right][F_{-1}(\eta_{4,5}) - F_{-1}(\theta_{4,5})] \tag{26.8}$$

where

$$\eta_{4,5} = (k_B T)^{-1}[E_{F26,5} - E_{45}], \quad \theta_{4,5} = \left[\eta_{4,5} - \frac{eV_D}{k_B T}\right]$$

and $E_{F26,5}$ is the Fermi energy in the present case which can be determined from

$$\frac{C_{ins}}{e}(V_T - V_{GS}) = \frac{g_v eB}{2\pi\hbar} \sum_{n=0}^{n_{\max}} \sum_{i=0}^{i_{\max}} [F_{-1}(\eta_{4,5}) + F_{-1}(\theta_{4,5})] \tag{26.9}$$

(d) The three-band Kane model

(i) Strong electric field limit

Using (21.12), (21.1), (21.2), (21.3) and (4.17) we get

$$\frac{J_D}{W} = \frac{e^2 B g_v}{\pi^2 \hbar^2} \sqrt{\frac{2}{m_c}} \sum_{n=0}^{n_{\max}} \sum_{i=0}^{i_{\max}} \left[\left[\sqrt{I_{11}(E_{47})}\right] [I'_{11}(E_{47})]^{-1}\right] \times [F_{-1}(\eta_{4,6}) - F_{-1}(\theta_{4,6})] \tag{26.10}$$

where

$$\eta_{4,6} = (k_B T)^{-1}[E_{F26,6} - E_{47}], \quad \theta_{4,6} = \left[\eta_{4,6} - \frac{eV_D}{k_B T}\right]$$

and $E_{F26,6}$ is the Fermi energy in the present case which can be determined from

$$\frac{C_{ins}}{e}(V_T - V_{GS}) = \frac{g_v e B}{2\pi\hbar} \sum_{n=0}^{n_{\max}} \sum_{i=0}^{i_{\max}} [F_{-1}(\eta_{4,6}) + F_{-1}(\theta_{4,6})] \tag{26.11}$$

(e) Two-band model of Kane (strong electric field)

Using (21.15), (21.1), (21.2), (21.3) and (4.20) we get

$$\frac{J_D}{W} = \frac{e^2 B g_v}{\pi^2 \hbar^2} \sqrt{\frac{2}{m_c}} \sum_{n=0}^{n_{\max}} \sum_{i=0}^{i_{\max}} \left[\left[\sqrt{E_{48}(1+\alpha E_{48})}\right] [1 + 2\alpha E_{48}]^{-1}\right] \times [F_{-1}(\eta_{4,7}) - F_{-1}(\theta_{4,7})] \tag{26.12}$$

where

$$\eta_{4,7} = (k_B T)^{-1}[E_{F26,7} - E_{48}], \quad \theta_{4,7} = \left[\eta_{4,7} - \frac{eV_D}{k_B T}\right]$$

and $E_{F26,7}$ is the Fermi energy in the present case which can be determined from

$$\frac{C_{ins}}{e}(V_T - V_{GS}) = \frac{g_v e B}{2\pi\hbar} \sum_{n=0}^{n_{\max}} \sum_{i=0}^{i_{\max}} [F_{-1}(\eta_{4,7}) + F_{-1}(\theta_{4,7})] \tag{26.13}$$

26.2.3 The $I_D - V_D$ equation in magneto MOSFETs of II–VI materials

Using (21.24), (21.1), (21.2), (21.3) and (4.23) we get

$$\frac{J_D}{W} = \frac{e^2 B g_v}{2\pi^2\hbar^2} \sum_{n=0}^{n_{\max}} \sum_{i=0}^{i_{\max}} \left[\left[\sqrt{C_0^2 + \frac{2E_{49}\hbar^2}{m_\perp^*}}\right][F_{-1}(\eta_{4,8}) - F_{-1}(\theta_{4,8})]\right] \tag{26.14}$$

where

$$\eta_{4,8} = (k_B T)^{-1}[E_{F26,8} - E_{49}], \quad \theta_{4,8} = \left[\eta_{4,8} - \frac{eV_D}{k_B T}\right]$$

and $E_{F26,8}$ is the Fermi energy in the present case which can be determined from

$$\frac{C_{ins}}{e}(V_T - V_{GS}) = \frac{g_v e B}{4\pi\hbar} \sum_{n=0}^{n_{\max}} \sum_{i=0}^{i_{\max}} [F_{-1}(\eta_{4,8}) + F_{-1}(\theta_{4,8})] \tag{26.15}$$

26.2.4 The $I_D - V_D$ equation in magneto MOSFETs of IV–VI materials

(i) Weak electric field limit

Using (21.26), (21.1), (21.2), (21.3) and (4.26) we get

$$\frac{J_D}{W} = \frac{e^2 B g_v}{\pi^2\hbar} \sqrt{\frac{2}{m_1}} \sum_{n=0}^{n_{\max}} \sum_{i=0}^{i_{\max}} \left[\left[\frac{\sqrt{E_{50}(1+\alpha E_{50})}}{(1+2\alpha E_{50})}\right] \times [F_{-1}(\eta_{4,9}) - F_{-1}(\theta_{4,9})]\right] \tag{26.16}$$

where

$$\eta_{4,9} = (k_B T)^{-1}[E_{F26,9} - E_{50}], \quad \theta_{4,9} = \left[\eta_{4,9} - \frac{eV_D}{k_B T}\right]$$

and $E_{F26,9}$ is the Fermi energy in the present case which can be determined from

$$\frac{C_{ins}}{e}(V_T - V_{GS}) = \frac{g_v e B}{2\pi\hbar} \sum_{n=0}^{n_{\max}} \sum_{i=0}^{i_{\max}} [F_{-1}(\eta_{4,9}) + F_{-1}(\theta_{4,9})] \tag{26.17}$$

(ii) Strong electric field limit

Using (21.26), (21.1), (21.2), (21.3) and (4.29) we get

$$\frac{J_D}{W} = \frac{e^2 B g_v}{\pi^2 \hbar} \sqrt{\frac{2}{m_1}} \sum_{n=0}^{n_{\max}} \sum_{i=0}^{i_{\max}} \left[\left[\frac{\sqrt{E_{51}(1+\alpha E_{51})}}{(1+2\alpha E_{51})} \right] \times [F_{-1}(\eta_{4,10}) - F_{-1}(\theta_{4,10})] \right] \quad (26.18)$$

where

$$\eta_{4,10} = (k_B T)^{-1}[E_{F26,10} - E_{51}], \quad \theta_{4,10} = \left[\eta_{4,10} - \frac{eV_D}{k_B T} \right]$$

and $E_{F26,10}$ is the Fermi energy in the present case which can be determined from

$$\frac{C_{ins}}{e}(V_T - V_{GS}) = \frac{g_v e B}{2\pi\hbar} \sum_{n=0}^{n_{\max}} \sum_{i=0}^{i_{\max}} [F_{-1}(\eta_{4,10}) + F_{-1}(\theta_{4,10})] \quad (26.19)$$

26.2.5 *The $I_D - V_D$ equation in magneto MOSFETs of stressed Kane type materials*

(i) Weak electric field limit

Using (21.32), (21.1), (21.2), (21.3) and (4.32) we get

$$\frac{J_D}{W} = \frac{e^2 B g_v}{\pi^2 \hbar^2} \sum_{n=0}^{n_{\max}} \sum_{i=0}^{i_{\max}} [[\bar{a}_0(E_{52})]']^{-1} [F_{-1}(\eta_{4,10}) - F_{-1}(\theta_{4,10})] \quad (26.20)$$

where

$$\eta_{4,10} = (k_B T)^{-1}[E_{F26,10} - E_{52}], \quad \theta_{4,10} = \left[\eta_{4,10} - \frac{eV_D}{k_B T} \right]$$

and $E_{F26,10}$ is the Fermi energy in the present case which can be determined from

$$\frac{C_{ins}}{e}(V_T - V_{GS}) = \frac{g_v e B}{2\pi\hbar} \sum_{n=0}^{n_{\max}} \sum_{i=0}^{i_{\max}} [F_{-1}(\eta_{4,10}) + F_{-1}(\theta_{4,10})] \quad (26.21)$$

(ii) Strong electric field limit

Using (21.32), (21.1), (21.2), (21.3) and (4.35) we get

$$\frac{J_D}{W} = \frac{e^2 B g_v}{\pi^2 \hbar^2} \sum_{n=0}^{n_{\max}} \sum_{i=0}^{i_{\max}} [[\bar{a}_0(E_{53})]']^{-1} [F_{-1}(\eta_{4,11}) - F_{-1}(\theta_{4,11})] \qquad (26.22)$$

where

$$\eta_{4,11} = (k_B T)^{-1} [E_{F26,11} - E_{53}], \quad \theta_{4,11} = \left[\eta_{4,11} - \frac{eV_D}{k_B T}\right]$$

and $E_{F26,11}$ is the Fermi energy in the present case which can be determined from

$$\frac{C_{ins}}{e}(V_T - V_{GS}) = \frac{g_v e B}{2\pi\hbar} \sum_{n=0}^{n_{\max}} \sum_{i=0}^{i_{\max}} [F_{-1}(\eta_{4,11}) + F_{-1}(\theta_{4,11})] \qquad (26.23)$$

26.2.6 *The $I_D - V_D$ equation in magneto MOSFETs of bismuth telluride*

(i) Weak electric field limit

Using (21.37), (21.1), (21.2), (21.3) and (4.38), we get

$$\frac{J_D}{W} = \frac{e^2 B g_v}{\pi^2 \hbar^2} 2\sqrt{I_1} \sum_{n=0}^{n_{\max}} \sum_{i=0}^{i_{\max}} \left[\left[\frac{\sqrt{E_{54}(1+\alpha E_{54})}}{(1+2\alpha E_{54})}\right] \times [F_{-1}(\eta_{4,11}) - F_{-1}(\theta_{4,11})]\right] \qquad (26.24)$$

where

$$\eta_{4,11} = (k_B T)^{-1} [E_{F26,11} - E_{54}], \quad \theta_{4,11} = \left[\eta_{4,11} - \frac{eV_D}{k_B T}\right]$$

and $E_{F26,11}$ is the Fermi energy in the present case which can be determined from

$$\frac{C_{ins}}{e}(V_T - V_{GS}) = \frac{g_v e B}{2\pi\hbar} \sum_{n=0}^{n_{\max}} \sum_{i=0}^{i_{\max}} [F_{-1}(\eta_{4,11}) + F_{-1}(\theta_{4,11})] \qquad (26.25)$$

(ii) Strong electric field limit

Using (21.37), (21.1), (21.2), (21.3) and (4.41), we get

$$\frac{J_D}{W} = \frac{e^2 B g_v}{\pi^2 \hbar^2} 2\sqrt{I_1} \sum_{n=0}^{n_{\max}} \sum_{i=0}^{i_{\max}} \left[\left[\frac{\sqrt{E_{55}(1+\alpha E_{55})}}{(1+2\alpha E_{55})}\right] \times [F_{-1}(\eta_{4,12}) - F_{-1}(\theta_{4,12})]\right] \tag{26.26}$$

where

$$\eta_{4,12} = (k_B T)^{-1}[E_{F26,12} - E_{55}], \quad \theta_{4,12} = \left[\eta_{4,12} - \frac{eV_D}{k_B T}\right]$$

and $E_{F26,12}$ is the Fermi energy in the present case which can be determined from

$$\frac{C_{ins}}{e}(V_T - V_{GS}) = \frac{g_v e B}{2\pi\hbar} \sum_{n=0}^{n_{\max}} \sum_{i=0}^{i_{\max}} [F_{-1}(\eta_{4,12}) + F_{-1}(\theta_{4,12})] \tag{26.27}$$

26.2.7 *The $I_D - V_D$ equation in magneto MOSFETs of germanium*

(a) Model of Cardona *et al.*

Using (21.42), (21.1), (21.2), (21.3) and (4.44), we get

$$\frac{J_D}{W} = \frac{e^2 B g_v}{\pi^2 \hbar} \sqrt{\frac{2}{m_\perp^*}} \sum_{n=0}^{n_{\max}} \sum_{i=0}^{i_{\max}} \left[\left[\frac{\sqrt{E_{57}(1+\alpha E_{57})}}{(1+2\alpha E_{57})}\right] \times [F_{-1}(\eta_{4,13}) - F_{-1}(\theta_{4,13})]\right] \tag{26.28}$$

where

$$\eta_{4,13} = (k_B T)^{-1}[E_{F26,13} - E_{57}], \quad \theta_{4,13} = \left[\eta_{4,13} - \frac{eV_D}{k_B T}\right]$$

and $E_{F26,13}$ is the Fermi energy in the present case which can be determined from

$$\frac{C_{ins}}{e}(V_T - V_{GS}) = \frac{g_v e B}{2\pi\hbar} \sum_{n=0}^{n_{\max}} \sum_{i=0}^{i_{\max}} [F_{-1}(\eta_{4,13}) + F_{-1}(\theta_{4,13})] \tag{26.29}$$

(b) Wang *et al.*

Using (22.41), (21.1), (21.2), (21.3) and (4.47), we get

$$\frac{J_D}{W} = \frac{e^2 B g_v}{\pi^2 \hbar} \sum_{n=0}^{n_{\max}} \sum_{i=0}^{i_{\max}} [[\omega'_{222}(E_{59})]^{-1} [F_{-1}(\eta_{4,14}) - F_{-1}(\theta_{4,14})]] \tag{26.30}$$

where

$$\eta_{4,14} = (k_B T)^{-1} [E_{F26,14} - E_{59}], \quad \theta_{4,14} = \left[\eta_{4,14} - \frac{eV_D}{k_B T}\right]$$

and $E_{F26,14}$ is the Fermi energy in the present case which can be determined from

$$\frac{C_{ins}}{e}(V_T - V_{GS}) = \frac{g_v e B}{2\pi\hbar} \sum_{n=0}^{n_{\max}} \sum_{i=0}^{i_{\max}} [F_{-1}(\eta_{4,14}) + F_{-1}(\theta_{4,14})] \tag{26.31}$$

26.2.8 *The $I_D - V_D$ equation in magneto MOSFETs of gallium arsenide*

(a) Rossler model

(i) Weak electric field limit

Using (21.45), (21.1), (21.2), (21.3) and (4.50), we get

$$\frac{J_D}{W} = \frac{e^2 B g_v}{2\pi^2 \hbar^2} \sum_{n=0}^{n_{\max}} \sum_{i=0}^{i_{\max}} \left[\left[\frac{\sqrt{\gamma_{4,\pm}(E_{60})}}{\gamma'_{4,\pm}(E_{60})}\right] [F_{-1}(\eta_{4,15}) - F_{-1}(\theta_{4,15})]\right] \tag{26.32}$$

where

$$\eta_{4,15} = (k_B T)^{-1} [E_{F26,15} - E_{60}], \quad \theta_{4,15} = \left[\eta_{4,15} - \frac{eV_D}{k_B T}\right]$$

and $E_{F26,15}$ is the Fermi energy in the present case which can be determined from

$$\frac{C_{ins}}{e}(V_T - V_{GS}) = \frac{g_v e B}{4\pi\hbar} \sum_{n=0}^{n_{\max}} \sum_{i=0}^{i_{\max}} [F_{-1}(\eta_{4,15}) + F_{-1}(\theta_{4,15})] \tag{26.33}$$

(ii) Strong electric field limit

Using (21.45), (21.1), (21.2), (21.3) and (4.53), we get

$$\frac{J_D}{W} = \frac{e^2 B g_v}{2\pi^2\hbar^2} \sum_{n=0}^{n_{\max}} \sum_{i=0}^{i_{\max}} \left[\left[\frac{\sqrt{\gamma_{4,\pm}(E_{61})}}{\gamma'_{4,\pm}(E_{61})} \right] [F_{-1}(\eta_{4,16}) - F_{-1}(\theta_{4,16})] \right] \quad (26.34)$$

where

$$\eta_{4,16} = (k_B T)^{-1}[E_{F26,16} - E_{61}], \quad \theta_{4,16} = \left[\eta_{4,16} - \frac{eV_D}{k_B T} \right]$$

and $E_{F26,16}$ is the Fermi energy in the present case which can be determined from

$$\frac{C_{ins}}{e}(V_T - V_{GS}) = \frac{g_v e B}{4\pi\hbar} \sum_{n=0}^{n_{\max}} \sum_{i=0}^{i_{\max}} [F_{-1}(\eta_{4,16}) + F_{-1}(\theta_{4,16})] \quad (26.35)$$

(b) Model of Stillman *et al.*

(i) Weak electric field limit

Using (21.50), (21.1), (21.2), (21.3) and (4.56), we get

$$\frac{J_D}{W} = \frac{e^2 B g_v}{\pi^2\hbar^2} \sum_{n=0}^{n_{\max}} \sum_{i=0}^{i_{\max}} \left[\left[\frac{\sqrt{\gamma_5(E_{62})}}{\gamma'_5(E_{62})} \right] [F_{-1}(\eta_{4,16}) - F_{-1}(\theta_{4,16})] \right] \quad (26.36)$$

where

$$\eta_{4,16} = (k_B T)^{-1}[E_{F26,16} - E_{62}], \quad \theta_{4,16} = \left[\eta_{4,16} - \frac{eV_D}{k_B T} \right]$$

and $E_{F26,16}$ is the Fermi energy in the present case which can be determined from

$$\frac{C_{ins}}{e}(V_T - V_{GS}) = \frac{g_v e B}{2\pi\hbar} \sum_{n=0}^{n_{\max}} \sum_{i=0}^{i_{\max}} [F_{-1}(\eta_{4,16}) + F_{-1}(\theta_{4,16})] \quad (26.37)$$

(ii) Strong electric field limit

Using (21.50), (21.1), (21.2), (21.3) and (4.59), we get

$$\frac{J_D}{W} = \frac{e^2 B g_v}{\pi^2 \hbar^2} \sum_{n=0}^{n_{\max}} \sum_{i=0}^{i_{\max}} \left[\left[\frac{\sqrt{\gamma_5(E_{63})}}{\gamma_5'(E_{63})} \right] [F_{-1}(\eta_{4,17}) - F_{-1}(\theta_{4,17})] \right] \quad (26.38)$$

where

$$\eta_{4,17} = (k_B T)^{-1} [E_{F26,17} - E_{63}], \quad \theta_{4,17} = \left[\eta_{4,17} - \frac{eV_D}{k_B T} \right]$$

and $E_{F26,17}$ is the Fermi energy in the present case which can be determined from

$$\frac{C_{ins}}{e}(V_T - V_{GS}) = \frac{g_v e B}{2\pi\hbar} \sum_{n=0}^{n_{\max}} \sum_{i=0}^{i_{\max}} [F_{-1}(\eta_{4,17}) + F_{-1}(\theta_{4,17})] \quad (26.39)$$

26.2.9 *The $I_D - V_D$ equation in magneto MOSFETs of gallium antimonide*

(a) The model of Sailer *et al.*

(i) Weak electric field limit

Using (21.54), (21.1), (21.2), (21.3) and (4.62), we get

$$\frac{J_D}{W} = \frac{e^2 B g_v}{2\pi^2 \hbar^2} \sum_{n=0}^{n_{\max}} \sum_{i=0}^{i_{\max}} \left[\left[\frac{\sqrt{\gamma_6(E_{64})}}{\gamma_6'(E_{64})} \right] [F_{-1}(\eta_{4,18}) - F_{-1}(\theta_{4,18})] \right] \quad (26.40)$$

where

$$\eta_{4,18} = (k_B T)^{-1} [E_{F26,18} - E_{64}], \quad \theta_{4,18} = \left[\eta_{4,18} - \frac{eV_D}{k_B T} \right]$$

and $E_{F26,18}$ is the Fermi energy in the present case which can be determined from

$$\frac{C_{ins}}{e}(V_T - V_{GS}) = \frac{g_v e B}{4\pi\hbar} \sum_{n=0}^{n_{\max}} \sum_{i=0}^{i_{\max}} [F_{-1}(\eta_{4,18}) + F_{-1}(\theta_{4,18})] \quad (26.41)$$

(ii) Strong electric field limit

Using (21.54), (21.1), (21.2), (21.3) and (4.65), we get

$$\frac{J_D}{W} = \frac{e^2 B g_v}{2\pi^2\hbar^2} \sum_{n=0}^{n_{\max}} \sum_{i=0}^{i_{\max}} \left[\left[\frac{\sqrt{\gamma_6(E_{65})}}{\gamma_6'(E_{65})} \right] [F_{-1}(\eta_{4,19}) - F_{-1}(\theta_{4,19})] \right] \quad (26.42)$$

where

$$\eta_{4,19} = (k_B T)^{-1}[E_{F26,19} - E_{65}], \quad \theta_{4,19} = \left[\eta_{4,19} - \frac{eV_D}{k_B T}\right]$$

and $E_{F26,19}$ is the Fermi energy in the present case which can be determined from

$$\frac{C_{ins}}{e}(V_T - V_{GS}) = \frac{g_v eB}{4\pi\hbar} \sum_{n=0}^{n_{\max}} \sum_{i=0}^{i_{\max}} [F_{-1}(\eta_{4,19}) + F_{-1}(\theta_{4,19})] \quad (26.43)$$

(b) The model of Mathur *et al.*

(i) Weak electric field limit

Using (21.59), (21.1), (21.2), (21.3) and (4.68), we get

$$\frac{J_D}{W} = \frac{e^2 B g_v}{\pi^2\hbar^2} \sum_{n=0}^{n_{\max}} \sum_{i=0}^{i_{\max}} \left[\left[\frac{\sqrt{\gamma_7(E_{66})}}{\gamma_7'(E_{66})} \right] [F_{-1}(\eta_{4,20}) - F_{-1}(\theta_{4,20})] \right] \quad (26.44)$$

where

$$\eta_{4,20} = (k_B T)^{-1}[E_{F26,20} - E_{66}], \quad \theta_{4,20} = \left[\eta_{4,20} - \frac{eV_D}{k_B T}\right]$$

and $E_{F26,20}$ is the Fermi energy in the present case which can be determined from

$$\frac{C_{ins}}{e}(V_T - V_{GS}) = \frac{g_v eB}{2\pi\hbar} \sum_{n=0}^{n_{\max}} \sum_{i=0}^{i_{\max}} [F_{-1}(\eta_{4,20}) + F_{-1}(\theta_{4,20})] \quad (26.45)$$

(ii) Strong electric field limit

Using (21.59), (21.1), (21.2), (21.3) and (4.71), we get

$$\frac{J_D}{W} = \frac{e^2 B g_v}{\pi^2 \hbar^2} \sum_{n=0}^{n_{\max}} \sum_{i=0}^{i_{\max}} \left[\left[\frac{\sqrt{\gamma_7(E_{67})}}{\gamma_7'(E_{67})} \right] [F_{-1}(\eta_{4,21}) - F_{-1}(\theta_{4,21})] \right] \tag{26.46}$$

where

$$\eta_{4,21} = (k_B T)^{-1} [E_{F26,21} - E_{67}], \quad \theta_{4,21} = \left[\eta_{4,21} - \frac{eV_D}{k_B T} \right]$$

and $E_{F26,21}$ is the Fermi energy in the present case which can be determined from

$$\frac{C_{ins}}{e}(V_T - V_{GS}) = \frac{g_v e B}{2\pi\hbar} \sum_{n=0}^{n_{\max}} \sum_{i=0}^{i_{\max}} [F_{-1}(\eta_{4,21}) + F_{-1}(\theta_{4,21})] \tag{26.47}$$

(c) The model of Zhang

(i) Weak electric field limit

Using (21.64), (21.1), (21.2), (21.3) and (4.74), we get

$$\frac{J_D}{W} = \frac{e^2 B g_v}{\pi^2 \hbar^2} \sum_{n=0}^{n_{\max}} \sum_{i=0}^{i_{\max}} \left[\left[\frac{\sqrt{\gamma_8(E_{68})}}{\gamma_8'(E_{68})} \right] [F_{-1}(\eta_{4,22}) - F_{-1}(\theta_{4,22})] \right] \tag{26.48}$$

where

$$\eta_{4,22} = (k_B T)^{-1} [E_{F26,22} - E_{68}], \quad \theta_{4,22} = \left[\eta_{4,22} - \frac{eV_D}{k_B T} \right]$$

and $E_{F26,22}$ is the Fermi energy in the present case which can be determined from

$$\frac{C_{ins}}{e}(V_T - V_{GS}) = \frac{g_v e B}{2\pi\hbar} \sum_{n=0}^{n_{\max}} \sum_{i=0}^{i_{\max}} [F_{-1}(\eta_{4,22}) + F_{-1}(\theta_{4,22})] \tag{26.49}$$

(ii) Strong electric field limit

Using (21.64), (21.1), (21.2), (21.3) and (4.77), we get

$$\frac{J_D}{W} = \frac{e^2 B g_v}{\pi^2 \hbar^2} \sum_{n=0}^{n_{\max}} \sum_{i=0}^{i_{\max}} \left[\left[\frac{\sqrt{\gamma_8(E_{69})}}{\gamma_8'(E_{69})} \right] [F_{-1}(\eta_{4,23}) - F_{-1}(\theta_{4,23})] \right] \quad (26.50)$$

where

$$\eta_{4,23} = (k_B T)^{-1}[E_{F26,23} - E_{69}], \quad \theta_{4,23} = \left[\eta_{4,23} - \frac{eV_D}{k_B T} \right]$$

and $E_{F26,23}$ is the Fermi energy in the present case which can be determined from

$$\frac{C_{ins}}{e}(V_T - V_{GS}) = \frac{g_v e B}{2\pi\hbar} \sum_{n=0}^{n_{\max}} \sum_{i=0}^{i_{\max}} [F_{-1}(\eta_{4,23}) + F_{-1}(\theta_{4,23})] \quad (26.51)$$

26.2.10 *The $I_D - V_D$ equation in magneto MOSFETs of indium antimonide*

(a) The model of Palik *et al.*

(i) Weak electric field limit

Using (21.69), (21.1), (21.2), (21.3) and (4.80), we get

$$\frac{J_D}{W} = \frac{2e^2 B g_v}{\pi^2 \hbar^2} \sum_{n=0}^{n_{\max}} \sum_{i=0}^{i_{\max}} \left[\left[\frac{\sqrt{\gamma_9(E_{70})}}{\gamma_9'(E_{70})} \right] [F_{-1}(\eta_{4,24}) - F_{-1}(\theta_{4,24})] \right] \quad (26.52)$$

where

$$\eta_{4,24} = (k_B T)^{-1}[E_{F26,24} - E_{70}], \quad \theta_{4,24} = \left[\eta_{4,24} - \frac{eV_D}{k_B T} \right]$$

and $E_{F26,24}$ is the Fermi energy in the present case which can be determined from

$$\frac{C_{ins}}{e}(V_T - V_{GS}) = \frac{g_v e B}{2\pi\hbar} \sum_{n=0}^{n_{\max}} \sum_{i=0}^{i_{\max}} [F_{-1}(\eta_{4,24}) + F_{-1}(\theta_{4,24})] \quad (26.53)$$

(ii) Strong electric field limit

Using (21.69), (21.1), (21.2), (21.3) and (4.83), we get

$$\frac{J_D}{W} = \frac{2e^2 B g_v}{\pi^2 \hbar^2} \sum_{n=0}^{n_{\max}} \sum_{i=0}^{i_{\max}} \left[\left[\frac{\sqrt{\gamma_9(E_{71})}}{\gamma_9'(E_{71})} \right] [F_{-1}(\eta_{4,25}) - F_{-1}(\theta_{4,25})] \right] \quad (26.54)$$

where

$$\eta_{4,25} = (k_B T)^{-1} [E_{F26,25} - E_{71}], \quad \theta_{4,25} = \left[\eta_{4,25} - \frac{eV_D}{k_B T} \right]$$

and $E_{F26,25}$ is the Fermi energy in the present case which can be determined from

$$\frac{C_{ins}}{e}(V_T - V_{GS}) = \frac{g_v e B}{2\pi\hbar} \sum_{n=0}^{n_{\max}} \sum_{i=0}^{i_{\max}} [F_{-1}(\eta_{4,25}) + F_{-1}(\theta_{4,25})] \quad (26.55)$$

(b) The model of Johnson *et al.*

(i) Weak electric field limit

Using (21.74), (21.1), (21.2), (21.3) and (4.86), we get

$$\frac{J_D}{W} = \frac{2e^2 B g_v}{\pi^2 \hbar^2} \sum_{n=0}^{n_{\max}} \sum_{i=0}^{i_{\max}} \left[\left[\frac{\sqrt{\gamma_{10}(E_{72})}}{\gamma_{10}'(E_{72})} \right] [F_{-1}(\eta_{4,26}) - F_{-1}(\theta_{4,26})] \right] \quad (26.56)$$

where

$$\eta_{4,26} = (k_B T)^{-1} [E_{F26,26} - E_{72}], \quad \theta_{4,26} = \left[\eta_{4,26} - \frac{eV_D}{k_B T} \right]$$

and $E_{F26,26}$ is the Fermi energy in the present case which can be determined from

$$\frac{C_{ins}}{e}(V_T - V_{GS}) = \frac{g_v e B}{2\pi\hbar} \sum_{n=0}^{n_{\max}} \sum_{i=0}^{i_{\max}} [F_{-1}(\eta_{4,26}) + F_{-1}(\theta_{4,26})] \quad (26.57)$$

(ii) Strong electric field limit

Using (21.74), (21.1), (21.2), (21.3) and (4.89), we get

$$\frac{J_D}{W} = \frac{2e^2 B g_v}{\pi^2 \hbar^2} \sum_{n=0}^{n_{\max}} \sum_{i=0}^{i_{\max}} \left[\left[\frac{\sqrt{\gamma_{10}(E_{73})}}{\gamma'_{10}(E_{73})} \right] [F_{-1}(\eta_{4,27}) - F_{-1}(\theta_{4,27})] \right] \quad (26.58)$$

where

$$\eta_{4,27} = (k_B T)^{-1}[E_{F26,27} - E_{73}], \quad \theta_{4,27} = \left[\eta_{4,27} - \frac{eV_D}{k_B T} \right]$$

and $E_{F26,27}$ is the Fermi energy in the present cas which can be determined from

$$\frac{C_{ins}}{e}(V_T - V_{GS}) = \frac{g_v e B}{2\pi\hbar} \sum_{n=0}^{n_{\max}} \sum_{i=0}^{i_{\max}} [F_{-1}(\eta_{4,27}) + F_{-1}(\theta_{4,27})] \quad (26.59)$$

(c) The model of Agafonov *et al.*

(i) Weak electric field limit

Using (21.79), (21.1), (21.2), (21.3) and (4.92), we get

$$\frac{J_D}{W} = \frac{2e^2 B g_v}{\pi^2 \hbar^2} \sum_{n=0}^{n_{\max}} \sum_{i=0}^{i_{\max}} \left[\left[\frac{\sqrt{\gamma_{11}(E_{74})}}{\gamma'_{11}(E_{74})} \right] [F_{-1}(\eta_{4,28}) - F_{-1}(\theta_{4,28})] \right] \quad (26.60)$$

where

$$\eta_{4,28} = (k_B T)^{-1}[E_{F26,28} - E_{74}], \quad \theta_{4,28} = \left[\eta_{4,28} - \frac{eV_D}{k_B T} \right]$$

and $E_{F26,28}$ is the Fermi energy in the present case which can be determined from

$$\frac{C_{ins}}{e}(V_T - V_{GS}) = \frac{g_v e B}{2\pi\hbar} \sum_{n=0}^{n_{\max}} \sum_{i=0}^{i_{\max}} [F_{-1}(\eta_{4,28}) + F_{-1}(\theta_{4,28})] \quad (26.61)$$

(ii) Strong electric field limit

Using (21.79), (21.1), (21.2), (21.3) and (4.95), we get

$$\frac{J_D}{W} = \frac{2e^2 B g_v}{\pi^2 \hbar^2} \sum_{n=0}^{n_{\max}} \sum_{i=0}^{i_{\max}} \left[\left[\frac{\sqrt{\gamma_{11}(E_{75})}}{\gamma'_{11}(E_{75})} \right] [F_{-1}(\eta_{4,29}) - F_{-1}(\theta_{4,29})] \right] \quad (26.62)$$

where

$$\eta_{4,29} = (k_B T)^{-1} [E_{F26,29} - E_{75}], \quad \theta_{4,29} = \left[\eta_{4,29} - \frac{eV_D}{k_B T} \right]$$

and $E_{F26,29}$ is the Fermi energy in the present case which can be determined from

$$\frac{C_{ins}}{e}(V_T - V_{GS}) = \frac{g_v eB}{2\pi\hbar} \sum_{n=0}^{n_{\max}} \sum_{i=0}^{i_{\max}} [F_{-1}(\eta_{4,29}) + F_{-1}(\theta_{4,29})] \quad (26.63)$$

26.2.11 *The $I_D - V_D$ equation in magneto MOSFETs of tellurium*

Using (22.28), (21.1), (21.2), (21.3) and (4.98), we get

$$\frac{J_D}{W} = \frac{e^2 B g_v}{\pi^2 \hbar^2} \sum_{n=0}^{n_{\max}} \sum_{i=0}^{i_{\max}} \left[\left[\sqrt{\psi_4^2 + 4\psi_2 E_{76}} \right] [F_{-1}(\eta_{4,30}) - F_{-1}(\theta_{4,30})] \right] \quad (26.64)$$

where

$$\eta_{4,30} = (k_B T)^{-1} [E_{F26,30} - E_{76}], \quad \theta_{4,30} = \left[\eta_{4,30} - \frac{eV_D}{k_B T} \right]$$

and $E_{F26,30}$ is the Fermi energy in the present case which can be determined from

$$\frac{C_{ins}}{e}(V_T - V_{GS}) = \frac{g_v eB}{4\pi\hbar} \sum_{n=0}^{n_{\max}} \sum_{i=0}^{i_{\max}} [F_{-1}(\eta_{4,30}) + F_{-1}(\theta_{4,30})] \quad (26.65)$$

26.3 Summary and Conclusions

In this chapter, in Section 26.2.1, we have formulated the $I_D - V_D$ equations for ballistic magneto MOSFETs made of nonlinear optical materials for both weak and strong electric field limits as given by (26.1) and (26.4), respectively and for the computation of which we need the corresponding expression of the 2D electron statistics for both the limits which are given in (26.3) and (26.5a), respectively. The $I_D - V_D$ equations for magneto

MOSFETs made of III–V, ternary and quaternary materials have been investigated in Section 26.2.2 under weak electric field limit in accordance with the three-band model of Kane, two-band model of Kane together with parabolic energy bands, are given by (26.4), (26.6b) and (26.8) and the corresponding 2D electron statistics are expressed through (26.5a), (26.7) and (26.9), respectively. Besides the $I_D - V_D$ equations in this case for strong electric field limit in accordance with the three-band model of Kane and two-band model of Kane are given by (26.10) and (26.12) and the corresponding 2D electron statistics are expressed through (26.11) and (26.13) respectively. In Section 26.2.3, we have studied the $I_D - V_D$ equation for ballistic magneto MOSFETs of II–VI materials which is given by (26.14) and the corresponding 2D electron statistics has been represented in (26.15). In Section 26.2.4, we have studied the $I_D - V_D$ equation in magneto MOSFETs of IV–VI materials for both weak and strong electric field limits which are given by (26.16) and (26.18), respectively and the corresponding 2D electron statistics have been represented in (26.17) and (26.19), respectively. In Section 26.2.5, we have investigated the $I_D - V_D$ equation in ballistic magneto MOSFETs of stressed Kane type materials for both weak and strong electric field limits which are given by (26.20) and (26.22), respectively and the corresponding 2D electron statistics have been represented in (26.21) and (26.23), respectively. In Section 21.2.6, we have studied the $I_D - V_D$ equation in ballistic magneto MOSFETs of bismuth telluride for both weak and strong electric field limits which are given by (26.24) and (26.26), respectively and the corresponding 2D electron statistics have been represented in (26.25) and (26.27), respectively. In Section 21.2.7, we have studied the $I_D - V_D$ equation in ballistic magneto MOSFETs of germanium in accordance with the models of Cardona *et al.* and Wang *et al.* which are given by (26.28) and (26.30) and the corresponding 2D electron statistics has been represented in (26.29) and (26.31), respectively. In Section 26.2.8, we have investigated the $I_D - V_D$ equation in magneto MOSFETs of gallium arsenide in accordance with the models of Rossler and Stillman *et al.* for weak electric field limits which are given by (26.32) and (26.36) and the corresponding 2D electron statistics has been represented in (26.33) and (26.37), respectively. In the same section, we have also investigated the $I_D - V_D$ equation in this case in accordance with the said models for strong electric field limits which are given by (26.34) and (26.38) and the corresponding 2D electron statistics has been represented in (26.35) and (26.39), respectively. In Section 26.2.9, we have studied the $I_D - V_D$ equation in ballistic magneto

MOSFETs of gallium antimonide in accordance with the models of Sailer, Mathur *et al.* and Zhang. for weak electric field limits which are given by (26.40), (26.44) and (26.48) and the corresponding 2D electron statistics has been represented in (26.41), (26.45) and (26.49), respectively. In the same section, we have also investigated the $I_D - V_D$ equation in this case in accordance with the said models for strong electric field limits which are given by (26.42), (26.46) and (26.50) and the corresponding 2D electron statistics has been represented in (26.43), (26.47) and (26.51), respectively. In Section 26.2.10, we have studied the $I_D - V_D$ equation in ballistic magneto MOSFETs of indium antimonide in accordance with the models of Palik, Johnson *et al.* and Agafonov *et al.* for weak electric field limits which are given by (26.52), (26.56) and (26.60) and the corresponding 2D electron statistics has been represented in (26.53), (26.57) and (26.51), respectively. In the same section, we have also investigated the $I_D - V_D$ equation in this case in accordance with the said models for strong electric field limits which are given by (26.54), (26.58) and (26.62) and the corresponding 2D electron statistics has been represented in (26.55), (26.59) and (26.63), respectively. In Section 26.2.3, we have studied the $I_D - V_D$ equation for ballistic magneto MOSFETs of tellurium which is given by (26.64) and the corresponding 2D electron statistics has been represented in (26.65).

We have not given the graphs of the $I_D - V_D$ equations for ballistic magneto MOSFETs made of many materials as discussed here since we feel that our readers will perform the intricate numerical computation to obtain the oscillatory plots for better assessment and joy of understanding. It is the band structure that changes in a fundamental way and consequently all the physical properties of all the electronic materials changes radically leading to new physical concepts.

26.4 Open Research Problem

(R26.1) Derive $I_D - V_D$ equations for ballistic Magneto MOSFETs made of all the materials whose DRs are given in (R1.1). Draw the corresponding graphs and explain the results.

Chapter 27

The Generalized Distribution Functions in Heavily Doped Materials

Thoughts become words and words become actions. Actions become habits and habits become character. Character becomes destiny.

27.1 Introduction

The concept of distribution functions of the carriers in different materials occupies a singular position in the arena of materials science in general for not only the characterization of different systems but also for the study of the transport features in different technologically important materials [1,2]. Fermi and Dirac (FD)(both independently) formulated the FD statistics for electrons [3,4] whereas the Bose–Einstein (BE) statistics appeared in the literature in 1924 for bosons [5] together with the fact that both of them get simplified into well-known Maxwell–Boltzmann statistics under certain limiting condition. The FD statistics is generally, applicable for studying the transport features of degenerate electronic materials, because in this case, the band structure prevails [6] and the Fermi energy (E_F) is above the edge of the conduction band (CB) and within the conduction band. Also, we know by Pauli exclusion principle, the occupational probability of an electron at an energy state is (1/2) at E_F [7]. The simple FD function cannot explain the existence of band tailing phenomena observed in the case of heavily doped systems at low temperatures where the quantum effects become prominent. Therefore, some modifications must be expected in the mathematical form of the well-known FD distribution function. It might be noted that in the case of disordered materials, the conduction band penetrates within the forbidden band (FB) and the carriers do exist in the tail region. We know that for a semiconductor with heavily doped conditions, the doping levels must satisfy the inequality $(a_D \cdot Ni^{1/3}) > 1$, where a_D is the Bohr radius and Ni is the doping concentration of the carriers, and in this case, the doping condition should satisfy the inequality [9–11] $\{(0.02 \leq [a_D \cdot Ni^{1/3}] < 1.0)\}$ [8–12]. As we know that the normal FD distribution function cannot explain the occupational probability of the carriers is the tail–zone (TZ), because it is valid for the band gap zone (BZ) only. Also, in general, the FB zone is known as imaginary band (IB). The TZ lies in the forbidden zone (FBZ). Therefore we might conclude that for the TZ region, the probability of the distribution of the carriers must be different from the normal FD distribution function. In this appendix, an attempt is made in the theoretical background to derive the generalized FD and BE in a very simplified manner and we also note various important conclusions which emerge from it. This new probability of distribution in the TZ region is termed as "Extended and Modified distribution functions. Sections 27.2, 27.3 and 27.4 contain the theoretical background, the result and discussions and 23 open research problems respectively in this context.

27.2 Theoretical Background

Let us assume that the probability of occupation of the energy level $\bar{E}$ by an electron is $F(\bar{E})$. Let us also assume that there are two initial energy levels namely $\bar{E}_1$ and $\bar{E}_2$ together with two final energy levels $\bar{E}_3$ and $\bar{E}_4$, respectively as shown in Fig. 27.1.

The probability of forward transmission of an electron in the initial energy level $\bar{E}_1$ to the final energy level $\bar{E}_3$ and an electron in the initial energy level $\bar{E}_2$ to the final energy level $\bar{E}_4$ is proportional to

$$F(\bar{E}_1)[1 - F(\bar{E}_3)]F(\bar{E}_2)[1 - F(\bar{E}_4)]$$

Besides, the probability of reverse transmission is similarly proportional to

$$F(\bar{E}_3)[1 - F(\bar{E}_1)]F(\bar{E}_4)[1 - F(\bar{E}_2)]$$

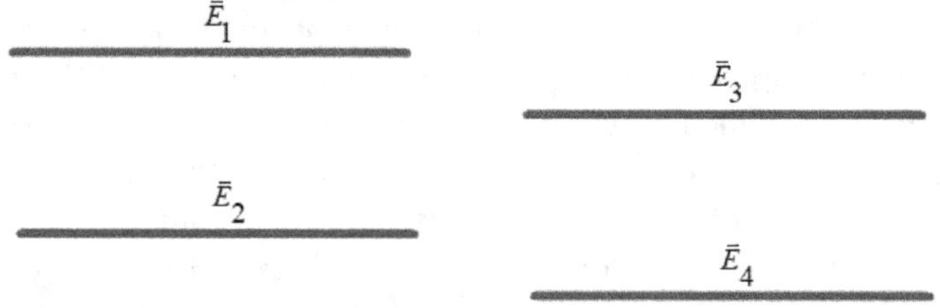

Fig. 27.1. This figure exhibits the probability of forward transmission of an electron in the initial energy level $\bar{E}_1$ to the final energy level $\bar{E}_3$ and an electron in the initial energy level $\bar{E}_2$ to the final energy level $\bar{E}_4$ respectively.

The application of the principle of detailed balance of Einstein leads to the result

$$F(\bar{E}_1)[1 - F(\bar{E}_3)]F(\bar{E}_2)[1 - F(\bar{E}_4)] = F(\bar{E}_3)[1 - F(\bar{E}_1)]F(\bar{E}_4)[1 - F(\bar{E}_2)] \tag{27.1}$$

Dividing both sides by $[F(\bar{E}_1)F(\bar{E}_2)F(\bar{E}_3)F(\bar{E}_4)]$ and assuming $[F(\bar{E}_1)F(\bar{E}_2)F(\bar{E}_3)F(\bar{E}_4)] \neq 0$ we get

$$\left[\frac{1}{F(\bar{E}_1)} - 1\right]\left[\frac{1}{F(\bar{E}_2)} - 1\right] = \left[\frac{1}{F(\bar{E}_3)} - 1\right]\left[\frac{1}{F(\bar{E}_4)} - 1\right] \tag{27.2}$$

The energy conservation rule gives us

$$\bar{E}_1 + \bar{E}_2 = \bar{E}_3 + \bar{E}_4 \tag{27.3}$$

Let us assume

$$\frac{1}{F(\bar{E})} - 1 = A + \exp\left(\frac{\bar{E} - E_F}{k_B T}\right) \tag{27.4}$$

where A is a constant. Using (27.2), (27.3) and (27.4) we get

$$\left[A + \exp\left(\frac{\bar{E}_1 - E_F}{k_B T}\right)\right]\left[A + \exp\left(\frac{\bar{E}_2 - E_F}{k_B T}\right)\right]$$
$$= \left[A + \exp\left(\frac{\bar{E}_3 - E_F}{k_B T}\right)\right]\left[A + \exp\left(\frac{\bar{E}_4 - E_F}{k_B T}\right)\right] \tag{27.5}$$

From (5) we can write

$$\left[\exp\left(\frac{\bar{E}_1}{k_B T}\right) + \exp\left(\frac{\bar{E}_2}{k_B T}\right)\right]^{-1} = \left[\exp\left(\frac{\bar{E}_3}{k_B T}\right) + \exp\left(\frac{\bar{E}_4}{k_B T}\right)\right]^{-1} \tag{27.6}$$

If (27.6) is satisfied by (27.4) then the $F(\bar{E})$ as given by (27.4) is a solution. Now for heavily doped nanomaterials which are being discussed in this chapter is a low-temperature phenomenon, i.e., $T \to 0$ and under this condition, (27.6) is being satisfied by (27.4).

Thus, the modified FD function can be written from (27.4) as

$$F(\bar{E}) = [1 + A + \exp(y)]^{-1} \tag{27.7}$$

where $y = ((\bar{E} - E_F)/(k_B T))$.

The conclusions from (27.7) are written as follows:

1. When $A = 0$, we get the ordinary FD function from (27.7) as

$$F(\bar{E}) = [1 + \exp((\bar{E} - E_F)/(k_B T))]^{-1} \tag{27.8}$$

2. When $A = -1$, from (27.7) we can write that the $F(\bar{E}) = C\exp(-\bar{E}/k_BT)$, where $C = \exp(E_F/k_BT)$ which the well-known Maxwell–Boltzmann distribution.
3. The substitutions $A = 0$ and $\bar{E} = E_F$ together with $A = -4$ and $\bar{E} = E_F$ in (27.7) generate the well-known Pauli's exclusion principle as $(\pm 1/2)$.
4. For $A = i - 1(i = \sqrt{-1})$ and $\bar{E} = E_F$ together with $A = -3 + i$ and $\bar{E} = E_F$ in (27.7) lead to complex value of the Pauli's spin in the tail zone as $1/2(\pm 1 \mp i)$, respectively.
5. The substitution $A = 1 + B_0 i$ (where B_0 is a real constant) leads to the modified and extended complex Fermi Dirac function as

$$F_c(\bar{E}) = [1 + B_0 i + \exp(y)]^{-1} \tag{27.9}$$

From (27.9) we can write

$$Re[F(\bar{E})] = [1 + \exp(y)][B_0^2 + (1 + \exp y)^2]^{-2} \tag{27.10}$$

and

$$Im[F(\bar{E})] = -B_0[B_0^2 + (1 + \exp y)^2]^{-2} \tag{27.11}$$

A few important points can also be noted in this context:

6. We know the concept of "spin g-factor" is very important for semiconductors; while studying the electron energy ($\bar{E}_5$) in the presence of a magnetic field B, applied to a semiconductor along z-axis which in this case assumes the form

$$\bar{E}_5 = \pm(g/2)[\hbar\omega_B] \tag{27.12}$$

where $\hbar\omega_B$ is the cyclotron resonance frequency of electron in this case.

Equation (27.12), exhibits the fact that it has only magnitude but no phase value.

The electron energy in the presence of the complex spin value $\pm (1/2)(1 - i)$ in the tail zone can be written as

$$\bar{E}_6 = \pm(g/2)(1 - i)[\hbar\omega_B] \tag{27.13}$$

from (27.13), we observe that the electron energy ($\bar{E}_6$) as the presence of a magnetic field, has both magnitude and phase value,

Thus, the magnitude is given by

$$|\bar{E}_6| = \frac{g}{2}\hbar\omega_B(\sqrt{2}) = 0.707\left[\frac{g}{2}\hbar\omega_c\right] \approx 71\% \text{ of } (g\hbar\omega_B) \tag{27.14}$$

and the phase value can be written as

$$\phi_{EX} = \tan^{-1}\left\{-\frac{\left[\pm\frac{g}{2}\hbar\omega_B\right]}{\left\{\pm\frac{g}{2}\hbar\omega_B\right\}}\right\} = \pm\frac{\pi}{4} \tag{27.15}$$

In the conventional case of Pauli's value is 50% and the phase value $\phi_c = 0^0$. When an electron rotates in its orbit around the nucleus, the average electron energy is about 71% of its maximum value. The phase value ϕ_{EX}, in this case is $\pm(\pi/4)$.

For Boson, Pauli's exclusion principle is not obeyed. Replacing (-1) in (27.2) and (27.4) with $(+1)$ we get

$$\left[\frac{1}{F(\bar{E}_1)}+1\right]\left[\frac{1}{F(\bar{E}_2)}+1\right] = \left[\frac{1}{F(\bar{E}_3)}+1\right]\left[\frac{1}{F(\bar{E}_4)}+1\right] \tag{27.16}$$

$$\frac{1}{F(\bar{E})}+1 = A + \exp\left(\frac{\bar{E}-E_F}{k_B T}\right) \tag{27.17}$$

Following the method as given above we note that the (27.17) is a solution of Equation (27.16). Substituting $A = 0$, we note that

$$F(\bar{E}) = \left[\exp\left(\frac{\bar{E}-E_F}{k_B T}\right) - 1\right]^{-1} \tag{27.18}$$

which is the well-known BE statistics.

27.3 Result and Discussions

Figures (27.2)–(27.4) exhibit the ordinary FD function, real and imaginary parts of the extended FD functions as given by (27.8), (27.10) and (27.11), respectively where for figures 27.3 and 27.4 we have taken $B_0 = 1$ for the purpose of numerical computations. Graphs 27.3 and 27.4 are new and should be used in dealing with all types of carrier properties in the TZ of HD quantum material.

In this appendix, we have derived the modified and extended form of (FD) distribution function for the carriers in the (TZ) region, while the carriers penetrate with is the (FBZ) in the form of tails. These carriers are free carriers as well as the carriers have finite effective masses. The transport properties of these carriers are not the same as those of the carriers in the normal semiconductors bands (27.2). In this chapter, we have derived the modified FD distribution function as given by (27.7) which is valid for carriers in the (TZ).

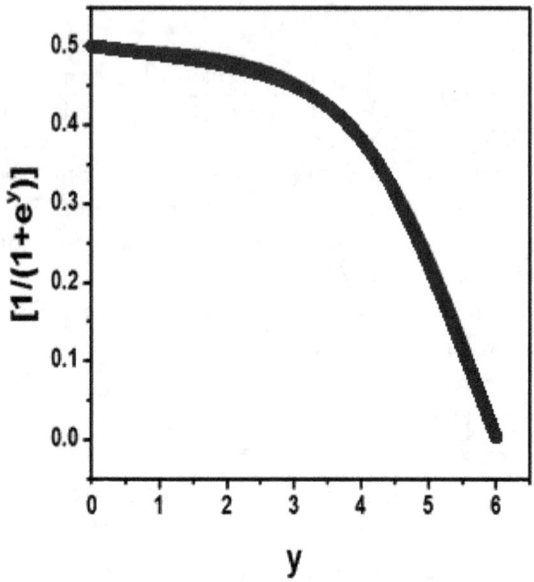

Fig. 27.2. Plot of the normal FD probability distribution factor (Equation (27.8)) as a function of y.

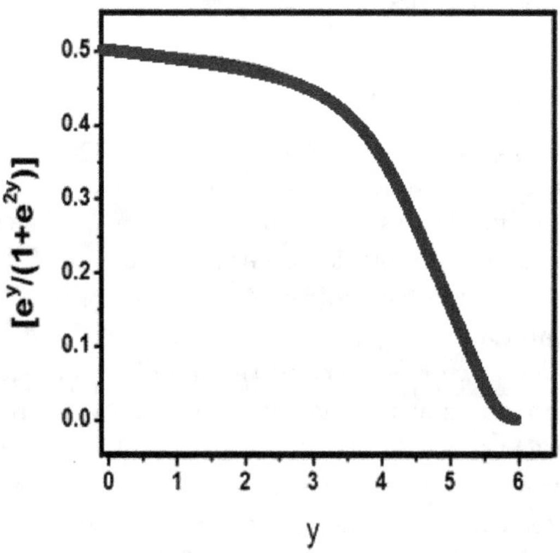

Fig. 27.3. Plot of the real part of the extended and modified FD probability distribution factor as given by Equation (27.10) as a function of y.

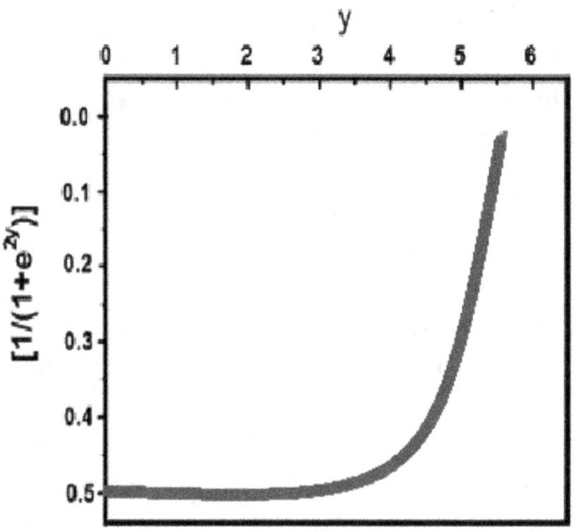

Fig. 27.4. Plot of the imaginary part of the extended and modified FD probability distribution factor as given by (27.11) as a function of y.

The validity of the generalized equation (27.7) for which we have given a very simplified derivation can easily be assessed from six special cases as given above where the first three lead to well-known results and the last three generate new concepts.

27.4 Open Research Problems

(R.27.1) Derive generalized distribution functions in degenerate materials in the presence of strong light waves and intense arbitrarily oriented electric fields respectively. Derive the dispersion relations and the density of state functions. Draw the corresponding graphs and explain the results.

(R.27.2) Derive the corresponding Boltzmann transport equation in the presence of time-dependent quantizing magnetic field.

(R.27.3) On the basis of (R.27.1), derive the corresponding ES in the said cases in the presence of time-dependent electric field. Draw the corresponding graphs and explain the results.

(R.27.4) Formulate the Burstein Moss shift using (R.27.1) in the presence of time-dependent quantizing non-uniform magnetic field. Draw the corresponding graphs and explain the results.

(R.27.5) Derive the Debye screening length using (R.27.1) in the presence of time-dependent non-uniform electric field. Draw the corresponding graphs and explain the results.

(R.27.6) Derive the diffusion coefficient of the minority carriers on the basis of (R.27.1) in the presence of time-dependent quantizing magnetic field and strain respectively. Draw the corresponding graphs and explain the results.

(R.27.7) Formulate nonlinear optical response on the basis of (R.27.1) in the presence of time-dependent quantizing magnetic field and non-uniform strain. Draw the corresponding graphs and explain the results.

(R.27.8) Formulate the third-order nonlinear optical susceptibility in this case on the basis of (R.27.1) in the presence of time-dependent quantizing magnetic field. Draw the corresponding graphs and explain the results.

(R.27.8) Formulate the generalized Raman gain in this case on the basis of (R.27.1) in the presence of time-dependent quantizing magnetic field. Draw the corresponding graphs and explain the results.

(R.27.9) Formulate the plasma frequency in this case on the basis of (R.27.1) in the presence of time-dependent quantizing magnetic field. Draw the corresponding graphs and explain the results.

(R.27.10) Formulate the activity coefficient in this case on the basis of (R.27.1) in the presence of time-dependent quantizing magnetic field. Draw the corresponding graphs and explain the results

(R.27.11) Formulate the magneto-thermal effect in quantized structures in this case on the basis of (R.27.1). Draw the corresponding graphs and explain the results.

(R.27.12) Formulate the normalized Hall coefficient in this case on the basis of (R.27.1) in the presence of time-dependent non-uniform quantizing magnetic field. Draw the corresponding graphs and explain the results.

(R.27.13) Formulate the reflection coefficient in this case on the basis of (R.27.1) in the presence of time-dependent quantizing electric field. Draw the corresponding graphs and explain the results.

(R.27.14) Formulate the heat capacity in this case on the basis of (R.27.1) in the presence of time-dependent quantizing magnetic field. Draw the corresponding graphs and explain the results.

(R.27.15) Formulate the magnetic susceptibilities in this case on the basis of (R.27.1) in the presence of time-dependent quantizing magnetic field. Draw the corresponding graphs and explain the results.

(R.27.16) Formulate the Faraday rotation in this case on the basis of (R.27.1) in the presence of time-dependent quantizing magnetic field. Draw the corresponding graphs and explain the results.

(R.27.17) Formulate the Fowler–Nordheim field emission in this case on the basis of (R.27.1) in the presence of time-dependent quantizing magnetic field. Draw the corresponding graphs and explain the results.

(R.27.18) Formulate the optical effective mass in this case on the basis of (R.27.1) in the presence of time-dependent quantizing magnetic field. Draw the corresponding graphs and explain the results.

(R.27.19) Formulate the Einstein's photoemission in this case on the basis of (R.27.1) in the presence of time-dependent quantizing magnetic field. Draw the corresponding graphs and explain the results.

(R.27.20) Formulate the Righi-Leduc coefficient in this case on the basis of (R.27.1) in the presence of time-dependent quantizing magnetic field. Draw the corresponding graphs and explain the results.

(R.27.21) Formulate the electric susceptibility in this case on the basis of (R.27.1) in the presence of time-dependent quantizing magnetic field. Draw the corresponding graphs and explain the results.

(R.27.22) Formulate the thermoelectric power in this case on the basis of (R.27.1) in the presence of time-dependent quantizing magnetic field. Draw the corresponding graphs and explain the results.

(R.27.23) Derive the electron mobilities for various scattering mechanisms in this case.

Dear Readers, we earnestly hope that you and only you can convert this cozy cottage of Quantum Capacitance in Quantized Transistors into a skyscraper, not only by solving the open research problems but also by extending the same in many other new directions, by inventing new research problems and new formalisms, both theoretical and experimental. The Quantum Capacitance in Quantized Transistors is the dual dance of symmetry reduction of wave vector space of the charge carriers of the materials and advanced quantum mechanics. Now we realize in the real sense of the term that the quantum phenomena in quantized materials is really a sea in itself.

Let us recall the famous words of the Prince of Mathematics, Carl Friedrich Gauss (Prince of Mathematics — Carl Friedrich Gauss, by M.B.W. Tent, A.K. Peters Ltd. Wellesley, Massachusetts, 2006) "Nothing has been done, if something remains to be done." Make an honest academic commitment to delve deep into this topic in deference to this supreme mathematician.

Move with creativity and enjoy your work. Good Luck!

References

1. O. Modelung, *Introduction to Solid state Theory*, Springer Series in Solid State Sciences, Vol. 2 (Springer, Germany 1978).
2. B. R. Nag, *Electron Transport in The Compound Semiconductors*, Springer Series In Solid State Science, Vol. 11 (Springer, Germany 1980).
3. E. Fermi, *Z. Phys.* **36**, 902 (1926).
4. P. A. M. Dirac, *Proc. Roy. Soc.* **112**, 661 (1926).
5. S. N. Bose, *Z. Phys.* **26**, 178 (1924).
6. B. I. Shaklovskii and A. L. Efros, *Electronic Properties of Doped Semiconductors*, Vol. 45 (Springer-Verlag, Berlin, New York, 1984).
7. W. Pauli, *Z. Physik.* **31**, 765 (1925).
8. B. K. Choudhuri, B. N. Mondal and P. K. Chakraborty, *Pramana J. Phys.* **90**, 18 (2019).
9. P. K. Chakraborty, B. N. Mondal and B. K. Choudhury, *Open Access Jo. Math. Theoreti. Phys.* **2**, 22, (2019).
10. P. K. Chakraborty and J. C. Biswas, *Jo. Appl. Phys.* **82**, 7, (1997).
11. P. K. Chakraborty and B. N. Mondal, *Quart. Phys. Rev.* **3**, 1, (2017).
12. C. Kittle, *Elementary Statistical Mechanics* (Dover Publications, Inc, New York, 1980).

Subject Index

C

F

G

H

I

K

L

M

P

Q

R

T

U

V

W

Materials Index

Series on the Foundations of Natural Science and Technology

(Continued from page ii)

Vol. 7 *Magneto Thermoelectric Power in Heavily Doped Quantized Structures*
by K. P. Ghatak

Vol. 6 *Nano-Engineering in Science and Technology: An Introduction to the World of Nano-Design*
by M. Rieth

Vol. 5 *Grasping Reality: An Interpretation-Realistic Epistemology*
by H. Lenk

Vol. 4 *What is Life? Scientific Approaches and Philosophical Positions*
by H.-P. Dürr, F.-A. Popp and W. Schommers

Vol. 3 *The Visible and the Invisible: Matter and Mind in Physics*
by W. Schommers

Vol. 2 *Symbols, Pictures and Quantum Reality: On the Theoretical Foundations of the Physical Universe*
by W. Schommers

Vol. 1 *Space and Time, Matter and Mind: The Relationship between Reality and Space-Time*
by W. Schommers

Forthcoming:

Vol. 10 *Space-Time Reality: Mach's Principle, Relativity and Novel Aspects*
by W. Schommers

Vol. 11 *The Origin of Mind: Interplay between Matter and Mind*
by W. Schommers

www.ingramcontent.com/pod-product-compliance
Lightning Source LLC
LaVergne TN
LVHW010321220926
848484LV00004BB/819